Characterization of Solid Surfaces

Characterization of Solid Surfaces

Edited by

Philip F. Kane and Graydon B. Larrabee

Materials Characterization Laboratory
Texas Instruments Incorporated
Dallas, Texas

PLENUM PRESS • NEW YORK - LONDON

Library of Congress Cataloging in Publication Data

Kane, Philip F 1920-
Characterization of solid surfaces.

Includes bibliographical references.
1. Solids. 2. Surfaces (Technology). 3. Metallography. 4. Surface chemistry. I. Larrabee, Graydon B., 1932- joint author. II. Title.
QC176.K33 541'.3453 73-84000
ISBN 0-306-30752-9

A Division of Plenum Publishing Corporation
227 West 17th Street, New York, N.Y. 10011

United Kingdom edition published by Plenum Press, London
A Division of Plenum Publishing Company, Ltd.
4a Lower John Street, London W1R 3PD, England

Printed in the United States of America

CONTENTS

Chapter 12

X-RAY FLUORESCENCE ANALYSIS

John V. Gilfrich

Chapter 13

SURFACE CHARACTERIZATION BY ELECTRON SPECTROSCOPY FOR CHEMICAL ANALYSIS (ESCA)

Shirley H. Hercules and David M. Hercules

Chapter 14

RESONANCE METHODS

D. Haneman

Chapter 17
ACCELERATOR MICROBEAM TECHNIQUES
T. B. Pierce

Chapter 18
ELECTRON PROBE MICROANALYSIS
Gudrun A. Hutchins

Chapter 21
MASS SPECTROMETRY
J. M. McCrea

Chapter 22

IMPURITY-MOVEMENT PROBLEMS IN ANALYSIS METHODS USING PARTICLE BOMBARDMENT

Daniel V. McCaughan and R. A. Kushner

Chapter 23

SURFACE COMPOSITION BY ANALYSIS OF NEUTRAL AND ION IMPACT RADIATION

C. W. White, D. L. Simms, and N. H. Tolk

LIST OF CONTRIBUTORS

William L. Baun
Air Force Materials Laboratory
Wright Patterson AFB

K. H. Beckmann
Philips Forschungslaboratorium

Chuan C. Chang
Bell Laboratories

R. D. Dobrott
Texas Instruments Incorporated

John V. Gilfrich
Naval Research Laboratory

A. C. Hall

D. Haneman
University of New South Wales

N. J. Harrick
Harrick Scientific Corporation

David M. Hercules
University of Georgia

Shirley H. Hercules
University of Georgia

Melvin C. Hobson, Jr.
Virginia Institute of Scientific Research

Gudrun A. Hutchins
Sprague Electric Company

Om Johari
IIT Research Institute

P. F. Kane
Texas Instruments Incorporated

Joseph A. Keenan
Texas Instruments Incorporated

R. A. Kushner
Bell Laboratories

Campbell Laird
University of Pennsylvania

Graydon B. Larrabee
Texas Instruments Incorporated

D. M. MacArthur
Bell Laboratories

W. D. Mackintosh
Atomic Energy of Canada Limited

J. M. McCrea
Indiana University of Pennsylvania

Walter C. McCrone
Walter C. McCrone Associates, Inc.

Daniel V. McCaughan
Bell Laboratories

T. B. Pierce
U.K. Atomic Energy Research Establishment

A. V. Samudra
La Salle Steel Company

J. L. Seeley
Colorado State University

D. L. Simms
Bell Laboratories

R. K. Skogerboe
Colorado State University

N. H. Tolk
Bell Laboratories

C. W. White
Bell Laboratories

David John Whitehouse
Rank Precision Industries Limited

INTRODUCTION

Until comparatively recently, trace analysis techniques were in general directed toward the determination of impurities in bulk materials. Methods were developed for very high relative sensitivity, and the values determined were average values. Sampling procedures were devised which eliminated the so-called sampling error. However, in the last decade or so, a number of developments have shown that, for many purposes, the distribution of defects within a material can confer important new properties on the material. Perhaps the most striking example of this is given by semiconductors; a whole new industry has emerged in barely twenty years based entirely on the controlled distribution of defects within what a few years before would have been regarded as a pure, homogeneous crystal. Other examples exist in biochemistry, metallurgy, polymers and, of course, catalysis. In addition to this recognition of the importance of distribution, there has also been a growing awareness that physical defects are as important as chemical defects. (We are, of course, using the word defect to imply some discontinuity in the material, and not in any derogatory sense.) This broadening of the field of interest led the Materials Advisory Board[1] to recommend a new definition for the discipline, "Materials Characterization," to encompass this wider concept of the determination of the structure and composition of materials.

In characterizing a material, perhaps the most important special area of interest is the surface. A surface, by definition, is an interface, a marked discontinuity from one material to another. Since no change in nature is ever instantaneous, there is a finite depth for any real surface, despite the mathematical definition, and in characterizing a surface one must at some point consider just what this depth is. As editors, we have avoided any statement of opinion and left the decision to our contributors. In practice, of course, the surface depth is defined by the technique, and results must be interpreted on this basis. From one chapter to another, the surface being examined may vary from an atomic layer to as much as 50 μ, and this should be borne in mind in considering the applications of a particular technique. However, this may not be inconsistent since the effect of the surface, that is the property being applied, will itself be a function dependent on depth, varying from application to application.

The chapters are divided, somewhat arbitrarily, into two sections, physical and chemical; this was the editors' choice and certainly open to argument, but we felt this did have the virtue of emphasizing the two subsets of materials characterization, structural and compositional. With two exceptions, the chapters are intended as broad reviews of particular techniques. In the confines of a single volume, it is of course not possible to give a great deal of depth to any one of these and, at the editors' request, the authors have given a minimum of mathematical theory and have emphasized the applications of the techniques to real surfaces in real-world problems. The intention is to acquaint the scientist working in the general field of materials research with the potential of the techniques available to him in surface studies. Should he decide to pursue one of these techniques further, adequate references should make a deeper study a relatively easy step.

The two chapters not devoted primarily to techniques (Chapters 19 and 22) are by way of being commentaries on the interactions of particles with matter, Chapter 19 largely dealing with electrons, Chapter 22 with ions. Both may stimulate some second thoughts on the several applications of particle beams to the analysis of surface material.

Two techniques which we had intended to include are unfortunately absent: We are acutely aware that electron diffraction is one of the most important methods for determining the structure of surfaces and, inferentially, their chemical compositions and has been for many years. Since it is a well-established technique, most scientists will have at least a nodding acquaintance with reflection electron diffraction (or RHEED, for reflection high-energy electron diffraction, as it is now often referred to) as used on the electron microscope. Although RHEED is still important, low-energy electron diffraction (LEED) has become more important over the past decade or so. An excellent review of both techniques has been given recently by Estrup and McCrae.[2] The other technique we were unable to include is low-energy ion scattering spectrometry (ISS). This is a deceptively simple method in which a beam of ions is rebounded from the surface atoms by an elastic collision, the loss of energy of these primary ions being related directly to the masses of the atoms struck. It is similar to the Rutherford scattering described in Chapter 16, but is carried out with low-energy ions. Whereas Rutherford scattering typically uses energies of 1–2 MeV, ISS uses energies below 5 keV. This greatly reduces the depth of interaction, making it a more truly surface effect and, at the same time, simplifies the equipment requirements. On the other hand, cross-section measurements are well defined and readily available for Rutherford scattering, but are still somewhat sketchy for low-energy ions. Much remains to be done in the interpretation of the spectra, particularly from a quantitative aspect, but it does appear to have considerable potential. A useful review has been given by Smith,[3] who has been instrumental in introducing this technique.

Our hope was, of course, that this volume would be a survey of the current status of surface characterization. In looking ahead, the evolving surface-characterization techniques are, without doubt, the probe techniques. This trend was noted by Kane and Larrabee[4] in a review of trace-analysis techniques for solids in which these techniques were shown developing from electron microscopy, x-ray fluorescence, and charged-particle scattering, all of which were in existence in the 1950's. Following a proliferation of methods through the 1960's, there has been a definite trend toward combining and consolidating various instruments, e.g., Auger electron spectroscopy (AES) and scanning electron microscopy (SEM).

Another way of looking at the current probe techniques used in surface characterization is shown in Fig. A. The modes of excitation and emission can best be categorized in terms of phonons ($h\nu$), electrons (e^-), and ions. Phonons include all wavelengths of electromagnetic radiation, but for these probe techniques are restricted to x rays and ultraviolet light. From this table it can be seen that ion- and electron-bombardment probe techniques generate the most data.

In attempting to envision what techniques and instrumentation will be available in the 1980 time frame, it is of considerable help to look at the excitation and detection systems. The concept of surface characterization technique consolidation becomes more readily apparent from Fig. B. Note that all excitation systems generate fluorescent x rays, and therefore a Si(Li) nondispersive detector would be one of the detectors. A secondary-electron detector would allow SEM operation for both electron and ion excitation. The incorporation of a cylindrical mirror analyzer would allow the analytical system to obtain Auger spectra. An ion mass analyzer would allow IMMA operation, and thus three-dimensional characterization would be possible from the sputtering capabilities of the ion beam. Light optics for both excitation and detec-

Emission \ Excitation	$h\nu$	e^-	ions
$h\nu$	XRF	EMP	IEX SCANIIR
e^-	ESCA PES AES	SEM AES EM	SEM
ions			IMMA SIMS ISS RUTHERFORD

AES—Auger electron spectroscopy
EM—electron microscopy
EMP—electron microprobe
ESCA—electron spectroscopy for chemical analysis
IEX—ion excited X-ray fluorescence and ion impact radiation
IMMA—ion microprobe mass analyzer
ISS—ion scattering spectroscopy
PES—photoelectron spectroscopy
RUTHERFORD—Rutherford scattering
SCANIIR—surface composition by analysis of neutral and ion impact radiation
SEM—secondary electron microscopy
SIMS—secondary ion mass spectroscopy
XRF—x-ray fluorescence

Fig. A. Summary of probe techniques for surface analysis.

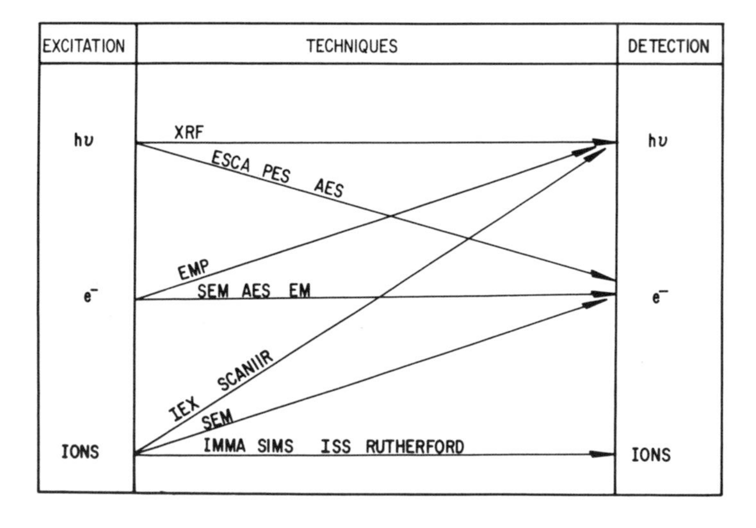

Fig. B. Excitation and detection systems used in surface analysis.

tion of emitted light would allow photoelectron spectroscopy (PES or ESCA) and SCANIIR characterization. Thus an instrument with an excitation beam of ions and electrons with multiple detection systems will evolve in the next decade.

The very nature of an instrument as complex as described above would dictate the use of a computer for both instrument operation and data collection. Minicomputers are available at the time of this writing which are capable of performing all of the described functions yet cost less than $3000 for the basic 4096-word machine. Even with a million-word disk and 24,576 words of memory, the current cost would be less than $20,000. This is a small part of the cost of even today's sophisticated characterization tools. By 1980, the computer cost will have dropped to well under $10,000 and every characterization instrument will be controlled by minicomputers. Thus, data acquisition will become far more sophisticated with direct reduction of spectral data using complex mathematical data smoothing such as fast Fourier transforms.

Surface-characterization technology has seen a renaissance in the early 1970's. It is appropriate that this volume, which is a comprehensive review of the analysis of surfaces, should be written and published at the crest of this resurgence. The mid and late 1970's will see continued development of sophisticated surface characterization, with a real understanding of the role of surfaces in controlling the properties of materials emerging as a direct result of this new technology.

BIBLIOGRAPHY

1. Characterization of Materials, Publication MAB-229-M, National Academy of Sciences, Washington, D.C. (March, 1967).
2. P. J. Estrup and E. G. McCrae, *Surface Sci.* **25**, 1 (1971).
3. D. P. Smith, *Surface Sci.* **25**, 171 (1971).
4. P. F. Kane and G. B. Larrabee, *Ann. Rev. Mater. Sci.* **2**, 33 (1972).

I

PHYSICAL CHARACTERISTICS

1

LIGHT MICROSCOPY

Walter C. McCrone

Walter C. McCrone Associates, Inc.
Chicago, Illinois

1. INTRODUCTION

The use of the light microscope for the study of surfaces has decreased considerably since the commercial introduction of scanning electron microscopes (SEM). A look at Chapter 5 on the SEM will quickly prove that this change is at least partly justified. These instruments will resolve detail one-tenth as large (20 nm = 0.02 μm) as that resolved by the light microscope, and the in-focus depth of field for the SEM is 100–300 times that of the light microscope.

There are other advantages of the SEM, including ease of sample preparation, elemental analysis by energy-dispersive x-ray analyzer, and, usually, excellent specimen contrast. How, then, do we justify a chapter on the light microscope? One important reason, perhaps, is the cost of an SEM, 10 to 50 times the cost of an adequate light microscope. There will always be laboratories unable to justify the high cost of an SEM.

Secondly, there are many routine surface examinations easily performed by light optics that do not justify use of the SEM. There are, at least, a few surface characterization problems that the SEM cannot do: surfaces of materials unstable under high vacuum or high-energy electron bombardment; samples too bulky for the SEM sample compartment; and finally, samples requiring manipulation on the surface during examination and vertical resolution of detail below 250 μm.

Hopefully, when some of the excitement of using an SEM wears off, the microscopist will be able to decide objectively whether his sample requires a stereobinocular or mono-objective light microscope or the SEM. Often, of course, both light microscopes and the SEM will be used. The stereobinocular microscope will be needed, if only to quickly decide what areas to study or to examine the pertinent areas in terms of the total sample and total problem. Even SEM examination should begin at low magnification and never be increased more than necessary.

There are accessories for the light microscope that greatly enhance its ability to resolve detail, differentiate different compositions, or increase contrast. Any microscopist who has attempted to observe thin coatings on paper, e.g., ink lines, with the SEM soon goes back to the light microscope. The Nomarski interference contrast system on a reflected light microscope gives excellent rendition of surface detail for metals, ceramics, polymers, or biological tissue. Multiple-beam interferometry will easily resolve differences in vertical dimensions as small as 1.5 μm. The SEM is 10 times better than the light microscope in horizontal resolution but 20 times worse in vertical resolution.

To the microscopist, characterization of a surface means: topography, elemental composition, and solid-state structure. He usually studies all three by what is usually termed morphological analysis, i.e., shape characteristics. Surface geometry or topography is obviously a matter of morphology, although the light microscopist may have to enhance contrast of transparent, colorless surfaces like paper or ceramics by a surface treatment, e.g., an evaporated-metal coating or by a special examination technique, e.g., interferometry.

Elemental composition determination is also often possible by study of morphology although, again perhaps, made easier by surface etching, staining, or examination by polarized light. Finally, solid-state structure is also often apparent morphologically, e.g., the various phases in steels (austenite, ferrite, pearlite, sorbite, troostite, and bainite) are recognizable by a metallographer.

When micromorphological studies fail, the microscopist then proceeds to the SEM for topography; to the electron microprobe, photoelectron spectroscopy (ESCA), or the SEM for elemental analysis; and to backreflection x-ray diffraction for solid-state structure.

2. TOPOGRAPHY

Let us examine first the microscopical techniques and instrumentation useful for the examination of microscopic topography. As indicated before, the topography (as well as the chemical and physical composition) of a surface greatly affects wear, friction, reflectivity, catalysis, and a host of other properties. Many techniques can be used to study surfaces, but most begin with visual examination supplemented by increasing magnification of the light microscope. Straightforward microscopy may be supplemented by either sample-preparation techniques or by use of specialized microscope accessories.

2.1. Darkfield Illumination

There are two general methods of observing surfaces: darkfield and brightfield. Each of these, however, can be obtained with transmitted light from a substage condenser and with reflected light from above the preparation (Fig. 1-1). For brightfield top lighting, the microscope objective itself must act as condenser for the illuminating beam. For darkfield transmitted light, the condenser numerical aperture (NA) must exceed the NA of the objective—and a central cone of the condenser illuminating beam, equal in angle to the maximum objective angular aperture, must be opaqued.

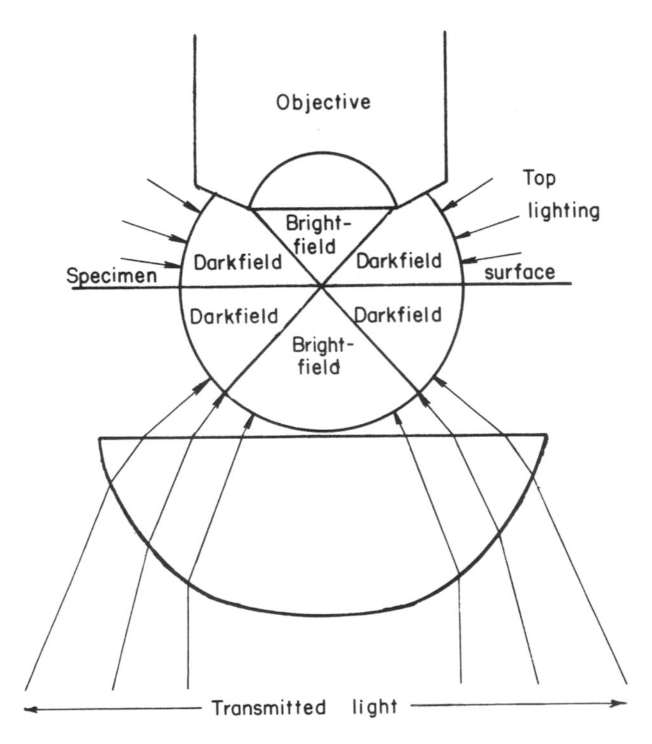

Fig. 1-1. Schematic representation of darkfield and brightfield illumination, both transmitted from beneath and reflected from above the preparation.

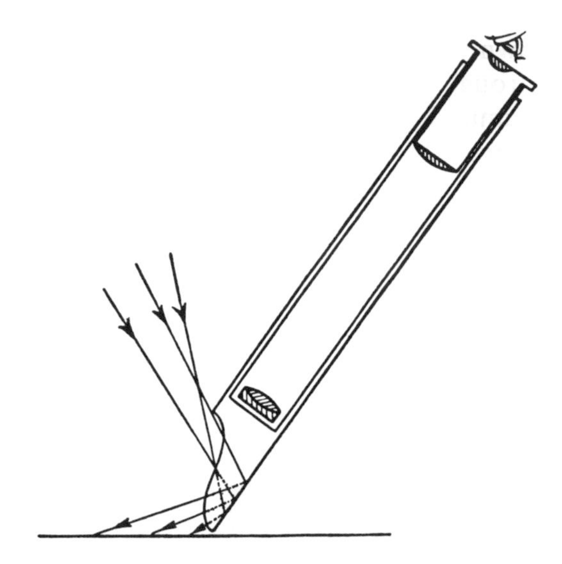

Fig. 1-2. Toolmaker's microscope.

2.1.1. Toolmaker's Microscope

The simplest compound microscope is undoubtedly the toolmaker's microscope. Shaped like a fountain pen and about the same size, it usually magnifies 60–100 ×. It is normally hand-held over any surface to be examined (Fig. 1-2). It is useful for casual examination of any surface at low power.

2.1.2. Stereobinocular Microscope

The stereobinocular microscope is an arrangement of two separate compound microscopes, one for each eye, looking at the same area of an object. Because each eye views the object from a different angle, separated by about 14°, a stereoimage is obtained. The physical difficulty of orienting two high-power objectives close enough together for both to observe the same object limits the NA to about 0.15 and the magnification to about 200 ×.

The erect image is, however, an advantage, and solution of most surface problems starts with the "stereo." There is ample working distance between the objective and preparation, and the illumination is flexible. Many stereos permit transmitted illumination and some permit brightfield top lighting. At worst, one can shine a light down one bodytube and observe the bright-light image with the second bodytube. This is more satisfactorily accomplished with a trinocular head which utilizes a beam splitter rather than beam switcher. In other words, the lamp can be shown down the third, or camera, bodytube with simultaneous vision through the binocular bodytubes.

The usual illumination for a stereo is furnished by one or two spotlights directed downward onto the area to be examined. The angle between the beam and the surface is very important; generally a small angle is best for contrast and relief (Fig. 1-3). Low relief and low contrast are obtained by arranging that the light strike the surface from as many angles as possible. Most stereo manufacturers furnish such illuminators. An equivalent effect can be obtained by surrounding the object with a paper cylinder (2 inches in diameter and 2 inches high is convenient) and aiming two lamps from opposite sides onto the cylinder. Very diffuse illumination is thereby furnished to the object.

a

b

Fig. 1-3. Effect of lighting angle on contrast and relief: (a) grazing angle, (b) large angle.

TABLE 1-1. Working Distance Resolution and Depth of Field as a Function of Objective Numerical Aperture

Microscope	NA	Resolution, μm	Working distance, mm	Depth of field, μm
Stereobinocular	0.14	2.2	50–90	15
Mono-objective	0.25	1.22	7	8
	0.50	0.61	1.3	2
	0.65	0.47	0.7	1
	0.85	0.36	0.5	1

2.1.3. Mono-Objective Microscope

2.1.3.1. Unilateral Oblique Illumination

The resolution of a stereobinocular microscope is only 2 μm, twenty times larger than the limit for a mono-objective microscope. Unfortunately, increased resolution is paid for by a smaller working distance and a smaller depth of field (Table 1-1). It becomes more difficult, as a result, to reflect light from a surface, using side spotlights, as the objective NA increases. In any case, the angle between the light rays and the surface must decrease rapidly as the NA increases and the working distance decreases. The surface should be uncovered, i.e., no cover slip, and consequently all objectives having NA >0.25 should be corrected for uncovered preparations (metallographic objectives).

a. Spotlights. Side spotlights work very well for 16-mm 10 × objectives (Fig. 1-4) and surprisingly well for higher-power objectives, including 4-mm 45 × objectives. Apparently enough light is reflected back and forth between the bottom lens of the objective and the preparation so that a portion reaches the area in the field of view. This latter kind of illumination has to be recognized when using high-power transmitted light to avoid misinterpretation of structures. Thin metal flakes often look transparent because of unsuspected light reflected from the top surface. This sometimes occurs with no intentional top light at all if a nearly full condenser NA allows light to hit the front objective surface and be reflected back into the preparation.

The spot illuminators must be capable of concentrating all of their light onto a small spot a few millimeters in diameter (Fig. 1-5a) in order to reflect sufficient light from the specimen for visual examination. The amount of reflected light is usually measured through the microscope for photomicrography by placing a small piece of white bond paper in the plane of the preparation. An exposure sufficient to render the paper

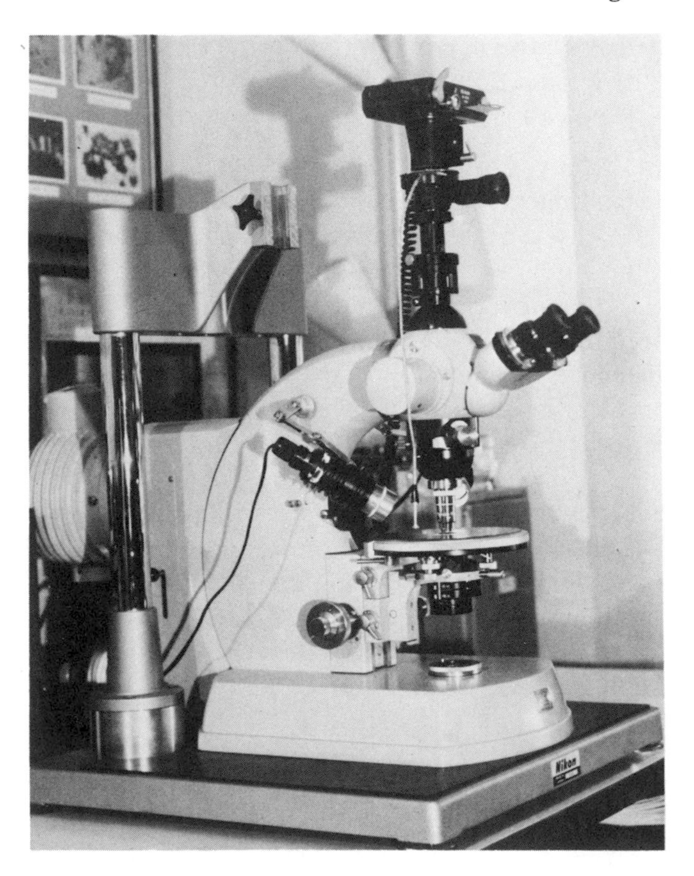

Fig. 1-4. Zeiss Universal microscope with two opposing spotlights for top lighting. These two lights are operated by a foot switch.

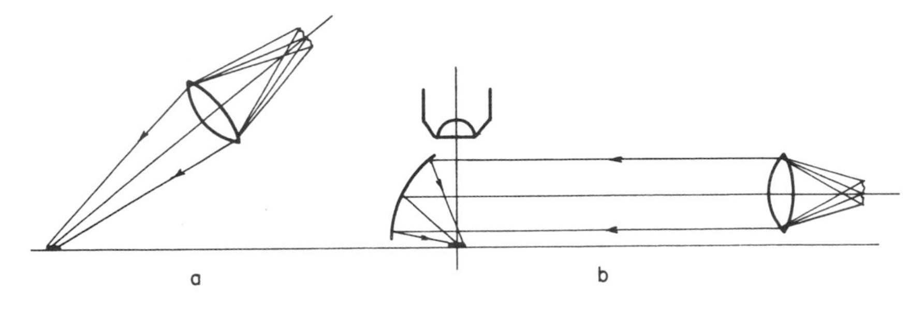

Fig. 1-5. Unilateral oblique top lighting by (a) a spot light and (b) a reflector.

just saturated as white will render properly the various colors and shades of gray in any preparation replacing the paper.

b. Reflectors. A variation on the use of spotlights when the lamp itself gets in the way is to move it (them) to one side and direct the light beam to a small mirror (or mirrors) near the objective. Small mirrors like those used by a dentist are excellent for this purpose. If concave (Fig. 1-5b), the mirror can help concentrate the light.

One of the major variables when using unilateral illumination is the angle between the illuminating beam and the surface. Figure 1-3 shows what a difference this angle can make in the appearance of a surface.

The problem of directing light onto the specimen by spotlight or reflector because the working distance is too small may be eased by using reflection objectives. Both 8-mm, 20× and 4-mm, 40–45× reflecting objectives are available with working distances as great or greater than that of the refracting 16-mm, 10× objective, i.e., 5–6 mm. Very adequate darkfield top lighting can be achieved with reflecting 20–45× objectives using either spotlights or reflectors.

2.1.3.2. Annular (Ring) Illumination

Another way to "flatten" a surface is to use several spot illuminators (or reflectors) or fully annular illumination. There are various ways of accomplishing this.

a. Light-Tent Illuminator. A simple way is through use of a ping-pong ball as proposed by Albertson.[1] He cut a ping-pong ball in half and then cut a small circular hole in the center of one hemisphere. This "light-tent" was placed over the preparation equator-side down and the objective was centered through the north pole opening. At least two spotlights were then focused on the outside of the hemisphere. A soft diffuse illumination thereby pervaded the specimen area, lighting highlights and generally flattening the object (Fig. 1-6).

b. Lieberkuhn Illuminator. A simple circular concave mirror fitting around the objective has also been used for annular top light. Introduced many years ago by Lieberkuhn, it is still a useful device (Fig. 1-7). To use it, the substage condenser must be removed so that parallel light coming up around the specimen is reflected by the mirror back onto the specimen.

c. Silverman Ring Illuminator. Still available from E. Leitz, it is a circular array of small light sources, again fitting around the objective so that light is directed downward from all sides onto the surface.

Fig. 1-6. Effect of diffuseness of illumination (photomicrograph taken with a light-tent; compare with Fig. 1-3).

d. Fluorescent Ring Illuminator. Another variation on this theme is a ring fashioned from fluorescent tubing of the size used in advertising signs. Aristo Grid Lamp Products, Inc.* makes such a lamp. Any neon-sign maker can also produce one of these with an appropriate transformer and a green-fluorescing gas. The tubing should be a 1–1.5 inch (i.d.) circle.

e. Ultropak. E. Leitz produces vertical illuminators for darkfield illumination (Fig. 1-8) under the name Ultropak. They are available in all magnifications from 3.8 × to 75 × ; the latter is an oil-immersion objective. All Ultropak objectives utilize an annular mirror above the objective surrounding the objective optical path, and at a 45° angle, to reflect light from an external illuminator down around the objective to the specimen. The low-power Ultropaks have an annular lens around

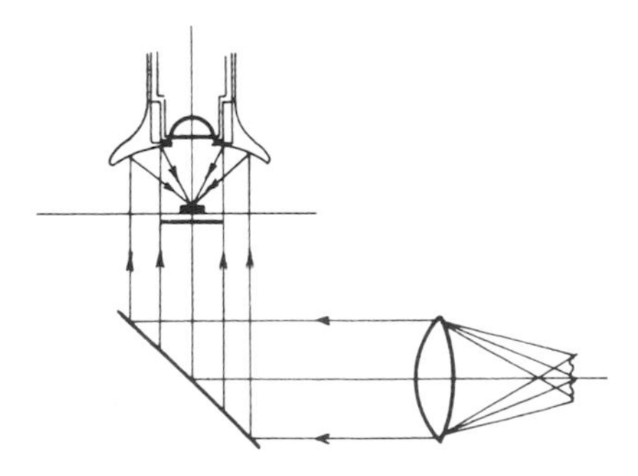

Fig. 1-7. Lieberkuhn top light illuminator.

* 65 Harbor Road, Port Washington, New York 11050.

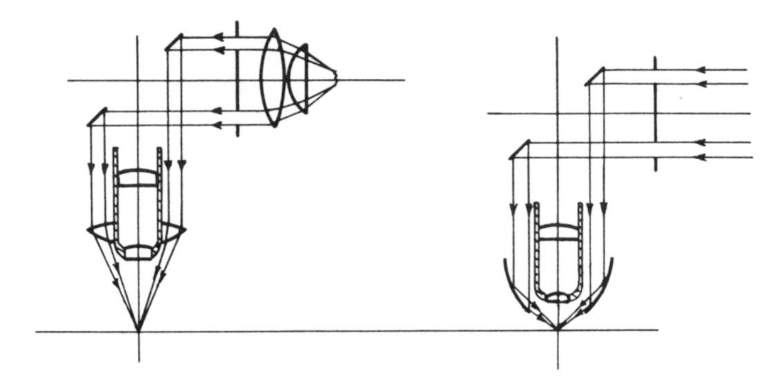

Fig. 1-8. Leitz Ultropak objectives.

the objective to focus the annular light rays onto the specimen plane. The higher-power Ultropaks use an annular mirror instead of a lens to focus the incident light.

2.2. Bright-Field Illumination

2.2.1. Epi-Condenser

A convenient transition from darkfield to brightfield illumination is the Epi-system. Used by several manufacturers, it is a vertical illuminator with a double-mirror system (Fig. 1-9). In one configuration, the Epi-Condenser is like the Ultropak, a darkfield vertical illuminator (Fig. 1-9a). However, by sliding a half-silvered mirror into position in the objective light path some light is reflected down through the objective onto the specimen. Some light reflected from the specimen under either of these conditions (central or annular mirror) re-enters the objective and some of this proceeds through the imaging process to the eye (Fig. 1-9b).

The annular mirror is, of course, a darkfield system; scratches on a polished metal surface, for example, appear white on a dark field. The central mirror, on the other hand, is a brightfield system, and scratches on a polished metal appear dark on a bright field.

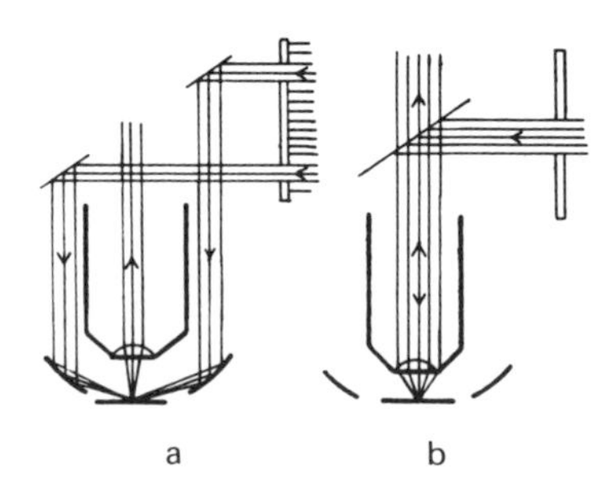

Fig. 1-9. Epi-condenser system: (a) darkfield arrangement, (b) brightfield arrangement.

There are other vertical illuminators capable only of brightfield illumination. These are simply objectives with a thin glass plate at 45° just above the top lens of the objective and with a side tube for an illuminating beam.

2.2.2. Half-Reflecting Mirror beneath the Objective

A crude but effective brightfield vertical illumination effect for, at least, low-power objectives can be achieved by supporting a small fragment of clean cover slip between the specimen and the objective. It must be small enough to fit entirely within the working distance of the objective and large enough to cover the entire light path (Fig. 1-10). The cover-slip fragment is held in a small flattened bead of modeling clay either on the specimen or, better, on the objective itself. A 10 × 16-mm objective is the highest power lens that can be used for this purpose without serious deterioration of the image because the cover-slip thickness when traversed at a 45° angle is far too thick (0.24 mm for a 0.17-mm cover slip) for good imaging. The 16-mm objective must, of course, be corrected for use with a cover slip. Metallographic objectives are corrected for zero cover-slip thickness.

2.2.3. Stereobinocular Microscope

Another simple way to obtain brightfield vertical illumination is to deliver light to the specimen through one bodytube of a stereobinocular microscope (see Section 2.1.2).

Another interesting system of top lighting utilizes the 2 × auxiliary magnifier furnished as an accessory for the Wild M-5 stereobinocular microscope. This lens has a concave front face (facing the specimen). It therefore acts like a Lieberkuhn reflector (Section 2.1.3.2.*b*) and light

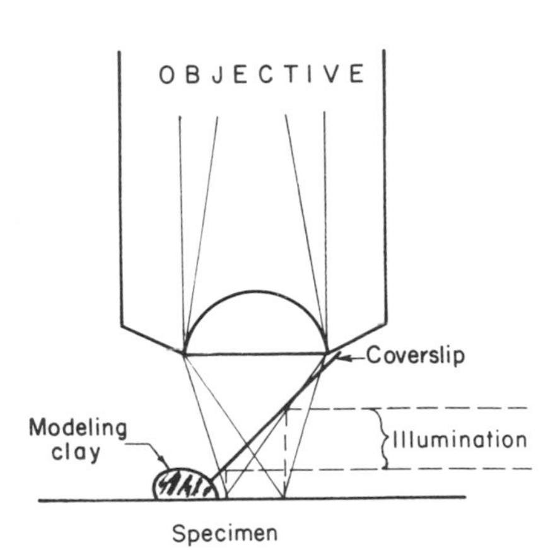

Fig. 1-10. Use of a cover-slip fragment to simulate a brightfield vertical illuminator.

transmitted from the substage mirror is reflected from the large-diameter front lens back onto the specimen. This gives adequate light for brightfield surface study.

2.2.4. Phase Contrast

When surface detail is not readily visible because contrast is low, phase contrast is a useful means of enhancing contrast. Phase contrast enhances optical path differences and, since surface detail generally involves differences in optical path (differences in height), these differences can be made more apparent to the eye by phase contrast.

2.2.5. Interference Microscopy

Reflected-light phase contrast is not used nearly as much as it was before the introduction of reflected-light Nomarski interference contrast (NIC). The latter also enhances optical-path differences but without the halos that mar the phase-contrast images. Figure 1-11 shows a metal surface by NIC. Although limited in resolving power compared to that of the light microscope, on which it is based, NIC gives fully as much information (and as pretty micrographs) as the scanning electron microscope.

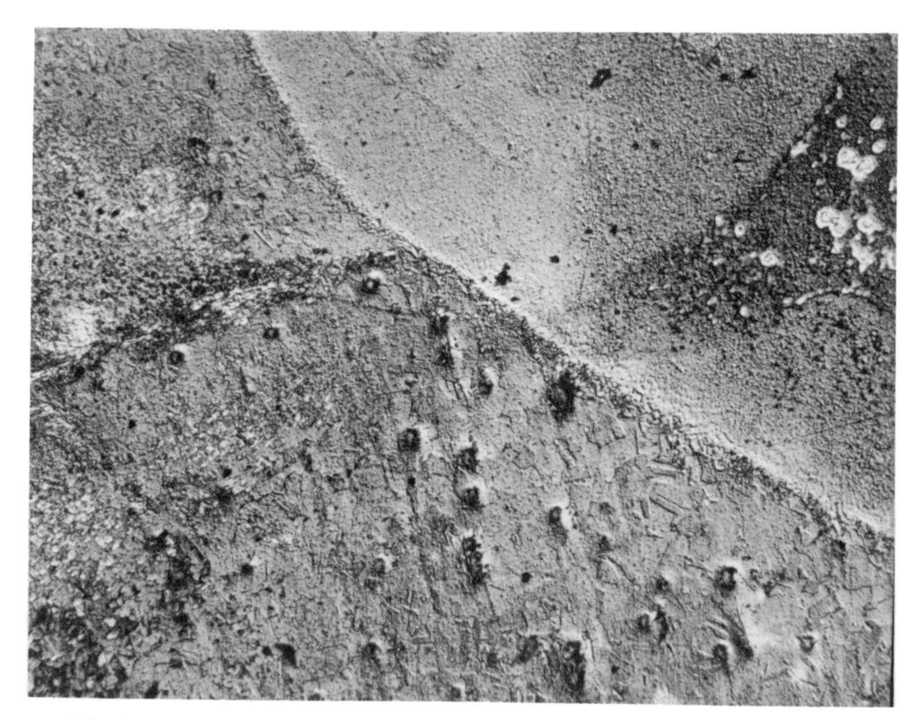

Fig. 1-11. A metal surface taken with reflected Nomarski interference contrast.

Fig. 1-12. The McArthur microscope.

2.2.6. McArthur Microscope

There are a few specialized microscopes for surface study. One of the most interesting of these is the McArthur microscope (Fig. 1-12). It is fully portable and hand-held. It rests on the surface to be examined and gives a standard light-microscope brightfield image with normal light-microscope resolution. It can, of course, be used on surfaces not observable with conventional light microscopes. It can also be converted quickly to a transmitted light microscope, and phase, fluorescence, and polarized-light accessories are easily added. Photomicrographic and video cameras are easily attached, and the results are indistinguishable from those obtained from standard size, weight, and shape microscopes.

2.2.7. Metallograph

The metallograph is a one-purpose instrument designed to give the best possible visual and photomicrographic image of an opaque surface. It is, nearly always, an inverted microscope system allowing easy access to a large unencumbered stage. All optics are below this stage and corrected for uncovered preparations. Generally, either brightfield or darkfield illumination is possible. Rigid construction helps to minimize vibration and metallographers find it almost indispensable for the routine examination of polished metal surfaces. It can, of course, be used to examine surfaces of any materials.

There seems to be a tendency to abandon the metallograph, as such, for the "universal" microscope with its greater flexibility for the addition of accessories such as Nomarski interference contrast.

It is an advantage to be able to have a single microscope–photomicrographic camera system with highest quality accessories for all types of surface study and for photomicrography at any stage. All major microscope manufacturers produce such equipment. The price approaches that of a low-priced SEM.

2.3. Special Methods

This section will discuss specialized instruments for surface study and methods of surface treatment used to enhance contrast and detail, or make microscopical examination easier.

2.3.1. Hole-Examining Microscope

There is an increasing demand, especially in the microelectronics industry, for borehole inspection microscopes. The problem is to see the whole bore at one time with good image quality. The problem has been solved by Schindl[2] with a spherical reflecting surface below the borehole which produces an intermediate image of the borehole. A conventional microscope enlarges this image. A special combination of incident and transmitted light illumination makes it possible to see not only the borehole itself, but also the upper and lower surfaces of the plate. Figure 1-13 shows the instrument itself and Fig. 1-14 shows the layout of the instrument.

2.3.2. Tolansky's High-Resolution Profile Microscope

Tolansky[3] first proposed projection of a straight line, e.g., a fine wire, obliquely onto an irregular surface. The image of the line delineates the surface irregularities; it can be observed visually or photographed over any part of the specimen. His system used an oil-immersion objective with brightfield vertical illuminator. A mirror covering a

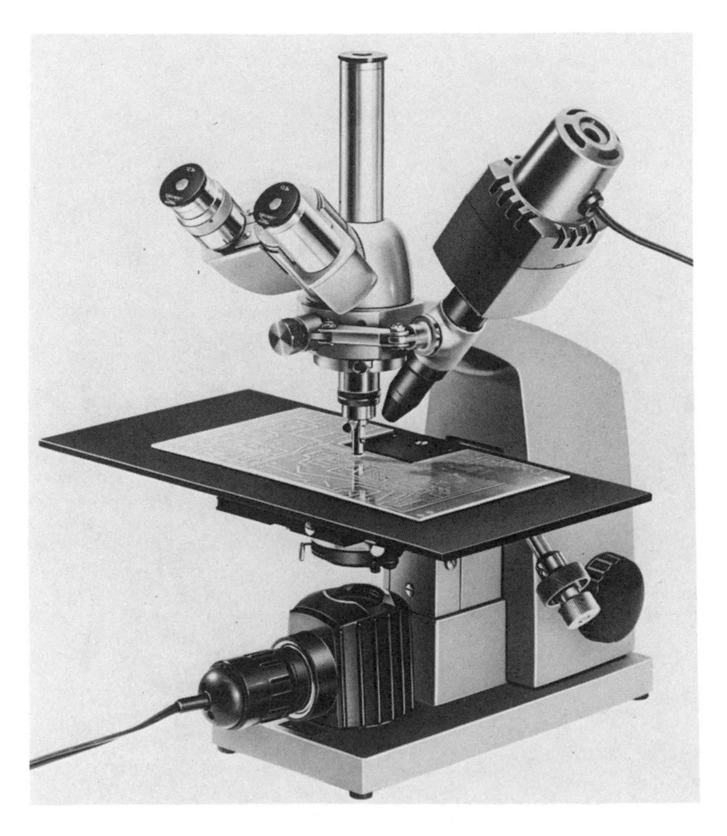

Fig. 1-13. The Reichert borehole-inspection microscope. (Photograph courtesy of C. Reichert, Vienna, Austria.)

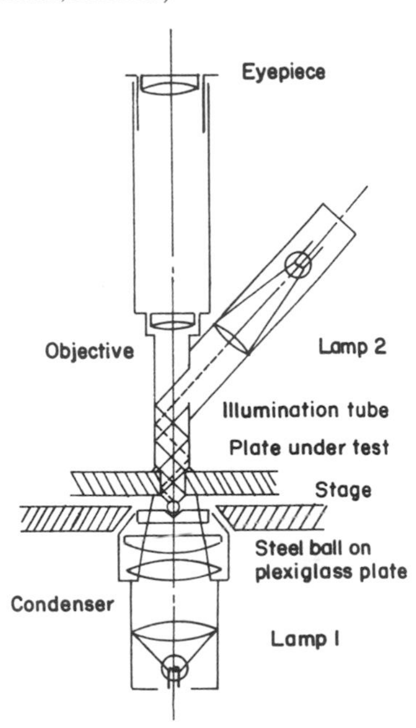

Fig. 1-14. A schematic drawing of the Reichert borehole-inspection microscope.

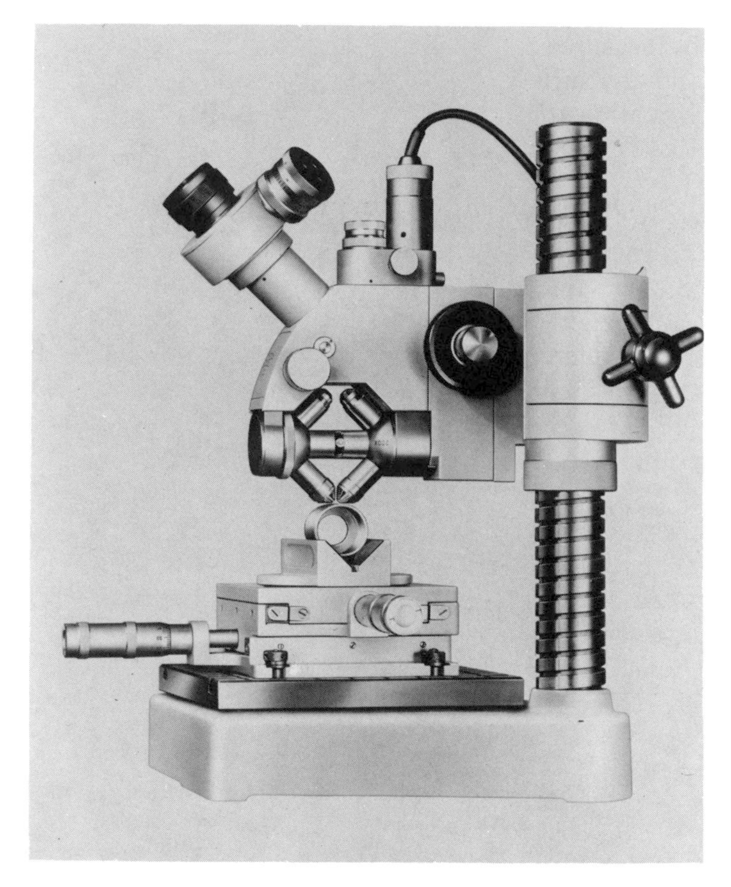

Fig. 1-15. C. Zeiss "Light Section Microscope." (Photograph courtesy of Carl Zeiss, Inc., New York.)

small sector of the light path reflects an image of a fine wire (placed in the field diaphragm plane) obliquely onto the specimen.

Carl Zeiss manufactures a somewhat different version useful at lower magnification (Fig. 1-15). Their Light Section Microscope uses two separate objectives 90° apart, one to project the image of a fine wire onto the specimen and the other to observe the projected image of the wire on the specimen. A revolving nosepiece with magnifications of 200 × and 400 × permits measurement of roughness from 1–400 μm. An eyepiece micrometer with a measuring drum scale gives direct readings of roughness in micrometers.

2.3.3. Interferometry

The most sensitive measure of surface topography is undoubtedly interferometry. Differences in height as small as 15 Å (1.5 μm) can be accurately measured by this means. It can be used on any reflecting surface, and any surface can be made more reflecting by evaporating a thin metal coating at normal incidence over that surface. These methods are fully discussed in Chapter 2.

2.3.4. Stereophotomicrography

Surface detail can be registered photographically by taking two pictures of the same field of view at slight differences in angle. The differences in height will appear normally at angle differences of about 14°. The same differences can be exaggerated by using larger angles, or minimized with smaller angles, between the viewing beams.

Stereophotomicrographs can be taken directly using a stereobinocular microscope using a stereocamera such as the Stereo-Realist.® They can also be taken with a single-lens (preferably reflex) 35-mm camera by taking two shots, one down each tube. Use of the stereobinocular microscope limits the resolution, of course, to about 2 μm and it can be used only for coarse detail at low magnification.

For higher-magnification stereophotomicrography, a monoobjective microscope is used in conjunction with a tilting stage[4] (Fig. 1-16). A reversal of the stereoscopic effect can occur if the two

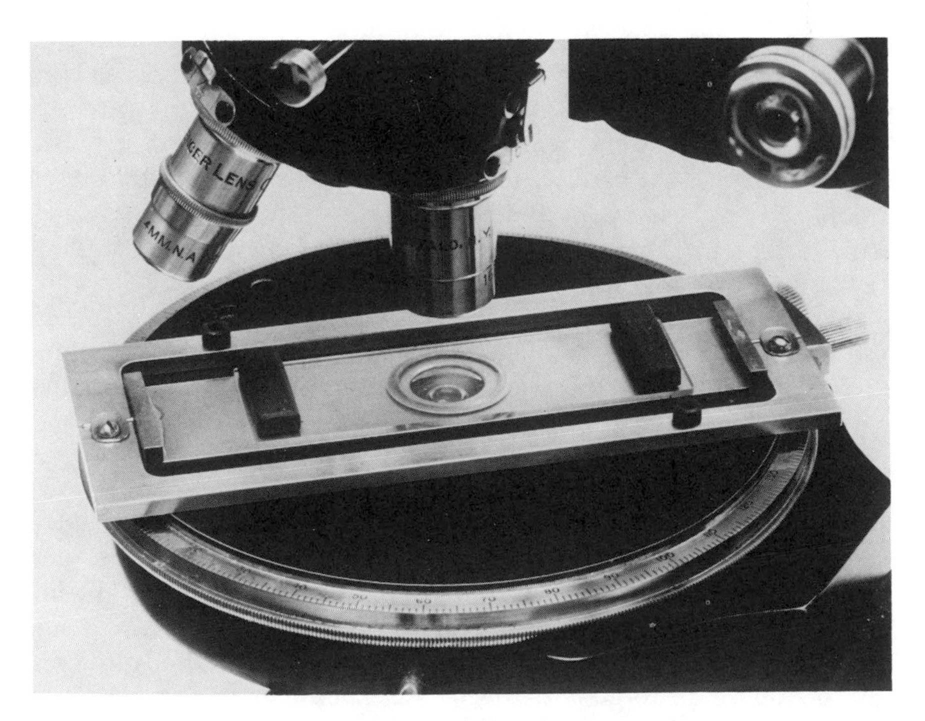

Fig. 1-16. Tilting stage for stereophotomicrography.

pictures are reversed left for right. It is often a problem to be certain that the observed stereo effect is the correct one; therefore, we sometimes place an object of known elevation in the field of view. For example, a line might be scribed across a corner of the field or a tiny droplet of oil might be placed within the field of view. If the stereopair is then registered so that a known bump is observed as a bump rather than a depression, all other detail in the field of view will be properly represented.

One of the most troublesome problems in surface study by microscopy, light or electron, is to be sure that bumps appear as bumps and holes as holes. Anyone who has seen photographs of lunar landscapes knows the craters often look turned inside out and that the proper appearance can be restored by inverting the picture. The same inversion occurs with photomicrographs (Fig. 1-17) and can be controlled by making certain the picture is mounted the right way up.

2.3.5. Metal Shadowing

We pass now to sample treatment procedures used to enhance contrast. There is one kind of surface difficult to study and virtually impossible to photograph by light microscopy. This is the surface of any transparent, colorless (or nearly colorless) multicomponent substance, e.g., paper, particle-filled polymers, porcelains, and some ceramics. So much light penetrates the surface only to be refracted and reflected back to the observer that the surface itself is lost in glare.

This problem is very easily solved, however, by evaporating a thin film of metal onto the surface. The metal (usually aluminum, chromium, or gold) may be evaporated under vacuum in straight lines at any angle to the surface, from grazing to normal incidence. An angle of about 30° is often used; under these conditions the heights of surface elevations can be calculated from shadow lengths. Normal incidence is usually used for photomicrography of, say, a paper surface or when the surface is to be studied by another viewing procedure such as interferometry or Tolansky's surface-profile microscope.[3]

2.3.6. Replication

Transparent film replicas of opaque surfaces can be studied by transmission light microscopy. This leads to the possibility of using transmission phase contrast or interferometry and the best possible optics, e.g., oil-immersion planapochromats to study surface detail. In addition to these obvious advantages, replication is almost the only way to study hard-to-get-at surfaces, e.g., the inner ball-race of a ball bearing. Finally, replication not only gives a faithful rendition of the

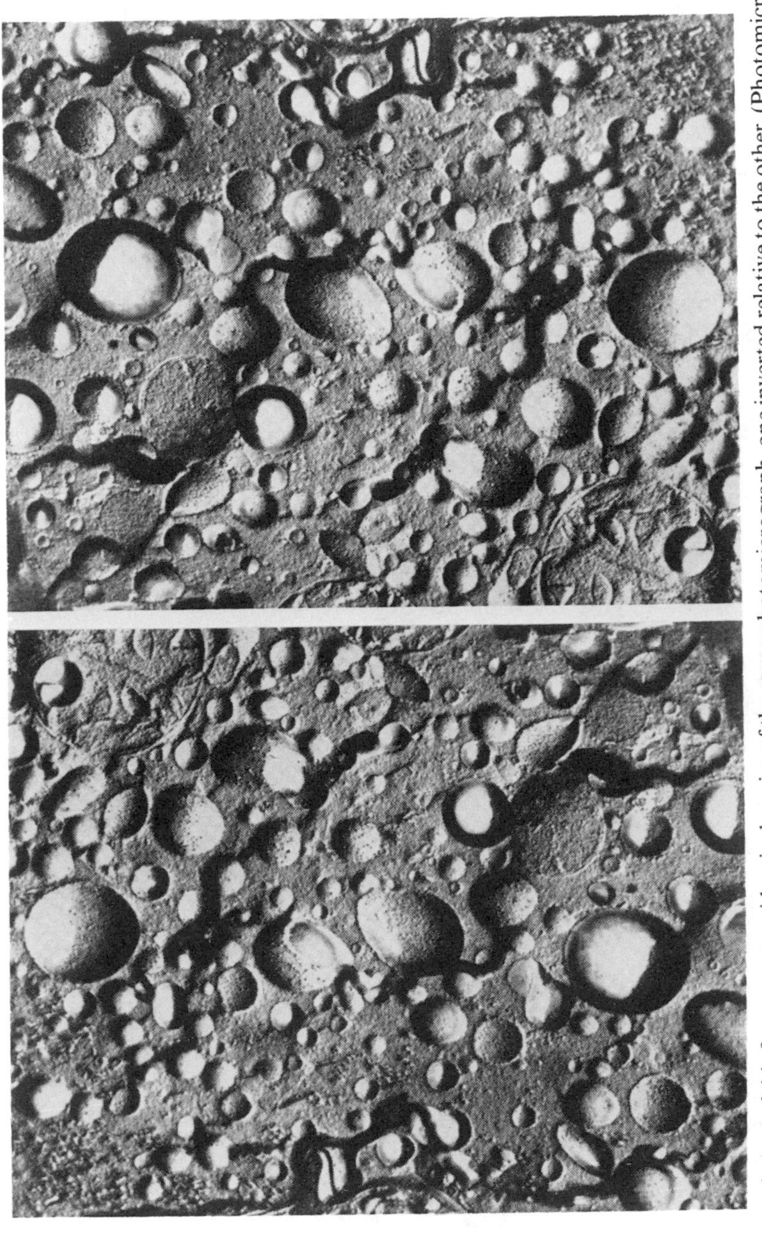

Fig. 1-17. The two halves of this figure are two identical copies of the same photomicrograph, one inverted relative to the other. (Photomicrograph courtesy of Praktische Metallographie.)

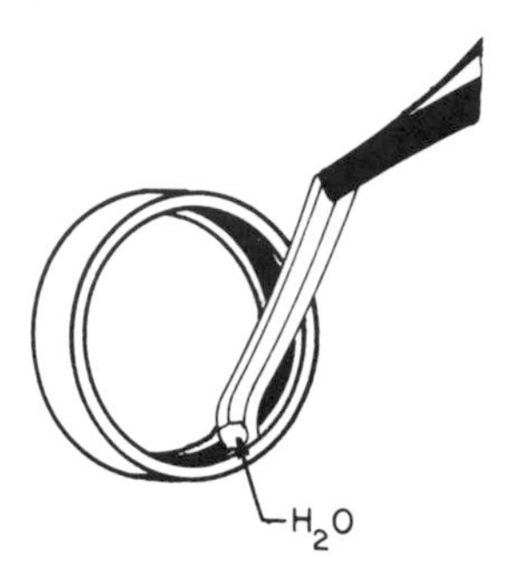

Fig. 1-18. Steps in replication of hard-to-get-at surfaces (here an inner ball-race of a ball bearing) using collodion.

surface detail but often picks off from the surface individual particles or coatings that may be related to the problem under investigation, e.g., scoring of the ball-race. The position of the particles relative to the surface geometry is also preserved by replication.

The best procedure for replication (if one objective is to remove surface debris with the replica) is coating with collodion solution followed by stripping of the dried replica (Fig. 1-18). The surface debris will become imbedded in the collodion, held securely in place on drying, and removed on stripping off the film. The replica can then be flattened on a microscope slide so that an entire curved surface can be viewed microscopically in one focal plane (Fig. 1-19).

2.3.7. Cross Sections

A direct way of examining a surface profile is to make a cross section and turn the surface up on edge for microscopical study. This

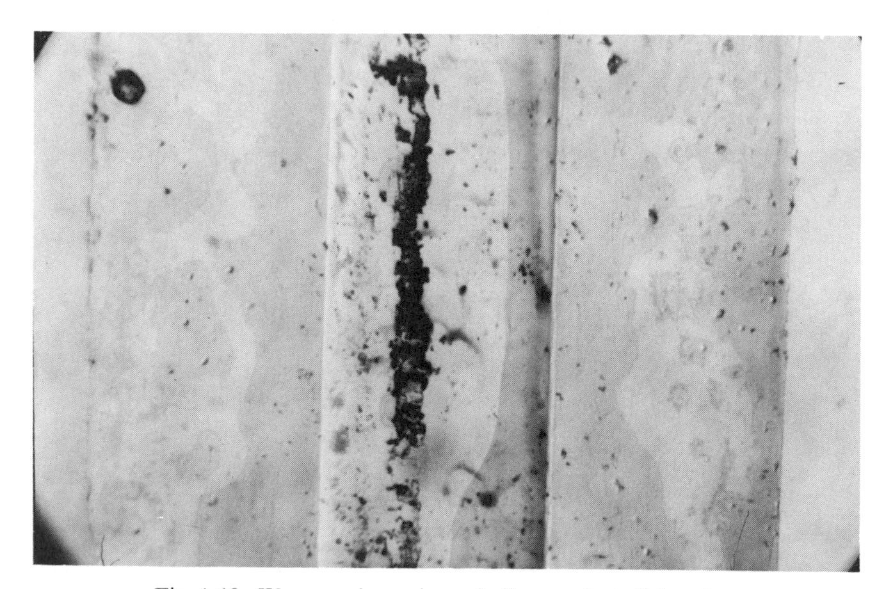

Fig. 1-19. Wear marks on inner ball-race of a ball-bearing.

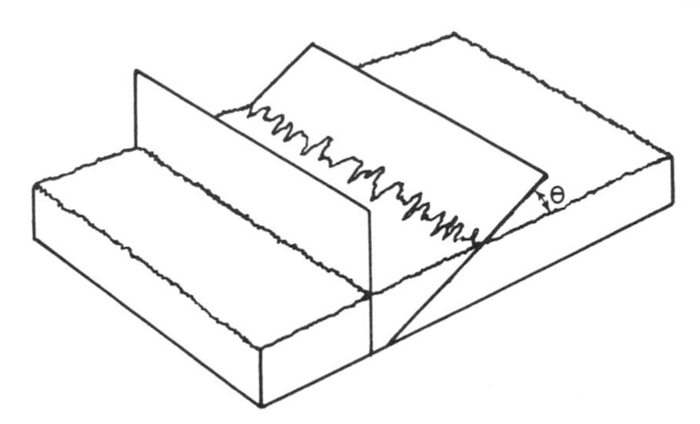

Fig. 1-20. Cross sections of a surface: (a) normal incidence and (b) a taper cross section at an angle θ to the surface.

usually involves mounting the piece in a metallographic mount, then grinding and polishing down to the desired section. If the surface is soft, e.g., copper, silver, polymers, etc., it can be protected during polishing by electroplating or evaporating a thin coat of nickel metal over the surface before mounting and sectioning.

An interesting variation of this sectioning procedure is to make the section at an angle other than normal to the surface. This has the effect of magnifying the heights of elevations; an angle of 5.7° from the surface gives a 10 × magnification in the vertical direction (Fig. 1-20). Moore[5] terms this a "taper-section." It is not, of course, restricted to cross sections of metals.

3. CHEMICAL COMPOSITION AND SOLID-STATE STRUCTURE

3.1. Morphological Analysis

Characterization of a surface includes not only topography but chemical composition and solid-state structure. An experienced microscopist can identify many microscopic objects in the same way all of us identify macroscopic objects, that is, by shape, size, surface detail, color, luster, etc. Unfortunately, it is usually as difficult to describe a surface in descriptive terms as it is to describe, say, another human being so that someone else would recognize that surface or that human being. Descriptive terms we have found useful for surfaces, if not usually for humans, include: angular, cemented, cracked, cratered, dimpled, drusy, laminar, orange-peel, pitted, porous, reticulated, smooth, striated, and valleyed.

A microscopist should recognize a paper as rag, mechanical coniferous, chemical nonconiferous, or as combinations of these by studying the paper surface. He should recognize surface features on small particles such as scales on wool, crossover marks on silk, striations on viscose rayon, twin bands on calcite, melt and crystal patterns on micrometeorites, lamellar steps on mica, fibrous structure of incinerated wood sawdust, etc. The nature of the surface in these, and for many other substances, helps to identify that substance.

3.2. Reflectance Measurements

Measurements of reflectance on polished surfaces can be used to calculate the refractive indices of transparent substances and to give specific reflectance data for opaque substances. The methods are discussed in detail by Cameron.(6) Reflectance and microhardness data are tabulated by Bowie and Taylor(7) in a system for mineral identification.

3.3. Surface Treatment Followed by Morphological Analysis

3.3.1. Polished and Etched Sections

The most sophisticated approach to morphological identification of substances is probably the use of metallography to identify metals. Polished and etched metal sections show details of structure a metallographer can use to identify the metal or alloy and to detail the heat treatment or other metallurgical processing that metal has undergone. Similar procedures of polishing and etching can be applied to other substances, e.g., explosives.(8) One can tell quickly whether the explosive was cast, pressed, or extruded, whether it is a single component or mixture, and, with training, what the components are. Pre-etching with dilute acetone helps to bring out the structure. The solid-state structure of metals, e.g., α or β brass, or explosives, e.g., α or β HMX, can be recognized on etched surfaces.

3.3.2. Staining Surfaces

Staining a surface, either chemically or optically, helps to differentiate different parts of a composite surface and to identify the various phases.

3.3.2.1. Chemical Stains

A variety of stains are available for diverse surfaces. Mineral sections can be etched with hydrofluoric acid then stained with $Na_3Co(NO_2)_6$ to differentiate quartz (unetched), feldspars (etched but unstained), and potassium feldspars (etched and stained yellow). Paper surfaces can be stained with Herzberg stain(9) to differentiate lignified cellulose [e.g., straw, manila, and mechanical wood pulp (yellow or yellow-green)] from purified cellulose [e.g., chemical wood pulp, bleached straw, or manila (blue or blue-violet)] or pure cellulose [e.g., cotton or linen (wine red)]. J. Isings[(10)] uses fluorescent dyes for selective staining of polymers. K. Kato[(11)] selectively stains unsaturated elastomers with osmium tetroxide.

3.3.2.2. Optical Staining—Parlodion Films

There are a variety of ways of building up thin films on surfaces. The thin film will then show interference colors if the right thickness is attained. Furthermore, the thickness may vary from grain to grain or composition to composition and provide a color guide for phase identification. One of the simplest procedures was developed by Staub and McCall.[(12)] They applied a drop or two of 0.75% Parlodion in amyl acetate over the polished surface and allowed the specimen to drain and dry in a vertical position. This method gives different thicknesses of film over the different phases thus leading to differentiation by differences in interference colors. The procedure works best for metals and opaque ores.

3.3.2.3. Optical Staining—Anodizing

Any metal that can be anodized will generally develop oxide coatings over each grain of a thickness related to composition and orientation of that grain. The transparent oxide films then show interference colors differentiating one grain from another. A colorful paper by M. L. Picklesimer[(13)] illustrates this technique for alloys of zirconium and niobium. His paper refers to other older ways of developing thin-film colors on metals both electrolytically and by heating.

3.3.2.4. Optical Staining—Evaporated Films

Allmand and Houseman,[(14)] among others, have vacuum-evaporated thin films of highly refractive transparent substances, e.g., TiO_2 or ZnS, onto a polished, etched metal surface. The film builds up to different thicknesses over different solid phases and each is then differentiated by the interference colors. Again, a colorful paper on this technique has appeared in *Microscope*.[(14)]

Techniques of this kind are often necessary to develop sufficient contrast in a specimen so that automatic image analysis can be used to make quantitative measurements of composition and grain size.

3.3.2.5. Decoration of Structures

The above optical and chemical staining procedures might well be called decoration of structures, but a final unique decoration technique delineates magnetic domains. To do this, McKeehan and Elmore[15,16] place over the surface a drop of a dilute suspension of submicrometer particles of magnetic iron oxide, Fe_3O_4. The oxide particles migrate to the surface and form a pattern accurately mapping the domain structure.

4. Conclusion

This chapter should show that the light microscope is the starting point for the study of surfaces. Although one of the least expensive research tools, the light microscope furnishes a direct approach to study of surface topography. Many questions of composition and such questions as number density of magnetic domains are most easily answered by light microscopy.

5. ACKNOWLEDGMENT

The author would like to acknowledge with grateful thanks the literature survey made by Jaqueline Smid of Walter C. McCrone Associates, Inc.

6. REFERENCES

1. C. E. Albertson, A light-tent for photomicrography, *Microscope* **14**, 253–256 (1964).
2. K. P. Schindl, A new bore-hole inspection microscope, *Microscope* **20**, 51–56 (1972).
3. S. Tolansky, A high resolution surface profile microscope, *Nature* **169**, 445–446 (1952).
4. W. C. McCrone, Stereophotomicrography using the tilting stage, *Microscope* **14**, 429–439 (1965).
5. A. J. W. Moore, A refined metallographic technique for the examination of surface contours and surface structure of metals—Taper sections, *Metallurgia* **38**, 71 (1948).
6. E. N. Cameron, *Ore Microscopy*, John Wiley & Sons, New York (1961).
7. S. H. U. Bowie and K. Taylor, A system of ore mineral identification, *Min. Mag.* **99**, 265–277, 337–345 (1958).
8. F. E. McGrath, A metallographic preparation procedure for RDX dispersed in TNT, *Metallography* **1**, 341–347 (1969).
9. J. B. Calkin, Microscopical examination of paper, *Paper Trade J.* **99**, 267 (1934).
10. J. Isings, in *Encyclopedia of Microscopy* (G. L. Clark, ed.) p. 390, Reinhold, New York (1961).
11. K. Kato, The osmium tetroxide procedure for light and electron microscopy of ABS plastics, *Polymer Eng. Sci.* **7**, 38–39 (1967).
12. R. E. Staub and J. L. McCall, Increasing the microscopic contrast of phases with similar reflectivities, *Metallography* **1**, 153–155 (1968).
13. M. L. Picklesimer, Anodizing for controlled microstructural contrast by color, *Microscope* **15**, 472–479 (1967).
14. T. R. Allmand and D. H. Houseman, Thin film interference—A new method for identification of non-metallic inclusions, *Microscope* **18,** 11–23 (1970).
15. L. W. McKeehan and W. C. Elmore, Surface magnetization in ferromagnetic crystals, *Phys. Rev.* **46**, 226–228 (1934).
16. W. C. Elmore, Ferromagnetic colloid for studying magnetic structures, *Phys. Rev.* **54**, 309–310 (1938).

2

MULTIPLE-BEAM INTERFEROMETRY

A. C. Hall

Dallas, Texas

1. INTRODUCTION

Several of the techniques that now are well established in the science and technology of surface studies are due essentially to the efforts of a single school. So it is with multiple-beam interferometry, whose application to the study of microtopography stems from the work of Tolansky and his associates.[1] The method is unique in its sensitivity and simplicity, providing as it does enormous vertical magnification by the use of only the simplest standard optical components. But multiple-beam interferometry is not yet a widely used tool in surface science, probably because a high degree of experimental skill is needed to fully exploit its potential. This outline is intended as an introduction to the method, which is briefly compared and contrasted with alternative techniques.

2. FORMATION OF FRINGES

The colors of soap bubbles seen in sunlight are very familiar visual phenomena caused by optical interference effects of a kind generally exhibited by transparent films whose thickness is only a few wavelengths. Such thin films may have refractive indices higher than those of the adjoining media, for instance, oil slicks on wet asphalt pavements; or may have lower indices, for instance, the air films trapped between elements of compound lenses. Colored fringes are readily seen in either case, and arise by interference of beams reflected from the upper and lower film boundaries. Because these films are very thin, the resulting differences in path length or, equivalently, angular phase must be treated coherently. The amplitudes of vibrations which differ by an integral number of wavelengths are mutually reinforced, while those which differ by an odd number of half-wavelengths, that is, which are 180° out of phase, are mutually weakened. In any interference pattern, the order of interference p at a given point is defined as

$$p = \delta/2\pi = \Delta L/\lambda_0 \tag{1}$$

where λ_0 is the wavelength, and ΔL and δ are the differences in path length and phase between interfering beams.* It may be mentioned in passing that if the given film is not very thin, colored fringes do not appear. First of all, except at angles of incidence close to zero, rays reflected from the upper and lower film boundaries are sufficiently displaced so that the pupil of the eye does not receive both. Second, as seen from the above definition, the orders of interference are very high. Thousands of wavelengths coincide at a given maximum, precluding chromatic effects. From these rather qualitative considerations, it is clear that fringe visibility is enhanced when (1) the film which causes interference is very thin, and (2) the light source is monochromatic.

Consider the fringe system formed upon illumination of the air gap between glass optical flats by a parallel monochromatic beam at normal incidence. At points of contact between the surfaces there is a minimum of reflected intensity. This is the well-known black spot at the center of Newton's rings, caused by interference between beams internally reflected from the lower surface of the upper plate and those externally reflected from the surface of the lower plate. Although the path difference between these points is minute compared with the wavelength, there is a 180° phase shift at external reflection, causing mutual cancellation of the reflected beams. As the intersurface distance increases to a separation of $\lambda/4$ (one quarter-wavelength), one of the beams traverses this twice, and so lags the other by an additional 180°, giving rise to an intensity maximum. The new minimum occurs where the surfaces are separated by a half-wavelength, and so on. Obviously, the separation between fringes, and therefore the number of fringes visible, depends upon the steepness of the contours or, if both surfaces are flat, upon the angle between them. Such systems are called Fizeau fringes, or fringes or equal thickness. Clearly, since one measures fringe

* For an introduction to the physics of interference and reflection, see the work by Jenkins and White.[2]

displacement relative to the separation of an order, meaningful observations demand the presence of at least two fringes, although too many fringes result in very close spacing so that fringe displacement cannot be measured accurately. Also, if interference fringes are to represent surface contours at all faithfully, they must be narrow—a very small fraction of an order. The familiar chromatic and monochromatic fringes already mentioned cannot satisfy this requirement, whereas multiple-beam fringes can. With transparent media there is, in addition to the fringe system formed in reflection, a complementary transmitted set. The former consists of dark lines on a bright field, and the latter of bright lines on a dark field. For reasons to be mentioned later, transmission fringes frequently are of low visibility. For several reasons, the study of microtopography generally uses only reflection fringes.

2.1. Airy's Treatment

Some of the points touched on above are clarified in an analysis due to Airy.[3] Figure 2-1 shows a collimated light beam of amplitude E_i and wavelength λ, incident at angle i upon a thin layer with refractive index n_1 and thickness d, lying in a medium of unit refractive index. As indicated, at every point on both surfaces incident plane waves divide into reflected and transmitted waves. Thus reflection and transmission at the layer are both represented by amplitude series, each member of which is generated from the preceding one by an internal reflection and double traversal of the layer. This means that every term has a phase lag δ relative to the preceding one:

$$\delta = 2 \cdot \frac{2\pi}{\lambda} n_1 d \cos i \tag{2}$$

The ratios of reflected and transmitted amplitudes to incident amplitude are given by the well-known Fresnel coefficients, (r, t), whose

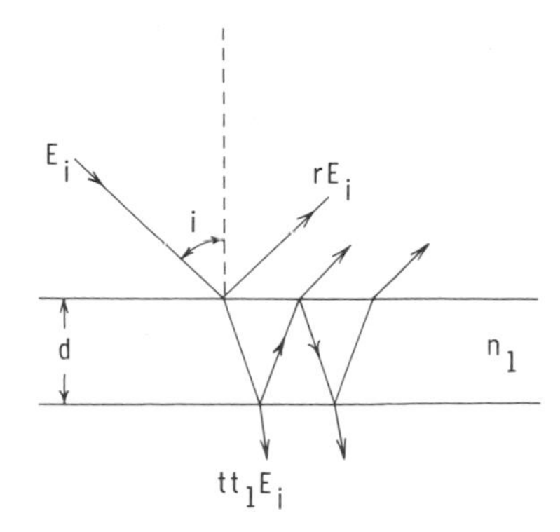

Fig. 2-1. Formation of multiple-beam fringes in reflection and transmission.

magnitudes depend upon n_1 and i. Except for $i = 0$ and $\pi/2$, they also depend upon whether the electric vector of the incident beam is parallel or perpendicular to the plane of incidence.

If (r, t) and (r_1, t_1) are the reflection and transmission coefficients for waves directed, respectively, into and out of the layer, then the reflected series is

$$rE_i,\ tt_1r_1E_i\,e^{i\delta},\ tt_1r_1{}^3E_i\,e^{i2\delta},\ldots,\ tt_1r_1{}^{(2x-3)}E_i\,e^{i(x-1)\delta}$$

and the transmitted series is

$$tt_1E_i,\ tt_1r_1{}^2E_i\,e^{i\delta},\ tt_1r_1{}^4E_i\,e^{i2\delta},\ldots,\ tt_1r_1{}^{2(x-1)}E_i\,e^{i(x-1)\delta}$$

Examining the members of these series, one notes that where the reflection coefficients are much smaller than the transmission coefficients, as in the case of the air/glass boundary, the first two members of the reflected series are of similar magnitude, while the third is smaller by a factor of several hundreds. The latter may be neglected in comparison with the first two, and the observed pattern is said to result from two-beam interference. By contrast, it is clear that in the transmitted series the first beam, which has come straight through, is far less attenuated than the second, which has suffered two internal reflections. Thus, even where the two differ in phase by 180°, the second beam applies only a small perturbation to the intensity of the direct beam. The resulting fringe system is pallid by comparison with that due to reflection, and, as previously noted, sometimes escapes even close examination.

Returning to the series representing the reflected wave, and superposing the first x reflected waves, the resultant is

$$\begin{aligned} E_r(x) &= [r + tt_1r_1\,e^{i\delta}(1 + r_1{}^2\,e^{i\delta} + \cdots + r_1{}^{2(x-2)}\,e^{i(x-2)\delta})]E_i \\ &= \left\{r + \left(\frac{1 - r_1{}^{2(x-1)}\,e^{i(x-1)\delta}}{1 - r_1{}^2\,e^{i\delta}}\right)tt_1r_1\,e^{i\delta}\right\}E_i \end{aligned} \tag{3}$$

In the theory of reflection[2] it is known that $r = -r_1$, so that as $x \to \infty$,

$$E_r = \frac{-r_1[1 - (r_1{}^2 + tt_1)\,e^{i\delta}]E_i}{1 - r_1{}^2\,e^{i\delta}} \tag{4}$$

From other results of reflection theory:

$$tt_1 = T, \qquad r^2 = r_1{}^2 = R, \qquad R + T = 1 \tag{5}$$

where R and T are the reflectance and transmittance; i.e., the ratios of energy reflected and transmitted to incident energy, respectively, all referred to unit surface area. Thus,

$$E_r = \frac{(1 - e^{i\delta})\sqrt{R}}{1 - R\,e^{i\delta}} E_i \tag{6}$$

To obtain intensity, the amplitude is multiplied by its complete conjugate so that

$$I_r = \frac{(2 - 2\cos\delta)RI_i}{1 + R^2 - 2R\cos\delta} = \frac{4R\sin^2\delta/2}{(1 - R)^2 + 4\sin^2(\delta/2)} I_i \tag{7}$$

By treating the transmitted amplitude similarly, its intensity is found to be

$$I_t = \frac{T^2}{(1 - R)^2 + 4R\sin^2(\delta/2)} I_i \tag{8}$$

Defining a parameter $F = 4R/(1 - R)^2$,

$$\frac{I_r}{I_i} = \frac{F\sin^2(\delta/2)}{1 + F\sin^2(\delta/2)} \quad \text{and} \quad \frac{I_t}{I_i} = \frac{1}{1 + F\sin^2(\delta/2)} \tag{9}$$

Figure 2-2 shows I_r/I_i as a function of δ for various values of R. In principle, R can be anything between 0 and 1, which corresponds to

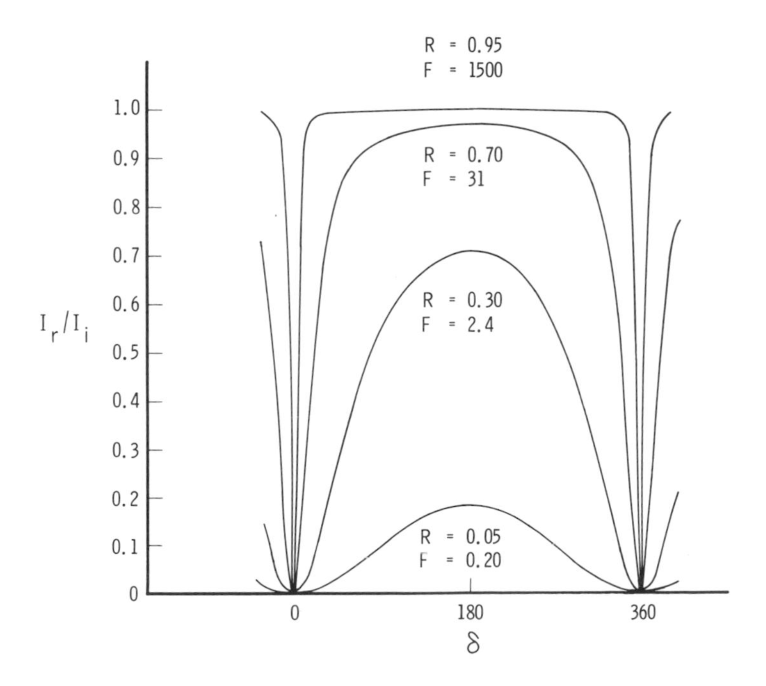

Fig. 2-2. Dependence of multiple-beam bandwidth on surface reflectivity.

$0 < F < \infty$. At a low value of R, say 0.05, characteristic of glass surfaces, it is seen that the reflected intensity is always low, and varies sinusoidally between 0 and its maximum. Such behavior typifies oscillations comprising two harmonic components of similar amplitude and identical frequency, thus corroborating the observation that at low reflectance only two beams need to be considered. Two-beam interference patterns, consisting of equally broad bright and dark fringes, can reveal large-scale surface contours but are of no great importance in microtopography.

As R increases, the ratios of reflected intensity approach asymptotically toward unity everywhere except in the near vicinity of the minima which are zero, thus generating a pattern of narrow dark fringes on a bright field. The complementary transmitted pattern consists of a dark field crossed by narrow bright fringes, and is illustrated by Fig. 2-2 if I_t/I_i replaces I_r/I_i on the ordinate, which is renumbered 1 to 0 in the positive direction.

In comparing fringes, use is made of their relative half-widths, i.e., width at one-half amplitude. For large values of F, it can be shown that the ratio of half-width to order separation is $(2/\pi)\sqrt{F}$, or about $0.63\sqrt{F}$. Thus, the very narrow fringes shown in Fig. 2-2 have a half-width of only 1/60 order. Such very fine fringes obviously satisfy one previously mentioned requirement for application to microtopography.

2.2. Intensity of Fringes

The preceding development is intended to present a clear, intuitively appealing picture without aiming at full analytical rigor. A complete treatment would entail mathematical elaboration and detail quite beyond the stated objective. Instead, further refinements are treated qualitatively so as to convey a notion of the experimental requirements. The theoretical part may be found in the literature cited.[1,4]

In reviewing the physical basis of fringe formation, a significant omission occurred in writing the equation $R + T = 1$, which implies that all incident light is either reflected or transmitted. Strictly speaking, light absorption must generally be allowed for, i.e., $R + T + A = 1$. The effect of any substantial light absorption is to reduce fringe visibility very greatly, though preserving fringe shape, which depends upon R. Incident light that is not reflected is either absorbed or transmitted. Therefore, if absorption occurs, the transmitted beam is correspondingly weakened. But examining the series which represents the reflected and transmitted amplitudes, it is seen that the former consists of a leading term rE_i, and that the remaining terms, apart from a constant factor, are identical with the transmitted series. If the transmitted series is weakened by absorption, all terms in the reflected series except the first are also weakened. The net result is that destructive interference can no longer reduce the intensity to zero and fringe contrast is lost. If absorption becomes large enough, fringes may vanish altogether.

2.3. Nonparallel Reflecting Surfaces

Figure 2-1 was used to illustrate the situation described by the Airy formula, in which the two reflecting surfaces are parallel so that every reflected beam differs in phase from its immediate neighbor in the same degree. But usually in the study of microtopography, the adjacent surfaces are not parallel, and it is required to know whether the phase condition can be approximated closely enough to permit formation of highly sharpened interference fringes. Moreover, since there are multiple internal reflections, it is clear that some of the rays arriving at the given point have traversed a considerable lateral distance so that the resultant local amplitude does not depend only upon the local properties of the film, and the beams are said to be linearly displaced. Fortunately, both effects are minimized when the distance between the test surface and the reference surface does not exceed a few wavelengths. It is in fact an essential experimental requirement that these surfaces be effectively in contact.[5]

At a point where the separation is d for a layer of refractive index n_1 between plane surfaces at an angle α, illumination by a beam of wavelength λ gives rise to p transmitted rays, leading to sharpened fringes provided that

$$n_1 d \ll 3\lambda/8p^3\alpha^2$$

Since at least two fringes should be visible, α cannot be made arbitrarily small, and if, as is usual, fringes are to be examined microscopically, then α must increase as magnification increases. With reference to beam displacement, it may be shown that the intensity depends upon film properties over lateral distances of the order of $2p^2 d\alpha$. The number of terms, p, increases rapidly with reflectivity, e.g., $R = 0.80$, 0.93, and 0.95 correspond, respectively, to $p = 34$, 120, 180.

2.4. Beam Divergence

Another practical difficulty ensues from failure to illuminate by a perfectly parallel, normal beam which results in impairment of fringe quality. For most purposes, a beam divergence of 15′ of arc is acceptable. Lastly, fringe broadening is caused by finite line width in a supposedly monochromatic source. However, if d is kept small, a low-pressure mercury arc can be used effectively. Some of these problems were studied quantitatively by Kinosita[6] for a sixth-order Fizeau fringe between surfaces of $R = 0.90$, $T = 0.02$, at $\alpha = 10^{-2}$ rad. The reflected intensity at the fringe minima is not 0, but $0.02I_i$, and occurs at order 6.01, that is, away from the wedge apex. The fringe is asymmetrical, has secondary maxima, and a half-width greater than indicated by Airy's distribution.

To summarize: For surfaces with high reflectivity and low absorption, very narrow fringes of width not exceeding 1/50 of an order are obtainable. At best, fringe displacements can be determined to about 1/5 of fringe width, so that fringe displacements of 1/250 of an order are measurable. The most commonly used wavelength is the 5461 Å mercury line, therefore surface features corresponding to vertical distances of 10 Å are resolved. Horizontal resolution is much less, due to beam displacement, and ranges from about 2 to 5 or 6 wavelengths. Whether failure to exhibit an Airy distribution of intensity is due to the phase effect, beam displacement, beam divergence, polychromaticity, or some combination of these, it can be minimized by reducing d. Although surface separation is not absolutely at the observer's disposal, it can often be limited to about 10^{-3} mm, which generally ensures adequate fringe quality.

2.5. Interference Contrast

There is an interesting and useful variation of the Fizeau fringe method, due to Tolansky and Wilcock.[7] Provided that the test surface is relatively flat, it can be made effectively parallel to the reference surface so as to expand a given fringe until it covers the field of observation. For values of d corresponding to the half-intensity as shown, for instance, by the curve of Fig. 2-2, the intensity becomes extremely sensitive to small changes in d. By means of this technique of interference contrast, height variations as small as 2 Å can be resolved. It has been used by Heavens[8] to examine surface irregularities on optical glass and ordinary glass plate.

2.6. Fringes of Equal Chromatic Order

Fringes of constant thickness (Fizeau fringes), as outlined in Section 2.3 are formed by suitable illumination of film wedges. Illumination of thin parallel-sided films by normally incident polychromatic light produces fringes of equal chromatic order (FECO), so called by Tolansky, who first described them in multiple-beam interference. Such fringes are loci of points for which d/λ is constant, so that if d is changed, the fringes shift to other wavelengths. In this case a carbon arc, for instance, can serve as light source, and the reflection fringes—again fine, dark lines on a bright background—are resolved when an image of the thin film is focused on the entrance slit of a spectrograph. The overall topography is obtained by scanning the surface so that all regions are sequentially imaged on the spectrograph slit. Use of a spectrograph involves experimental complexity beyond what is required for the study of Fizeau fringes, yet by comparison with them, fringes of equal chromatic order offer certain advantages:

1. Since the surfaces may be exactly parallel, there is no fringe broadening due to phase lag, and the Airy intensity distribution is observed.
2. Although, because of spectrographic dispersion, fringe spacing is nonuniform, the fringes are replicas showing the effect of interference at a given point of the film, which is imaged at the entrance slit.
3. Fringe spacing depends upon the separation d of the parallel surfaces, therefore, it is feasible to observe fringes under high magnification.
4. The sense of curvature of topographic features, whose determination in Fizeau fringes can be rather complicated, is immediately evident. Protuberances and pits on the surface bend the fringes convex and concave, respectively, to the high end of the spectrum.

Fringes of equal chromatic order have been used by Scott *et al.*[9] to measure film thickness down 15 ± 5 Å, and to evaluate the effects of surface polish on the results. Koehler and White[10] demonstrated the influence of surface roughness on the half-widths of fringes of equal chromatic order.

3. EXPERIMENTAL TECHNIQUE AND SURFACE COATING

Figure 2-3 shows schematically the simplicity of the arrangements used for interference microscopy: S is a light source, essentially monochromatic for Fizeau fringes, or polychromatic for fringes of equal chromatic order; C is a collimator from which light passes to a beam splitter and falls perpendicularly upon the reference surface R; the test surface T is inclined at a small angle for Fizeau fringes, or effectively parallel for fringes of equal chromatic order or for interference contrast; M is a magnifying system or spectrograph. For transmission fringes the beam-splitter is removed, and the observation system aligned to receive light directly from T, which must, of course, transmit a certain amount of light.

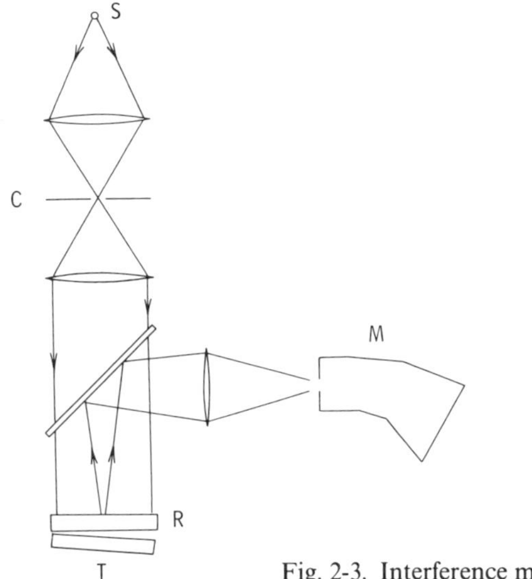

Fig. 2-3. Interference microscope.

The power of multiple-beam interferometry in the study of microtopography derives from the optical and mechanical properties of silver films deposited in a vacuum by thermal evaporation. Below 500 Å, reflectivity is deficient, and at about 700 Å absorption is excessive, so these are the effective limits on film thickness. Surprisingly, such silver films accurately reproduce on their surfaces features of molecular thickness buried beneath them. This topographical fidelity is mandatory for the film carried by the test surface. Moreover, a freshly evaporated silver film must be used, since aged films develop increased absorption. On reference flats it is possible to use dielectric multilayers instead of silver coatings. These have the high reflectivity and low absorption that are required for highly sharpened fringes, but are difficult to prepare and do not reproduce surface contours well. Also, they are intensely chromatic, and their high reflectivity is confined to a narrow wavelength band which must, of course, correspond to the spectrum of the light source.

The reference surface must be relatively flat over the area of the test surface that is to be examined. But since fringes are nearly always viewed under magnification, the reference surface needs to be locally very smooth rather than highly planar overall. Ordinary plate glass is usually flat within a few hundredths of a wavelength over the distances involved. Tolansky has shown that fire-polished glass is particularly well suited to serve as a reference surface, being essentially free of secondary structure such as polish marks.

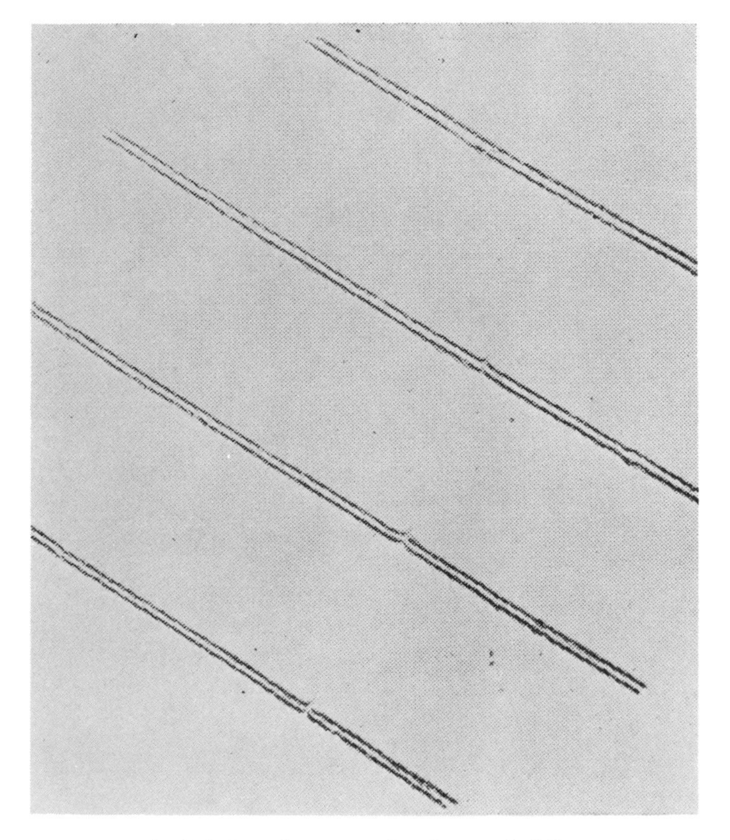

Fig. 2-4. Interferogram of a glass optical flat.

4. APPLICATIONS

Figure 2-4 is a multiple-beam interferogram obtained using a high-grade optical flat. The fringes are paired because the source is a mercury arc emitting the yellow lines at 5770 and 5790 Å. Fringe width is near 1/60 order, and much smaller than the doublet spacing, which increases as the spacing *d* increases. The fringes are linear and parallel, showing the absence of large-scale surface features. Numerous small excursions, many less than a fringe width in amplitude, prove that microtopographical irregularities are present. Vertical intervals of about 2500 Å correspond to fringe spacing of 2 cm, for magnifications of $8 \cdot 10^4 \times$. In contrast to this topographically simple surface is the surface of a diamond, semipolished after sawing. The contours of the saw marks are strikingly reproduced in Fig. 2-5.

An interferogram with crossed fringes and high interference contrast is shown in Fig. 2-6. A relatively flat, natural octahedral face

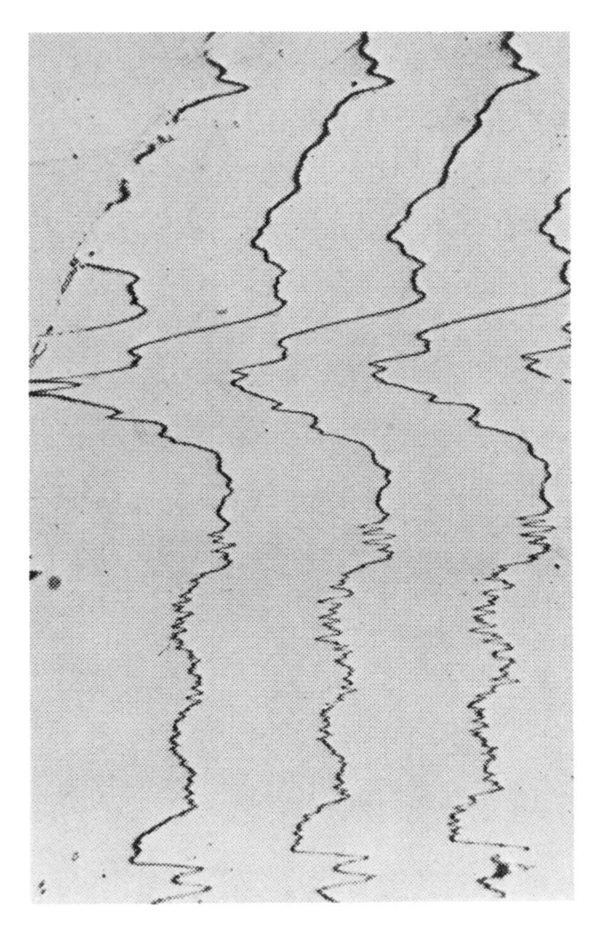

Fig. 2-5. Interferogram showing saw cuts on a diamond surface.

of diamond was arranged at a very small wedge angle with the reference flat leading, as previously outlined, to conspicuous variations in reflected intensity from areas of different spacing *d*. Then the dispersion was increased, to produce a series of triplet fringes from the two yellow lines and the green line of mercury. Finally, the dispersion was changed so as to generate a fringe system approximately orthogonal to the previous one. Successive exposures of a single negative produce the highly informative final picture. Large-scale surface structures show up as areas of varying reflected intensity. Such distinct regions are interconnected by the fringe network, which clearly reveals their relative altitude and orientation.

Figure 2-7 is again of an optical flat, this time producing fringes of equal chromatic order at a magnification of 142,000 ×. The fringes are

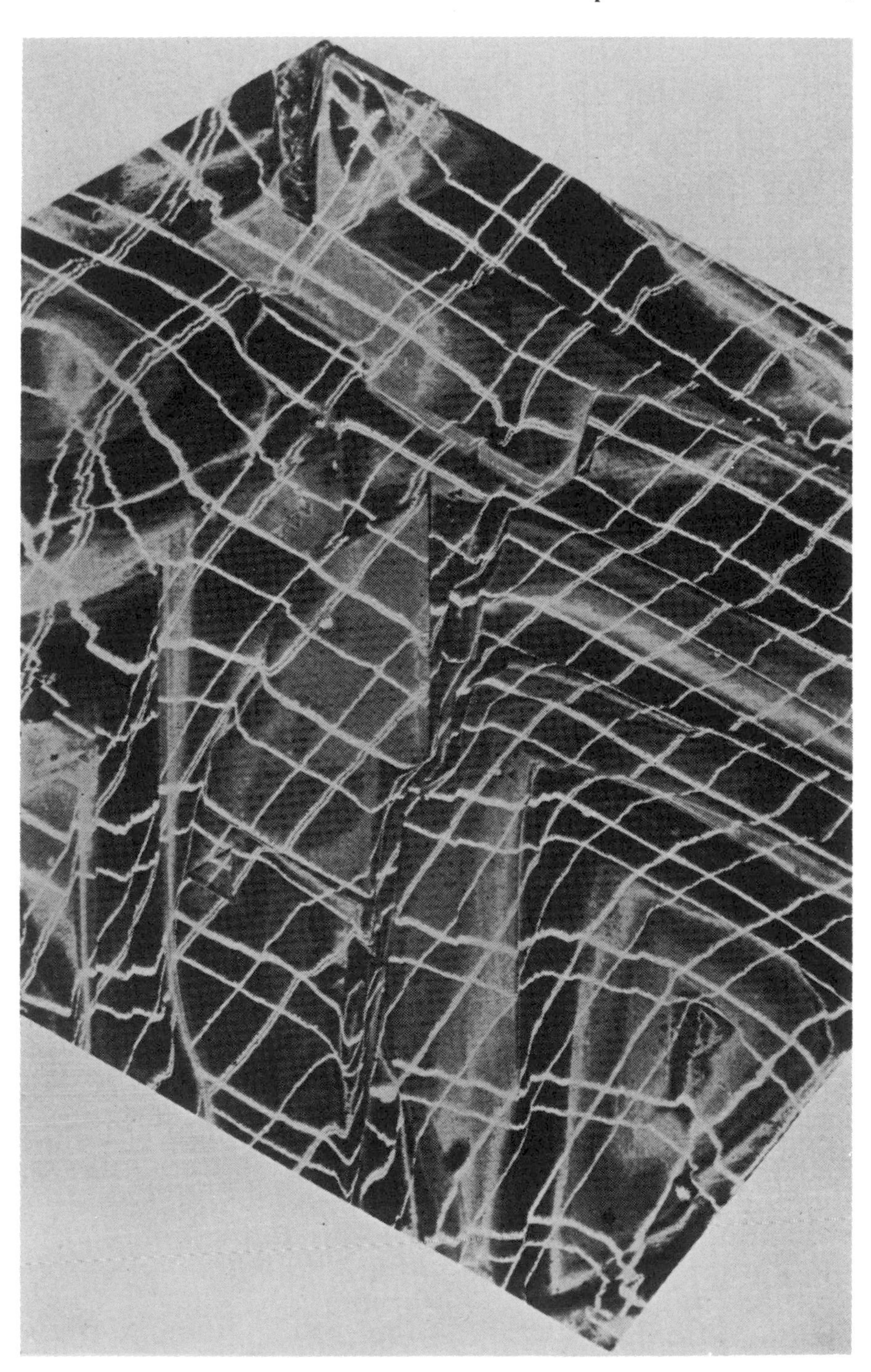

Fig. 2-6. Natural surface of diamond with crossed fringes and high interference contrast.

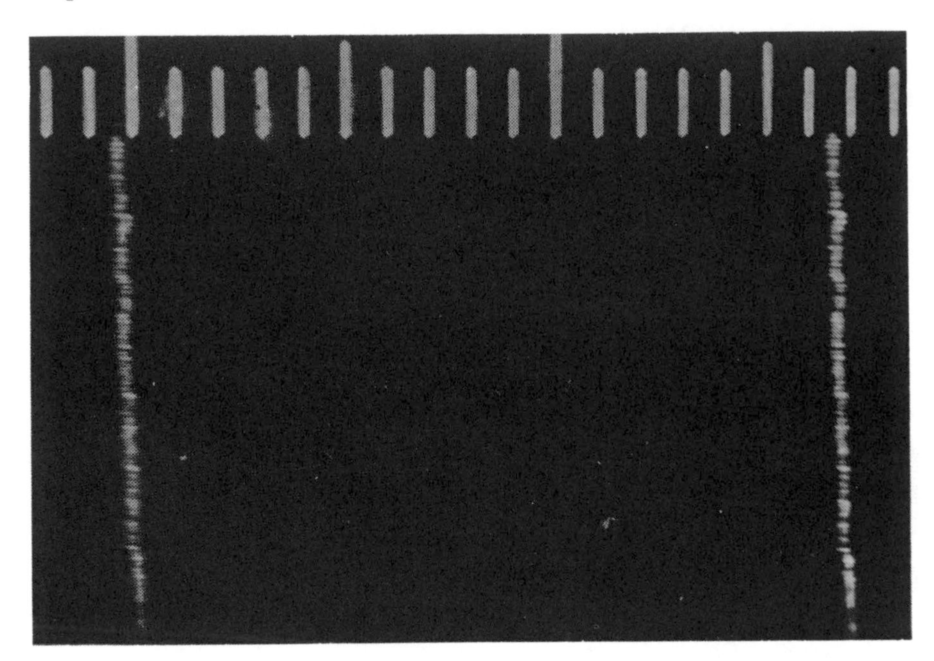

Fig. 2-7. Fringes of equal chromatic order on a glass optical flat.

alike, and about 1/50 order wide. Because reflectivity decreases with wavelength, there is fringe broadening toward shorter wavelengths. A 1-mm surface length is covered, and microstructure is apparent.

Other properties of FECO are shown in the interferogram of Fig. 2-8, a growth feature on synthetic quartz. Its height happens to be somewhat less than a half-wavelength, so a Fizeau fringe system would be ambiguous. The profile of a section across the width of the structure shows the expected fringe broadening from red (right) through blue. As the wavelength decreases, the fringe occupies an increasing fraction of an order since the elevation, which is just over 2×10^3 Å, naturally is constant. It is evident that the feature is a growth rather than a pit, since the fringes are convex toward the violet.

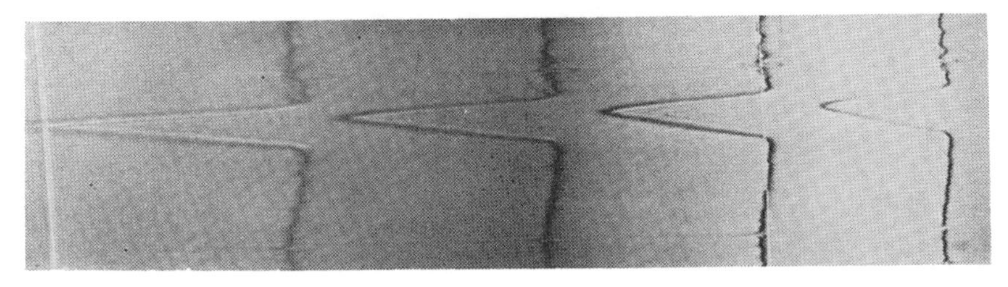

Fig. 2-8. Fringes of equal chromatic order; a growth feature on synthetic quartz.

5. COMPARISON WITH OTHER METHODS

Although highly useful in measuring thickness of thin films and in studies of microtopography, the methods of multiple-beam interferometry have certain obvious limitations. Delicate biological specimens are inaccessible, as are any other structures which cannot withstand the rigors attending deposition of thermally evaporated silver films. Also, it is impossible to follow the course of changes, such as crystal growth, occurring in real time. In these respects, ellipsometry and optical microscopy are superior, and the former technique offers vertical resolution generally superior to that obtained in multiple-beam interferometry. But ellipsometry requires a far greater investment in equipment, quantitative results must be derived via lengthy computations, and although ellipsometry can yield surface optical constants, it cannot give comprehensive microtopographical information. Stylus instruments, for example, the Rank–Taylor–Hobson Talystep (see Chapter 3) can give vertical resolution down to 25 Å, and surface contours are depicted directly on a strip chart. Because the stylus has an extremely small radius, test surfaces are placed under great pressure, and soft materials are subject to damage. The cost of stylus devices typically exceeds that of multiple-beam interference apparatus.

Comparisons between multiple-beam interferometry and electron microscopy are almost beside the point, in view of the enormous cost difference between them. The latter method is much superior in horizontal resolution, and scanning instruments convey an impression of three-dimensionality quite surpassing even interferograms that use crossed fringes and high interference contrast, but in vertical resolution multiple-beam interferometry is superior by a factor of several hundreds.

A newly developed instrument[11] uses a field emission probe to scan specimen surfaces. Changes in voltage between probe and surface due to variations in their relative spacing are annulled by a piezoelectric ceramic element which acts to maintain constant spacing. Surface profile is deduced directly from the piezovoltage. This method is said to give vertical resolution only slightly inferior to that of multiple-beam interferometry, and horizontal resolution that is somewhat superior. Moreover, it is probable that secondary-electron and photon emission will allow additional information to be derived by the instrument. Naturally, the apparatus is rather complex, requiring high vacuum and electrically conducting specimens. It has been called the "topografiner" by its developers, but it may be hoped that a less awkward name, for instance, microtopograph, will be adopted.

The study of solid surfaces comprises a much wider range of topics than just microtopography, which has very little to do with individual atoms, molecules, valence states, energy bands, and so forth. Yet microtopography constitutes a vigorous and indispensable branch of surface science. Recent advances in electronic and related technology have produced several new and powerful methods for the study of surface profiles. These do not supersede but rather complement multiple-beam interferometry, for no other technique of microtopography can compare with it if maximum experimental accuracy, simplicity, and beauty are to be achieved with minimum investment in laboratory space and equipment.

6. ACKNOWLEDGMENT

The author thanks Clarendon Press, Oxford, for permission to use Figs. 2-4 through 2-8, which were taken from S. Tolansky, *Multiple-Beam Interferometry of Surfaces and Films*.[1]

7. REFERENCES

1. S. Tolansky, *Multiple-Beam Interferometry of Surfaces and Films*, Clarendon Press, Oxford (1948).
2. F. A. Jenkins and H. E. White, *Fundamentals of Optics*, McGraw-Hill, New York (1957).
3. G. B. Airy, *Phil. Mag.* **2**, 20 (1833).
4. M. Born and E. Wolf, *Principles of Optics*, Pergamon Press, London (1964).
5. J. Brossel, *Proc. Phys. Soc.* (*London*) **59**, 224 (1947).
6. K. Kinosita, *J. Phys. Soc. Japan* **8**, 219 (1953).
7. S. Tolansky and W. L. Wilcock, *Nature* **157**, 583 (1946).
8. O. S. Heavens, *Proc. Phys. Soc.* (*London*) **64B**, 419 (1951).
9. G. D. Scott, T. A. McLauchlan, and R. S. Sennett, *J. Appl. Phys.* **21**, 843 (1950).
10. W. F. Koehler and W. C. White, *J. Opt. Soc. Am.* **45,** 1011 (1955).
11. R. Young, J. Ward, and F. Scire, *Rev. Sci. Instr.* **43** (7), 999 (1972).

3

STYLUS TECHNIQUES

David John Whitehouse

Rank Precision Industries Ltd.
Metrology Division
Leicester, England

1. INTRODUCTION

Stylus techniques were developed mainly to meet the need for quantitative assessment of the surfaces of manufactured components. The method received impetus because of the need for accurate methods of quality control during the war, in particular, because of the need for controlling the quality of surfaces of components manufactured by subcontractors. Since then, the uses of stylus techniques have increased manyfold. Reduction of the dimensional tolerances of parts has been achieved by improvements in manufacturing processes; this in turn has meant that the surface texture plays a more significant role in the determination of how well a component will behave when in operation. Exactly the same can be said for the functional importance of the roundness and straightness of parts. The necessity for accurate and fast methods of surface metrology incorporating the measurement of surface texture, out-of-roundness, etc. has long been apparent.

To get an idea of the nature of the problems involved in designing instruments for surface metrology, consider the measurement of surface texture. The range of sizes of the marks typically produced is 1000:1 in their amplitudes and 100:1 in their spacings. This when taken together with the complex component shapes often met with in practice makes comprehensive instrumentation difficult.

The two most natural ways of assessing surface texture are by eye and fingernail. Optical techniques are the successors to the eye, and stylus methods are the successors to the fingernail. This does not mean that the eye and fingernail are still not used today, they are, although only for quick assessment in the workshop or prior to a more accurate evaluation. Obviously, in the same way that it is difficult to reproduce by optical means the properties of the human eye, it is correspondingly difficult to reproduce the properties of the fingernail. Many factors are involved: for instance, the frictional coefficient between the nail and the workpiece is important as well as the roughness height itself. Indeed, because of this difficulty, the evolution of stylus instruments has taken two paths, the most important being that in which the stylus is used to accurately follow the geometry of the surface. However, purely frictional devices have also been made.

The main reasons why stylus types of instruments have found favor in the workshop, the laboratory, and inspection room rather than the optical methods are due to their ease of operation, and robustness, as well as the convenient form of the output and the fact that there is no need to specially prepare the specimens for measurement. One other advantage makes use of one of the criticisms sometimes leveled at stylus instruments, which is that the stylus exerts some mechanical force on the surface. Although on some very soft materials this can be important, in the vast majority of cases this is not so and, furthermore, the mechanical force on the stylus helps remove extraneous matter from the surface, for instance, debris, coolant, etc. often found in workshop environments.

Since the early development of stylus instruments, several variations have been designed, each having a different capability to suit some particular need. Because of the large number of these, a description of each is impossible; therefore, in the next section a breakdown of a simple surface-texture measuring instrument will be given. More detailed accounts can be found in the literature.[1]

2. BASIC INSTRUMENT

This comprises a pickup, mechanical system, electronic unit, and a metering and/or recording device. When being used, a sharply pointed stylus is rested lightly on the surface and is carefully traversed across it. The up and down movements of the stylus relative to a suitable mechanical datum are magnified and a meter value taken, or a record made (this recording is usually referred to as a profile graph).

2.1. Stylus and Pickup

The stylus is generally a pyramidal or conical diamond with a flat or rounded tip. The American Standard for Surface Texture, B46, calls for a 60° cone with a radius of 0.0005 inch (12.5 μ). In special cases this is modified to 0.0001 inch (2.5 μ) at the tip. In Britain a 90° pyramidal form with a flat or rounded tip not more than 0.0001 inch (2.5 μ) in width is generally preferred, but comparative instruments with tips of 0.0005 inch in radius have been introduced and tips of less than 0.000004 inch (0.1 μ) are being used for research.

Special styli having angles finer than those already mentioned have also been used. In the pyramidal type of stylus tip, the dimension normally referred to is the one being moved across the machine marks or lay of the texture. The dimension parallel to the lay is sometimes larger, for instance, up to 0.0003 inch (7.5 μ). This is acceptable because most finishing processes, such as grinding or diamond turning, result in texture which is anisotropic in character. An advantage of having the longer dimension of the stylus parallel to the lay is that it gives more mechanical strength to the stylus and therefore helps to make the instrument more robust.

A force is intentionally applied to the stylus in order to maintain its contact with the surface during the traverse. For a typical instrument, this force is about 100 mg, but in certain circumstances can be made less than 1 mg. A guide to the maximum permissible load on the stylus is given by a formula in the American Standard, which is that the maximum force in grams is 0.00001 times the tip radius in microinches squared. This ensures that, for most materials, there is no mechanical damage to the surface.

The stylus is usually connected via a bar of light, stiff material, such as Dural, to a transducer which may be either along the same axis as the stylus movement or perpendicular to it. In either case, it is essential that to a high degree of accuracy the movement of the armature in the transducer has only one degree of freedom.

The transducer element connected to the stylus may be the armature of an inductor in an inductance bridge, it may be a plate of a capacitor in a capacitance bridge, or it may be an optical system in which two photocells connected differentially are illuminated by light controlled by a mask on the end of the stylus beam. Alternatively, it may be in the form of a piezoelectric crystal such as barium titanate whose dynamic characteristic responds to the rate of movement of the stylus rather than to its absolute position. In any event, the mass of the stylus, the connection to the transducer, and the moving part of the transducer is, as a rule, kept as low as possible in order to keep the resonant frequency high and the dynamic forces low.

Whereas the moving part of the transducer is connected to the stylus, the fixed part of the transducer is connected by some mechanical means to the reference surface. Consequently, any movement of the moving part of the transducer relative to the stator is a direct result of the movement of the stylus, due to the test surface, relative to the reference surface.

This mechanical movement causes the generation of a current or voltage in the transducer of value proportional, or nominally proportional, to the extent of the movement. Because the movements of the stylus are very small, of the order of tens of microinches, the magnitude of the electrical signal is correspondingly small. In the case of a voltage generator, it may be of the order of microvolts.

2.2. Mechanical Arrangement

The mechanical datum from which the surface is measured should ideally conform with the nominal shape of the surface. Furthermore, this reference surface should be longer than the surface to be measured and should have errors over the length of the test surface which are small compared with the texture on the test surface. These requirements become progressively harder to meet if the shape of the test surface is complex or if its deviations are small.

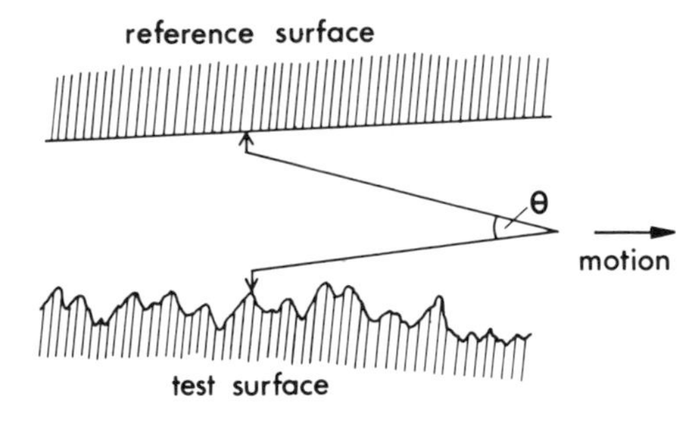

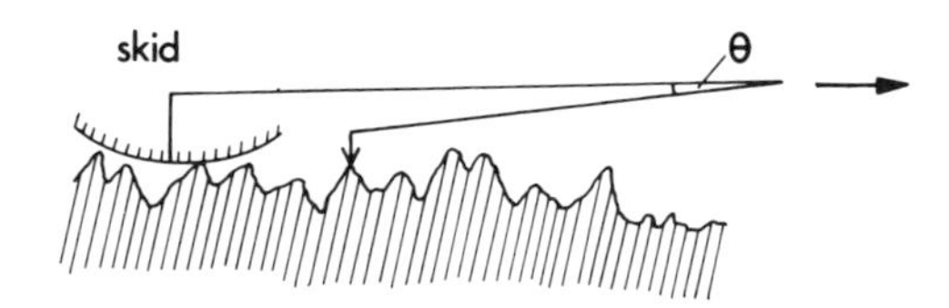

Fig. 3-1. Diagrammatic representation of stylus instrument for measuring surface texture: (a) accurate method incorporating a smooth reference surface of nominally the same shape as the test surface; (b) approximate method in which the reference surface is replaced by a blunt foot or shoe riding across the test surface.

Usually, the reference is a smooth surface shaped according to that of the test surface, but it can be a reference made up of a set of linkages which provide a reference movement. Under both of these circumstances the pickup arrangement can be compared with a caliper that is pulled between the test and reference surfaces (Fig. 3-1a). For practical purposes, the use of a smooth datum can sometimes be somewhat relaxed by using a skid or self-aligning shoe which is large in size compared with the stylus tip (Fig. 3-1b). The assumption made here is that the up and down movements of the skid during its traverse will normally be much smaller than those of the stylus. This is a condition which is usually met in practice providing that the crest spacings are considerably smaller than the skid radius and the skid-radius separation.[2] In some instruments, the skid is alongside the stylus; or even two skids may be used. In general, the use of a skid can considerably reduce the setting-up problem.

The length of surface usually included in one measurement varies between 0.4 and 2.0 inches (10–50 mm). The speed of traverse of the stylus across the surface can take many values; however, a typical value is about 0.05 inch/sec when a meter reading is being taken, and between 5 and 25 times slower when a profile graph is required. For some special applications, instruments having much slower recording speeds have been devised, permitting horizontal magnifications of 2000 and over to be achieved.

When the gearbox drags the stylus across the surface, the motion is usually linear in time, although other motions have been developed from time to time.

2.3. Electronic Unit

The input to the electronic unit is the electrical signal generated by the movement of the stylus. This signal may be voltage or current, depending on the type of transducer. The type of transducer from the electrical point of view generally falls into two classes. One is the carrier-modulated device in which the amplitude of an alternating current of high frequency is controlled by the position of the stylus relative to the datum. The first instrument of this kind was the Talysurf made by Taylor, Taylor, and Hobson in 1940. Other instruments of the same nature are the Proficorder, the Hommel instruments, and the Perth-O-Meter. Current- or potential-generating devices in which a current or potential is generated according to the rate of change of the position of the stylus as it is displaced are also being manufactured. Some commercial examples are the Profilometer, the Surfagage, Microtest, and Surtronic instruments; the first instrument of this kind, the Profilometer, designed by Abbott, appeared in 1936, and derivatives continue to be sold by the Micrometrical Division of the Bendix Corporation. In the modulated-carrier type of instrument, the carrier frequency is typically a few kilocycles per second. The position of the stylus controls the carrier amplitude, i.e., the depth of modulation. Movement of the stylus generates a current of which the highest frequency is much lower than that of the carrier. The signal thus derived is then amplified. After amplification, the carrier frequency is suppressed leaving a current signal whose amplitude is proportional to the position of the stylus at any time. In this way a faithful reproduction of the vertical movement of the stylus is obtained which is independent of the crest spacings on the surface.

Generating devices such as the moving coil type used in the Profilometer produce signals proportional to both the amplitude and frequency of the stylus movement. Corrective circuits are introduced to ensure a flat frequency characteristic over the roughness band.

In those instruments using piezoelectric crystals for the transducer, the output is proportional more to the movement of the stylus than to its position, especially at low frequencies. Both of these latter types of transducers inherently introduce a long-wavelength attenuation. How this is achieved in the carrier type of instrument will be seen presently.

So far, only the electrical signal from the transducer and its amplification have been considered. The next problem is how to process this signal in order to get a measure of the roughness. This is achieved either graphically, by metering, or by computing. The theory behind this will be explained in Section 3. In the following section only the hardware will be described.

2.4. Recording and Metering Units

The earliest method of assessment of the signal was to output the signal onto a recorder and evaluate the signal graphically. (Because of the inherent differentiation of the signal produced by the generator type of instrument, these instruments are not really suitable for use with graphical assessment.)

The usual method of assessment is to pass the signal from the amplifier of a carrier-type instrument through an electrical filter which cuts out the long wavelengths, rectify the signal, and then input it into a meter which reads average or peak values for a given length of time. In the earlier days of surface metrology, meters approximating the rms value were sometimes used.

Recent additions to the processing hardware have been the data logger[3] and the computer. In the former case, the profile signal is digitized and put out onto paper tape, magnetic tape, or cards. This paper tape then represents a record of the profile just as valid as a profile graph from a recorder. Even more recent has been the use of computers directly linked to the instrument.[4,5] The use of such sophisticated tools is at present restricted to research, but it seems evident that before long computers will be used extensively in more practical situations.

Simple instruments like the one just described are in use throughout the world. Instruments like this for the measurement of texture fall into roughly two cost categories. First, the typical workshop instrument selling at about $250 and then the inspection instrument which is probably over ten times more expensive. Obviously the extent of the metering and recording facilities of an instrument will depend upon the purpose for which the instrument has been designed. For instance, in the workshop a recorder is not usually required, whereas, it is usually an integral part of an inspection instrument.

3. THEORETICAL BACKGROUND

3.1. Reference Lines

The method of assessment considered here will be specifically related to the assessment of surface texture. There are other situations involving stylus instruments, for instance, in the assessment of roundness and step height, where different techniques are used.[17]

It is generally not sufficient to merely get an electrical signal out of an instrument and measure it. The complete waveform will often include extraneous wavelengths that are irrelevant to the texture, which arise from imperfect leveling of the specimen relative to the instrument table, and errors due to chatter, etc., caused by imperfect machining. These have to be eliminated so that the signal from the surface contains only the elements relevant to the texture to be measured. Although the signal has been obtained from the surface with respect to the mechanical datum, this does not ensure that it is in a form suitable for processing. The effect of the mechanical datum has been to keep the signal level to within the range of the transducer throughout the traverse. A further operation is required to separate the roughness signal from the profile. This operation in effect establishes a reference line within the profile itself from which the roughness can be measured. This is usually done by one of two methods: that is, by either drawing procedures on the chart or by filters in the electronic unit.

In the method adopted by the International Standard for the Profile Method, R468, the profile to be assessed is split, in its simplest form, into a sampling length, i.e., the minimum length over which the roughness value is considered to be significant. A range of sampling lengths has been selected by international agreement to be

0.0003 0.01 0.03 0.1 0.3 inch

or

0.08 0.25 0.8 2.5 8.0 mm

Once the sampling length required to select the texture to be measured has been decided upon, it is usual, in the case of repetitive waveforms, to use the nearest preferred value of sample length. Determining the sample length for the surface is not difficult and can be done approximately by eye. It is the shortest sample which will just include all the wavelengths of the roughness to be measured. If too long a sample is used, then extraneous long wavelengths can affect the reading.

Once the sample length is decided upon and a region of the chart isolated a line is put through the sample such that it lies in the general direction of the surface. The line is such that the area enclosed between the profile and the line is the same above as it is below. This definition has been specified also as a least-squares best-fit line. In order to get some good statistical reliability, it is usual to take five consecutive samples. This question of reliability is fundamental and should be taken into account. (Most instruments measure a profile about five sample lengths long in one traverse.)

The other practical method is that of electrical filtering. Here the filter in the electronic unit effectively looks continuously only at a length of the surface equal to the graphical sample length. This means that the meter (or filter) cutoff is nominally equal to the sampling length. By international agreement, the filter used is a standard 2CR filter. This nominally achieves electrically the same effect as that of the graphical procedure on the profile graph; it attenuates those long wavelengths likely to be irrelevant to the assessment of the roughness.

These two methods—graphical and electrical—of filtering the waveform have been simultaneously adopted for practical convenience, since it is difficult (but not impossible) to implement the electrical filtering action as a mathematical procedure[6] suitable for implementation on the graph. It is also difficult to filter the electrical signal by means of the best-fit mean line normally achieved graphically. These two accepted methods usually give similar results, but if there are any differences, they are now described as "method divergence."

Other methods of filtering have been suggested, one of which is the use of a wave filter having optimum characteristics[7] called the phase-corrected filter. This filter has certain advantages over other methods and is used in research work in a number of countries. However, it is expensive to implement and usually requires the use of a computer.

It is important to realize that a defined degree of filtering may sometimes be required at the short-wavelength as well as at the long-wavelength end of the spectrum, e.g., in the measurement of the slopes on the surface or the curvature at the peaks. Any short wavelengths due, for example, to mechanical vibration must be excluded from the signal before the assessment can be made.[8] The stylus tip itself will provide some filtering action by virtue of its finite tip dimension and sometimes by the slope of its flanks. Although the short-wavelength components are important in some applications, they are not very important when the average roughness value (R_a) or any integrating type of parameter is being measured.

3.2. Parameters

After a reference line is established within the profile from which to assess the roughness the next problem is to decide how to provide a quantitative evaluation of what has been measured.

A single profile graph contains many independent bits of information. What the inspector requires is simply one bit: Is the part good or not? The problem is to condense the information in the profile into as small a number of bits of information as possible without throwing out anything of significance. This is not an easy task.

What has been used often in the past has been simply the R_a value (the British *CLA* and American *AA* value). This R_a value is the average value of the modulus of the deviations of the profile from the mean line. Being of an integrated nature, it tends to be reliable. In Germany and elsewhere in Europe, peak values have been used which, although less reliable, are easier to measure from a graph.

Fundamentally, in order to specify the information in a single-dimensional random waveform (such as a surface profile graph), two things are needed; the amplitude probability density function and the autocorrelation function. The former, commonly called the height distribution, is a measure of the frequency with which the profile equals a given level; the autocorrelation function is a measure of the dependence of one part of a profile on another. Hence the height distribution is a function from which the distribution of peak heights can be determined, and the autocorrelation function (or alternatively, its Fourier transform, the power spectral density) is a measure of the spacings on the surface. All methods of assessment in the past have been estimates of either one or the other or a mixture of these functions; for instance, the R_a value is an estimate of the height distribution.

Estimates similar to the R_a have been used, for instance, the rms value, the maximum peak-to-valley value, the average peak height value and associated parameters, the bearing-ratio curve, and many others.[1] To some extent they are all saying the same thing. Similarly the peak count, average wavelength,[8] and high-spot count are all estimates of the autocorrelation length or power spectrum.

Hybrid parameters have also been used in attempts to give a more meaningful number to put on a drawing. Some examples are the average slope of the flanks, the curvature at the peaks, or the curvature at the valleys. In fact, there is such a profusion of different ways of numbering surfaces that there is a great danger of confusion arising in workshops. Instruments which purport to measure a multitude of parameters can be more of a hindrance than a help.

This whole problem of assessment of surfaces brings up the subject of classification or typology.

3.3. Typology

For a number of years, people have been attempting to help control the functioning of parts by specifying the surface. Usually this has involved a statement of the manufacturing process together with a numerical value such as the R_a value to control it. This method has the advantage that both the geometrical and physical properties of the surface and surface layers have been defined to some extent, i.e., the surface integrity is defined. However, for some applications, the geometric specification of amplitude (R_a) needs to be augmented by a wavelength-conscious parameter.[8] An example of this is in the sheet steel industry.[9] There has also been pressure to remove the manufacturing process from the specification. Two reasons for this are that production engineers want freedom to use any available machine plant and even using the same machining process different properties can be introduced unless there is a good deal of control over the process itself.[10] The problem of specifying surfaces without reference to the process is difficult; whether it will be possible remains to be seen.

There appear to be two ways of tackling this problem: The first is to try to glean more information from the profile itself, and the second is to measure directly some of the physical properties of the surface skin and then to combine these with a geometric measure.

Ideally, if there has to be a control of surface properties which allows the maximum amount of flexibility for the production engineer, then the tests should be carried out immediately prior to assembly. This implies that any successful test will be able to make a judgment of surface performance independent of its manufacturing history. This means, in typological terms, that all that needs to be known about the surface can be deduced from its "as received by the inspector" state.

A purely geometrical start has been made in this direction by considering not only the scale of size of the surface in the horizontal and vertical directions as determined by the two parameters mentioned previously, but also the shape of the profile. Breaking this last statement down into its constituents means that information can be obtained from the shape of the height distribution and the shape of the autocorrelation function. It is this classification according to shape that can give certain information about the type of process that may be needed for satisfactory behavior, and hence, indirectly, about the physical properties of the skin. So far, this approach is in its infancy and as yet no comprehensive treatment has evolved. However, there have been a few attempts to develop schemes. How successful any method based on this approach will be will not be clear for a good many years.

One of the problems of classification by shape is the complexity likely to be involved. The best that can be achieved is some sort of elementary code. One of the first serious attempts was by Peklenik,[11] who classified the shape of the autocorrelation function into five groups depending on the relationship between the periodic and random components. In order to quantify this coding, he also introduced two spacing parameters—one to determine the wavelength of any periodicity present, the correlation wavelength, and the other to specify the bandwidth of the random element, the correlation length. His grouping is as follows:

1. Steady value or cosinusoidal
2. Random plus periodic
3. Random multiplied by periodic
4. Complex combination of groups 1 and 2
5. Random (exponential decay)

The fundamental assumption is that the type of process determines the shape of autocorrelation function (which seems to be quite a reasonable proposition). Certainly the shape of the correlation function can detect the presence of chatter and hence, to some extent, abusive machining.

One problem with this method is that the reliable measurement of autocorrelation is difficult. Whitehouse[12] alleviated this problem by relaxing the grouping on the autocorrelation function and incorporating with it a shape classification based on the skew of the height distribution. Thus nine basic groups emerge, determined by whether the autocorrelation is mostly random (R), periodic (P), or mixed (M), and whether the skew of the height distribution is positive ($+$), negative ($-$),

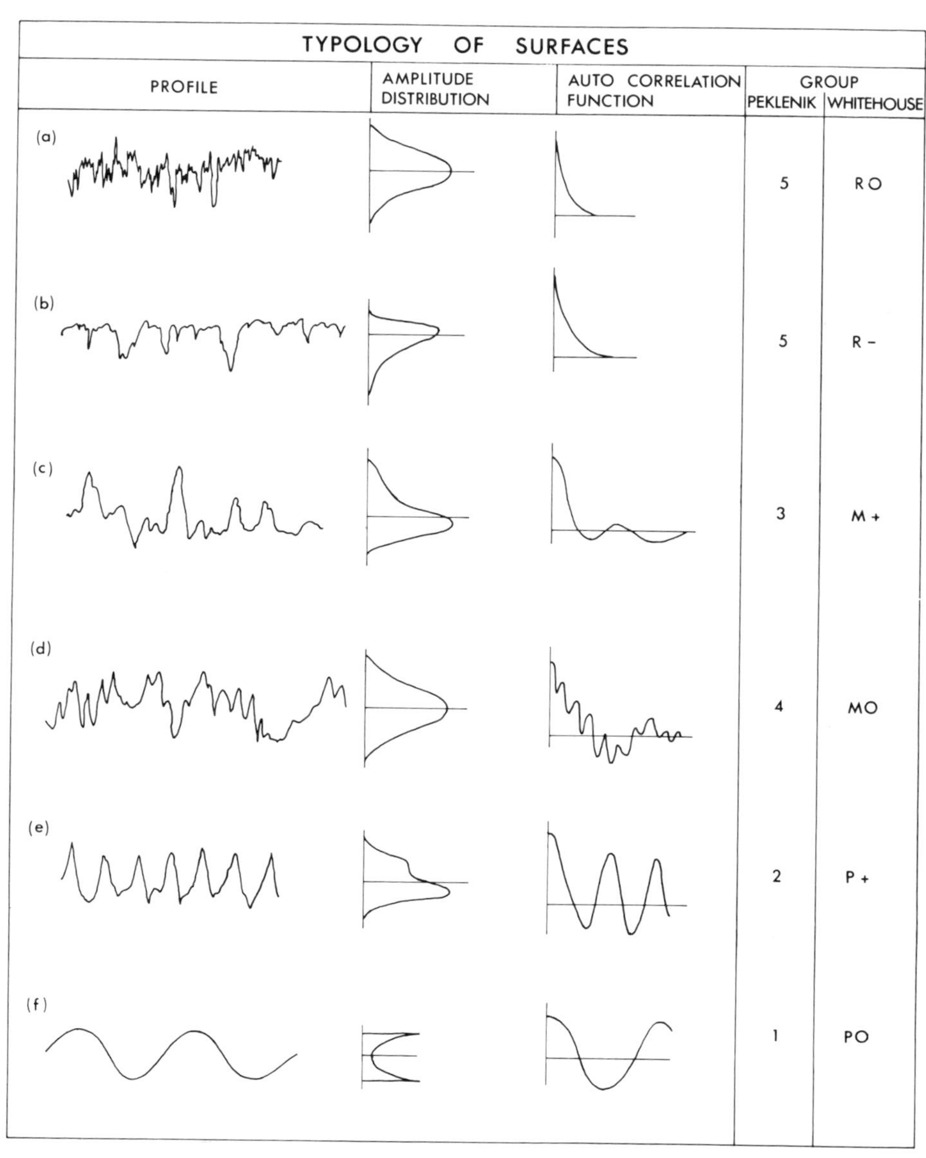

Fig. 3-2. Two methods of surface characterization using the amplitude distribution and the autocorrelation function: (a) skew zero, simple random, typical of a ground surface, (b) skew negative, simple random, typical of a lapped surface, (c) skew positive, modulated random, typical of a bead-blasted surface, (d) skew zero, complex correlation, typical of a milled surface, (e) skew positive, random plus periodic, typical of a turned surface, (f) deterministic, etched or ruled standards.

or symmetrical (0). This classification is not nearly so difficult to assess. Figure 3-2 shows a few representative surfaces coded according to these two classifications. As size parameters, the $R\hat{a}$ value and the average wavelength[8] are used in the Whitehouse method, whereas the correlation length and period are used in the Peklenik method.[11] Another classification system based on a series of the moments of the height distribution has been proposed by Al-Salihi,[13] but high-order moments are difficult to measure, so again the practicality is suspect. Other methods based only on the shape classification of the height distribution have also been tried.[14]

Finally, in the same vein, Myers[15] has proposed a classification based on a series of the profile derivatives, incorporating the rms height, slope, and second derivative. Other methods of typology have been described in the literature.[12]

Turning now to a classification which includes physical quantities, the most notable so far has been due to Greenwood and Williamson.[16] They proposed a plasticity index ϕ which is determined from the composite elastic modulus E' of the surfaces, the hardness H, the rms value of the peak height σ, and the mean radius of the peaks β:

$$\phi = E'/H(\sigma/\beta)^{1/2}$$

The plasticity index is important because it attempts to answer the question, in problems of contact, of whether the peaks will deform elastically or be crushed. In other words, ϕ is a number that can be attached to a surface which is predictive of its behavior under contact conditions. For $\phi < 0.6$, the deformation will be elastic up to quite high loads, whereas for $\phi > 1$, deformations will be plastic at a touch. It is important also because it brings together geometrical and physical properties in a way which attempts to solve a particular functional problem. It may be that this is the way things will go in the future, but more work needs to be done.

One of the difficulties of the approach is determining whether surface or bulk physical properties should be used, and another is how to define peaks. In Section 5.4 some results are shown of measurements of both geometrical and physical properties taken with a stylus instrument.

4. NATURE OF THE INFORMATION REVEALED BY STYLUS TECHNIQUES

The information revealed by a stylus instrument, as with any other technique, has to be understood for what it is, and the nature and extent of any limitations must be taken into account.

In the first place, the dimension of the tip itself affects what is seen on the profile graph. Since the tip is not infinitely sharp, it has the effect of a mechanical low-pass filter, i.e., it reduces in amplitude all wavelengths of a size equal to and smaller than that of the tip. The rate of attenuation depends upon the shape of the tip. For a flat tip the attenuation is severe, i.e., none of the smaller wavelengths are seen, whereas for a spherical tip the attenuation is more gradual. In practical instruments the tips may start off by having either a formal radius or a nominal flat; with continued use, however, they usually end up with a spherical type of tip having a long radius of curvature. A knowledge of the tip dimension is useful in determining the maximum rate at which the surface needs to be digitized in computational applications.

Another limitation sometimes but not often imposed by the stylus results from the angle of the stylus, which is typically 90° or 60°. Angles larger than these cannot be picked up by the stylus method. Obviously, no re-entrant features of the surface can ever be revealed either.

The slope and dimensional limitations are not serious for most practical surfaces. Peaks having these separations and slopes are usually of small amplitude and hence functionally insignificant.

The mechanical filtering nature of the stylus is amplitude- and wavelength-dependent. This means that peaks will appear to be blunter and the valleys sharper than they really are. These differences are generally small, however, and only need to be taken into consideration in a few research programs.

Critics of the stylus method often point to the finite force exerted by the stylus and complain of serious surface damage. Although it is true that a mark is sometimes left on the softer metals, this need not be cause for alarm. To put the criticism into perspective, one would hardly suggest that a mountain could not be surveyed if the grass gets trampled down. This is the relative scale of size. Most optical pictures of damage have had to have been taken at high magnification in order to see the track; this fact alone results in a very limited field of view. The resulting photograph merely corresponds to looking at a small area on the side of the mountain!

A point that needs to be considered when dealing with manufactured surfaces and their appearances, as revealed by stylus-type instruments, is that because of the difference in the magnitude of the spacings on the surface as opposed to the height of the asperities, it is usually necessary to have a horizontal magnification on the graph different from the vertical one. This has the effect of causing distortion, and in fact can give misleading impressions of how a surface really is. For instance, in the extreme case of a surface having a 45° angle and a magnification ratio in the horizontal and vertical directions of 1:10, this will make a 45° angle appear to be about 85° on the graph, i.e., almost perpendicular. Hence this sort of distortion produced by different magnifications should always be considered when interpreting visually from the graph.

Styli used in the measurement of roundness and straightness are intentionally made larger to filter out the roughness. They are usually of a hatchet shape, made out of tungsten carbide, and have radii of about 0.5 inch in one direction and 0.05 inch in the other. Also, they usually work at greater pressures than those used for texture measurements. For more details of the effect of the stylus in roundness measurement and methods of assessment, refer to Reason.[17]

Finally, the great benefit of the stylus method despite the limitations described is that it will give an output which has a very high correlation to the true geometry of the surface, virtually independent of surface contamination.

Two recent exercises comparing the results of stylus methods with those of optical and electro-optical methods have produced some interesting results. King *et al.*[18] at the National Physical Laboratory have carried out an exhaustive exercise on the measurement of thin-film thickness by comparing a stylus instrument,* with two highly sophisticated interference methods. The correlation between the three sets of results was very high indeed, and the stylus method was much more practical than the others. Another exercise[19] has been carried out on surface-texture measurement comparing optical, scanning electron microscope, and stylus methods. These results show that the scanning electron microscope does not give a true surface representation suitable for quantitative evaluation unless the stereoscopic effect is utilized—not a very convenient method.

5. APPLICATION OF STYLUS TECHNIQUES

5.1. Introduction

Before considering some of the applications of stylus techniques, the methods of calibration will be touched upon.

The stylus technique, in common with any other, is only as good as its method of calibration. During the last 20 years or so, methods of calibration have been evolved which guarantee the usefulness of the technique. To start with, the pickup and amplifiers are usually calibrated by means of gauge blocks.[20] Small height differences between two gauge blocks can be very accurately reduced by scaling down with

* Talystep, Rank Precision Industries Limited, Metrology Division, Leicester, England.

a lever. The gauge blocks are calibrated with reference to the wavelength of light. Hence, any graph can be calibrated with reference to the wavelength of light. The meters and filters are calibrated by means of reference standards which are usually in the form of a set of ruled lines of known height, spacing, and waveform. The stylus is made to track across these standards and the meter reading is adjusted to the standard value. Nowadays, the absolute value of the standard can be calculated from the profile signal using computers, the filtering action being taken very accurately into account.(6) Thus it can be seen that both the profilegraph magnification and the meter values can be directly traced back to the wavelength of light.

Another point relevant to the application of stylus techniques is ways of measurement when the surface is inaccessible. Under such circumstances, replication methods have to be used.

The usual replication material consists of a resinous type of material, generally made from a powder and a liquid mixed in precise proportions, which is poured onto the cleaned surface and then allowed to set. In some cases, a release agent is put onto the surface to allow easy removal of the replica.

The earliest replication process was by Faxfilm.(2) This method appeared to be a cellulose acetate strip softened with a solvent and then pressed against the surface. Other examples are Cristic, Technovit, and the encapsulated dispenser kit by Rank Precision Industries.

The fidelity of these techniques is usually good for measuring small areas, where reproduction to within a few microinches is possible. It should always be remembered, however, that the replica is a negative of the surface.

The usefulness of stylus instruments in the quality control of manufactured components, particularly in the fields of surface texture and roundness, is well known and has been well documented.(12) Detailed descriptions of all the many applications would be too involved to go into here. It must suffice to record that in practically all branches of industry, but particularly in the aeroengine, motorcar and associated industries, many successful uses have been found.

Instead of the well-known applications, some examples will now be given of applications of the stylus technique which are not quite so obvious. These few topics will serve to indicate how the technique is evolving.

5.2. Two-Dimensional and Accurate-Location Profilometry

The very nature of the stylus technique dictates that only one cross section of the surface is measured at any one time. Generally the information obtained from such a profile is sufficient for engineering needs. However, as a general rule if there is any doubt as to the reliability of the information, more than one track should be taken and statistical limits put on the expected variations in such cases. To get average behavior such as the R_a value reasonably accurately, a few tracks, say three, need to be taken; but this number has to be increased considerably if freak events, for instance surface cracks, are being investigated. In fact, an increasing number of workers requiring quantitative information over an area are turning to stylus methods, and instruments have been devised which have this facility built in.*

The real need for quantitative information at present is in research into contact phenomena. Here a further constraint has to be applied: the positioning in height and spacing between the tracks has to be exactly known.[21]

Williamson has mapped out whole surfaces using the technique of accurate location and then, making use of the convenient output form of the stylus instrument, has performed contact experiments within a computer. Putting two surfaces together in this way has enabled information to be obtained about the size and spacing of contact regions hitherto impossible to obtain because of the inaccessibility of the region. Figure 3-3 is a computer-simulated map of the contact areas between two bead-blasted aluminum surfaces at four progressively smaller separations (as shown by the solid, dash, dot—dash, and dot boundaries). Work of this sort has revealed that contacts do not as a rule occur at the tips of the peaks, but are more likely to occur on the shoulders.

Fig. 3-3. Map of bead-blasted aluminum surface showing the areas of contact at four separations. The solid boundaries correspond to a separation between mean planes of 2.5 standard deviations; the dash, dot-dash, and dot boundaries to 2.0, 1.5, and 1.0 standard deviations, respectively.

* Micro-topographer 200, Gould-Cleavite, Inc., Gauging and Control Division, Elmont, California.

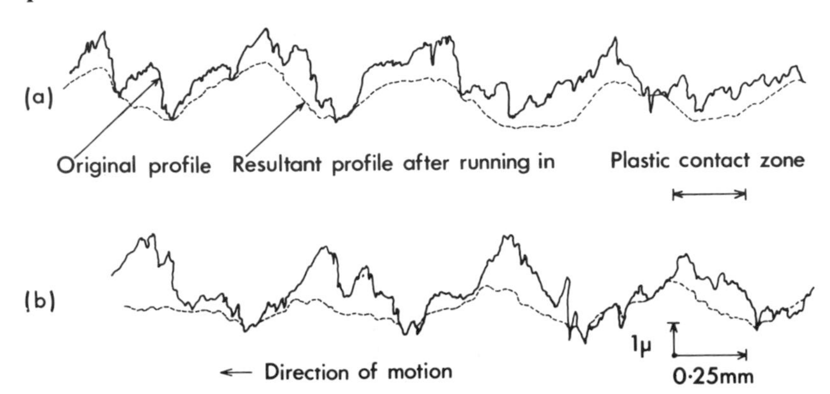

Fig. 3-4. Effect of the wear process on surface profiles: (a) in the center of the wear track, and (b) at the edge of the wear track.

It seems probable that work on the complete three-dimensional analysis of complex shapes including roundness, straightness, surface texture, and, ultimately, cylindricity will be investigated in this sort of way.

Another application of relocation profilometry is in the investigation of the changes of the surface topography during the time that the surface is being used.[19] In Fig. 3-4a, two profiles are shown. These are of a wear experiment carried out in a crossed-cylinders machine; the upper line shows the original profile and the lower (dotted) line shows how the surface has degenerated during running in. By data logging both profiles *in situ*, the most significant geometrical parameters can be found. In Fig. 3-4b a graph is shown which compares the original profile in the center of the wear track with a run-in (dotted) profile at the side of the track. Notice how the metal has flowed to the side. This, in this particular case, is due to the geometry of the wear apparatus. Experiments such as this can be carried out with a surety of quantitative results.

5.3. The Measurement of Step Height and Very Fine Surface Texture

The development of new instruments with very high vertical sensitivity enabling gains of 10^6 to be achieved, together with a corresponding reduction in the stylus forces by almost two orders of magnitude, has considerably extended the uses of stylus instruments in the research fields. Great advances have been made in recent years in the accurate measurement of the thickness of thin films.[18] Figure 3-5 shows a few typical examples. In Fig. 3-5a, the effect of exposure and bleaching on a high-quality photographic emulsion is shown. The precise amount of

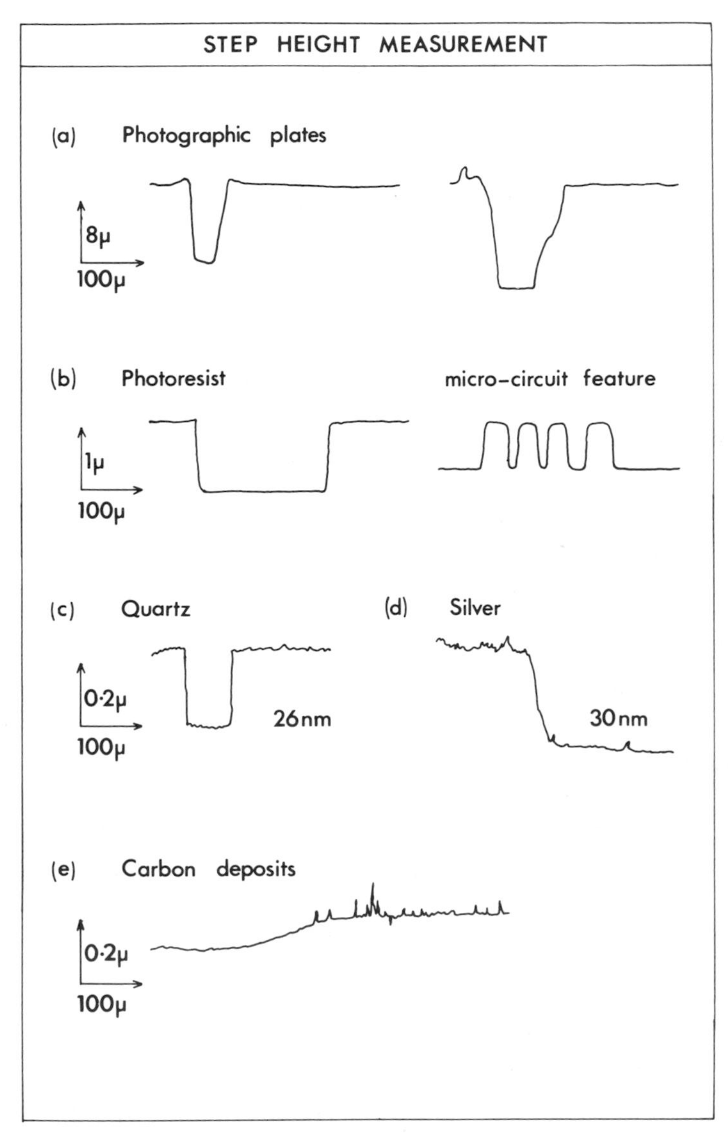

Fig. 3-5. Thin-film measurement using a 10-μ stylus tip: (a) photographic-emulsion thickness with and without exposure, (b) thickness of photoresist before and after development of feature, (c, d, e) measurement of thickness of thin films of varied hardness.

swelling can be measured. Figure 3-5b shows the effect of development on a photoresist substance, KTFR, deposited on a silicon dioxide substrate. The first picture shows the depth of a line drawn through the resist by a razor blade; the second picture shows the height of a micro-circuit feature on the processed slice. The reduction in the height of the feature relative to the line is due to losses of resist during development. Similar measurements can be made on the edges of coatings used in integrated circuits and also to investigate the growth of epitaxial layers.

So far, the layers in these examples have been of the order of 10 and 1 μ, respectively. Much smaller layers can also be measured, for instance, quartz on glass, 26 nm, (Fig. 3-5c); silver, 30 nm, (Fig. 3-5d); and carbon, 100 nm, (Fig. 3-5e). Films of metals as varied in hardness as chromium and gold have also been satisfactorily measured. As a rule of thumb, if the stylus has a tip of about 2.5 μ, then no breakdown of the film under loads of the order of tens of milligrams occurs in practice, even for the soft metals.

Advances have also been made in the measurement of ultrafine surface finish mainly due to the development of sensitive instruments and also to new techniques for the making of sharp styli.[22] Figure 3-6 shows some examples. Figure 3-6a shows how the texture of an emulsion on a photographic plate changes with exposure. This type of measurement is important in research into the texture of holograms (and in particular, phase holograms) which rely on this surface detail for their holographic properties. Spacings of a micron in the texture, and heights of much less than a micron are not uncommon in the best-quality holograms.

Another case where a knowledge of fine texture is useful is in the polishing of optical glass. Figure 3-6b shows two profile graphs. The first shows the texture on a lens near its center as measured with a stylus with a 0.1-μ tip. The other shows a profile near to the edge, illustrating a greatly increased roughness and, by inference, a lack of final polishing.

The apparently rough piece of metal shown in the next profile (Fig. 3-6c) is in fact an extremely finely polished gage block. This measure of roughness has been used as a quality-control factor.

The quality of carbon replication in electron microscopy (Fig. 3-6d) has also been investigated. The first profile shows a surprising amount of roughness even though an electron beam has been used to heat the source. Not surprising is that the roughness is considerably less than that obtained when using an electric arc, as shown in the second profile.

The final profile in this figure (Fig. 3-6e) shows how smooth a piece of cleaved topaz is. Using this technique it has been possible to measure the edges of dislocations and, hence, assess the lattice spacing.

Figure 3-7 shows the result of an experiment that has been carried out with an instrument having a sharp stylus, a high vertical magnification, a high horizontal magnification, and a relocation stage.

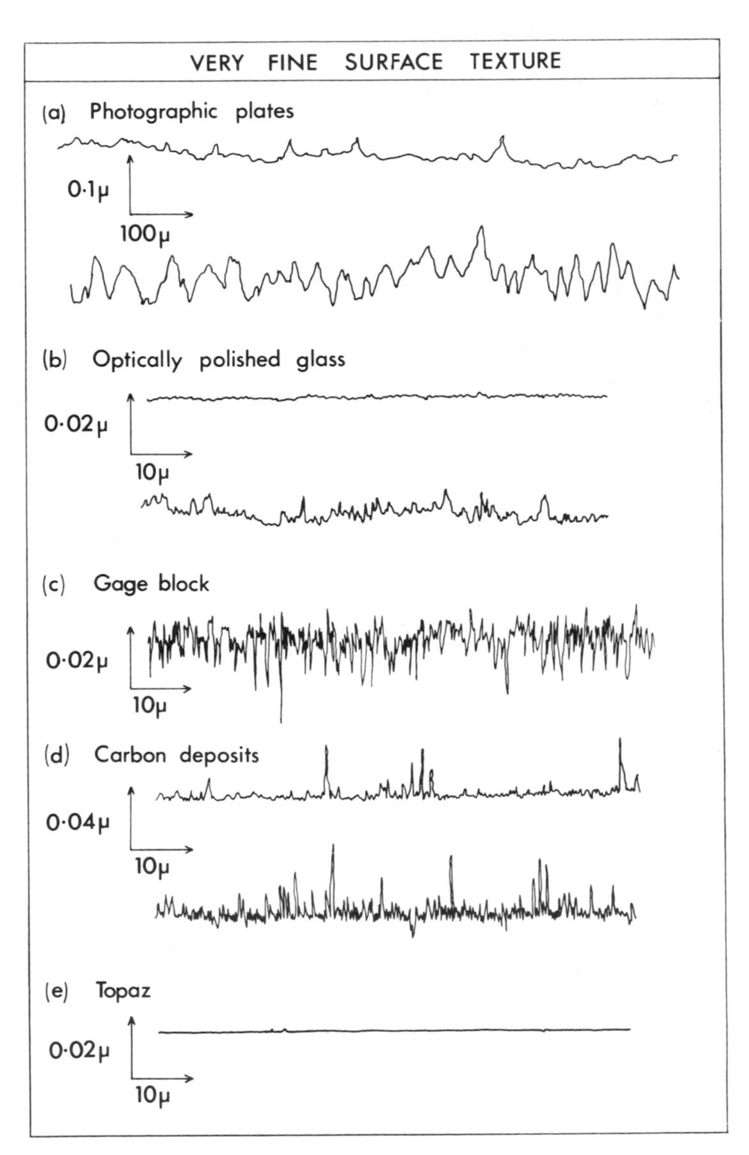

Fig. 3-6. Practical measurement of fine texture using light loads: (a) photographic plate with and without exposure using 10-μ tip, (b) well-polished and badly-polished glass using a 0.1-μ tip, (c) gauge block using a 0.1-μ tip, (d) comparison of texture produced on carbon film using electron beam source and electric arc using a 0.1-μ tip, and (e) molecular smoothness of cleaved topaz using a 0.1-μ tip.

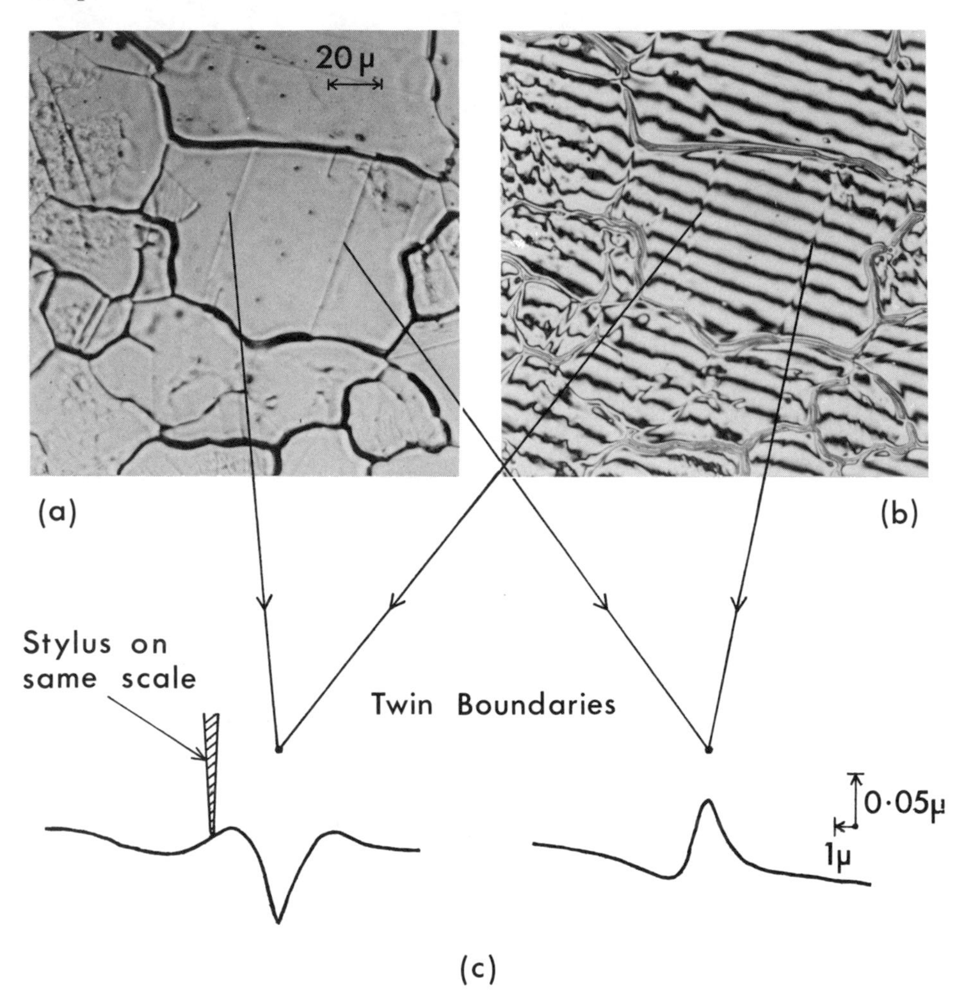

Fig. 3-7. Methods of measuring the profile of grain and twin boundaries in alloy steel to determine the surface energy: (a) microscopic, (b) interferometric, and (c) stylus.

The experiment was carried out by Nicholas[23] at U.K.A.E.A., Harwell. The research required very detailed measurement of grooves and ridges produced by thermal annealing of alloy steels, in particular, those ridges produced where grain and twin boundaries intersected the surface of the steel. It was necessary to make detailed comparisons between the profiles of the boundaries in order to measure the surface energy between the surface and grain boundary.

The figure shows a comparison of the results obtained using optical, interferometric, and stylus methods. Notice that in the interferometrical method, although the boundaries are clearly visible, they do not have sufficient detail to be able to make detailed measurements.

Unambiguous results having the accuracy required can only be obtained with the stylus method.

5.4. Physical Properties

Scientific areas other than those of a purely geometrical nature can also be studied using stylus techniques. For instance, the physical properties of surface layers have been investigated down to the order of depth of the surface texture; there has been a growing feeling in engineering research that the properties of the surface skin may be as important if not more important than those of the bulk material.

Work of a tentative nature has been carried out in a number of places. The following is an example of some of the work on this subject being carried out in the Rank Precision Industries Metrology Research Laboratories. A stylus instrument having a very sharp stylus of known angle and tip dimension is used. This instrument also has a means for continuously varying the load on the stylus from 1 to 300 mg. Figure 3-8a shows how the experiment is carried out. The stylus is tracked across the surface to a point of interest. A preload of 1 or 2 mg is then applied. This corresponds to the null position of the stylus shown in the figure. A further known load is applied, the stylus indents the surface, and finally the load is released. The permanent plastic flow and the elastic recovery in the skin (and instrument) can be found from the chart. Knowing the geometry of the diamond at the very tip and knowing the recovered indentation depth, the pyramidal area can be worked out and, hence, an equivalent Vickers hardness of the topmost skin of the surface. An estimate can also be made of the elastic recovery in the skin layers, but this has to be first isolated from that in the instrument.

A typical result on copper is shown in Fig. 3-8b. There are two graphs, one with the copper electrochemically polished, and the other with the copper mechanically polished. As one would expect, the mechanical polishing introduces a degree of work hardening. However, there is an apparent increase in the measured hardness for small loads which is common to both curves. This interesting phenomenon has been observed elsewhere,[24] and current thinking attributes it to a "size" effect. In any event, it may be of considerable importance in engineering.

The bringing together in one instrument of the measurement of physical and geometrical properties is illustrated in Fig. 3-8c. Here a profile is shown before and after indentation. Since only one instrument is used for the experiment and the position of the specimen or stylus is not changed, an excellent register is possible between the two profiles, thus enabling an investigation to be made of pile-up and metal flow during indentation.

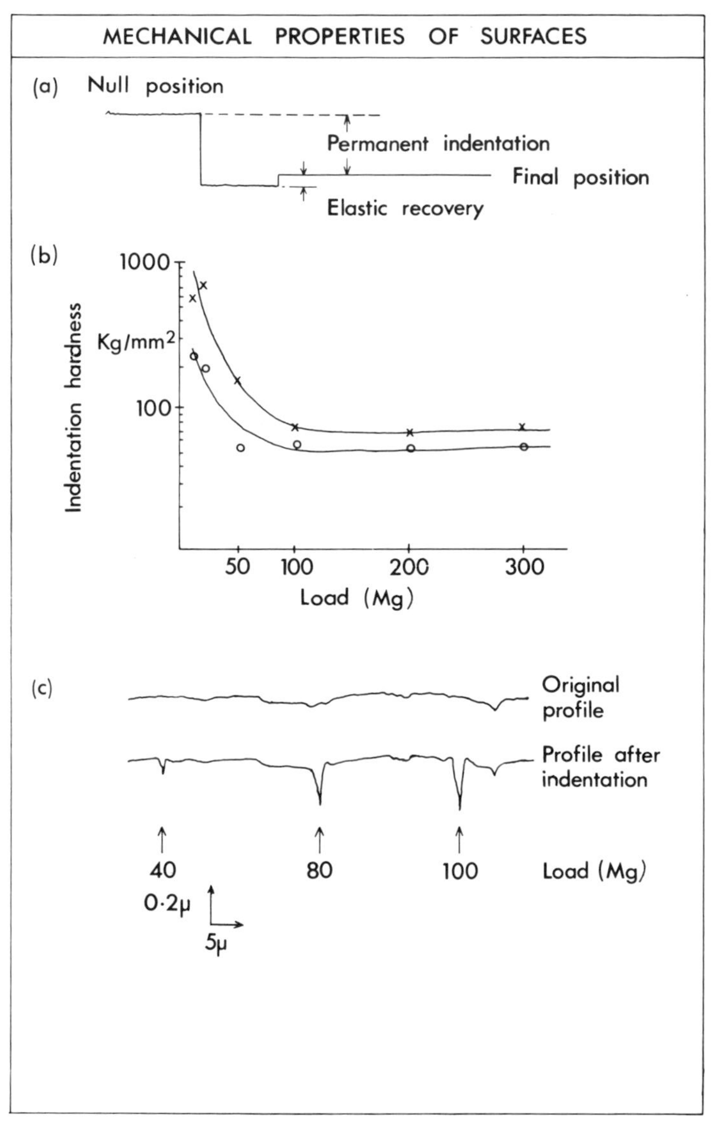

Fig. 3-8. Use of a stylus instrument to measure the geometrical and physical properties of the surface skin using a known stylus tip and a known stylus force: (a) typical trace from which the measurement is taken, (b) apparent reduction in measured hardness with load, (c) tracing of surface profile before and after indentation.

6. CONCLUSIONS

This brief note has been intended to give a background in the use of stylus instruments for the measurement and classification of surfaces. Emphasis has been placed on the measurement of texture rather than roundness, etc. Because of the extensive coverage in the literature of the many instances in engineering where the technique has proved to be useful, some examples of a research nature have been demonstrated. It is hoped that not only the advantages, but also some of the disadvantages, of stylus methods have been brought out.

Finally, there is little doubt that, taking all into account, the stylus technique seems to be rapidly developing in sophistication, in the number and variety of applications, and, last but not least, as an economic tool. There is no reason to believe that this trend will change in the future.

7. ACKNOWLEDGMENTS

I would like to thank Dr. J. B. P. Williamson and the Institution of Mechanical Engineers for permission to publish Fig. 3-3 and the Director of U.K.A.E.A., Harwell for permission to publish Fig. 3-7. I should also like to acknowledge the contribution of Mr. J. Jungles to the results shown in Figs. 3-5 and 3-6, and to Mr. R. T. Hunt for the results shown in Fig. 3-8. Finally, I would like to thank the Directors of Rank Precision Industries for permission to publish this work.

8. REFERENCES

1. R. E. Reason, *The Measurement of Surface Texture, Modern Workshop Technology*, Part 2, Macmillan, London (1970).
2. R. E. Reason, M. R. Hopkins, and R. I. Garrod, 1944 Report on the Measurement of surface finish by stylus methods, The Rank Organization (1963).
3. R. E. Reason, Le calcul automatique des critères des profils de surfaces, *Automatisme IX* **5**, 177–178 (1964).
4. J. E. Torjusen, On-line statistical analysis of surface roughness by means of a small digital computer, Institute for Maskinteknisk Fabrikkdrift og Vertoymaskiner Trondheim, Norway (1968).
5. D. Kinsey and D. G. Chetwynd, *Some Aspects of the Application of Digital Computers to the On-line Measurement of Surface Metrology*, Imeko, VI, Dresden (1973).
6. D. J. Whitehouse and R. E. Reason, *The Equation of the Mean Line of Surface Texture Found by an Electric Wave Filter*, Rank Taylor Hobson, Leicester, England (1965).
7. D. J. Whitehouse, Improved type of wavefilter for use in surface finish measurement, *Proc. Inst. Mech. Engrs. (London)* **182**, Part 3K, 306–318 (1967–1968).
8. R. C. Spragg and D. J. Whitehouse, A new unified approach to surface metrology, *Proc. Inst. Mech. Engrs. (London)* **185**, 47/71, 697–707 (1970–1971).
9. R. D. Butler and J. R. Pope, Surface roughness and lubrication in sheet metal working, *Proc. Inst. Mech. Engrs. (London)* **182**, Part 3K, 162–170 (1967–1968).
10. J. F. Kahles and M. Field. Surface integrity—A new requirement for surfaces generated by material—Removal Methods, *Proc. Inst. Mech. Engrs. (London)* **182**, Part 3K, 31–45 (1967–1968).
11. J. Peklenik, Investigation of the surface typology, *C.I.R.P. Annals* **15,** No. 4, 381–385 (1967).

12. D. J. Whitehouse, Typology of manufactured surfaces, *C.I.R.P. Annals* **15**, No. 4, 417–431 (1971).
13. T. Al-Salihi, Ph.D. Thesis, Univ. of Birmingham (1967).
14. M. Pesante, Determination of surface roughness typology by means of amplitude density curves, *C.I.R.P. Annals* **12**, No. 2, 61–68 (1963).
15. N. O. Myers, Characterization of surface roughness, *Wear* **5**, 182–189 (1962).
16. J. A. Greenwood and J. B. P. Williamson, Contact of nominally flat surfaces, *Proc. Roy. Soc. (London)* **A295**, 300–319 (1966).
17. R. E. Reason, Report on the measurement of roundness, Rank Precision Industries (1966).
18. R. J. King, M. J. Downs, P. B. Clapham, K. W. Raine, and S. P. Talim, A comparison of methods for accurate film thickness measurement, *J. Phys. E.* **5**, 445–449 (1972).
19. D. J. Whitehouse, Modern methods of assessing the quality and function of surface texture, *Soc. Manuf. Engrs.* **1Q72–206** (1972).
20. R. C. Spragg, Accurate calibration of surface texture and roundness measuring instruments, *Proc. Inst. Mech. Engrs. (London)* **182**, Part 3K, 397–405 (1967–1968).
21. J. B. P. Williamson, Microtopography of surfaces, *Proc. Inst. Mech. Engrs. (London)* **182**, Part 3K, 21–30 (1967–1968).
22. J. Jungles and D. J. Whitehouse, An investigation of the shape and dimensions of some diamond styli, *J. Phys. E.* **3**, 437–440 (1970).
23. Application Report 26, Talystep in the laboratory, Rank Precision Industries (1968).
24. N. Gane and J. M. Cox, The micro-hardness of metals at very low loads, *Phil. Mag.* **22**, 881–891 (1970).

4

ELECTRON MICROSCOPY

Campbell Laird

School of Metallurgy and Materials Science
University of Pennsylvania
Philadelphia, Pennsylvania

1. INTRODUCTION

There is hardly a field in materials science where the physical nature of the surface is not an important feature. For example, in fatigue fracture, cracks nucleate at the surfaces of materials and the rate at which they nucleate is greatly influenced by the detailed topography of the surfaces. In the field of thin-film devices, the manufacturing tendency has been to reduce the size of electronic components. Surface-to-volume ratios are now exceedingly high. Young points out that we are not far from the point where we can anticipate devices employing single layers of atoms.[(1)] However, the device industry, which presently employs films in the 10- to 100-Å range, suffers very high failure rates due to surface imperfections, stacking-fault intersections, voids in the films, thermally induced pits, and multiple steps. As a result of these deficiencies, large resources have been employed to control the imperfections by close control of processing variables. In other areas, elaborate polishing, cleaning, and smoothing techniques have been developed in an effort to eliminate the variability associated with surfaces. However, none of these efforts can improve upon a detailed knowledge of the actual surface topography.

The purpose of this chapter is to describe how transmission electron microscopy has been, or can be, applied to the study of surfaces. The transmission microscope is similar to the ordinary optical microscope in that it simultaneously illuminates the whole specimen area and employs Gaussian optics to generate the image. This is the only type of electron microscopical instrument to be considered here. A comparative review of the capability of all kinds of topographic measurers has been given by Young,[1] and the flying-spot and other types of instruments are treated in detail elsewhere in this volume.[2] However, it is worth pointing out briefly the advantages and disadvantages of the transmission microscope with respect to the scanning microscope, its most serious competitor, at least in terms of numbers. Unlike the transmission microscope, the scanner illuminates only one spot on the specimen at a time and forms its image sequentially. The transmission microscope (as is generally true of types that employ Gaussian optics) has greater resolving power than an equivalent scanner, and it spreads the illumination over the whole specimen rather than concentrating it in one high-density spot. In consequence, the scanner must employ a much smaller beam current than the transmission microscope and, in my experience, causes much less overall specimen damage than the transmission microscope in highly susceptible materials such as polymers. On the other hand, the transmission microscope, working with metals and regular accelerating voltages (100–150 kV), and equipped with a good decontamination device, can operate virtually *ad infinitum* without serious deterioration of the area under observation. The same is hardly likely in the case of a scanning instrument, unless it also is equipped with a good decontamination device.

Flying-spot instruments permit point-by-point analysis of surface properties. At first sight it would appear that transmission microscopes, illuminating the whole sample, would not be capable of such application. In general this is so. However, a new transmission microscope, named EMMA 4, has been developed with combined transmission microscope and probe capability by the introduction of a "minilens" in the illumination system.[3,4] This instrument should be considered a special case of microprobe analysis, also treated in this volume.[5] EMMA 4 has demonstrated considerable power in a number of applications and could easily be applied to surfaces, but it will not be further considered here because our primary emphasis is on topography.

A great advantage of the scanning instrument is its ability to deal with bulk specimens. Unfortunately, nonconducting samples have to be given a light coating of metal, typically gold; otherwise, charging effects will seriously impair the resolution of the image. Transmission microscopes are not subject to this limitation and the techniques to be described here apply universally to all materials. Such a statement is, of course, "theoretical" because numerous practical problems beset the preparation of all kinds of materials for observation in the transmission microscope.

In the transmission microscope, the electrons that form the image must pass through the specimen; thus the specimen thickness is limited to a few thousand angstroms, or to a few microns for the fortunate owners of a high-voltage instrument. If one is to study the surfaces of solids, two approaches are possible. In one approach, a replica of the surface can be made—for example, a carbon replica may be made by vacuum-depositing a 100- to 1000-Å film on the surface—and be carefully removed by some etching technique and then mounted in the microscope. The image obtained from such a replica does represent the surface topography, but it is frequently subject to distortion and artifacts and is often difficult to interpret. Moreover, the process of replication seriously cuts down the resolution ultimately obtainable with the instrument. In the other approach, it is necessary to plate a suitable material onto the surface of interest and then to section a thin slice normal to that surface. The section is then mounted for observation in the microscope and it permits one to observe the surface in profile. The resolving power of the instrument can be fully exploited by this method (the profile method) and it has the additional advantage of revealing the surface topography in relation to the underlying structure of the material. Alternatively, to obtain a bird's-eye view of a surface rather than the one-dimensional profile of a section, it would be feasible, for many materials, to protect the surface of interest with a nonconducting layer and then to electropolish the sample from the "other side." Local thinning near the surface would enable an observer to see rather large areas of the surface if the nonconducting layer could be dissolved away by washing in a suitable solvent.

The aim of this volume is too broad to permit detailed description of any one kind of instrument or of the theory by which it is employed. Since many excellent books have been written on the microscope itself,[6–15] on methods of preparing specimens,[7,11,14–16] and on the theory of contrast,[9,11–13] only a very brief description of contrast principles and specimen-preparation methods will be given here. The main thrust will be to describe methods and applications where replication and sectioning techniques have been successfully employed to study surfaces, with the aim of illustrating the scope of the instrument, the resolution obtained, and the limitations of the methods.

2. CONTRAST THEORY

The problem we now have to solve is how to interpret the electron images obtained by the two approaches available for studying surfaces:

the replication and profile methods. Since the electrons pass through the samples, the images formed from them are going to be strongly affected by the interaction of the electrons with the material of the sample. The atomic spacings of most materials and the wavelengths of the electrons obtained from the accelerating voltages employed are suitable for diffraction effects to occur. Both elastic and inelastic scattering is possible, with the inelastic type dominating in thick samples (>2000 Å for 100 kV electrons) and the elastic type more important for readily usable thicknesses. Many different types of inelastic scattering occur,[11,13] plasmon losses, phonon interactions, bremsstrahlung radiation, and so forth. The net effect is that some of the incident electrons are deflected from the collimated, axially parallel beam focused on the specimen by the illumination system. As shown in Fig. 4-1, these de-

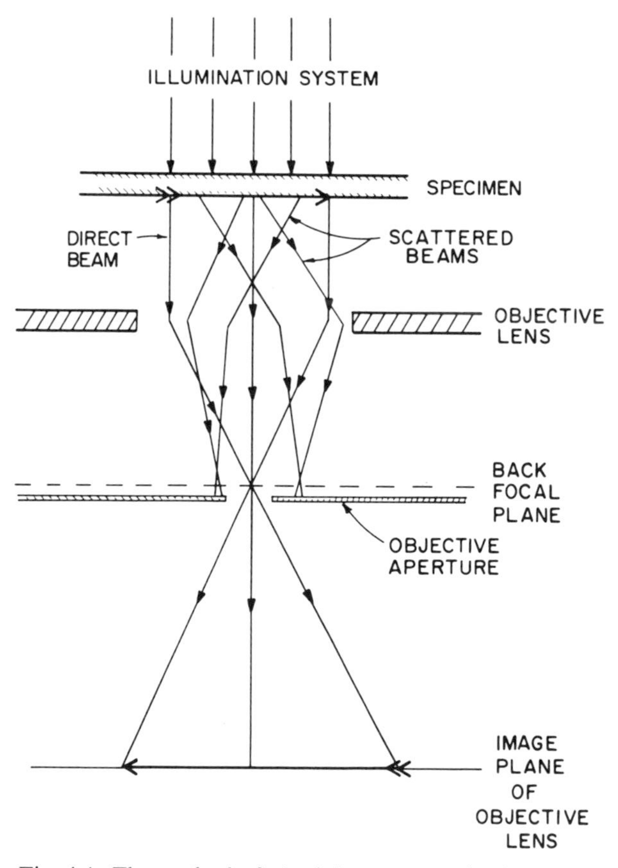

Fig. 4-1. The method of obtaining contrast in the transmission microscope by inserting an objective aperture in the back focal plane and stopping off the scattered beams focused on that plane.

flected beams are focused at different points in the back focal plane of the objective lens. To obtain contrast in the image, an objective aperture is inserted in the back focal plane to block the scattered beams and to permit only the direct beam to form an image in the projection lens system of the microscope. This image is called the bright-field image and its details are determined by the extent to which scattering has occurred in different regions of the specimen. Alternatively, one can form a dark-field image by shifting the objective aperture laterally so as to block the direct beam and to permit only one of the scattered beams to pass into the image system of the microscope. The different information contained in the bright- and dark-field images can be employed to determine many details about the imperfections contained within the specimen or at its surface.(11–13)

Although this method of obtaining contrast is quite general, the scattering processes involved are going to vary widely for different materials, and it is convenient to discriminate between those which occur in the two approaches employable for studying surfaces. In the replication method, most replicas are essentially amorphous. The diffraction of electrons from replicas is therefore going to be different from the type which occurs in profile sections which are more likely to be crystalline. In replicas, the diffraction patterns (i.e., the distribution of electron intensity in the back focal plane) are rather hazy affairs with a fairly high intensity scattered at a Bragg angle corresponding to the most populous interatomic spacing (see the schematic in Fig. 4-2a). Since the structure is generally uniform, intensity distributions in the electron images are also uniform unless the thickness of the replica varies. The scattering amplitude must increase where the thickness of the replica is greater, due to a greater probability of scattering events. Therefore such regions will appear darker in relation to the background in a bright-field image. The way by which the topography of a surface affects contrast is shown in Figs. 4-2b–d. Clearly, if a uniformly thick replica is taken from a surface, a step will give a region two or more times as thick as the average, and inclined regions will have their effective thickness increased by the secant of the angle of inclination. Heidenreich has worked out in detail the contrast to be expected from such specimens.(9)

It usually happens that the materials used for replication, such as carbon, are so transparent to electrons that small thickness variations produce no observable contrast. It is usual, therefore, to enhance contrast by shadowing the replica with a heavy metal, which produces marked variations in contrast. In addition, the shadows help to bring out height differences in the specimen and open the way to obtain

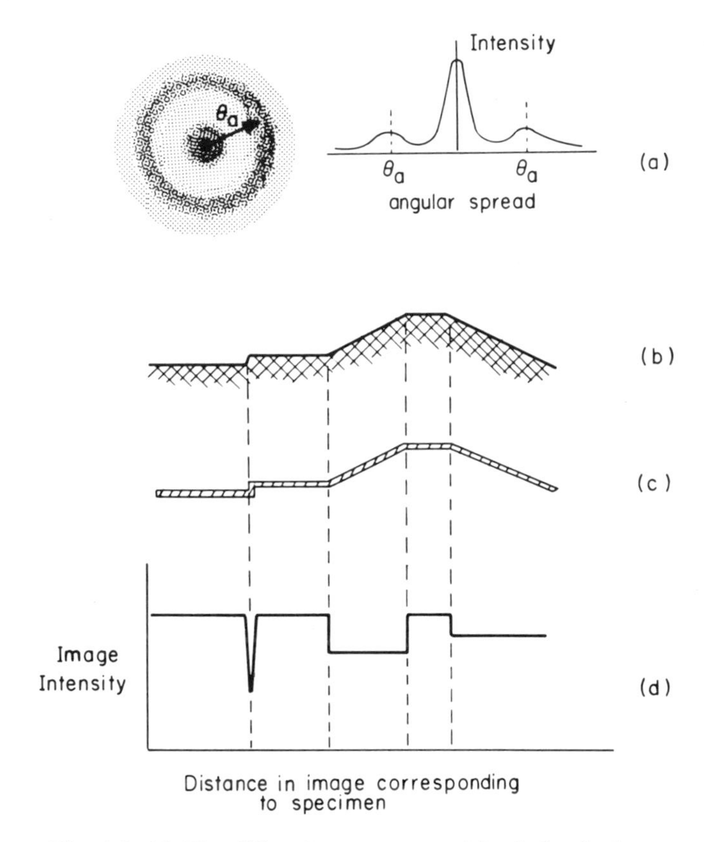

Fig. 4-2. (a) The diffraction pattern and back focal plane intensity plot obtained from an amorphous substance; (b) a surface profile to be studied by replication; (c) the replica, uniform in thickness, taken from the surface in (b); and (d) the intensity plot observed in a Gaussian image formed from the replica in (c).

quantitative information about the surface topography via stereomicrometry.

For profile specimens, the ordered nature of the crystals will give rise to marked elastic scattering of the incident beam. If the specimen is monocrystalline, the diffraction pattern will be a spot pattern, readily identifiable by the techniques described in much more detail elsewhere.(11) Since the theory of electron diffraction is well understood,(9,11,13) detailed quantitative information can be obtained from the specimen by tilting it *in seriatim* to different orientations and exciting a variety of Bragg reflections.(9–13) This information can be obtained both about the crystallography of the specimen and the defects within it. Although both the specimen to be observed in profile and the material plated on the surface to preserve its outline can be uniform in thickness, the contrast is entirely different from that observed in a replica. Small deflections inevitably occur in the ordered atomic arrangements of the crystals, either by elastic strains (either gross or localized to defects) or by variations in composition. Since electron wavelengths are generally quite small in relation to atomic spacings, Bragg's law indicates that the scattering angle of the diffracted beams should also be small, $\sim 1°$. Consequently, even small elastic strains in a specimen can locally distort the crystal away from the Bragg condition and cause large variations in scattering intensity and, therefore, in image contrast.

Interpretation of the Gaussian image therefore requires a detailed knowledge of diffraction theory. The kinematical theory of diffraction, familiar from x-ray theory, is not suitable for electron microscopy because this theory requires only a small proportion of the incident electrons to be scattered. The most interesting effects, however, are observed at crystal orientations near Bragg reflections, where large intensities are scattered and the kinematical theory no longer applies. Instead, observers turn to the dynamical theory of diffraction, formulated either in terms of wave mechanics or in those of wave optics. The wave mechanical formulation is much more convenient and the mathematics employed are relatively simple.[11–13] In this theory, allowance is made for large scattering intensities. Further, if the crystal orientation satisfies the Bragg condition for the direct beam, then it will also satisfy it for the diffracted beam (see Fig. 4-3). Consequently, the diffracted beam will be doubly diffracted back into the direct beam. If the crystal is thick enough, and it usually is, multiple reflections will occur. Dynamical theory takes such reflections into account, and also two other important factors, as follows: the electrons undergo an interaction with the potential of the crystal which varies periodically with the atomic array. Electrons between atom rows experience a higher potential energy than electrons in the neighborhood of the atoms. Since the electrons are given a fixed amount of energy by the electron gun, those electrons with higher potential energy will have a lower kinetic energy and, therefore, a different wavelength. Consequently, the electron beams within the crystal beat together. This interference is normally observed as fringe patterns at inclined interfaces within the specimen (grain boundaries, twins, stacking faults, precipitates, and any other planar defects). It permits an observer to see such interfaces if he tilts the specimen close to the Bragg condition where the interference effects are optimized. Finally, the dynamical theory takes into account absorption losses of electrons, which constitute another fruitful source of informative imaging phenomena.

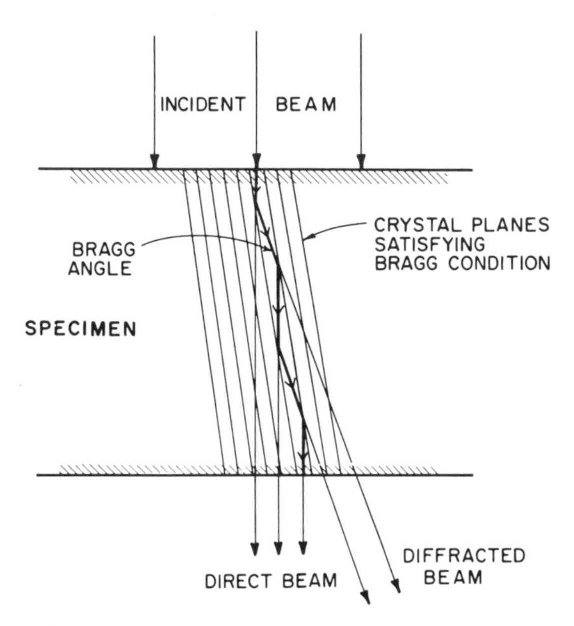

Fig. 4-3. The geometry of direct and diffracted electron beams illustrating the possibility of multiple diffraction.

The changes thus caused in the electron images of crystalline specimens allow us to study the nature of surfaces in profile specimens and also in those specimens where the surface of interest is also one of the surfaces of the thin object film. Of course, the details are quite complicated but they are thoroughly documented.[9,11–13]

3. TECHNIQUES

Replication techniques have been developed to a considerable degree of sophistication, comprising both one-stage and two-stage methods, and make use of a wide variety of replicating materials, depending on the application.[16] Plastic replicas have a serious resolution limitation in that the molecule of the plastic itself may be larger than the resolving power of the instrument; the aggregate of the replica can interfere, then, with the fine details of the surface of interest. Consequently, shadowed carbon replicas, having much better resolution, are used almost exclusively in the most exacting work; they are generally prepared by a two-stage method. In this method, the surface of interest is coated with a Formvar (or equivalent) film by flooding with a 2% solution of the plastic in chloroform and draining the excess away. When the film is dry, it is scored into squares, backed with Scotch Tape, and stripped for evaporation of carbon onto the surface structure of the film. Then, 100–200 Å of carbon are deposited, along with a lighter deposit of shadowing metal, which can be applied before, during, or after the evaporation of the carbon. Finally, the plastic film is dissolved by washing it in a solvent such as acetone, and the carbon replicas are netted out for mounting in the microscope. The stages of this replication process are illustrated schematically in Fig. 4-4. Although the necessary procedures and artistic hints are supplied in the standard texts,[14–16] considerable practice is required to obtain good results, and the details vary greatly in relation to the craft of the practitioner. The most important factor seems to be the thickness of the plastic film. The smoother the surface to be replicated, the thinner the plastic film should be, with the main objective of preventing the disintegration of the carbon replica during washing of the plastic. Before the plastic dissolves, it expands; the strains transmitted to the carbon replica can be destructive. For very rough surfaces, such as fractures, the corrugations in the carbon replica strengthen it, and thicker plastic coatings can be used.

While a carefully prepared replica should accurately reflect horizontal distances in the surface of interest, greater doubts may exist about how well vertical dimensions are represented. Gross features in the vertical direction will collapse and rumple, of course, but from a wide variety of experiences with powder samples, surfaces with slip steps, and fracture surfaces, reasonable confidence has been obtained

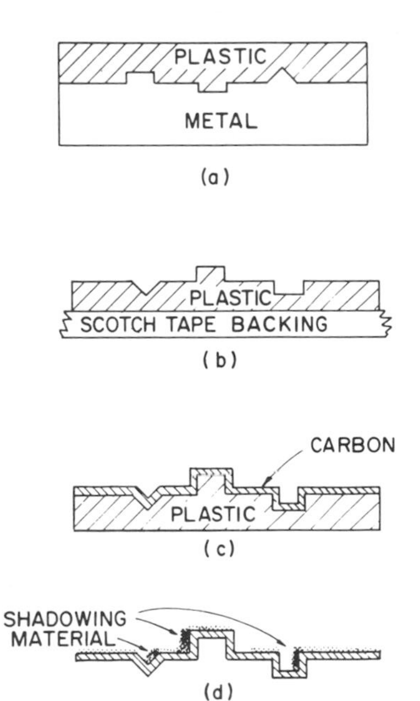

Fig. 4-4. The two-stage (plastic/carbon) replication method for solid surfaces: (a) surface covered with a layer of plastic, (b) the backed plastic replica after stripping and (c) coating with carbon, and (d) the final carbon replica after removal of the plastic by washing.

that, over short distances, the vertical features of a surface are accurately replicated and retained in the microscope. Thus stereomicroscopy, permitting the surface to be viewed and measured directly in three dimensions, is regarded as a valid technique.

The value of routine stereoscopic viewing in understanding complicated surfaces does not appear to be widely appreciated. The depth definition afforded by the technique gives a clearer overall view and shows up otherwise-hidden details, just as aerial stereophotographs are used to spot camouflage.[17] Thomas,[17] and others,[18,19] have pointed out how easy it is to obtain stereo pictures as follows: two different views of the same specimen area are required at different tilts. Depending on the amount of stereo effect desired at the final viewing magnification, the relative angle of tilting is 10–20°. By tilting the replica equal angles about its horizontal position, the stereo view will appear to lie horizontally. For depth measurements, the angle of tilt between the two views Φ and the magnification M should be accurately known. The height differences Δh between vertically spaced points in the image can be measured from the parallax Δp between the corresponding pairs of image points in the two views of the stereo pair, using the relation $\Delta h = \Delta p/[2M \sin(\Phi/2)]$.[17] The parallax is measured by a stereomicrometer which is placed on the photographs. Full details about how to use such an instrument and how to position and orient the micrographs with respect to the stereoviewer can be obtained from the various manufacturers' handbooks.[20] Complete surface profiles can be obtained by such techniques, and contour maps can be drawn. However, a very large number of measurements is required to obtain contours, and the procedure is exceedingly tedious. In spite of the reluctance an observer might accordingly feel, a few examples can be found in the literature.[21] Useful hints and general experience may be gained from stereo applications in scanning microscopy and reflection electron microscopy.[22–25]

Frequently, it is important to learn whether surface features are depressed or elevated with respect to the average surface level. Although a decision can sometimes be made unambiguously by study of the shadow morphology, the correct perspective can be obtained from stereopairs if the relative sense of tilt between the two views is known.[17] If the micrographs are not properly viewed, elevations and depressions will appear to interchange. In Fig. 4-5, a bump on a surface, replicated in a two-stage replica as a depression, is used to show the relationship between three-dimensional space as interpreted through the eyes of an observer, and a pair of stereophotographs. Each eye views one picture of the stereopair and the mind corrects the two images into a three-

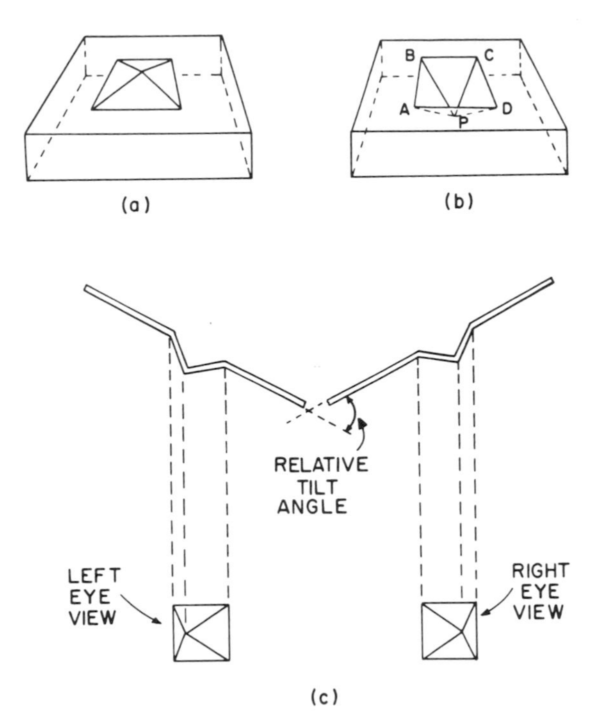

Fig. 4-5. The correlation between a three-dimensional object as viewed by human eyes, and via the formation of a stereopair: (a) the surface containing a bump, (b) the plastic replica taken of the surface, and (c) the second-stage carbon replica viewed in stereo. The left eye would see the APB facet as being contracted with respect to the CPD facet, and *vice versa* for the right eye. (Adapted from Thomas.[17])

dimensional image. Clearly, if the carbon replica were tilted in the opposite sense, the depression would appear differently.

Methods of preparing transmission specimens of metals and other materials are also widely described.[7,11,16] Essentially they consist of obtaining a uniformly thick ($\sim$0.005 inch) foil and thinning subsequently by electropolishing until the foil perforates. If polishing is then immediately stopped, the edges of the holes are usually thin enough to transmit electrons for observation, and small specimens can be cut from the edges for mounting in the microscope. To apply the profile method for observing a surface in section, a rather more controlled method must be used to perforate a section in the exact location of the surface. Sections for subsequent thinning can be cut by spark machining,[16] and an accurately located hole obtained either by jet polishing or by ion thinning

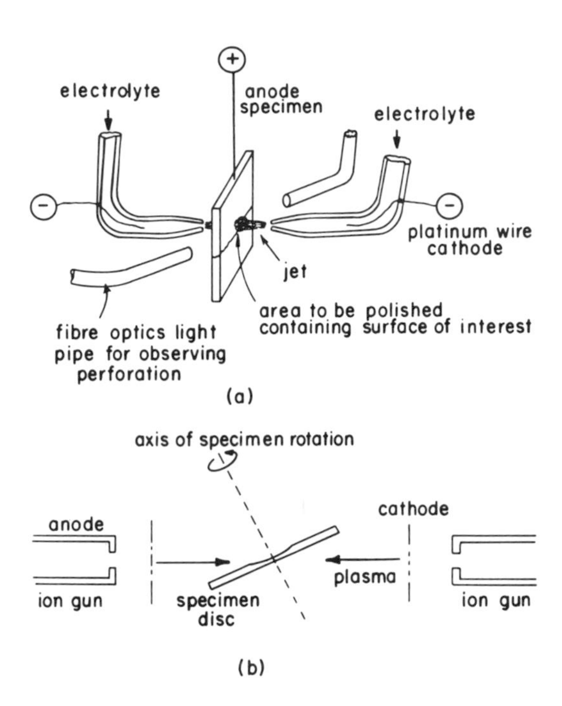

Fig. 4-6. Methods of thinning electron microscope specimens at a precisely located spot containing a surface of interest: (a) jet electropolishing, (b) ion thinning.

(see Fig. 4-6).* The latter method has been relatively rare until recently, but its advantages are now widely appreciated. A typical apparatus consists of opposing ion guns which bombard both sides of a specimen with 5- to 10-kV argon ions *in vacuo*. Atoms are knocked off the surface by a sputtering process and the surfaces are smoothed by inclining the specimen to the axis of the beams and rotating it. Since the ion guns can be focused to a spot approximately 1 mm square, the accuracy required to locate a perforation exactly on a surface/protective-layer interface is easily obtained. Ion damage to the sample can also be minimized by inclining the specimen, and the method will work for virtually any material. Numerous references to special methods of jet polishing and ion machining with designs for jigs to hold specimens can be found in journals such as *Review of Scientific Instruments* and *Metallography*.

* Previous reference to sectioning in the profile method mentioned cutting at right angles to the surface. The advantages of sectioning at a glancing angle to the surface should be pointed out, however, because a "taper magnification" is made possible by such a cut. Moreover, the electron microscopic foil specimen would also contain a large area of the surface inclined to the electron beam, rather than parallel to it, permitting direct examination of its topography.

4. TRANSMISSION MICROSCOPY APPLIED TO THE STUDY OF SURFACES

4.1. Replica Method

4.1.1. Solid–Vapor Surfaces

The literature dealing with the application of transmission microscopy to solid–vapor surfaces is enormous. The purpose of this chapter is not to review that material, however, but to show what the transmission microscope can do. A few recent references illustrating different aspects of the replica method have accordingly been selected for detailed discussion. This selection has also been made so as to include studies of different kinds of material.

In metals, frequent occasions for studying surface topography are provided by the field of deformation, where slip-band morphologies yield valuable information about the deformation mechanisms involved. A typical example is provided by the work of Desvaux and Charsley.[26,27] These investigators observed and measured slip-line densities on the surfaces of both copper and Cu–Al alloy single crystals. They were interested in how the slip-step height and density varied with strain in the Stage I regime of deformation (i.e., at small strains). They were also interested in how strain reversal took place when, after a small prestrain in tension, the specimen was given a small compressive strain. One hypothesis that they investigated was that dislocations generated in the prestrain would simply reverse their paths during compression. If this were correct, then it would follow that at least some of the slip steps formed during the tensile prestrain should become smaller, disappear, or change sign during compression. On account of the small strains involved, the slip lines were very faint and extreme demands for both horizontal and vertical resolution were made on the workers.

To study this problem, Desvaux and Charsley used two-stage Pt–C replicas (first stage, Formvar) taken from the surface region where the Burgers vector was normal to the direction of the slip lines in the single crystals. It was necessary to look at the same part of the gauge length after each tensile strain increment to avoid strain inhomogeneities affecting the results of slip-line density. Observation of the same surface region (2 × 2 mm) was effected by making two faint scratches 3 mm apart on the relevant part of the surface. These were visible on the shadowed replica and the area between them was cut out and observed after each strain increment. To select the same area for the reverse straining experiments, a trapezium-shaped area was electropolished on the crystal (see Fig. 4-7), the rest of the surface being protected by

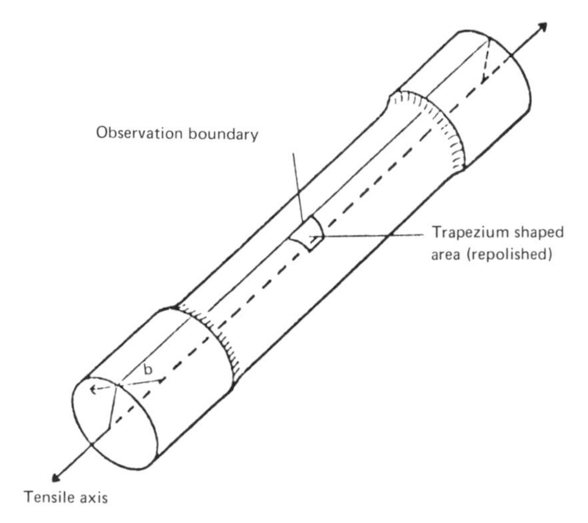

Fig. 4-7. A view of the specimen gauge length employed by Charsley and Desvaux.[26,27] (From Fig. 2 in Reference 26.)

nonconducting lacquer. The long side of the trapezium lay across the slip lines. This region could be seen on the replicas and slip-line counts were made on both sides of the boundary line in order to discriminate between the new lines forming in compression and changes occurring in the slip lines formed initially in tension.

A typical result is shown in Fig 4-8 (from Charsley and Desvaux[27]). The specimen shown in this figure was of Cu–Al; it was prestrained by 0.008 in tension and, after a boundary was polished parallel to the tensile axis, was compressed to a strain of -0.012. Replicas of the boundary revealed that new slip lines had formed during compression in the vicinity of the tensile slip clusters. Furthermore, many compressive lines joined onto the tensile slip lines at the boundary. The right-hand side of the micrograph shows the initial tensile lines, not removed during the polish. The left-hand side (polished part) contains blunted traces of the tensile lines present before polishing and a curved compressive line (B) which joins the tensile line (A) at the base of the step. Charsley and Desvaux thus concluded that the slip components in tension and compression are shown resolved. Furthermore, since no white slip trace can be seen along A, it is clear that A has partially unslipped during compression; the magnitude of reverse slip was determined from a measurement of the width of B and a knowledge of geometrical slip factors.

As another example of the interesting slip effects revealed by this technique, Fig. 4-9 shows a cross-slip reaction on an unpolished part

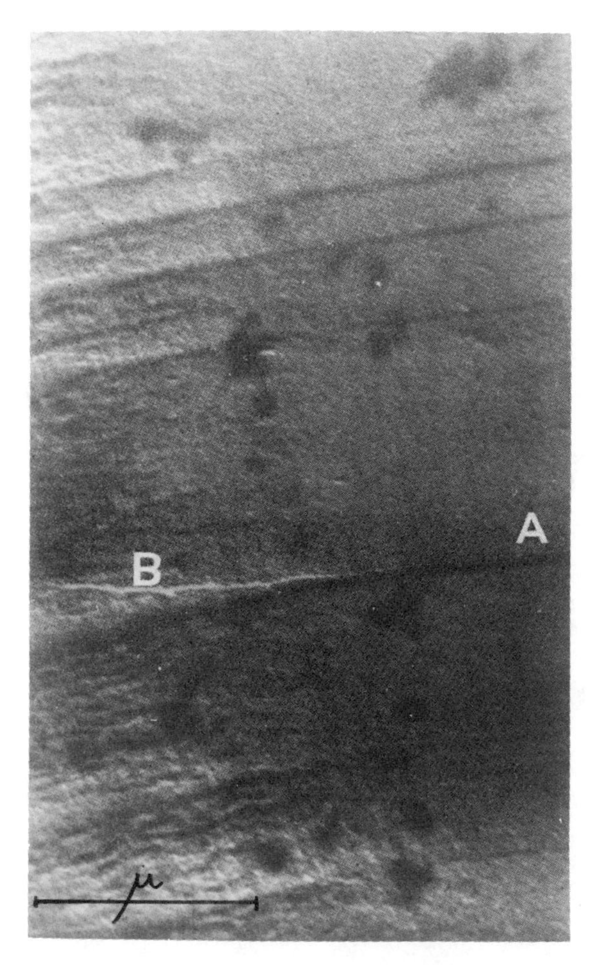

Fig. 4-8. Partial reverse slip, revealed by replication, in a Cu–Al crystal after one cycle of tension and compression. (Courtesy of Charsley and Desvaux.[27])

of the specimen shown in Fig. 4-8, but after an additional cycle of tension and compression. A cross-slip step connects a compressive (P_1) and tensile slip line (P_2) together, which both end abruptly at the cross-slip step. This result has useful implications for mechanisms of fatigue crack nucleation,[28] and Charsley and Desvaux were able to substantiate the dislocation reversal hypothesis, never before verified directly.[27]

An important step in this investigation was the development of the high-resolution replica technique which enabled Charsley and Desvaux to measure the incremental change in slip-step heights following a strain increment.[29] This technique, which was applied to Cu–Al alloy, employs the trapezium-shaped electropolished region described above

Fig. 4-9. Cross slip joining tensile and compressive slip lines, P_2 and P_1, respectively, on the surface of the crystal shown in Fig. 4-8. (Courtesy of Charsley and Desvaux.[27])

and shown in Fig. 4-8. A schematic view of the boundary between the electropolished region and protected surface is shown in Fig. 4-10. The desired result is the amount of glide G represented by the width W of δ, measured normal to the general trace of the slip lines at the point T. By trigonometrical manipulation,[29]

$$G = W/\sin \chi \sin \beta(\cot^2 \chi \cot^2 \beta - 1)^{1/2}$$

where χ is the angle between the Burgers vector of the dislocations emerging in the slip band and the surface and β is the angle between the

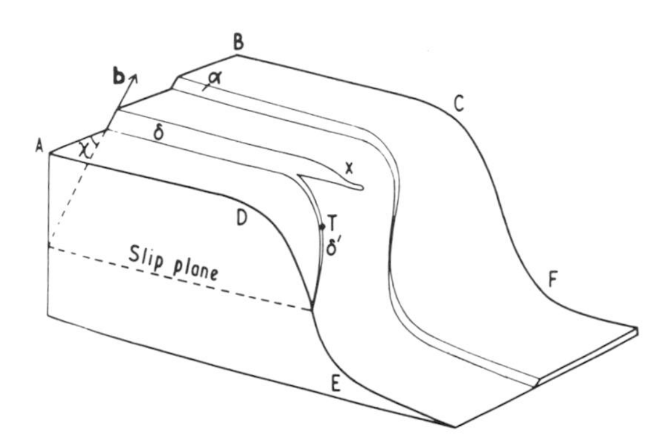

Fig. 4-10. Schematic view of the boundary shown in Fig. 4-8, with an incremental slip line δ' and a slip line α which first appeared after the strain increment. (Courtesy of Desvaux.[29])

tangent to δ' at T and the general trace of the slip bands. Desvaux felt that accurate measurement of the step height, i.e., W, begins only when the step is over 100 Å high, and gave the following typical data: $W \simeq 0.04 \pm 0.01$ cm, where $\beta = 27°$, at a magnification of 30,000 × ; $\chi = 47°$.[29] Thus $G = 255 \pm 65$ Å. By very careful shadowing, he was able to reduce the error to 30 Å, the limit of resolution; there is not much foreseeable improvement which can be made to the absolute resolution, at least by the use of *regular* replicas.

The vertical resolution obtained by Desvaux,[29] although very good for that kind of application, cannot be regarded as the ultimate in replication, however. In 1958, Bassett[30] developed a remarkable, if specialized, method for observing single- or multiple-atom steps on ionic crystal surfaces. He found that gold atoms, lightly evaporated onto cleaved rock salt to give a mean thickness of 10 Å, formed nuclei which collected along the edges of steps on the crystal surface. When a carbon replica was subsequently deposited, the gold nuclei were incorporated into the replica which could be removed for observation in the electron microscope by gently lowering the crystal into water. The gold nuclei appeared as step decorations in the microscope image, and the process was thus named the "gold-decoration" technique. It would be impossible to see such steps by a regular replication technique, for the contrast would be too small. The steps do have to be 30 Å apart to be visible, of course. When slip steps crossed cleavage steps in the decorated surfaces, Bassett noted that they were deflected by an amount that gave the total height of the cleavage steps. By counting the number of slip steps and assuming that each was of atomic dimensions, he was able to arrive at an independent measure of the cleavage slip

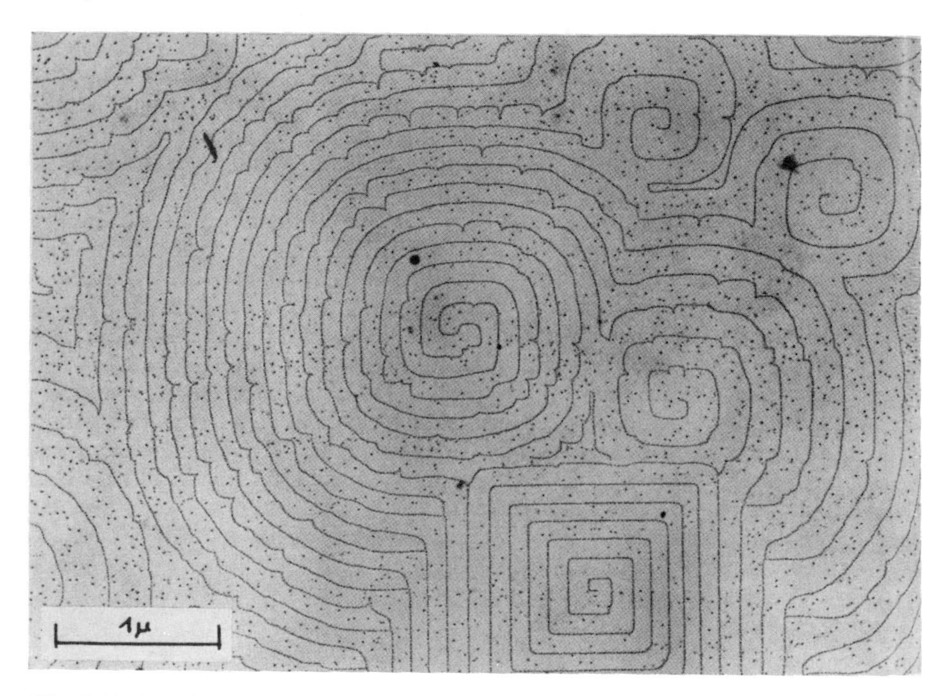

Fig. 4-11. Atomic steps on a rock salt crystal, revealed by the gold-decoration technique. Emergent screw dislocations, singly or in various combinations, are thought to give rise to the spirals of different morphology. (Courtesy of Bethge.[31])

height. Thus he concluded that monatomic steps could be revealed by the gold-decoration technique. The typical appearance of such decorated steps is shown in Fig. 4-11. Many useful observations of such decorated surfaces have been reviewed by Bethge.[31]

Bethge[32] and Allpress and Sanders[33] have extended the decoration technique to metals. The chief problem in work with metals is preparing a surface in which there are so few steps that their spacing is much greater than the size of the decorating nuclei. Allpress and Sanders, for example, working on silver, overcame this problem by thermally etching relatively smooth facets. Simple planes, which appear perfectly flat in regular shadowed replicas, were found to contain much interesting detail when examined by the decoration technique. Specimens with $\{111\}$ surfaces, thermally etched overnight, showed natural steps decorated by filaments of gold, clearly more than one atom high, and often divided into narrower steps decorated by separated nuclei (see *C* in Fig. 4-12). Many of the lines of nuclei terminated in the simple plane and were considered by Allpress and Sanders to be monatomic steps running from the points of emergence of dislocations. Slip steps were recognized because they appeared on the simple $\{111\}$ planes as straight lines along close-packed directions. Figure 4-13 illustrates the decoration of an area where extensive slip was introduced by cutting the specimen prior to decorating. The excellent clarity of the decoration on this surface was achieved by a heating pretreatment followed by decoration at a substrate temperature of 400°C. This kind of subtlety might well be explored for decoration in other metals besides the noble ones.

4.1.2. Solid–Solid or Solid–Liquid Interfaces

It seems unlikely that replication methods can be applied to solid–solid interfaces. However, J. M. Finney[34] made an interesting discovery during a routine microfractographic investigation of a fatigue fracture. In that case, a fatigue crack had run along the rolling direction in a commercial aluminum alloy and had taken an intergranular path, passing along the matrix–precipitate interfaces at grain-boundary precipitates. These interfaces, examined by a regular two-stage replica, were found to contain interesting growth spirals (see Fig. 4-14). This observation was especially interesting in that the growth kinetics of such grain-boundary precipitates have never been interpreted by a ledge mechanism.[35] The possibilities of the technique are presently

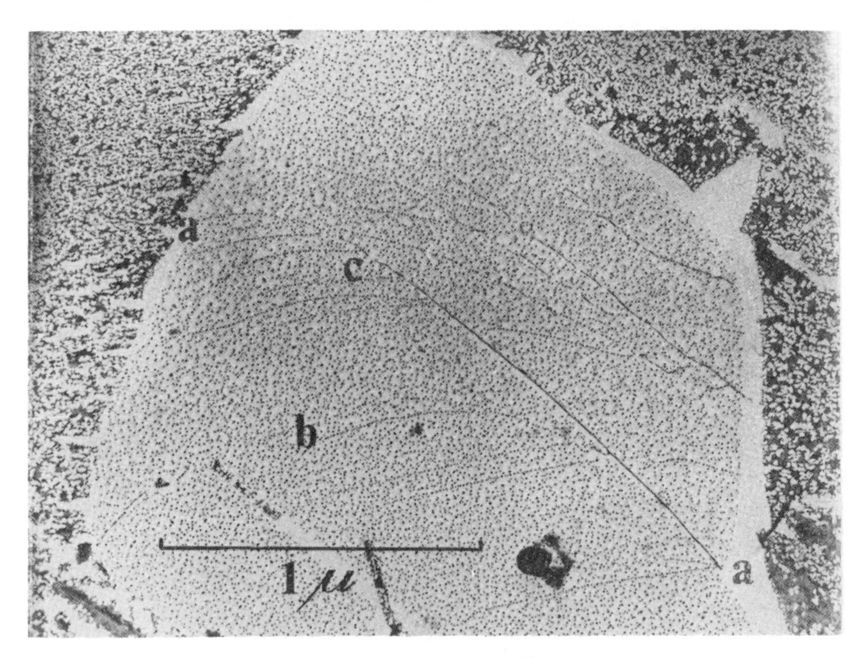

Fig. 4-12. Natural steps on a {111} plane of thermally etched silver, revealed by gold decoration (10-Å layer.) (Courtesy of Allpress and Sanders.[34])

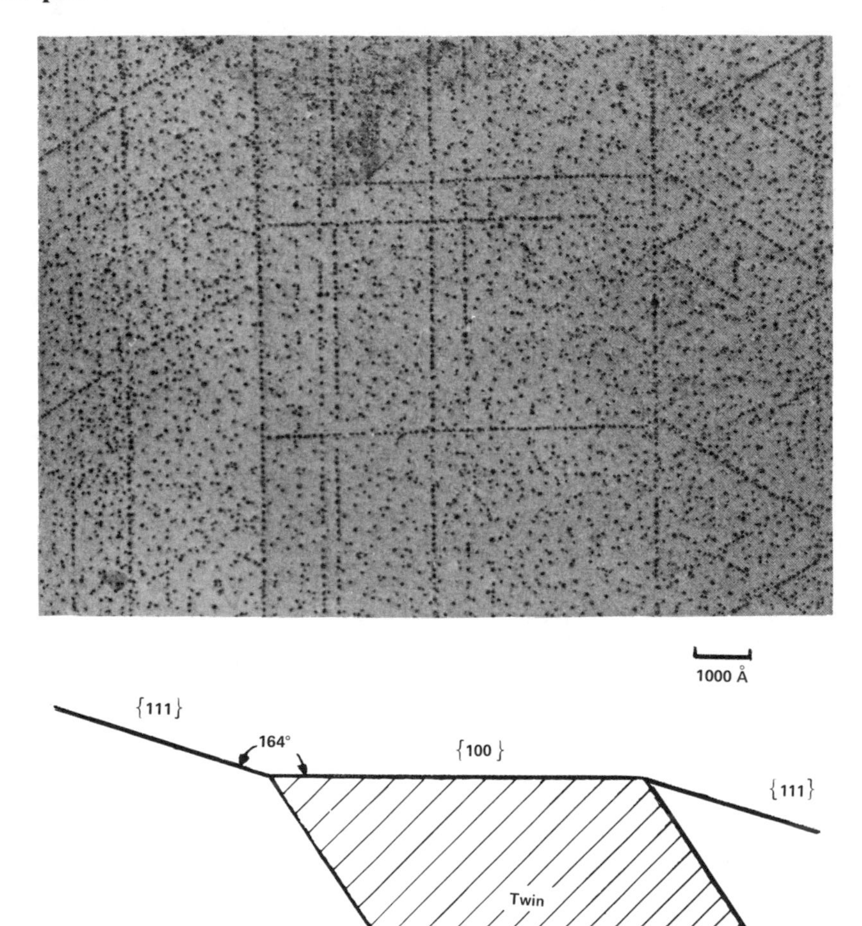

Fig. 4-13. Decoration of slip steps on {111} and {100} planes (10 Å of gold). The parent crystal, having a {111} surface, is traversed by a narrow twin whose surface has etched to a {100} plane. (Courtesy of Allpress and Sanders.[33])

being explored for studies of phase transformations in this author's laboratory.

4.2. Profile Method

4.2.1. Solid–Vapor Surfaces

In theory, there is no limit to the kinds of problem which may be attacked by the profile method, involving direct transmission microscopy. Those best known to this author are in the field of mechanical behavior.

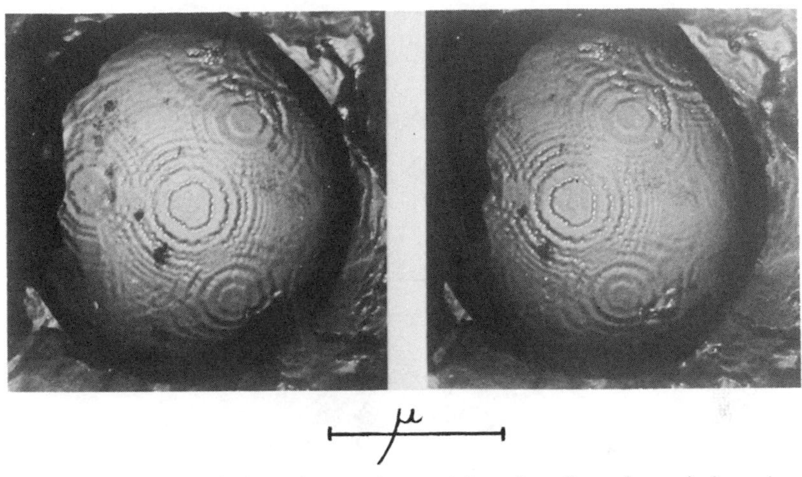

Fig. 4-14. Growth spirals at the matrix–precipitate interface of a grain-boundary precipitate in 2024 Al alloy. (Courtesy of J. M. Finney.[34])

As an example of the technique, the elegance of the observations by Lukáš and Klesnil *et al.*[36–39] calls for their description here. These investigators have been interested in the dislocation structures near the surfaces of fatigued metals and in the relation between these structures and the surface relief. As is well known, fatigue cracks nucleate in intense slip bands at the surface (known as "persistent" slip bands). Both the profile of the bands and the structure underlying them control the process of crack nucleation. In order to study these features, Lukáš and Klesnil fatigued single crystals of different metals (having different slip modes) with the orientation shown in Fig. 4-15. After the fatigue tests, they plated their crystals and sectioned them for examination in the electron microscope; they thus observed that the persistent slip bands in copper (and metals of similar slip mode) were underlain with planar zones, at most a few microns thick, containing dislocation cells arranged in a honeycomb fashion, i.e., with the axes of the cylindrical cells normal to the plane of the zone (see Fig. 4-15). An electron micrograph of a typical surface section is shown in Fig. 4-16. In this case, the metal within the band extruded from the surface and the section shows the topography of the extrusion very well. Note the dislocation-free zone which lies immediately beneath the surface and the different kind of dislocation structure adjacent to the persistent slip band. It is further important to note that the features shown here are "real." For example, fearing that the dislocation-free zone might be an artifact of specimen preparation, Lukáš and Klesnil carried out the following experiment.[39] After fatiguing the crystal, but before depositing the protective copper layer, they briefly electropolished the surface to a depth greater than the

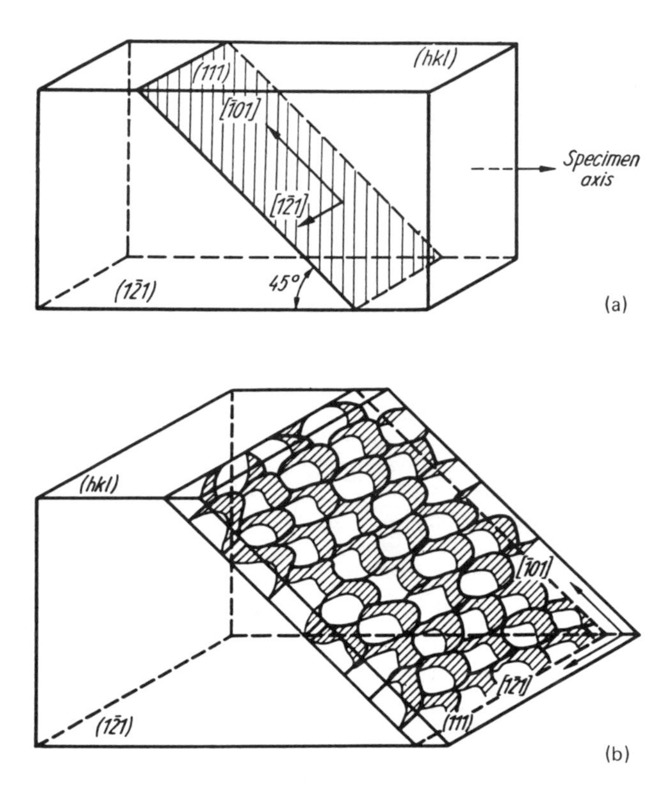

Fig. 4-15. (a) The orientation of the single crystals used by Lukáš and Klesnil for their fatigue studies; (b) a space model of the persistent slip bands they observed in copper. (Courtesy of Lukáš *et al.*[37])

dislocation-free layer previously observed. They then performed the deposition and the rest of the sectioning; electron microscopic examination showed that the dislocation structures now persisted right up to the surface, proving the validity of the dislocation-free layer. From the results of similar experiments,[39] these observations are regarded with great confidence.

The surface relief associated with the fatigue of a different kind of material, one with a more planar mode of slip, is shown in Fig. 4-17. Unlike in copper, regular persistent slip bands are not observed. Instead, the surface relief, related to interior slip planes containing dislocations, is confined to closely spaced slip lines with small offsets. By observations such as these, Lukáš and Klesnil obtained useful insights into the structure sensitivity of fatigue.

The results of an investigation by Swann,[40] using the same technique but with a different aim, demonstrate an important limitation

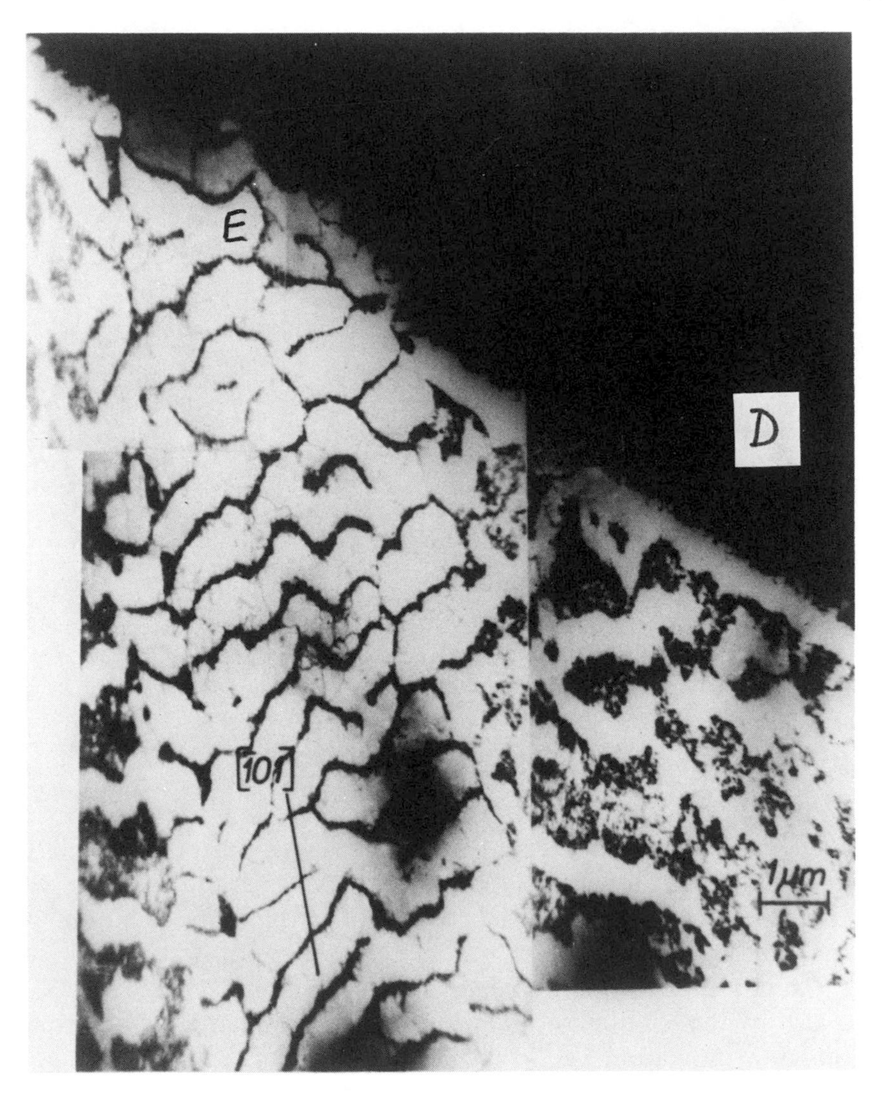

Fig. 4-16. An extrusion on the (*hkl*) section of a copper monocrystal stressed as shown in Fig. 4-15 in relation to the underlying dislocation structures. Section $(1\bar{2}1)$, D = electrodeposit, E = extrusion. (Courtesy of Lukáš *et al.*[37])

of the profile method. Swann was concerned with checking a hypothesis concerning surface effects in deformation. In this hypothesis, Kramer and Demer[41,42] proposed that a region of higher dislocation density develops at the surface during deformation. However, as Swann showed,[40] the dislocation structures near the surface did not differ significantly from those in the bulk of the specimen. The evidence for this conclusion is shown in Fig. 4-18. Also clearly visible in Fig. 4-18 is the

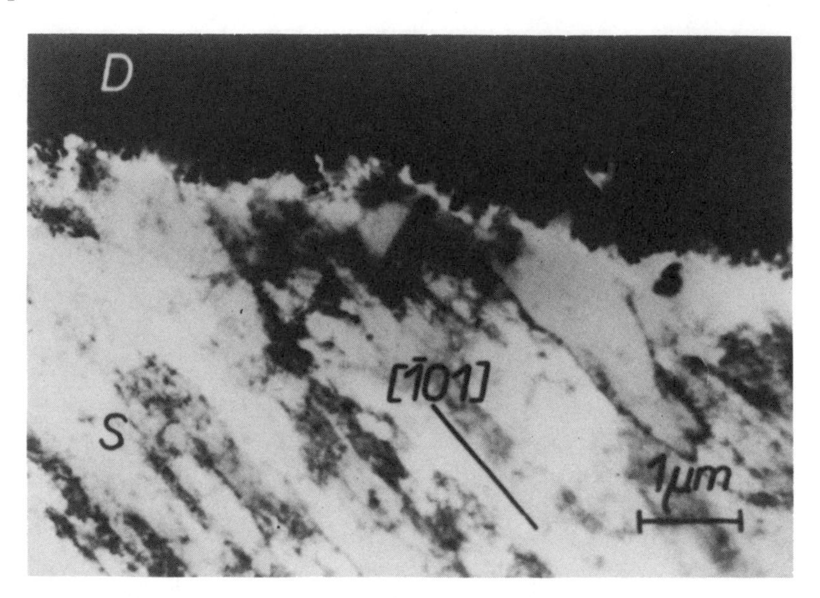

Fig. 4-17. The surface relief on the (*hkl*) surface and underlying dislocation structures observed in Cu-31%Zn fatigued at a stress amplitude of ± 8.4 kg/mm^2. The single crystal was oriented in the manner shown in Fig. 4-15. In this figure, D = electrodeposit, s = specimen. (Courtesy of Lukáš and Klesnil.[38])

structure of the specimen–deposit interface. The dislocation arrays created by the epitaxial deposition of the copper protective layer, although of little importance to Swann, could seriously obscure the phenomena of interest in other kinds of surface investigation. The epitaxial nature of the deposit was demonstrated to Swann by the fact that the contrast does not change across the boundary (see Fig. 4-18) and also by the continuity of annealing twin boundaries at other parts of the same interface.

Many applications of the profile method can be of use in surface problems of technological interest. For example, the structure of machined surfaces can be studied in this way. Turley[43] has recently used a modification of the profile method for measuring the strains produced in the deformed layers beneath machined surfaces of 70:30 brass. Since a detailed knowledge of the relationship between dislocation substructures and strain in deformed brass was available to him, he related the dislocation structures adjacent to the machined surfaces to similar structures for which the corresponding strains were known.

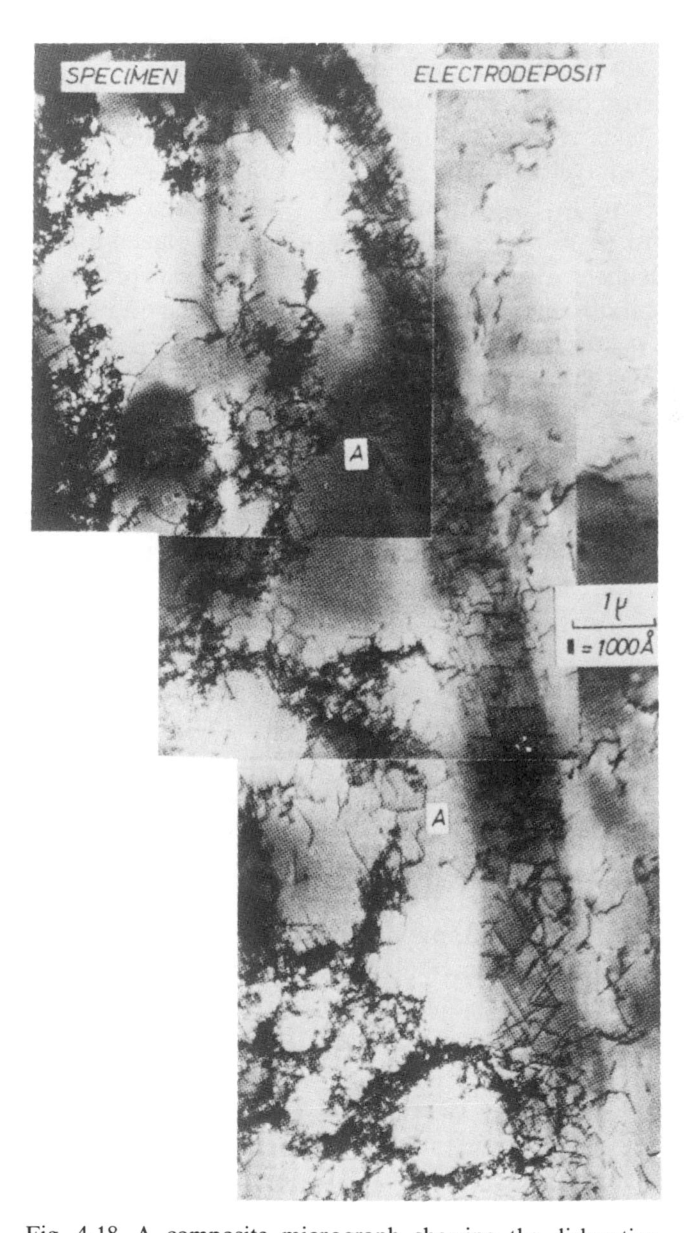

Fig. 4-18. A composite micrograph showing the dislocation distribution near the surface of 15% deformed copper. Some decomposition of cell walls appears to have occurred at the places marked *A*. The hatched region running down the center of the micrograph is the inclined surface of the specimen. (Courtesy of Swann.[40])

4.2.2. Solid–Solid Interfaces

Although solid–vapor interfaces are of most interest in this volume, electron microscopy is ideal for studying solid–solid interfaces, and reference to such studies should accordingly be made. During the last 10 years, there have been a large number of interface studies. This work has been reviewed by Aaronson *et al.*[35] and in two more recent conferences yet to be published.* Emphasis has been placed on the observation of misfit-dislocation structures. It is perhaps surprising that it was not until a long time after the development of electron microscopy that the method was used to observe misfit dislocations. This was probably due to the fact that the conditions for imaging such dislocations are somewhat more stringent than those for dislocations wholly contained within a single phase. Moreover, the misfit-dislocation spacing (for a single phase) can be equal to, or less than, the width of dislocation images. In such circumstances, Thölen[44] has shown that dislocations and Moiré fringes appear much alike. In early studies there was confusion between the Moiré fringes expected of misfitting phases and the associated dislocations. Most of these observations were therefore executed deliberately and with great effort. As a result of improvements in theory,[11–13] and the consequent increasing confidence of experimenters, these difficulties have been overcome and the structures of solid–solid interfaces are now documented in great detail.

As an example of this work, a hitherto unreported study of coherency breakdown in the η precipitate in Al–Au alloy[45] is briefly described below. Another important feature of electron microscopes, the ability to employ accessories such as heating and cooling stages, is also illustrated in this example. Many previous attempts had been made to observe coherency-breakdown mechanisms by direct observation in the electron microscope hot stage, but a variety of difficulties were encountered. For example, in the Al–Cu system, where θ' was the precipitate of interest, the equilibrium phase θ nucleated at the surfaces of the thin foils, causing the θ' precipitates to dissolve before interesting misfit-dislocation reactions could occur.† In Al–Ag, the processes occurred so slowly as to be beyond the patience of the interested observer. The following example is the first, to this author's knowledge, that can be regarded as an unqualified success. The η precipitate in Al-rich Al–Au alloys has a face-centered-cubic structure orientated with one of the cube planes parallel to $\{001\}$ of the α matrix. In addition,

* IBM Conference, Summer 1971; and Iron and Steel Institute Conference, London, November 1971.

† The solid–solid interface processes which occurred during dissolution at the precipitate interfaces did prove, however, to be of great interest.[46]

the η precipitate has a plate morphology with $\{001\}$ habit plane. The misfit between the α and η phases is relatively large, 4.83%, which probably explains why viable coherency-breakdown processes can be observed in the hot stage. A coherent η precipitate, viewed in plan, appears as an octagonal plate, the edges of which are parallel to $\langle 110 \rangle$ and $\langle 100 \rangle$, the edges parallel to $\langle 110 \rangle$ dominating (see Fig. 4-19a). On heating this precipitate in the hot stage of an electron microscope, Sankaran and Laird[45] observed misfit dislocations to nucleate at those edges of the plate which lay parallel to $\langle 110 \rangle$.* These dislocations (Burgers vector $(a/2)\langle 110 \rangle$) subsequently glided inward until they intersected near the corners of the plates (Fig. 4-19b) and reacted to form pairs of $(a/2)\langle 100 \rangle$ dislocations in edge configuration.† With continued isothermal annealing, the sections of $(a/2)\langle 110 \rangle$ dislocations attached to the $(a/2)\langle 100 \rangle$ were annihilated in favor of the $(a/2)\langle 100 \rangle$ type. Eventually all the $(a/2)\langle 110 \rangle$ dislocations were consumed and the fully semicoherent plate was entirely covered with orthogonal arrays of $(a/2)$ $\langle 100 \rangle$ dislocations. Some of these complicated details are illustrated in Fig. 4-19, but a complete description would be too lengthy to include here. These results, which are to be published elsewhere, have explained a number of perplexing difficulties concerning the energetics of dislocations on $\{100\}$ interfaces and the mechanisms of their formation.[35] It is important to note that, at high temperature, the dislocations nucleated and reacted too quickly to permit their detailed analysis by standard contrast methods. The investigators therefore adopted a quench–analysis–up–quench technique[45] in which a foil of interest was heated until a particular stage of the breakdown occurred. The foil was then quenched to room temperature by switching off the hot-stage power supply and the imperfections were analyzed to any extent necessary. The foil was subsequently heated again until the reactions progressed further, was again quenched and analyzed, and so forth.

Another application of hot-stage microscopy which has yielded exciting results concerning solid–solid interfaces is in diffusion studies.[47] The imperfections which develop at the interface of the substances comprising a diffusion couple, and which affect the kinetics of diffusion, can be studied directly to good effect.

Of course, since the "environment" of the electron microscope can be controlled to a large degree, the physical nature of surface–vapor

* This is not the only mechanism by which misfit dislocations were observed to nucleate, but it is one of the more important ones.

† A reader knowledgeable in the dislocation theory of single-phase materials may question the occurrence of dislocations with Burgers vectors of $(a/2)\langle 100 \rangle$. However, in this case the evidence is rather strong.

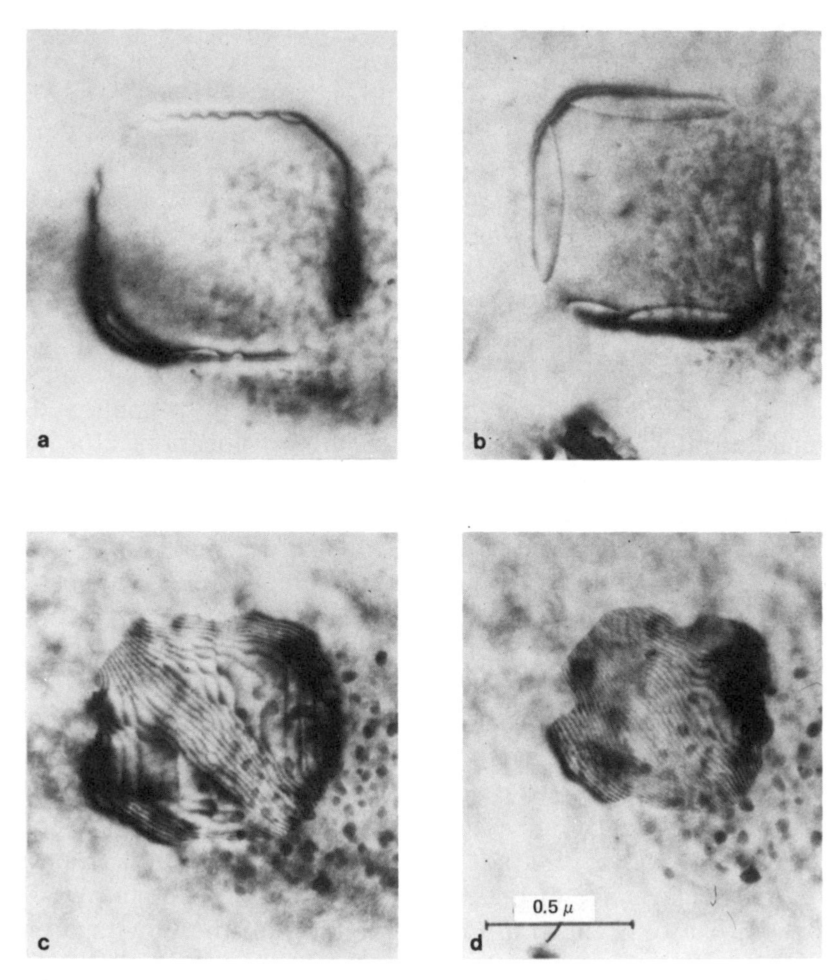

Fig. 4-19. The process of coherency breakdown on the broad faces of an η plate in Al–Au alloy: (a) the initially coherent octagonal plate showing mismatch dislocation loops forming at the edges. Zone axis [001], $g = 200$, $w > 0$; (b) the first misfit dislocation to join up from the loops having Burgers vectors of $(a/2)\langle 110\rangle$ parallel to the plane of the plate and lying in edge configuration; $g = 020$, $w > 0$; (c) the same plate after many $(a/2)\ \langle 110\rangle$ type dislocations have formed and reacted to produce bands of $(a/2)\langle 100\rangle$ dislocations running across the mid-regions of the plates; $g = 200$, $w > 0$ (one set of $(a/2)\langle 010\rangle$ dislocations is invisible); (d) the fully semicoherent η plate covered with $(a/2)\,[100]$ and $(a/2)\,[010]$ dislocations; $g = 200$, $w > 0$.

reactions or the effect of environment on solid–solid interactions also can be studied directly via electron microscopy with or without use of a hot stage to stimulate the kinetics of the reactions. A typical example in this respect is the recent study of sintering by Easterling and Thölen.[48]

In this work, a controversial subject was examined: do dislocations play a role in the initial stages of growth in the conjunctive necks between particles? The stresses generated at these necks were thought to reach values much greater than the yield stress of the material. The sintering process had accordingly been related to creep since Weertman had suggested that neck growth could occur by dislocation climb.[49] Since there had been no direct evidence of plastic flow in metal compacts (without external loading), the purpose of Easterling and Thölen's work was to investigate the early stages of neck growth, i.e., when necks are small, the stresses are highest. Accordingly creep processes should be exaggerated.

A typical result of this work is shown in Fig. 4-20. The structure of ~550 Å fcc (Fe–Ni) particles is seen before and after sintering. Clearly the particles are initially free from line defects (Fig. 4-20a), and none are observed after coalescence (Fig. 4-20b) in spite of the fact that grain boundaries frequently developed (e.g., at *C*). Measurements showed a gradual decrease in the distance between particle centers, and pores between particles closed up (e.g., in the group of particles marked *B*). Faceting is quite pronounced. On the basis of these results, Easterling and Thölen concluded that the maximum shear stresses generated in sintering are too small to nucleate fresh dislocations, and thus neck growth is controlled solely by diffusional processes.[48]

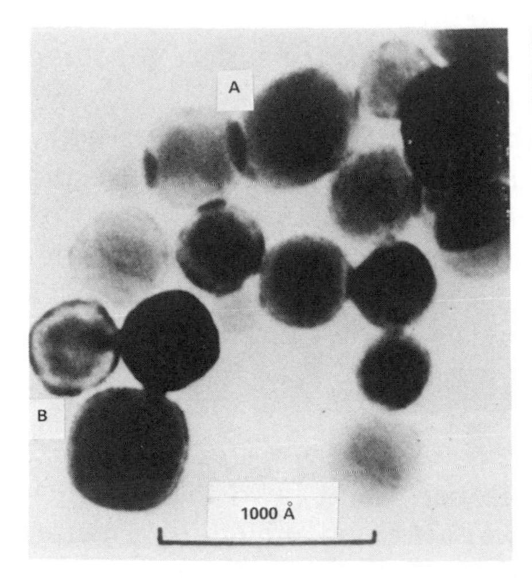

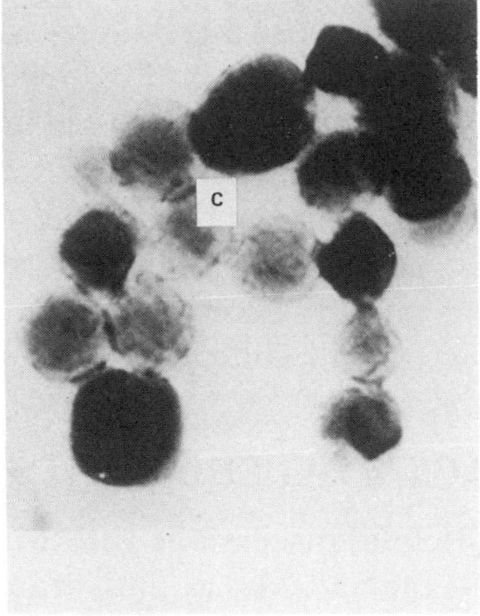

Fig. 4-20. Fcc (Fe–Ni) powder: (a) unsintered, and (b) sintered for 60 min at 800°C. The particles are slightly overlapped at *A*. (Courtesy of Easterling and Thölen.[48])

5. CONCLUSIONS

The conclusions of this review may be briefly summarized as follows:

1. The contrast theory and techniques of transmission electron microscopy can easily be applied to the study of surfaces and interfaces, and are generally adequate for purposes of interpretation.
2. Both replication methods and the profile method with actual materials of interest have been applied to solving problems connected with surfaces. In regular replication methods, vertical resolution of 100 Å in conjunction with a horizontal resolution of about 30 Å seems as much as can be expected from the present state of the art. However, monatomic step heights can be resolved if the "decoration" technique is employed in conjunction with carbon replication. Of course, the steps have to be separated by the normal 30-Å resolution to be observable.
3. In the profile method, the full resolving power of the electron microscope can be employed, but the surface of interest is vulnerable to the methods of preparing thin sections. This technique has the advantage that sandwich specimens or compacts can be examined in three dimensions as well as at high resolution because of the large depth of field of the instrument.
4. Techniques such as stereomicroscopy can be employed both on replicas and profile sections for measuring surface topography, but at the cost of tedious labor.
5. Although a considerable literature in surface studies with the microscope already exists, relatively few fields have been studied in great depth and the opportunities for useful application of the instrument are enormous, especially when consideration is given to accessories such as heating and cooling stages, and also to high-voltage microscopy.

6. ACKNOWLEDGMENTS

Helpful discussions with Mr. Robert White (who, along with Mrs. D. Doane and Mr. R. Sankaran, was also kind enough to criticize the manuscript) are gratefully acknowledged; likewise, we are grateful for the support of ARPA contract DAHC-15-67-C-0215 which enabled the author and his co-workers to carry out the unpublished work mentioned in this review.

7. REFERENCES

1. R. D. Young, Surface microtopography, *Phys. Today* **24**, 42–48 (1971).
2. O. Johari, in *Characterization of Solid Surfaces* (P. F. Kane and G. B. Larrabee, eds.) Scanning microscopy, Chapter 5, Plenum Press, New York (1974).
3. C. J. Cooke and P. Duncumb, Performance analysis of a combined electron microscope and electron probe microanalyser, EMMA, in *Fifth International Congress on X-ray Optics and Microanalysis* (G. Mollenstedt and K. H. Gaukler, eds.) pp. 245–247, Springer-Verlag, Berlin (1969).
4. M. H. Jacobs, Microstructural studies with a combined electron microscope and electron probe microanalyser (EMMA-3) *Proc. 25th Anniv. Meeting EMAG Inst. Phys.* (1971).
5. G. A. Hutchins, Electron probe microanalysis, in *Characterization of Solid Surfaces* (P. F. Kane and G. B. Larrabee, eds.) Chapter 18, Plenum Press, New York (1974).
6. C. Klemperer, *Electron Optics*, Cambridge University Press, London (1953).
7. G. Thomas, *Transmission Electron Microscopy of Metals*, J. Wiley and Sons, New York (1962).
8. R. Haine and V. E. Cosslett, *The Electron Microscope, The Present State of the Art*, Interscience, New York (1961).
9. R. D. Heidenreich, *Fundamentals of Transmission Electron Microscopy*, Interscience, New York (1964).
10. P. Grivet, *Electron Optics*, Pergamon, London (1965).
11. P. B. Hirsch, A. Howie, R. B. Nicholson, D. W. Pashley, and M. J. Whelan, *Electron Microscopy of Thin Crystals*, Butterworths, London (1965).
12. S. Amelinckx, *The Direct Observation of Dislocations*, Academic Press, London (1964).
13. S. Amelinckx (ed.), *Modern Diffraction and Imaging Techniques in Material Science*, North-Holland, Amsterdam (1970).
14. C. E. Hall, *Introduction to Electron Microscopy*, 2nd edition, McGraw-Hill, New York (1966).
15. R. W. Wyckoff, *Electron Microscopy, Technique, and Applications*, Interscience, New York (1949).
16. D. Kay, *Techniques for Electron Microscopy*, Blackwell Scientific Publications, Oxford (1961).
17. L. E. Thomas, Course notes in electron microscopy, Univ. of Pennsylvania (1971).
18. A. Howie, in *Techniques for Electron Microscopy* (D. Kay, ed.) pp. 438–440, F. A. Davis Co., Philadelphia (1965).
19. J. F. Nankivell, The theory of electron stereo microscopy, *Optik* **20**, 171–198 (1963).
20. *Handbook for Wild ST4 Mirror Stereoscope*, P2 307e, Wild, Heerbrugg (1967).
21. A. Boyde, Observations on enamel and dentine by surface electron microscopy, *J. Roy. Microscop. Soc.* **86**, 359–365 (1967).
22. A. Boyde, Practical problems and methods in three-dimensional analysis of scanning electron microscopy images, in *Proceedings of the Third Annual Scanning Electron Microscopy Symposium*, pp. 107–112, IIT Research Institute, Chicago (1970).
23. D. E. Bradley, J. S. Halliday, and W. Hirst, Stereoscopic reflection electron microscopy, *Proc. Phys. Soc. (London)* **69**, 484–486 (1956).
24. J. S. Halliday, Reflection electron microscopy, in *Techniques for Electron Microscopy*, pp. 306–324, Blackwell Scientific Publications, Oxford (1961).
25. S. J. Jones and A. Boyde, Experimental studies on the interpretation of bone surfaces studied with SEM, in *Proceedings of the Third Scanning Electron Microscopy Symposium*, pp. 195–200, IIT Research Institute, Chicago (1970).

26. M. P. E. Desvaux and P. Charsley, Slip lines on pure copper deformed in tension and compression, *Mater. Sci. Eng.* **4**, 221–230 (1969).
27. P. Charsley and M. P. E. Desvaux, The behavior of Cu–12% Al under simple reversed stresses, *Mater. Sci. Eng.* **4**, 211–220 (1969).
28. C. Laird and D. J. Duquette, Mechanisms of fatigue crack nucleation, in *Proceedings of Corrosion Fatigue Conference, Storrs, Conn.*, pp. 88–117, N.A.C.E., Houston, Texas (1972).
29. M. P. E. Desvaux, A replica technique for measuring incremental slip step heights, *J. Sci. Instr., Ser. 2*, **1**, 558–560 (1968).
30. G. A. Bassett, A new technique for decoration of cleavage and slip steps on ionic crystal surfaces, *Phil. Mag.* **3**, 1042–1045 (1958).
31. H. Bethge, Oberflächenstrukturen und kristalbaufehler in elektronenmikroskopischen bild, untersucht am NaCl, *Phys. Stat. Sol.* **2**, 3–27 (1962).
32. H. Bethge, Electron microscopic studies of surface structures and some relations to surface phenomena, *Surf. Sci.* **3**, 33–41 (1964).
33. J. G. Allpress and J. V. Sanders, Decoration of facets on silver, *Phil. Mag.* **9**, 645–658 (1964).
34. J. M. Finney, Univ. of Pennsylvania, Philadelphia, and ARL, Melbourne, Australia, private communication (1970).
35. H. I. Aaronson, C. Laird, and K. R. Kinsman, Mechanisms of diffusional growth of precipitate crystals, in *Phase Transformations*, ASM, pp. 313–390 (1970).
36. P. Lukáš, M. Klesnil, J. Krejčí, and P. Ryš, Substructure of persistent slip bands in cyclically deformed copper, *Phys. Stat. Sol.* **15**, 71–82 (1966).
37. P. Lukáš, M. Klesnil, and J. Krejčí, Dislocations and persistent slip bands in copper single crystals fatigued at low stress amplitude, *Phys. Stat. Sol.* **27**, 545–558 (1968).
38. P. Lukáš and M. Klesnil, Dislocation structures in fatigued Cu–Zn single crystals, *Phys. Stat. Sol.* **37**, 833–842 (1970).
39. P. Lukáš and M. Klesnil, Fatigue damage and resultant dislocation substructures, in *Corrosion Fatigue Conference, Storrs, Conn.*, pp. 118–132, N.A.C.E., Houston, Texas (1972).
40. P. R. Swann, The dislocation distribution near the surface of deformed copper, *Acta Met.* **14**, 900–903 (1966).
41. I. R. Kramer and L. J. Demer, The effect of surface removal on the plastic behavior of Al single crystals, *Trans. AIME* **221**, 780–786 (1961).
42. I. R. Kramer, The effect of surface removal on the plastic flow characteristics of metals, *Trans. AIME* **227**, 1003–1010 (1963).
43. D. M. Turley, Dislocation substructures and strain distributions beneath machined surfaces of 70/30 brass, *J. Inst. Metals* **99**, 271–277 (1971).
44. A. R. Thölen, On the ambiguity between Moiré fringes and the electron diffraction contrast from closely spaced dislocations, *Phys. Stat. Sol.* (*a*) **2**, 537–550 (1970).
45. R. Sankaran and C. Laird, Studies of precipitate morphology and growth kinetics, unpublished work (1971).
46. C. Laird and H. I. Aaronson, Direct observations of the thinning of θ' plates in Al–4% Cu by lateral movement of ledges, *J. Inst. Met.* **96**, 222 (1968).
47. K. Shinohara, Behavior of misfit dislocations during interdiffusion, Ph.D. Thesis, Ohio State Univ. (1972).
48. K. E. Easterling and A. R. Thölen, A study of sintering using hot-stage electron microscopy, *Met. Sci. J.* **4**, 130–135 (1970).
49. J. Weertman, Steady-state creep of crystals, *J. Appl. Phys.* **28**, 1185–1189 (1957).

5

SCANNING ELECTRON MICROSCOPY

Om Johari

IIT Research Institute
Chicago, Illinois

and

A. V. Samudra

LaSalle Steel Company
Hammond, Indiana

1. INTRODUCTION

A detailed examination of materials is vital to any investigation relating to the processing properties and behavior of materials. Characterization includes all information relating to topographical features, morphology, habit and distribution, identification of differences based on chemistry, crystal structure, physical properties, and subsurface features, among others.

Before the advent of the scanning electron microscope (SEM) several tools such as the optical microscope, the transmission electron microscope, the electron microprobe analyzer, and x-ray fluorescence were employed to accomplish partial characterization; this information was then combined for a fuller description of materials. Each of these tools has proficiency in one particular aspect and complements the information obtainable with other instruments. These bits of information, though occasionally valuable, are limited because of the inherent limitations of each method such as the invariably cumbersome specimen preparation, specialized techniques of observation, and interpretation of the results. It is no wonder that a versatile tool such as the SEM has literally taken the scientific community by storm. The SEM has so many material-characterization capabilities that it can singly be considered the ideal tool for total material characterization.(1)

2. COMPARISON OF SEM WITH OTHER CONVENTIONAL TOOLS

The SEM is often described as bridging the gap between the optical microscope and the transmission electron microscope, although the currently emerging new development, the transmission scanning electron microscope (TSEM), approaches the resolution and magnifications obtainable by the TEM. The SEM has a magnification of 3–100,000 ×, a resolution of about 200–250 Å, and a depth of field at least 300 times or more that of the light microscope—all of which result in the characteristic photographs of dramatic three-dimensional quality. Because of the large depth of focus and large working distance, the SEM permits direct examination of rough conductive samples at all magnifications without additional preparation.

3. PRINCIPLE OF SEM CONSTRUCTION AND OPERATION

The principle of the SEM as used in its most common mode, the emissive mode, is illustrated schematically in Fig. 5-1. Electrons from a

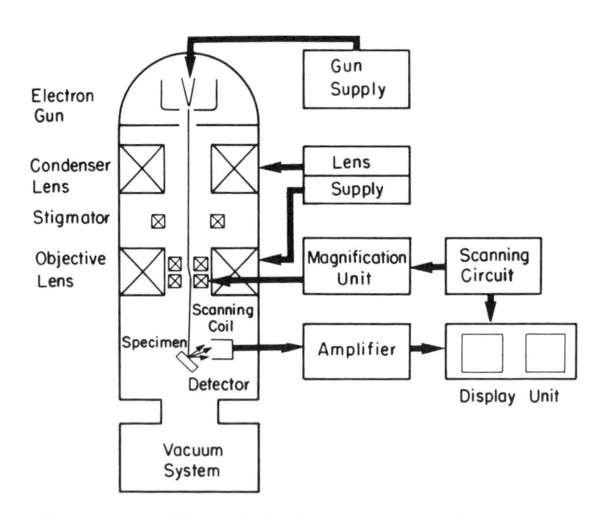

Fig. 5-1. Simplified block diagram of scanning electron microscope.

filament are accelerated by a voltage commonly in the range of 1–30 kV and directed down the center of an electron optical column consisting of two or three magnetic lenses. These lenses cause a fine electron beam to be focused onto the specimen surface. Scanning coils placed before the final lens cause the electron spot to be scanned across the specimen surface in the form of a square raster, similar to that on a television screen. The currents passing through the scanning coils are made to pass through the corresponding deflection coils of a cathode ray tube (CRT) so as to produce a similar but larger raster on the viewing screen in a synchronous fashion.

The electron beam incident on the specimen surface causes various phenomena, of which the emission of secondary electrons is the most commonly used. The emitted electrons strike the collector and the resulting current is amplified and used to modulate the brightness of the CRT. The times associated with the emission and collection of the secondary electrons are negligibly small compared with the times associated with the scanning of the incident electron beam across the specimen surface. Hence, there is a one-to-one correspondence between the number of secondary electrons collected from any particular point on the specimen surface and the brightness of the analogous point on the CRT screen. Consequently, an image of the surface is progressively built up on the screen.

The SEM has no imaging lenses in the true sense of the word. The image magnification is determined solely by the ratio of the sizes of the rasters on the CRT screen and on the specimen surface. In order to increase the magnification, it is only necessary to reduce the currents in the SEM scanning coils. For example, if the image on the CRT screen is 10 cm across, magnifications of 100×, 1000×, and 10,000× are obtained by scanning the specimen 1 mm, 0.1 mm, and 0.01 mm across, respectively. One consequence of this is that high magnifications are easy to obtain with the SEM, while very low magnifications are difficult. Thus, for a magnification of 10× it would be necessary to scan a specimen approximately 10 mm across, and this presents difficulties because of the large deflection angles required. For instance, the electron beam may strike the lens pole pieces or aperture, and at the extremes of the scan, linearity may not be maintained.

The completely different operation of the SEM compared to most other microscopes is possible because there are no imaging lenses, and any signal that arises from the action of the incident electron beam (reflected electron, transmitted electrons, emitted light, etc.) can be used to form an image on the CRT screen. Figure 5-2 shows various signals available. It is only necessary that the signal be collected sufficiently

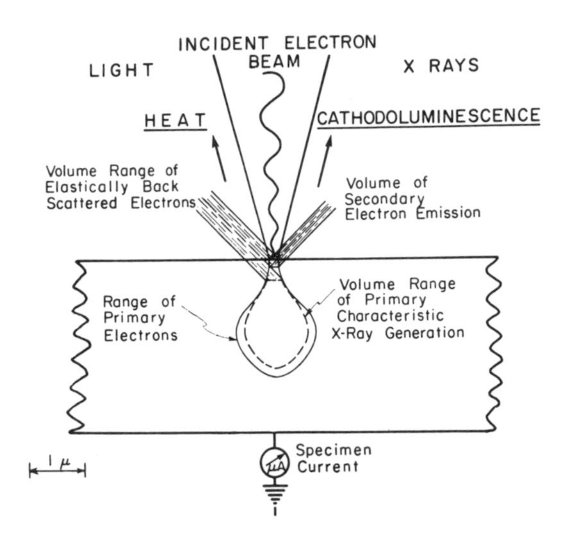

Fig. 5-2. Various signals available.

rapidly and converted, if necessary, into an electric current. It also follows that several such signals can be dealt with at the same time, and hence several different types of information can be simultaneously displayed on adjacent CRT's or electronically superimposed on the same CRT.

Another major difference is that each particular point in the image is recorded sequentially in time. Thus, a complete image is not seen until the whole raster has been scanned. This has practical disadvantages in that focusing and recording tend to be more tedious and special electronic amplifiers are required. However, it has important advantages in that the signal is in a convenient form to be "operated" on in a variety of ways, thereby enabling considerably more information to be obtained. Such operations include electronic signal processing, energy analysis, and wavelength separation; all of these operations can be applied to each point of the image in turn, and the resulting information can be appropriately displayed. Such procedures are not usually possible with other types of microscopes because all points in the image are recorded simultaneously.

Unlike other instruments, the SEM allows several different modes of operation, each corresponding to the collection of a different signal (Fig. 5-2) arising from the incident electron beam interacting with the specimen. The main modes are:

1. Emissive (secondary electrons)
2. Reflective (backscattered electrons)

3. Absorptive (leakage current)
4. Transmission
5. Beam-induced conductivity
6. Cathodoluminescence
7. X-ray
8. Auger electron

Since the intensity of signal in different modes is affected differently by the material features or characteristics, a varied type of information is obtained in the SEM.

4. MATERIAL CHARACTERIZATION POSSIBILITIES

The ideal setup for characterization should include the following features:

1. Ability to detect specific signals.
2. Separation or clear understanding of different factors contributing to contrast in that signal.
3. A rapid method for quantifying the information available in that signal.

The following sections are meant to describe the state of the SEM today and what developments can be expected in the near future.

4.1. Quantitative Topography of Surfaces

The most important contribution to the contrast in the case of the secondary-electron mode and the backscattered-electron mode of operation is the surface topography. In practice, the secondary-electron mode is nearly always preferred, because the resolution is better, the signal is greater, and even details of specimen areas not in the direct line of sight of the collector are visible.

The topographical information seen in a single SEM photograph, at best, affords only a qualitative interpretation. For an unambiguous interpretation one must use a stereopair. Stereoviewing permits recognizing image artifacts (due to specimen charging, etc.) and quantitatively calculating height and depth. Thus it is possible to arrive at volume, surface area, and other geometrical parameters.

A stereopair is obtained by taking two photographs of the same area at two tilt angles, usually differing by 6–10°. The techniques of analysis are well known and have been used extensively in aerial photography and surveying.

The third dimension is calculated from parallax measurement either with a stereoscope or by measuring the difference in separation between two identical points on the two photographs of the pair. If p is this parallax, α the angle of tilt between the two photographs, and M the magnification, then the height difference is $Z = (p/2M)(\sin \alpha/2)$. Development of this formulation assumes a parallel-beam situation, which may not be the case at low magnifications (for a detailed review, see Howell and Boyde[2] and Boyde[3]). In terms of depth resolution, for a stereopair at 20,000 × with a tilt angle difference of 10° and parallax p measured with a stereoscope as 0.1 mm, Z (approximately 290 Å) is about the same as the resolution of the SEM in the secondary mode (about 200 Å).

The accuracy of stereocalculations depends on the accuracy of parallax measurement and on the accuracy of tilt angle and magnification calibration. Boyde[2,3] has shown that the tilt angles obtained from the graduation on SEM stages may be in error by as much as 1°. For accurate work, one must obtain appropriate corrections in tilt angles by calibrating the stages. Also, corrections for distortions in the CRT should be made.

While the need for stereoviewing is obvious, procedures to obtain stereopairs by pressing buttons on the SEM are not yet available. It would be desirable to have a system for stereopairs whereby sample tilting could be effected by a beam deflection system and thus avoid the time-consuming process of physically tilting and relocating the desired sample area for the second photograph. Such a system was recently described by Dinnis,[4] but the aberrations introduced by the additional deflection coil limit the operations of the SEM at higher magnifications. The interest in such a system is immense, and it should be perfected in the near future. Such a system will offer the added possibility of simultaneously taking two photographs and storing data from them on some form of electronic data-processing system so that, with appropriate logic, it should be possible to directly obtain topographical maps and, with suitable machines, three-dimensional models at any magnification.

Rotation and tilt motions available on the SEM allow many views of the same area to be obtained. The degrees of freedom now available permit viewing of five of the six faces of a cube (all faces except the one on which the cube is resting). Although the specimen is usually tilted about an axis perpendicular to the detector, some models now permit the specimen to be tilted about an axis parallel to the detector. The latter motion facilitates the taking of the stereopairs, since the intensity of identical features does not change much and identification of the same features in two photographs for automatic treatment should be easier.

The treatment of stereopairs from rough or smooth samples is identical, except that good photographs are difficult to obtain from very smooth samples. The contrast in secondary and backscattered electron images depends heavily on topography of the sample (assuming no atomic number effects are interfering), and, for smooth surfaces, features are hard to find and focus. In such cases, particularly at low magnification, the use of backscattered electrons has many advantages in obtaining stereopairs. In backscattered images, only those electrons having a straight-line path reach the detector, so that images have very high contrast due to shadow effects. Any small topographical changes are thus readily recognized. At higher magnifications, secondary images can be used, if necessary, to provide better resolution.

Since only a fraction of all scattered electrons are collected in the backscattered mode, brighter beams have to be used. In practice, this means increasing the diameter of the beam and deteriorating the spatial resolution. The position of the detector influences resolution also. For an SEM where the secondary detector is used as a backscattered detector (by turning off the collection potential), the detector is located on a side, and collection efficiency is poor compared to when the detector is directly above and looking down on the sample. For an SEM with a heated tungsten filament source, the resolution for the two collector locations is, respectively, 1000–1500 Å and 300–500 Å. In practice, easily installable solid-state detectors for high-resolution backscattered images are readily available.

To use the SEM in characterizing surfaces that have undergone superfinishing operations, use of backscattered electrons is essential. Similar considerations apply in examining polished and etched sections for studying microstructures.

On microscopes with the backscattered-electron detector on the side, one can use higher beam currents (larger beam diameter) at lower magnifications while still working the SEM in the secondary mode. Since a secondary image always contains some backscattered signal, increasing the primary beam intensity and reducing the signal amplification in the detector system (so as not to saturate the signal) will increase the backscattered proportion of the image-forming signal, and smooth surfaces can be readily examined. With practice, the primary beam current (through the lens currents) and the detector power supply can be so manipulated at each magnification that images with the best resolution and information content are obtained.[5] As previously mentioned, stereopairs can be taken under appropriate operating conditions for quantitative topographic analysis.

4.2. Atomic-Number Contrast

Contrast can result from specimens if they contain elements with differences in atomic number Z. Higher-Z elements have more electrons and hence give rise to a greater number of backscattered and smaller number of absorbed electrons than lower-Z elements. This simple explanation assumes that all other factors contributing to contrast are equivalent from the low- and high-Z regions. Ideally this would be the case when the surface is flat and uniformly conducting so that geometrical and charging contributions to the atomic-number contrast is reduced to practically zero. Despite these factors, however, this mode of information is usable only for relatively large atomic-number differences between neighboring regions. This property has been used for identification of certain phases by Price and Johnson.(6) Pfefferkorn and Blaschke(7) have shown that by coating unetched metallurgical samples with a very thin layer of evaporated gold, the microstructure can be brought out. The phases with higher average Z generate more backscattered electrons which, while traveling through the thin gold layer, generate more secondary electrons. Conversely, phases with lower average Z generate fewer backscattered electrons. The resulting contrast differences can then be used to separate the microstructural constituents.

4.3. Surface Magnetic and Electric Fields, Voltage Contrast, and Beam-Induced Conductivity

If magnetic fields occur close to the specimen surface in small local areas, such as where magnetic domain walls meet the surface, then the secondary emitted electrons from these areas will undergo small deflections as they leave the surface, and so the angular-distribution curve will be slightly tilted. Hence, if the emissive mode is used, these magnetic field areas will be revealed in the image by differences in contrast. Surface electric fields and voltage distributions produce analogous contrast effects. Such contrast effects are very small with backscattered electrons. These methods have the advantage that no magnetic particles need be deposited on the specimen, and hence the changes in the position of the walls or the strengths of the fields, as caused by applying an external magnetic field, can be continuously followed.

The electron beam can produce many electron–hole pairs in semiconductor specimens. If an electrical field is applied across a p–n junction of a semiconductor specimen, it will cause motion of these carriers which, in turn, results in observable contrast changes in the specimen. A detailed discussion of various phenomena related to electric and magnetic contrast, has been provided by Thornton.(8)

5. VARIETY OF ANALYSES AND INTERPRETATIONS OF SEM IMAGES

5.1. Stereologic Analysis

Many mathematical methods based on "stereology" have been developed to treat light-microscopy images for quantitative information.[9] Similar methods can be used, with care, in some SEM applications[10–12] such as size measurements for particulate matter, blood cells, and second phases on smooth surfaces. The intensity of secondary or backscattered electron signals can be used if no interfering topographical effects are present. Suitable discriminator and logic units can be set up to identify and store information from any phase, based on the contrast due to atomic-number effects only. If necessary, one can use two signals—for example, the intensity gray level from the secondary electron image combined with the presence or absence of a certain element as indicated by x-ray spectroscopy—with the logic and discrimination systems. Initial operator judgment in selecting appropriate signals, discriminator settings, and logic are very important in obtaining meaningful quantitative data. Such methods for examining polished microstructural sections and for particles embedded in a solid matrix and revealed on a polished face have been developed and tried by Dorfler and Russ[10] and by Braggins *et al.*[12] (see Fig. 5-3).

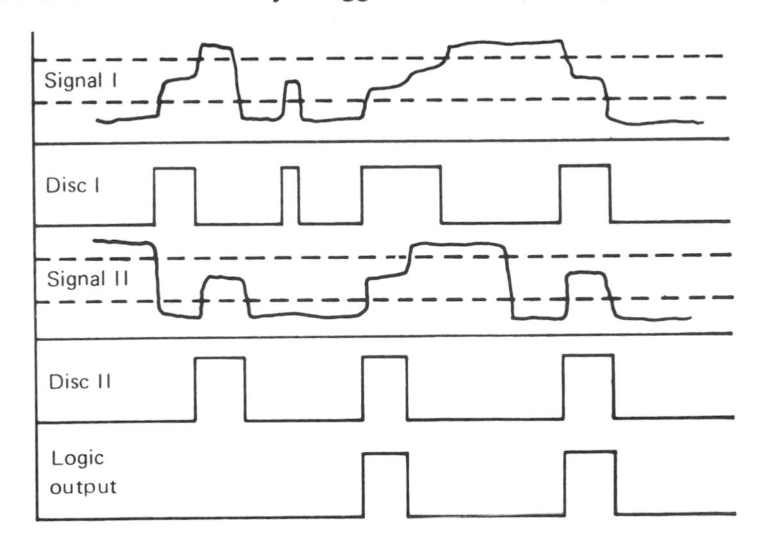

Fig. 5-3. Use of two signals from the SEM to discriminate phases present in the sample.[10] Each signal is processed by two discriminators optimally set for discrimination of a particular phase. In some cases outputs are caused by mixture of two phases; coincidence circuit uses AND gate to provide unique digital indication of particular phase selected.

5.2. Transmission Scanning Electron Microscopy

Although most commercial SEM's are used to study surface features, signals transmitted through thin samples can be collected by a suitable detector placed below the sample, and thus SEM can be used in the transmission mode (TSEM). Comparison of the TSEM with a conventional transmission electron microscope (TEM) shows that the two microscopes are equivalent, so that information obtained from a TEM theoretically can also be obtained in a TSEM.[13] Dark-field and diffraction work is possible for metallurgical and materials science applications, although in commercial SEM's the resolution (approximating the beam diameter—about 50–75 Å, depending on the thickness, primary kV, and other parameters) is much poorer than that of a TEM, about 2–5 Å. The main advantage is in biological work: (1) Thicker samples can be examined in the TSEM than in the TEM, because the scattered electrons are collected point-by-point and amplified, unlike TEM where primarily coherent electrons are collected and the image is magnified through an electron optical system; in other words, TSEM is similar to a built-in image intensifier. (2) Examination at low magnifications is possible; and since surface examination can also be made, correlation with other results, such as those from various types of optical microscopy, is easily possible. (3) Since the SEM's can run at a much lower voltage (5–25 kV, typically), specimen damage is reduced compared to a TEM (typically run at 50–100 kV). (4) Since the specimen is not physically placed in the field of the final lens, much larger specimens can be used for TSEM than for TEM.

Counting dispersed particles and obtaining their size distribution pose different problems. Particles must be so dispersed that aggregates are broken up. To obtain exact particle counts (not counting the same particle twice), a delay line that stores a scan line for comparison to the next scan line is used. Thus, the continuous boundary of a single isolated particle is followed, and the particle is counted at the point where the last line intersects it. Pattern recognition circuits or the ratio of the particle area to the perimeter as a function of particle size can be used to separate particles when they may be touching. Errors caused by features intersecting the images are avoided by additional logic circuits. Magnification has to be carefully selected so that fine particles are crossed with enough scan lines, yet spurious details from the substrate are not picked up. To some extent, this discrimination may again be obtained by using x rays or some other suitable second signal to separate substrate effects or to eliminate or detect the presence of different types of particles.

For these two cases, smooth surfaces and particle distribution, techniques are well developed. However, rough-surface analysis by these techniques is severely limited. In rough surfaces, the intensity gray levels depend heavily on topography, some features may be masked by others, and use of an additional signal may similarly suffer. In a way, future developments in automatic SEM stereometric analysis are tied in with automatic photogrammetric analysis of stereopairs, for which rapid image-analyzing techniques are not yet available.

Interpretation of TSEM images presents some unique problems because three-dimensional features from larger thicknesses project on a two-dimensional photograph. Perhaps correlations with thinner sections will be useful in working out these interpretation problems.

Specially built TSEM's with a field-emission source and an ion-pumped vacuum system have been used by Crewe to obtain point resolutions of 5 Å and to resolve atoms of uranium.[14]

Some contrast mechanisms are unique to TSEM.[13,14] These arise because of the way in which the information is presented—the TSEM yields a signal which builds up point-by-point as compared to TEM where the whole signal is available simultaneously. These signals can be collected and treated by advanced image-improvement techniques. Energy analyzers can be used to detect energy spectra from single-atom excitation or collective plasma losses. Treatment of elastically scattered electrons can provide information about mass per unit area or total mass of the object under examination. The ratio of elastic to inelastic signal is proportional to atomic number and has been used to provide contrast from heavy atoms.[14] Thus a variety of information about the sample can be obtained in the TSEM operation.

5.3. X-Ray Spectroscopy

Use of x-ray spectroscopy tremendously enhances the analytical value of the SEM in material characterization by providing chemical analysis of the sample along with surface topography.

Characteristic x rays emitted under the effect of the electron beam provide information about the nature and amount of elements present in the volume excited by the primary beam. Energy-dispersive (ED) x-ray attachments, consisting of a lithium-drifted silicon crystal, a multichannel analyzer, and necessary electronics, are finding increasing use on many SEM's. Wavelength diffractometers (WD), used with electron probe microanalyzers, are also available as an accessory on the SEM. Depending on the need and amount of funds available, one or both of these systems up to any level of sophistication can be provided.

A brief description of the two x-ray detection methods is warranted before comparing them. In the WD method, a crystal of a known spacing d separates x rays according to Bragg's law, $n\lambda = 2d \sin \theta$, so that at a diffraction angle of θ (collection angle of 2θ), x rays of specific wavelengths are detected. To cover the whole range, the diffractometers are usually equipped with many crystals. Even then, considerable time is needed to obtain an overall spectrum of all elements present. The resolution of the crystal in separating x rays of different wavelengths is very good (of the order of 10 eV), but its efficiency is very poor. To improve the collection efficiency, curved-crystal fully focusing diffractometers are used.

For nondispersive (ED) spectrometers, the energy of an incoming x-ray photon is converted into an electric pulse in a lithium-drifted silicon crystal. A bias voltage applied to the crystal collects this charge, which is proportional to the energy of the x ray. This pulse is amplified, converted to a voltage pulse, and fed into a multichannel analyzer. The analyzer sorts out the pulses according to their energy and stores them in the memory of the correct channel. The resulting spectrum can be displayed on a CRT, plotted on a chart, or printed out numerically.

5.3.1. Energy Resolution and Detectability Limit

The best resolution available on the ED systems (about 150–170 eV) is much poorer than the resolution of the WD systems (about 10 eV). In practice, however, the poorer energy resolution of the ED system is not a severe limitation because most elements above fluorine (atomic number 9) are readily detected. ED analyzers have been used to detect elements down to oxygen, but problems are encountered when elements whose x-ray energies are too close together coexist in a sample. For example, in the low-Z region, the K lines of boron, carbon, nitrogen, and oxygen lie between 190 and 530 eV. These elements are usually present with other elements of higher atomic number so that L and M lines of the higher-Z elements interfere with the absolute detection of these elements of low atomic number. Thus, this is the major factor in light-element detection. Other factors limiting light-element detection are window thickness and electronic noise. Better detectors and electronics (lower noise) and thinner window or windowless detectors may lower the detectability limit to ranges comparable to WD systems up to boron. Detailed considerations of various aspects of low-Z-element detection have been discussed by Russ.[15,16]

5.3.2. Beam Current and Spatial Resolution

Since the WD collection efficiency is much poorer, beam currents required to generate satisfactory x-ray statistics are much higher than for ED systems. ED systems work satisfactorily at the best resolution condition of the SEM, that is, current about 10^{-11} A and beam diameter about 50–75 Å. On the other hand, typical currents for WD systems are 10^{-6}–10^{-8} A at a beam diameter of 1000–3000 Å. Brighter source guns do provide a better current density, and brighter beams with lower diameters are possible. One would think that the lower the beam diameter, the better the spatial resolution, but this is not always the case because the spatial resolution is a function of the beam penetration, which depends on the voltage and the properties of the sample under examination.

For a 25-kV beam, the penetration may be as much as 1 μ, so that in essence, whatever the size of the beam, one is analyzing a depth of about 1 μ. Lower voltages may be used, but the potential is determined by the x-ray absorption edges of the elements to be detected. A finer beam is useful, however, in detection of particles. If particles were put on a suitable substrate so that x rays from the substrate do not interfere with the x rays from the specimen, particles approaching beam diameter could be analyzed using x-ray spectroscopy.

5.3.3. Analysis Time and Sensitivity

The ED system has a much larger acceptance angle than the WD system, and since all x rays are simultaneously analyzed and more x rays are collected, normally 0.5–1 min is all the time required to obtain a spectrum from a given region of a sample in the ED system, compared to about 15 min in a WD system. The comparative rapidity coupled with satisfactory resolution are the most attractive features of the ED system. The signal-to-noise ratio for WD spectrometers is much higher than for the ED systems, but in a detailed analysis Sutfin and Ogilvie[17] showed that the sensitivity of the two systems is comparable and, for x-ray spectra of low intensity, ED systems are preferable.

5.3.4. Geometry of the Samples

Fully focusing WD spectrometers cannot be easily used to examine rough surfaces because the intensity drops off appreciably for a sample position above or below the point of focus.[17] Semifocusing WD spectrometers do not suffer from this limitation, but larger beam diameters have to be used to compensate for the intensity loss caused by the semifocusing. No such limitation exists for ED systems. One must be careful, however, that the area being impacted by the primary beam is in a direct, straight-line path to the ED detector, or erroneous analysis can be obtained.[18] In this respect, stereoexamination may be very helpful to understand the geometry of the area under examination relative to the detector position.

5.3.5. Quantitative Work

For smooth surfaces, quantitative work is possible with both the ED and WD spectrometers. Since the WD methods have been in use for nearly a decade, the correction procedures are well developed and many theories known.[19] The same theories and corrections are applicable to the ED method, although not as extensive literature has been developed. Examples of quantitative work on some alloy steels have been recently published by Berkey[20] and show that quantitative analysis can be as readily carried out with ED systems as with WD systems.

In summary, line scan, point analysis, and x-ray photographs are readily obtained with either system. All things considered, it is safe to predict that except for some highly specialized cases requiring accurate quantitative analyses of elements in small concentrations or of low atomic numbers, the ED detectors will be used more and more.

5.4. Auger Electron Spectroscopy

Many recent articles indicate that there is considerable interest in Auger electron spectroscopy (AES). (For excellent reviews see MacDonald,[21] Chang,[22] and Chapter 20 of this volume.) The technique is most powerful, providing analysis of the first few atom layers (10 Å or less) on the surface of the sample. Many applications include detection of very low concentrations of impurity segregation, which critically affect mechanical properties of some body-centered-cubic metals and alloys such as phosphorus in tungsten, and arsenic, antimony, and tin in steels.[23]

Auger electron analysis with the SEM requires appropriate energy-analyzing equipment, but more importantly a much better vacuum system than is currently available in most instruments. Since AES analyzes only the first few surface layers, the samples must be free from any surface film, and the vacuum system should be such that no contamination layer is allowed to build up lest one analyzes the contaminants and not the surface layer of interest. Since very few SEM's with such high vacuums (10^{-8} Torr or better) are in use, very limited AES work has been done on the SEM. As mentioned in the beginning, a better vacuum system must be cleaner and also faster. Thus the future of AES in SEM is closely tied to developments in vacuum systems.

Nevertheless, the potentials of this technique are tremendous. Adsorbed and absorbed surface layers can be removed either by high vacuum alone or by low-energy (less than 1 keV) ion bombardment.

Techniques of deriving qualitative and quantitative information from Auger spectra are similar to those used by LEED and other non-SEM researchers. Peak differentiation and comparison with the standards have been used to clearly define peaks and obtain quantitative information. The required energy-analysis equipment is available as standard commercial items. Standard Auger spectra are available for all elements for various Auger transitions.[22] Overlapping spectra from two elements may create some problems of elemental separation, but with high-resolution energy-analysis equipment, procedures similar to those used with x rays can be employed to obtain elemental separation.

Auger electron production is a very inefficient process—approximately 10^3–10^5 primary electrons are required to produce one Auger electron. This means that much higher beam currents are required, causing spatial resolutions to be poorer than 1 μ. In this respect, brighter guns offer a distinct advantage in considerably improving the spatial resolution. Another important parameter is accelerating voltage—typically 500–1000 V acceleration is used (SEM's may require a special power supply)—although useful voltage is dependent on the ionization voltage for the particular level in the atom. Higher accelerating voltages will deteriorate the spatial resolution, besides reducing the extent of ionization, because the incident electron does not spend sufficient time in the neighborhood of the atom for appreciable interaction.

Examination of nonconductive samples in the SEM creates problems because of specimen charging. While use of low accelerating voltages may eliminate charging effects as far as Auger analysis is concerned,[24] the SEM resolution will be unsatisfactory unless 5–10 kV can be used. This requires coating of samples. These are some of the problems that will have to be faced as Auger electron spectroscopy finds increasing applications with the SEM.

5.5. Cathodoluminescence

Under the influence of electron bombardment, some solid materials will cathodoluminesce, resulting in radiation in the visible or near-infrared region of the electromagnetic spectrum. Cathodoluminescence (CL) provides another signal that can be used for analytical purposes. A number of mechanisms limit the spatial resolution when using this signal (for details, see Thornton[8]). For opaque materials that absorb the luminescent radiation, resolution may approach beam diameter; otherwise it depends on the penetration and scattering of the primary beam. To obtain an appreciable signal, however, large beam diameters (typically 1000 Å or more) are used, and thus spatial resolution is already poor from the start.

Besides many applications in the area of semiconductors (GaAs, AlAs, AlP, ZnS), shift in CL spectra has been used to detect the presence of trace elements of impurities in certain materials. For example, see Remond *et al.*,[25] and for a review of CL spectra, see Muir *et al.*[26] For SEM examination, nonconductive samples have to be made conductive by depositing a carbon or metallic evaporated film. Even with these inherent difficulties, however, the technique is applicable in certain specialized cases.

5.6. Diffraction in the SEM

Since Coates[27] first detected the pseudo-Kikuchi effect,* rapid strides have been made in the past two years in the development of diffraction techniques and information content in the SEM.[29,30] While the techniques as they are now developed are applicable on surfaces with no interfering topographical effects, areas down to 2 μ in diameter can be analyzed. The spatial resolution should decrease further with the use of brighter guns, so that even smaller areas could be analyzed to provide the diffraction information. The patterns obtained resemble Kikuchi line patterns obtained in transmission electron microscopy and have been called Selected Area Electron Channeling Patterns (SAECP).

To obtain SAECP's, the primary electron beam, impinging at a point on the specimen, is made to rock at that point so that it views the specimen at a number of incident angles. The angle of incidence can be made to vary in a suitable manner, either mechanically[27] or electrically.[29] Modifications on some commercial SEM's are available so that channeling-pattern conditions are readily obtained.

The image obtained by this rocking action of the beam at the point of impingement contains bright and dark lines in a geometrical pattern because of anomalous-absorption or channeling effects. Either backscattered- or absorbed-electron images are used—usually backscattered images are preferred. The analysis of these patterns is similar to that used with the Kikuchi diffraction patterns of the TEM.

The SAECP can be used to obtain the crystal structure of the area under examination. The spacing of the pairs of bright and dark lines is related to the *d* spacings of the planes in the crystal. This plus the angle measurements between different lines can be analyzed to provide crystal structure data. For ease and comparison to find orientation of unknown regions, one can construct an ECP map—by tilting and rotating the sample and obtaining SAECP's from each region. The

* See also Coates.[28]

geometry of the pattern, therefore, can be used to obtain a variety of crystal structure data: (1) lattice parameters for the crystal and, with the help of this information and standard data books, identification of phases; (2) orientation relationships for polycrystalline samples and twinned regions, phase-transformation studies, and so forth; (3) orientation determination of individual regions and, through these, studies of preferred orientation and textures; and (4) epitaxial growth studies to obtain orientation of the films relative to the substrate.

The second aspect of SAECP's that can provide valuable information is the quality of the patterns. Perfect crystals give sharp patterns, but any type of crystal imperfection causes the patterns to be diffused (similar to x-ray diffraction line broadening). The diffuseness of the line patterns can be measured and quantitatively related to the nature of crystallographic imperfections present in the samples. Thus, the pattern sharpness (or width, or resolution) has been used as a criterion to determine ion or electron radiation damage, extent of deformation, and dislocation density in a given region of a deformed sample stress distribution ahead of a progressing crack, progress of recovery and recrystallization during annealing of metals, and determination of the extent of deformation away from a fractured or deformed region of the sample in a number of metals and alloys.[30,31]

The applications of these techniques in material studies are numerous, and as the techniques become increasingly routine, their usefulness should increase considerably also. The main advantage in comparison to x-ray diffraction is that here diffraction analyses are done on areas down to micron range, compared to about millimeter range in x-ray diffraction.

Future improvements will include enhancement of the SAECP quality and reduction in the area from which they are obtained. Besides brighter guns, these improvements can be made using energy analysis, dynamic focusing corrections, and so forth.[32] As more and more work is done using these techniques, it is only a matter of time before computer programs to analyze the patterns and more accurate and perhaps automated procedures to determine line width become available. Some experiments have also been done in the SEM with Kossel patterns.[32]

5.7. Examination of Internal Structure

The study of internal structure through SEM studies of thin films has already been covered. Ion etching can be used to remove surface layers selectively and expose the inner structure of material. The features of the structure so brought out depend on the ion-etching parameters (such as the nature of the gas and ion energy) and on the properties of materials being etched (for example, certain crystallographic planes etch preferentially with respect to other planes). Ion etching can be done both inside and outside the SEM, or by providing a vacuum connection chamber. Although some ion-etching experiments have been done both ways, it is not yet clearly resolved whether one is viewing the true inner structure of the material or the structure of the artifacts introduced by ion etching.

In material sciences, chemical etching has been used to bring out the inner structure of polished surfaces. Although this work has been traditionally done on a light microscope, it can be shown that the advantages offered by the SEM in traditional metallography are many: large depth of focus allows clear understanding of the role of the etchant and identification of etching debris; higher resolution provides more details; chemical and crystallographic information is possible through x-ray spectroscopy and channeling patterns, respectively.

A new approach, found very useful in studying the inner details of microstructure, is the study of carefully preserved fracture surfaces. Since the mechanical properties of materials are controlled by the nature and distribution of other phases such as precipitates, inclusions, and porosities, these features are best brought out on a fracture face. A propagating crack will go through all inhomogeneities in its vicinity, so that a fracture face will present the nature and distribution of those microstructural constituents that affect the mechanical properties. Since no sample preparation is involved, dangers of second phases dropping out or porosities getting filled in during polishing or other preparation do not exist. The technique of examining microstructures through fractures has been successfully used in metals, ceramics, plastics, and in interfacial studies.

5.8. Examining Samples in Controlled Atmospheres

An electron optical system has to operate under vacuum; otherwise, the electrons will collide with the molecules in the chamber and lose all their energy. This has been a limitation in examining samples not stable under vacuum conditions, particularly biological materials. Many specimen-fixing and preparative procedures are used by biologists in preparing their samples for SEM examination; no single method works for all situations. Ideally one would like to examine the specimens in their natural state.

One approach has been described by Lane.[33] In a specially built stage, a partial pressure of water or any other gaseous phase is maintained, so that there is an equilibrium with the specimen. Using this system, he was able to examine a drop of water and the sublimation of ice. This method, obviously, is severe on the vacuum system and necessitates frequent cleaning. The approach requires further experimentation.

In another approach, a certain amount of gas is let into the specimen chamber to cause dynamic reactions such as oxidation or reduction that then would be directly observed. Even at the highest scanning speed available in the SEM (in the TV mode), the reaction was too fast for detailed studies.[34] Nitrogen-gas purge has been used to lower the contamination on the sample[35]; whether it would be possible to similarly use a continuous jet of another gas to maintain the sample stability remains to be proven.

Examination of samples outside the vacuum has also been tried. Cowley,[36] using a 600-kV transmission scanning electron microscope, allows the electrons to leave the vacuum chamber through a very thin plastic window. These high-energy electrons go through the sample and impinge on a collector placed on the other side of the sample. The distance traveled by electrons in air is kept as short as possible. This is, of course, a highly specialized system, but similar approaches have been tried with commercial scanning microscopes. Such systems can only work in transmission, because for surface scanning microscopy, the sample would have to be so placed that the field of the collecting detector would not interfere with the primary beam. This would require too much space and would cause the electrons to be lost by collisions with atmospheric molecules.

Finally, recent availability of temperature-controlled stages allows examination of samples at low temperatures where moisture or other liquid phases can be retained because of their very low vapor pressure. Results reported by Echlin[37] show that a resolution of 700 Å could be obtained with this stage. With this approach, care has to be taken to avoid ice-crystal damage and sample heating during examination, leading to a loss of moisture. The sample is frozen in Freon first, and then stored in liquid nitrogen. The examination is made on the cold stage, which is maintained at $-170°C$, after the ice crystals have been removed by heating the sample briefly to $-100°C$. Various fixative methods can be combined with this technique, although some problems are encountered with specimen charging if the accelerating potential is above 5 kV.

Thus thc area of examination of specimens in their natural state is still a matter of further research into refinements of these approaches or development of new techniques.

A wide variety of stages can be constructed so that specimens can be heated, cooled, elongated, compressed, twisted, fatigued, fractured, coated, sublimed, bombarded by ions, chemically reacted, electrically stressed, magnetized, etc. Observations may be made either at intervals or continuously.

6. ADDITIONAL METHODS FOR OBTAINING MORE INFORMATION

6.1. Signal Processing

Because the signal collected is a continuously varying function of time, electronic processing can be performed on the signal before it is used to produce the image, and so considerable additional information is made available.

The contrast can be increased overall by backing off the dc level and increasing the amplification. These enable an increase in the signal-to-noise ratio and hence improve the relative contrast.

In a second version, the contrast can be changed in a variable manner by taking the nth root ($n = 2, 3, 4, \ldots$) of the signal, and increasing the amplification. This variable γ control increases weak contrast in low-signal regions and decreases strong contrast in high-signal regions.

The contrast can be increased and features sharpened by differentiating the signal with respect to time. This can be performed using simple resistance–capacitance circuits with suitable time constants.

Derivative processing of signals can bring out small variations in intensity, and thus provide additional details and ease of interpretation. In this technique,[38] high-frequency components undergo preferential amplification over the low-frequency components. This enhances the contrast by bringing out details in low-contrast areas and suppressing the extreme contrast conditions, thereby improving the discrimination of details. This is particularly useful in stereoexamination, because fine details in dark holes and bright crests become readily recognizable. Examination of rough samples at low magnifications and smooth samples with not easily recognizable features should benefit immensely from such a system.

6.2. Image Processing

Photographs can be processed through holographic techniques to obtain enhanced resolution if the electron optics and operating conditions of the microscope are known.[39] Optical-diffractometer and convolution-camera techniques have been used to derive valuable information in studying the arrangement of clay particles in consolidated and unconsolidated soil samples.[40]

Color images themselves have no real meaning in SEM, but can be used in some instances. Hayes and Pawley[41] have used color images to code different gray levels, to highlight special features, and to make interpretations easier. Two or more modes of SEM images (secondary, backscattered, or x ray) or operating conditions (kV, contrast) may be presented by different colors on the same photograph for ease of comparison and to bring out additional information.

6.3. Beam Chopping

In this technique, a pulsed incident electron beam is used, instead of a steady beam. The main applications of this technique have been either with regard to improving the signal-to-noise ratio or for stroboscopic observations.

6.4. Contrast Separation

It is often advantageous to be able to separate the various contrasts that occur in SEM images. Surface topography and atomic-number contrast have been separated by using two symmetrical collectors, one on either side of the specimen. The images formed by subtracting the two signals show mainly topography contrast, while images formed by adding the two signals show mainly atomic-number contrast.

Voltage contrast has been separated from a combination of surface topography and atomic-number contrast by a beam-chopping technique.[42] The voltage is supplied to the semiconductor specimen during the even-number pulses, but not during the odd-number pulses. Signals corresponding to each set of pulses are separately extracted electronically and subtracted. The resulting image shows only the voltage contrast.

6.5. Energy Analysis

In both the emissive and reflective modes, the electrons collected have a range of energies, and separating the electrons in a narrow energy distribution can offer some advantage. If the signal intensity is plotted as a function of energy, the shape of the curve, position of peak, etc., allow information to be deduced regarding the specimen such as the identity of the material and the potential of the surface. The manner in which the image contrast changes when electrons of different energies are collected provides information about contrast mechanisms. If only the high-energy electrons are used to form the image, spreading effects in the specimen are almost completely eliminated and high-resolution images can be obtained.

6.6. Low-Loss Image

The principle of low-loss image in the backscattered electron mode consists of rejecting the slower backscattered electrons by means of an energy filter.[43] The low-loss signal is obtained from a very thin surface layer (typically 50–200 Å), and the resolution is 170 Å or better. This image cannot show details that are shielded from the detector.

6.7. Low-Voltage Operation

A new technique using a low-voltage electron beam is receiving attention of research workers in those cases where specimen charging or irradiation damage has to be avoided. The difficulties in examining nonconducting materials in SEM are due to the buildup of static charge, causing image distortions. In normal SEM practice, such materials are coated with a thin layer of vapor-deposited metal to render them electrically conducting. Another advantage with low-voltage operation is the decrease in contamination.

7. FUTURE WITH BRIGHTER ELECTRON SOURCES

The spatial resolution (the smallest feature that can be examined) will be a function of, among other things, the smallest spot to which the incident beam can be focused, consistent with enough intensity that an adequate collectible signal is obtained in a reasonable time and that problems associated with too weak a signal or an unfavorable signal-to-noise ratio are avoided. Other major parameters affecting the spatial resolution for a particular collected signal are the characteristics of the electron optical system, the primary accelerating voltage, and the properties of the material under examination. However, it is obvious that the brighter the electron source, the greater are the chances of better spatial resolution. Table 5-1 (updated from Broers[44]) compares important parameters for various electron sources. Nearly all of the commercial scanning electron microscopes now in use have a tungsten-filament source. Lanthanum hexaboride guns are now available as an accessory on some models, and a few manufacturers have recently announced SEM's with field-emission sources.

In principle it should be possible to modify an SEM to use the most suitable electron source, but, as Table 5-1 indicates, vacuum is a major limitation. Much of the rapidity of the present SEM is due to the ease and readiness with which the specimens can be changed. A better vacuum does not necessarily have to be slower, but it will definitely require many modifications in the vacuum system to make it adaptable for a brighter source. Some features in modifying a conventional SEM

TABLE 5-1. Brighter Sources Need Better Vacuums[a]

Cathode Type	Probe Diameter (not resolution), d_{min}	Brightness, A/cm^2-sr	Typical life for given brightness, hr	Temperature, °K	Typical satisfactory vacuum level, Torr
Tungsten hairpin (0.005 inch)	50 Å, 15-mm focal length. Surface or transmission microscopy.	7×10^4	30	2850	5×10^{-4}
LaB_6 rod Schottky emission (1-μ tip diameter)	25 Å (diffraction-aberration limited), 15-mm focal length. Surface or transmission.	6×10^6	~100	2000	1×10^{-6}
Tungsten field emission tip	5 Å (diffraction-spherical aberration limited), 0.6-mm focal length. Transmission only.	2×10^9	No limit; but reconditioning is required approximately every hour.	Ambient	1×10^{-9} or better

[a] After Broers,[44] with supplementary data.

to a LaB_6 gun are described by Brandis *et al.*[35] A cleaner and better vacuum offers many advantages (reducing specimen contamination, cutting down frequent column cleaning, longer electron-source life, and improved possibility of Auger electron spectroscopy, among others), but it must be weighed against the slower operation due to longer specimen-exchange times and the added cost. Ideally a cleaner, better, and faster vacuum system would be desirable.

One approach to overcoming these difficulties involves the use of differential pumping systems, an ion pumping system for the gun region, differential pumping apertures, and, for the specimen chamber, a conventional mechanical diffusion pump system fitted with suitable decontamination devices so that the partial pressure of contaminating hydrocarbons is lowered to the range of 10^{-9}–10^{-19} Torr. However, such systems are not yet readily available for present commercial microscopes.

8. CONCLUSION

While the first SEM was built in 1938, commercial units became available only in 1965. Since then, the SEM has already established itself as an immensely useful tool. Quantitative topographical studies and rapid x-ray spectroscopic analysis are now in regular use. Many new areas of characterization are being developed to a stage where they will soon be routinely available to the researcher.

Thus, total material characterization with a single instrument can be foreseen. The challenge is to wed the techniques described here—and possibly others—to the same instrument and still retain its basic characteristics of simplicity and rapidity.

9. REFERENCES

1. O. Johari, Total materials characterization with the scanning electron microscope, *Res./Develop.* **22**(7), 12–20 (1971).
2. P. G. T. Howell and A. Boyde, in *Scanning Electron Microscopy/1972* (O. Johari and I. Corvin, eds.) pp. 233–240, IIT Research Institute, Chicago (1972).
3. A. Boyde, in *Scanning Electron Microscopy/1970* (O. Johari, ed.) pp. 105–112, IIT Research Institute, Chicago (1970).
4. A. R. Dinnis, in *Scanning Electron Microscopy/1971* (O. Johari and I. Corvin, eds.) pp. 41–48, IIT Research Institute, Chicago (1971).
5. O. Johari, I. Corvin, R. F. Dragen, and N. M. Parikh, in *Scanning Electron Microscopy/1969* (O. Johari, ed.) pp. 277–284, IIT Research Institute, Chicago (1969).
6. C. W. Price and D. W. Johnson, in *Scanning Electron Microscopy/1971* (O. Johari and I. Corvin, eds.) pp. 145–152, IIT Research Institute, Chicago (1971).
7. G. E. Pfefferkorn and R. Blaschke, Staubuntersuchungen mit hilfe des raster-elektronenmikroskops stereoscan, *Staub-Reinhalt. Luft* **27**, 30–33 (1967).
8. P. R. Thornton, *Scanning Electron Microscopy*, Chapman and Hall, London (1968).
9. H. Elias, *Stereology*, Springer-Verlag, Berlin (1967).
10. G. Dorfler and J. C. Russ, in *Scanning Electron Microscopy/1970* (O. Johari, ed.) pp. 65–72, IIT Research Institute, Chicago (1970).
11. E. W. White, H. Gorz, G. G. Johnson, Jr., and R. E. McMillan, in *Scanning Electron Microscopy/1970* (O. Johari, ed.) pp. 57–64, IIT Research Institute, Chicago (1970).
12. D. W. Braggins, A. M. Gardner, and D. W. Gibbard, in *Scanning Electron Microscopy/1971* (O. Johari and I. Corvin, eds.) pp. 393–400, IIT Research Institute, Chicago (1971).
13. E. Zeitler, in *Scanning Electron Microscopy/1971* (O. Johari and I. Corvin, eds.) pp. 25–32, IIT Research Institute, Chicago (1971).
14. A. V. Crewe, High-resolution scanning microscopy of biological specimens, *Ber. Bunsen Ges. Phys. Chem.* **74**, 1181–1187 (1970).
15. J. C. Russ, in *Scanning Electron Microscopy/1971* (O. Johari and I. Corvin, eds.) pp. 65–72, IIT Research Institute, Chicago (1971).
16. J. C. Russ and A. Kabaya, in *Scanning Electron Microscopy/1969* (O. Johari, ed.) pp. 57–64, IIT Research Institute, Chicago (1969).
17. L. V. Sutfin and R. E. Ogilvie, in *Scanning Electron Microscopy/1970* (O. Johari, ed.) pp. 17–24, IIT Research Institute, Chicago (1970).
18. O. Johari, in *Scanning Electron Microscopy/1971* (O. Johari and I. Corvin, eds.) pp. 529–536, IIT Research Institute, Chicago (1971).
19. W. Reuther, Electron probe microanalysis, *Surface Sci.* **25**, 80–119 (1971).
20. E. Berkey and G. A. Whitlow, in *Scanning Electron Microscopy/1971* (O. Johari and I. Corvin, eds.) pp. 73–80, IIT Research Institute, Chicago (1971).
21. N. C. MacDonald, in *Scanning Electron Microscopy/1971* (O. Johari and I. Corvin, eds.) pp. 89–96, IIT Research Institute, Chicago (1971).

22. C. C. Chang, Auger electron spectroscopy, *Surface Sci.* **25**, 53–59 (1971).
23. N. C. MacDonald, H. L. Marcus, and P. W. Palmberg, in *Scanning Electron Microscopy/1970* (O. Johari, ed.) pp. 25–32, IIT Research Institute, Chicago (1970).
24. R. L. Weber, Auger electron spectroscopy for thin film analysis, *Res./Develop.* **23**(10), 22–28 (1972).
25. G. Remond, S. Kimoto, and H. Okuzumi, in *Scanning Electron Microscopy/1970* (O. Johari, ed.) pp. 33–40, IIT Research Institute, Chicago (1970).
26. M. D. Muir, P. R. Grant, G. Hubbard, and J. Mundel, in *Scanning Electron Microscopy/1971* (O. Johari and I. Corvin, eds.) pp. 401–408, IIT Research Institute, Chicago (1971).
27. D. G. Coates, Kikuchi live reflection patterns observed with the scanning electron microscope, *Phil. Mag.* **6**, 1179–1184 (1967).
28. D. G. Coates, in *Scanning Electron Microscopy/1969* (O. Johari, ed.) pp. 27–40, IIT Research Institute, Chicago (1969).
29. G. R. Booker, in *Scanning Electron Microscopy/1970* (O. Johari, ed.) pp. 489–496, IIT Research Institute, Chicago (1970).
30. G. R. Booker, in *Scanning Electron Microscopy/1971* (O. Johari and I. Corvin, eds.) pp. 465–472, IIT Research Institute, Chicago (1971).
31. R. Stickler, C. W. Hughes, and G. R. Booker, in *Scanning Electron Microscopy/1971* (O. Johari and I. Corvin, eds.) pp. 473–480, IIT Research Institute, Chicago (1971).
32. D. C. Joy, G. R. Booker, E. O. Fearon, and M. Bevis, in *Scanning Electron Microscopy/1971* (O. Johari and I. Corvin, eds.) pp. 497–504, IIT Research Institute, Chicago (1971).
33. W. C. Lane, in *Scanning Electron Microscopy/1970* (O. Johari, ed.) pp. 41–48, IIT Research Institute, Chicago (1970).
34. N. Ujiiye, S. Kimoto, and Y. Kawasaki, in *Scanning Electron Microscopy/1971* (O. Johari and I. Corvin, eds.) pp. 97–104, IIT Research Institute, Chicago (1971).
35. E. K. Brandis, F. W. Anderson, and R. Hoover, in *Scanning Electron Microscopy/1971* (O. Johari and I. Corvin, eds.) pp. 505–510, IIT Research Institute, Chicago (1971).
36. J. M. Cowley, D. J. Smith, and G. A. Sussex, in *Scanning Electron Microscopy/1970* (O. Johari, ed.) pp. 9–16, IIT Research Institute, Chicago (1970).
37. P. Echlin, in *Scanning Electron Microscopy/1971* (O. Johari and I. Corvin, eds.) pp. 225–232, IIT Research Institute, Chicago (1971).
38. A. Boyde, in *Scanning Electron Microscopy/1971* (O. Johari and I. Corvin, eds.) pp. 1–8, IIT Research Institute, Chicago (1971).
39. G. A. Stoke, M. Halioua, A. J. Saffir, and D. J. Evins, in *Scanning Electron Microscopy/1971* (O. Johari and I. Corvin, eds.) pp. 57–64, IIT Research Institute, Chicago (1971).
40. N. K. Tovey, in *Scanning Electron Microscopy/1971* (O. Johari and I. Corvin, eds.) pp. 49–56, IIT Research Institute, Chicago (1971).
41. T. L. Hayes, R. M. Glaeser, and J. B. Pawley, in *Proceedings of the 29th Annual EMSA Meeting*, pp. 410–411, Claitor Publishing House, Baton Rouge (1969).
42. C. W. Oatley, Isolation of potential contrast in the scanning electron microscope, *J. Sci. Instr.* **2**, 742–744 (1969).
43. O. C. Wells, in *Scanning Electron Microscopy/1972* (O. Johari and I. Corvin, eds.) pp. 169–176, IIT Research Institute, Chicago (1972).
44. A. N. Broers, in *Scanning Electron Microscopy/1970* (O. Johari, ed.) pp. 1–8, IIT Research Institute, Chicago (1970).

6

FIELD ION MICROSCOPY

Philip F. Kane

Materials Characterization Laboratory
Texas Instruments, Inc.
Dallas, Texas

1. INTRODUCTION

Perhaps the most fascinating of all the techniques available for the study of surfaces is the field ion microscope inasmuch as it is the only technique capable of giving an image of the atoms themselves. It was developed from the earlier field emission microscope and has recently been given another dimension by the addition of a time-of-flight mass spectrometer. This latter combination has been described as the atom probe and it is well-named since it can locate and identify individual atoms in the sample surface.

It is seldom that a technique can be attributed so uniquely to one scientist as can this one. Throughout its evolution, from the first development of the field emission microscope in 1936 to its present-day form as the atom-probe field ion microscope, the acknowledged leader has been Erwin W. Müller, who initially worked in Germany but for the past twenty years has worked at Pennsylvania State University. A review of field ion microscopy is essentially a review of the work of Müller and his co-workers. For a more complete and authoritative survey, the reader is referred to the book by Müller and Tsong[1] which deals in detail with field ion microscopy. The development of the atom probe is reviewed in a later paper by Müller.[2]

2. FIELD EMISSION AND THE FIELD EMISSION MICROSCOPE

When a metal filament is raised to a high temperature, electrons are emitted from the surface and, under the influence of an electric field, travel from the cathode filament towards the anode. This phenomenon, thermionic emission, is widely used as the electron source in radio tubes and in many electronic instruments such as electron microscopes. It is characterized by voltages of the order of 50–100 kV and high currents which generate temperatures over 1500°K. Figure 6-1a represents the energy diagram for a metal with no external field applied. Here W_i represents the Fermi energy and Φ is the work function of the surface. What this implies is that electrons will be retained in the surface by electrostatic forces unless their kinetic energy is greater than $W_i + \Phi$. Application of an external potential, i.e., if the metal is made a cathode, reduces the energy required to surmount the electrostatic potential barrier leading to increased thermionic emission.

If the field is high enough, as in Fig. 6-1b, another mechanism becomes possible. Under the very high potential slope, in the range of 10 MV/cm, the distance AB may become comparable to the wavelength

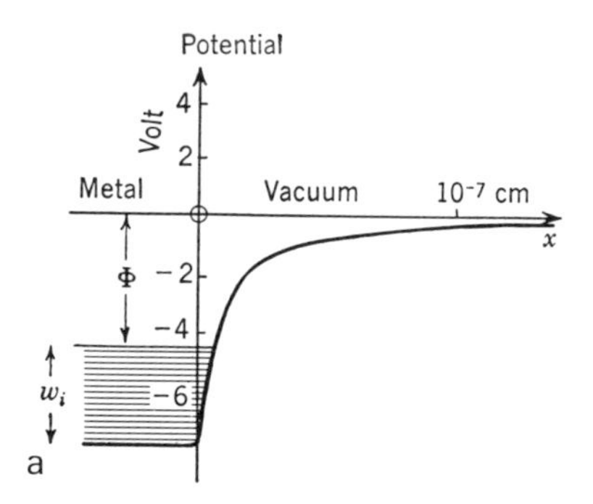

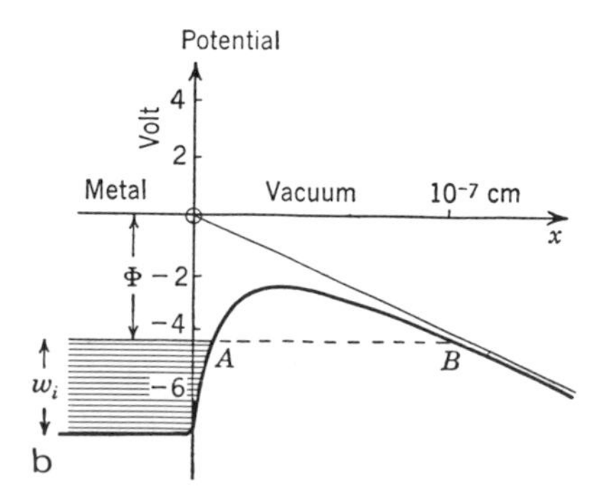

Fig. 6-1. (a) Potential on a metal surface with no external field. (b) Potential on a metal surface with an external field of 40 MV/cm.[3]

of the electrons and the phenomenon of tunneling occurs; i.e., electrons are not reflected by the potential barrier but drift through, arriving at B with zero velocity perpendicular to the metal, from which point they are rapidly accelerated away from the surface by the high field. This mechanism does not require high temperatures. It is this low-temperature emission under very high potential fields that is termed field emission.

A much more detailed mathematical treatment of this subject has been given by Good and Müller.[4]

The field emission microscope was developed by Müller[5] in 1936 and is shown schematically in Fig. 6-2. It is basically a simple device although considerable practical difficulties arise in preparing the tip, which is also the sample. To achieve the very high potential necessary for field emission, the tip must be of very small radius, of the order of tenths of a micron. To allow a reproduction of the emissivity to be registered by the fluorescent screen, the tip must be hemispherical. This is probably the most difficult to achieve; usually the wire sample is etched in a suitable solvent (e.g., molten sodium nitrite for tungsten) and heat-polished in a vacuum. This last step consists of heating the tip in the apparatus by passing a current through the loop supporting

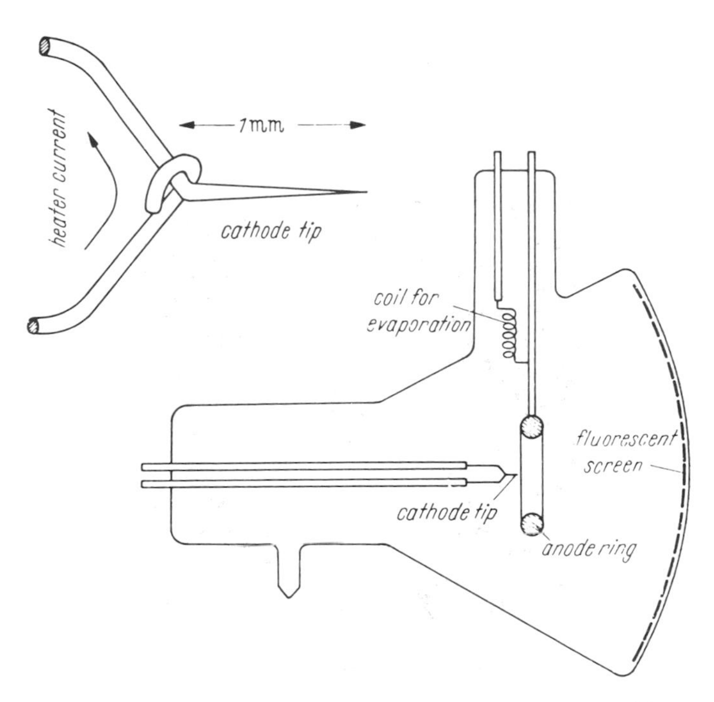

Fig. 6-2. Schematic diagram of field emission microscope.[4]

the tip. The surface atoms migrate to give a smooth surface at the atomic scale. With a tip of about 1–2000 Å in radius, a few thousand volts is sufficient to generate field emission.

The emission of electrons is dependent, as was shown in Fig. 6-1, on the applied potential and the work function. The applied potential will vary with the curvature of the surface, but with a carefully prepared tip, the major effect will be due to the work function. This, in turn, is dependent on such factors as the crystallographic orientation and the presence of other atoms in the lattice, i.e., on both structure and composition. Since the electrons travel essentially radially from the tip, under the influence of the strong radial electronic field, and if both tip and screen are hemispherical, then a work-function distribution map of the surface will be reproduced on the screen. With tips of the size suggested and a tip to screen distance of a few centimeters, then this map will be magnified about 10^6 times.

A typical field emission micrograph is shown in Fig. 6-3 which is of the emission from a clean tungsten tip. The low-index directions show up as small dark regions surrounded by larger areas containing the other possible indices. It may be compared with the usual stereographic projection for a cubic crystal (Fig. 6-4) which assists in identifying the particular planes in the crystal. The temperature of the tip can be determined either pyrometrically or by infrared emission or, at lower temperatures, by the change in resistance of a wire loop held on the cathode heater loop. Micrographs have been made down to liquid-helium temperatures where the surface atoms are frozen in.

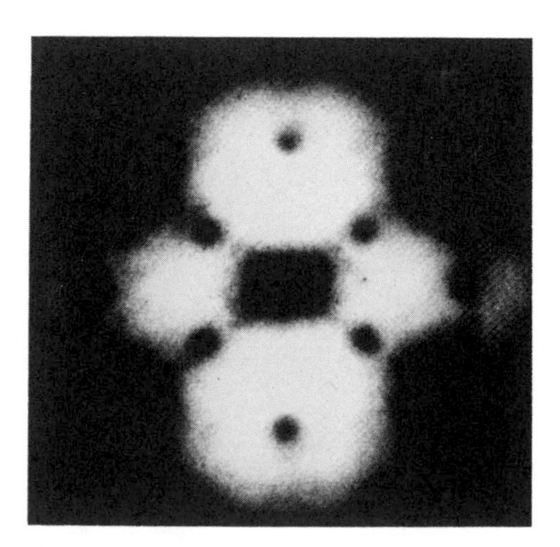

Fig. 6-3. Field emission pattern of clean tungsten ($r = 2.5 \times 10^{-5}$ cm, $V = 3500$ V, $i = 10^{-5}$ A).[4]

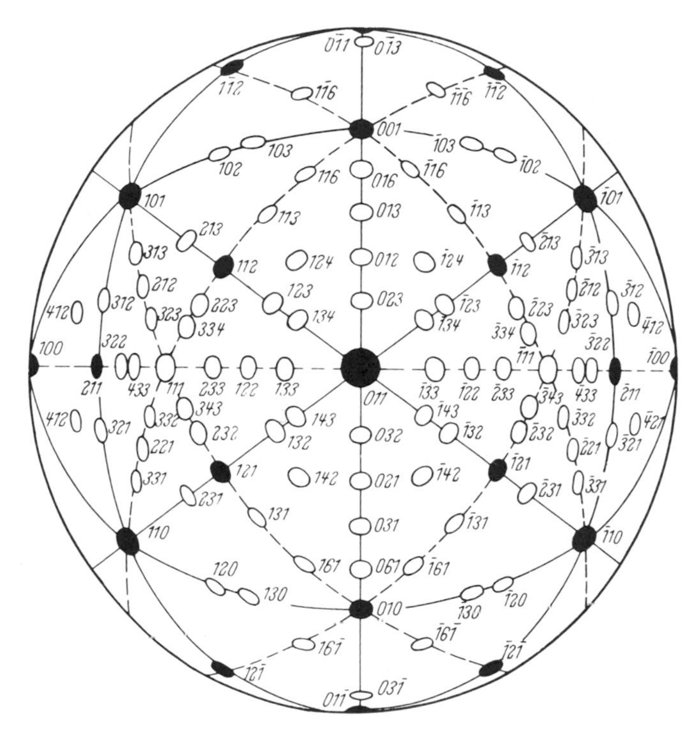

Fig. 6-4. Crystallographic map for the field emission pattern of a cubic crystal.[4]

Considerable work was done with the field emission microscope on adsorption effects on metal surfaces, in particular, the effect of oxide formation. Figure 6-5 is a series of micrographs taken from a tungsten tip and the effects at various stages of adsorption and desorption can be studied. The heating cycles were of two-minute duration, and it is interesting to note the similarity between *a* and *i*. The desorption leading to this return to a comparatively thin film is characterized by effects which cause decreases in the emission at differing planes; *a* and *i*, for example, show decreases from the (123) planes. Many studies of this type have been made on a number of metals; in addition, studies of several adsorbed molecules have also been made, among them various organic molecules. Detailed reviews of these applications have been given by Müller.[3,4]

3. FIELD ION EMISSION

Field emission microscopy under the best conditions can give a resolution of about 20 Å. It can be shown[4] that the resolution can be significantly improved by the use of particles larger than the electron;

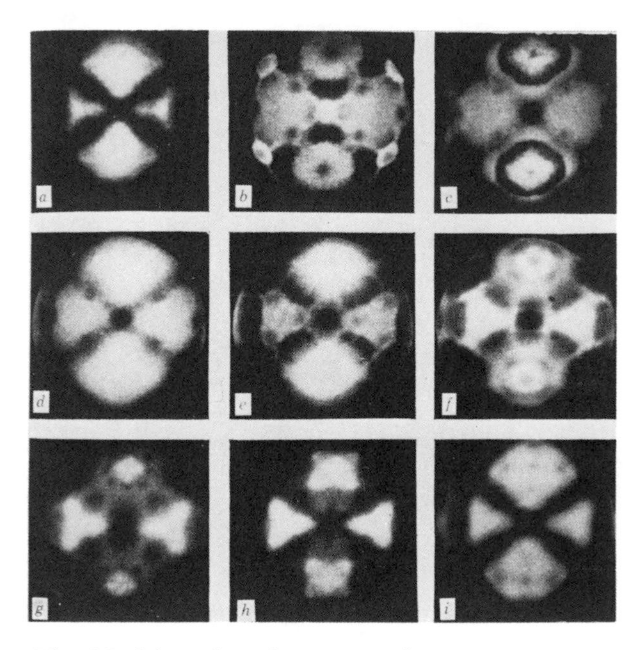

Fig. 6-5. Adsorption of an oxygen layer on a tungsten tip:[3] (a) small amount of oxygen adsorbed (10^{-7} mm for 5 sec), (b) surface saturated with oxygen (10^{-3} mm), (c) desorption after 2 min annealing in high vacuum at 600°C, (d) heated at 700°C, (e) heated at 1100°C, (f) heated at 1200°C, (g) heated at 1350°C, (h) heated at 1600°C, (i) heated at 1800°C.

this led Müller,[6] in 1951, to the use of hydrogen ions as the imaging particles. The tip was made positive instead of negative and a low pressure of hydrogen was leaked into the microscope. The gas molecules close to areas of high energy were ionized and traveled linearly to the screen where a fluorescent image formed. The resolution achieved approached the atomic scale, about 3–6 Å, and individual rows of atoms in the surface could be detected.

Field ionization is analogous to field emission inasmuch as a tunneling mechanism is believed to occur. Figure 6-6a is a representation of the energy levels of an electron in the vicinity of its parent ion. Under the influence of a strong field, the energy diagram may take the shape shown in Fig. 6-6b and a tunneling mechanism becomes possible in which the electron may drift from A to B by exactly the same phenomenon which allows electrons to leave the surface in field emission (Fig. 6-1). In the vicinity of a positively charged conductor, the situation can be represented by Fig. 6-6c. In fact, X_c is a fairly critical distance. If X_c is too small, the energy level of the electron may be below the Fermi level

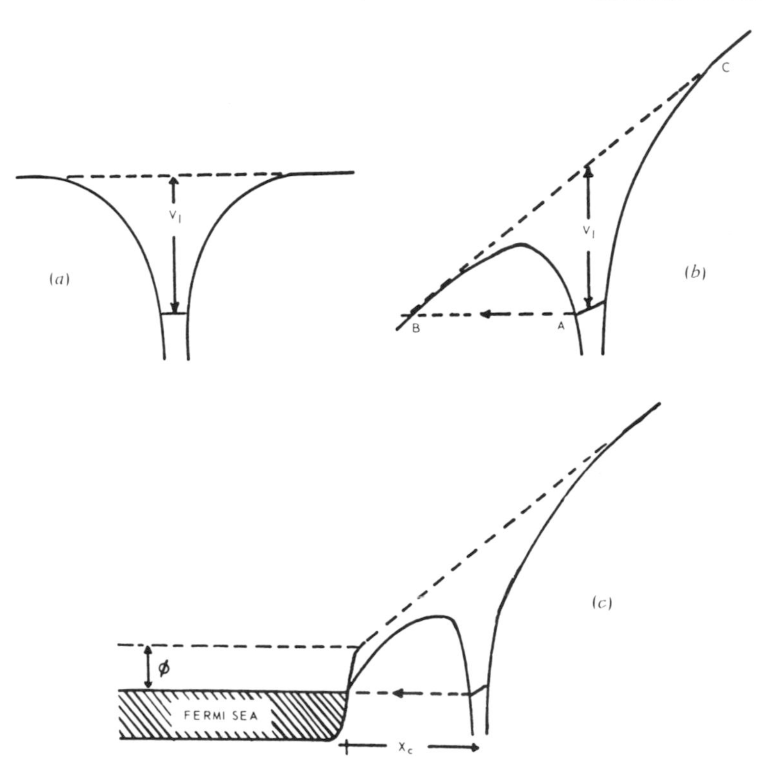

Fig. 6-6. Potential diagrams for an electron in field ionization: (a) the potential due to the parent ion in zero field, and (b) the potential due to the parent ion in a field BC, and (c) the potential in an applied field near the surface of a metal of work function ϕ.[7]

of the conductor, in which case the electron may have only a low probability of finding a vacancy. On the other hand, near the surface, another effect tends to raise the probability of tunneling. This is the so-called image effect in which a dipole tends to form between the gas atom and one in the surface. This tends to lower the potential at B making tunneling more likely. The net result of these opposing effects is to make the ionization probability highest in the region between 5 and 10 Å from the surface. Of course, once ionized the ion is accelerated rapidly away from the surface under the influence of the high field.

In the first application by Müller,[6] the emitter was at room temperature. With this condition, molecules in the region 5–10 Å from the surface will have a lateral velocity component proportional to this temperature, i.e., to T_{gas}. However, only a small number of the ions will originate from molecules approaching the tip. Far more will enter this zone by rebound, and their temperatures will be that of the tip, i.e., T_{tip}.

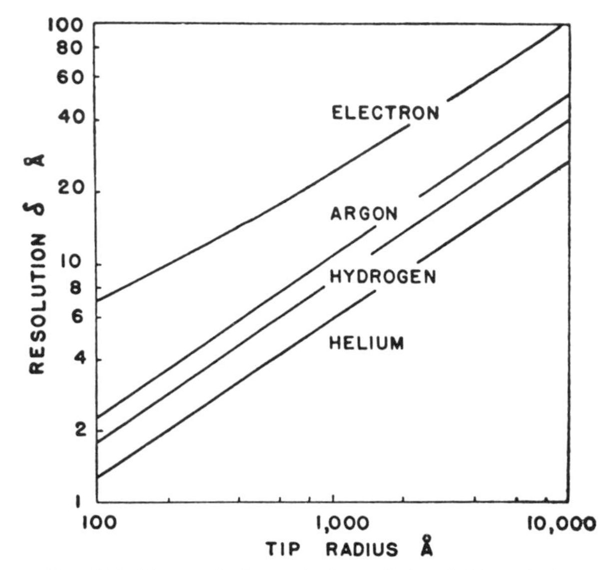

Fig. 6-7. Theoretical resolution of the field emission microscope, for electrons and for helium, hydrogen, and argon ions.[9]

Obviously, if the tip is at room temperature, this fact is of no consequence, but if we make $T_{tip} \ll T_{gas}$, then the lateral velocity component can be significantly reduced for many of the ions. This lateral component is an important factor in the image resolution and the introduction of a cryogenic tip by Müller[8] substantially improves this property.

Although hydrogen was used initially as the imaging gas, the elucidation of the mechanism indicated that, because of the dipole attraction referred to previously, the resolution of the image is dependent, among other things, on the polarizability of the gas. Studies by Müller and Bahadur[9] confirmed this hypothesis and showed that helium, with its low polarizability, gave the best resolution. The theoretical resolutions for helium, hydrogen, and argon ions are given in Fig. 6-7 and compared with that for electrons, i.e., to the resolution of the field emission microscope.

3.1. Field Evaporation

In preparing the specimen tips for field emission microscopy, the necessary hemispherical shape was achieved by heating the etched tip by means of an auxiliary resistance heater. At higher temperatures, the atoms migrated to form an atomically smooth surface. However, the increased resolution of the field ion microscope placed even more stringent requirements on the shape of this tip. Fortunately, the phenomenon of field evaporation had been discovered at just about the same time as the field ion microscope was designed, and this enabled tips to be prepared which were essentially identical to the bulk material.

The mechanism of field evaporation has been dealt with elsewhere (see, for example, Brandon[10]). Briefly, the positive potential of the specimen is increased to the point at which surface atoms are ionized and ejected from the tip. Since those atoms at protrusions will be subjected to the highest localized field, the effect will be to "polish" the tip to an atomically "smooth" surface.

Continued application of this high field will result in the stripping of atomic layers from the tip. This allows an investigation of the sample in depth.

3.2. Field Ion Microscope

Since most investigators in this field have built their own instruments, there are a considerable number of variations on the basic theme. However, the instruments generally incorporate a cryogenic sample holder cooled by liquid hydrogen or liquid helium, a high vacuum system capable of at least 10^{-7} Torr, a high-voltage power supply capable of supplying up to 50 kV, and an auxiliary circuit for superimposing an additional voltage pulse in order to perform the field evaporation.

The phosphor screen employed in both the field ion and field emission microscopes is a device with a very low light level, and this requires a darkened room and a period of adaptation to view it. Various methods have been used to improve this (for example, McLane *et al.*[11] used an external image intensifier), but perhaps the most useful recent innovation was the use of the channel-plate image converter, first applied to the field ion microscope by Turner *et al.*[12] This resulted in a gain of about 10^3 to be obtained; the secondary electron image giving a relatively bright display on the phosphor screen.

A typical commercial instrument (although without the channel-plate converter) is shown in Fig. 6-8. It includes a liquid-hydrogen cryostat tip, a source of helium and/or neon for the imaging gas, a high vacuum system, and a high-voltage power supply with a pulser. Like most of the more recent designs, this is a metal system which can be baked out. It doubles as a field emission microscope by a simple reversal of the field.

3.3. Applications of the Field Ion Microscope

Reference to Fig. 6-7 will indicate that the field ion microscope should show an improvement in resolution of about fivefold over the field emission microscope, bringing it into the 2- to 4-Å range. This, of

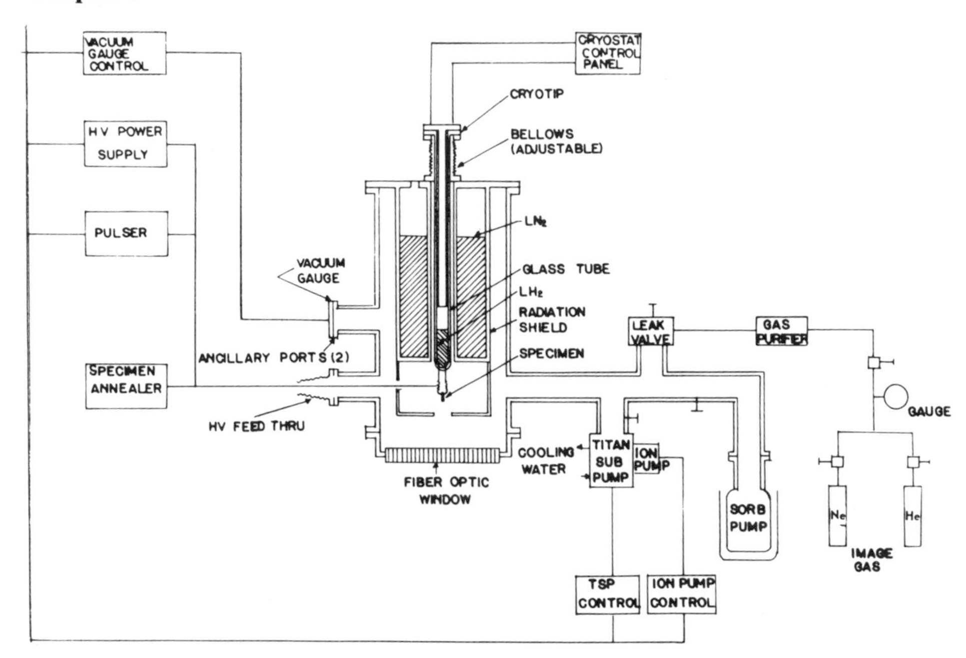

Fig. 6-8. Functional diagram of combined field ion and field emission microscope with UHV system, voltage pulser, and specimen heater. (Courtesy of Jackson and Church Electronics.)

course, is within the range of atomic distances, and it is this unique ability to "see" individual atoms that makes this instrument such a valuable tool for surface studies. Müller and Tsong[1] devoted a considerable portion of their book to a review of the many possible applications, and reference should be made to this work for a more comprehensive presentation. A few examples will be given here to try to indicate the potential of this technique as it applies to surface studies. Of course, the technique always looks at the surface of the sample, but in fact much of the work has been aimed at studies that are essentially concerned with bulk structure. For example, many of the metallurgical studies have been directed towards the crystal structure of alloys. Like the example given for field emission microscopy (Fig. 6-3), the technique is capable of revealing much information on the positions of the lattice planes and the defects associated with them, but the studies themselves are not really surface studies.

Again, as in the case of the field emission microscope, a considerable amount of effort has gone into studies of corrosion and oxidation of metals. Figure 6-9 shows a tungsten surface after exposure to oxygen, and it is interesting to compare the resolution of this micrograph with that of Fig. 6-5. Considerable basic information has been obtained from

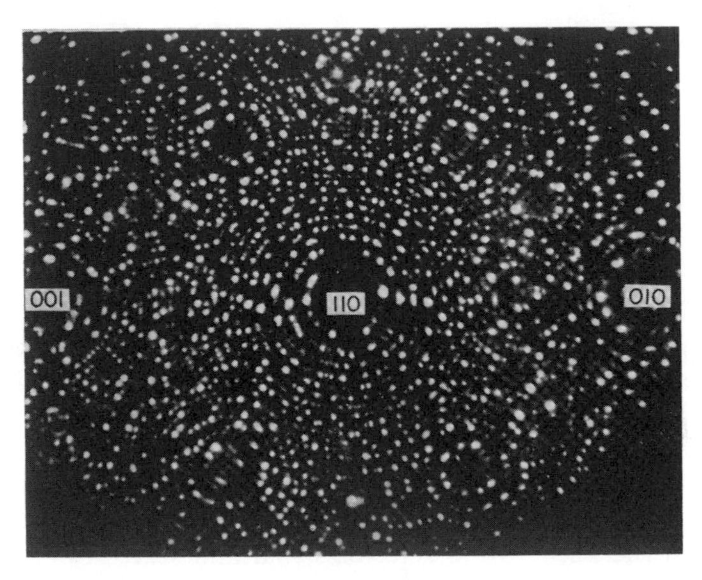

Fig. 6-9. Corroded tungsten surfacc after exposure to 10^{-4} Torr oxygen for a few seconds at 78°K (imaged at 10^{-3} kV, 21°K).[1]

this and other metals, e.g., Pt, Re, on the mechanism of oxygen adsorption and desorption, adding significantly to the understanding of oxide film formation. Other gases, e.g., water, carbon monoxide, and nitrogen, have also received attention.

Another fruitful field of investigation has been the nucleation and deposition of metals, mostly by evaporation. Bassett[13] has recently reviewed this application. He points out that field ion microscopy is the only technique which can verify some aspects of nucleation theory. However, it must be used in conjunction with some other technique, e.g., LEED, to aid in interpreting the ion micrographs. Vapor deposition can, of course, be carried out within the instrument and this has consequently received most attention. Some work has been done, however, on electrodeposited films. The electrolytic growth of platinum on iridium and tungsten was studied by Rendulic and Müller.[14] They found that a more stable film was obtained with ac current than with dc. In a later reference,[1] Müller also added that addition of, for example, coumarin to the bath gave smoother deposits with smaller crystallite size and fewer defect types.

Since the introduction of the field emission microscope with its potential for resolving atoms, there has been considerable interest in its application to organic molecules. If one could "see" the atoms making up the molecule, one could avoid a great deal of the effort currently expended in deductive chemistry toward elucidating the structure of complicated biochemical molecules such as RNA. As early as 1950, Müller[15] obtained images from copper phthalocyanine molecules, but subsequent progress has been difficult. To date, the original objective of an atomic map of an organic molecule has not been satisfactorily achieved. However, this could be a useful technique in the study of adsorption and desorption of organic molecules from surfaces.

4. THE ATOM PROBE

In 1968, Müller *et al.*[16] added a new dimension to the field ion microscope by combining it with a time-of-flight mass spectrometer. A hole was drilled through the screen large enough to cover the image of one atom (about 1–2 mm in diameter) and the drift tube of the mass spectrometer placed in line behind this. The image of the atom of interest is adjusted over the probe hole. A field-evaporation pulse ejects the atom from the surface into the mass spectrometer; the flight is timed from the pulse to the detector signal and serves to identify the atom.

4.1. Atom-Probe Field Ion Microscope

The more recent versions of the atom probe incorporate a channel-plate converter and improved electronics; a design due to Brenner and McKinney[17] is shown in Fig. 6-10. Since the channel-plate converter must be viewed from the back, a 45° mirror is added to allow the operator

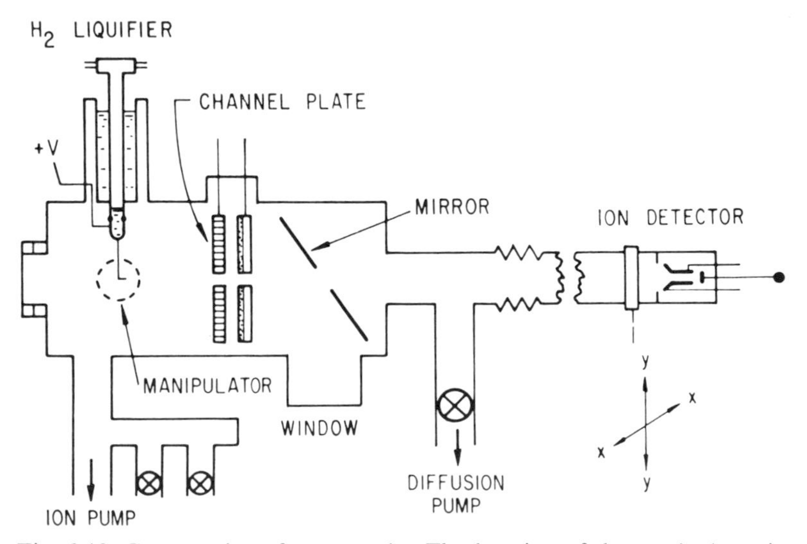

Fig. 6-10. Construction of atom probe. The location of the manipulator is indicated by dashed circle. The tip is attached to the liquid-hydrogen cold finger and high-voltage line by flexible braid. The image is viewed on the phosphor screen adjacent to channel plate.[12]

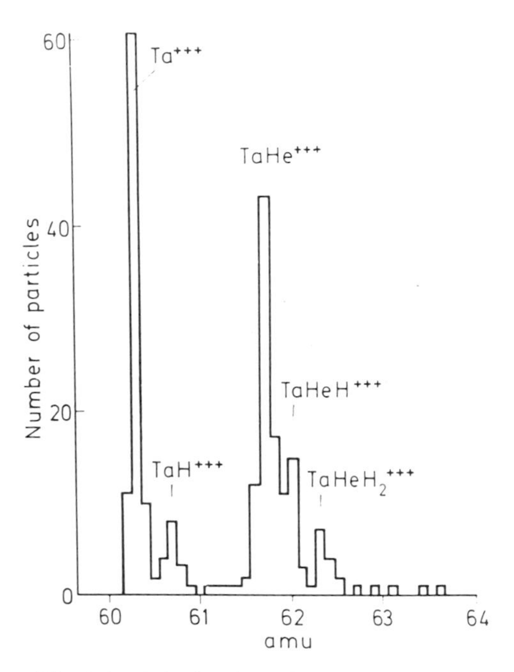

Fig. 6-11. Distribution of 232 triply charged ions from Ta evaporating in 10^{-4} Torr He at 78°K. The background H_2 pressure is below 10^{-8} Torr.[2]

to view the screen. Otherwise, the instrument is similar to that first designed by Müller *et al.*[16] The detector is a channeltron electron multiplier.

4.2. Applications of the Atom Probe

The major role of the atom probe so far has been in the elucidation of the reactions occurring in the field ion microscope between, for example, the imaging gas and the sample surface. A typical spectrum of this type is shown in Fig. 6-11; a number of molecular ions are seen arising from the adsorption of both hydrogen and helium on the tantalum surface.

A study of a ferrous alloy by Brenner and McKinney[17] gave good agreement with the expected composition for the molybdenum concentration (4.9%). Similar results were obtained for Fe–3.8% Au, Pt–20% Ni, and Fe–1.55% Cu.

The potential of this technique has hardly been tapped as yet, and there are many possible applications to metallurgy, e.g., grain-boundary studies, which have not been undertaken although they undoubtedly will be in the near future. It is a unique research tool in its combination of both compositional and structural information at the atomic level.

5. CONCLUSION

The field ion microscope is the only technique capable of imaging atoms in the surface of the sample. It can actually map the distribution of these atoms in their crystallographic habit and the addition of the time-of-flight mass spectrometer to form the atom probe can identify these atoms. It is a research tool in the sense that the surfaces examined are not "real." They must be formed in the machine and consequently are artificial; however, the reaction of these surfaces with gases and vapor-deposited metals can give much valuable information on the probable nature of "real" surfaces and the formation of thin films.

This instrument is restricted to conductors but, within this limitation, it is capable of giving unique information on both the structure and composition of surfaces.

6. REFERENCES

1. E. W. Müller and T. T. Tsong, *Field Ion Microscopy*, Elsevier, New York (1969).
2. E. W. Müller, *Naturwiss.* **57**, 222–230 (1970).
3. E. W. Müller, Field emission microscopy, in *Physical Methods in Chemical Analysis* (W. G. Berl, ed.) Vol. III, pp. 135–182, Academic Press, New York (1956).
4. R. H. Good and E. W. Müller, Field emission, in *Encyclopedia of Physics* (S. Flügge, ed.) Vol. 21, pp. 176–231, Springer-Verlag, Berlin (1956).
5. E. W. Müller, *Phys. Z.* **37**, 838–841 (1936).
6. E. W. Müller, *Z. Physik* **131**, 136–142 (1951).
7. M. J. Southon, in *Field Ion Microscopy* (J. J. Hren and S. Raganathan, eds.) pp. 6–27, Plenum Press, New York (1968).
8. E. W. Müller, *J. Appl. Phys.* **27**, 474–476 (1956).
9. E. W. Müller and K. Bahadur, *Phys. Rev.* **102**, 624–631 (1956).
10. D. G. Brandon, Field evaporation, in *Field Ion Microscopy* (J. J. Hren and S. Ranganathan, eds.) pp. 28–52, Plenum Press, New York (1968).
11. S. B. McLane, E. W. Müller, and O. Nishikawa, *Rev. Sci. Inst.* **35**, 1297–1302 (1964).
12. P. J. Turner, P. Cartwright, M. J. Southon, A. Van Oostrom, and B. W. Manley, *J. Phys. E.* **2**, 731–733 (1969).
14. K. D. Rendulic and E. W. Müller, *J. Appl. Phys.* **38**, 550–553 (1967).
15. E. W. Müller, *Naturwiss.* **37**, 333–334 (1950).
16. E. W. Müller, J. A. Panitz, and S. B. McLane, *Rev. Sci. Instr.* **39**, 83–86 (1968).
17. S. S. Brenner and J. T. McKinney, *Surface Sci.* **23**, 88–111 (1970).

7

X-RAY DIFFRACTION METHODS

Robert D. Dobrott

Materials Characterization Laboratory
Texas Instruments, Inc.
Dallas, Texas

1. INTRODUCTION

The application of x-ray diffraction to surface characterization requires some consideration of the definition of the material surface. If it is the present-day surface definition of one, five, or twenty monolayers that the ion-scattering, Auger, or ESCA techniques see, then x-ray diffraction probably has no place in surface characterization. However, by returning to the definition used a decade ago, where the first 5–10 K Å of a bulk material or any thin film deposited upon a substrate was considered the material surface, then x-ray diffraction does have an important application.

This latter surface definition will be accepted since, to observe any diffraction effects, the wavelength of the x rays must be of the same order as or smaller than the interplanar spacings. For real materials, this restricts the technique to the use of the harder, more penetrating x rays. A second restriction arises from the fact that to observe any diffraction, the volume of material must contain a sufficient number of scattering centers for interference effects. Consequently, the depth considered as surface throughout this chapter will be limited by either absorption effects or the total film thickness if different from the substrate.

2. THEORETICAL

2.1. Intensities

Many of the applications of x-ray diffraction will depend on some diffraction-intensity model for their interpretation. In this section, it will be assumed that the reader is familiar with the basic theory of the x-ray diffraction mechanism. A review of this theory has been given by Cullity.[1]

Darwin[2] formulated the basic expressions for diffraction intensity for perfect and ideally imperfect (mosaic) crystals early in the twentieth century. These basic wave mechanical expressions have established the foundations for most of the diffraction experiments since that time. Darwin's basic intensity expressions are, for a perfect crystal,

$$I_p = I_0 N |F| B \tag{1}$$

and, for a mosaic crystal,

$$I_m = I_0 N^2 |F|^2 B'/\mu \tag{2}$$

where I_0 is the intensity of the primary beam, N is the number of unit cells, F is the crystal-structure factor, B and B' are a group of parameters not needed for this discussion, and μ is the linear absorption coefficient for the material. These expressions indicate that I_p will be much smaller than I_m since in Eq. (1), N and F are to the first power and in Eq. (2) N and F are squared. Also, μ does not appear in the case of the perfect crystal suggesting some other effect (primary extinction) is limiting the intensity. Most crystals do not approach either the perfect-crystal or the mosaic-crystal model, but lie somewhere in between. However, in the last decade, elemental semiconductor crystals have closely approached the perfect-crystal model. This intensity difference is the basis for all the perfection methods to be discussed later.

2.2. Absorption

The maximum depth that the x rays will explore in a bulk medium can be calculated by using the mosaic intensity expression which has its limitation in μ, the linear absorption coefficient. Assuming the material to be a pure absorber, the transmitted intensity can be expressed as

$$I = I_0 \exp(-\mu t) \tag{3}$$

where t is the thickness of the absorber. From this expression and a suitable model, the maximum diffraction surface thickness can be calculated.

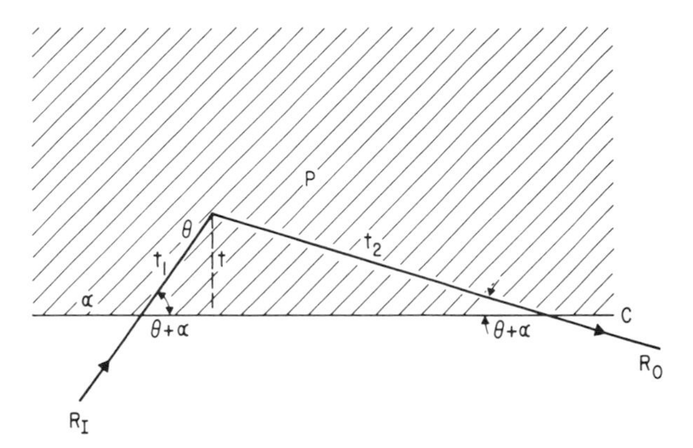

Fig. 7-1. Model for x-ray path length in an absorbing crystal, where θ is Bragg angle, α is the angle between crystallographic planes P and specimen surface C, t_1 is path length of incident ray R_I in the interior of the crystal, t_2 is path length of diffracted ray R_0 in the interior, and t is the effective surface depth.

Figure 7-1 shows a model suitable for Bragg-type diffraction. The crystallographic planes are intersected by the primary x-ray beam at the Bragg angle θ and these planes and the sample surface intersect with angle α. As shown (Fig. 7-1), the actual path of the x rays is $t_1 + t_2$; by plane geometry this can be expressed as a function of surface thickness t by $t \csc(\theta - \alpha) + t \csc(\theta + \alpha)$. The absorption expression, Eq. (3), can then be transformed to

$$\ln I/I_0 = -\mu t[\csc(\theta - \alpha) + \csc(\theta + \alpha)] \tag{4}$$

The linear absorption coefficient can be calculated from the relationship

$$\mu = \rho \sum_i P_i \mu_{mi} \tag{5}$$

where ρ is the density of the material, P_i is the weight fraction of component i, and μ_{mi} is the mass absorption coefficient[3] for element i.

Figure 7-2 is a graphical summary of surface depths for Cr K_α radiation as a function of θ and α assuming that, when $I/I_0 = 0.01$, the diffracted intensity is too weak to be useful.

2.3. Primary Extinction

The x-ray surface depth in the case of the perfect crystal is not as easily calculated. For the perfect crystal, the diffracted intensity is limited by primary extinction of the incident beam.[4] Primary extinction occurs when all crystallographic planes are completely parallel. The parallelism of the planes allows the reflected beam to undergo a second

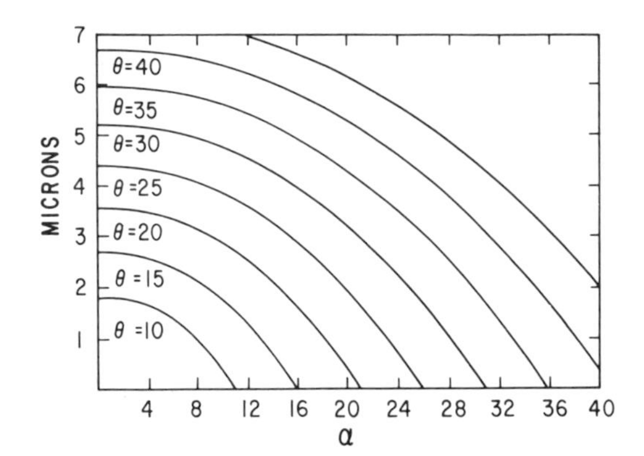

Fig. 7-2. Maximum surface-depth penetration of Cr K_α radiation in GaAs as a function of Bragg angle θ and the angle α between the diffracting planes and the specimen surface.

and third reflection in the crystal. Since the phase change on reflection is $\pi/2$, the total phase change after the second reflection will be π and the direction returned to will be the direction of the incident beam. Consequently, this twice-reflected beam will result in a diminished amplitude of the incident beam in the interior of the crystal.

2.4. Secondary Extinction

If the crystallographic planes are almost but not quite parallel, the phase change for the twice-reflected beam is not exactly π and the direction is not exactly returned to the direction of the incident beam; however, the overall intensity (not amplitude) is still reduced. This latter effect is known as secondary extinction.

Hence, if material to be examined is mosaic, the surface depth can be calculated by applying the pure-absorber relationship. As we approach the perfect crystal state where primary and secondary extinction dominate, all that can be readily deduced is that the surface depth is probably considerably smaller than the absorber model predicts.

3. POLYCRYSTALLINE METHODS

In the last decade, the use of very thin polycrystalline films has become extremely important, especially in the semiconductor industry. The thickness of interest of these films normally ranges from 1000 to 10,000 Å. In this section we shall regard the entire thin film as the surface of our sample.

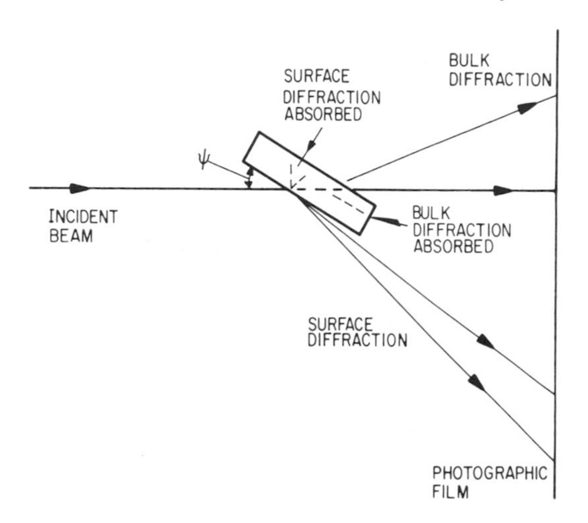

Fig. 7-3. Low-angle Laue geometry.[5]

3.1. Laue Method

Walker[5] introduced a low-angle Laue technique in which grain sizes and preferred orientation can be inferred. The geometry of this method is shown in Fig. 7-3. This method is applicable to thin films since the low incident angle ψ gives a much longer path length for the beam in the film. By changing the angle ψ, a depth study of the film can be made. If the substrate is sufficiently thin, the diffracted rays from the film and film substrate interface will be recorded on the lower half of the photographic film whereas the information from the back of the substrate will be recorded on the upper half. Hence, if the substrate is single-crystalline, the Laue spots can be compared and information on how the film is affecting the substrate can be obtained.

Figure 7-4 is a low-angle Laue photograph of two 12-kÅ Al films on Si substrates. The Debye rings diffracted from Al films reveal the preferred orientation. Since the thickness of the film is known, the grain size can be calculated by counting the spots on the Debye ring, and using the relationship[6]

$$G = \left(\frac{\pi d p D^3 \cos \theta}{8NR} \right)^{1/3} \tag{6}$$

where d is the film thickness, p the multiplicity of the reflection, θ the Bragg angle, D the diameter of the beam, N the number of spots on the Debye ring, and R the distance from the specimen to the film. The value of D^3/R is a constant for each camera and is best determined empirically by determining the grain size of a standard film by electron

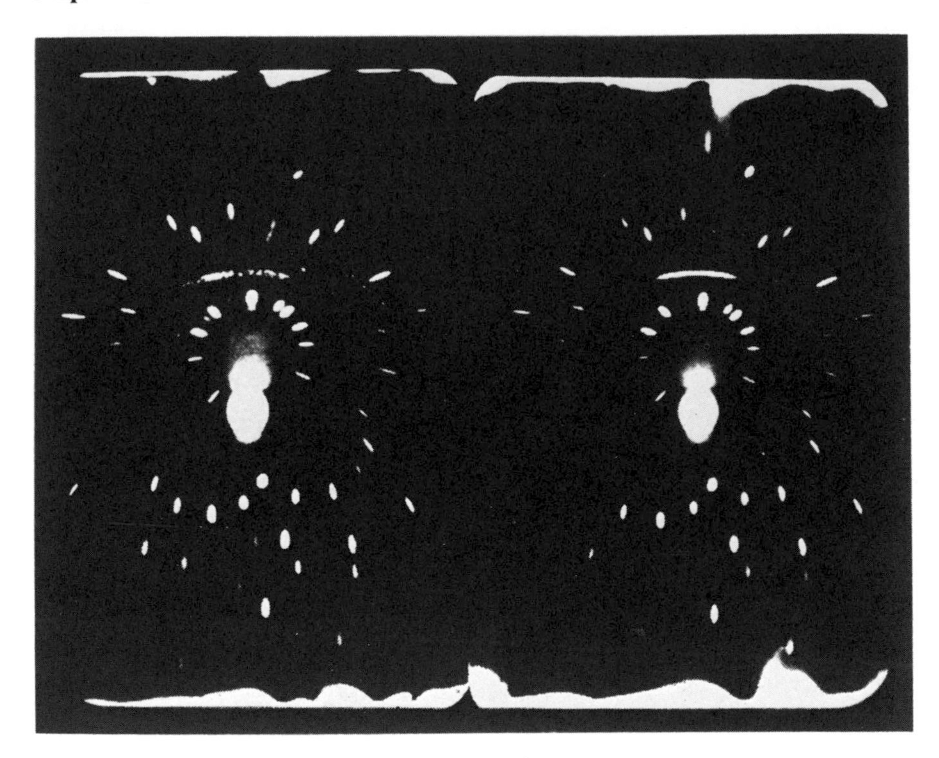

Fig. 7-4. Low-angle Laue photograph of 12-kÅ Al on Si showing the difference in grain size between two films.[5]

microscope techniques. This method for grain size is good down to about 2000 Å.

3.2. Diffractometer Methods

3.2.1. Bragg–Brentano

Walker[5] also reported a Bragg–Brentano geometry x-ray diffactometer study of the two 12-kÅ Al films. Figure 7-5 illustrates the Bragg–Brentano diffraction geometry. With this geometry, care must be taken in the data interpretation since, due to the high diffracting power of single crystals, diffractions of extraneous radiations from tube contaminants can give peaks comparable in intensity to the thin-film peaks. In most cases, these peaks can be eliminated or reduced by misorienting the sample in the ω direction of the Bragg–Brentano geometry. Figure 7-6 is a series of diffractometer traces of the 12-kÅ Al film illustrating the effect of misorientation on the substrate peaks.

After obtaining the diffraction pattern of the film and separating the peaks due to film from those due to substrate, a peak-shift analysis will give information on residual strain and a combination of peak-shift[7] and peak-broadening data will give information on microstrain, faulting probability, and particle size.[8] In order to ensure that there is no geometrical broadening or shift of the diffractometer peaks, many corrections must be made. A Rachinger[9] correction is made for $K\alpha_1$–$K\alpha_2$ separation and a Stokes[10] or Laue integral breadth[11] correction is made which allows for geometrical effects. To separate the combined effects on broadening of nonuniform strain, effective particle size, and faulting, there is a choice between Fourier analysis[8] or the Laue integral breadth method.[7] The Fourier method has the advantage of giving distribution of particle size and nonuniform strain whereas the Laue integral breadth method is much easier to use if no computer is available. The reader is referred to Walker[5] for a detailed description of these calculations.

3.2.2. Seeman–Bohlin

A second approach to the diffractometer studies of polycrystalline thin films has been described by Feder and Berry.[12] In this arrangement a modified Seeman–Bohlin[13–14] arrangement is used to obtain the

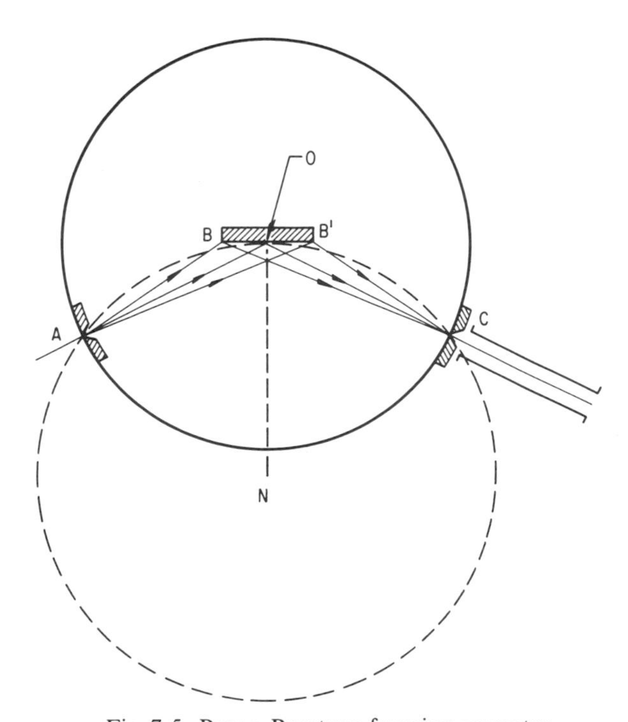

Fig. 7-5. Bragg–Brentano focusing geometry.

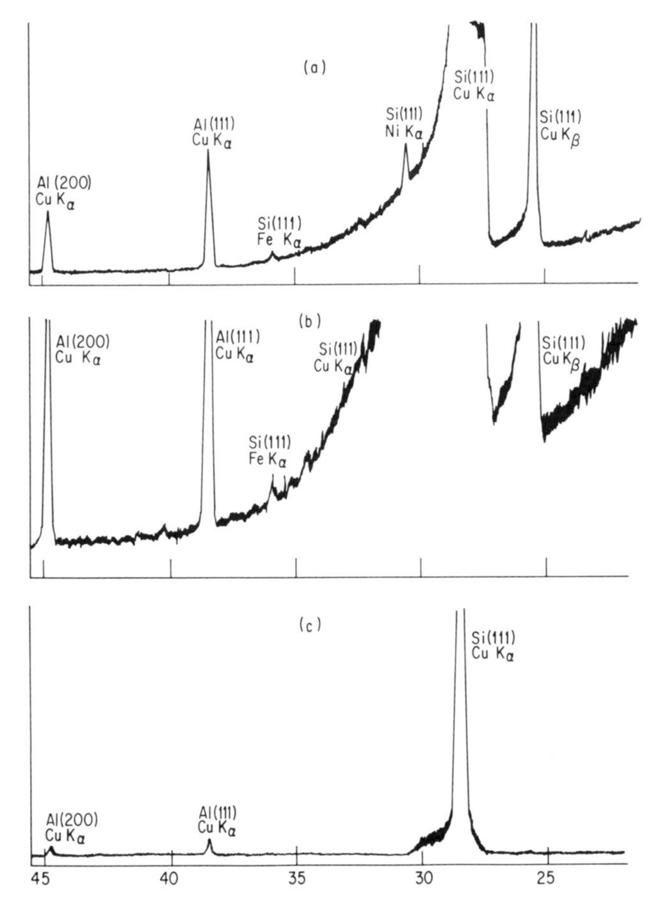

Fig. 7-6. Bragg–Brentano diffractometer traces of a thin Al film on single-crystal Si under various experimental conditions: (a) scale 1×10^5 cpm, (b) scale 2×10^4 cpm, (c) substrate off axis by ω rotation of 1°, scale 5×10^5 cpm.[5]

diffraction pattern. The geometry of this apparatus is shown in Fig. 7-7. A specimen S placed on the circumference of a circle of radius R is irradiated by a divergent x-ray beam from a monochromatic line source located at f on the circle. It has been demonstrated that the diffracted x rays also focus on the circumference of the circle.[15,16] The crystal monochromator shown at the left of Fig. 7-7 also employs the Seeman–Bohlin geometry, and the focus point f must be common to both the monochromator and diffractometer circles.

This arrangement differs from the Bragg–Brentano arrangement in that the x-ray beam impinges at a near-grazing angle of incidence with the specimen and remains stationary. Thus, the path length can be many times larger than the specimen film thickness t which can be much smaller than the volume needed for diffraction. The effective path length is given by $t\csc\gamma$, where γ is the angle of incidence. Thus, for example, a γ of 5° yields a path length of $12t$. Figure 7-7 also illustrates the requirement imposed upon the scanning mechanism used to drive the x-ray counter and slit assembly C. To maintain the correct attitude of this assembly, such that its axis always points directly to the specimen, the primary rotation around the center of the diffractometer must be accompanied by a secondary or α rotation of the counter about the traveling axis P, such that the condition $\alpha = \beta/2$ is always maintained.

To minimize the counting times needed for thin-film samples, Feder and Berry[12] used a pyrolytic graphite monochromator crystal and a helium chamber to minimize the air absorption of the diffracted beam. A data-acquisition control system coupled to their diffractometer is shown in block form in Fig. 7-8. The operation of the diffractometer was automated so that a preset step scan could be executed automatically, with the output collected on punched cards, paper tapc, or directly fed into a computer. The data were then immediately ready for Fourier analysis of each diffraction peak and resulted in (a) the lattice parameter (from each diffraction peak), (b) the rms value of inhomogeneous strain, and (c) the average crystallite size of the sample.

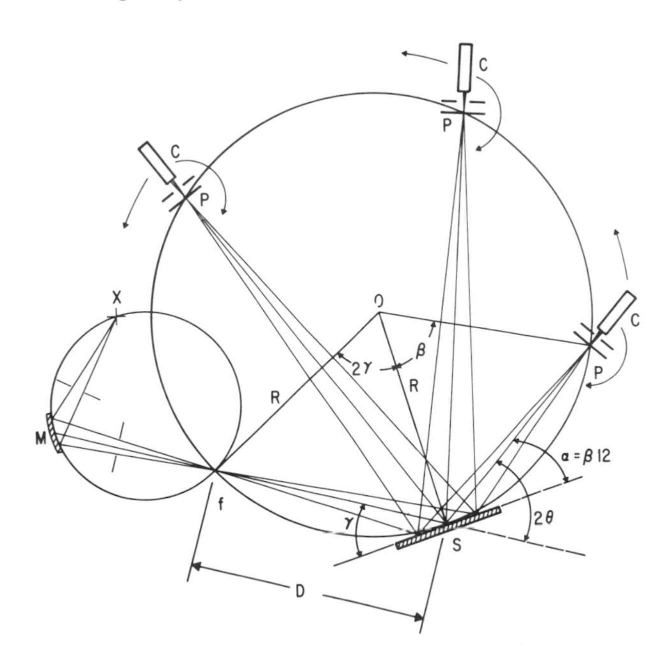

Fig. 7-7. Seeman–Bohlin diffractometry geometry with an incident-beam monochromator.[12]

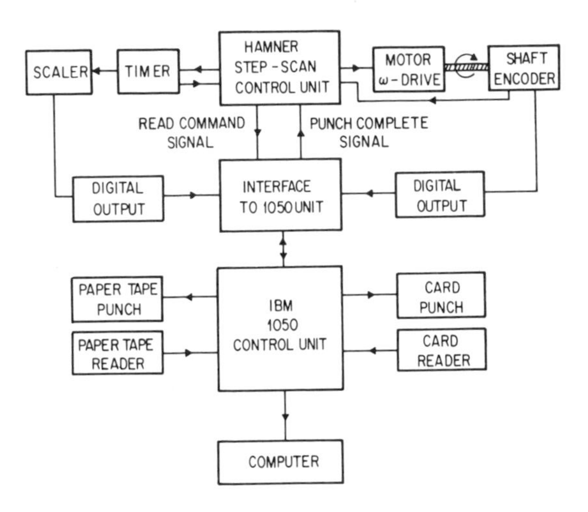

Fig. 7-8. Block diagram of Seeman–Bohlin diffractometer control and data-acquisition system.[12]

Figure 7-9 demonstrates the sensitivity of this arrangement for very thin films; it shows the diffraction pattern of a copper film only 150 Å thick. The scanning was performed in steps of 0.2° (4θ), using a counting time of 75 sec/step. Even though this film was only about 50 atomic layers, 5 distinct diffraction peaks (each corresponding to different crystallite orientations) were resolved. The broadening observed is consistent with a particle size of less than 100 Å.

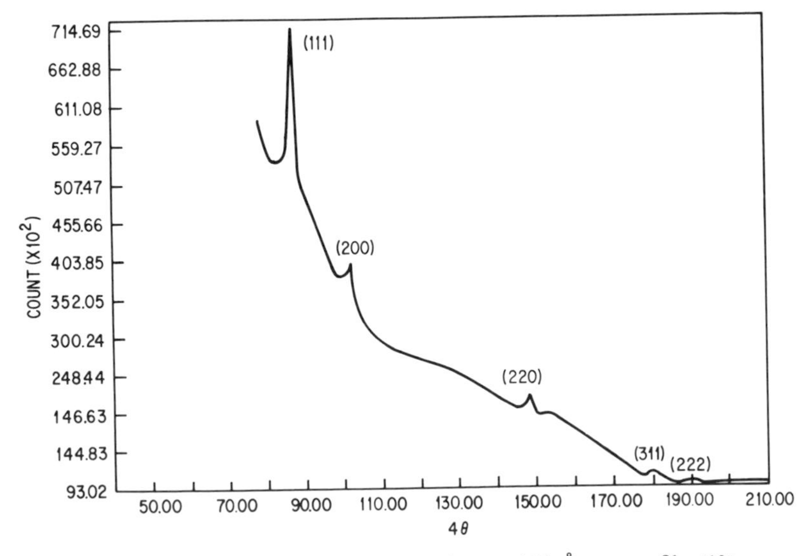

Fig. 7-9. The diffraction pattern from a 150-Å copper film.[12]

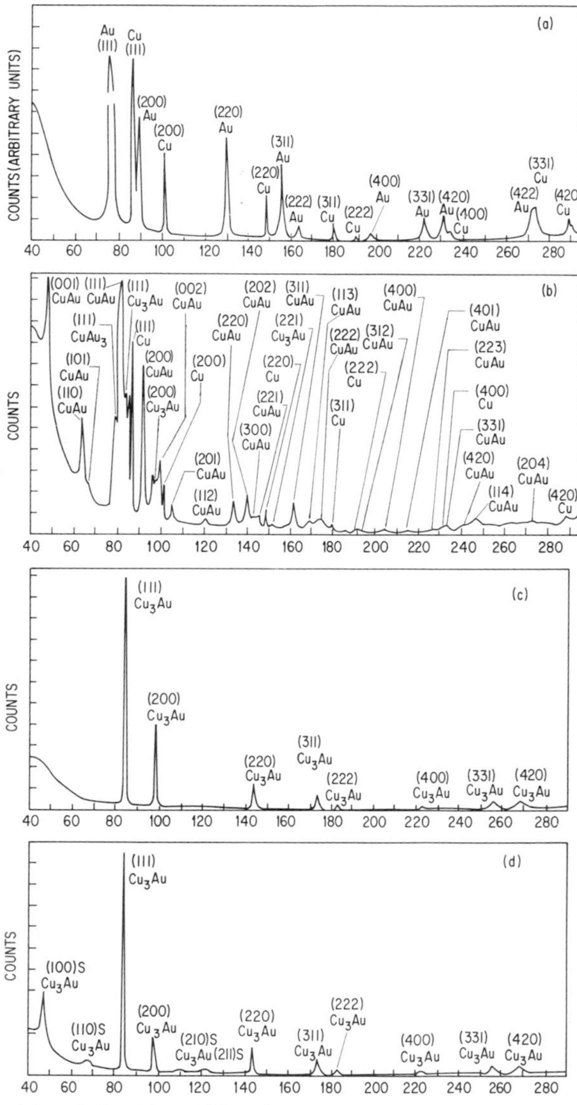

Fig. 7-10. The diffraction pattern from a Cu–Au composite film after the following sequential treatments: (a) as prepared, (b) after 1 hr at 290°C, (c) after 1 hr at 450°C, (d) after 1 hr at 290°C.[12]

This arrangement is especially suited to the study of multicomponent or multilayer films. Figure 7-10 illustrates this application of the technique to a 6000-Å copper film under a 2000-Å gold film on a glass substrate after various heat treatments. Figure 7-10a, taken before any heat treatment, shows all the diffraction patterns corresponding to either pure Cu or pure Au as indicated. Figure 7-10b shows the dramatic change in the diffraction pattern after a 1-hr anneal at 290°C. Most of the peaks can now be indexed to the CuAu or Cu_3Au intermetallics with some unalloyed Cu. Figure 7-10c is the pattern after a 1-hr anneal at 420°C and is completely indexed as the Cu_3Au stoichiometric composition. Since the critical temperature for long-range ordering of Cu_3Au is 390°C, the sample was treated again for 1 hour at 290°C. Figure 7-10d illustrates that this latter treatment did produce superlattice lines. These diffraction patterns demonstrate the potential of this type of diffractometry to the study of very thin films.

4. SINGLE-CRYSTAL METHODS

4.1. Nucleation and Texture

Brine and Young[18] have reported a method for determining the texture of thin films deposited on suitable substrates. They employed a single-crystal orienter[19] with a scintillation detector. The sample to be examined was mounted on a goniometer head attached to the orienter. This allowed rotation of the sample about the two mutually perpendicular axes of the instrument as well as about the diffractometer axis. These motions allow a complete hemisphere in reciprocal space to be explored. The film thicknesses studied by Brine and Young ranged from 500 to 1500 Å. Since the films in this range are relatively nonabsorbing, the incident x-ray beam penetrates to the substrate and the x-ray reflections from the substrate are observed together with the x-ray reflections from the film itself. Therefore, the orientation of the film crystallites with respect to the substrate may be deduced directly from the measured angular positions of the sample at which the reflections occur.

4.2. Crystal Polarity

Most of the Group III–Group V compounds crystallize in a zincblende type of structure. This structure is noncentrosymmetric and contains a polar axis. Because of this axis, the faces on opposite sides of the crystal may exhibit opposite polarity. Figure 7-11 is a schematic of GaAs viewed perpendicular to the (111) direction. The polarity of the (111) and $(\bar{1}\bar{1}\bar{1})$ planes are readily apparent in the illustration.

Anomalous scattering which results when the incident x-ray energy is just below the absorption edge can be used to distinguish the Ga face (A) from the As face (B). The anomalous scattering effect can be described as a departure from Friedel's law[20] for noncentrocymmetric crystals. Friedel's law states that it is impossible to determine by x-ray diffraction whether a crystal has a center of symmetry; that is, the *hkl* intensity

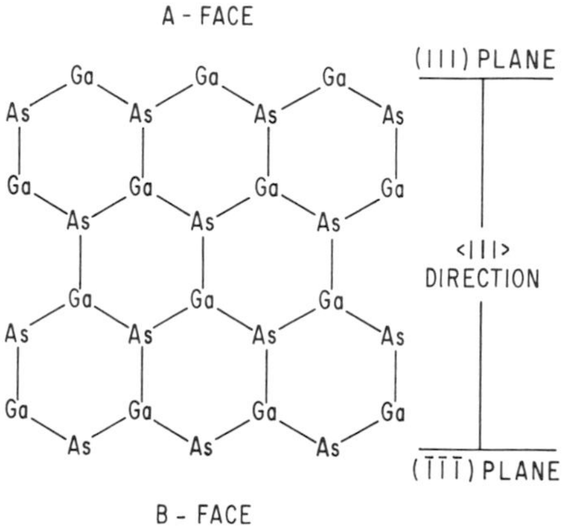

Fig. 7-11. Crystal structure of GaAs illustrating the (A)Ga face and (B)As face.

equals the $\bar{h}\bar{k}\bar{l}$ intensity. The breakdown of this law occurs when the wavelength is near the absorption edge of one type of atom, since the wave reflected from this atom is advanced slightly more than 90° with respect to normal scattering, whereas the wave itself is advanced exactly 90°.

The effect of this on the amplitude of the combined scattered wave is illustrated by the vector diagrams shown in Fig. 7-12 for a two-atom noncentrosymmetric structure. First, in Fig. 7-12a the vectors are shown as they would be if the scattering were normal for both atoms. The vector **A** from atom A is $\pm\theta$-advanced compared to vector **B** from atom B; for both $+\theta$ and $-\theta$ advance, however, the resultant vectors

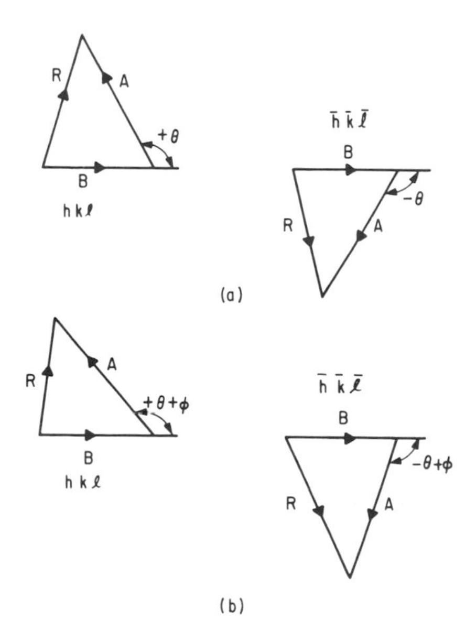

Fig. 7-12. Vector diagrams representing the amplitude summation of two reflected beams: (a) normal scattering, and (b) anomalous scattering.

R and **R̄** are identical. Figure 7-12b is the corresponding vector diagram when the wavelength is just below the absorption edge for atom *A*. Two differences are apparent: (1) the vector **A** is smaller due to the high absorption effect and (2) the wave is advanced by an extra amount ϕ which does not change sign in the *hkl* to $\bar{h}\bar{k}\bar{l}$ reflections. The result of this extra phase advance ϕ is that the resultant vector **R** for the *hkl* reflection is smaller than the resultant vector **R̄** for $\bar{h}\bar{k}\bar{l}$ reflection.

Figure 7-13 is a trace of the diffraction pattern from opposite sides of (111)-oriented GaAs crystal.[25] The trace was made on a standard powder diffractometer using the white radiation from a tungsten source. The (111) and $(\bar{1}\bar{1}\bar{1})$ faces can be easily identified at a glance. The upper trace exhibits a drop in intensity nearly equal in magnitude at both the Ga edge and As edge, while the lower trace has a large decrease at the Ga edge with only a small decrease at the As edge. With this information the $(\bar{1}\bar{1}\bar{1})$ As face can be assigned to the upper trace and the (111) Ga face to the lower trace. These (111) and $(\bar{1}\bar{1}\bar{1})$ traces are typical for GaAs and each exhibits characteristics sufficient to identify the polarity.

This polarity analysis can be carried out on any noncentrosymmetric crystal provided (a) faces can be prepared which exhibit sufficient polarity, (b) a strong reflection is available, and (c) there are edges in a convenient wavelength region.

This anomalous scattering phenomenon has proven to be extremely valuable in many other crystal structure determinations.[21–25]

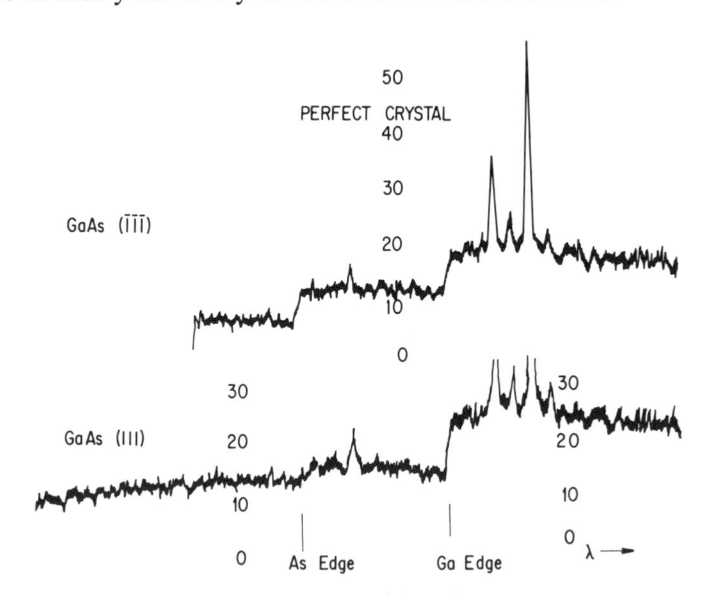

Fig. 7-13. Spectrometer traces of (111) reflection of GaAs crystal. The polarity of the faces are evident at a glance.[25]

4.3. Crystal Perfection

4.3.1. Classical Method

The classical method for determining crystal perfection is to measure the total integrated intensity diffracted by the crystal oriented near the Bragg angle and the angular width of the Bragg peak. The x-ray diffraction theory previously discussed relates both the integrated intensity and angular width to the crystal perfection. This technique has been used in many ways to determine gross crystal perfection.(26) Batterman(27) and Chandrasekaran(28) have used this method (double-crystal rocking curve) for determining dislocation densities between about 10^4 and 10^7 cm^{-2}. Since the equipment necessary for the measurements is very elaborate and data are difficult to interpret, this classical method has been almost completely replaced by the x-ray topographic methods.

4.3.2. X-Ray Topography

Most real crystals can be regarded as ideally perfect crystals with local perturbations of the perfect lattice in the neighborhood of a defect. A technique which photographically records the intensity differences between the perfect regions and defect regions is called x-ray topography. This technique has a resolution of 1–10 μ which for low-dislocation-density material results in a defect distribution map. For high dislocation densities, $>10^5$, x-ray topography cannot resolve individual dislocations but is still useful for studying the size, locations, and relative orientation of grains, the existence, position, and habit of twins, and the direction and magnitude of crystal deformation.

4.3.2.1. Berg–Barrett Method

The most surface-sensitive x-ray topographic method was originally devised by Berg(29,30) and was modified by Barrett.(31) The technique was later refined by Newkirk(32) for geometrical resolutions of about 1 μ. The Berg–Barrett–Newkirk scheme is shown in Fig. 7-14. This method is relatively simple; it consists of holding the specimen S in a

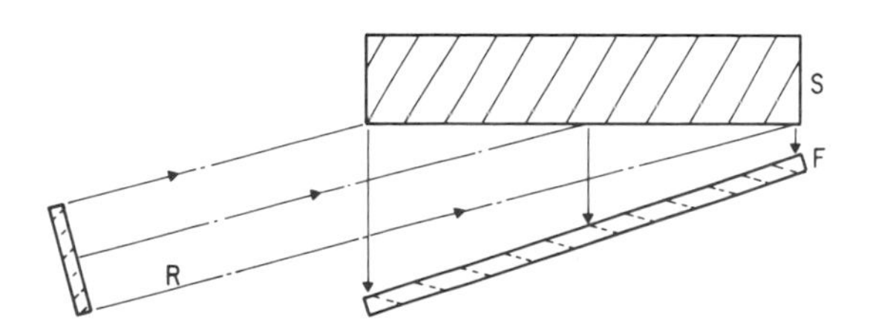

Fig. 7-14. Berg–Barrett geometry.

roughly collimated x-ray beam R so that a set of *hkl* planes makes the proper Bragg angle for diffraction. The diffraction image is recorded on a high-resolution fine-grained photographic plate F which is held as close as possible to the specimen without intercepting the incident beam.

Figure 7-15 shows a series of topographs taken by this method. Figure 7-15a shows a light micrograph of a cleaved and etched (100) face of a lithium fluoride crystal which had been compressed to produce slip on a few {110} planes. Figures 7-15b–d are images of the same crystal surface as seen by x-ray diffraction topography using three different sets of *hkl* planes. Note that in Figs. 7-15b and 7-15c, only some of the dislocations are visible while, in Fig. 7-15d, all are visible. This dependency of contrast visibility on the diffraction vector allows the identification of the type and direction of dislocations from a set of *hkl* topographs. In symbolic form, the requirement for the visibility of a dislocation is

$$(\mathbf{g} \cdot \mathbf{b}) \neq 0 \tag{7}$$

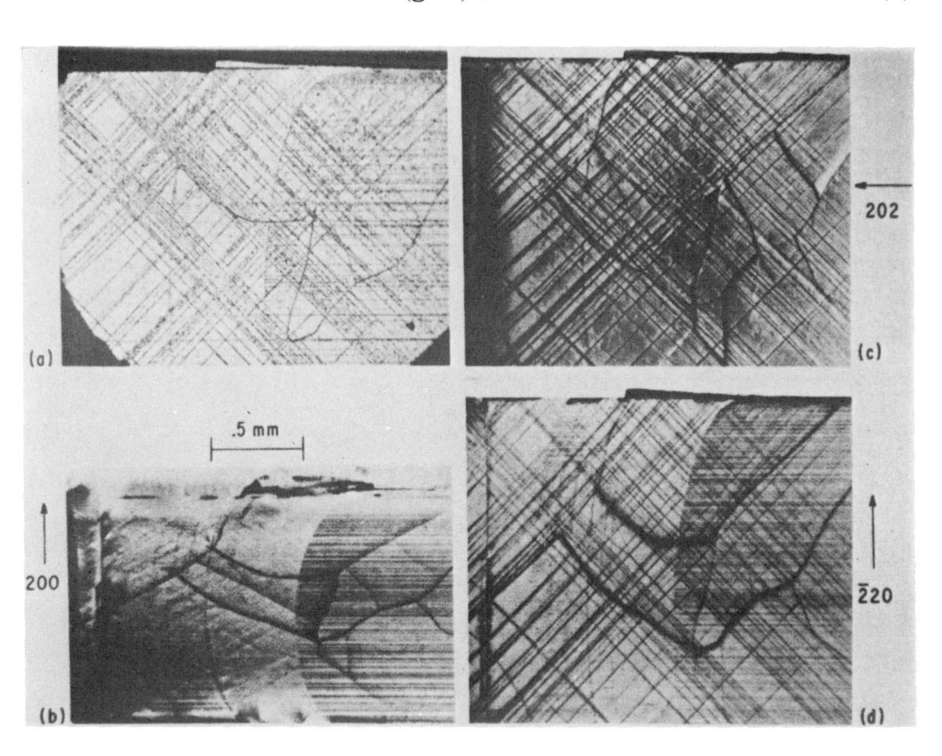

Fig. 7-15. (a) Light micrograph and (b–d) diffraction micrographs of single crystal LiF. Berg-Barrett method (b) 200, (c) 202 and (d) $\bar{2}20$ reflections. Some grain boundaries and dislocations (diagonal straight lines) can be seen. From the contrast variation of the dislocations with the various *hkl* reflections, the directions of the Burgers vectors can be determined to be $\langle 110 \rangle$.[32]

where **g** is the diffraction vector of the topograph and **b** is the Burgers vector of the dislocation.

4.3.2.2. Lang Limited-Projection Method

Lang[33] developed a transmission high-resolution method which consists of irradiating a nonabsorbing crystal slab with a slit-collimated beam of monochromatic x rays and recording the transmitted x rays on a fine-grained photographic plate. In order to record the image of the entire slab, the specimen and photographic plate are translated as shown in Fig. 7-16. In this arrangement, the specimen *C* and the film *F* are translated in direction *T*, keeping the receiving slit, S_1, S_2, or S_3, fixed. If the topograph is taken with receiving slit positioned as S_1, then the normal Lang topograph results with the defects throughout the entire crystal thickness recorded. If slit arrangement S_2 or S_3 is selected, then a limited-projection topograph of the crystal is obtained with either entrance or exit surface features enhanced. Lang[34] developed this technique to block such surfacc features from the topographs since heavy surface damage obscures the defects in the interior of the crystal.

For nearly perfect crystals where anomalous transmission[35] of x rays is observed, a similar technique has been used for the detection of local surface-related strain fields.[36] This method is applicable to highly absorbing crystals and is extremely sensitive to small strain fields such as those produced by evaporating a small metal dot on the surface of an otherwise perfect crystal.

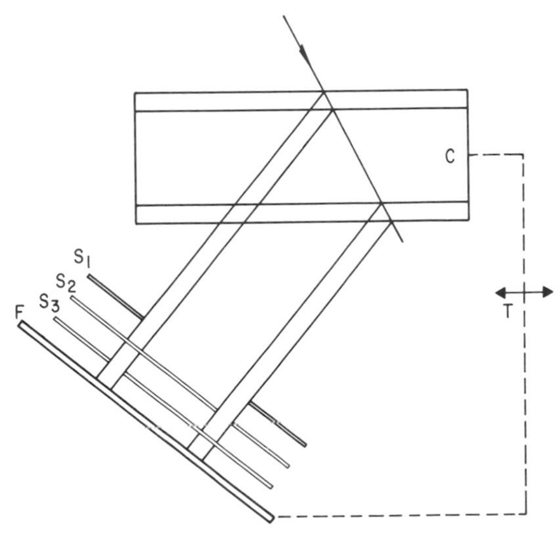

Fig. 7-16. Lang limited-projection geometry. Selection of slit S_2 or S_3 enhances either entrance or exit surface of the crystal *C*.

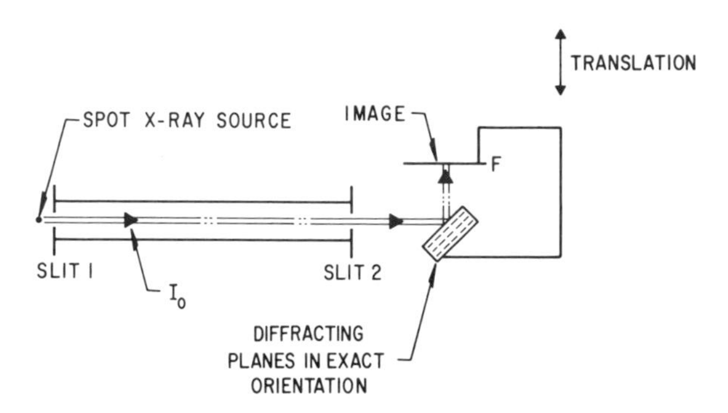

Fig. 7-17. Scanning-reflection geometry.[37]

4.3.2.3. Scanning-Reflection and Composition Methods

The scanning-reflection method was first described by Howard and Dobrott.[37] The method consists of combining the larger-area capability of the Lang method with the surface sensitivity and resolution of the Berg–Barrett method. The experimental arrangement is shown in Fig. 7-17. The substrate crystal C in the diffracting position is coupled to a film holder F mounted perpendicular to the diffracted beam D. The incident-beam slits are adjusted until only the K_{α_1} component of the characteristic radiation can diffract. The crystal is adjusted until the ribbon-shaped incident beam makes the Bragg angle with the selected *hkl* planes and the diffraction image is recorded on the fine-grain photographic plate F. The crystal and plate are translated to record the entire surface diffraction topograph.

Figure 7-18 is a scanning-reflection (404) topograph of a (111) GaAs slice with an optically damage-free surface. The topograph readily reveals the unremoved polishing scratches still in the surface of this slice. The region marked with the arrow is a micromisorientation of the bulk crystal; this feature also was not apparent by visual examination.

Compositional x-ray topography[38] is an extension of the scanning-reflection method. It is a very powerful tool for studying the perfection of heteroepitaxial films deposited on a suitable substrate when there is a difference in the lattice parameter. If the film is thin, then the film and substrate surface can be examined separately by judicious choice of the diffraction angle. Only the volume element which has the desired lattice parameter will diffract to form the image.

Figure 7-19 illustrates the diffraction geometry for a three-layer GaAs–GaP system. It is apparent from this diagram that the substrate surface or either film layer can be studied by selection of θ, θ_1, or θ_2 Bragg angle.

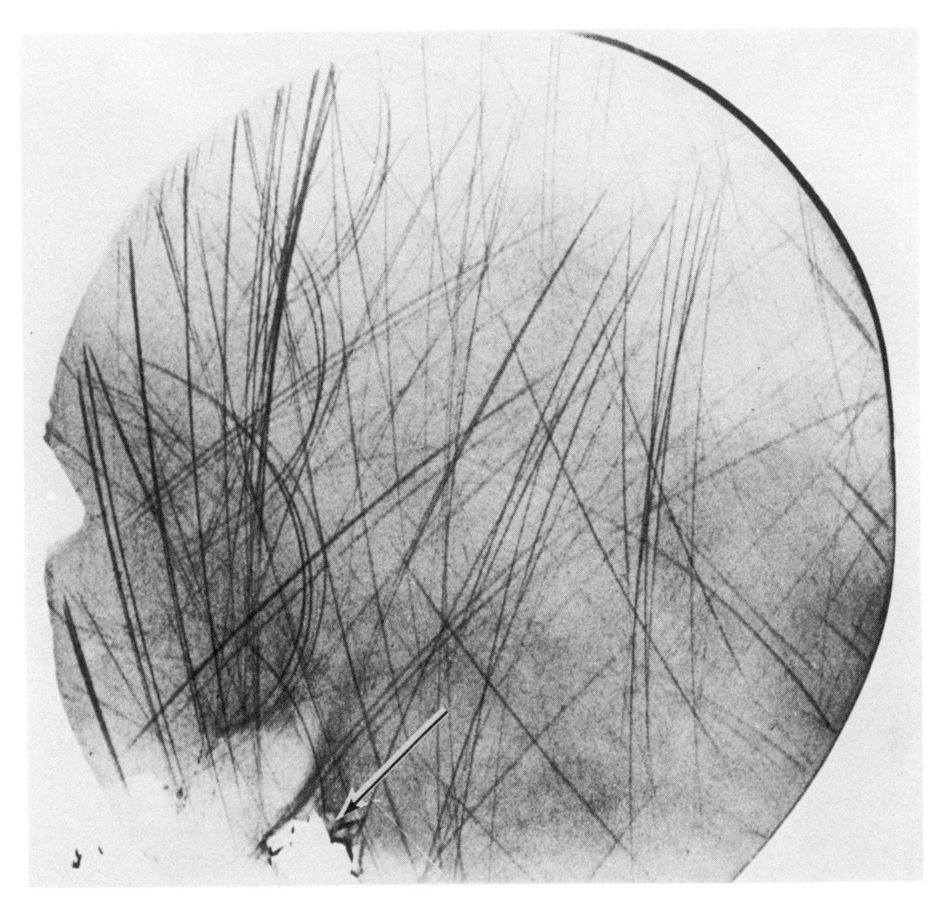

Fig. 7-18. Scanning-reflection topograph showing residual polishing damage in GaAs slice.[37]

Figure 7-20 is an x-ray θ–2θ diffractometer trace of a three-component system. This trace by itself can be used to study overall composition homogeneity of heterofilms and to determine their precise lattice parameters.[39] In compositional topography, it is used for Bragg-angle selection.

Figure 7-21a is a (440) scanning-reflection topograph taken at the Bragg angle of 50.4° which corresponds to the GaAs-substrate layer. The triangular null-contrast regions result from total absorption of the x rays by the external layers. Figures 7-21b and 7-21c are (440) topographs taken at Bragg angles of 50.8° and 51.2° corresponding to compositions of $GaAs_{0.85}P_{0.15}$ and $GaA_{0.67}P_{0.33}$, respectively. From intensity considerations, the 15% GaP layer is identified as the initial deposit. This series of topographs shows that the initial deposit of the

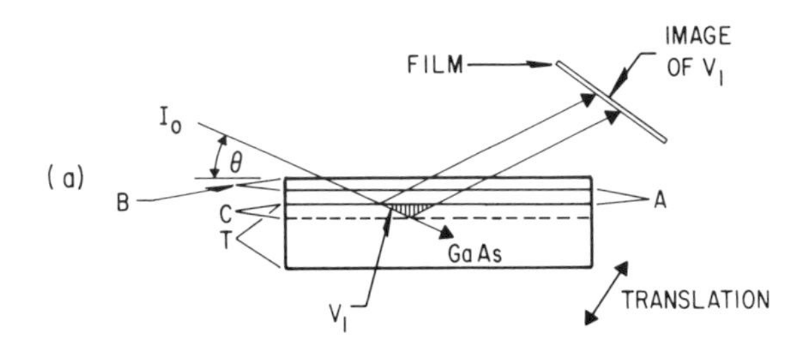

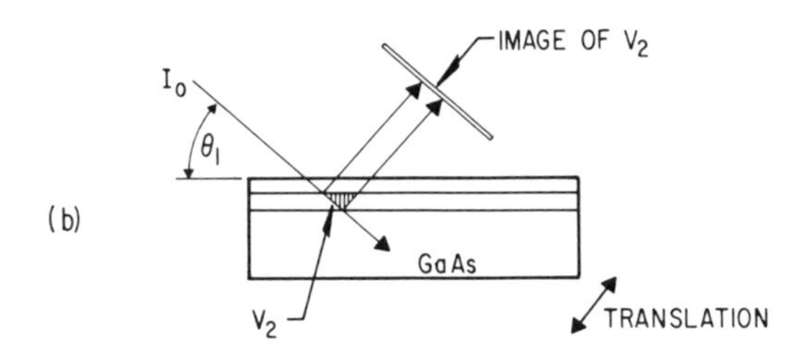

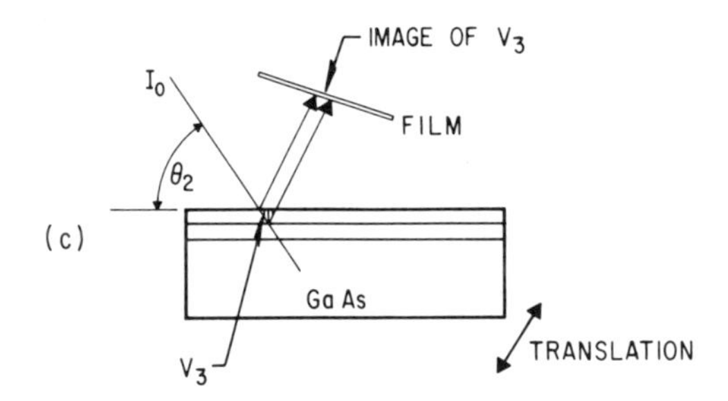

Fig. 7-19. Diagram for compositional x-ray topography. Each layer can be diffracted separately by utilizing the diffraction angle which corresponds to the lattice parameter (composition) of that layer.[38]

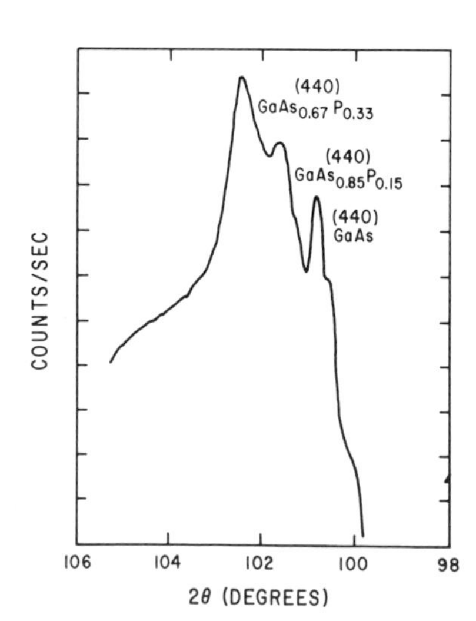

Fig. 7-20. Diffractometer trace of a three-layer Ga (As, P) system.[38]

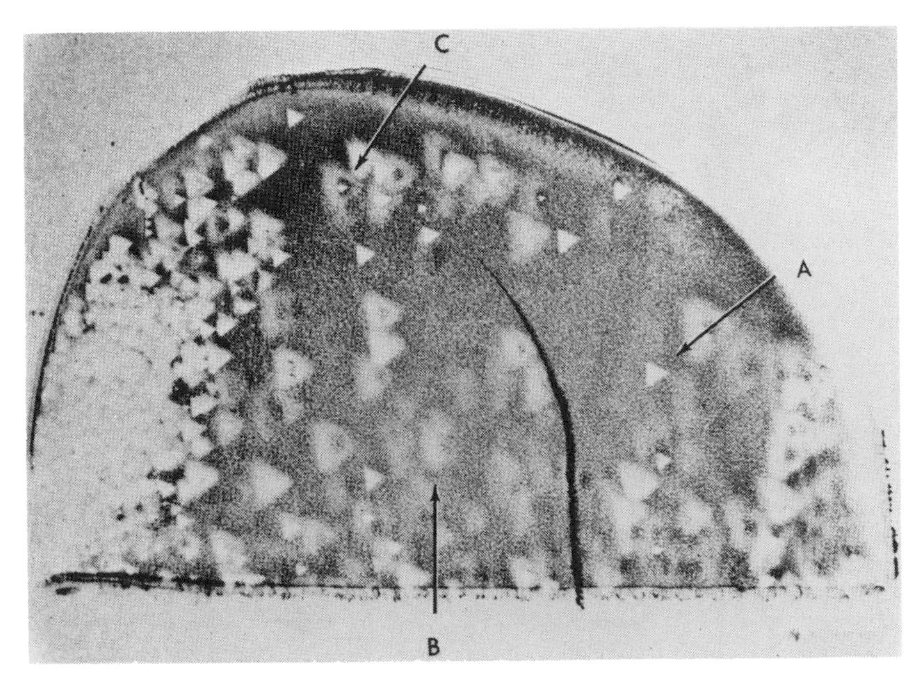

Fig. 7-21a. The (440) topograph of the GaAs substrate which reveals the hillocks as triangular regions of null contrast. The dark line traversing the image is a spurious reflection.[38]

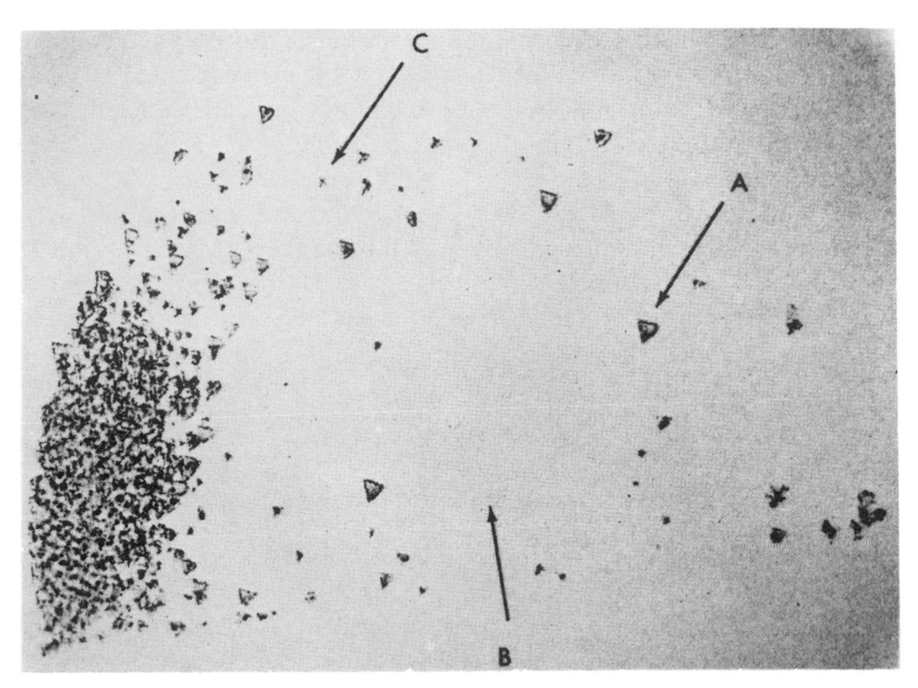

Fig. 7-21b. The (440) topograph of the $GaAs_{0.85}P_{0.15}$ layer; only the hillocks are in diffracting position.[38]

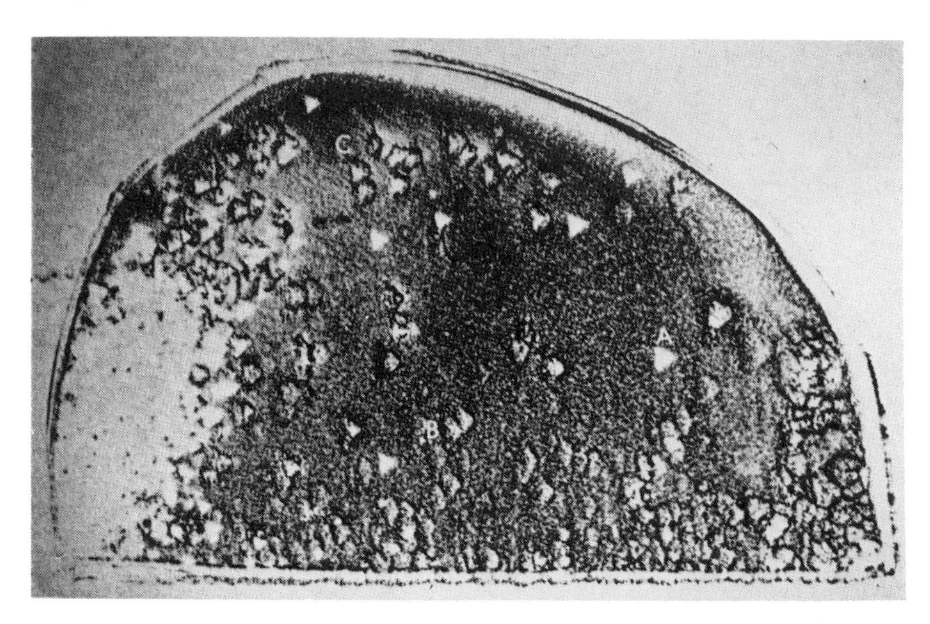

Fig. 7-21c. The (440) topograph of the $GaAs_{0.67}P_{0.33}$ layer; the dark contrast results from local cracks in the deposit.[38]

15% GaP composition formed a high density of triangular hillocks slightly misorientated with respect to the matrix of the film (null contrast in Fig. 7-21b except for hillocks) and that the top layer (33% GaP) formed around these hillocks and frequently overgrew the defects. The intense darkening seen in Fig. 7-21c indicates a high density of local cracks in the deposit.

This method has also been applied to planar compositional gradients in the GaAs–InAs system, axial changes in composition of GaAs–GaP by studying a cleaved edge perpendicular to the deposit, stress-relief mechanisms of deposits of germanium and silicon substrates, and deposits of (Ga, In) As into holes etched into GaAs substrates.[38]

4.3.2.4. Laue Backreflection

The simplest of all the x-ray topographic methods is based on the Laue backreflection technique. The method is extremely surface sensitive, especially for highly absorbing materials since most of the information is derived from the longer-wavelength components of a polychromatic source. Many different geometrical arrangements using polychromatic radiation have been described,[40–44] but the Laue technique developed by Swink[45] is of considerable interest.

The geometry of the Laue method is shown in Fig. 7-22. The polychromatic x rays from a point source P are collimated to a 6- to

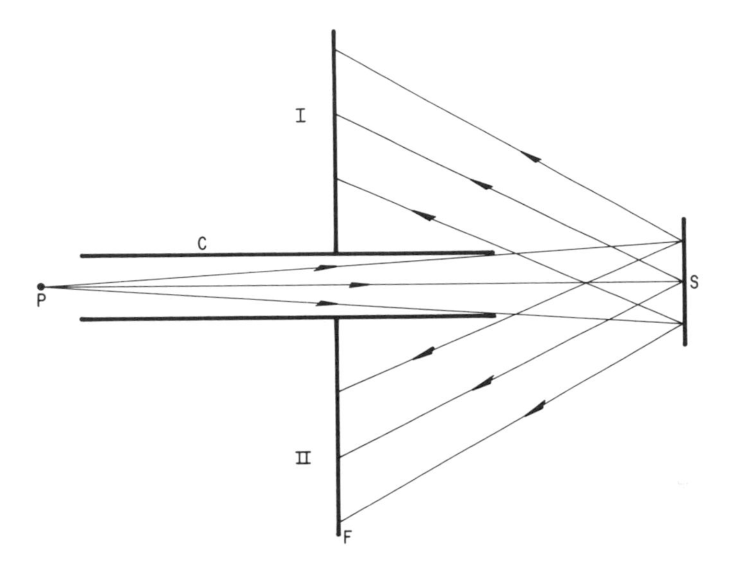

Fig. 7-22. Laue backreflection topography geometry.

12-mm beam, *C*, and impinge upon the specimen *S* at a 90° incident angle. The diffracted images *I* are then formed by backreflection in the same manner as in pinhole-orientation applications.[1] Each set of *hkl* planes which diffracts will have a characteristic backreflection angle which results in separation of the *hkl* diffraction images on the film plane. Any grain tilts, twins, or crystal deformations can be regarded as having an independent set of *hkl* planes and, as such, will be recorded on different portions of the film. Dislocation aggregates which do not constitute a grain mismatch sufficient to be separated from the more perfect parts will appear as intensity contrasts in each image similar to the monochromatic topographs discussed above.

Figure 7-23a is a Laue backreflection topograph of a HgCdTe crystal. For comparison, the high-resolution scanning-reflection topograph is shown in Fig. 7-23b. Both topographs show discrete areas of nearly constant density separated by sharp boundaries from neighboring areas of different density. In the scanning-reflection topograph image (Fig. 7-23b), the contrast observed can result from three possible crystal faults, (1) the grains are all the same composition (lattice constants) but have slightly different orientations, (2) the grains have the same orientation but different compositions, and (3) the grains have widely different dislocation densities. Many more scanning-reflection topographs from different *hkl* planes would be necessary to identify the correct fault. Analysis of the single Laue topograph reveals that postulate (1) can be the only correct one. Postulates (2) and (3) are false since in other areas

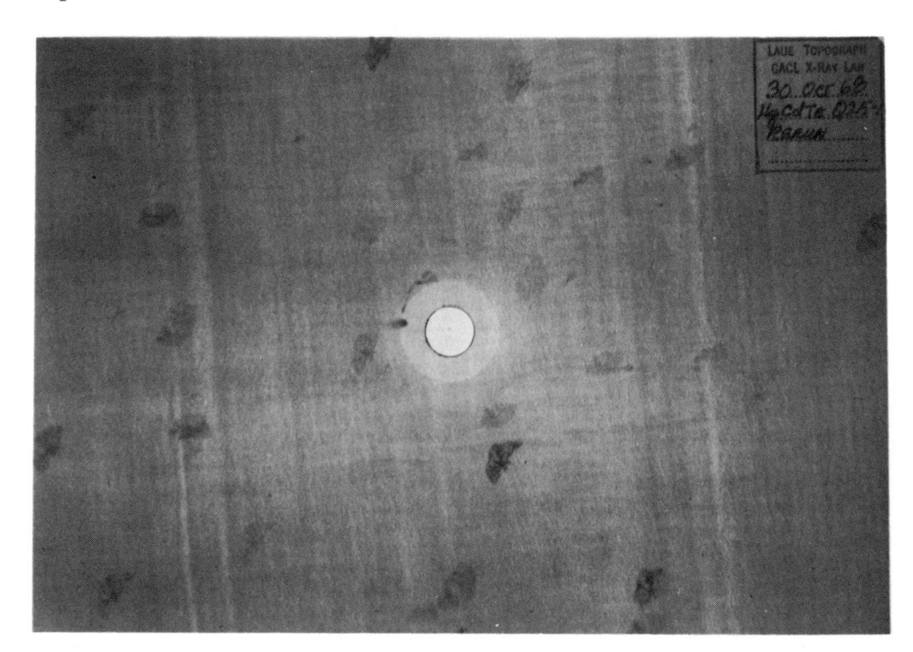

Fig. 7-23a. Laue backreflection topograph of HgCdTe crystal.

Fig. 7-23b. Scanning-reflection topograph of the same HgCdTe crystal used for Fig. 7-23a.

on the film individual grains are recorded. This conclusion is reached since the position of the diffraction image in Laue topography is not dependent upon lattice constant in the cubic system and a strictly dislocation-density mechanism would not result in any individual grain images.

Figure 7-24 shows backreflection Laue topographs of a thin Ga(As, P) film illustrating an ordered and a disordered array of dislocation domains. These topographs are excellent examples of pure dislocation-intensity topographs without any extra reflections recorded which would arise from grain tilts.

Laue topographs as normally recorded require only a one-hour exposure time without any tedious alignment of the crystal to the x-ray beam. This simplicity renders the technique very powerful in the early stages of crystal-growth procedures and for observation of the induced damage in crystal processing. The Berg–Barrett exposure time is much faster, about 5 minutes, but it does require critical alignment and demands working extremely close to the x-ray beam. The Lang and scanning-reflection techniques require very precise alignments of the crystal and slits as well as extremely long exposure times, 3–24 hours, but both techniques do yield images of very high resolution over very large areas. An extremely good review of x-ray topographic techniques and applications has been recently compiled by Meieran.[36]

5. SPECULAR AND ANOMALOUS SURFACE REFLECTION OF X RAYS

The measurement of thin-film thickness using the angular spacing of the intensity maxima of specularly reflected x rays was first presented by Kiessig.[46] The Kiessig expression which related the thickness to these maxima was

$$N = R + (2t/\lambda)(\theta_n^2 - \theta_c^2)^{1/2} \tag{8}$$

where N is the order number of the fringe, R is an unknown phase constant, t is the thickness of film, θ_n the angular position of the fringe, and θ_c is the critical angle for total reflection.

In 1963, Yoneda[47] reported an anomalous surface reflection of x rays for a glancing angle of incidence greater than the critical angle. Warren and Clark[48] described an interpretation for this anomalous reflection using a model which used small-angle scattering by "dirt" on the surface of a mirror. Guentert[39] expanded the dirty-surface model in terms of small-angle scattering arising from irregularities at the surface and/or a film–substrate interface.

(a)

(b)

Fig. 7-24, Laue topograph of GaAsP illustrating (a) an ordered and (b) a disordered array of dislocation networks.

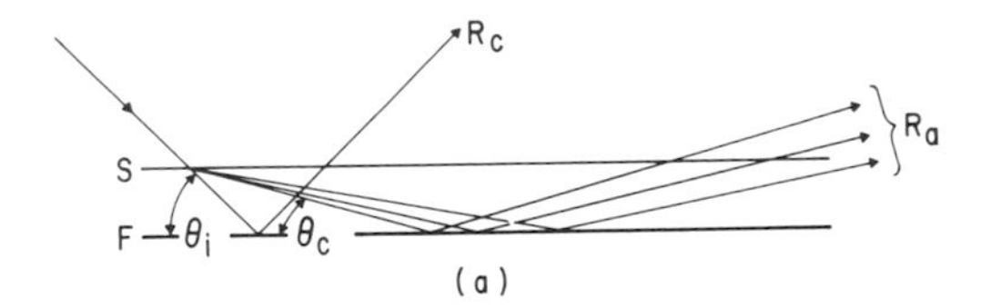

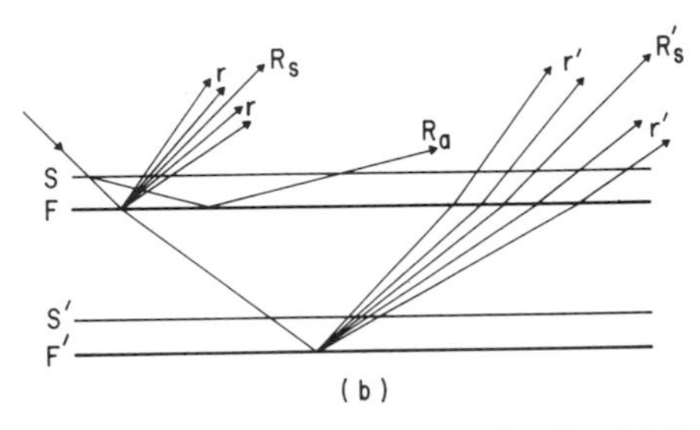

Fig. 7-25. Illustration of x-ray reflection (a) from a single surface F with scattering layer S, and (b) from a thin film. The radiation components are specularly reflected rays (R_s, R_s^1), anomalously reflected rays (R_a), and general small-angle scattering (r, r^1).[49]

The Guentert model, which encompasses both specular and anomalous surface reflection, is shown schematically in Fig. 7-25a. The reflecting face F is considered ideally smooth but covered with a thin sheet of scattering substance, S. The primary beam, incident at angle $\theta_i > \theta_c$, travels through S, is partially reflected from F, and the reflected radiation returns to the outside as the specular beam R_s. On its way toward F, the primary beam undergoes a certain amount of small-angle scattering in S. Rays scattered into the incident angle, $\theta_i \leq \theta_c$, are totally reflected from F and return as the anomalous beam R_a. The scattered substance S could be dirt, an oxide, an absorbed layer, or irregular projections in the bulk material.

Figure 7-25b shows the radiation components for a corresponding model of a thin film with surface F, interface F', and associated scattering layers S and S'. In this case, the total reflection restricts its occurrence to cases of reflection from optically less-dense media. For x rays, the optical density is expressed by $n = 1 - \delta$ where δ is proportional to the physical density ρ. The small-angle-scattered rays r and r' may actually originate anywhere within the scattering layers S and S'.

Figure 7-26 is a schematic of the x-ray reflectometer used by Guentert.[50] Monochromatic radiation of low divergence is obtained by passing the target radiation through an anomalously transmitting

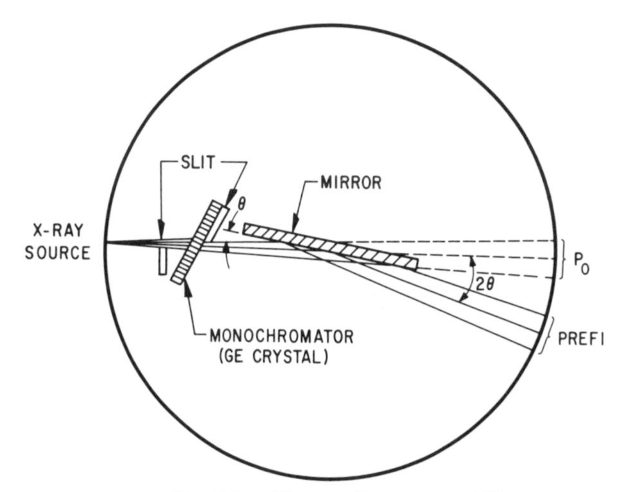

Fig. 7-26. X-ray reflectometer.[49]

germanium crystal.[51] The K_{α_1} and K_{α_2} components are sufficiently separated by the monochromator that their intensity ratio can be controlled by the monochromator slit. A scintillation counter with pulse-height discrimination was used as a detector.

Figure 7-27 is a total-reflection curve from a nickel film using the standard θ–2θ motion with both an open detector window (upper trace)

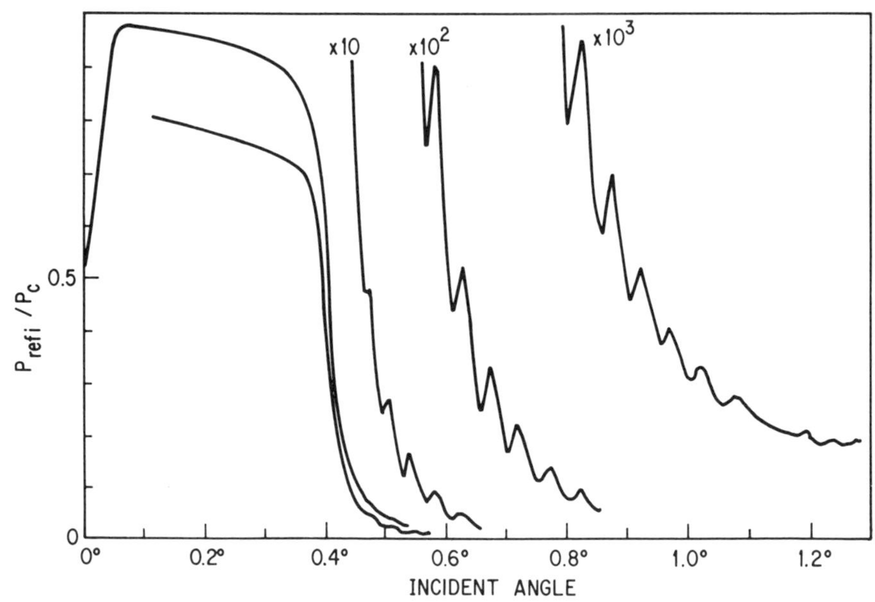

Fig. 7-27. Total-reflection curve with Kiessig pattern from a 755-A nickel film measured with open detector window (upper) and with 0.05° detector slit (lower). Cu K_{α_1} radiation.[49]

and with a 0.05° detector slit (lower trace). The data were recorded by step scanning with fixed time periods and 2θ intervals of 0.005°. The reflected intensity P_{refl}, plotted as P_{refl}/P_0 *vs.* θ_i, starts low since about half the incident beam is blocked by the specimen in this zero portion. The critical angle for this film is taken as 0.405°, the point at which the Kiessig interference begins. Then using the Kiessig relationship, the analysis of the fringe pattern gives a thickness of 755 Å for the film.

Figure 7-28 shows the angular distribution of x rays from the same nickel film in which $\theta = \theta_i$ (the angular positions of successive maxima of the Kiessig pattern). The traces show an overall intensity drop with increasing θ_i and the expected high-angle interference pattern. Sauro *et al.*[52] postulates that some structure should be observed in the region

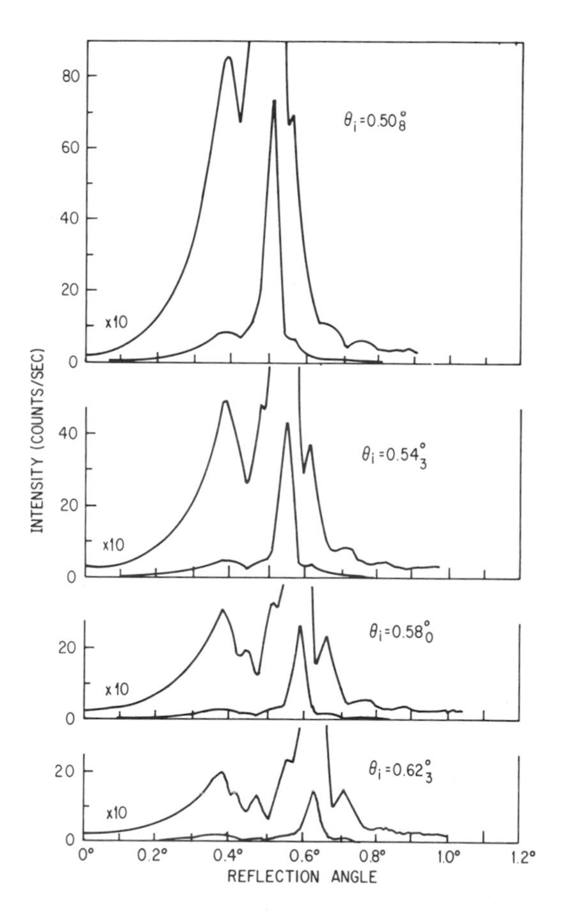

Fig. 7-28. Angular distribution of x rays reflected from a 755-Å nickel film when θ_i is set to successive Kiessig maxima. Cu K_{α_1} radiation.[49]

between the critical angle and the specular reflection as well. These traces do indicate structure in this region as well as a strong peak near 0.40° which is independent of the incident angle. The position and shape of this peak identified it as resulting from the anomalous surface-reflection effect. The intensity of this anomalous reflection increases with increasing surface roughness. This phenomenon could prove to be extremely powerful for determining surface smoothness and possible surface contaminants on the angstrom scale.

Kapp and Wainfain[53] have studied barium stearate multilayer films using this technique and obtained a double-layer spacing of 50.3 Å, in excellent agreement with other types of experimental measurements.

6. SUMMARY

The use of x-ray diffraction methods for surface characterization has been described. The surface depth of a material is limited either by an absorption mechanism in the case of bulk homogeneous materials or by the total film thickness in the case of thin heterogeneous films.

A technique for polycrystalline films based on powder diffractometry using a Seeman–Bohlin geometry is used to give an essentially complete diffraction pattern from metallic films as thin as 150 Å. Standard Laue techniques and Bragg–Brentano diffractometry have been shown to be applicable to films about 1 μ thick. These techniques are used for the determination of particle size, nonuniform strain, faulting, and lattice parameters. An extremely interesting application is in intermetallic phase determination in multilayer films. This type of information can be very important in the area of reaction kinetics.

Several x-ray topographic methods for the characterization of surface perfection of single-crystalline materials are available. The Berg–Barrett and scanning-reflection methods using the Bragg back-reflection geometry are capable of high resolution and can map the distribution of individual dislocations on the more perfect specimens ($< 10^5$ dislocations/cm^2). The Lang limited-projection method enhances the contrast due to surface features by spatial diffraction volume selection. The backreflection Laue topographic method reveals all the information needed for gross perfection determination with a single exposure.

Specular and anomalous surface-reflection techniques are more surface sensitive than any of the other x-ray methods. The interference pattern of the specular reflection can be used to calculate the thickness of extremely thin films. The anomalous-reflection effect has its source from small-angle scattering of surface or interface irregularities which makes this phenomenon a useful tool for measurement of surface smoothness on the atomic scale.

7. REFERENCES

1. B. D. Cullity, *Elements of X-ray Diffraction*, Addison-Wesley, Reading, Mass. (1956).
2. C. G. Darwin, *Phil. Mag.* **43**, 800 (1922).
3. *International Tables for X-ray Crystallography*, Vol. III, p. 157, Kynoch Press, Birmingham (1962).
4. R. W. James, *Solid-State Physics*, No. 15, p. 53, Academic Press, New York (1963).
5. G. A. Walker, *J. Vac. Sci. Tech.* **1**, 465 (1970).
6. A. Taylor, *X-ray Metallography*, p. 667, John Wiley & Sons, New York (1961).
7. C. N. J. Wagner, in *Local Atomic Arrangements Studied by X-Ray Diffraction* (J. B. Cohen and T. E. Hilliard, eds.) Chapter 6, Gordon and Breach, New York (1965).
8. B. E. Warren, *Progr. Metal Phys.* **8**, 147 (1959).
9. W. A. Rachinger, *J. Sci. Instr.* **25**, 254 (1948).
10. A. R. Stokes, *Proc. Phys. Soc. (London)* **61**, 382 (1948).
11. C. N. J. Wagner and E. N. Aqua, *Adv. X-Ray Anal.* **7**, 46 (1964).
12. R. Feder and B. S. Berry, *J. Appl. Cryst.* **3**, 372 (1970).
13. G. Wasserman and J. Wiewiorowsky, *Z. Metallk.* **44**, 567 (1963).
14. A. Segmuller, *Z. Metallk.* **48**, 448 (1957).
15. W. Parrish and M. Mack, *Acta Cryst.* **23**, 687 (1967).
16. M. Mack and W. Parrish, *Acta Cryst.* **23**, 693 (1967).
17. D. A. Brine and R. A. Young, Vacuum Technology Transactions: Proceedings of the Seventh National Symposium of the American Vacuum Society, pp. 250–259, Pergamon Press, New York (1963).
18. D. A. Brine and R. A. Young, *Phil. Mag.* **8**, 651 (1963).
19. T. C. Furnas, Jr., *Single Crystal Orienter Manual*, General Electric, Milwaukee (1957).
20. G. Freidel, *Compt. Rend.* **157**, 1533 (1913).
21. R. W. James, *Optical Principles of X-Ray Diffraction*, Bell, London (1950).
22. W. H. Zachariasen, *Theory of X-Ray Diffraction in Crystals*, John Wiley & Sons, New York (1945).
23. C. W. Bunn, *Chemical Crystallography*, Clarendon, Oxford (1961).
24. E. P. Warekois and P. H. Metzger, *J. Appl. Phys.* **30**, 960 (1959).
25. H. Cole and N. R. Stemple, *J. Appl. Phys.* **33**, 2227 (1962).
26. P. B. Hirsch, *Proc. Metal Phys.* **6**, 236 (1956).
27. B. W. Batterman, *J. Appl. Phys.* **30**, 508 (1959).
28. K. S. Chandrasekaran, *Acta Cryst.* **12**, 916 (1959).
29. W. Berg, *Naturwiss.* **19**, 391 (1931).
30. W. Berg, *Z. Krist.* **89**, 286 (1934).
31. C. S. Barrett, *Trans. AIME* **161**, 15 (1945).
32. J. B. Newkirk, *Trans. AIME* **215**, 483 (1959).
33. A. R. Lang, *Acta Met.* **5**, 358 (1957).
34. A. R. Lang, *Brit. J. Appl. Phys.* **14**, 904 (1963).
35. G. Borrman, W. Hartwig, and H. Irmler, *Z. Naturforsch.* **13A**, 423 (1958).
36. E. S. Meieran, *Siemens Rev.* **XXXVII**, 39 (1970).
37. J. K. Howard and R. D. Dobrott, *Appl. Phys. Letters* **1**, 101 (1965).
38. J. K. Howard and R. D. Dobrott, *J. Elec. Chem. Soc.* **113**, 567 (1966).

39. E. W. Williams, R. H. Cox, R. D. Dobrott, and C. E. Jones, *Electrochem. Tech.* **4**, 479 (1966).
40. A. Guinier and J. Tennevin, *J. Acta Cryst.* **2**, 133 (1949).
41. L. G. Shultz, *Trans. AIME* **201**, 1082 (1954).
42. C. T. Wei and P. A. Beck, *J. Appl. Phys.* **12**, 1508 (1956).
43. R. A. Coyle, A. M. Marshall, J. H. Auld, and N. A. McKinnon, *Brit. J. Appl. Phys.* **8**, 79 (1957).
44. P. J. Holmes, *J. Appl. Phys.* **6**, 180 (1955).
45. L. N. Swink and M. J. Brau, *Met. Trans.* **1**, 629 (1970).
46. H. Kiessig, *Ann. Physik.* **10**, 715 (1931).
47. Y. Yoneda, *Phys. Rev.* **131**, 2010 (1963).
48. B. E. Warren and J. S. Clark, *J. Appl. Phys.* **36**, 324 (1965).
49. O. J. Guentert, *Phys. Rev.* **138**, A732 (1965).
50. R. L. Mozzi and O. J. Guentert, *Rev. Sci. Instr.* **35**, 75 (1964).
51. G. Borrmann, *Physik Z.* **42**, 157 (1941).
52. J. Sauro, I. Fankuchen, and N. Wainfan, *Phys. Rev.* **132**, 1544 (1963).
53. D. S. Kapp and N. Wainfain, *Phys. Rev.* **138**, A1490 (1965).

II

CHEMICAL CHARACTERIZATION

8

ELECTROCHEMICAL TECHNIQUES

D. M. MacArthur

Bell Laboratories
Murray Hill, New Jersey

1. INTRODUCTION

Electrode processes, by definition, occur at an interface. It follows that electrochemical measurements on electrodes are affected by surface conditions and tell something about the state of the surface. Electrochemical measurements, however, require an electrolyte on one side of the interface, and high potential gradients across this region may occur ($\approx 10^6$ V/cm). The surface can be materially different from that in equilibrium at the solid–gas interface as a result of rearrangement of surface atoms and adsorption from solution of ions or molecules in the electrolyte. There have been few attempts, for this reason, to use electrochemical methods to obtain knowledge about the solid–gas interface. There are many practical situations where a knowledge of the solid–electrolyte interface is desired, however, and it is in these situations where the techniques discussed in this chapter may be of value.

This book is concerned with solid surfaces. There is a great deal of information available about the mercury–electrolyte interface, but it is outside the scope of this book and will not be discussed here.

Traditional electrochemistry (or more properly, that part of it which may be described as voltammetry) has been concerned with the three variables—potential, current, and time. There are only a limited number of combinations of these variables. In general, they have not been sufficient to provide unambiguous answers concerning happenings at the surface, and increasing attention is being paid to optical techniques* such as ellipsometry and specular and internal-reflection spectroscopy combined with the traditional voltammetric methods. As a method for characterizing surfaces, these approaches appear to be primarily dependent on optical methods and are properly left for other chapters of this book. There is, however, much that can be learned from the traditional electrochemical techniques. These include determination of the true surface area, the cleanliness of the surface, the presence of surface films and adsorbed layers, the surface composition, and, for semiconductors, the presence of surface states and the space charge.

The various electrochemical methods that have been used to study solid surfaces in contact with electrolytes will be reviewed in the first section of this chapter; the application of these methods to surface studies will be illustrated in the second section; and a brief description of circuits, cell design, and reference electrodes will be given in the third section. All the voltammetric methods have the advantage of requiring relatively simple and inexpensive equipment to do experiments that, only rarely, are time consuming. A previous review article[1] has been concerned with electrochemical techniques for the characterization of clean surfaces in solution.

2. METHODS

2.1. Reaction Kinetics

The rate of a simple electrochemical reaction without specific adsorption and mass transport effects can be written[2] in the form

$$i = nFA\left[C_0 k_f \exp\left(-\frac{\alpha nF}{RT}E\right) - C_R k_b \exp\left(\frac{(1-\alpha)nF}{RT}E\right)\right] \tag{1}$$

(The symbols are defined in the glossary at the end of this chapter.) At equilibrium the current is zero and the above expression can be written as

$$E_e = \frac{RT}{nF}\ln\frac{C_0 k_f}{C_R k_b} \tag{2}$$

Sufficiently far from equilibrium (roughly 0.12/nV) the reverse reaction of Eq. (1) can be neglected and for a reduction reaction the rate may be written as

$$i_c = nFAC_0 k_f \exp\left(-\frac{\alpha nF}{RT}E\right) \tag{3}$$

* The interested reader may find a detailed review of these techniques applied in electrochemistry in Vol. 9 of the series, *Advances in Electrochemistry and Electrochemical Engineering* (R. H. Muller, ed.) pages 61–166, Wiley–Interscience, New York, 1973.

It is customary to write the exchange current as

$$i_o = nFAC_0k_f \exp\left(-\frac{\alpha nF}{RT}E_e\right) \tag{4}$$

Thus for a cathodic reaction (reduction)

$$i_c = i_o \exp\left[-\frac{\alpha nF}{RT}(E - E_e)\right] \tag{5}$$

and for an anodic reaction (oxidation)

$$i_a = i_o \exp\left[\frac{(1 - \alpha)nF}{RT}(E - E_e)\right] \tag{6}$$

These equations are in the form of the familiar Tafel relationship ($\log i$ linear with E). The exchange current (i_o) and transfer coefficient (α) are usually obtained from a plot of log current density *vs.* potential. The presence of impurities on the surface has at least two effects—a variation of apparent current density with coverage and a change in potential of the double layer. (The double layer is described below.) These effects are evident in the value obtained for the exchange current.

This method can be a useful one since the exchange current is often very sensitive to the cleanliness of the surface. In addition, the exchange current is sensitive to the exposed surface area of the electrode and may be used to obtain the actual surface area. To ensure that the reaction kinetics are not controlled by diffusion effects, the solution should be stirred. Measurements should be taken over at least a decade of current values. They may be made using either the current or the potential as the dependent variable. The reactions studied are usually either hydrogen or (on oxide electrodes) oxygen evolution from aqueous solutions on the assumption that these reactions do not alter the surface in the same manner as would oxidation of an electrode or reduction of metal cations. The potentials at which these reactions may be studied, however, are a long way from the potential of zero charge (defined in Section 2.3) for the electrode, and it must be assumed that the large potential gradient across the interface could alter the surface of the electrode to an unknown extent.

2.2. Electrode Potential

In most cases when an electrode is immersed in an electrolyte, a stationary potential is observed after a period of time. Unless the electrode and electrolyte have been chosen specifically to provide a reversible system it is usually not an equilibrium potential [Eq. (2)] that is observed but a mixed or corrosion potential.

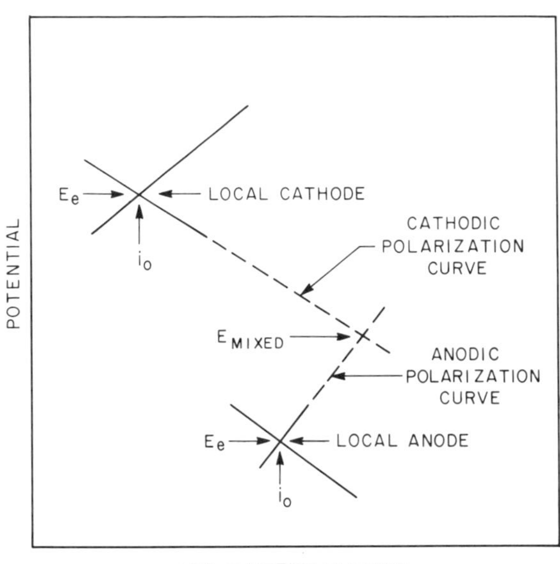

Fig. 8-1. Schematic to illustrate the origin of the mixed potential on an electrode.

There are localized areas on the surface of the electrode where an oxidation reaction occurs (the anode) and a reduction reaction occurs (the cathode), and the potential attained is that at which the rate of the oxidation reaction is equal to the rate of the reduction reaction. This is illustrated in Fig. 8-1 where Tafel lines have been drawn for the reactions occurring at the localized anode and cathode. The mixed potential is determined both by the kinetics of the respective reactions and the areas where these reactions occur. The presence of an impurity changes either or both of these factors, and in most cases will alter the observed potential. The effect may be a complex one and it is difficult to predict whether the potential will be shifted in the anodic or cathodic direction. However, measuring the potential can be a useful indication of a soiled electrode if its potential can be compared to that of a clean electrode in the same electrolyte.

The method is a very simple one requiring only a very-high-impedance electrometer and a reference electrode. It does suffer from being nonquantitative and variable since different impurities may cause different potential changes or no change at all.

2.3. Impedance Measurements

A redistribution of charge carriers, which behaves like an electrical capacitor, occurs at the phase boundary of a solid in an electrolyte.[3]

On the solution side, the distance of the redistribution from the surface depends on the electrolyte concentration and, in dilute electrolytes, extends up to 1 μ into the solution. For descriptive purposes, the charge redistribution on the electrolyte side is divided into the inner Helmholtz layer, which is the locus of the electrical center of adsorbed unsolvated ions on the surface; the outer Helmholtz layer, which is the locus of the electrical center of solvated ions in contact with the surface; and the diffuse or Gouy–Chapman layer which extends into the solution. This double layer on the electrolyte side may be considered as two capacitors in series, and the electrode capacitance may be written as

$$\frac{1}{C} = \frac{1}{C_i} + \frac{1}{C_{\text{Gouy}}} \tag{7}$$

The capacitance of the Gouy layer is given by[4]

$$C_{\text{Gouy}} = \left(\frac{Z^2 e_o{}^2 C_o \varepsilon}{2\pi k T}\right)^{1/2} \cosh\left[\frac{Z e_o(\varphi_2 - \varphi_E)}{2kT}\right] \tag{8}$$

This function has a minimum for $\varphi_2 - \varphi_E = 0$, and the potential where this occurs is called the potential of zero charge. The value of C_{Gouy} depends on the electrolyte concentration. The inner capacitance C_i, on the other hand, is relatively insensitive to electrolyte concentration in the absence of specific adsorption, but depends on electrode potential. When specific adsorption does occur it is reflected in the value of C_i.

On the solid side, the extent of the redistribution depends on the concentration of the mobile charge carriers. In a metal, where the density of the charge carriers is high, a very thin layer of electrical charge accumulates on the solid side of the interface. If the solid is a semiconductor, the charge may extend some distance from the surface producing a field within the solid, and the conduction and valence bands are bent upward for a positive charge accumulation and downward for a negative charge accumulation in the usual convention. A sequence of different space-charge layers is described[5] ranging from strong inversion (i.e., majority carrier at the surface is of opposite sign to that of the bulk) through the flat-band condition, to the accumulation layer.

The space-charge layers in the solid and in the electrolyte change rapidly with the applied potential, but the charge associated with surface states has a relatively long relaxation time. Thus a determination of the capacitance of an electrode at several frequencies can provide useful information about the surface. The capacitances arising from the charge accumulation are in series combination[6]

$$\frac{1}{C} = \frac{1}{C_S} + \frac{1}{C_i} + \frac{1}{C_{\text{Gouy}}} \tag{9}$$

where the capacitance of the space-charge layer and surface states in a semiconductor has been included in the total capacitance. For even moderately concentrated solutions ($>M/10$)

$$C_{\text{Gouy}} \gg C_S \tag{10}$$

The capacitance of the semiconductor is described as comprising two capacitances in parallel, that associated with the space charge C_{SC} and that associated with surface states C_{SS}. Thus $C_S = C_{SC} + C_{SS}$. At low frequencies, when the relaxation time of the surface states is not exceeded,

$$\frac{1}{C} = \frac{1}{C_S} + \frac{1}{C_i} \tag{11}$$

For most semiconductor electrodes,

$$C_{SC} \gg C_{SS} \tag{12}$$

At higher frequencies, up to about 10^4 Hz, the relaxation time of the surface states is exceeded, and

$$C \approx C_{SC}$$

The capacitance of the space-charge layer is given by[7]

$$C = \frac{e^2 n_i^2 \mathscr{L}^2}{2kTq_{sc}^{1/2}} \left\{ \lambda \left[\exp\left(e\frac{\varphi_b - \varphi_s}{kT} \right) - 1 \right] + \lambda^{-1} \left[\exp\left(-e\frac{\varphi_b - \varphi_s}{kT} \right) - 1 \right] \right\} \tag{13}$$

For the frequently encountered condition of an exhaustion or weak inversion layer,[5]

$$\frac{1}{C^2} = \frac{2}{\varepsilon\rho}(\varphi_b - \varphi_s) \tag{14}$$

These equations provide a means of obtaining the surface potential of the semiconductor.

For metal electrodes in moderately concentrated solutions

$$C \approx C_i \tag{15}$$

and C_i is very sensitive to the presence of impurities.

A wide range of techniques has been employed to measure electrode capacitance. The simplest uses ac bridge methods employing Wheatstone bridges or, for larger capacitances, transformer bridges. Transient methods employing current-step or potential-step functions have also been used. It is the differential capacitance

$$C = \frac{\Delta q}{\Delta E} \tag{16}$$

which is obtained. One of the problems with transient impedance measurements is the need to use an electrical analog for analysis of the measurements. For a metal electrode, the simplest analog (Fig. 8-2) is that of a capacitor and resistor in parallel to represent the interface and a series resistance representing all other resistive components of the circuit. Analysis of the time transients gives

$$\ln\frac{dE}{dt} = \ln\frac{I}{C} - \frac{t}{R_pC} \tag{17}$$

at constant current, where E is measured between points A and B, and

$$\ln\frac{dI}{dt} = \ln\frac{1}{EC} - \frac{t}{R_pC} \tag{18}$$

at constant voltage, where E is held constant between A and B. More complex electrical analogs have been used to represent the electrode impedance, but none has found wide acceptance.

In the current-step method, a constant current is instantaneously applied to the electrode and its potential–time behavior is observed on an oscilloscope. Figure 8-3 schematically shows the potential–time behavior for a current step and the current–time behavior for a potential step. The current may be consumed in Faradaic reactions and in charging of the double layer. The charging of the double layer, if it is done in a time period in which the capacitance does not change, follows the above exponential time–potential relationship.

The simultaneous occurrence of a Faradaic reaction may give anomalously high capacitances. It may also show up as a plateau in the potential–time curve (Fig. 8-3). If the reaction is that of reduction or oxidation of an impurity on the surface, a quantitative measure of the impurity may be obtained through the integral $\int_{t_1}^{t_2} i\,dt$ and a knowledge of the number of electrons consumed per reacting species.

In the potential-step method the electrode is allowed to equilibrate either at its open-circuit potential or at some adjusted potential and

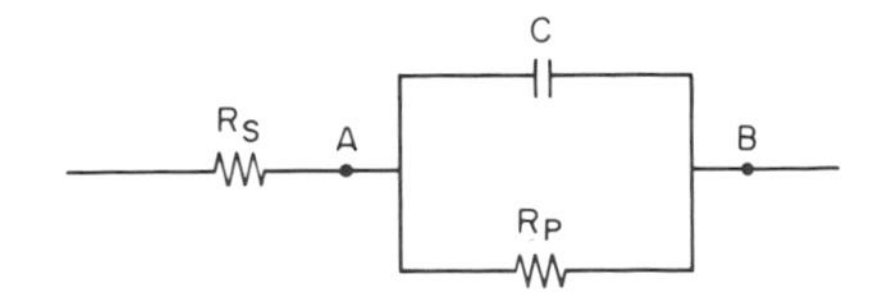

Fig. 8-2. A simple electrical analog of an electrode–electrolyte interface.

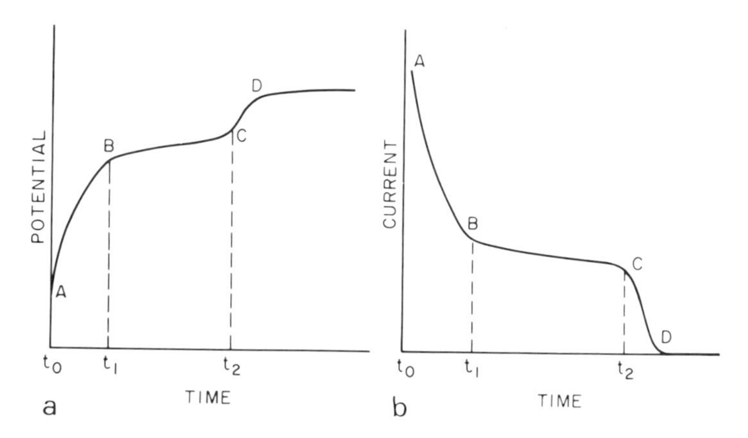

Fig. 8-3. (a) Schematic of the potential–time transient at constant current of an electrode with a surface electroactive agent. (b) The current–time transient at constant potential for a situation similar to (a).

then is instantaneously switched to another predetermined potential. The current to the electrode is observed on an oscilloscope as a function of time (Fig. 8-3). As with the current-step method, the charging of the double layer is observed first, and an oxidation or reduction current plateau is observed next if some Faradaic reaction can occur at that applied potential. An adsorbed species, if it is electrochemically active, can usually be quantitatively measured by determining $\int_{t_1}^{t_2} i\,dt$. The potential-step method has an advantage over the current-step method in that it is easier to study a potential range where no Faradaic reactions occur.

Both the current-step and potential-step functions may be used in a repetitive mode. If sufficient time is allowed between sufficiently small steps, then equilibrium may be achieved on each cycle and the above equations hold. If equilibrium is not achieved, the equations must be modified to take this into account.

The necessary switching that goes on using these methods frequently introduces extraneous transients in the circuitry. Several approaches[8,9] have been used to solve this problem. It is also frequently necessary to compensate for the IR component of the measured variable. This may be done electronically.[10] Electronic circuits[11,12] for directly obtaining the interfacial capacitance have also recently been described.

2.4. Potential Sweep

In the potential-sweep method the potential of an electrode is changed in a predetermined manner as a function of time. A linear function is most frequently used. The current flowing in the electrode is monitored. For surface characterization, a single sweep moving either in the anodic or cathodic direction and then back to the starting point is most useful. At slow sweep speeds (<10 mV/sec), a steady-state condition is approached and the current is determined by reactions at the electrode occurring at the immediate potential. At high sweep speeds (>100 mV/sec) the current is a measure of the state of the electrode at the commencement of the sweep. A limit to the sweep speed is reached when the current consumed in charging the double layer becomes relatively large.

The method can be used to quantitatively determine the presence of electroactive surface contaminants. It can also be used to determine the surface composition. Figure 8-4 shows potential scans of a Pt electrode, a Au electrode, and a 65Pt–35Au alloy electrode. The curve for the alloy electrode is seen to be a composite curve made up of the individual Pt and Au contributions. This curve is cited[13] as evidence that the surface of the alloy is composed of discrete Pt and Au phases.

A great deal of information can be obtained from potential-sweep measurements. Current peaks have different shapes depending on the reaction occurring. If the reacting species is an adsorbed species or a surface film, the current rises to a maximum and then falls to essentially zero as the quantity of reactants, which is limited, is consumed. The number of coulombs in the peak is a measure of the quantity of the reactant. It should be independent of the sweep rate over a wide range of velocities, but the range in which this occurs is dependent on the kinetics

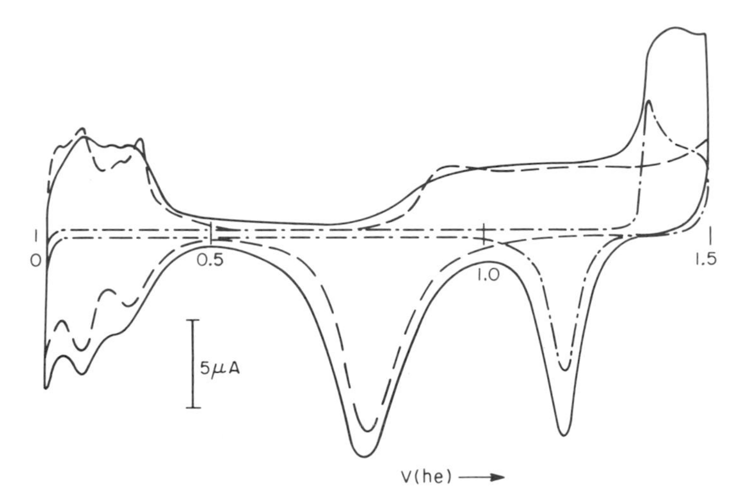

Fig. 8-4. Current–potential curves for Pt(– – – –), Au(– · – ·), and a 65Pt–35 Au alloy(—) in 1M H_2SO_4. (Reproduced by permission of the publisher.[13])

of the reaction. The potential-sweep technique has an advantage over the potential-step technique (which is essentially a very fast sweep) in that a slower-speed chart recorder can be substituted for the oscilloscope, and, also, double-layer charging currents are small.

2.5. Charge Step-Potential Decay

In this method a quantity of charge (Δq) is injected into an electrode over a short time interval and the potential of the electrode is observed on an oscilloscope as a function of time. The objective is to obtain the differential capacitance [Eq.(16)]. In principle this is obtained by dividing the quantity of injected charge by the change in potential (ΔE), but obtaining the value of ΔE and making corrections for any possible Faradaic process can be difficult. If the potential increment is kept small (a few millivolts) the potential relaxation curve for a Faradaic process is given by[14]

$$\Delta E = \Delta E_{t_o} \exp - \left(\frac{i_o}{C} \frac{nF}{RT} t \right) \tag{18}$$

This equation provides ΔE at time zero (the end of the charge-injection period) and a means of obtaining the exchange current i_o. Thus both capacitance and the kinetic parameter i_o can be obtained in this type of experiment.

2.6. Other Methods

A unique method[15] to determine the point of zero charge (pzc) for a solid electrode (which cannot be determined by electrocapillary methods) uses the bending of a thin-film metal electrode deposited on a glass strip. The radius of curvature of the electrode is plotted as a function of potential to obtain the pzc. The change in length of a ribbon electrode measured with an extensometer has also been used.[16]

3. APPLICATIONS

The characterization of solid surfaces includes such things as the presence of films or adsorbed layers on the surface, the composition of the surface, and the presence of surface contaminants. Also included for semiconductors are surface states and space charge. Selected examples are given here of the application of the various electrochemical techniques to such a characterization. The list is not comprehensive—a few examples have been chosen from the recent literature to provide a means whereby the interested reader can pursue the subject.

There appear to be no studies where the primary interest was the physical character (e.g., stress, hardness) of the surface. This would appear to be an area worthy of examination.

3.1. Films and Adsorbed Layers

The potential-step method has been used[17] to measure the amount of surface oxide formed on Pt and Au at various potentials. Electronic integration was used to obtain the charge capacity of the oxide. A plot of charge capacity *vs.* time was then used to extrapolate to zero time corresponding to the commencement of the potential step, and the quantity of surface oxide initially present on the electrode was calculated.

The application of a square-wave voltage to obtain surface capacitance of Pt, Ag, Cu, Al, and Ta in 1 N H_2SO_4 has been used.[18] An analysis of the potential–time traces using the simple electrical analog circuit (Fig. 8-2) was made for various assumptions of the values for the series and parallel resistors, and the capacitance deduced. The analysis has been carried further by others[19] who clearly point out that the technique is intended for surface-area measurements when no Faradaic process occurs.

The potential-sweep technique has been applied[20] to study the formation of platinum oxide at the metal–gas interface. This is one of very few examples where electrochemical techniques have been used to study a surface before it contacts the electrolyte. The platinum was exposed to a gas stream and then transferred to an electrolytic cell where the oxide formed during exposure to the gas was reduced by applying a potential sweep. Integration of the area under the curve provided a quantitative measure of the oxide. In another work[21] examining oxygen-containing films on Rh, Pd, and Au electrodes, potential sweeps were used to determine the stoichiometry of chemisorbed oxygen and the real surface area of the electrode.

The "poisoning" of the surface of an oxygen fuel-cell electrode by adsorption of cadmium ions has been studied[22] by a fast potential-sweep technique. In this work the presence of the cadmium was detected by a reduction of the hydrogen wave on the platinum fuel-cell electrode and an increase in a cadmium oxidation peak at about 1.0 V (reversible hydrogen reference). By making assumptions about the oxidation of the adsorbed cadmium, it was possible to deduce a fractional coverage of the surface by cadmium. The current-step technique has also been used[23] to determine the exposed platinum surface area in fuel-cell electrodes.

Using the method of charge step-potential decay, the double-layer capacitance of Pt in H_2SO_4 was determined.[9] A high-speed switching technique to prevent saturation of the oscilloscope amplifier during charge injection was described. The capacitance measurements by this technique were in agreement with those determined by a constant current step.

3.2. Surface Composition

The potential-sweep method has been used[24,25] to measure hydrogen adsorption on heterogeneous Pt–Au alloys. It was shown that the current–potential curves in 1 N H_2SO_4 were the sum of the curves for the pure metals adjusted for the fraction of the surface occupied by each metal. This is also illustrated in Fig. 8-4. In later work,[26] the same technique was used to identify Au islands on copper-plated wire surfaces; however, it was not possible in this instance to obtain a quantitative measure of the surface coverage. Using a similar technique, the surfaces of homogeneous Pt–Au alloys are found to be composed of two distinct phases at the temperature of measurement, irrespective of the bulk composition.[27] These separate phases were too small to be detected by x-ray diffraction analysis.

The potential-sweep method has also been used[28] for surface analysis of Fe–Al–Mn alloys. In this case, the currents for oxidation of the metallic components of the alloy (slow scan) were used for the analysis. It was found that sweeps in two different solutions provided data for quantitative analysis of the Mn and Al contents within about 1% mean deviation.

A steady-state potentiostatic passivation technique (constant potential) has been used to determine passivation currents on single Ni crystals[29] and Pt crystals[30] of different orientations. These and other types of electrochemical measurements on single crystals are an area that has been largely neglected by electrochemists. It appears reasonable that application of the techniques described here could provide useful information about their surfaces.

3.3. Surface Contaminants

The study of surface contaminants by electrochemical techniques has been pursued primarily by electroplaters because of the observation that such factors as electroplate adherence, uniformity, porosity, and stress are sensitive to substrate surface contamination. A study[31] of reaction kinetics for copper deposition on clean and soiled copper cathodes has been made. The transfer coefficient α was smaller for soiled electrodes than for clean electrodes, but the exchange current did not follow a recognizable trend. Some progress is being made in an effort[32,33] to classify and understand the role of surface-active agents in the electrodeposition of copper using constant-current transients. One of the problems with using an electrodeposition reaction to study surface contamination is that the substrate surface soon becomes covered and one is no longer studying the surface of interest. Reaction-kinetics measurements have been made[34] for the hydrogen reaction on clean and soiled copper electrodes to avoid this problem.

Potential–time[35,36] studies of clean and contaminated Pt electrodes have been used to determine the degree of surface contamination of materials for electron-device fabrication. The procedure was intended as a simple and rapid method of determining the cleanliness of materials before assembly.

3.4. Semiconductor Electrodes: Surface State and Space Charge

The usual approach to characterizing the surfaces of semiconductors is to measure the capacitance using conventional ac bridges or potential-step and current-step transients. The $1/C^2 - \Delta E$ relationship [Eq.(14)] is frequently observed, the so-called Mott-Schottky case, which includes the exhaustion and weak inversion situations. The space charge may then be calculated. The existence of surface states is usually postulated for any unexpected departure from the usual relationships. Reaction-kinetics measurements have also frequently led to the postulation of surface recombination–generation centers. The physical picture of surface states (and indeed the need to invoke them at all[37]) appears to be in a state of flux at the present time. Gerischer[38] has described surface states as those that interact with the semiconductor electronic system and would exclude ionic surface charges resulting from adsorption. The role of adsorbed ions in the inner Helmholtz layer in the charge transfer reactions, however, continues to be a matter of some discussion. Further description of processes at the semiconductor–electrolyte interface may be found in the literature.[6,40,41] The study of semiconductor surfaces has been severely hampered by the fact that only a few semiconductors have even a moderate potential range where Faradaic reactions do not occur. One of the more attractive in this respect is ZnO.

Impedance measurements[42] of the single-crystal-ZnO–electrolyte interface result in the proposal of surface states which are identified as due to the crystal itself and not the electrolyte. Reaction kinetics at the ZnO interface for reduction of various ions in solution have been studied.[43] A maximum for the electron transfer rate is found for energy levels in the outer Helmholtz layer just below the conduction-band minimum. These results are discussed on the basis of surface-state capture theory.

The role of surface states in the charge transfer reaction at the CdS and CdSe interface has been examined[44] using reaction-kinetics and capacitance methods. It has been shown that surface states control the electrochemical reactions. Measurements[45] of the interface capacitance of heavily doped n^+ silicon in 5% HF solution have been made using a square-wave generator. The surface generation of minority carriers is proposed as the rate-limiting step in anodic dissolution of the n^+ silicon. It is apparent from this short description that characterization of the surfaces of semiconductors has more often than not resulted in vague references to surface states.

The picture of the semiconductor–electrolyte interface, for those semiconductors that are entirely covalent, involves an adsorbed oxide, as for germanium, $Ge{-}OH + H_2O \rightleftarrows Ge{-}O^- + H_3O^+$. This reaction is used to explain the observed relationship $dE/dpH = 0.06$ v, which is found for germanium. It is much more difficult to provide such an argument for the more-ionic semiconductors at the present time. More experimental work appears to be required.

The earlier literature contains examples of the determination of the space-charge layer. A good example is the impedance measurements on germanium electrodes that have been made[46] using a small ac voltage superimposed on an applied dc voltage. The results are interpreted in terms of an equivalent circuit and absolute values of the space-charge capacity at high frequencies are obtained.

Reaction-kinetics, measurements[47] of the oxygen-evolution reaction on potassium tantalate anodes and on semiconductive TiO_2[48] have been made. Capacitance, using a current-step technique, was also obtained. The data were interpreted in terms of electron tunneling from surface oxide ions to the conduction band.

There are few examples where electrochemical techniques other than impedance or reaction-kinetics measurements have been made on semiconductor electrodes. It would appear that potential sweep techniques could be useful. They have been used[49] on semiconducting tin oxide electrodes, and a region has been identified where oxidation and reduction of surface states occurs. These states appear to be related to specific adsorption of chloride ion.

4. Cell Design and Reference Electrodes

The characterization of soiled surfaces by electrochemical techniques usually requires a comparison to a clean surface. Attaining a clean surface is not easy and may require extensive purification of reagents and working in a controlled environment under the most rigorous conditions.[50] For most work, however, more modest efforts may be sufficient. Attention should be paid to the electrolyte as well as the electrodes. In most cases it is necessary to use doubly or triply distilled water, reagents of the highest purity, and pre-electrolysis of the electrolyte between inert electrodes. Oxygen may be removed by bubbling nitrogen or helium through the solution, and the gas trains should have means for purification of the gas. Such procedures are described in many of the references at the end of this chapter. It is highly desirable to have the solid electrode which is to be studied (here called the working electrode) in a form in which it can be easily removed from the cell for cleaning. A problem is encountered in mounting the working electrode. An all-glass cell with a glass support for the electrode is desirable, but of the common electrodes only platinum can be sealed in glass without minute cracks forming, and in this case only in soft glass with some difficulty. Teflon is frequently satisfactory as a mounting medium, but a tight seal to the electrode is difficult. Cast epoxy resins which receive careful attention during the curing cycle so that electroactive agents are not evolved in solution can also be used. Temperature cycles should be avoided since the differing degrees of expansion between the electrode and the mounting material invariably cause leaks. It is desirable to use double-walled glass electrolytic cells so that circulating water can be used for temperature control.

When currents are small, IR potential gradients with moderately concentrated electrolytes are no problem, and the auxiliary electrode (here called the counter electrode) may be located where reaction products from it will not contaminate the solution around the working electrode. An H cell (Fig. 8-5) can be used with a sintered glass disk separating the two compartments. When larger current densities such as those that occur during some transient techniques are encountered, IR potential gradients may be significant and it is then necessary to place the counter and working electrodes in the same compartment with due consideration of the current distribution. A third electrode, the reference electrode, should be located near the working electrode and if compensation for IR potentials is not made electronically in this part of the circuit, the Luggin probe of the reference electrode (Fig. 8-5) should be brought very close to the surface of the working electrode. The most commonly used reference electrode is the saturated calomel electrode. This is suitable in most acid and neutral solutions, but for strongly alkaline solutions the Hg/HgO reference may be used. The hydrogen electrode may be used through the pH range 1–14. These reference electrodes and others are described in detail elsewhere.[51] Thought should be given to minimizing liquid-junction potentials. For a uni-univalent electrolyte, this is accomplished using KCl or KNO_3

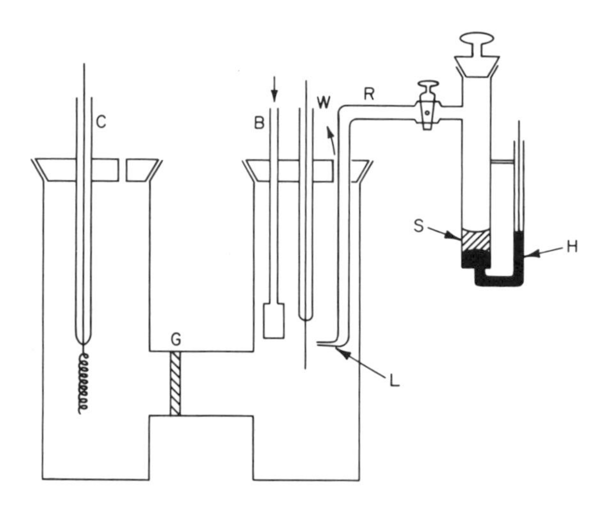

Fig. 8-5. Diagram of an electrolytic cell, where *C* is the counter electrode; *G*, the glass frit; *B*, the gas bubbler; *W*, the working electrode; *L*, the Luggin capillary; *R*, the reference electrode; *S*, the calomel mix, and *H*, the mercury contact.

salt bridges. Care should be taken to keep contamination of the electrolyte by the reference electrode to a minimum. The simplest method is to use a stopcock as shown in Fig. 8-5. For very careful work, however, this may not be sufficient, in which case a salt bridge to an intermediate reservoir may be used. To avoid the use of lubricants, Teflon stopcocks in glass are satisfactory. A high-impedance voltage follower in the reference circuit allows use of the reference with the stopcock closed.

A relatively simple control circuit primarily for potential-sweep experiments but with provision for potential-step and current-step transients is shown in Fig. 8-6. A detailed discussion of operational amplifiers in electrochemistry is available in the literature.[52] For potential sweeps, switch *A* is closed and switch *B* is in position 1. Function generator 1 which has provision for constant potential is used to set the starting potential. Function generator 2 is used to provide the sweep function. The potential step is provided by switching from function generator 1 to function generator 2, with switch *A* closed and switch *B* in position 1. The current step is provided by closing switch C or D, with switch *A* open and switch *B* in position 2.

5. SUMMARY

An electrochemical approach to characterizing surfaces appears to be useful only where information about the solid–electrolyte interface is desired. The few attempts to characterize the solid–gas interface by

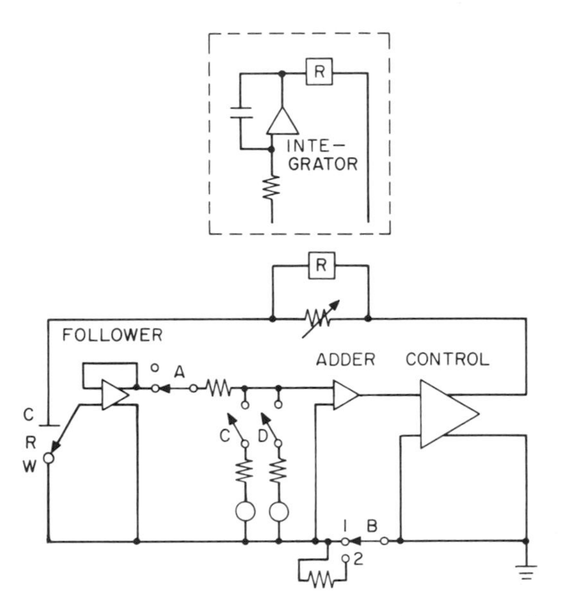

Fig. 8-6. A simple circuit for potential-sweep and current- or potential-transient experiments, where C is the counter; R, the reference; W, the working electrodes, and R, the recorder or oscilloscope.

subsequently immersing the solid into an electrolyte and using an electrochemical technique have suffered from the reasonable supposition that the surface is altered on immersion.

Of the techniques that have been employed for surface characterization, impedance, and potential-sweep measurements appear to be the most useful. Impedance measurements have been very useful in semiconductor electrochemistry to determine the presence of surface states and to measure the space-charge layer. The potential-sweep technique is finding an application in determining the surface composition of metal and metal-alloy electrodes. There does not appear to be any instance, however, where the conclusions about the surface composition deduced by the potential-sweep technique have been confirmed by another independent technique while the electrode is still immersed in the electrolyte. This is a very difficult thing to do. Finally, mention should be made of the application of electrochemical techniques to the measurement of adsorbed species at the surface. These adsorbed species frequently play an important role in catalytic reactions and the techniques mentioned here, particularly impedance measurements, are finding an application in the studies of surface catalysis.

6. GLOSSARY OF TERMS

A	Area
C	Capacitance
C_{SC}	Capacitance of space-charge layer in a semiconductor
C_S	Capacitance of the space-charge layer and surface states in a semiconductor
C_{SS}	Capacitance of the surface states
C_{Gouy}	Capacitance of the Gouy layer
C_o	Bulk concentration of an ion
C_0	Concentration of a reducible ion
C_R	Concentration of an oxidizable ion
E	Electrode potential
E_e	Equilibrium open-circuit potential
e	Charge on an electron
F	Faraday
I	Current density
i	Current
i_o	Exchange current
k	Boltzmann constant
k_f	Heterogeneous rate constant in the forward direction
k_b	Heterogeneous rate constant in the backward direction
n	Number of electrons
n_i	Intrinsic electron concentration in a semiconductor
R	Gas constant
T	Temperature
t	Time
z	Charge on an ion
ε	Permittivity
$\mathscr{L}$	Debye length $\left[= \left(\frac{\varepsilon kT}{8\pi n_i e^2} \right)^{1/2} \right]$
φ_2	Potential at the outer Helmholtz plane
φ_E	Potential in the bulk electrolyte
$\varphi_b - \varphi_s$	Potential difference between the bulk and surface of a semiconductor
ρ	Charge density
λ	ratio of intrinsic electron concentration to the bulk electron concentration in a semiconductor.

7. ACKNOWLEDGMENTS

The author would like to acknowledge the assistance of A. R. Pierce in the literature search for this chapter; and P. J. Boddy, J. D. E. McIntyre, and Y. Okinaka for many useful discussions.

8. REFERENCES

1. S. Srinivasan and P. N. Sawyer, in *Clean Surfaces: Their Preparation and Characterization for Interfacial Studies* (George Goldfinger, ed.) pp. 195–218, Marcel Dekker, New York (1970).
2. P. Delahay, *New Instrumental Methods in Electrochemistry*, Interscience, New York (1954).
3. J. O'M. Bockris and A. K. N. Reddy, *Modern Electrochemistry*, Vol. 2, Plenum Press, New York (1970).
4. R. Parsons, in *Modern Aspects in Electrochemistry*, Vol. 1 (J. O'M. Bockris, ed.) pp. 103–180, Academic Press, New York (1954).
5. H. U. Harten, The semiconductor–electrolyte interface, *Electrochim. Acta* **13**, 1255–1261 (1968).
6. M. Green, in *Modern Aspects of Electrochemistry*, Vol. 2 (J. O'M. Bockris, ed.) pp. 343–407, Academic Press, New York (1959).
7. V. A. Myamlin and Y. V. Pleskov, *Electrochemistry of Semiconductors*, Plenum Press, New York (1967).
8. S. Bruckenstein and B. Miller, Circuit for transient—free current—potential control conversion, *J. Electrochem. Soc.* **117**, 1040–1044 (1970).
9. D. R. Flinn, M. Rosen, and S. Schuldiner, Double layer capacitance on platinum, *Collection Czechoslov. Chem. Commun.* **36**, 454–463 (1971).
10. A. A. Pilla, R. B. Roe, and C. C. Herrmann, High speed non-faradaic resistance compensation in potentiostatic techniques, *J. Electrochem. Soc.* **116**, 1105–1112 (1969).
11. N. Tshernikovski and E. Gileadi, New techniques for double layer capacitance measurements at solid metal electrodes, *Electrochim. Acta* **16**, 579–584 (1971).
12. D. E. Aspnes, A capacitive divider technique for fast interface capacitance measurement, *J. Electrochem. Soc.* **116**, 585–591 (1969).
13. R. Woods, The surface composition of platinum–gold alloys, *Electrochim. Acta* **16**, 655–659 (1971).
14. P. Delahay, Coulostatic method for the kinetic study of fast electrode processes, *J. Phys. Chem.* **66**, 2204–2207 (1962).
15. R. A. Fredlein, A. Damjanovic, and J. O'M. Bockris, Differential surface tension measurements at thin solid metal electrodes, *Surface Sci.* **25**, 261–264 (1971).
16. T. R. Beck, "Electrocapillary curves" of solid metals measured by extensometer instrumentation, *J. Phys. Chem.* **73**, 466–468 (1969).
17. D. E. Icenhower, H. B. Urback, and J. H. Harrison, Use of the potential step method to measure surface oxides, *J. Electrochem. Soc.* **117**, 1500–1506 (1970).
18. J. J. McMullen and N. Hackerman, Capacities of solid metal-solution interfaces, *J. Electrochem. Soc.* **106**, 341–346 (1959).
19. R. G. Barradas and E. M. L. Valeriote, On the electrical analog circuit for the study of metal–solution interfaces by the square wave technique for capacitance measurements, *J. Electrochem. Soc.* **117**, 650–651 (1970).
20. K. Sasaki and Y. Nishigakiuchi, Cathodic reduction of adsorbed oxygen on platinum, *Electrochim. Acta* **16**, 1099 1106 (1971).
21. D. A. J. Rand and R. Woods, The nature of adsorbed oxygen on rhodium, palladium, and gold electrodes, *J. Electroanal. Chem.* **31**, 29–38 (1971).
22. S. Gilman, Adsorption of cadmium ions from KOH solution at a platinum electrode, *J. Electrochem. Soc.* **118**, 1953–1957 (1971).
23. J. F. Connolly, R. J. Flannery, and G. Aronowitz, Electrochemical measurement of the available surface area of carbon supported platinum, *J. Electrochem. Soc.* **113**, 577–580 (1966).

24. M. W. Breiter, Hydrogen adsorption on heterogeneous platinum–gold alloys in sulphuric acid solution, *Trans. Faraday Soc.* **61**, 749–754 (1965).
25. M. W. Breiter, Electrochemical characterization of the surface composition of heterogeneous platinum–gold alloys, *J. Phys. Chem.* **69**, 901–904 (1965).
26. M. W. Breiter and F. E. Luborsky, Identification of gold islands on copper plated wire surfaces by cyclic voltammetry, *J. Electrochem. Soc.* **118**, 867–869 (1971).
27. R. Woods, Electrolytically co-deposited platinum–gold electrodes and their electrocatalytic activity for acetate ion oxidation, *Electrochim. Acta* **14**, 533–540 (1969).
28. G. Todd and G. A. Wild, Novel technique for surface analysis of solid metallic specimens using selected anodic current–voltage characteristics, *Anal. Chem.* **43**, 476–480 (1971).
29. R. M. Latanision and H. Opperhauser, On the Passivation of Nickel Monocrystal Surfaces, RIAS Tech Report 71-16c (August 1971), paper presented at N.A.C.E. Corrosion Research Conference, Chicago (March 23, 1971).
30. S. Schuldiner, M. Rosen, and D. R. Flinn, Comparative activity of (111), (100), (110), and polycrystalline platinum electrodes in H_2 saturated 1 M H_2SO_4 under potentiostatic control, *J. Electrochem. Soc.* **117**, 1251–1259 (1970).
31. L. Karasyk and H. B. Linford, Electrode kinetic parameters for copper deposition on clean and soiled copper cathodes, *J. Electrochem. Soc.* **110**, 895–904 (1963).
32. H. Schneider, A. J. Sukava, and W. J. Newby, Cathode overpotential and surface active additives in the electrodeposition of copper, *J. Electrochem. Soc.* **112**, 568–570 (1965).
33. A. J. Sukava, H. Schneider, D. J. McKenney, and A. T. McGregor, Cathode overpotential and surface active additives in the electrodeposition of copper, *J. Electrochem. Soc.* **112**, 571–573 (1965).
34. M. A. Farrell and H. B. Linford, Cleaning and preparation of metals prior to electroplating, *Plating* **53**, 1110–1114 (1966).
35. D. O. Feder and E. S. Jacob, Electrode potential: A tool for the control of materials and processes in electron device fabrication, *ASTM STP* **No. 300**, 53–66 (1961).
36. D. G. Schimmel, Detection of inorganic contamination on surfaces by an EMF measurement, *ASTM STP* **No. 300**, 46–52 (1961).
37. H. Gobrecht and R. Blaser, On the mechanism of surface recombination at semiconductor electrodes, *Electrochim. Acta* **13**, 1285–1292 (1968).
38. H. Gerischer, in discussion of the paper, Charge transfer processes at semiconductor–electrolyte interface in connection with problems of catalysis, *Surface Sci.* **18**, 97–122 (1969).
39. P. J. Boddy, Impedance measurements at the semiconductor–electrolyte interface, *Surface Sci.* **13**, 52–59 (1969).
40. P. J. Boddy, The structure of the semiconductor–electrolyte interface, *J. Electroanal. Chem.* **10**, 199–244 (1965).
41. H. Gerischer, Charge transfer processes at semiconductor–electrolyte interfaces in connection with problems in catalysis, *Surface Sci.* **18**, 97–122 (1969).
42. W. P. Gomes and F. Cardon, Surface states at the single crystal zinc oxide–electrolyte interface, *Ber. Bunsenges Phys. Chem.* **74**, 431–436 (1970).
43. S. R. Morrison, Electron capture by ions at the ZnO–solution interface, *Surface Sci.* **15**, 363–379 (1969).
44. V. A. Tyagai and G. Ya. Kolbasov, The contribution of surface states to the charge transport process across CdS, CdSe–electrolyte interface, *Surface Sci.* **28**, 423–436 (1971).

45. R. L. Meek, n^+ silicon–electrolyte interface capacitance, *Surface Sci.* **25**, 526–536 (1971).
46. R. Memming, On the interpretation of the impedance of the semiconductor–electrolyte interface, *Philips Res. Repts.* **19**, 323–332 (1964).
47. P. J. Boddy, D. Kahng, and Y. S. Chen, Oxygen evolution on potassium tantalate anodes, *Electrochim. Acta* **13**, 1311–1328 (1968).
48. P. J. Boddy, Oxygen evolution on semiconducting TiO_2, *J. Electrochem. Soc.* **115**, 199–203 (1968).
49. H. A. Laitinen, C. A. Vincent, and T. M. Bednarski, Behavior of tin oxide semiconducting electrodes under conditions of linear potential scan, *J. Electrochem. Soc.* **115**, 1024–1028 (1968).
50. S. Schuldiner, B. J. Piersma, and T. B. Warner, Potential of a platinum electrode at low partial pressures of hydrogen and oxygen, *J. Electrochem. Soc.* **113**, 573–577 (1966).
51. D. J. G. Ives and G. J. Janz, *Reference Electrodes*, Academic Press, New York (1961).
52. W. M. Schwarz and I. Shain, Generalized circuits for electroanalytical instrumentation, *Anal. Chem.* **35**, 1770–1778 (1963).

9

EMISSION SPECTROMETRY

J. L. Seeley and R. K. Skogerboe

Department of Chemistry
Colorado State University
Fort Collins, Colorado

1. DESCRIPTION OF THE TECHNIQUE

Emission spectroscopy has played a very distinctive and important role in the development of analytical chemistry. This has derived, in part, from the general flexibility of the technique; but is due largely to the fact that simultaneous multielement analyses can be readily accomplished on limited sample sizes. Excitation of atomic species to produce characteristic emission spectra forms the fundamental basis of the technique. Of primary interest in this chapter will be the use of electrical discharges for excitation.

In the simplest operational configuration, a discharge is produced between two electrodes, at least one of which is composed of, or contains, the sample to be analyzed. Electrons from the cathode are accelerated through the arc gap under a potential gradient between the electrodes, resulting in the generation of heat and the release of positive ions at the anode. In this manner some portion of the sample is vaporized by the discharge, the vaporized species are dissociated, and the resultant atoms and ions are excited either thermally or by collisions of the first and second kind. The light emitted by the excited species is dispersed into its component wavelengths with a grating or prism spectrograph and is detected with either a photographic plate or a photomultiplier readout system, i.e., a direct reader. The essential components of an emission spectroscopic analytical system are, therefore, a source for vaporization and excitation, a wavelength-dispersing unit, and an appropriate means for detecting and measuring the atomic spectral intensities. The excitation source must generally fulfill the dual role of vaporization and excitation of the sample material.

The intensity of a spectral line is determined by the Boltzmann equation:

$$I_{\lambda i} \propto N_i = N_0 g_i \exp(-E_i/kT) \tag{1}$$

where $I_{\lambda i}$ is the spectral line intensity at wavelength λ due to emission from the ith excited state of a particular atom, N_i is the number of those atoms in that excited state during the period of measurement, N_0 is the excitation plasma population of those atoms in the ground (unexcited) state, g_i is the statistical weight factor for transition of the atom to the ith state, E_i is the excitation energy required to reach that state, k is the Boltzmann constant, and T is the (effective) absolute temperature of the excitation medium. A difficulty associated with such intensity measurements originates from the fact that a number of experimental factors may cause variation in the plasma temperature with concomitant variations in the intensities. Thus, most intensity measurements rely heavily on the use of the internal-standard principle. It can be shown that the ratio of the intensities of two atomic lines will be determined by

$$\frac{I_{\lambda i}}{I_{\lambda j}} \propto \frac{N_i}{N_j} = \frac{{}_iN_0 g_i}{{}_jN_0 g_j} \exp[(E_i - E_j)/kT] \tag{2}$$

where j denotes a particular excited state of the internal standard element emitting radiation at λ_j. In essence, if $E_i = E_j$, the exponential term goes to unity and temperature can vary without effect on the intensity ratio. To apply the internal-standard method, the excitation potentials are matched as closely as possible to minimize the effects of temperature variations. The internal-standard element may be added to the samples such that its concentration (${}_jN_0$) from sample-to-sample is constant, or an element present in the samples at relatively constant concentrations may be selected. Figure 9-1 illustrates another requirement that must be imposed. The curves shown are time–vaporization curves for three different elements (A, B, and C) contained in a sample subjected to a particular set of vaporization–excitation conditions. It may be readily noted that elements A and B exhibit parallel vaporization behaviors, while that for C is quite different. Recognizing that changes in the discharge temperature will be reflected by changes in vaporization behavior, it is important that the behavior of the internal-standard element be matched to that of the analytical element. If the intensity

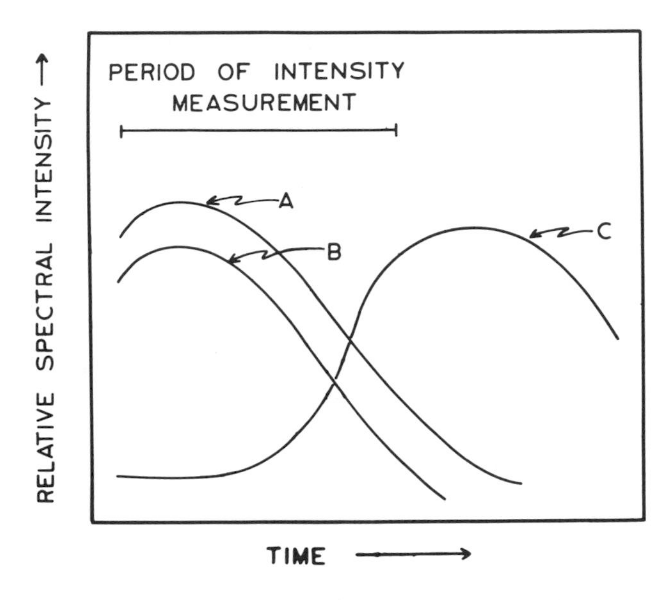

Fig. 9.1. Vaporization–excitation curves for three elements.

measurements were made over the period indicated in Fig. 9-1, *A* would serve as a good internal standard for *B* (and *vice versa*), but *C* would probably not adequately reflect changes which occur during this period and would be a poor choice as an internal standard for *A* or *B*. Other less critical considerations in selecting an internal standard are listed by Nachtrieb,[1] and Ahrens and Taylor.[2]

The foregoing discussion has been emphasized because of its pertinence to the application of emission spectroscopy to surface analysis. More extensive information about types of excitation sources, wavelength dispersion systems, and measurement of spectral intensities is readily available in a number of texts including those cited above. Additional descriptive information will be presented where appropriate in following sections.

2. THE PROBLEM OF SURFACE ANALYSIS

Spatial resolution is perhaps the principal connotation associated with the term surface analysis. While some might argue that the primary impetus of surface characterization is that of analyzing thin films, examination of the literature indicates that the characterization of segregates is also often considered as a form of surface analysis. In either case, spatial resolution is still the prevailing consideration and we shall define the analysis problem on this basis.

Surface or thin films may be divided into the following descriptive categories: alien surface contamination, intentional surface contamination, and sputtered films.[3,4] The distinction between the latter two categories derives from differences in the mode of deposition. The thickness of any of these films may range from a few angstroms to several hundred micrometers. The degree of uniformity of alien and intentional surface contaminations may be generally expected to be quite variable, while sputtered films are expected to be fairly uniform. In any case, the analysis problem may assume several distinct aspects.[4] In some cases, the establishment of the identity of the principal components of the film is the only requirement imposed on the analyst. For the more stringent case, the analyst must establish the variation in the composition of the film in the x, y, and z dimensions, or determine only the variations in the z dimension of thickness. Differentiation between the constituents of the surface film and those of the substrate is almost always required. Identification of the constituents of a segregate or an inclusion contained in a particular sample is a similar problem. Again, determination of the segregate composition and the compositional variation in three dimensions is frequently of interest. In this instance, however, the spatial resolution required of the analysis system may be much more restrictive in the x and y dimensions than would necessarily be typical of thin-films analysis.

In view of the physical dimensions of thin films and segregates, and considering the general analytical requirements outlined above, there is one overriding requirement imposed on the analytical technique utilized—high absolute sensitivity. For example, if a sample of 1 μm depth is to be analyzed and the xy sampling area is 2 mm^2, this represents a sampling volume of 2×10^{-6} cm^3. For a material with a density of 5 g/cm^3, a total of 1×10^{-5} g is sampled. At an impurity level of 1 ppm, the analytical technique must be capable of determining 1×10^{-11} g if that impurity is to be characterized. Optical emission spectrography offers a reasonable analysis capability at these levels. An examination of the compilation of detection limits by Skogerboe and Morrison[5] indicates that nearly 40 elements can be determined at levels between 0.01 and 10 ng. The fact that these can be determined simultaneously is a matter of prime interest for many surface-analysis problems.

Accepting the fact that emission spectrometry offers sensitivity adequate for many such problems, it must be admitted that there are difficulties in the use of this technique for surface analyses which originate from the spatial-resolution and the quantitation requirements. A number of specialized approaches have been developed to overcome these limitations, and these will be discussed below.

3. EMISSION SPECTROMETRIC METHODS OF SURFACE ANALYSIS

Although emission methods have been used extensively for the solution of a wide variety of analytical problems, their actual utilization for surface analysis may be described as limited. Categorically, one may consider the analytical approaches used as either direct or indirect. The former category involves the utilization of specific types of sampling-excitation systems which will be discussed below. The indirect methods involve film recovery or removal.

3.1. Indirect Methods

Several types of chemical and mechanical techniques for film removal have been discussed by Arcus.[6] The approach may be illustrated by considering the analysis of tantalum films on glass substrates as carried out by Nohe.[7,8] Since tantalum is soluble only in hydrofluoric acid or concentrated strong bases, and as both of these etch glass, there is a strong possibility that impurities detected would originate from the glass. To avoid chemical removal, the film was thermally oxidized at 600°C to flakes which were removed by scraping with a platinum spatula. The flakes were intimately mixed with graphite powder and placed in graphite spectroscopic electrodes for analysis by dc arc excitation.[7,8] Quantitation of the impurities in the tantalum film was accomplished using multielement standards prepared in a graphite matrix with appropriate amounts of pure tantalum oxide added. It was possible to determine 22 elements at concentrations of 10 ppm and below, referred to the tantalum weight.

Runge and Bryan[9] used a tedious, but effective, drilling technique to remove microconstituents (segregates) from metals for identification. Microdrills as small as 0.00025 inches in diameter were used under a microscope to remove inclusions weighing as little as 2 μg. The drillings were transferred to pointed spectroscopic electrodes surfaced with Duco cement to aid the transfer. The constituents were qualitatively identified using arc excitation with a fast spectrograph and high-speed photographic plates.

Although the authors are certain that chemical removal methods can be used with optical emission spectroscopy as a means of thin-films analysis, there do not appear to be many publications discussing such applications. Perhaps the requirement of chemical selectivity serves as an explanation. If the constituents of the surface film are to be distinguished from those of the substrate, it would be necessary to selectively dissolve away either the thin film or the substrate. Given the ability to selectively remove the film, emission analysis could be carried out by a number of solution or solution-residue techniques for which it would be relatively easy to prepare standards.[5] Similarly, if the substrate were selectively removed, the film could be as easily analyzed as the solid.

3.2. Direct Analyses

Because direct analyses are less subject to the contamination problems characteristic of indirect methods and because they offer a better opportunity for obtaining spatially resolved concentration information, they are much more attractive to the surface analyst. Direct methods of analysis are summarized by the following examples.

3.2.1. Spark Methods

Medium- and high-voltage spark vaporization–excitation methods may actually be broadly classified as surface analysis methods under a variety of circumstances. Even a rigorous high-voltage spark to a metallic surface will vaporize a milligram of material or less when sustained for a fairly normal exposure period of one minute. By adjustment of the spark parameters, the area sampled in the xy dimension may be varied from approximately 1 mm^2 to perhaps 1 cm^2. Typically the depth of sampling will be less than 1 mm and, again, proper selection of the spark parameters will allow a fair degree of latitude. Strasheim and Blum,[10] for example, have shown that a medium voltage spark to an aluminum surface produces sampling at depths ranging from 40 to 120 μ, depending on the inductance in the discharge circuit. By systematically varying the inductance, the radius of the spark channel can also be changed from approximately 0.1 to 2.0 mm.[10] Moreover, Strasheim and Blum[10] present photographs which dramatically illustrate the effect of changing the resistance and capacitance of the circuit on the crater dimensions. Thus, the spark method offers the opportunity to couple a fair degree of spatial resolution with the multielement sensitivity of emission spectroscopy. Some of the capabilities and limitations may be deduced from the following examples.

A spectrographic microanalysis method utilizing a point-to-plane spark was developed by Hurwitz[11] for the study of diffusion in the solid state. The spark device, which was arranged to permit traversing the sample surface at selected rates below the counter electrode, is schematically illustrated in Fig. 9-2. In order to relate the concentrations of species vaporized and excited along the path by the spark, the photographic plate was moved to obtain a continuous exposure which was referenced to the speed of sample traverse. By adjusting the rate of traverse and the rate with which the photoplate was moved, control of

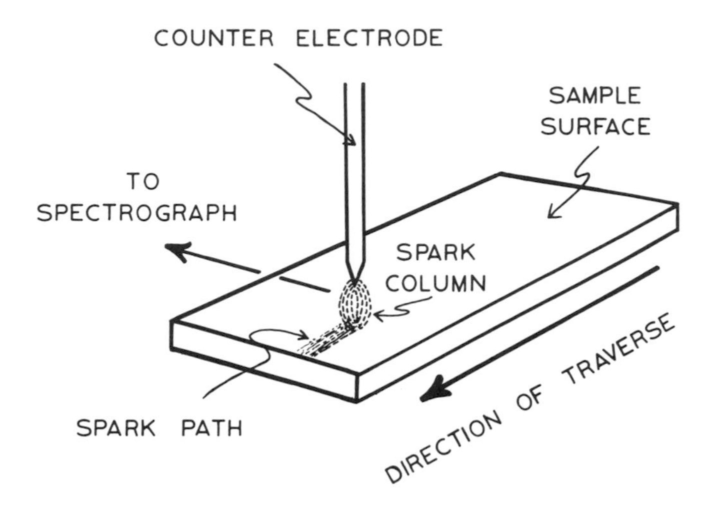

Fig. 9-2. Illustration of a system for two-dimensional surface analysis.[11]

the spatial definition was obtained within reasonable limits.[11,12] Control of the width of the spark path and the depth of surface penetration was possible through the adjustment of the spark parameters and the rate of traverse. Others have shown that the spark could be confined to a small area by passing it through a 0.02-mm capillary opening in a quartz tube.[13,14]

To illustrate the capabilities of this approach, Hurwitz[11] clamped two steel blocks to form a distinct surface boundary and recorded traverse data under a variety of conditions. Figure 9-3 shows the results obtained by scanning across this interface, progressing from a block with a nickel concentration of 4.1% to one with a concentration of 1.1%, where the length of the spectrograph slit was the only parameter varied.[11] These results distinctly imply that concentration changes over a lineal distance on the order of 0.1–0.2 mm can be resolved. This system has also been used to identify the constituents of segregates in steel samples with good success.[11] Similarly, Bryan and Nahstoll[13] and Murray *et al.*[14] have used a spark method constricted to small areas for the microanalysis of segregates.

The presence of hydrogen-containing compounds as surface contaminants is the main source of porosity or weld defects in the welding of aluminum.[15] In order to determine the level of contamination that contributed to the degree of weld-bead porosity, Loseke *et al.*[15] used a controlled atmosphere chamber to occlude air and a spark technique to determine surface hydrogen. Calibration of the technique was accomplished by exposure of the surface to absolute

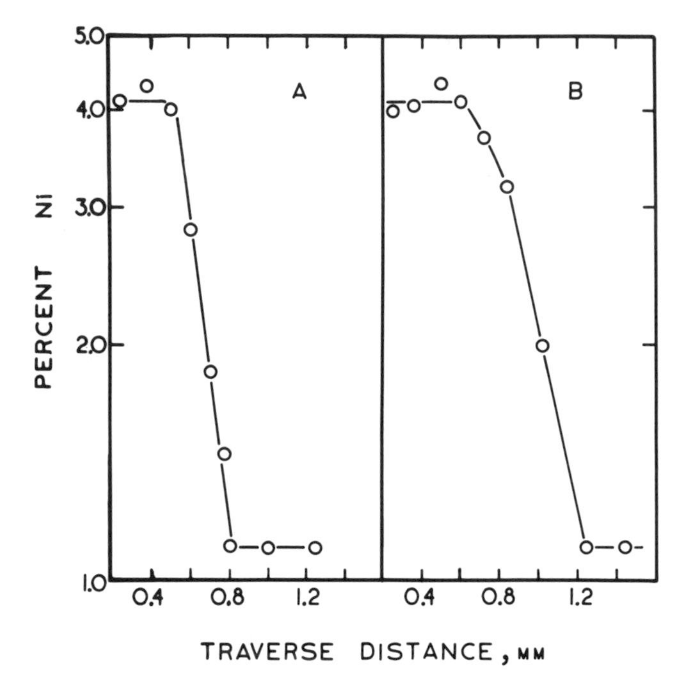

Fig. 9-3. Concentrations for traverses at different spectrographic slit length: (a) slit length = 0.6 mm, (b) slit length = 1.1 mm. (Reproduced from Hurwitz[11] by courtesy of the author and the Optical Society of America.)

humidities. Hydrogen concentrations as low as 10 ppm could be determined and the method was judged to provide a reliable means of monitoring surface contamination by hydrogen-containing compounds.[15]

While these examples indicate the potential of emission spectroscopy for direct surface analysis, it must be admitted that quantitation of the results in terms of concentrations and in terms of spatial definition represent fairly complex problems. Among the various techniques used for surface analysis, these problems are not unique and, in fact, relative concentration–spatial profiles are frequently sufficient for the solution of the problem at hand. In such cases, the spark techniques can play an important role in surface analysis.

3.2.2. Laser Vaporization Techniques

The use of laser vaporization and excitation techniques in emission spectroscopy has been extensively reviewed by Baldwin.[16] The vaporization and excitation characteristics of pulsed lasers have been studied by Rasberry, Scribner, and Margoshes[17] and by Scott and Strasheim.[18]

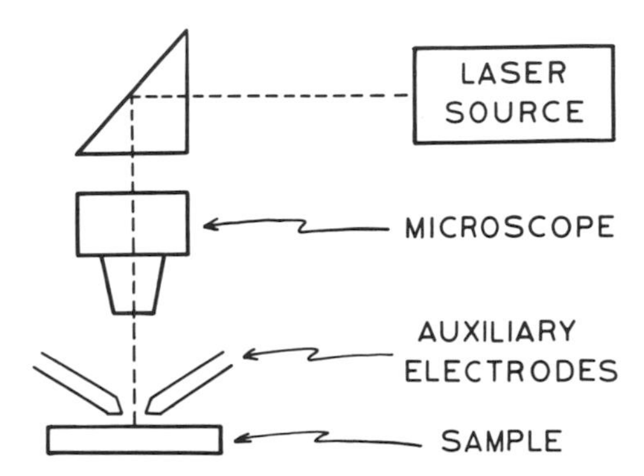

Fig. 9.4. Schematic illustration of the laser microprobe.

The system most widely used for analysis is illustrated schematically in Fig. 9-4. By focusing the laser energy incident on the sample surface through a microscope, control of the volume sampled can be achieved, and analysis of localized areas down to approximately 5–10 μm in diameter is possible.[17–24] In this laser microprobe system, a pulsed laser with a pulse duration in the range of tens of nanoseconds is fired through the microscope and thence to the sample surface. The Q-switched lasers used are generally capable of emitting peak powers in the 10^7 to 10^8-W range, so they are capable of vaporizing the vast majority of materials. Some characteristic radiation is emitted by the vaporized material, but its intensity is generally low enough to preclude detection of all but the major constituents of the material. Thus, pointed electrodes (usually graphite) are placed in opposition across the path of the vapor plume. These electrodes are charged to a high potential by use of a capacitor system such that when the vapor plume is formed, the electrode gap becomes conductive and the energy dissipated in a spark pulse across the gap serves to excite atoms in the plume. By using this secondary means for excitation, the sensitivity of this spectrographic technique is enhanced considerably.

An example of the type of surface crater produced by the laser probe is shown in Fig. 9-5. The laser pulse tends to leave a ridge of

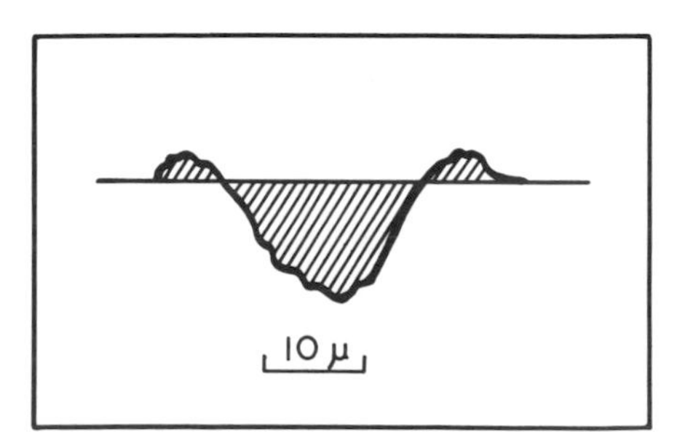

Fig. 9-5. Example of a laser crater.

TABLE 9-1. Crater Depth Produced by a 10^9-W/cm^2 Laser Pulse of 44-nsec Duration[22]

Material	Crater depth, μ
Stainless steel	1.1
Brass	2.5
Aluminum	3.6
Copper	2.2
Nickel	1.2

ablated material around the crater, but it is apparent that localized sampling is possible. Successive pulses to the same point may be used to sample at greater depths, but some caution should be exercised because selective vaporization may occur.[24] In examining the effects due to absorption of laser radiation, Ready[22] observed the penetration depths given in Table 9-1. The crater depth obtained may be controlled by varying the power of the laser pulse, the magnification of the microscope, or by defocusing the microscope to some extent, as illustrated in Table 9-2.[17] Recognizing these experimental possibilities, the data in Tables 9-1 and 9-2 verify that surface, thin-film, and surface-segregate analyses are possible. Because a relatively small amount of material is vaporized by the laser (in the microgram range), it is generally possible to determine constituents at concentration levels in the range of 100–1000 ppm. Some elements can be determined at lower levels in specific matrices using a photomultiplier readout system.[20] Although the laser probe is destructive, it is capable of providing spatially defined multielement analyses at levels of practical significance. On this basis it can be regarded as a competitor of the electron and ion microprobe systems (see Chapters 18 and 21). One advantage of practical significance is derived from the fact that samples need not be conductive. Example applications of the

TABLE 9-2. Crater Dimensions as a Function of Laser Energy[17]

Energy, J	Depth, μ
0.16	15
0.32	16
0.48	27
0.64	36
0.80	48
0.96	59

technique are contained in references cited above[16–24] and others contained therein.

3.2.3. Ion Bombardment Techniques

The use of ion bombardment systems for analysis will be discussed in Chapter 21. The interaction of an energetic ion with a solid target is always followed by changes in the crystal lattice of the solid, secondary emission of different types of particles, and emission of light. While most analytical interest has been placed on secondary particle emission, the emission of light has stirred less interest but does show analytical potential.[25–30] A system for such measurements offering fairly general capabilities was described by Sawatsky and Kay.[27] Rare-gas ions from a duoplasmatron are focused onto the target inside a Faraday cage. Some sputtered particles pass through a slit in the target region and are excited by a low-energy electron beam; the light emitted is analyzed with a grating spectrometer. According to the authors,[27] the system can be used to monitor sputtering yield *in situ* of individual species ejected from the target or it may be used to characterize the particles sputtered from the surface by ion bombardment. Very good quantitative results were obtained for Cu[27] and Al, Ag, and Mo.[28] Considering the surface-analysis characteristics associated with ion bombardment techniques in general, one may predict that emission analysis methods based on this approach will emerge in the near future.

4. SUMMARY

It is difficult to conclude exactly what the future role of emission spectroscopy will be in the area of surface characterization. Although we have contended above (at least by implication) that emission methods can serve as powerful means for surface analysis, we must admit that there is a paucity of good examples in the literature to demonstrate this. Emission spectrography is a multielement analysis technique capable of high absolute sensitivity for a diverse range of elements. A number of vaporization–excitation media commonly used are actually surface sampling systems. Quantitation of these systems in either concentration or spatial terms is obviously subject to some difficulty, but no more so than many of the other techniques discussed in this text. When these attractive features of emission spectroscopy are considered together with others of less importance, it must still be maintained that the technique is a potentially significant tool for surface characterization that can compete very favorably with other techniques. The combination of some of the more recent developments in instrumentation with some imaginative research will surely prove this.

5. REFERENCES

1. N. H. Nachtrieb, *Spectrochemical Analysis*, New York (1950).
2. L. H. Ahrens and S. R. Taylor, *Spectrochemical Analysis*, Addison-Wesley, Reading, Mass. (1961).
3. D. L. Malm, in *Physical Measurement and Analysis of Thin Films* (E. M. Murt and W. G. Guldner, eds.) Plenum Press, New York (1969).
4. R. K. Skogerboe, in *Inorganic Mass Spectrometry* (A. J. Ahearne, ed.) Academic Press, New York (1972).
5. R. K. Skogerboe and G. H. Morrison, in *Treatise on Analytical Chemistry* (I. M. Kolthoff and P. J. Elving, eds.) Part 1, Vol. 9, John Wiley & Sons, New York (1971).
6. A. A. Arcus, Emission spectrographic analysis of thin films and residues, paper presented at the International Conference on Spectroscopy, College Park, Maryland (1962).
7. J. D. Nohe, *Appl. Spectry.* **21**, 364 (1967).
8. J. D. Nohe, in *Physical Measurement and Analysis of Thin Films* (E. M. Murt and W. G. Guldner, eds.) Plenum Press, New York (1969).
9. E. F. Runge and F. R. Bryan, *Appl. Spectry.* **13**, 74 (1959).
10. A. Strasheim and F. Blum, *Spectrochim. Acta* **26B**, 685 (1971).
11. J. K. Hurwitz, *J. Opt. Soc. Am.* **42**, 484 (1952).
12. J. Convey and J. H. Oldfield, *J. Iron Steel Inst.* (*London*) **152 No. II**, 473P (1945).
13. F. R. Bryan and G. A. Nahstoll, *J. Opt. Soc. Am.* **37**, 311 (1947).
14. W. M. Murray, Jr., B. Gettys, and S. E. Q. Ashley, *J. Opt. Soc. Am.* **31**, 433 (1941).
15. W. A. Loseke, E. L. Grove, G. Morks, and L. A. White, *Appl. Spectry.* **24**, 206 (1970).
16. J. M. Baldwin, Bibliography of laser publications of interest to emission spectroscopists, U.S.A.E.C. Reports IN-1219, IN-1262 (1968).
17. S. D. Rasberry, B. F. Scribner, and M. Margoshes, *Appl. Opt.* **6**, 81 (1967).
18. R. H. Scott and A. Strasheim, *Spectrochim. Acta* **25B**, 311 (1970).
19. W. H. Blackburn, Y. J. A. Pelletier, and W. H. Dennen, *Appl. Spectry.* **22**, 278 (1968).
20. E. S. Beatrice, I. Harding-Barlow, and D. Glick, *Appl. Spectry.* **23**, 257 (1969).
21. M. S. W. Webb, *Anal. Chim. Acta* **43**, 351 (1968).
22. J. F. Ready, *J. Appl. Phys.* **36**, 462 (1965).
23. K. Vogel and P. Backlund, *J. Appl. Phys.* **36**, 3697 (1965).
24. J. M. Baldwin, *Appl. Spectry.* **24**, 429 (1970).
25. W. Steinmann, *Phys. Rev. Letters* **5**, 470 (1960).
26. R. A. Ferrell and E. A. Stern, *Am. J. Phys.* **30**, 810 (1962).
27. E. Sawatzky and E. Kay, *Rev. Sci. Instrum.* **37**, 1324 (1966).
28. I. Terzic and B. Perovic, *Surface Sci.* **21**, 86 (1970).
29. R. V. Stuart and G. K. Wehner, *J. Appl. Phys.* **35**, 1819 (1964).
30. W. C. Krey, *J. Appl. Phys.* **35**, 3575 (1964).

10

INTERNAL REFLECTION SPECTROSCOPY

N. J. Harrick

Harrick Scientific Corporation
Ossining, New York

and

K. H. Beckmann

Philips Forschungslaboratorium
Hamburg, Germany

1. INTRODUCTION

Reflection of light is a surface phenomenon—it is strongly dependent on the nature of the surface and can therefore be used to study surfaces. If the surface is flat and smooth, the nature of the reflection is called specular, i.e., mirrorlike, and obeys the simple law that the angle of incidence equals the angle of reflection.

When the light approaches an interface from a medium which is optically less dense to one which is optically more dense, the reflection is called external reflection, and its dependence on polarization and angle of incidence is shown by the solid lines of Fig. 10-1 [for the interface water ($n = 1.33$) → germanium ($n = 4$)]. The reflectivity may be modified by the presence of a thin film on the surface, especially for parallel polarization near Brewster's angle. An example of this is shown in Fig. 10-2 where 500 Å of SiO_2 film on the Al surface is undetectable near normal incidence but exhibits about 30% reflection loss at large angles of

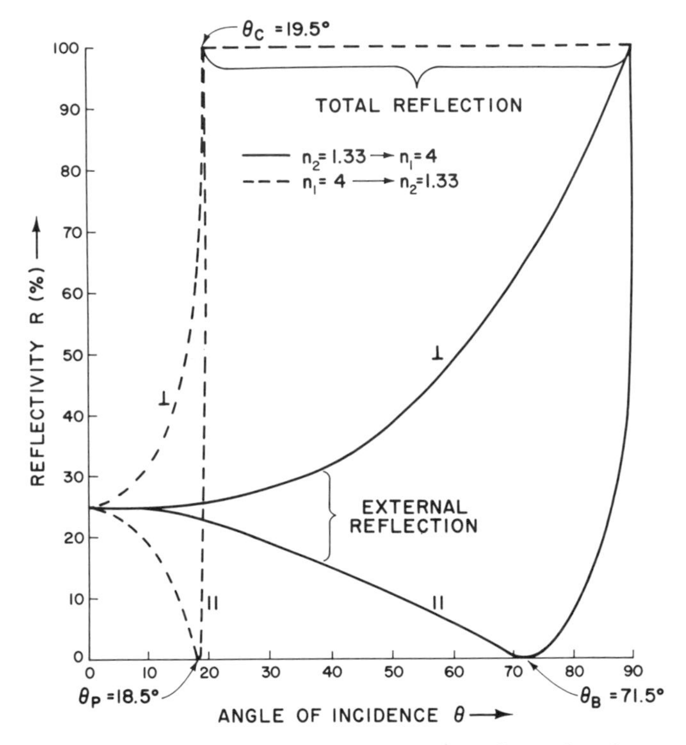

Fig. 10-1. Reflectivity *vs.* angle of incidence for an interface between media with indices, $n_1 = 4$ and $n_2 = 1.33$, for light polarized perpendicular, $R_\perp$, and parallel, $R_\parallel$, to the plane of incidence for external reflection (solid lines) and internal reflection (dashed lines). Angles θ_c, θ_B, and θ_p are the critical, Brewster's, and principal angles, respectively.

incidence. The reason the film is undetectable at normal incidence is that the incoming and reflected light waves interact to set up a standing wave with a node (zero electric field) at the reflecting surface for external reflection.

External reflection has been highly neglected as a method of studying surfaces but is now being more widely used. It has been used to detect the presence of monolayer films adsorbed on the surface. For a clean surface, reflection measurements can also yield information about the change in optical constants of the material resulting from the change in atomic bonding as one moves from the surface to within the bulk. Details of some of these studies can be found in a number of published papers.(1)

When the light propagates toward the interface from a medium which is optically more dense to one which is less dense, the reflection is called internal reflection and behaves much the same as external reflection

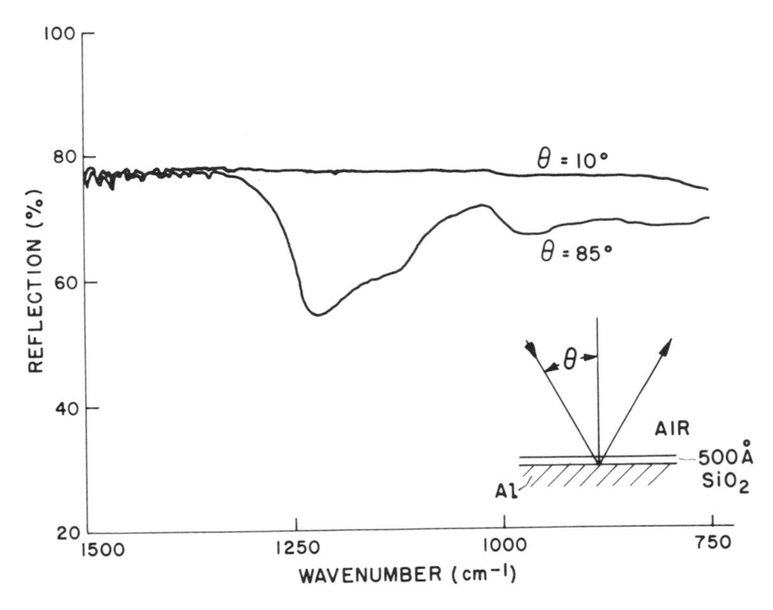

Fig. 10-2. Spectra showing the power of specular external reflection near Brewster's angle for the study of thin films on metal surfaces. The SiO_2 film which is undetectable at 10° gives a signal at 85° of 30%.

from normal incidence until the critical angle is reached, as shown by the dashed curves of Fig. 10-1. Beyond the critical angle, there is total reflection and the interface acts as a perfect mirror. There are many common examples of total internal reflection: The hypotenuse of a prism or the side of a glass of water suddenly appears silvered if viewed at certain angles. Perhaps the most common example of total reflection that occurs in nature is the mirage (Fig. 10-3). Here the layer of warm air adjacent to a hot road or sand surface is optically less dense (has a lower refractive index) than the cooler air above it and for large angles

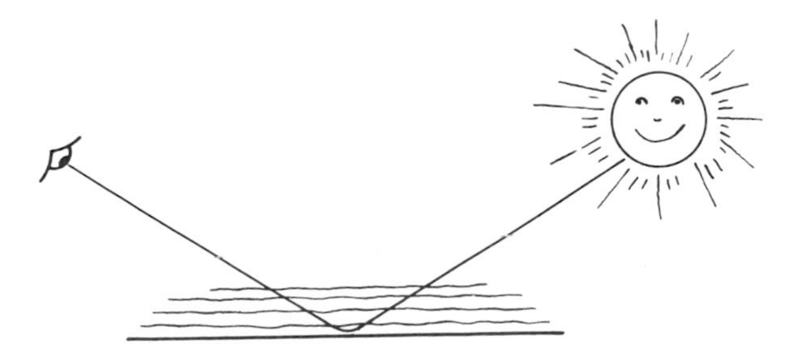

Fig. 10-3. A mirage, an example of total internal reflection found in nature. Light at grazing incidence is totally reflected by the less-dense warm air near the road surface.

of incidence the condition for total reflection is satisfied. Thus, when looking at the road, one sees the sky which gives the impression of the presence of water in the distance. Some aspects of total internal reflection and its application to studies of surfaces and thin films will be discussed in this chapter; details, however, can be found in other works[(2)] devoted to this subject.

In general, surfaces are not as well behaved as shown by the curves in Fig. 10-1 because they are not perfectly smooth, and therefore in many practical cases we have what is called diffuse reflectance. These cases include reflection of light from terrain, fabrics, papers, etc. There is no rigorous solution to a general diffuse-reflectance case since the reflection is a combination of external specular reflection, internal reflection, transmission, and scattering, and is therefore dependent on particle size, shape, refractive index, absorption coefficient, polarization, etc. A further difficulty in making measurements is that a white scatterer for the infrared, the fingerprint region for optical spectroscopy, has not been available. There are materials, however, such as KRS-5, which are reasonably inert and are nonabsorbing over a wide spectral range and thus might be used for this purpose. In spite of these difficulties, empirical methods have been developed to treat certain cases. Also, as pointed out, this method is important because of the many cases in nature where we have diffuse reflectance and, furthermore, some samples can only be obtained in granular form. Details of diffuse reflectance can be found in other works[(3)] that deal with this subject.

2. INTERNAL REFLECTION SPECTROSCOPY (IRS)

Internal reflection spectroscopy is a method of recording spectra by introducing light into an optically transparent medium at angles above the critical angle and measuring the intensity of the emerging radiation after it has suffered one or often hundreds of internal reflections. Although similar methods had been employed on occasions many years ago, e.g., by Taylor *et al.* in *Studies in Refractive Index* (see N. J. Harrick,[(2)] p. 10), the modern beginning of its current wide use was its application to demonstrate surface-state relaxation phenomena associated with the effect of metal-to-semiconductor contact potential on semiconductor surface-barrier height. Some results of this experiment (see N. J. Harrick[(4)]) are shown in Fig. 10-4. [Similar relaxation phenomena have been reported by Rzhanov and Sinyukov.[(8)]] In this case the results are obtained from a study of the absorption of infrared radiation by electrons and holes in the semiconductor space-charge region. The technique of internal reflection was subsequently developed to study adsorbed molecules, or thin films, on the surface and materials of lower

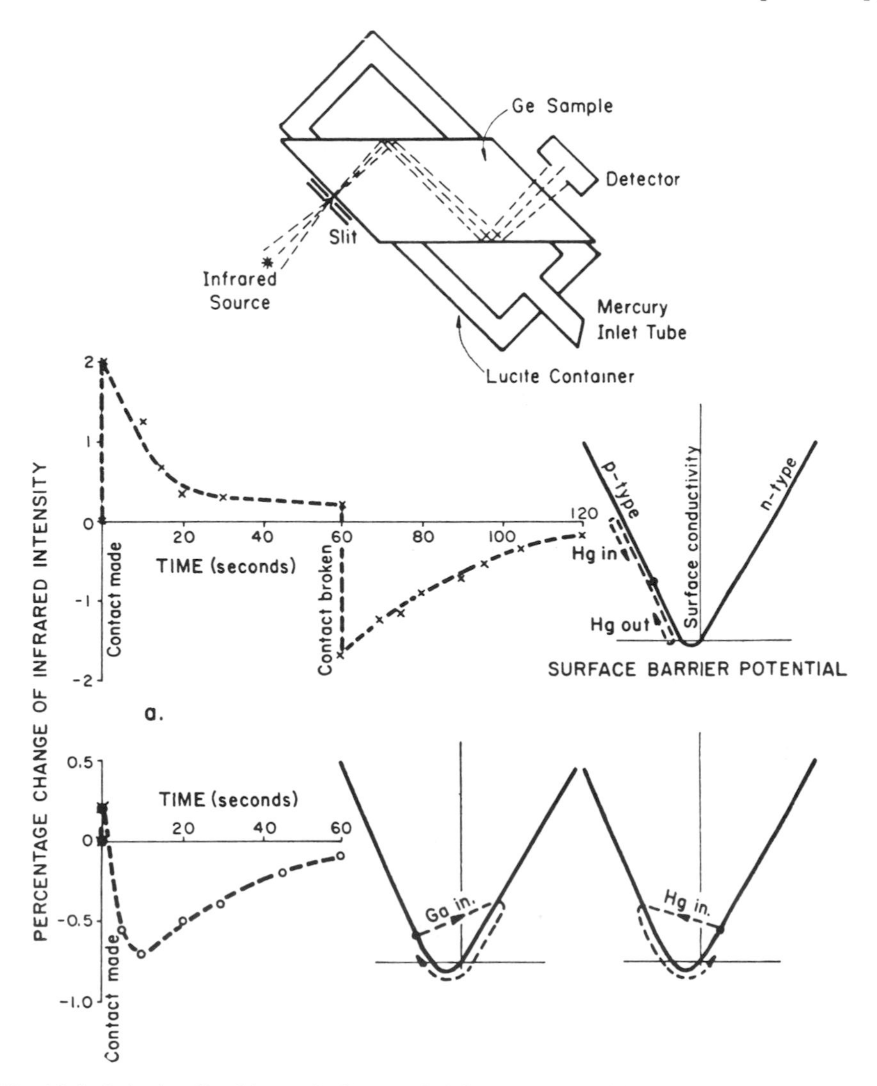

Fig. 10-4. Bringing liquid metals (Hg or Ga) into contact with Ge changes the surface-barrier height, and hence the free-carrier density in the semiconductor space-charge region, as observed from study of change in absorption of infrared by electrons and holes. Relaxation phenomena are associated with transition of electrons and holes from space-charge region to surface states and *vice versa*.

refractive index outside of the surface[5] as well as the surface itself (surface states).[6] This technique is thus a very useful one since not only can it be used for the study of phenomena inside the surface but also the surface itself and species outside of the surface.

It is at first puzzling that species outside of the surface can be studied by means of a light beam which is reflected from the inside of the surface.

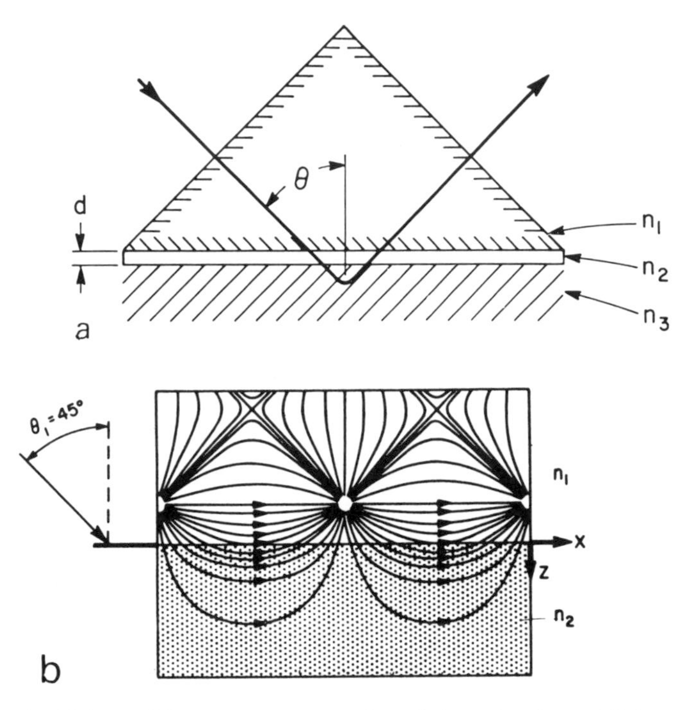

Fig. 10-5. (a) Schematic representation of the path of ray of light—total internal reflection. The ray penetrates a fraction of a wavelength (d_p) beyond the reflecting interface. Medium 1 is the optically transparent IRE, medium 2 is the sample material, the surrounding medium 3 is usually air. (b) Two-dimensional wave pattern near surface for total reflection. The density of the lines indicates the magnitude of the Poynting vector. The whole pattern is moving to the right with the velocity determined by medium 1.

The reason for this is now widely known—it is the fascinating phenomenon discovered by Newton that occurs for total internal reflection, namely, the electromagnetic field associated with the reflected light beam penetrates a certain distance beyond the reflecting interface, and this penetrating field can interact with a medium brought to within a distance from the surface comparable to this penetration depth (Fig. 10-5). Advantage has been taken of this possible interaction in a wide variety of applications including inkless fingerprinting, coupling light into thin films, modulation of light, deflection of light, construction of beam splitters, optical filters, cold mirrors, and recording of spectra of media on or outside the surface.

A partial list of applications of internal reflection techniques to the study of surfaces follows: (some of which will be discussed in more detail in the latter part of this chapter).

1. Spectra of holes and electrons in semiconductor space-charge regions: These measurements give, in addition to monitoring surface potential, information on the mobility of carriers in the space-charge region.
2. Relaxation effects associated with surface states: Such effects can be readily investigated by measuring the absorption of the light by the free carriers in the space-charge region.[4]
3. Optical spectrum of surface states: It was originally predicted, and later confirmed by observation,[6] that absorption edges rather than absorption bands should be measured. More recently, absorption bands for surface states on germanium have been observed.[7–9]
4. Quantization of energy levels in the space-charge region.
5. Properties of thin oxide films: The oxide, lattice, and hydroxyl bands as well as those of other species incorporated in the oxide can be detected.
6. Adsorbed molecules on the semiconductor surface: In addition to the detection of the adsorbed species it is possible to obtain information on the nature of the surface bonding in cases such as naturally adsorbed species, complexes formed on the surface due to etching, ion implantation, and photoenhanced adsorption and desorption.
7. Thin films on a surface: In studies of thin films placed on the surface it is possible to identify configurations under compression, reaction between thin films, and reactions between film and surface.[10]
8. Study of electrode reactions: In these studies the semiconductor itself may act as the electrode, or a thin conducting film of SnO_2, InO_2, Pt, or Au may be placed on the surface of the internal reflection plate.
9. Determination of optical constants of thin films.

3. INTERNAL REFLECTION ELEMENTS (IRE)

The heart of the internal reflection technique is the internal reflection element. A few geometries for IRE's of fixed and variable angles of incidence are shown in Fig. 10-6. The simplest is the single-pass (SP) plate which may be a trapezoid (T) or parallelepiped (P). Here the light is introduced into an aperture at one end, propagates down the length via multiple reflections $[N = (l/t)\tan\theta]$, and exits via an aperture at the other end. The double-pass (DP) and double-sampling (DS) geometries have entrance and exit apertures at the same end. Since they require only one window, they are attractive for use in vacuum systems or

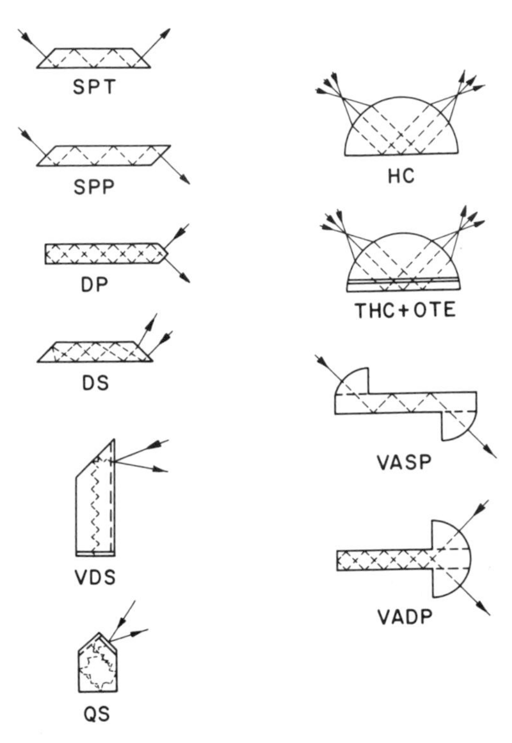

Fig. 10-6. Geometries of internal reflection elements (IRE's) commonly used for internal reflection spectroscopy. The group on left are fixed-angle elements while group on right are variable-angle elements.

Dewars and, since one end is free, they are useful for dipping into liquids and powders to record their spectra. The double-pass plate has an aperture only half the size of the single-pass plate. In double-sampling geometry the entrance and exit beams use the same aperture. The other bevel is metallized so that light cannot escape. Separation of the beams is achieved by cocking the second bevel so that the return beam propagates at a slightly different angle of incidence and is refracted upon exiting. In the quadruple-sample (QS) geometry, the light, in addition to traversing the length of the plate twice, zigzags across the width and therefore strikes each point on the surface four times, thus it has four times the sensitivity of the single-pass plate.

Variable-angle internal reflection elements are useful for changing spectral contrast, investigating layers of various thicknesses (by changing depth of penetration), and are particularly suited for measuring optical constants. Shown here are the simple hemicylinder (HC) and truncated hemicylinder (THC). The removable substrate (e.g., optically transparent electrode) is optically contacted to the THC to make a full hemicylinder. The multiple-reflection variable-angle internal reflection elements consist of quarter rounds optically contacted to a plate to form variable-angle single- or double-pass plates.

A wide variety of optical materials have been used to make internal reflection elements. These include Ge, Si, uv quartz, uv sapphire, MgO, KRS-5, AgCl, AgBr, ZnS, ZnO, ZnSe, CdTe, As_2S_3, As_2Se_3, Ge glass ($Ge_{35}Se_{50}As_{15}$), NaCl, and KBr. The demands on the purity of the material are quite stringent since the path length in the material often exceeds 15 cm and therefore weak absorption arising from impurities, free carriers, and weak lattice bands, often undetectable in conventional transmission measurements through windows, can pose a severe problem for internal reflection spectroscopy. For example, Ge and Si which are considered acceptable as window materials out to 23 and 8.5 μ, respectively, cut off rather sharply for internal reflection applications at 11 and 6.5 μ due to the presence of weak lattice bands. Similarly, Ge can be used as an optical window at temperatures above 250°C, however, it cannot be used above 50°C for internal reflection applications because of absorption losses resulting from the increased density of free carriers.

When many reflections are employed, the requirements on surface finish and maintenance of geometrical tolerances also become more stringent. These demands have in a number of cases led to the development of new techniques for polishing a number of materials. For surface studies, single-crystal-oriented material which has been chemically polished to remove mechanical damage should be employed. Oriented surfaces should also be employed if the IRE is cycled to high temperatures since thermal etching and slippage along crystalline planes can lead to a degradation of surface finish.

Careful surface preparation and cleanliness is imperative in surface studies. Standard cleaning techniques often do not remove small traces of contamination which affect the surface properties and produce spurious absorption bands, especially when the techniques are capable of detecting submonolayer films. A technique that has been found very useful in the last stages of surface preparation is ion bombardment or plasma cleaning. Ion bombardment in internal reflection studies was first employed by Becker and Gobeli[11] in the production of SiH_4 via hydrogen ion bombardment of a silicon IRE and subsequent removal of such films by argon bombardment. Their results, shown in Fig.10-7, reveal band shifts arising from use of higher-energy ions, indicating a change in the nature of bonding as ions penetrate deeper into the silicon bulk.

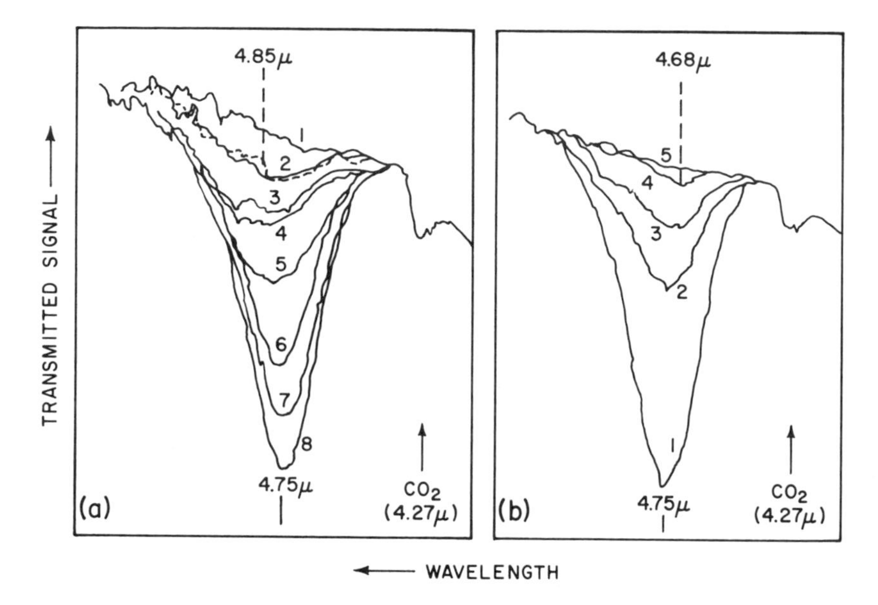

Fig. 10-7. Internal reflection spectra of a silicon surface, $N = 200$, $\theta = 45°$. (a) Curve 1 is an initially clean Si surface after sputtering in argon and annealing. The dashed curve nearly superimposed on curve 2 shows the maximum absorption peak resulting from exposure of silicon to atomic hydrogen. Curves 2–8 are due to bombardment with hydrogen ions of increasing energy and for longer times. Note the shift in the absorption peak. (b) Series of spectra showing decrease in strength of silicon hydride band as the silicon is bombarded with argon ions for various lengths of sputtering. Curve 1(b) shows a thick surface layer [same as curve 8(a)]. Note the band shift for curve 4(b) relative to curve 2(a).

A simple device that has been found useful in removing organic contaminants is the small Plasma Cleaner shown in Fig. 10-8. The instrument consists of an electrodeless gas-discharge apparatus. It is only necessary to adjust the pressure (via a roughing pump) to about 0.1 Torr and expose the substrate to the gas discharge for a few seconds. The resulting surface is both clean and sterile. Figure 10-9 shows the removal of hydrocarbons and photoresist with the plasma cleaner.

4. INSTRUMENTATION

Most commercial spectrometers are generally well suited only for simple transmission measurements. Because of the limited sampling space and means provided for mounting, compromises must be made in developing accessories for internal reflection, specular reflection, and other applications. The sampling space of most spectrometers is located in the undispersed beam and, since high-energy sources are used, overheating of the sample often occurs. Furthermore, few designs have taken advantage of all of the latest developments in electronics, e.g.,

Fig. 10-8. The Plasma Cleaner is a compact (vol. = 0.5 ft^3 and wt = 12 lb), table-model, electrodeless, radio-frequency, gas-discharge apparatus. The vacuum envelope (1 or 3 inch diameter) has an O-ring quick-disconnect seal to the cover plate for ease of access to the vacuum chamber. The pump is connected to an outlet at the back side of the vacuum chamber. The needle valve can be used to break the vacuum gently, to control the pressure, or to admit a special gas for the plasma discharge. The power level in the rf coil can be changed by the selector switch.

phase-sensitive lock-in amplifiers, to yield sensitivity often required in surface studies. Since many hundreds of such spectrometers are out in the field, however, there is a demand for accessories and these are available from a number of firms.

One attachment (VRA) employing double-sampling internal reflection plates was developed initially for surface studies as shown in Fig. 10-10. In this accessory, reflection plates of any length can be employed without altering the optics, and vacuum chambers or Dewars having only one optical window can be employed. The optical path projects the beam focus away from the instrument thereby relaxing the restrictions of small sample compartments. Figure 10-10 shows the VRA with a bakeable (to 400°C) ultrahigh vacuum chamber constructed especially for surface studies. Techniques have been developed for brazing a wide variety of windows (SiO_2, Al_2O_3, Ge, Si, MgO, CdTe, ZnS, etc.) thus eliminating the need for gaskets.

In addition to its use for internal reflection, the VRA can be employed for many other applications. As an example, Fig. 10-11 shows the optical layout and photograph of the VRA with a retro-mirror

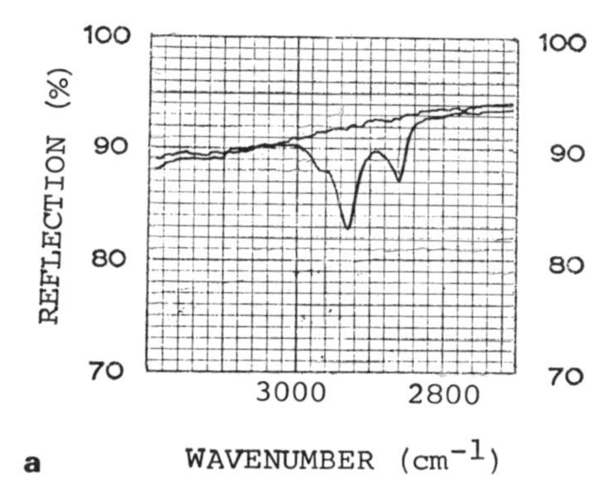

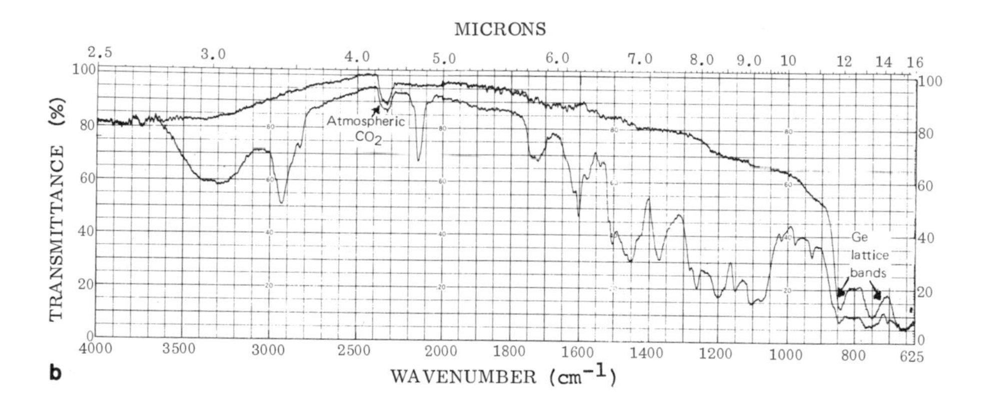

Fig. 10-9. (a) Removal of hydrocarbons from the surface of a silicon internal plate ($N = 60$, $\theta = 45°$). (*Lower trace*) The C–H band representing 10% absorption. (*Upper trace*) The C–H band—completely erased after a 1-minute exposure to air plasma. (b) Internal reflection spectra ($\theta_{av} = 45°$, $N = 20$) of a Ge surface before and after plasma cleaning. Lower trace shows Ge having surface coated with a thick (about 1 μ) film of photoresist (AZ111). Upper trace shows the same surface after 15 minutes of O_2 plasma cleaning indicating Ge is restored to its original organic-free condition with photoresist stripped off the surface.

accessory (RMA) for specular reflection. Once this system is aligned for one angle it remains in alignment for all others. Results taken with the VRA and RMA together have already been shown in Fig. 10-2, and indicate the enhancement in sensitivity obtained by employing large angles of incidence in the study of thin films via external reflection.

For many applications, commercially available spectrometers with well-designed attachments yield satisfactory performance. However, where complete flexibility and high sensitivity are required advantage should be taken of recently developed instrumentation. Now such systems can be readily put together since good monochromators and

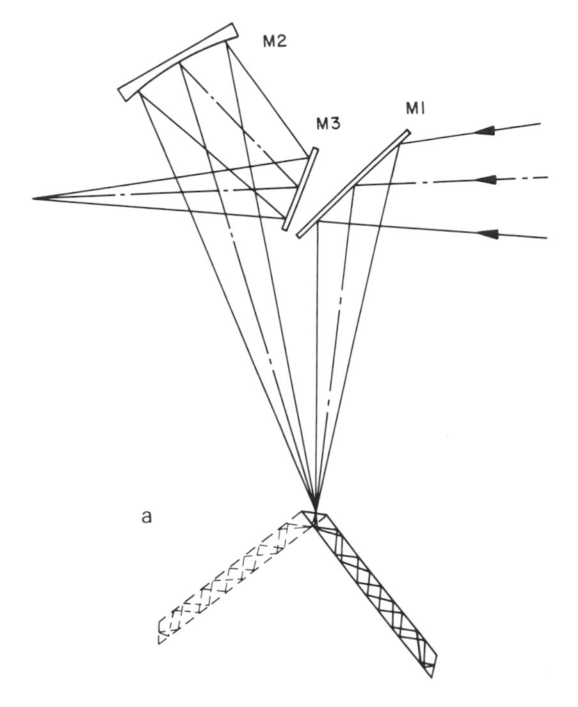

Fig. 10-10. The versatile reflection attachment (VRA) for internal reflection spectroscopy and specular reflection. The optical layout relaxes restrictions on the sampling compartment and permits use of Dewars and vacuum chambers having only one optical window.

electronic systems are available from a number of firms. For example, reflectivity changes of a silicon surface of 1 part in 10^7 were measured in the study of transition involving surface states[6] by employing lock-in amplifiers (before they were commercially available). An additional

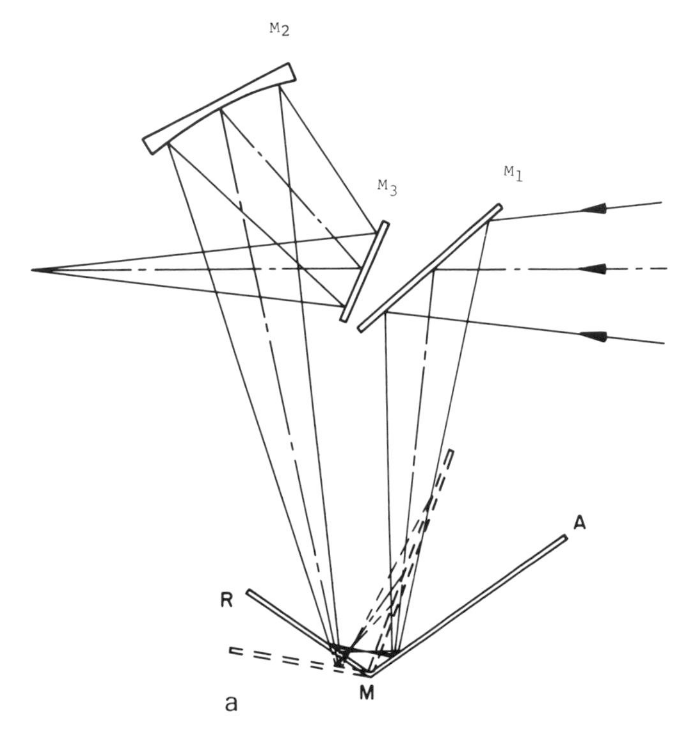

Fig. 10-11. VRA with retro-mirror accessory (RMA) for variable-angle specular reflection. Once the VRA + RMA is aligned for one angle of incidence, it remains in alignment for all others.

feature which adds to flexibility for almost any measurement in transmission, external reflection, or internal reflection is the possibility of changing the angle of incidence. The ideal instrument would thus be a double-beam goniometer spectrometer. The optical layout of such an instrument is shown in Fig.10-12. The source optics are mounted on an arm which pivots at A at the rate of 2θ. The internal reflection elements for the sample and reference beams, as well as the beam chopper and recombiner, are mounted on a platform which is coupled to the 2θ arm and which also pivots about A, but at the rate of θ. This system remains in alignment for all angles of incidence. The beam splitter and recombiner are driven synchronously and must move symmetrically relative to AB at rate of $\tan\theta$ as the angle of incidence is changed. Better than 0.01° mechanical accuracy can be obtained on both the θ and 2θ arms. With the ability to control the angle of incidence and the location of the beam focus relative to A, a high degree of flexibility is achieved.

By chopping the light beam simultaneously at two frequencies—one via a butterfly mirror, which allows the beam alternately to strike sample point A or reference point B, and a second much faster on–off chopper—and use of lock-in amplifiers and ratiometer,[12] the signals can be processed to display I/I_o or $\Delta I/I_o$ without slit drives or optical nulls.

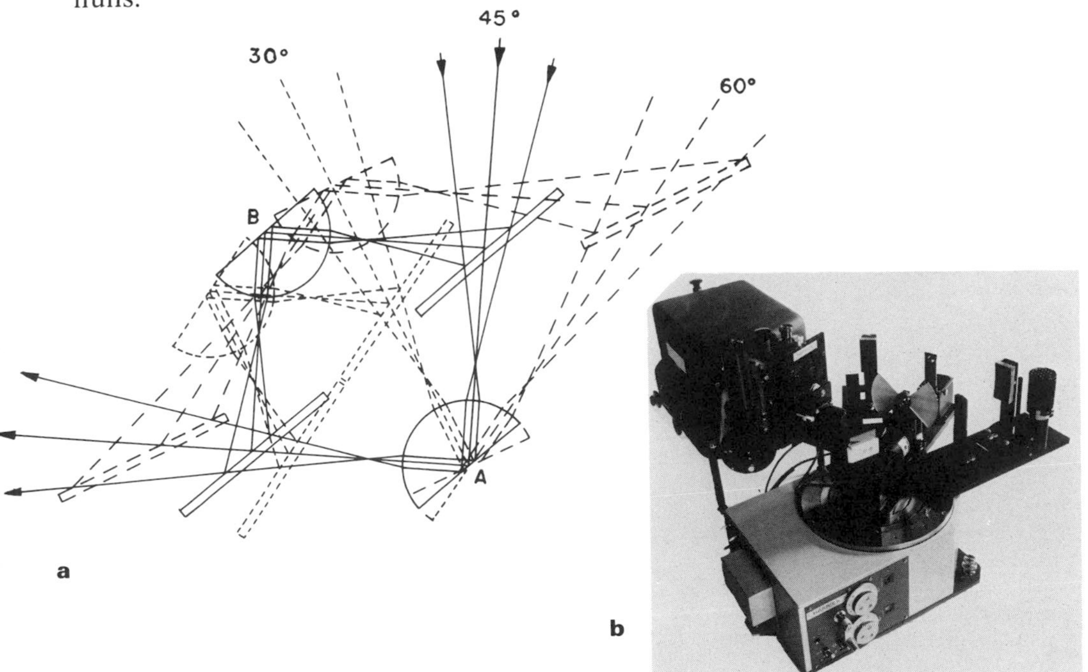

Fig. 10-12. Double-beam goniometer spectrometer for internal reflection, external reflection, or transmission. Optical layout (a) shows alignment with hemicylinder and that angle of incidence can be changed continuously without misalignment. Other IRE's that can be employed include variable angle double-pass plates, double sampling plates or single-pass plates.

5. PRINCIPLES

In transmission spectroscopy the only parameter that can be changed is sample thickness. This makes the technique foolproof but gives it little flexibility. For internal reflection spectroscopy, on the other hand, the parameters that can be changed and that affect the character of the spectrum include the refractive index of the IRE, the angle of incidence, and the polarization of the light. The strength of interaction of the standing wave with the absorber near the totally reflecting surface and its dependence on all of these parameters is expressed exactly by Fresnel's equations which unfortunately, in most cases, can be solved only with the aid of a computer. For most practical cases for either thick or thin films when the absorption coefficient is not too high, the first-order approximation gives the strength of interaction to a high degree of accuracy. Furthermore, these expressions are quite simple and give insight into the nature of the interaction permitting one to judiciously select the parameters to optimize the interaction. These first-order expressions for thick and thin films have been discussed in detail elsewhere.[2] We will review only the case for a thin-film absorber.

For transmission, neglecting reflection losses, the transmitted energy is related to the incident energy, absorption coefficient, and sample thickness as follows:

$$I = I_o e^{-\alpha d} \tag{1}$$

$$= I_o(1 - \alpha d) \qquad \text{for} \qquad \alpha d \ll 1 \tag{2}$$

Similarly, for reflection

$$R = (1 - \alpha d_e) \tag{3}$$

where d_e is defined as the effective film thickness and is the film thickness required to give the same spectral contrast in the transmission measurement as is obtained in the reflection measurement. It will become evident that d_e may be much greater or much less than the actual film thickness. For multiple reflection,

$$R^N = (1 - \alpha d_e)^N \tag{4}$$

$$= 1 - N\alpha d_e \qquad \text{for} \qquad \alpha d_e \ll 1 \tag{5}$$

For a thin film on the reflecting surface (see Fig. 10-5)

$$d_{e\perp} = \frac{4n_{21} d \cos\theta}{1 - n_{31}{}^2} \tag{6}$$

$$d_{e\parallel} = \frac{4n_{21} d \cos\theta[(1 + n_{32}{}^4)\sin^2\theta - n_{31}{}^2]}{(1 - n_{31}{}^2)[(1 + n_{31}{}^2)\sin^2\theta - n_{31}{}^2]} \tag{7}$$

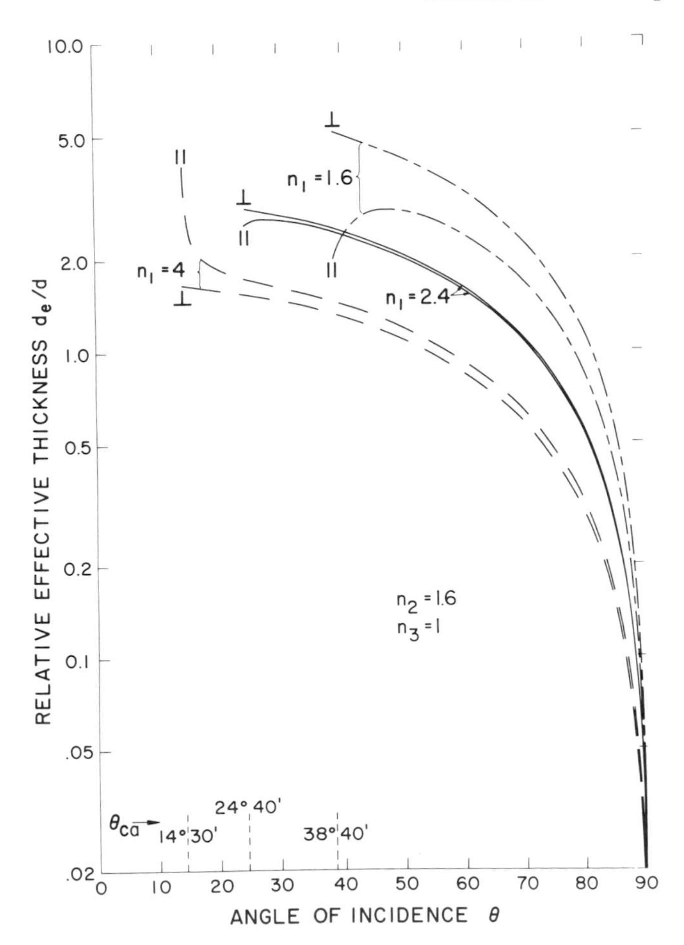

Fig. 10-13. Relative effective thickness d_e/d *vs.* angle of incidence for different refractive indices of the IRE (n_1), index of refraction of the film ($n_2 = 1.6$), and of the environment ($n_3 = 1$), (where d = film thickness). The critical angles θ_{ca} for the interface IRE–air are also indicated.

The possibility of changing the effective thickness by varying any of the parameters mentioned is quite evident and their effect on the character of the spectrum has been tested and discussed elsewhere.[(2)] Figure 10-13 shows these effects for a few interfaces.

There is still another parameter which can be varied for thin films and which has not been previously discussed. This parameter is the refractive index of medium 3. It is varied by immersing the IRE, usually in the form of a plate, into a medium of higher refractive index. The effect of immersion can readily be determined from the previous equations for the effective thicknesses of a thin absorbing film on a totally reflecting surface.[2] The change in spectral contrast resulting from an increase in the refractive index of medium 3 is shown in Fig. 10-14a for a few selected cases at an angle of incidence of 45°. To maintain critical reflection, the refractive index of the outer medium can be increased only until it reaches the critical index for total reflection, namely, $n_{3c} = n_1 \sin \theta_c$, or $n_3/n_1 = 0.707$ for $\theta = 45°$. In general, the effective thickness will increase as the refractive index of the surrounding medium is increased. The reason for this is that the electric field amplitudes tend to increase as the refractive index approaches the critical index. This is always the case for perpendicular polarization; at most, an increase in the effective thickness by a factor of almost 2 can be obtained. For parallel polarization, on the other hand, it will be noted that the effective thickness can

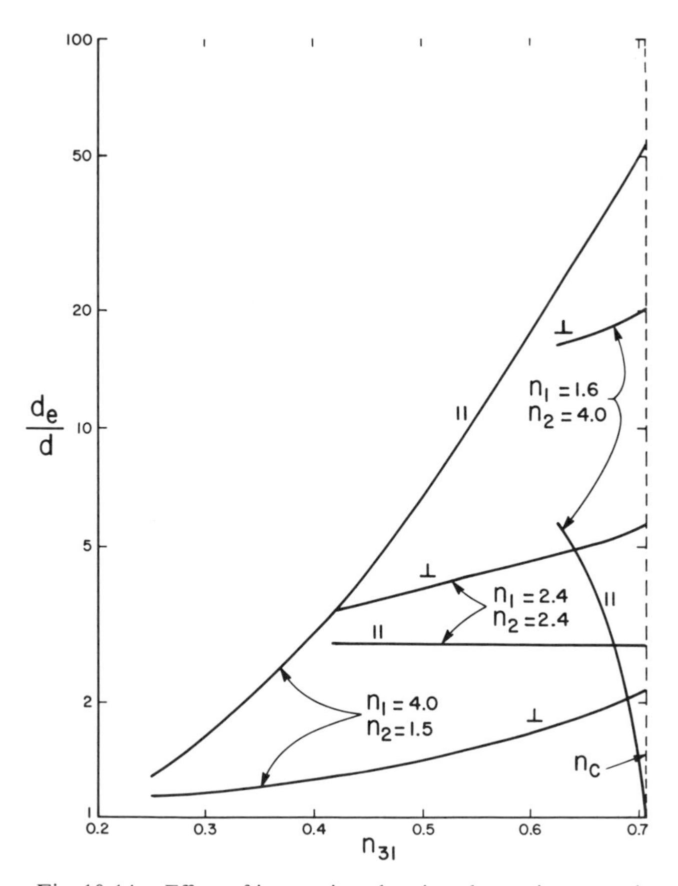

Fig. 10-14a. Effect of immersion showing change in spectral contrast as the refractive index of medium 3 is increased. Note that in general, spectral contrast increases, but in special cases may remain constant or even decrease.

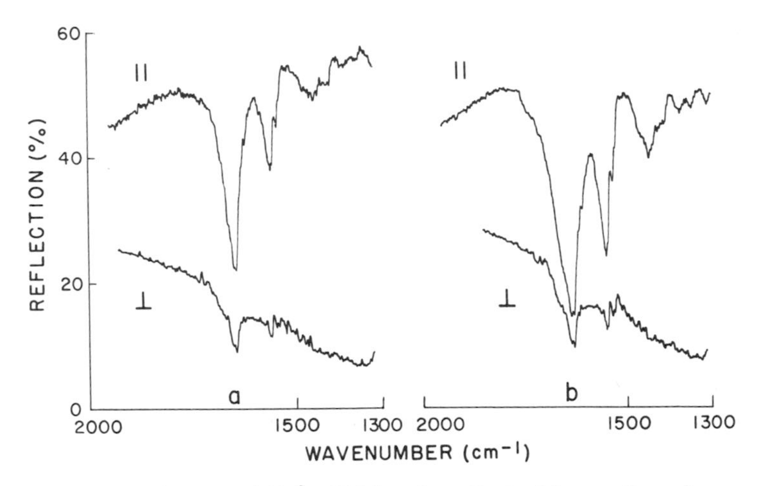

Fig. 10-14b. Spectra of 50 Å of Nylon 4 on Ge double-sampling plate; (a) before and (b) after bringing AgCl in contact with Nylon covered Ge plate. Note increase in spectral contrast with greater increase for parallel polarization.

increase sharply by more than an order of magnitude, remain unchanged, or even decrease as n approaches n_c. When $n_1 = n_2$ and the angle of incidence is 45°, $d_{e\parallel}$ is independent of the refractive index of medium 3. It can readily be understood that for other cases $d_{e\parallel}$ will either increase or decrease, depending on whether E_z or E_x predominates.

The variation of spectral contrast by immersion is quite practical for the visible spectral region where a wide range of optically transparent liquids are available and optical contact is thus readily achieved. For the infrared, on the other hand, there are a number of soft, solid materials to choose from, which, when pressed against the internal reflection plate, will give fairly good contact. These materials include selenium–sulfur mixtures where the refractive index can be varied between 2 and 3, AgCl ($n = 2$), AgBr ($n = 2.2$), and GaSe ($n = 2.2$).

Figure 10-14b shows the effect on spectral contrast resulting from pressing AgCl ($n = 2$) onto a Ge ($n = 4$) double-sampling plate having about 35 reflections at 45° and coated with a 100-Å film of Nylon 4. It should be noted that $d_{e\parallel}/d_{e\perp} = 1.2$ as predicted by theory. Also, as predicted by the above discussion, there is an increase in both $d_{e\parallel}$ and $d_{e\perp}$, and, furthermore, $d_{e\parallel}$ increases to a greater degree than does $d_{e\perp}$. The fact that the increase in d_e is not as large as theoretically predicted can be explained by lack of intimate contact between the AgCl and Ge.

These results further demonstrate the usefulness of the effective thickness expressions which permit one to judiciously select the parameters to optimize sensitivity, to make quantitative measurements, etc.

They also demonstrate that in certain cases there is still another parameter that can be varied which gives internal reflection still more versatility.

6. APPLICATIONS

As discussed in the preceding sections dealing with the description of internal reflection spectroscopy, this method can be used for various types of surface investigations; i.e., the region near the surface within the solid, as well as different kinds of interactions between the solid and the adjacent medium, can be studied. These include adsorption from the gas or liquid phase and chemical reactions between the solid and the surrounding medium leading to a second solid phase.

Semiconductors have found wide applications as IRE's. The magnitude of their index of refraction leads to a low critical angle making it possible to obtain a high number of total internal reflections for a given length of the element. In addition, the good infrared trans-

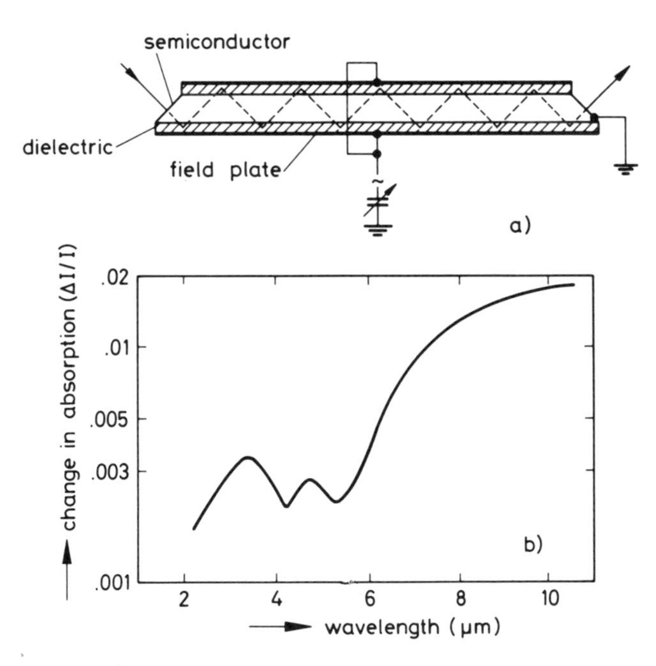

Fig. 10-15. Optical studies of field effect modulation of semiconductor surfaces by internal reflection spectroscopy. (a) Condenser system for field effect modulation of semiconductor or barrier height and hence, modulation of free-carrier density in space-charge region and population in surface states of a semiconductor. (b) Infrared absorption of free holes in the space-charge region of intrinsic Ge for a change in hole density of 10^{11} per cm^2. Angle of incidence = 45°, 25 reflections[2].

mitting properties allow extension of spectroscopic studies into the infrared region. A further advantage of these materials is that semiconductors are very well suited for studies of the electronic interaction between solids and the environments since the type and density of charge carriers at the surface can be controlled by applying appropriate voltages across the surface, and the carrier density can be measured optically by IRS. By employing modulation techniques, very high sensitivities can be achieved which correspond to absorption changes ΔI due to modulation as little as $10^{-7} I_0$, where I_0 is the incident intensity. Figure 10-15a shows an arrangement that has been employed for measuring surface effects by means of modulation spectroscopy. A semiconducting IRE as shown may be prepared of such dimensions that more than 200 reflections can be obtained. Instead of using a metal plate as a counter electrode for modulating the surface potential, the IRE may also be immersed into an electrolyte with a platinum electrode relatively remote from the IRE.[13] By application of a voltage between this semiconductor and the Pt electrode, a change of the voltage drop within the space-charge region of the semiconductor surface may be obtained.

The absorption spectrum of free holes in the surface space-charge region of an intrinsic-germanium IRE thus measured is shown in Fig. 10-15b. By applying a certain bias and a superimposed ac voltage between the counter electrode and the IRE, a periodic variation of the density of free holes at the surface is achieved. This produces a periodic variation of the intensity of the transmitted light, whose wavelength dependence is due to the absorption by holes in the valence band.

Investigations of this type have also been performed with silicon.[6] The silicon IRE's were thermally oxidized, the oxide film serving as a dielectric between the semiconductor IRE and the metal plate as counter electrode. By employing this technique, the absorption spectrum of free holes within the space-charge region could be measured, and from the analysis of the voltage dependence of this optical signal, information could be obtained on the relation between applied voltage and the bending of the energy bands at the surface. In addition to this, at wavelengths shorter than 4 μ, absorption changes appeared whose magnitude and sign were strongly dependent on the applied dc voltage. These changes were assigned to transitions of the charge carriers into or from the surface states. A broad distribution of surface states was deduced from the spectral behavior of the magnitude and phase of these changes, and in the range in which measurements were possible, the density of these states was found to decrease from the center of the band to zero at the upper edge of the valence band. The physical or chemical nature of these centers is not yet known.

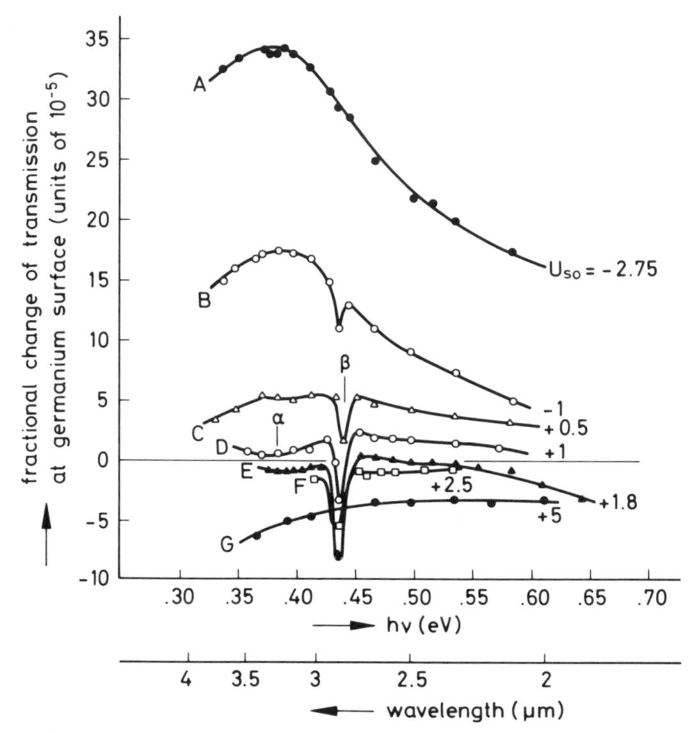

Fig. 10-16. Fractional change of transmission as a result of a change in the density of free charge carriers in the space-charge region and in the occupancy of the surface states.[7] The bending of the bands in units of kT is given at the right-hand side of each curve.

Discrete surface states have also been detected and their energetic position has been determined by means of IRS in combination with modulation spectroscopic techniques. Samoggia *et al.*[7] observed transitions from surface states into the conduction band in n-type germanium (see Fig. 10-16). The bending of the bands U_{SO} (surface potential) is given by the distance of the middle of the forbidden zone on the surface from the position of the Fermi level in units of kT.

Curve A was recorded with upward bending of the bands. The intensity change ($\Delta I/I_0$) due to the superimposed ac voltage corresponds to the absorption spectrum of the free holes. Curve B, which was obtained with less-pronounced upward bending of the bands, shows smaller changes in the hole density as a function of the ac voltage, as well as an additional absorption shifted in phase by 180° in a narrow energy range around 0.44 eV(2.8 μ). The absorption signal of the free holes in this case is due to the fact that the upward bending of the bands is stronger and the hole density on the surface is therefore greater in the positive than in the negative half-wave of the superimposed alternating voltage. The fact that the signal β is opposite in phase means that this absorption takes place in the negative half-wave. The restriction of this additional signal to a narrow energy range indicates that a discrete state is involved in the transition. It follows from the opposite phase that this signal is due to a transition from the occupied state into the conduction band. The transition reaches its maximum rate at a surface potential $U_{SO} = +1.8$. Thus at this band bending, the center coincides in its energy with the Fermi level. The center is therefore situated 0.045 eV above midgap. The broad selective absorption band denoted by α can be assigned to a continuum of surface states distributed about an energy of 3.5 kT above midgap. If the band is bent further downward, we finally obtain the absorption spectrum of the free electrons (curve G). Another absorption band opposite in phase to β should now occur as a result of transitions from the valence band into the same state. However, the authors believe that the quantum mechanical transition probabilities are too low for observation of this band. Rzhanov and Sinyukov[8,9] have observed a similar band at 2.72 μ on a Ge surface after heat treatment in vacuum. Their conclusion however is that although this band could be due to surface states, it is more likely due to absorption by adsorbed water molecules.

Shortly after the internal reflection technique was proposed as a means of obtaining spectra of adsorbed molecules and thin films, there was rather wide interest in its use for electrode-reaction studies. For the infrared one can employ a semiconductor (e.g., Ge) as an electrode. If a high conductivity is required to study fast reactions the semiconductor surfaces may be doped. Problems that arise are that Ge as an electrode makes the system complicated and that there is a lack of windows in the infrared region, i.e., the electrolytes are often absorbing in spectral regions of interest.

In the visible spectral region, one can use glass plates covered with thin, transparent, and electrically conducting films (such as tin oxide, indium oxide, gold, or platinum) as IRE's to monitor molecule or ion concentrations in electrolytes in contact with these transparent electrodes. The aim of such investigations is to study electrode reactions and to obtain information on intermediates and products of such reactions. Furthermore it is of interest to determine whether adsorption processes on the electrode surface occur during such reactions. In the visible and uv spectral regions, many organic molecules and radicals which are of maximum interest in this field have characteristic absorption bands. Furthermore, most solvents are transparent so that there is no interference from additional absorption bands.

As an example of the applicability of IRS in the study of species generated at electrodes, we will discuss some results of Prostak, Mark, and Hansen.[14] They used IRE's consisting of a glass plate covered with a gold film approximately 50 Å thick as an electrode to determine ion concentrations in the boundary layer of an electrolyte contacting the gold electrode. They also used a 0.0125 M solution of 4,7-dimethyl ferroin. At a potential of +0.379 V against calomel electrode, these molecules are known to be in the reduced form and appear colored, while at +0.94 V they are in an oxidized form and are practically colorless. The potential E was varied between these limits, and the absorbance of monochromatic light of 5120-Å wavelength was measured at each potential. The total change in absorbance between 0.5 and 0.94 V potential was given by ΔA_{max} and the changes between 0.5 V and the potential E were ΔA. According to the Nernst equation, the potential of this electrode is proportional to the change in the concentration of the molecules in the reduced form near the electrode. Therefore, it should follow that

$$E = E_0 + 2.3\frac{RT}{nF}\log\frac{\Delta A}{\Delta A_{max} - \Delta A} \tag{8}$$

From the results of their measurements, the authors plotted $\log[(\Delta A/(\Delta A_{max} - \Delta A)]$ *vs.* the measured potentials E. They found a linear dependence with a slope corresponding to 52 mV, whereas according to the Nernst equation it should be 59 mV. Therefore, this result can be regarded as a proof that IRS may be used for quantitative determinations in the study of electrode reactions.

Kuwana and his co-workers[15] used a glass plate covered with a thin film of transparent and conducting tin dioxide as an IRE and electrode. Their papers contain discussions of spectral changes due to variations in the optical constants of the electrode material and of the ionic double layer and the electrochemical generation of intermediates, as well as a study of fast kinetic rates of homogeneous chemical reactions for a variety of mechanisms. They obtained excellent results; however, they sometimes encountered problems due to interference phenomena resulting from multiple reflections of the light beam within the tin dioxide film. Memming and Möllers[16] observed that it was not only geometric optical phenomenon which led to disturbing interference effects. When using modulation techniques by applying ac voltages between the tin dioxide and a counter electrode there was also a modulation of the free-carrier density within the tin dioxide film. According to Drude's theory, this leads to a modulation of the refractive index. The disturbances by this latter effect together with the geometric optical-interference phenomenon resulted in intensity modulations of up to 10%. Memming and Möllers, however, reduced the modulation to below 0.01% by preparing films of inhomogeneous thickness.

In their paper,[16] Memming and Möllers exhibit very fine demonstrations of the potentialities of IRS for the study of electrode (surface) reactions. They investigated the reduction processes of benzyl-viologene

$$(V^{2+} = C_6H_5-CH_2-\overset{+}{N}\langle\bigcirc\rangle-\langle\dot{\bigcirc}\rangle N-CH_2-C_6H_5) \quad (9)$$

at tin dioxide electrodes in acetonitrile. This compound can be reduced in two steps ($V^{2+} + e^- \rightarrow \dot{V}^+$; $\dot{V}^+ + e^- \rightarrow V$). Both processes have markedly different redox potentials (standard potentials for $V^{2+}/\dot{V}^+ = -500$ mV, for $V^+/\dot{V} = -900$ mV). In Fig. 10-17a we see the absorption spectrum of a solution of the semireduced species of benzyl-viologene ($\dot{V}^+$) which was electrochemically reduced prior to the performance of the measurement, which occurred by transmission of light through this solution in a standard photometer. Figure 10-17b gives the spectrum obtained by IRS and periodic modulation of the electrode potential

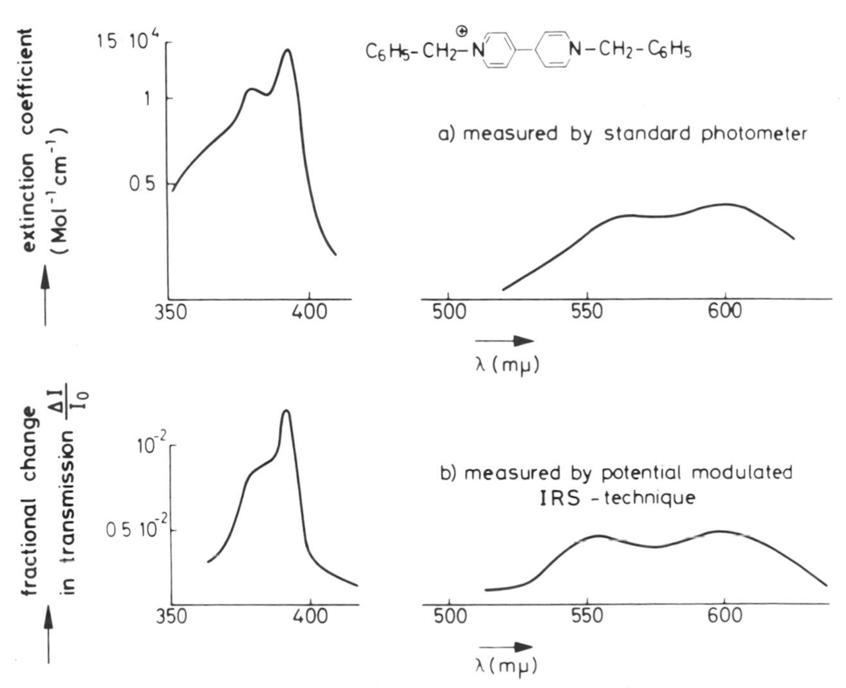

Fig. 10-17. (a) Absorption spectrum of semireduced benzyl-viologene in deaerated acetonitrile; (b) spectral distribution of the modulated absorbance.[16]

between +200 and −700 mV. Since the only absorption peak of V^{2+} in this range is situated at $\lambda = 260$ nm, there is no disturbance of the spectrum by the oxidized form. Comparison of Fig. 10-17a with 10-17b clearly shows that both spectra agree very well. If one takes into consideration that these reduction processes occur at the electrode surface, this experiment clearly demonstrates that the optical spectra of adsorbed molecules and molecules in the transition region on top of solid surfaces can be measured by this method. It is known from the literature that in the spectrum of Fig. 10-17a the peaks at 394 and 600 nm correspond to the monomer and those at 380 and 555 nm to the dimer.[17] With these facts, further insight into these electrode processes can be obtained from the following experiment. By variation of the angle of incidence at which the incident beam strikes the tin dioxide electrolyte interface, the number of reflections within the IRE and the depth of penetration into the rarer medium (the solution) will change. Above the critical angle, the depth of penetration decreases with increasing angle of incidence. In Fig. 10-18 two spectra, which were taken at angles of incidence of 60° (solid line) and 74° (dashed line) and at the same modulation poten-

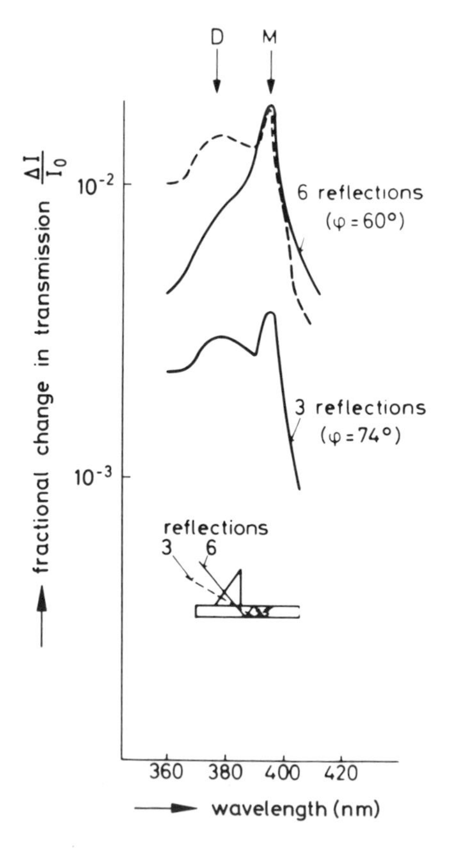

Fig. 10-18. Modulated absorption of semi-reduced benzyl-viologene in acetonitrile for different angles of incidence.[16]

tial, are compared at different modulation frequencies. The increase of the angle of incidence leads to a reduction of the depth of penetration and of the effective thickness by more than a factor of 1/10. The spectra were of different strengths because of the different number of reflections in the two cases. Hence they were normalized in the upper part of Fig. 10-18 so that the monomer peaks coincided. Figure 10-18 must be interpreted either that the dimer concentration is larger close to the electrode or that these dimers are adsorbed at the surface. Unfortunately it could not be proved by performing these measurements with polarized light of different planes of polarization that adsorption of the dye molecules occurred at the electrode surface, since the light intensities were too low in this case. It could be shown clearly with methylene blue, however, that such dye molecules were adsorbed at the electrode surface.[16] In addition, the authors were able to measure the signal due to variation of the concentration of the original species and that due to the potential modulation of the reaction product. They could also measure absorption signals by intermediates formed during such reactions.

When modulation techniques cannot be applied, the relatively large signal corresponding to the transmitted intensity as well as the variation of this signal with wavelength can be cancelled by using differential techniques.[2,18] In Fig. 10-19 such a setup is shown in which

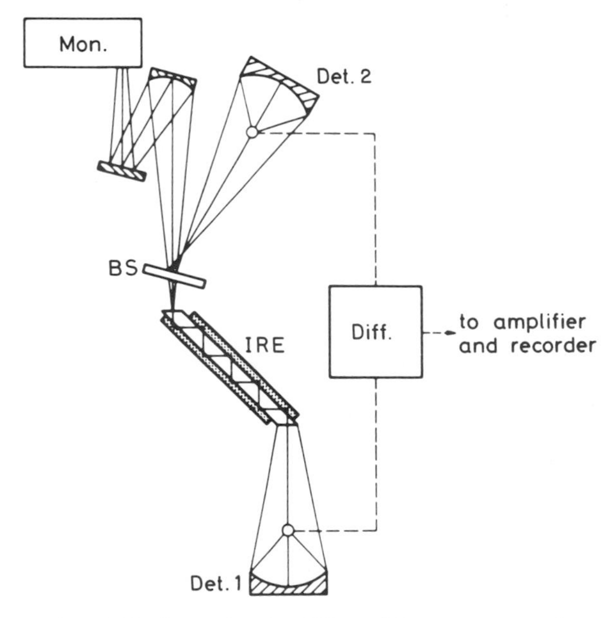

Fig. 10-19. Setup for the differential measurement by IRS of the infrared absorption spectrum of a thin film on a substrate serving as an IRE.[18]

the light reflected from a beam splitter is recorded by a thermocouple (Det. 2) which produces the reference signal I_0. This signal and the one from the sample channel, which is detected by Det. 1, are electrically attenuated so that the differential signal equals zero in absorption-free regions. Figure 10-20 shows as an example the wavelength dependence of the reference I_0 and that of sample signal I. In the differential spectrum $(I_0 - I)$, a marked deviation from the zero line in the middle of the graph is seen at the wavelength at which the thin film under study absorbs. Upon further electronic amplification of this differential signal, a pronounced absorption band $(I_0 - I)^*$ can be observed. The use of this method to obtain high sensitivity in the study of adsorbed O–H and C–H groups on a Si surface can be further demonstrated (see Harrick,[2] p. 181).

By means of this technique, very thin surface films on silicon wafers have been investigated. It is a well-known fact that silicon having been

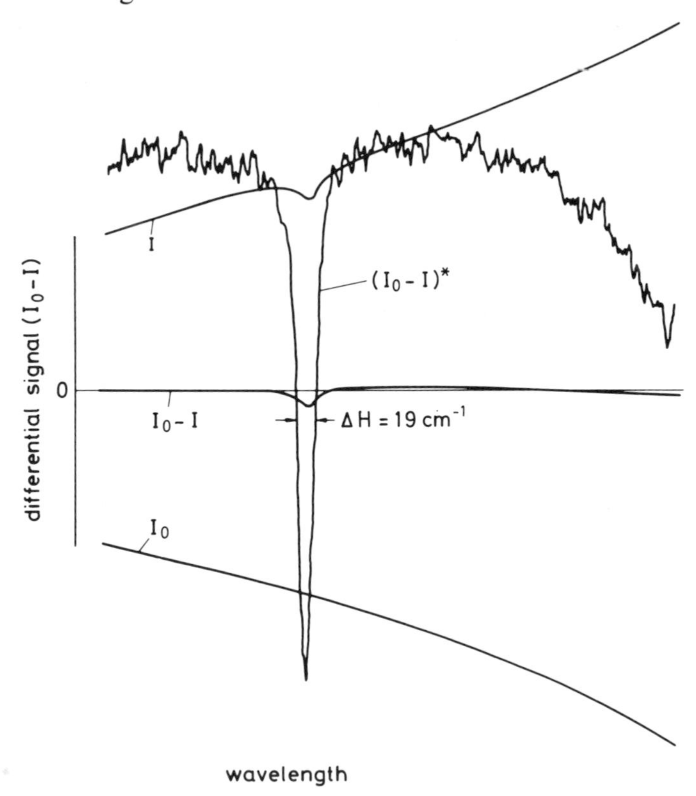

Fig. 10-20. Reference (I_0), signal (I), and differential spectra $(I_0 - I)$, and $(I_0 - I)^*$ of the SiH absorption band at 4.25-μ wavelength in an anodically produced SiO_2 film on silicon. These spectra were taken with the setup shown in Fig. 10-19 (25 reflections).

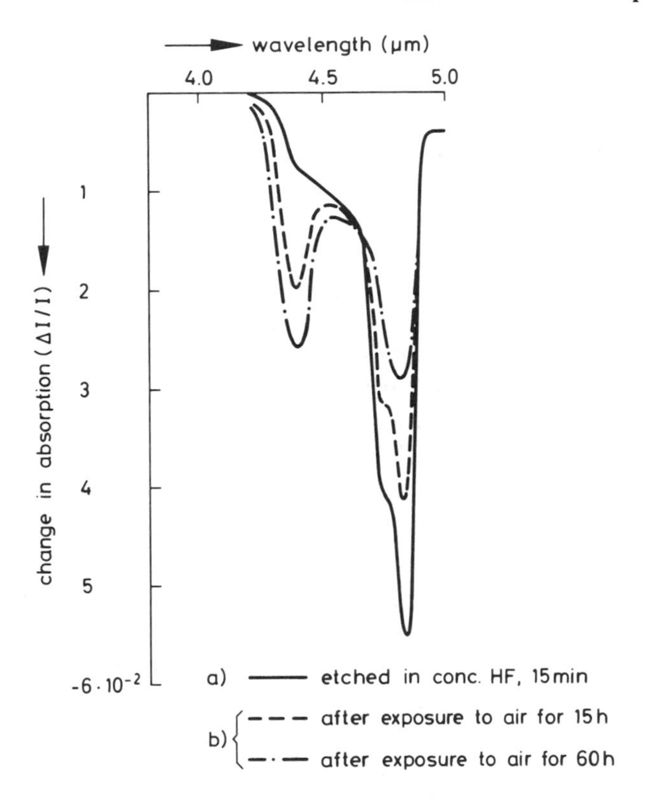

Fig. 10-21. (a) Spectral position of the absorption due to the SiH valence vibration in a polymerized silicon hydride film as produced by etching a silicon IRE in 10 N hydrofluoric acid; film thickness is approximately 20 Å. (b) Shift of this absorption band to shorter wavelength due to partial oxidation of the Si–Si bonds of this polymer (180 reflections, angle of incidence = 45°).

freshly etched in hydrofluoric acid is hydrophobic, whereas it turns hydrophilic upon long-time exposure of this wafer to room air. By using IRS, characteristic absorption bands in the wavelength region between 4 and 5 μ were observed (Fig. 10-21). According to former results,[19] these bands could easily be assigned to be SiH valence vibrations of silicon hydride groups which formed a kind of polymerized silene. After this wafer had been exposed to room air for some days, we observed a characteristic shift of these absorption bands towards shorter wavelengths which is due to the oxidation of the Si–Si bonds to Si–O–Si bonds.

The chemical nature of surface films produced on germanium in different etchants was also studied by this technique.[20]

The application of IRS has yielded some insight into the mechanism of the growth of native oxide films by anodic or thermal oxidation of silicon surfaces.[18] The infrared absorption spectra of SiH and OH groups incorporated into these oxide films have been measured in the wavelength range of 2.5–5.0 μ. From the strength of the absorption signal, quantitative determinations of the concentration of the groups under study within the film could be performed. By stepwise etching of the oxide layers and measuring the variation in the strength of absorption with reduced film thickness, the distribution of these groups as a function of the distance from the Si–SiO_2 interface could be determined.[18] From these distributions, an important role of water or hydroxyl in the oxidation process of silicon could be demonstrated.

An IRS study by Zolotarev, Veremei, and Gorbunova[21] of SiO_2 films deposited on Ge has shown the presence of an absorption band which is absent in transmission spectra. The explanation for this must be the presence of a longitudinal electric field associated with the standing wave at the reflecting interface for internal reflection which, of course, is absent for propagating waves (transmission spectra).

This section has given some examples of the application of internal reflection spectroscopy to the study of surfaces and thin films. Many other workers are applying these techniques to similar studies and thus additional publications should be anticipated. The potential application of internal reflection techniques as an investigative tool has been far from exhausted. As an example, thousandfold sensitivity enhancements[22] over conventional techniques have been obtained in the study of bulk materials and thin films via internal reflection fluorescence spectroscopy. Investigation is currently under way to evaluate its potential in Raman spectroscopy.

7. REFERENCES

1. S. A. Francis and A. H. Ellison, *J. Opt. Soc. Am.* **49**, 131 (1959); H. G. Tompkins and R. G. Greenler, *Surface Sci.* **28**, 194 (1971); G. W. Poling, *J. Colloid Interface Sci.* **34**, 365 (1970); D. J. (Messervey) Drmaj and K. E. Hayes, *J. Catalysis* **19**, 154 (1970); J. D. E. McIntyre and D. E. Aspnes, *Surface Sci.* **24**, 417 (1971); J. D. E. McIntyre, Specular reflection spectroscopy of the electrode–solution interphase, in *Advances in Electrochemistry and Electrochemical Engineering* (R. H. Muller, ed.) Vol. 9, Interscience, New York (1973).
2. N. J. Harrick, *Internal Reflection Spectroscopy*, Interscience, New York (1967).
3. W. W. Wendlandt and H. G. Hecht, *Reflectance Spectroscopy*, Interscience, New York (1966); W. W. Wendlandt (ed.), *Modern Aspects of Reflectance Spectroscopy*, Plenum Press, New York (1968); G. Kortüm, *Reflectance Spectroscopy*, Springer-Verlag, New York (1969).
4. N. J. Harrick, *J. Phys. Chem. Solids* **8**, 106 (1959). [Presented at International Conference on Semiconductors, Rochester, New York (1958).]

5. N. J. Harrick, *Phys. Rev. Letters* **4**, 224 (1960).
6. N. J. Harrick, *Phys. Rev.* **125**, 1165 (1962).
7. G. Samoggia, A. Nucciotti, and G. Chiarotti, *Phys. Rev.* **144**, 749 (1966).
8. A. V. Rzhanov and M. P. Sinyukov, *Sov. Phys., Semiconductors* **2**, 416 (1968).
9. A. V. Rzhanov and M. P. Sinyukov, *Sov. Phys., Semiconductors* **2**, 424 (1968).
10. L. H. Sharpe, *Proc. Chem. Soc.*, 461 (1961); G. I. Loeb and R. E. Baier, *J. Colloid Interface Sci.* **27**, 38 (1969); G. L. Haller and R. W. Rice, *J. Phys. Chem.* **74**, 4386 (1970); U. P. Fringeli, H. G. Müldner, Hs. H. Günthard, W. Gashce, and W. Leuzinger, The structure of lipids and proteins studied by attenuated total reflection (ATR) infrared spectroscopy. I. Oriented layers of tripalmitin, *Zeit. f. Naturfor.* **27**, 780 (1972); P. Fromherz, J. Peters, H. G. Müldner, and W. Otting, An infrared spectroscopic study on the lipid protein interaction in an artificial lamellar system, *Biochem. Biophys. Acta* (to be published).
11. G. E. Becker and G. W. Gobeli, Surface studies by spectral analysis of internally reflected infrared radiation: Hydrogen on silicon, *J. Chem. Phys.* **38**, 2942 (1963).
12. Donald Munroe, An Improved Electronic System for Double-Beam Optical Measurement, presented at the Optical Society of America Meeting, in San Francisco, California (October 1972); also appeared in Ithaco Data Sheet, IAN-32.
13. H. U. Harten, *Z. F. Naturforschg* **16a**, 459 (1961).
14. A. Prostak, H. B. Mark, and W. N. Hansen, *J. Phys. Chem.* **72**, 2576 (1968).
15. V. S. Srinivasan and T. Kuwana, *J. Phys. Chem.* **72**, 1144 (1968); N. Winograd and T. Kuwana, *Electroanal. Chem. Interf. Electrochem.* **23**, 333 (1969).
16. R. Memming and F. Möllers, *Symp. Faraday Soc. Nr.* **4**, 145 (1970).
17. W. M. Schwarz, Thesis, Univ. of Wisconsin, 1961.
18. K. H. Beckmann and N. J. Harrick, *J. Electrochem. Soc.* **118**, 614 (1971).
19. K. H. Beckmann, *Surface Sci.* **3**, 314 (1965).
20. K. H. Beckmann, *Surface Sci.* **5**, 187 (1966).
21. V. M. Zolotarev, V. A. Veremei, and T. A. Gorbunova, *Optics Spectry.* **31**, 40 (July 1971).
22. N. J. Harrick and G. Loeb, Internal reflection fluorescence, *Anal. Chem.* **45**, 687 (1973).

11

RADIOISOTOPE TECHNIQUES

Joseph A. Keenan and Graydon B. Larrabee

Materials Characterization Laboratory
Texas Instruments, Incorporated
Dallas, Texas

1. INTRODUCTION

A wide spectrum of radiochemical techniques can be employed for the characterization of surfaces. In broad terms these techniques can be separated into the two fields of tracer analysis and activation analysis.

In tracer analyses, radioactive isotopes of individual elements are introduced into a chemical system and their fate as the system reacts is followed by radioassay. In activation analyses, the entire sample is irradiated and the type and quantity of stable isotopes present prior to irradiation are deduced from a measure of the radioactive isotopes produced.

Detailed treatments of the use of Rutherford scattering and microbeam techniques for the characterization of surfaces are given in Chapters 16 and 17, respectively. This chapter will cover the use of radioactive tracer techniques and activation analysis. Finally, special treatment will be given to the use of autoradiography in radioassay of surfaces.

2. RADIOASSAY TECHNIQUES

All of the radiochemical techniques described in this chapter depend ultimately upon some method of radioassay. The radioactive decay process involves simple first-order kinetics as shown in Eq. (1). Consequently the problem in radioassay is simply to determine the disintegration rate.

$$DPS = \lambda N \tag{1}$$

where DPS is the number of disintegrations per second, N is the number of radioactive atoms present, and λ is the decay constant in units of seconds^{-1} ($\lambda = 0.693$/half-life in seconds).

The disintegration rate in Eq. (1) is actually the differential of the number of radioactive atoms present with respect to time. Integration of Eq. (1) leads to Eq. (2).

$$N_t = N_0 e^{-\lambda t} \tag{2}$$

where N_t is the number of atoms at time t, N_0 is the number of atoms at time zero, λ is the decay constant in seconds^{-1}, and t is the elapsed time in seconds. Since the disintegration rate is proportional to the number of atoms present, Eq. (1), the disintegration rate also changes exponentially with time. A plot of the logarithm of the disintegration rate *vs.* time should therefore be a straight line with slope λ. If several isotopes with different λ contribute to the observed activity it is often possible to resolve the composite curve into a series of straight lines. Each isotope may then be identified by its characteristic decay constant. Radiochemists, however, usually prefer to speak of the half-life which is proportional to the decay constant as was shown in Eq. (1).

The relationship between the number of events observed per unit time and the actual radioactive disintegration rate will depend upon the properties of the detector being used and the type of radiations being observed.

For all types of radioassay, the observed count rate is converted to a disintegration rate by some form of Eq. (3).

$$DPS = \frac{CPS}{\varepsilon(E) \cdot \text{branching ratio}} \tag{3}$$

where DPS is disintegrations per second, CPS is counts per second observed, $\varepsilon(E)$ is the detection efficiency at the energy of the detected radiation, and the branching ratio is the fraction of radioactive disintegrations which lead to a radiation of the type detected.

The branching ratio is a property of the radioactive isotope. The detection efficiency is a property of the radiation detector. It can generally be divided into an energy-dependent term which is an intrinsic property of the detection medium and a geometry factor which depends upon the solid angle subtended by the detector with respect to the source.

2.1. Types of Radioactive Decay

The familiar classifications of α, β, and γ radiations refer to processes which differ with respect to the radiations emitted and the nuclear

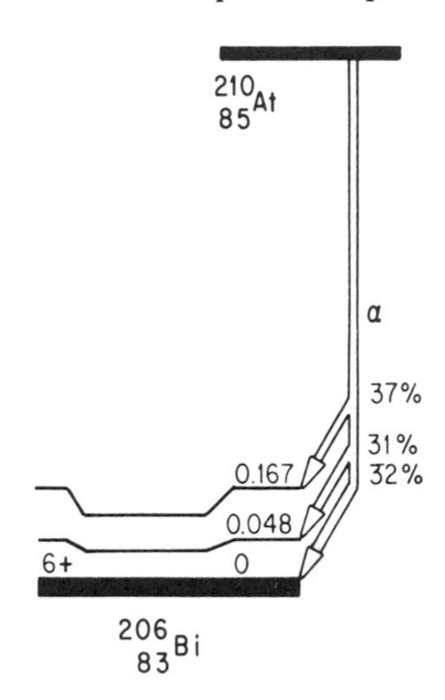

Fig. 11-1. Decay scheme of ^{210}At showing α emission.

transitions from which they arise. Both α and β particles result from the decay of an unstable parent nucleus to the ground state or an excited state of a daughter nucleus. Alpha particles, doubly charged He ions, carry away two protons and two neutrons. Thus the daughter nucleus is an isotope with a proton number which is 2 less than that of the parent, and a total of 4 mass units lighter. Figure 11-1 shows the decay scheme of ^{210}At to levels in ^{206}Bi. Three different α particles can be emitted here depending upon which branch of the decay scheme is followed for a given disintegration. For instance, 32% of the disintegrations result in a direct transition to the ground state of ^{206}Bi. Since the nuclear transitions in α decay are between discrete quantum states, the particles emitted have unique energies (e.g., 5.252 MeV for the transition to the ground state).

Examples of two kinds of β decay are shown in Fig. 11-2. The decay of ^{198}Au involves the emission of a negatively charged electron. The effect is to change one neutron to a proton, forming ^{198}Hg. In the decay of ^{64}Cu, the process is reversed. A proton is transformed to a neutron resulting in ^{64}Ni. As indicated in the figure, this second type of β decay occurs by either of two mechanisms. The letters *EC* indicate an electron capture. In this process an orbital electron is incorporated into the nucleus. The only emissions readily observed are x rays. The alternative is the formation of an electron–positron pair with emission of the positron.

Positrons are extremely unstable particles. They readily annihilate, forming pairs of 511-keV gamma rays which are emitted at 180° to one another. Consequently, positron emitters can be counted using gamma

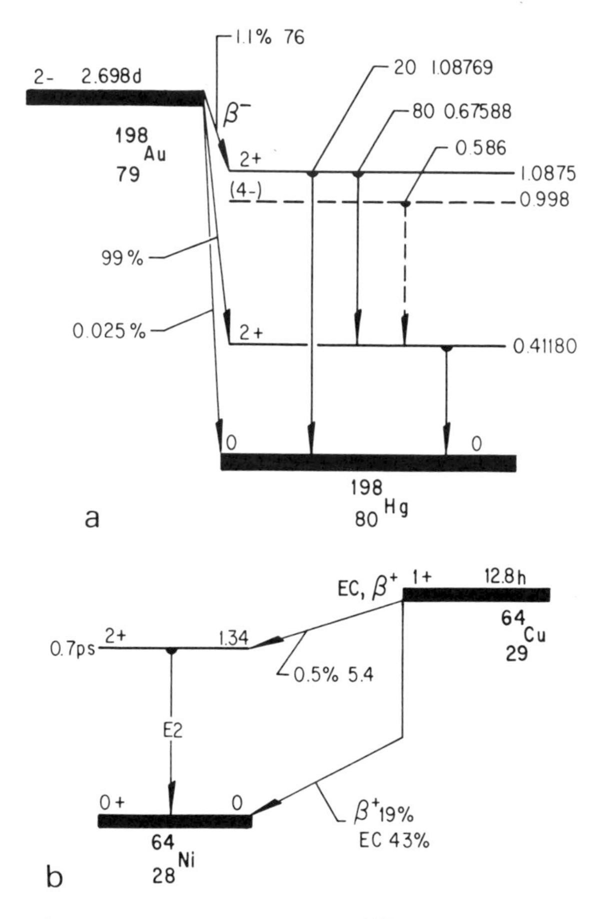

Fig. 11-2. Decay scheme of (a) ^{198}Au and (b) ^{64}Cu showing β emission.

detectors, while beta-minus emitters must be counted by detecting the emitted electrons.

From the point of view of counting β particles, the particles emitted are either positively or negatively charged electrons. Actually, however, β decay also involves the emission of neutrinos. The energy difference between the two quantum states in β decay is divided between the β particle and the neutrino in accordance with the law of conservation of momentum. Consequently, the energy spectrum of β particles is distributed over a wide range, from zero up to the total transition energy.

Figure 11-2 can also be used to illustrate the origin of γ rays. In the decay of ^{64}Cu, 0.5% of the β decays populate an excited state in ^{64}Ni at 1.34 MeV. The subsequent transition to the ground state is accomplished by γ emission. The γ ray or photon is emitted with a unique energy equal to the energy difference between the connected quantum states.

The majority of the isotopes encountered in the following sections decay by β–γ emission. Chemical separations are usually required for the assay of pure β emitters, both because of the polyenergetic distribution of the β spectrum and because a thin source must be counted in order to minimize absorption of β particles within the source. Isotopes which are pure β emitters are commonly identified on the basis of their half-lives.

The greater penetrating power of γ rays makes it possible to assay samples in the form of solids or solutions by γ-ray spectroscopy. Since γ rays are monoenergetic and specific for the emitting isotope, most elements can be identified without resorting to chemical separations or half-life measurements. Consequently γ-ray spectroscopy will normally be the most desirable technique for radioassay.

2.2. Types of Detectors

This section offers a brief discussion of some radiation detectors which seem to be the most useful for radiochemical analysis of surfaces. These detectors may be divided into three types: gas-filled detectors, scintillators, and semiconductor detectors. The use of film to produce autoradiograms is discussed in Section 5.

Gas-filled detectors are among the oldest of nuclear radiation detector types. A detailed discussion of their operation can be found in a book by Price.[1] A diagram of a gas counter is shown in Fig. 11-3.

Nuclear radiations enter the tube and ionize the gas. The charge collected on the electrode results in a pulse for each primary ionizing event. Gas-filled proportional counters are operated in a region of bias in which the size of the pulses produced is proportional to the energy of

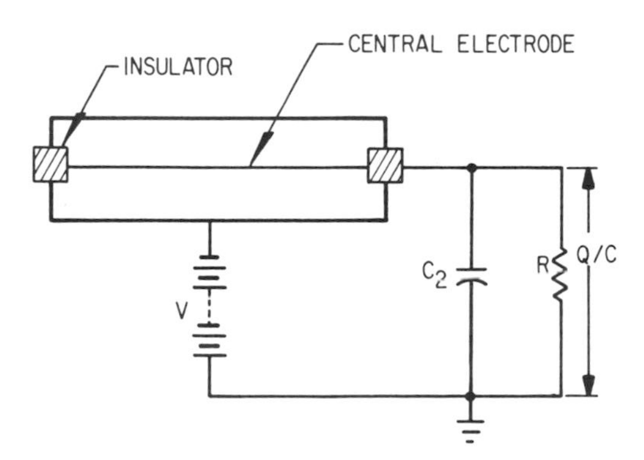

Fig. 11-3. Diagram of a gas counter.

the ionizing radiation. Energy resolution is therefore possible. In addition, dependence of pulse size on the primary ionization makes it possible to discriminate between radiation types. For instance, α particles can be counted easily in the presence of β particles. Since the detection mechanism involves the ionization of a gas, these detectors are most sensitive to charged particles such as α or β particles. Proportional counters are also used to detect very-low-energy photons such as soft x rays. However, they are poor γ-ray detectors.

Scintillation detectors made of NaI(Tl) have become a standard tool for the detection of γ rays. The principles of γ-ray spectrometry using NaI(Tl) detectors have been published by Heath.[2] A diagram of a NaI(Tl) scintillator and its associated photomultiplier tube is shown in Fig. 11-4.

Photons interact with the crystal to cause the emission of light (scintillation). The light strikes the cathode of the photomultiplier tube producing photoelectrons. An amplified pulse results at the photomultiplier anode. The pulse height observed is proportional to the energy deposited in the crystal by the photon. Scintillations can be induced in the crystal by charged particles as well as photons. Consequently, an absorber is normally used to permit only photons to enter the crystal.

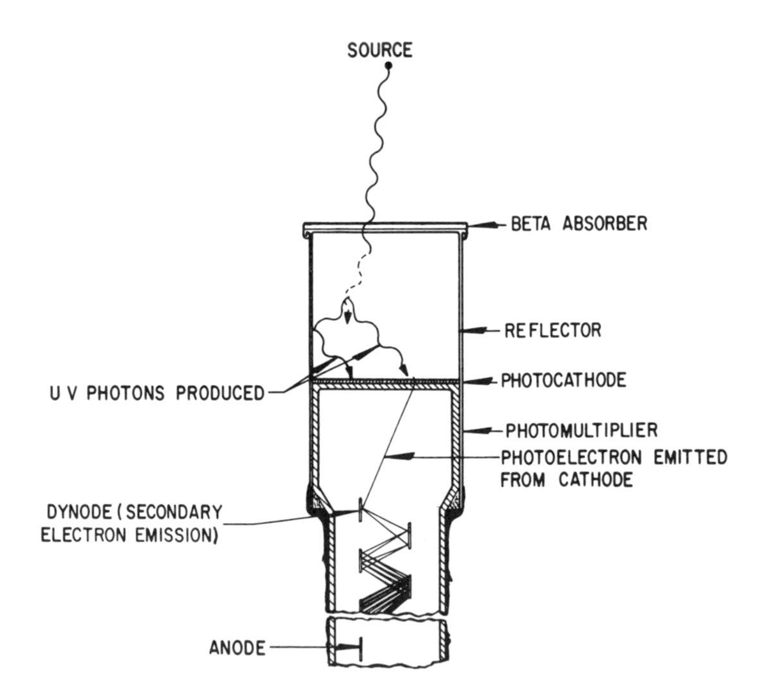

Fig. 11-4. Diagram of a NaI(Tl) scintillator with photomultiplier.

There are three basic interactions of photons with matter. Photoelectric interactions deposit all the photon energy in the crystal. The resulting pulse height is proportional to the photon's full energy. Compton scattering events deposit less than the full energy of the photon. This type of event can be envisioned as a billiard-ball-like collision between a photon and an electron with full conservation of momentum. If the scattered photon escapes the crystal, the resulting pulse height is less than "full energy." Particularly in a large detector, multiple Compton events may seem to produce a full-energy pulse.

Photons with more than 1022 keV of energy can produce electron–positron pairs. The positron usually annihilates within the crystal producing a pair of 511-keV gamma rays. If both annihilation γ rays escape without detection, the resulting pulse height represents an energy which is 1022 keV less than full energy (double escape). If one 511-keV photon is photoabsorbed, the pulse height is proportional to an energy 511 keV less than full energy (single escape). Of course, combinations of photoabsorption and Compton scattering can lead to a pulse height anywhere between double escape and full energy.

The relative probability of each type of interaction for NaI is shown as a function of energy in Fig. 11-5. A pulse-height distribution or spectrum of ^{24}Na is shown in Fig. 11-6. The full-energy, single-escape, and double-escape peaks for the 2753-keV photons can all be seen.

Semiconductor radiation detectors are the newest and most versatile type of detectors. All types of semiconductor detectors are basically diodes operated under reverse bias as shown in Fig. 11-7. The reverse bias depletes the high-field region of electron–hole pairs. Nuclear radiation striking this depleted region produces new electron–hole pairs. These are swept out by the bias, producing a pulse for each event. A detailed review of what is currently available in semiconductor detectors has been published by Muggleton.[5]

Briefly there are two means of producing a depletion region. A simple *p–n* junction will suffice although the depletion region produced in this manner will be very small, typically about 50 μ. Alternatively, the depletion region can be enlarged by the technique of lithium drifting. Depletion regions ranging from a few millimeters to nearly 2 cm in depth can be produced in this manner. Drifted detectors have the disadvantage that they must be operated under vacuum and cooled to liquid-nitrogen temperatures.

Junction devices such as silicon surface-barrier detectors are useful for detecting α particles and low-energy β particles. The small size of their depletion regions makes them relatively insensitive to high-energy β particles or γ rays.

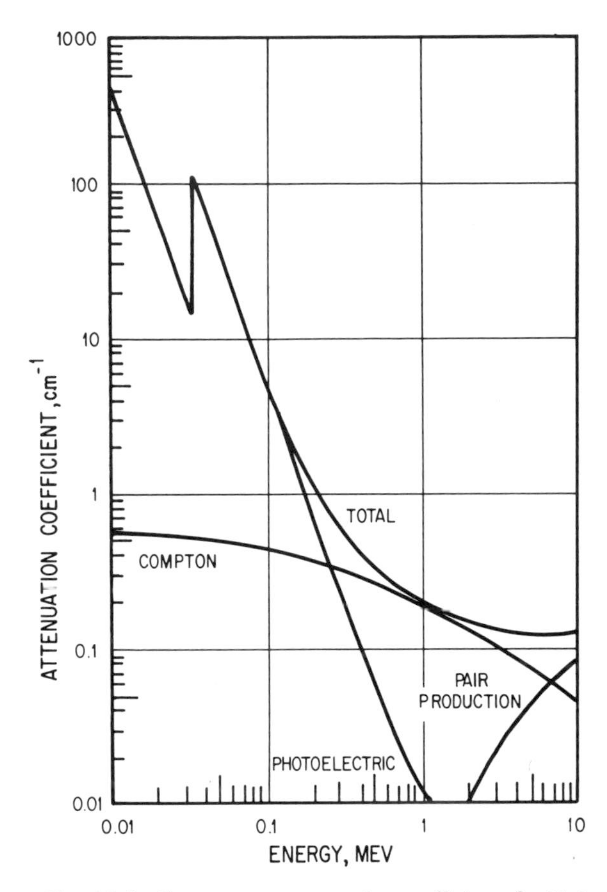

Fig. 11-5. Gamma-ray attenuation coefficients for NaI as a function of energy. (Reprinted from Harshaw Scintillation Phosphors,[3] data from G. R. White.[4])

Lithium-drifted-silicon detectors [Si(Li)] are very good β detectors. When they are shielded from betas, they are equally good x-ray detectors. Because of their greater mass, lithium-drifted-germanium detectors [Ge(Li)] are more efficient γ-ray detectors. This can be seen by comparing the relative probability of photoelectric absorption for Si, Ge, and NaI as shown in Fig. 11-8. The detection efficiency for a given detector [see Eq. (2)] is roughly proportional to the linear attenuation coefficient for photoelectric effect. Consequently, Fig. 11-8 suggests that NaI(Tl) gamma-ray detectors offer the greatest sensitivity in radioassay. In cases where interference between γ rays from various elements are not significant, this is correct. The value of Ge(Li) gamma-ray detectors depends more upon their energy resolution than their efficiency. The normal means of measuring energy resolution is to consider the full

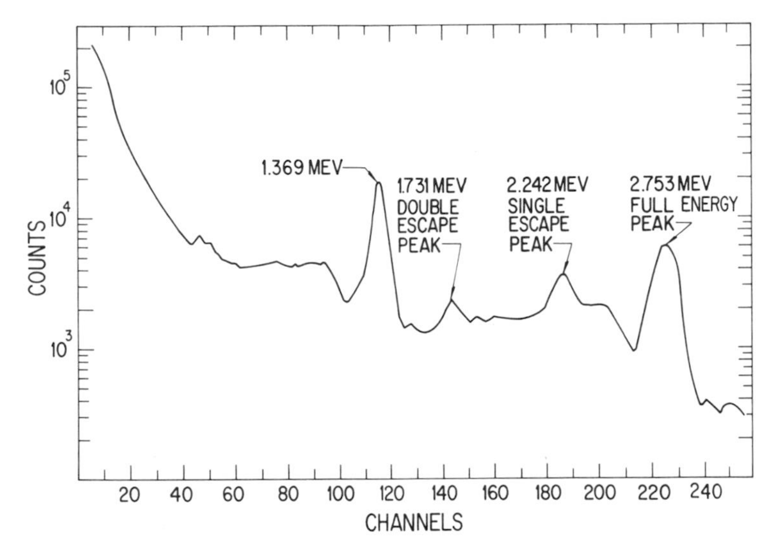

Fig. 11-6. Gamma spectrum of ^{24}Na using a NaI(Tl) detector.

width at half maximum (fwhm) of a full-energy peak. At 661 keV, a 7.6 × 7.6-cm NaI(Tl) detector may have fwhm = 46 keV while a good Ge(Li) system should easily reach fwhm = 2.5 keV. Figure 11-9 shows the dramatic difference between NaI(Tl) and Ge(Li) detection of a radioactive sample.

Thus for simple systems where γ–γ interferences are not significant, a NaI(Tl) detector is the better choice. But when interferences become significant, Ge(Li) detectors are superior. As increasingly larger Ge(Li)

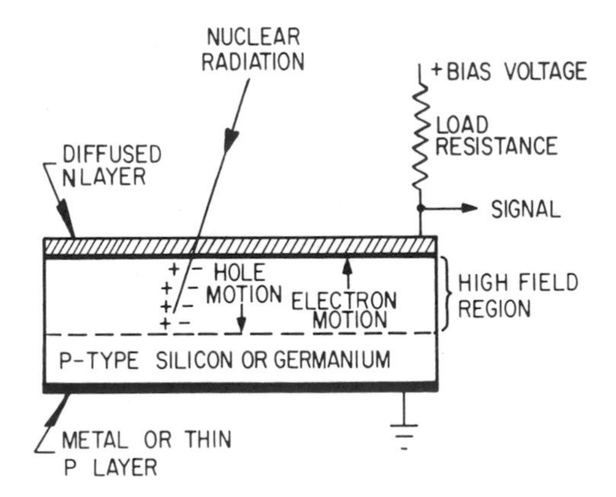

Fig. 11-7. Diagram of a semiconductor radiation detector.

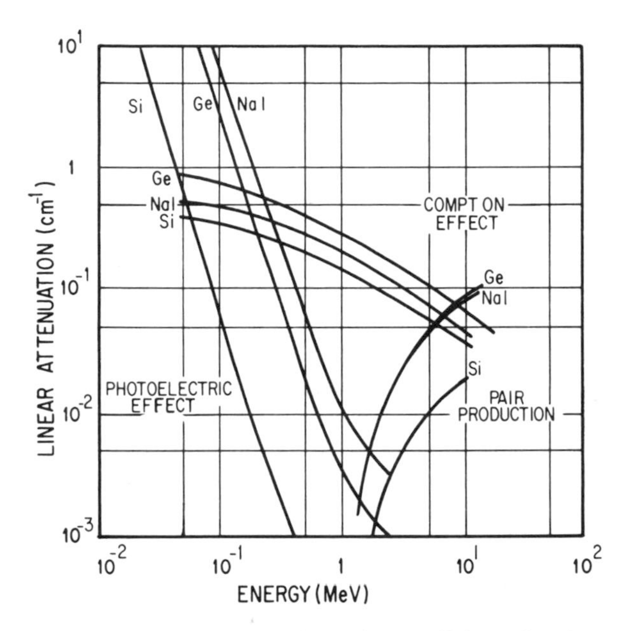

Fig. 11-8. Gamma-ray absorption coefficients for Ge, Si, and NaI as function of energy.

detectors become available, the reasons for considering NaI(Tl) at all become more economic rather than scientific. At the present time Ge(Li) detectors are more expensive than NaI(Tl) and they require more sophisticated data-acquisition systems.

2.3. Data Acquisition

All of the detectors described in the previous section respond to nuclear radiations with pulses. The pulse height is proportional to the

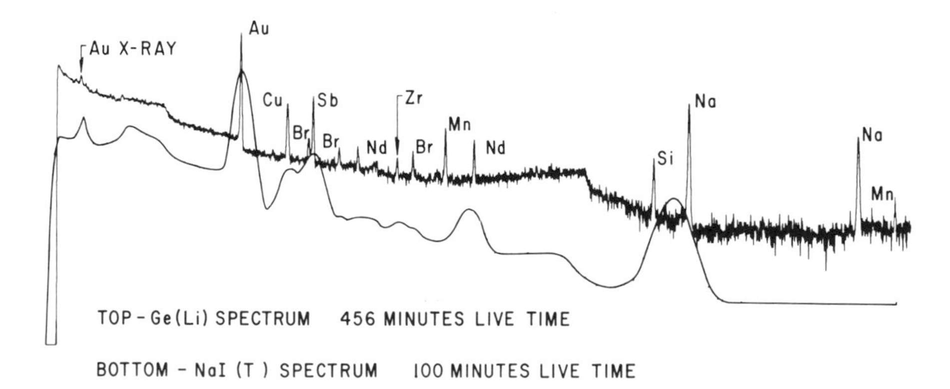

Fig. 11-9. A NaI(Tl) and a Ge(Li) detector spectrum of the same radioactive sample as Fig. 11-8. (After Keenan and Larrabee.[7])

energy deposited in the detector. The pulse rate or count rate is related to the disintegration rate by Eq. (2). In order to determine the count rate for a specific radiation, it is necessary to know the count rate as a function of pulse height. This process is referred to as pulse-height analysis. Generally the pulses emitted from radiation detectors must be shaped and amplified before pulse-height analysis. Particularly with semiconductor detectors, sophisticated amplifier systems may be required. This subject has been covered by Nowlin.[6] For the purposes of this work we shall assume that suitable amplification has been provided.

Pulse-height analysis can be performed by analog techniques such as a single-channel analyzer (SCA) with a timer-scaler or by digital means using a multichannel analyzer (MCA).

An SCA provides an upper- and lower-level discriminator, and all pulses which fall between these discriminator levels cause a logic pulse to be put out. The timer-scaler simply sums the number of logic pulses observed over a fixed time period. The discriminator levels can be used to focus upon a specific peak in a γ spectrum or to distinguish between high- and low-energy events on a proportional counter.

With the advent of NaI(Tl) scintillators for γ spectroscopy, multichannel pulse-height analysis became desirable. The pulse-height distributions shown in Figs. 11-6 and 11-9 were obtained by this technique. An MCA consists of two basic components; an analog to digital converter (ADC) and a memory. The ADC measures pulse heights and the memory stores a tally on the number of events observed at each pulse height.

The ability of an MCA to distinguish between different pulse heights will depend upon the number of channels it contains. If the amplifier system supplies pulses on a scale from 0 to 10 V, then a 500-channel ADC has resolution of 20 mV and a 4000-channel ADC has resolution of 2.5 mV. Because of the binary basis of digital electronics, ADC's usually come in sizes which are powers of 2, e.g., 512 channels and 4096 channels.

Normally it is desirable for the MCA memory to contain as many words as the ADC has channels. Nevertheless, useful information can be derived using a memory which is smaller than the ADC. The number of channels required depends upon the energy resolution of the detector. With NaI(Tl) detectors, 256 channels are normally sufficient to cover the region from 0 to 2 MeV. With high-quality Ge(Li) detectors at least 4096 channels are normally required. This increase in the number of data points per spectrum presents problems in data reduction. The usual solution is to turn to computer-based techniques.[7]

3. RADIOTRACER TECHNIQUES

The ready availability of radioactive isotopes, which started in the early 1950's with the introduction of nuclear reactors, has led to the extensive use of radiotracers in characterization techniques. The extension to surface studies is a natural one and these techniques have been widely applied to solid surfaces.

3.1. Principles

Radioactivity is a nuclear property and as such has a negligible effect upon the outer valence electrons of the atom. Since it is these outer electrons that control the chemical properties of the atom, a radioactive isotope, e.g., ^{59}Fe, behaves chemically in exactly the same manner as the stable isotopes of that element, e.g., ^{54}Fe, ^{56}Fe, ^{57}Fe, and ^{58}Fe. By utilizing this separability of nuclear and chemical properties, it is possible to trace the path of a particular element through a chemical system by following the radioactive isotopes of that element.

The ability to control the number of radioactive atoms in a given isotopic mixture provides a powerful quantitative tool for the researcher. This property is referred to as the specific activity and is generally described in units of activity per weight, e.g., counts per minute per microgram (cpm/μg) or millicuries per gram (mCi/g). The sensitivity, or minimum detectable quantity, can be controlled by varying the amount of radioactivity used in the experiment. This is accomplished by the controlled addition of the radioactive species. For example, the addition of 10^6 cpm of activity to 1 g or 1 mg of a stable species varies the sensitivity by a factor of 1000, yet the amount of activity is the same in either. Certain radioisotopes can be obtained "carrier-free", that is, essentially free of any stable isotopes. These "carrier-free" radioisotopes provide radiotracer sensitivity on the atomic scale. However, experimental problems are very difficult because of the few atoms in the system.

3.2. Techniques

The techniques used in the radiochemical characterization of a surface utilize all of the standard radiochemical technology. In brief, the process involves (1) adding the radioactive isotopes to the experiment and ensuring that chemical equilibration has been established between the active and stable isotopes, (2) conducting the experiment, and (3) performing a radioassay, e.g., β-, γ-, and α-counting, to locate and measure the amount of radioactivity at any point in the experiment.

Friedlander and Kennedy published their book, *Nuclear and Radiochemistry*, in 1955 with a second edition in 1964[8] and even today it remains a classic text on the techniques of radiochemistry. Overman and Clark[9] cover much of the experimental detail required for studies using radioisotopes.

3.3. Applications

Characterization of surfaces using radioisotopes generally involves observing the manner in which the radioactive species interact with the surface. The amount of radioactivity on the surface can be determined by counting techniques and the distribution by autoradiography (Section 5).

Schuerenkaemper[10] and Houstman and Medema[11] reported the use of ^{133}Xe and ^{85}Kr as adsorbates to determine surface areas of 0.003–1000 m^2/g. The qualitative observation of compositional differences at surfaces using adsorption and desorption of radioisotopes was used by Brandreth and Johnson.[12] Heinen and Larrabee[13] used radioactive ^{131}I to study the retention of organic photoresists on silicon surfaces. A sensitivity of 20 ng of residual photoresist on a slice with a 90% confidence error of $\pm 3.4\%$ was reported.

The reaction of radioactive ions from solutions on solid surfaces has been extensively used to understand semiconductor surfaces.[14–16] Memming[17] used ^{64}Cu and ^{198}Au to study the origin of fast surface states at the germanium–electrolyte interface. Haissinsky and Tuck[18] have reviewed the application of radioisotopes to a large number of surface properties including catalysis and electrochemistry.

4. ACTIVATION ANALYSIS

4.1. Principles

When a sample is irradiated with particles such as photons, neutrons, or ions, nuclear reactions induced by these particles produce one or more reaction products for each atomic species present. These products are specific for a particular type of parent isotope. The radioactive products can be identified on the basis of their unique radioactive properties such as half-life, type of radiation emitted, and the energy of that radiation. The identity of the parent nucleus can then be established from a knowledge of the nuclear reactions induced during irradiation.

In the case where the target atoms are not radioactive, the number of target atoms N originally present in the sample can be determined from Eq. (4):

$$N = \frac{DPS}{\sigma\phi(1 - e^{-\lambda t})(e^{-\lambda T})} \tag{4}$$

where DPS is disintegrations per second of radioactive reaction product,

σ is the activation cross section in cm^2, ϕ is the particle flux in particles/cm^2/sec, N is the number of target atoms, λ is the decay constant of the reaction product (0.693/half-life), t is the irradiation time, and T is the decay time.

The activation cross section σ for a given exciting particle is a property of the target atom. Generally, it will vary with the energy of the exciting particle. The decay constant λ is a property of the radioactive reaction product and is independent of the manner by which that product was formed. Therefore, the activity produced for specific elements can be maximized within practical limits by the choice of the energy and the type of exciting particle as well as by varying the flux, the irradiation time, and the decay time.

Many of the nuclear reactions used for activation analysis can be understood in terms of the compound-nucleus model.[19] This model assumes that when an incident particle penetrates to the nuclear surface of a target isotope, the incident particle is absorbed and a new nucleus is formed which is in an excited state. This compound nucleus is analogous to the "activated complex" used in absolute-reaction-rate theory.[20] The compound nucleus subsequently decomposes by the emission of charged particles, neutrons, or γ rays to form the reaction product. The type and number of emissions depends upon many factors, one of which is the excitation energy of the compound nucleus.

The energy deposited when thermal neutrons are absorbed into target nuclei is so low that generally only γ emission follows. The reaction for gold may be represented as

$$^{197}\text{Au}(n, \gamma)^{198}\text{Au}(t_{1/2} = 64.8 \text{ hr})$$

The product of an (n, γ) reaction therefore, is an isotope of the target nucleus, but 1 mass unit heavier.

4.2. Techniques

After irradiation, it is necessary to identify the reaction products and to determine their disintegration rates. The traditional method of activation analysis involves chemical separation of the various elements prior to radioassay. These separations aid in identifying the radioactive products and assure maximum sensitivity for all elements by eliminating interferences between similar radiations. Multielement analyses, however, can become very long, complex procedures. Analysis by instrumental means is therefore much more desirable.

The most widely used instrumental technique is γ-ray spectroscopy. The penetrating power of γ rays makes it possible to assay samples in the form of solids or solutions by this technique. The simplicity of source preparation for γ-ray spectroscopy, however, is often counterbalanced by the complexity of data reduction. Computer techniques are commonly used for this purpose. With low-resolution NaI(Tl) gamma-ray spectrometers, moderately complex γ-ray spectra can be unfolded using least-squares analyses such as the one developed by Helmer *et al.*[21] These analyses require a library of standard spectra for each element or isotope in the composite spectrum. With high-resolution Ge(Li) detectors, impurity concentrations can be derived from peak areas without reference to a library of experimentally observed standard spectra.[22]

In the general case, the particle flux, the activation cross section, the detection efficiency, and the branching ratio for the characteristic radiation must all be determined in order to convert an observed count rate to an impurity concentration. In practice the simplest and most accurate quantitative determinations are made by a direct comparison of the activity induced in the sample and in a standard. When the samples and standards are irradiated in the same particle flux and assayed on the same detector, Eqs. (2) and (3) reduce to a simple ratio involving only a correction for decay after irradiation.

$$\frac{CPS_{\text{sample}}}{CPS_{\text{standard}}} = \frac{N_{\text{sample}}\, e^{-\lambda T_1}}{N_{\text{standard}}\, e^{-\lambda T_2}} \tag{5}$$

where T_1 and T_2 are the times after irradiation at which the sample and standard were assayed.

As can be seen from Eqs. (2) and (3), the sensitivity which can be achieved in activation analysis depends upon the irradiation conditions and the efficiency of the detector used for radioassay. The activities which can be induced for specific irradiation conditions using 35-MeV photons, 14-MeV neutrons, 18-MeV ^{3}He, and thermal neutrons are given for a few elements in Table 11-1. For the purpose of comparison, the units are disintegrations per second per microgram of target material. For a given radiation detector it should be clear from Eqs. (2) and (3) that the limit of detectability can be expressed in terms of the minimum number of target atoms. The minimum detectable concentration therefore will vary depending upon the volume to be analyzed.

In order to analyze large samples, it is necessary to produce a uniform flux of exciting particles across the entire sample. Of the four irradiation methods compared in Table 11-1, thermal neutrons from a nuclear reactor are most readily applied to a wide variety of sample sizes. Moreover, except for the light elements, this method also has the best sensitivity. On the other hand, the use of charged particles such as

TABLE 11-1. Induced Activity in *DPS*/μg at the End of Irradiation

Element	35-MeV photons, 100 μA for 4 hr	14-MeV neutrons, 10^9n/cm^2/sec for 1 hour	18-MeV ^{3}He, 10 μA for one half-life	Thermal neutrons, 1.5×10^{13} n/cm^2/sec for 14 hr
Au	100	0.000056	—	864,000
Cu	2000	3.2	—	298,000
Na	0.09	0.0004	1238	101,000
As	10	0.038	—	229,000
Sb	100	2.27	—	112,000
Ag	2000	2.64	—	477
W	3	—	—	729,000
Si	30	4	1238	345,000
C	500	—	8256	—
O	600	3.4 (1 min)*	4661	—
N	800	0.214 (45 min)*	1247	—

* Irradiation time was less than 1 hr due to the short half-life of the product.

^{3}He has an inverse advantage. Since beams of charged particles can be highly focused, it is possible to determine the lateral distribution of impurities by successive irradiations and analysis of small spots. Photons can be obtained either from isotopic sources or from the bremsstrahlung resulting from the stopping of charged particles such as electrons. The latter was the case for the activities given in Table 11-1. Similarly, high-energy or fast neutrons can be obtained either from isotopic sources or from nuclear reactions induced by charged particles from an accelerator. Both photons and fast neutrons can be classified as intermediate in their sensitivity and their applicability to a variety of sample sizes.

Clearly the choice of irradiation conditions and detector systems should be aimed at maximizing sensitivity for the impurities of interest. At the same time, careful consideration must be given to the problem of interferences between induced activities. In many instances the reaction products which are of interest can be separated chemically after irradiation. Often, however, a suitable choice of irradiation conditions can be made to avoid the problem.

Many light elements ($Z \leq 14$) have the combination of small thermal-neutron cross sections and/or very short-lived radioactive (n, γ) products. Therefore, matrices of such elements can be readily analyzed for heavier impurities by neutron activation. One notable exception is sodium whose (n, γ) product (15-hr ^{24}Na) is relatively long-lived and is produced with a good cross section (0.53 b). Tang and Maletskos[23] have suggested a technique for reducing interferences from sodium activities by heterogeneous isotopic exchange.

As a positively charged projectile approaches a similarly charged target nucleus, the Coulomb repulsion gives rise to a potential barrier. The height of the barrier, V, for a target of proton number Z_1 and nuclear radius R_1 and a projectile with Z_2 and R_2 is given by Eq. (6):

$$V = 1.44 \frac{Z_1 Z_2}{R_1 + R_2} \text{ MeV} \tag{6}$$

The nuclear radius can be obtained from Eq. (7):

$$R = 1.6A^{1/3} \text{ fermi } (10^{-13} \text{ cm}) \tag{7}$$

where A is the mass number. It follows from Eq. (6) that the Coulomb barrier for a given target nucleus is about half as high for protons or deuterons as it is for ^3He or α particles. The barrier increases in height with increasing Z of the target nucleus. Charged particles may therefore be used to activate light elements in heavy matrices by adjusting the bombarding energy to stay below the Coulomb barrier for the elements of the matrix.

4.3. Applications

A silicon matrix is particularly well suited for activation analysis using thermal neutrons. The only radioactive (n, γ) product from the matrix is 2.6-hr ^{31}Si while the (n, γ) products of some of the most important impurities in silicon are 15.0-hr ^{24}Na, 12.9-hr ^{64}Cu, 64.8-hr ^{197}Au, 26.5-hr ^{76}As, 2.8-day ^{122}Sb, and 60.2-day ^{124}Sb. Therefore, considerable sensitivity for these and many other elements in silicon can be achieved by permitting the 2.6-hr ^{31}Si to decay after irradiation. Theoretical detection limits for impurities in silicon for a specific set of irradiation conditions, sample size, and efficiency in radioassay are given in Table 11-2.

In the study of surfaces by activation analysis it is usually a practical necessity to irradiate the entire sample and to distinguish between bulk and surface impurities by chemical or physical means following activation. Selective etching techniques are commonly used for this purpose.

Lunde[25] used neutron activation to investigate the impurities on the surfaces of polished silicon slices. The sensitivity required for such an analysis of course is inversely proportional to the thickness of the "surface layer" removed. Lunde augmented his sample size by analyzing etchants from the surfaces of ten silicon slices as one sample. He defined the "surface layer" by alternatively etching and assaying the samples. When the activity removed corresponded to the weight decrease in the

TABLE 11-2. Detection Limits for Impurities in Silicon Epitaxial Material Etched Off for Analysis Assuming a 2-inch Slice, Analyzing 10 μ*

Element	Atoms/cc	Element	Atoms/cc
Antimony 122	6.949E + 12	Neodymium 147	1.957E + 14
Antimony 124	1.759E + 14	Osmium 191	4.525E + 13
Arsenic 76	6.377E + 12	Palladium 109	5.105E + 13
Barium 131	1.084E + 16	Phosphorus 32 beta	3.389E + 13
Barium 133	6.971E + 17	Platinum 197 0.2 MeV	5.380E + 14
Bromine 82	1.338E + 13	Platinum 197 77 keV	1.286E + 14
Cadmium 115	1.854E + 15	Potassium 42	3.839E + 15
Cadmium 115M	2.982E + 16	Praseodymium 142	1.002E + 14
Calcium 45 beta	4.690E + 15	Rhenium 188	1.328E + 12
Cerium 141	1.862E + 14	Rubidium 86	2.364E + 15
Cerium 143	1.419E + 14	Ruthenium 97	7.149E + 14
Cesium 134	1.087E + 14	Ruthenium 103	3.421E + 14
Chromium 51	1.006E + 15	Samarium 153	3.270E + 11
Cobalt 60	7.392E + 14	Scandium 46	2.207E + 13
Copper 64	1.519E + 13	Selenium 75	1.555E + 15
Dysprosium 165	7.031E + 15	Silver 110M	7.782E + 14
Erbium 171	4.041E + 13	Sodium 24	7.254E + 13
Gadolinium 159	7.954E + 13	Strontium 85	4.938E + 16
Gallium 72	1.295E + 13	Sulfur 35 beta	3.174E + 15
Germanium 77	4.587E + 15	Tantalum 182 0.1 MeV	5.480E + 13
Gold 198	1.289E + 11	Tantalum 182 1.0 MeV	1.147E + 14
Hafnium 181	3.520E + 13	Tellurium 127	3.495E + 17
Holmium 166	1.318E + 12	Terbium 160	1.163E + 13
Indium 114	2.157E + 14	Thallium 204 beta	1.110E + 14
Indium 116	1.447E + 22	Thulium 170	4.120E + 13
Iridium 192	1.796E + 12	Tin 113	4.138E + 16
Iridium 194	1.094E + 12	Tungsten 187 0.1 MeV	4.985E + 12
Iron 59	1.542E + 17	Tungsten 187 0.7 MeV	4.988E + 12
Lanthanum 140	4.467E + 12	Ytterbium 169	1.061E + 13
Lutecium 177	3.973E + 12	Yttrium 90M	2.364E + 18
Manganese 56	2.698E + 15	Zinc 65	1.634E + 16
Mercury 197	2.492E + 13	Zinc 69M	8.500E + 14
Mercury 203	1.145E + 14	Zirconium 95	4.215E + 16
Molybdenum 99	1.426E + 15	Zirconium 97	1.510E + 16

* Sample counted on a 3 × 3 NaI spectrometer. 14.00 h irradiation at 1.50E + 13 neutrons/cm^2/sec. 36.00 h decay. (From Larrabee and Keenan.[24])

slices, all surface impurities were assumed to have been removed. A check was made for losses during etching by comparing the activity initially observed in the slices with the sum of the activity in the etchant and the residual activity in the slices.

TABLE 11-3. Concentrations of Three Impurities in Epitaxial Silicon Showing Distribution

	Concentration (atoms/cc)	
Impurity	Surface, 4μ	Epitaxial film, >4μ
Gold	1×10^{13}	1×10^{12}
Copper	7×10^{14}	6×10^{13}
Sodium	3×10^{15}	7×10^{13}

Larrabee and Keenan[24] used neutron-activation analysis and γ-ray spectroscopy for the analysis of epitaxial films on silicon slices. The conditions for ordinary analyses were those outlined in Table 11-2. Sufficient sensitivity for the analysis of individual slices was achieved through careful shielding of the detectors, background subtracting, and spectrum stripping. For the routine analysis of epitaxial films over many months, the first four microns were taken to represent the "surface layer." Results of the analyses of this surface layer and of the remainder of typical epitaxial films are shown in Table 11-3. A better feeling for what constitutes surface contamination on these films was achieved by profiling epitaxial layers. Two such profiles are shown in Figs. 11-10 and 11-11.

Successive etches ranging in depth from 0.5 to 4.0 μ were removed and analyzed individually in order to derive these profiles.

Osborne, Larrabee, and Harrap[26] analyzed silicon dioxide films on silicon for sodium. The distribution of sodium across such a film is shown in Fig. 11-12. Clearly, profiling is an effective means of distinguishing between surface and bulk impurities.

Butler and Wolicki[27] used the high Coulomb barrier of gold and platinum together with the low penetrating power of charged particles in matter to analyze the surface of gold and platinum foils for carbon, oxygen, and aluminum. They used 4.5-MeV ^{3}He particles for excitation by means of the nuclear reactions

$$^{12}C(^3He, \alpha)\,^{11}C$$

and

$$^{16}O(^3He, p)\,^{18}F$$

for carbon and oxygen. As can be seen from Eq. (6) the Coulomb barrier for gold and platinum with respect to ^{3}He is approximately 20 MeV. Consequently 4.5-MeV ^{3}He projectiles were unable to penetrate the Coulomb barrier sufficiently to react with the nuclei of the matrices.

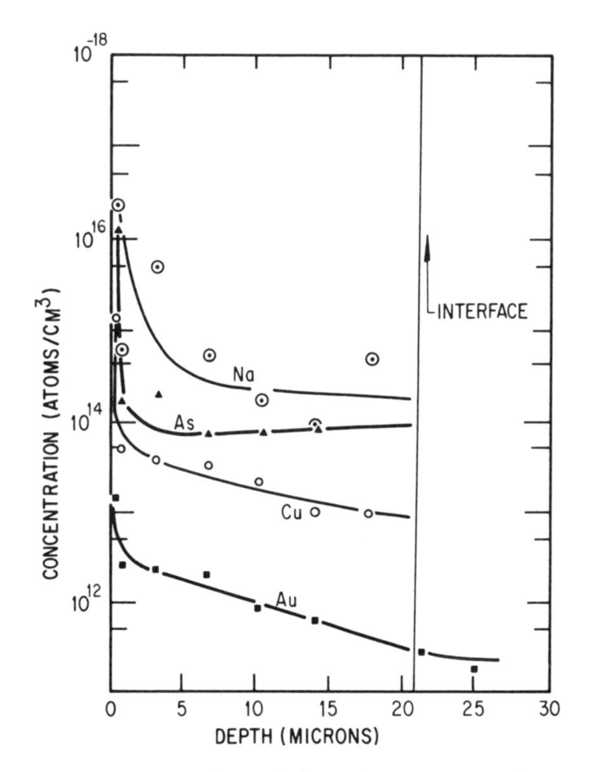

Fig. 11-10. Profile of impurity concentrations through a 20-μ silicon epitaxial film on a boron-doped substrate. (After Larrabee and Keenan.[24])

They could, however, penetrate the lower Coulomb barriers of nuclei such as carbon and oxygen which have fewer protons and therefore less nuclear charge. Since charged particles lose their energy rapidly in solids,[28] the incident ^{3}He particles retained sufficient energy to produce the desired reaction products only if they underwent nuclear reactions very near the surface.

The reaction products both emit positrons and the associated annihilation radiation (511 keV). Radioassay was accomplished by following the decay of 511-keV radiation and separating the composite decay curve into two components, ^{11}C($t_{1/2}$ = 20 min) and ^{18}F($t_{1/2}$ = 110 min). Concentrations of the target atoms were calculated using Eq. (5), the measured flux (beam current), the detection efficiency, and published reaction cross sections.

Aluminum on the surface of the same disks was determined by observing the reaction γ ray from the nuclear reaction

$$^{27}\mathrm{Al}(p, \gamma)\,^{28}\mathrm{Si}$$

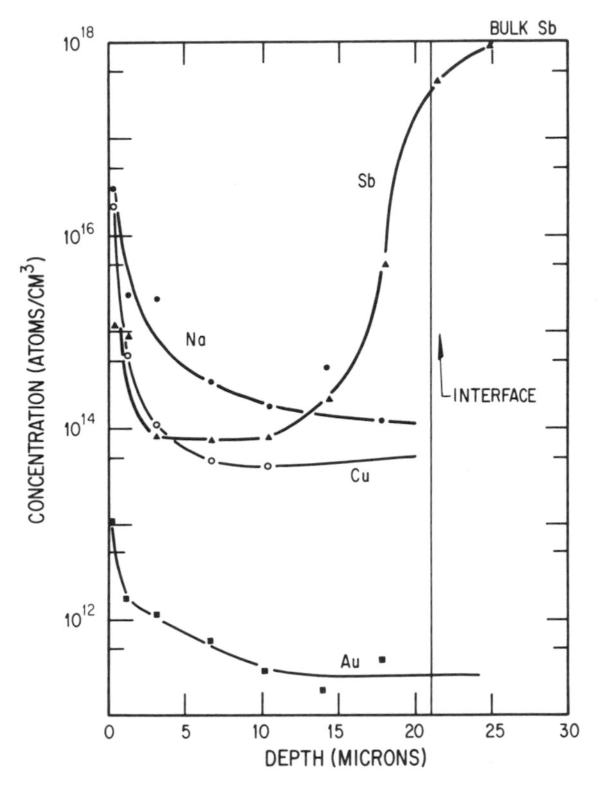

Fig. 11-11. Profile of impurity concentrations through a 20-μ silicon epitaxial film on an antimony-doped substrate. (After Larrabee and Keenan.[24])

A γ detector was mounted in close proximity to the target. The aluminum concentration was determined by comparing the γ intensity for the sample and for an aluminum foil of known thickness under the same irradiation conditions.

Barrandon and Albert[29] used tritium particles to detect oxygen on the surface of zirconium and aluminum; the reaction

$$^{16}O(^{3}H, n)\,^{18}F$$

can be induced by 2-MeV tritium particles without significant excitation of the matrix. The reaction product was identified on the basis of its half life by following the decay of γ-ray intensity at 511 keV.

5. AUTORADIOGRAPHY

Autoradiography is a surface-analysis technique that is used to complement the other radioisotope techniques. It can be used effectively in conjunction with radiotracer studies to obtain information on the

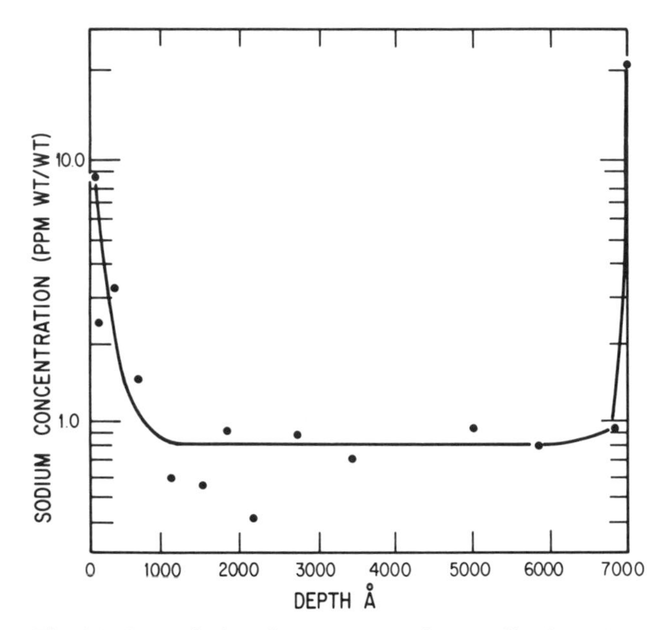

Fig. 11-12. Typical sodium-concentration profile through a 7000-Å silicon dioxide film. (After Osborne, Larrabee, and Harrap.[26])

topographic distribution of the radioactivity across the surface. This technique normally gives semiquantitative information. Under favorable conditions quantitative information on the radioactive species may be obtained. However, the adage "a picture is worth a thousand words" certainly holds for autoradiography. This aspect of the technique makes it a particularly useful method for surface analysis.

Henri Becquerel,[30] in 1896, discovered the blackening of photographic plates by crystals of a uranium salt, and in so doing provided the first observation of both radioactivity and autoradiography. The observations remained a curiosity until 1924 when Lacassagne and Lattes[31] used photographic emulsions to study the distribution of polonium (polonium is naturally radioactive) in biological samples. Autoradiography remained a tool for biological studies until the early 1950's when adequate quantities of a variety of radioactive species became available from atomic reactors. At this time other disciplines, e.g., metallurgy, could use radioactive tracers and autoradiography to solve a wide variety of surface problems.

5.1. Principles

Autoradiography is very simple in principle; however, as will be discussed later, the technology tends more to be an art with highly

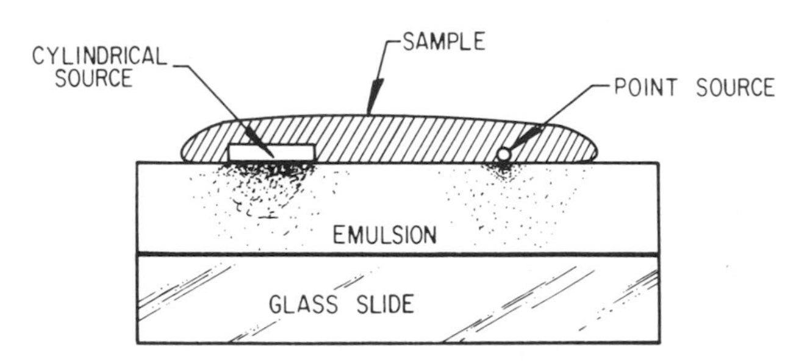

Fig. 11-13. Schematic illustrating the technique of autoradiography.

specialized emulsions and darkroom techniques. In principle, the technique is shown in Fig. 11-13 where the radioactive sample is brought into intimate contact with a photographic emulsion. The ionizing radiations from the specimen interact with the silver bromide in the emulsion. In the presence of a developing agent, the activated site catalyzes the complete conversion of the entire silver bromide crystal to metallic silver. Unexposed silver bromide crystals are dissolved out of the emulsion by the developer.

Since the entire technique depends upon an efficient interaction between the ionizing radiation and the photographic emulsion, it is apparent that the type of radiation (α, β, γ, etc.) will have a profound effect on both the fidelity of the autoradiograph as well as the techniques used to obtain optimum exposure.

Norris and Woodruff(32) calculated the rate of energy loss for α and β particles in a nuclear emulsion. Their results are shown in Fig. 11-14. Alpha particles appear to be superior to beta particles for autoradiographic work. This is because α particles are double charged and have a rest mass 8000 times that of the electron. Unfortunately, radioisotopes that are α emitters exist only for elements with very high mass, e.g., greater than 200. Normally, this part of the periodic table is of little interest for radiotracer work.

5.2. Techniques

Rogers(34) has published an excellent book on the wide spectrum of techniques employed in autoradiography. While Rogers' work is restricted to biological specimens, thc techniques are widely applicable to other materials.

There is always a compromise between resolution and exposure time. Unfortunately the two criteria are mutually exclusive for any given radioisotope since large-grain emulsions are required to minimize exposure times while very fine-grain emulsions are necessary to obtain maximum resolution. Wainwright *et al.*[35] prepared a nomogram to aid in the calculation of exposure time. Their work was based on the formula of Axelrod and Hamilton[36] that states that 2–10 × 10^6 beta particles per square centimeter striking the emulsion give a suitable exposure. These estimates must be adjusted for the energy of the particular β particle, the type of emulsion used, and the distribution of the radioactivity on the sample.

In most radiochemical laboratories, a log of each autoradiogram is kept, where sample and isotope identification, β counts per minute, type of emulsion, exposure time, and overall quality of the final autoradiogram are recorded. In a very short while a good working empirical relationship between exposure time and radioisotope activity is evolved.

Virtually all autoradiograms are made from β-emitting radioisotopes. It is obvious from Fig. 11-14 that the amount of interaction with the nuclear emulsion will be strongly dependent upon the energy of the particle. This is further complicated by the energy spectrum of β emission. Unlike α and γ emission, β particles are emitted in a continuous energy spectrum ranging from zero to the maximum energy or end-point energy (E_{max}). This latter value, E_{max}, is used to describe the β

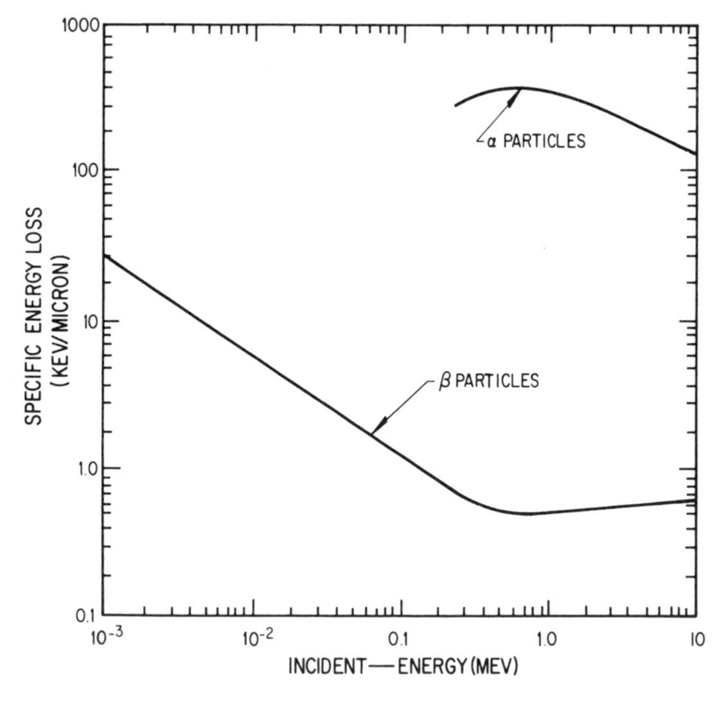

Fig. 11-14. Rate of energy loss in nuclear emulsion material. (After Norris and Woodruff.[32])

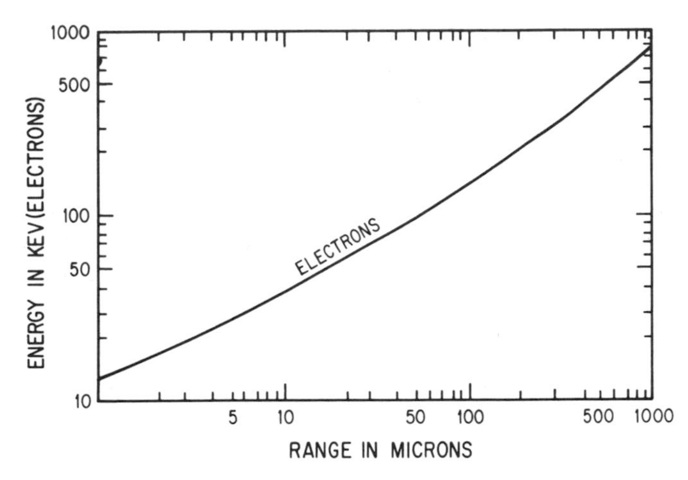

Fig. 11-15. Beta-particle range in nuclear-emulsion material. (After Hertz.[33])

emitted from a particular nuclide, e.g., ^{14}C, E_{max} of 0.156 MeV. From the standpoint of autoradiography there are very few β particles with E_{max}. In fact, the β spectrum is such that the majority of emissions will have energies near $1/3E_{max}$. However, published data showing the range of β particles in nuclear emulsions as a function of β energy normally refer to E_{max}. Such a plot is shown in Fig. 11-15. Using such data it can be seen that a particular emulsion can be chosen to maximize β interaction for each E_{max}.

Nuclear emulsions are generally transparent to γ rays. A small amount of interaction occurs through infrequent secondary electrons which ultimately only give rise to background in the autoradiogram. It is possible, however, to obtain γ autoradiograms by using very thick emulsions and optimizing the generation of secondary electrons by backing the nuclear emulsion with high-mass material such as lead.

A wide variety of emulsions are available, ranging from the highest-resolution nuclear emulsion through the lowest-resolution but highest-sensitivity x-ray emulsions. Osborne[37] described the use of dental x-ray films for a wide variety of semiconductor research problems. A comparison between dental x-ray films, x-ray films, and nuclear emulsions by Osborne[37] clearly showed that for resolutions in the 10-μ range the less-expensive and faster dental x-ray emulsions were quite adequate.

5.3. Applications

Verkerk[38] in 1956 was probably one of the earliest workers to exploit autoradiography as a surface-characterization technique.

Verkerk studied the adsorption of chlorine, using ^{34}Cl and ^{38}Cl, on aluminum oxide and showed that the oxide layer was preferentially attached on cracks and along grain boundaries. Resolutions in the 2–3-μ range were reported.

Kohn,[39] in an excellent review of the use of radioisotopes in metallurgical research, reviewed the application of autoradiography to a wide variety of wear and friction problems. Since both friction and wear involve the rubbing of surfaces, it is a fertile area for the use of radioisotopes and autoradiography.[40,41]

Gal, Gruzin, and Yudina[42] used autoradiography to study the surface diffusion of metals. Bacon,[43] studying the corrosion or surface oxidation of iron, used autoradiography to show localized points of attach at the surface. Autoradiographic studies of metals rendered passive by chromate solutions indicate that radiochromium is concentrated at the anodic sites of the surface.

6. SUMMARY

Radioisotope techniques offer two types of information in the characterization of surfaces. Activation analysis provides a means for determining the chemical impurities at or near the surface. When coupled with autoradiography, activation analysis can provide both quantitative data on concentration as well as topographic localization of the impurity sites.

Radiotracer techniques provide indirect characterization of the chemical nature and composition of the surface. They provide insight into the role of chemical or physical species at the surface during processes such as adsorption–desorption, corrosion, and catalytic activity. Here particularly, autoradiography is essential to understanding the radiotracer characterization of the surface.

7. REFERENCES

1. W. J. Price, *Nuclear Radiation Detection*, McGraw-Hill, New York (1958).
2. R. L. Heath, Scintillation spectrometry, **IDO** 16880 (1964).
3. *Harshaw Scintillation Phosphors*, p. 11, The Harshaw Chemical Company, Cleveland (1962).
4. G. R. White, National Bureau of Standards Circular 583 (1957).
5. A. H. F. Muggleton, *J. Phys. E* **5**, 390 (1972).
6. C. H. Nowlin, *IEEE Trans. Nucl. Sci.* NS (17), 226 (1970).
7. J. A. Keenan and G. B. Larrabee, *Chem. Inst.* **3** (2), 125 (1971).
8. G. Friedlander, J. W. Kennedy, and J. M. Miller, *Nuclear and Radiochemistry*, 2nd ed., John Wiley & Sons, New York (1964).
9. R. T. Overman and H. M. Clark, *Radioisotope Techniques*, McGraw-Hill, New York (1960).
10. A. Schuerenkaemper, *J. Phys. Chem.* **69**, 2300 (1965).

11. J. P. W. Houstman and J. Medema, *Ber. Banenges. Physik. Chem.* **70**, 489 (1966).
12. D. A. Brandreth and R. E. Johnson, *Science* **169**, 864 (1970).
13. K. G. Heinen and G. B. Larrabee, *Solid State Tech.* **12** (4), 44 (1969).
14. W. Kern, *RCA Rev.* **31**, 207 (1970).
15. G. B. Larrabee, *J. Electrochem. Soc.* **108**, 1130 (1961).
16. W. Gebauhr, *Z. Anal. Chem.* **245**, 209 (1969).
17. R. Memming, *Surface Science* **2**, 436 (1964).
18. M. Haissinsky and D. G. Tuck, *Nuclear Chemistry and its Applications*, Addison-Wesley, Reading, Mass. (1964).
19. G. Friedland, J. W. Kennedy, and J. M. Miller, *Radioisotopic Techniques*, Chapt. 10, p. 337, McGraw-Hill, New York (1964).
20. T. L. Hill, *An Introduction to Statistical Thermodynamics*, Chapt. II, pp. 189–200, Addison-Wesley, Reading, Mass. (1962).
21. R. G. Helmer, R. L. Heath, D. D. Metcalf, and G. A. Cazier, **IDO** 17015 (1964).
22. J. A. Keenan and G. B. Larrabee, *Chem. Instr.* **3** (2), 125–140 (1971).
23. C. Tang and C. J. Maletskos, *Science* **167** (3914), 52 (1970).
24. G. B. Larrabee and J. A. Keenan, *J. Electrochem. Soc.* **118** (8), 1352–1355 (1971).
25. G. Lunde, *Solid State Technol.* **13**, 61–69 (1970).
26. J. F. Osborne, G. B. Larrabee, and V. Harrap, *Anal. Chem.* **39** (10), 1144–1148 (1967).
27. J. W. Butler and E. A. Wolicki, *Modern Trends in Activation Analysis*, National Bureau of Standards Special Publication 312 (1969), Vol. II, pp. 794–801.
28. M. D. Tran and J. Tousset, *Modern Trends in Activation Analysis*, National Bureau of Standards Special Publication 312 (1969), Vol. II, pp. 754–767.
29. J. N. Barrandon and P. Albert, *Modern Trends in Activation Analysis*, National Bureau of Standards Special Publication 312 (1969), Vol. II, pp. 794–801.
30. H. Becquerel, *Compt. Rend.* **122**, 501–689 (1896).
31. A. Lacassagne and J. S. Lattes, *Bull. Histol. Appl. Tech. Microscop.* **1**, 279 (1924).
32. W. P. Norris and L. A. Woodruff, *Ann. Rev. Nucl. Sci.* **5**, 297 (1955).
33. R. H. Herz, *Nucleonics* **9** (3), 24 (1951).
34. A. W. Rogers, *Techniques of Autoradiography*, Elsevier, New York (1967).
35. W. W. Wainwright, E. C. Anderson, P. C. Hammer, and C. A. Leman, *Nucleonics* **12** (1), 19 (1954).
36. D. J. Axelrod and J. G. Hamilton, *Am. J. Pathol.* **23**, 389 (1947).
37. J. F. Osborne, *Int. J. Appl. Rad. Isotopes* **18**, 829 (1967).
38. B. Verkerk, *Nucleonics* **14** (7), 60 (1956).
39. A. Kohn, *Met. Rev.* **2** (10), 143 (1958).
40. J. T. Burwell and C. O. Strang, *J. Appl. Phys.* **23**, 18 (1952).
41. M. E. Merchant, M. Ernst, and Krabacher, *Trans. ASME* **75**, 549 (1953).
42. U. V. Gal, P. L. Grazin, and G. K. Yudina, *Fiz. Metal. i Metalloved.* **30** (5), 950 (1970).
43. C. G. Bacon, *Gen. Elec. Rev.* **52** (5), 7 (1949).

12

X-RAY FLUORESCENCE ANALYSIS

John V. Gilfrich

Naval Research Laboratory
Washington, D.C.

1. INTRODUCTION

X-ray fluorescence analysis has come a long way in the sixty years since Moseley began his classic experiments on the relation of x-ray wavelength to atomic number.[1] During this period several significant milestones were reached: In 1923 Coster and Von Hevesy confirmed the existence of element 72, Hafnium, from the x-ray spectra of Norwegian zircon;[2] in the 1930's concentrations of a fraction of a percent could be measured; in the mid-1940's the availability of high-powered sealed x-ray tubes, large single crystals, and geiger counters enabled Friedman and Birks to demonstrate a practical system for chemical analysis.[3] In the last twenty years, the technique has developed to a high degree of sophistication and is presently accepted as one of the most useful rapid and economical analytical methods available. The method is simple and straightforward with commercial equipment and a high vacuum is not required.

X-ray spectrochemical analysis is a surface technique in the sense that the x-rays being emitted, and therefore being measured, originate in a surface layer of finite thickness. Section 5 will discuss this criterion in detail, but first it is necessary to give a rather detailed description of the principles of x-ray generation and how these photons are used for chemical analysis. The prime purpose of this chapter is to explain the rudiments of the technique and how it might be applied to surface analysis. It is hoped that enough detail is given so the reader can judge whether x-ray fluorescence is applicable to his problem.

2. PRINCIPLES

2.1. Origin and Properties of X Rays

X rays are electromagnetic radiation of a specific wavelength, usually between about 0.1 and 250 Å, although these are not absolute limits. X rays result from the interaction of high-energy photons or particles (electrons, protons, α particles, etc.) with the atoms in a material. The x rays themselves, as all electromagnetic radiation, have energy equal to the product of Planck's constant, h, and the frequency, ν

$$E = h\nu \tag{1}$$

By substituting for all the constants in this equation, the wavelength, λ, and energy can be shown to have the simple relationship

$$\lambda = 12398/E \tag{2}$$

where λ is in angstroms and E is in electron volts. Thus a 2-Å x-ray photon has an energy of 6199 eV.

If the particle interacting with a target is an electron of sufficiently high energy, as in an x-ray tube, two kinds of x rays are produced: continuum and characteristic lines. The former are so called because the x rays are present as a continuous spectrum over some finite wavelength range. These x rays are also called bremsstrahlung (braking radiation) or white radiation (by analogy to the white light of the visible spectrum). The latter are named for the fact that their discrete wavelengths are characteristic of the element which makes up the target. No continuum is generated by photon excitation, and the quantity of continuum generated by particles heavier than electrons is smaller than that generated by electrons by approximately the square of their mass ratios.

2.1.1. Continuous Spectrum

If the electron striking the target in an x-ray tube loses all or part of its energy by interaction with the nucleus of the atom, the x-ray photon emitted has the same energy as that lost by the electron.[4] This process is sometimes referred to as deceleration (hence, "braking radiation"). This is a rather simplified but useful representation. The minimum wavelength present in the continuous spectrum can be calculated from Eq. (2) for the maximum energy of the electrons and occurs when this energy is entirely lost in one encounter. Because this "short-wavelength limit" (SWL) is a function only of the electron energy, it is independent of the atomic number, Z, of the target. The continuous spectrum has a

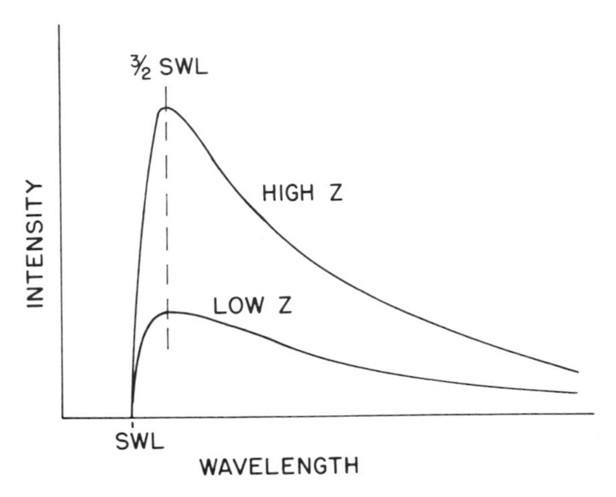

Fig. 12-1. Shape of the continuous spectrum.

characteristic shape as shown qualitatively in Fig. 12-1 where the peak wavelength is approximately equal to $\frac{3}{2}$ SWL. The intensity of the continuous spectrum is however strongly dependent on the atomic number of the target, varying approximately as Z. In x-ray fluorescence analysis, the continuous spectrum plays an important role. When the radiation from an x-ray tube is used to excite the sample, a large part (if not all) of the excitation is due to the continuum. Characteristic lines in the x-ray tube spectrum can also contribute to the excitation.

Protons incident on a target also create a continuous spectrum, but the intensity is lower than that for electrons by about the square of the ratio of the masses of the two particles, i.e., 1800^2, and it is therefore not easily observable.

2.1.2. Characteristic Lines

If, instead of interacting with the nucleus of the target atom, the incident electron knocks one of the electrons out of the inner shells of the atom creating a vacancy, this vacancy may be filled by an electron from the outer shells. When this happens there exists some probability that an x-ray photon will be emitted. This photon will have an energy related to the energies with which the involved electrons are bound to the nucleus of the atom. If the ejected electron is from the K shell (the innermost, most tightly bound), the photon which is emitted when this vacancy is filled from the L, M, or N shells is called a K-series x ray. Figure 12-2 is a simplified energy-level diagram showing the transitions involved in some K- and L-series emission lines.

Unlike the continuous spectrum, the characteristic spectrum is generated by electrons, protons, other particles, or x-ray photons

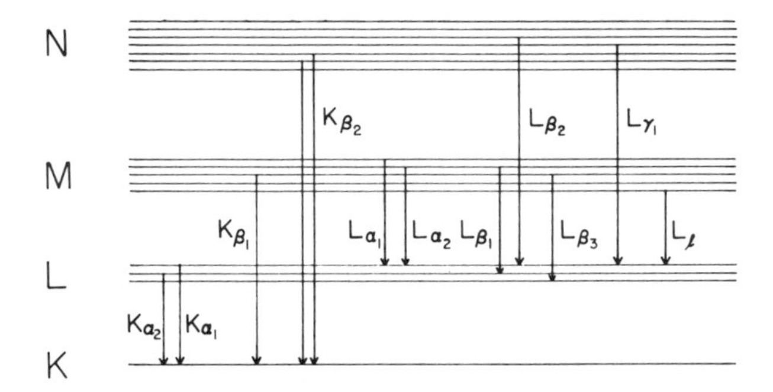

Fig. 12-2. Simplified energy level diagram.

provided the incident quantum has enough energy to eject an inner electron from the atom. Different quanta generate characteristic lines with varying degrees of efficiency, but they can all be used for chemical analysis in one way or another. In x-ray fluorescence analysis, the term "fluorescence" is used to indicate the excitation of photons by other photons. Therefore, we are talking about the use of an x-ray photon source such as an x-ray tube or a radioisotope to excite a sample to emit its own characteristic x rays. Because x rays do not generate a continuum, the background should be low (if the source contains a continuum which can be scattered into the measuring instrument) or virtually nonexistent (for an essentially monochromatic source like a K-capture radioisotope).

2.2. Interactions of X Rays with Matter

2.2.1. X-Ray Absorption

The absorption of electromagnetic radiation through materials is an exponential process. X rays are no exception. The relationship

$$I_T = I_0 \exp[-(\mu/\rho)\rho t] \tag{3}$$

where I_T is the transmitted intensity, I_0 is the incident intensity, (μ/ρ) is the mass-absorption coefficient, ρ is the density, and t is the thickness, permits the calculation of absorption if the various other parameters are available. The mass absorption coefficients of all elements for a wide range of x-ray wavelengths are available from a variety of sources, but unfortunately for some elements and over some wavelength ranges the values are uncertain to as much as 20%. Perhaps the best available compilation is that by Veigele,[5] extracted in a useful form for x-ray analysts by Birks.[6] Figure 12-3 is an example of the mass-absorption-coefficient curve for some elements. These coefficients are additive. Therefore the value for an alloy or a compound is simply the weight

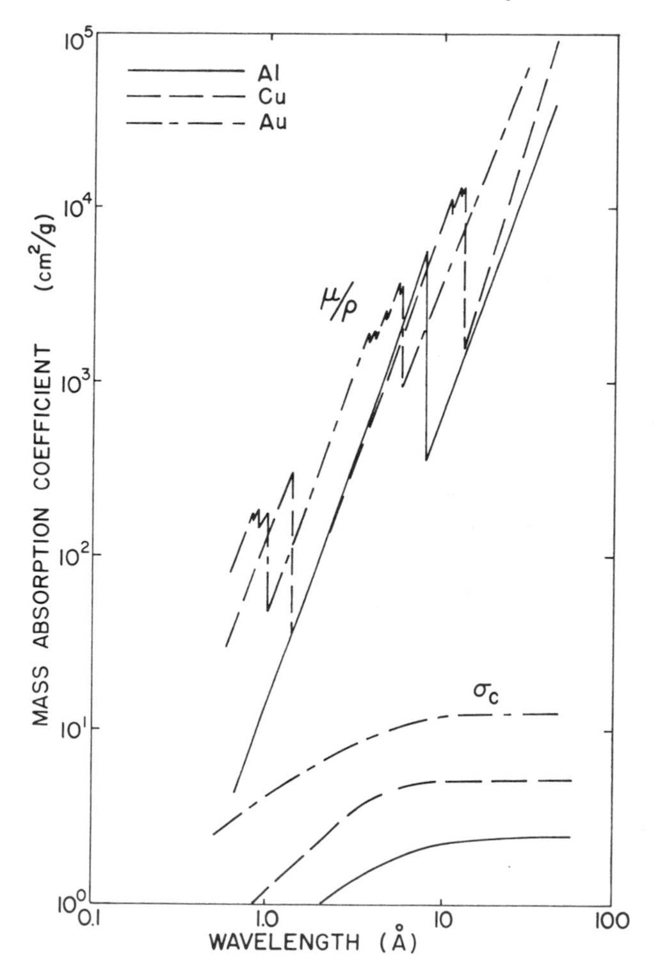

Fig. 12-3. Absorption and coherent-scattering coefficients for Al, Cu, and Au.

fraction of one element multiplied by its coefficient plus the weight fraction of the second element multiplied by its coefficient, and so forth. The product of the mass absorption coefficient and the density $[(\mu/\rho) \times \rho = \mu]$ is called the linear absorption coefficient and has units of cm^{-1}, so the thickness in the absorption equation must be in centimeters.

When an x-ray photon is photoelectrically absorbed in a material, a photoelectron is ejected. This photoelectron may not have enough energy to escape from the material because ordinarily it will have fairly low energy and therefore a short range. For example, if a Cu K_α photon (8.0 keV) is absorbed in Fe by knocking out an Fe K-shell electron (absorption edge energy = 7.1 keV), the energy of the photoelectron will be only 0.9 keV (8.0 − 7.1). If, however, the incident photon has a high energy relative to the absorption edge of the material, the photoelectron will have enough energy to generate some bremsstrahlung to contribute to the background in an x-ray fluorescence measurement. This is a small contribution however and would be difficult to measure.

2.2.2. X-Ray Scattering

The mass absorption coefficient is actually made up of three components: the photoelectric absorption τ, the coherent scattering σ_C, and the incoherent scattering coefficients σ_I. The photoelectric component is the predominant part (well over 90%) for the wavelength region of interest in x-ray fluorescence, as shown in Fig. 12-3. Also shown is the coherent scattering coefficient; the incoherent scattering coefficient is too small to be shown even on this logarithmic graph covering five orders of magnitude.

To all intents and purposes, the two different types of scattering can be pretty well ignored in x-ray fluorescence analysis except as they contribute to the background by scattering the primary radiation into the detector. Coherent and incoherent scattering of the continuum cannot be separated from one another because they are unresolvable. Because incoherent scattering causes a small increase in wavelength, however, it is wise to consider its effect on the characteristic lines. The change in wavelength of an x-ray line caused by incoherent scattering (also called Compton scattering) is a function of the angle through which the photon is scattered

$$\Delta\lambda = 0.0243\,(1 - \cos\phi) \tag{4}$$

where $\Delta\lambda$ is the change in wavelength and ϕ is the scattering angle. Thus, a photon incoherently scattered through 90° will have its wavelength increased by 0.0243 Å. If the scattering angle is 180°, $\Delta\lambda$ = 0.0486 Å. Incoherently scattered lines are usually not as sharp as the line itself. Brooks and Birks[7] have shown how incoherently scattered W L_α and L_β lines from the target of the x-ray tube can interfere with the x-ray analysis for Ta.

2.2.3. X-Ray Diffraction

Because the wavelengths of x rays are comparable to the distance between atoms in materials, x rays are used to study the structure of these materials. This widely used technique, called x-ray diffraction, is the foundation of the whole field of x-ray crystallography and is based on the Bragg equation

$$n\lambda = 2d\sin\theta \tag{5}$$

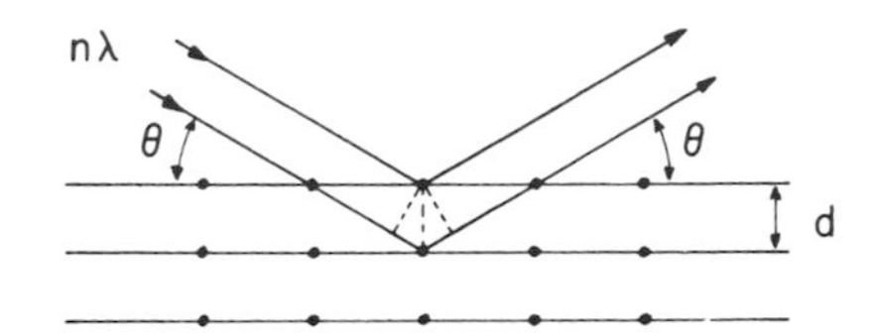

Fig. 12-4. Bragg diffraction.

where n is the order of diffraction, λ is the wavelength in angstroms, d is the spacing between the diffracting planes in angstroms, and θ is the angle between the incident beam (or the diffracted beam) and the crystal planes. This is shown diagrammatically in Fig. 12-4; the derivation of the equation can be found in any text on x-ray diffraction.[8]

X-ray diffraction, used for the study of crystal structure, uses x rays of known wavelength to determine the interplanar spacing in the materials. When crystal spectrometers are used in x-ray fluorescence analysis, a crystal of known spacing is employed to measure the x-ray wavelength. More will be said of this in Section 4.1.1.

2.3. Important Parameters

The mass absorption coefficient has been mentioned in Section 2.2.1. It is an important parameter in x-ray fluorescence analysis for several reasons. The photoelectric component is a measure of the probability that an incident photon will ionize the atom, beginning the process which may lead to the emission of an x-ray photon characteristic of the atom. When x rays are generated within a material, the absorption coefficient of the matrix determines the probability that the generated photon can escape from the depth at which it is generated. The efficiency of detectors is a function of the absorption coefficient of the detecting material.

The fluorescence yield ω is a measure of the probability that a photon will be emitted when a vacancy is created in one of the electron shells of the atom. This parameter varies with atomic number as shown in Fig. 12-5. The intensity of an emitted characteristic x ray can be considered in a very simplified way to be merely a function of the probability that a vacancy will be created (μ/ρ) and that an x ray will result from de-excitation (ω). The relatively low intensity of low-energy x-ray lines (K lines for low Z and L lines for intermediate Z) is a direct result of the low fluorescent yield.

3. EXCITATION FOR ANALYSIS

The first step in performing an x-ray analysis is to cause the sample to emit its characteristic x rays. As described in Section 2.1, this can be accomplished by irradiating the specimen with photons or particles of

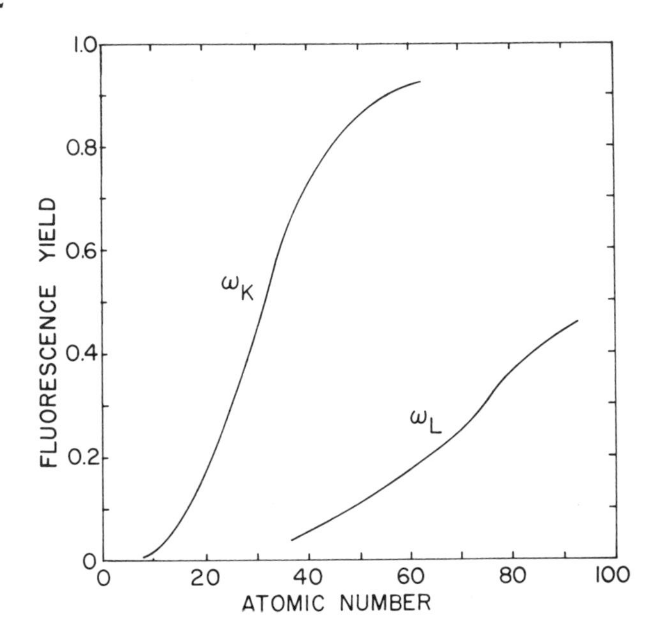

Fig. 12-5. Fluorescence yields.

sufficient energy to create an inner-shell vacancy. Electron excitation, by far the most efficient, is used in the electron microprobe and certain other types of instruments, but that is the subject of another chapter of this book. Other particles used for excitation such as protons, α particles, or heavier ions can also be used, but they are not properly the subject of this chapter on fluorescence since this involves only excitation of photons by photons.

If one considers the sources of photons which might be used for excitation, the x-ray tube might be the first to come to mind. And, of course, the x-ray tube is the most widely used source for x-ray fluorescence analysis. However, there are also a significant number of radioisotopes which emit either x rays or γ rays and these can be used as well as source–target assemblies where it might be desirable to use photon energies not available from the primary source.

3.1. X-Ray Tubes

The usual source of excitation for x-ray fluorescence analysis is the sealed x-ray tube. These are available with a variety of targets and window thicknesses in order to provide varying efficiency for different regions of the periodic table. Because the intensity of the continuum generated by the electron beam in the x-ray tube is a function of the atomic number of the target, a high atomic number is desirable. However, the characteristic lines have considerably more intensity than the continuum. If a target can be chosen so that its characteristic lines have the right energy to excite the element of interest, greatly improved sensitivity can result. Because an element is most efficiently excited by photons having energy only slightly higher than its absorption edge, the light-element *K* lines (or intermediate-element *L* lines) require that the x-ray tube have a thin window so that the low-energy photons can exit the tube and be available to excite the sample. Commercial x-ray tubes are available with targets ranging in atomic number from 24 (Cr) to 79 (Au) and with windows (usually Be) as thin as 0.125 mm. Not all atomic numbers can be made into targets for engineering reasons, e.g., thermal conductivity, but spectrographic tubes are available having targets of Cr, Cu, Mo, Rh, Ag, W, Pt, Au, and perhaps a few more. It has recently been suggested that a thin-window Rh target is the best single tube to use because its atomic number is high enough to generate an intense continuum, the energies of the *K* lines (19 and 22 keV) are high enough to provide considerable additional excitation for the intermediate elements, and the energies of the *L* lines (2.7–3.2 keV) are efficient for the light elements. The relative spectral distributions of several x-ray tubes have been measured[9] and Fig. 12-6 shows the spectrum of a Rh tube such as described above. Most x-ray spectrographic tubes are limited to 50- or 75-kV operation. Some years ago, it was thought that designing an

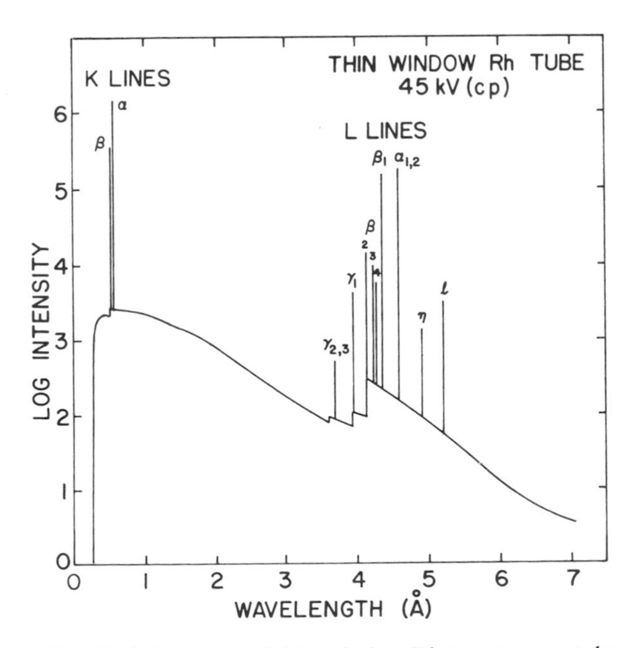

Fig. 12-6. Spectrum of thin-window Rh-target x-ray tube.

x-ray tube for 100 kV would improve the sensitivity for some of the high-intermediate elements because the K lines could be measured. However, the unavailability of analyzing crystals of small enough d spacing to diffract these short wavelengths to reasonable Bragg angles led to poor acceptance of the instruments designed to use the tubes. Also, the peak to scattered continuum was not very good because the measurements were being made in a wavelength region of intense continuum. For energy-dispersion measurements (see Section 4.2) there may be some revitalizing arguments for high incident x-ray energies.

An ultrathin-window x-ray tube operating at low kilovoltage and high current has been designed[10] for use in the ultrasoft x-ray region ($\lambda > 15$ Å), and was commercially produced for a time. Because the window is subject to radiation damage, which necessitates its replacement from time to time, the tube requires a continuously pumped vacuum and a means for protecting the fragile window when the sample chamber is opened to the atmosphere. The windows used on such a tube might be 25-μ Be, stretched polypropylene, or a nitrocellulose film. Because of the continuous vacuum pumping the tube is demountable and various target materials can be used.

3.2. Radioisotopes

Radioisotopes may emit x rays, γ rays, α particles, β rays, and various other kinds of radiation having energies sufficient to excite x rays. Choice of the isotopic source is based on the range of elements to be covered. The simplest type of radioisotope for practical use is the K-capture source. In these sources, the decay scheme involves the capture of a K-shell electron by the nucleus, creating a K-shell vacancy which is filled by an electron from one of the outer shells. This results in the emission of the K lines of the next lower atomic number. There are two usable isotopes whose nuclear decays involve essentially only the K-capture process: ^{55}Fe which emits Mn K lines and ^{109}Cd producing predominantly Ag K_α and K_β. Other sources emit both K-capture radiation and γ rays in varying amounts: ^{125}I emits Te K lines and 35-keV γ rays, ^{153}Gd produces Eu K lines and γ rays of 97 and 103 keV, and ^{57}Co generates Fe K lines and γ rays of 14, 122, and 136 keV. ^{241}Am is a pure γ-ray emitter, its photons having energies of 59.6 keV. The α-particle and β-ray sources are not ordinarily used. However, β rays do generate a continuum and sources of this type can be mixed with other materials to produce a source which emits this continuum. Two commonly used sources of this type are ^{3}H–Zr and ^{147}Pm–Al.

The one major disadvantage of radioisotope sources is low intensity. This makes them unusable with the crystal spectrometric technique of x-ray fluorescence analysis. They are, however, very stable and, as will be seen in Section 4.2, have some valuable utility for certain applications.

3.3. Source–Target Assemblies

It was mentioned in Section 3.1 that characteristic x rays are produced efficiently by other photons whose energy is close to, and only slightly higher than, the absorption edge of the element of interest. This has an interesting consequence for situations where it is desirable to excite some elements and not excite others. This can be accomplished for select cases by the use of a radioisotope, e.g., ^{55}Fe which emits Mn K_α and K_β. The Mn K lines have energies of 5.9 and 6.5 keV, respectively, and will therefore excite the K lines of elements below atomic number 24 (Cr) and L lines below atomic number ~ 57 (La). Cr K lines would not be excited very well because only Mn K_β has a high enough energy, but both K_α and K_β would contribute for V and below.

Because isotopic sources of this type are few, the use of an x-ray tube to excite a fluorescer to emit its characteristic lines is a very efficient method of obtaining a monochromatic source at any energy within the range of interest for x-ray fluorescence analysis. Actually, this is not a monochromatic source since if K lines are being excited there is no way to obtain only K_α or K_β; both will be present. However, β filters can be used (and frequently are in x-ray diffraction experiments) to remove the K_β component. These β filters are thin foils of an element whose K absorption edge lies between the K_α and K_β. If the L series is being excited, there may be a dozen or more lines present. The intensity from an x-ray tube–fluorescer combination is two or three orders of magnitude lower than the direct radiation from an x-ray tube. If this arrangement were used with a crystal spectrometer, long counting times would be required to achieve reasonable statistics in most cases. Radioisotopes can be used to excite fluorescers, as well. In this case, however, intensity becomes a real problem because of the undesirable radiation hazard inherent in using high-activity isotopic sources.

4. MEASUREMENT TECHNIQUES

There are basically two experimental methods of x-ray fluorescence analysis: wavelength dispersion and energy dispersion. In both cases an x-ray photon source is used to excite a sample which emits its own characteristic x rays. The wavelength or energy of these emitted photons is measured to determine what elements are present in the sample (qualitative analysis) and their intensity is a measure of how much of each element makes up the specimen (quantitative analysis). Practically, the recognition of the elements is simply the identification of the peaks present in the spectrum. Quantitative analysis, on the other hand, is a more complex problem because of what is called "matrix effect."

For many years, the matrix-effect problem was avoided by using comparison standards which were chemically analyzed samples similar in composition to the unknown. Because in many cases this necessitated maintaining a library of many standards, attempts have been made to solve the problem mathematically. In recent years there have been two different approaches used. Although a detailed description of these methods goes well beyond the scope of this chapter, mention is made of them because of their usefulness in practical analysis. The so-called "empirical-coefficient" or "regression-equation" method uses coefficients empirically determined from a set of regression equations to represent the matrix effect of one element on another.[11–13] The more general "fundamental-parameter"[11,12] method makes use of the spectral distribution for the x-ray tube, mass absorption coefficients, fluorescence yields, and various other parameters to calculate the composition from the relative intensity. The only standards that are necessary are the pure elements, although compound standards can be used. These problems, however, are more in the nature of data interpretation than measurement.

4.1. Wavelength Dispersion

The use of grating or crystal spectrometers to disperse the radiation is the classic technique for x-ray fluorescence analysis. Because the wavelength being diffracted by the dispersing device is a function of the measurable Bragg angle, the name "wavelength dispersion" has been given to this method. As was implied previously (Section 2.2.3), a crystal spectrometer can measure any x-ray wavelength provided that a crystal with a suitable interplanar spacing is available.

4.1.1. Flat-Crystal Spectrometers

The flat-crystal spectrometer is the most widely used instrument for making wavelength-dispersion x-ray fluorescence measurements. Early spectrometers were adaptations of powder diffractometers which could be used most readily with flat crystals. For analysis of bulk samples, this is an appropriate geometry and is schematically shown in Fig. 12-7. In this figure, an x-ray tube excites a few square centimeters of the sample. The specimen is located close to the window of the x-ray tube to provide as high an incident flux as possible. The emitted radiation is directed by a collimator to the analyzing crystal where it is diffracted to the detector set at the appropriate angle. The detector and crystal are coupled in a

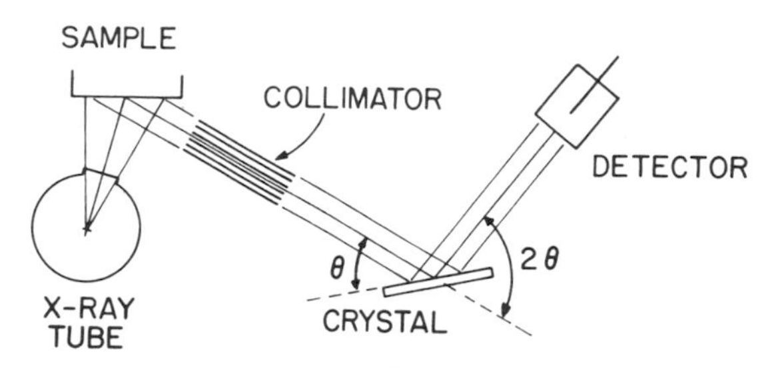

Fig. 12-7. Schematic of a flat-crystal x-ray spectrometer.

manner that provides twice as much angular movement to the detector as to the crystal. Thus, with the crystal set at the angle θ, the detector will be at 2θ to detect the diffracted intensity. The spectrometer may be fixed, to measure the intensity of a single x-ray line; it may be a scanning model, which can be manually or motor driven through a range of Bragg angles to measure a range of wavelengths; or it may be programed to move at a rapid rate of speed from one angle to another, stopping long enough at each position to measure the intensity. The modern multichannel simultaneous x-ray fluorescence instrument, designed for production analytical control, may have as many as 24 fixed spectrometers to measure that many elements at the same time; these spectrometers may use either flat or curved (focusing) crystals. These automatic machines may have programable scanning spectrometers incorporated also. The sequential analyzers ordinarily are of the programable scanning type. Research instruments are usually the manual or motor-driven type.

Most modern spectrometers are capable of being operated with an air atmosphere, a helium or other controlled gas atmosphere, or in a vacuum. In an air path, the analyst is limited to measuring x-ray wavelengths shorter than about 2.5 Å because longer wavelengths are severely attenuated in the air. For softer x rays ($\lambda > 2.5$ Å), a vacuum, hydrogen, or helium path is necessary. Of these options, the vacuum is the most desirable because a reproducible usable vacuum can be obtained by the use of a simple mechanical pump.

One of the important parameters of a spectrometer is the resolution. For a flat-crystal instrument, the ability to separate two x-ray lines close to one another in wavelength can be calculated simply by differentiating Eq. (4). Using only the first-order diffraction, the n term drops out.

$$\Delta\lambda = 2d \cos \theta \Delta\theta \qquad (6)$$

All of the terms here are the same as in Eq. (4) and it remains only to define $\Delta\theta$, the divergence of the system. The divergence of the system is a function of the collimator(s) and the crystal, but usually the collimator has the predominant effect. Most spectrometers have provision for a primary collimator between the sample and the crystal about 10 cm long with blade spacings of 0.01–0.1 cm. The divergence allowed by the collimator is merely arc tan (spacing/length) or 3.4 to 34 min of arc. The divergence of the crystal is the full width at half maximum of the rocking curve and will vary between about 10 sec of arc for a nearly perfect crystal like freshly cleaved potassium acid phthalate (KAP) to about 2 min of arc for a highly mosaic crystal like abraded and etched lithium fluoride (LiF). The graphite crystal has a rocking-curve width of about 0.4° and will not be considered here because it would not be used if resolution is of concern. In the simplest case, the combination of collimator divergence and crystal rocking curve by the rule of variance defines $\Delta\theta$. Only when the finest of the collimators and a crystal with a broad rocking curve are used does the crystal have significant effect on the divergence, e.g., for an abraded and etched LiF crystal, with a rocking curve of 2′, used with the 0.01 × 10-cm collimator having a divergence of 3.4′

$$\Delta\theta = (2^2 + 3.4^2)^{1/2} = 3.9'$$

The resolution (in angstroms) is not independent of the Bragg angle and will vary over a wide range for any given crystal–collimator combination, as shown in Table 12-1, for some typical examples. It should be obvious from this table that the best resolution occurs at large Bragg angles. Therefore, the best resolution is always obtained by using the smallest-spacing crystal consistent with the longest wavelength to be measured.

The reflectivity of the crystal used in the spectrometer is a critical factor in determining the intensity of the x-ray line being measured. There are three parameters commonly used to define the reflectivity of a crystal:[(14)] (1) W, the full width at half maximum of the rocking curve, mentioned above, (2) P, the peak diffraction coefficient, and (3) R, the integral reflection coefficient. The integral reflection coefficient is the integral intensity under the rocking curve divided by the intensity incident on the crystal, i.e., $\sim P \times W/I_0$, and is the factor by which crystals should be compared for use in an x-ray fluorescence spectrometer.

4.1.2. Curved-Crystal Spectrometers

If the source of x rays is small, considerably higher intensities can be obtained by focusing the radiation into the detector. If the diffracting planes of the crystal are cylindrically curved to a radius equal to the

TABLE 12-1. Resolution of Crystal Spectrometer

Collimator	0.01 × 10 cm, 3.4′			0.1 × 10 cm, 34′			0.01 × 10 cm, 3.4′		
Crystal	Abraded LiF, 2′ $2d = 4.03$ Å			EddT, 2′ $2d = 8.8$ Å			Cleaved KAP,[b] 10″ $2d = 26.6$ Å		
Resolution	3.9′, 0.011 rad			34′, 0.0100 rad			3.4′, 0.0010 rad		
λ, Å	1.04	2.29	3.74	2.29	5.36	8.31	7.11	11.9	23.0
E, keV	11.9	5.40	3.32	5.40	2.31	1.50	1.74	1.05	0.54
	Br K_α	Cr K_α	K K_α	Cr K_α	S K_α	Al K_α	Si K_α	Na K_α	O K_α
θ.	15°	35°	68°	15°	38°	71°	15°	27°	60°
$\cos\theta$	0.966	0.819	0.375	0.966	0.788	0.325	0.966	0.891	0.500
$\Delta\lambda$, Å[c]	0.0043	0.0036	0.0017	0.0850	0.0695	0.0286	0.0257	0.0237	0.0133
ΔE, eV[d]	49	8.5	1.5	200	30	5.1	6.3	2.0	0.31

[a] EddT = Ethylene diamine ditartrate.
[b] KAP = Potassium acid phthalate.
[c] $\Delta\lambda = 2d \cos\theta\Delta\theta$.
[d] $\Delta E = E\Delta\lambda/\lambda$.

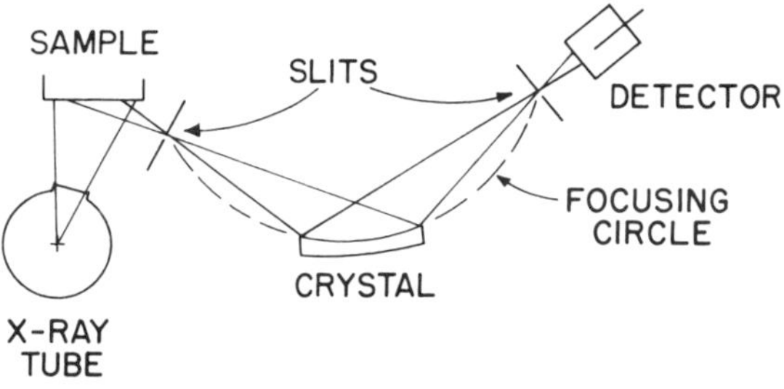

Fig. 12-8. Schematic of a curved-crystal x-ray spectrometer.

diameter of the focusing circle and the surface of the crystal is ground to a radius equal to the radius of the focusing circle, it is possible to diffract a point source to a line image. This type of x-ray optics is used in electron-probe microanalysis. It is also possible to bend the crystal to a logarithmic curve which is said to avoid the need for grinding in order to achieve focusing. In x-ray fluorescence of bulk samples, however, it is necessary to have a slit lying on the focusing circle to satisfy the conditions. For small amounts of sample such as might be contained in a capillary tube, it is possible to have a narrow line source which can take advantage of the increased intensity available with focusing optics. For the extended sample, the slit corresponds roughly to the collimator with flat-crystal optics and little, if any, increase in intensity is realized. The use of a secondary slit at the detector can decrease the amount of radiation scattered by the crystal which reaches the detector and thus improve the line-to-background ratio. Figure 12-8 is a schematic diagram of one type of curved-crystal spectrometer for use in x-ray fluorescence analysis.

4.1.3. Choice of Crystals

It is possible to specify the characteristics of the ideal crystal for a particular x-ray measurement. In practice the analyst is constrained by the limited number of available materials and it is necessary to choose among them. The first parameter of interest is the interplanar spacing. The $2d$ spacing must be large enough so that the wavelength of interest can be diffracted. The longest wavelength which can be diffracted is, in theory at least, equal to this spacing, because $\sin\theta$ in the Bragg equation cannot have a value larger than unity. In practice, it is impossible for a spectrometer to go beyond some θ value smaller than 90° and most instruments are limited to θ of about 70°. The longest-wavelength x rays which can be measured with a particular crystal is then equal to about $0.94 \times 2d$. On the other hand, if resolution is of any concern, the smallest-spacing crystal should be used consistent with the above criterion. If

TABLE 12-2. X-Ray Fluorescence Analyzer Crystals

Crystal	(*hkl*)	$2d$, Å	Fluorescing elements	λ, Å	R-value, rad
LiF	(220)	2.85	—	1.66	1×10^{-4}
LiF	(200)	4.03	—	2.10	3×10^{-4}
Graphite	(002)	6.71	—	2.75	2×10^{-3}
PET[a]	(002)	8.74	—	3.74	2×10^{-4}
EddT[b]	(020)	8.81	—	5.37	1×10^{-4}
ADP[c]	(101)	10.6	P	6.16	5×10^{-5}
Gypsum	(020)	15.2	Ca, S	7.13	7×10^{-5}
Mica	(002)	19.8	K, Al, Si	8.34	2×10^{-5}
KAP[d]	(100)	26.6	K	9.89	7×10^{-5}
OHM[e]	—	63.4	C	11.9	1×10^{-4}
Pb St[f]	—	100	C, Pb	18.3	2×10^{-4}

[a] PET = Pentaerythritol.
[b] EddT = Ethylene diamine ditartrate.
[c] ADP = Ammonium dihydrogen phosphate.
[d] KAP = Potassium acid phthalate.
[e] OHM = Octadecyl hydrogen maleate.
[f] Pb St = Lead stearate, a multilayer soap "pseudo-crystal."

possible, the crystal should not have any elements in it which can fluoresce in a measurable-wavelength region because this merely increases the background. The crystal should have as high an integral reflection coefficient as possible to diffract high intensity, and it obviously must be available in large enough blanks to intercept the x-ray beam being analyzed. Table 12-2 lists some of the crystals commonly used along with the parameters above for each.

4.1.4. Grating Spectrometers

Diffraction gratings can also be used to disperse soft x rays. Holliday[15] has described this technique in a lucid fashion. There is considerable evidence that the blazed gratings are more efficient than the Siegbahn type; both must be used at grazing incidence in this energy range. Figure 12-9 compares the two types of gratings. The grating equation is somewhat different than the Bragg equation, and when written for the blazed grating takes the form

$$n\lambda = 2a\beta(\phi + \beta) \tag{7}$$

where λ is the wavelength, n is the order, a is the grating constant, β is the blaze angle, and ϕ is the grazing angle of incidence. It can be seen that for any $n\lambda$ there is a unique value of ϕ, just as there is a distinct value of θ for any $n\lambda$ in the Bragg equation.

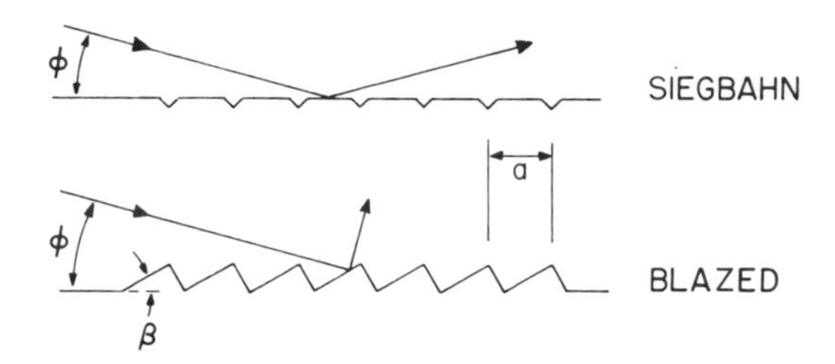

Fig. 12-9. Schematics of two types of gratings.

Grating spectrometers are not ordinarily used to any extent in x-ray fluorescence analysis and so will not be treated any further here. It did seem appropriate to include some mention of them for completeness because they are used by researchers in the soft x-ray region.

4.2. Energy Dispersion

The use of x-ray detectors whose output pulse amplitude is proportional to the energy of the incident photon provides a second technique for distinguishing x-ray energies and intensity. If the electronic counting equipment contains a module capable of discriminating the pulse amplitudes present in the signal, it is possible to measure the incident x rays. Such an electronic device is called a pulse-height analyzer. Let us suppose we have a detector which puts out a 1-V pulse when a 1-keV photon is detected, a 10-V pulse for a 10-keV photon, and so forth. The pulse-height analyzer must have the capability of varying the pulse amplitude which it will detect. This is commonly done by having an adjustable baseline E below which no pulses will be counted and an adjustable window ΔE which sets an upper limit at $E + \Delta E$. If an alloy of Cr and Ni is caused to emit its characteristic x rays, they will be Ni K_α (7.5 keV), Ni K_β (8.3 keV), Cr K_α (5.4 keV), and Cr K_β (5.9 keV). Because the K_α-to-K_β intensity ratio is about 5:1, the spectrum should look like Fig. 12-10. Since we have chosen to set the detector in this ideal case so that 1 keV gives 1 V, the horizontal axis can represent either photon energy in keV or pulse amplitude in volts. In practice, this ideal situation does not exist as explained in the next section.

4.2.1. Detectors

In order to make use of the energy dispersive technique, it has been said that we need a detector whose output is proportional to the energy of the photon incident upon it. There are several such devices available. Figure 12-11 contains schematic diagrams of the three available detectors. The scintillation detector is very efficient over a wide energy range, the

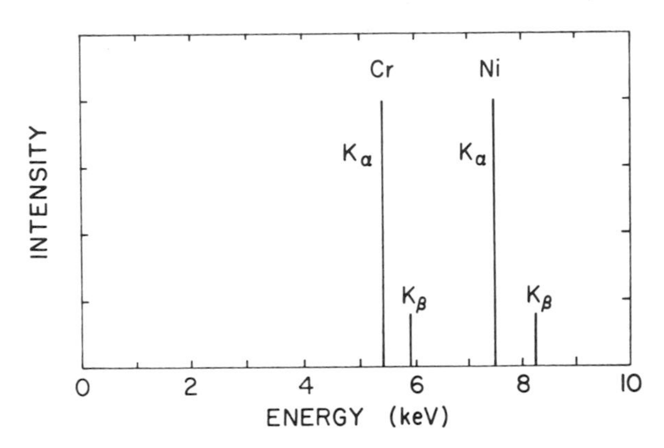

Fig. 12-10. Idealized energy spectrum of Cr and Ni *K* lines.

proportional counter can be used with various counting gases and thus its efficiency can be tailored to some extent, and the solid-state detector, the newest of the group, has very good energy resolution.

4.2.1.1. Proportional Counters

The proportional counter may be one of two types. The sealed proportional counter is filled with one of the counting gases: xenon, krypton, argon, or neon and has sealed windows for the radiation to enter and leave. The exit window is desirable in a proportional counter because any radiation of energy high enough to be incompletely absorbed in the gas would cause fluorescence of the wall of the detector if

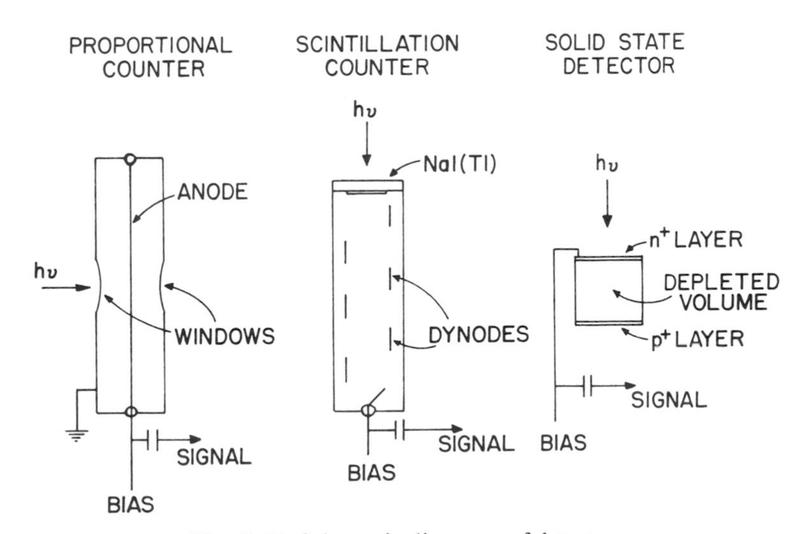

Fig. 12-11. Schematic diagrams of detectors.

it were not permitted to escape. These sealed detectors are used for radiation shorter than 2 or 3 Å. For radiation longer than this, it is necessary to have thin windows which are not leakproof. Since these long-wavelength detectors are frequently used in a vacuum, some means must be provided to replenish the counting gas. For these purposes a gas-flow proportional counter is used. The counting gas, which may be argon, helium, or methane, is flowed into the detector continually. The windows are 6-μ Mylar, stretched polypropylene, or thin nitrocellulose films supported by a highly transparent grid. Because these windows are relatively fragile, they must be replaceable.

The spectrum for Cr and Ni shown in Fig. 12-10 cannot be realized in practice because the statistical fluctuations create some pulses of more or less amplitude than the value calculated. Although it has been suggested[(16)] that the statistical treatment to determine the energy resolution of a detector is not realistic, it is a convenient way to demonstrate the principle and, in fact, does seem to hold rather well for proportional counters.

When a photon enters a proportional counter, the most likely interaction with the gas is knocking out an electron and imparting to it kinetic energy equal to the photon energy minus the ionization energy of the gas. This energetic electron will lose its energy by creating ion pairs in the process of ionizing other gas atoms. The most probable number of ion pairs is a function of the photon energy and the average ionization potential of the gas. The ionization potential which is used in this relationship is twice the value of the first ionization potential for the gas. The factor of two is an empirical number which corrects the first ionization potential to a value agreeing with experiment.

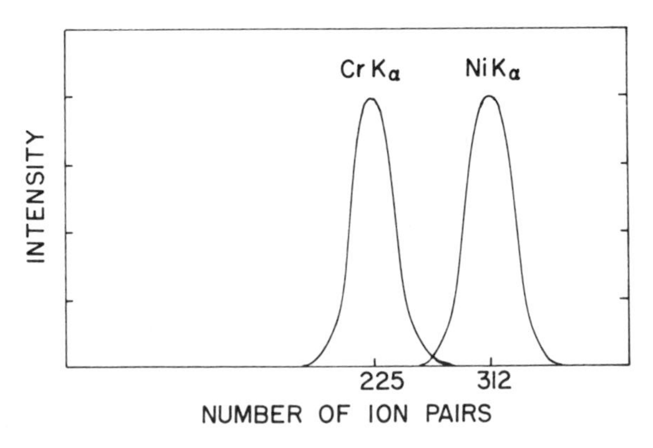

Fig. 12-12. Ion-pair distribution for Cr and Ni K_α in xenon.

Thus, if we return to our Cr–Ni alloy of Fig. 12-10 and ignore the K_β lines to simplify the consideration, we see that for a Cr K_α photon where the energy is 5.4 keV and a Xe counter where the first ionization potential is 12 eV, the number of ion pairs produced is 5400/(2 × 12), or 225. The statistical spread about this value is $(225)^{1/2}$ or 15. For the Ni K_α, photon energy is 7.5 keV and the number of ion pairs is 7500/(2 × 12), or 312, with a statistical variation of about 18. The pulse-amplitude distribution of these two energies would look like Fig. 12-12, with the horizontal axis expressed in number of ion pairs, although it could be pulse amplitude in volts (after amplification) or photon energy. A simple way to represent the resolution in percent (fwhm) is $236/\sigma$, which would be 16% for Cr K_α and 13% for Ni K_α.

4.2.1.2. Scintillation Counters

The scintillation counter is usually a thallium-activated sodium iodide crystal optically coupled to a photomultiplier tube. An incident x-ray photon generates a number of visible-light photons in the scintillator which are collected by the photocathode of the multiplier tube. Electrons are emitted and are multiplied in their passage through the stages of the tube so that a large pulse appears at the output. Although the energy used up in generating a visible-light photon in the scintillator is small (2 or 3 eV) compared to the 20 or 30 eV necessary to generate an ion pair in a proportional counter, the scintillation counter has poorer resolution by a factor of 2 or 3 because the photomultiplier converts light photons into electrons very inefficiently. In practice, the scintillation counter is only used in energy dispersion in conjunction with filters to provide improved energy selection. Its energy resolution is adequate to discriminate against second-order diffraction in a crystal spectrometer, however.

4.2.1.3. Solid-State Detectors

The appearance of the solid-state detector on the x-ray analytical scene[17] has created a tremendous interest in energy dispersion. Although this technique had been used previously with proportional counters, the limited resolution available required that overlapping peaks be unfolded mathematically. In the past seven years the solid-state detector has achieved a resolution of ~150 eV at Mn K_α (5.9 keV) which corresponds to 2.5% compared to about 15% for a proportional counter at this same energy. The solid-state detector as shown in Fig. 12-11 is a semiconductor (Si or Ge) which has been diffused with Li atoms to produce an intrinsic layer. Because the mobility of the Li atoms is fairly high in these materials, they must be maintained at liquid-nitrogen temperatures. The first stage of the associated preamplifier, a field-effect transistor, is usually fabricated on the back of the same semiconductor material in order to maintain the capacitance at a low value and is also kept cold to decrease thermal noise. The choice between Si and Ge is primarily one associated with the energy range of interest. In the x-ray analytical region Si is usually used, but if higher energy measurements are to be made (>25 or 30 keV) Ge has a much better efficiency. It is in this region that there may be some inspiration for using higher-voltage x-ray tubes since the Ge detectors are relatively efficient up to 100 keV or higher and they provide a means of measuring the *K*-series lines of the high atomic numbers.

4.2.2. Multichannel Analyzers

Earlier in Section 4.2 mention was made of a pulse-height analyzer as an electronic device with a baseline which could move a window across the pulse-amplitude distribution of a spectrum to determine what energies were present. This operation is fairly laborious and time-consuming. The electronic circuitry usually used in energy-dispersion x-ray analysis is a multichannel pulse-height analyzer. This instrument functions by sensing the amplitude of an incoming pulse, sorting it into the proper bin according to the energy calibration, and storing the information in its memory. After accumulating data for the appropriate time, it is possible to display the pulse-amplitude spectrum on a cathode-ray tube, plot it on an x-y recorder, or type it out on a printer. Figure 12-13 shows a typical spectrum. The sample was barium formate being ex-

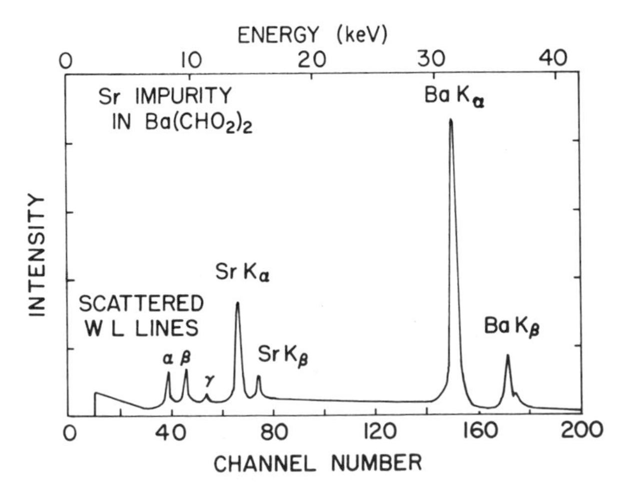

Fig. 12-13. Energy dispersion spectrum of $Ba(CHO_2)_2$ containing Sr impurity.

amined for strontium impurity; it was excited by a tungsten-target x-ray tube and measured with a Si(Li) detector. Mention must be made of the high background between the Sr and Ba peaks. A large part of this background is due to inefficient charge collection of the Ba radiation, a problem which is being minimized in new detectors (c. 1972).

4.2.3. Advantages and Limitations

In discussing the advantages and limitations of energy dispersion, the major emphasis will be on solid-state detectors because of their widespread use at this time. One of the major advantages of this technique is the large solid angle intercepted by the detector from any point on the sample. For example, a 1-cm^2 detector can be placed about 3 cm from a sample permitting it to accept $\frac{1}{10}$ of a steradian. Thus, very-low-activity radioisotopes or low-power x-ray tubes can be used to excite the specimen. The energy-dispersion spectrum is also recorded at one time, permitting the observer to note what elements are present within the energy range being recorded. Referring back to Fig. 12-13, it is immediately obvious that Sr is present at significant levels in this $Ba(CHO_2)_2$ sample and that no other elements are present to any appreciable extent. With the present-day availability of detectors with 150-eV resolution, there exists some difficulty examining samples containing many elements where some of them are immediate neighbors. Although 150-eV resolution is adequate to separate the K_α lines of immediate neighbors, it is not sufficient to separate the K_β line of atomic number Z from the K_α line of atomic number $(Z + 1)$ in the Z range from 17 (Cl)–26 (Fe). If the sample in question had appreciable quantities of Cr and Fe it would be very difficult to be certain whether or not a small amount of Mn was present. Mathematical techniques for unfolding spectra are available,[18] but the quantification of results obtained as the small difference between two large numbers is suspect. The solid-state detector has some minor limitations when attempting to measure low-energy x rays because of the window necessary to provide a high-vacuum environment for the detecting element, and further because of the thin dead layer on its front surface. The window thickness can be made as thin as perhaps 10 μ of Be but these are very fragile. The dead-layer thickness is the order of a micron. This is not necessarily a limitation of energy dispersion as a technique because flow proportional counters can have windows as thin as 2000 Å of nitrocellulose and their resolution approaches that of the Si(Li) detector at low energies. The large solid angle intercepted by the energy-dispersive detector has been mentioned as an advantage. It can also be a disadvantage because the detector sees all the radiation at the same time and suffers from a count-rate limitation in the range of 10,000–50,000 cps. Therefore, it is almost a necessity to use a relatively low-powered source.

4.3. Comparison of Wavelength and Energy Dispersion

It is difficult to consider all the facets of a comparison between two analytical techniques, but there are some facts which must be mentioned. The resolution of a solid-state detector used in energy dispersion is, at best, 150 eV (for a small-area detector at reasonably low count rate) for 5.9-keV x rays and varies approximately as the square root of the photon energy (plus the noise contribution). The resolution of a crystal spectrometer varies from less than 1 eV to several tens of eV over the usual operating range. Energy dispersion intercepts a solid angle several hundred times as large as the crystal spectrometer collimator (a mixed blessing perhaps). Radioisotopes of moderate or low activity, source–target assemblies, and low-power x-ray tubes all have sufficient intensity to be used as sources for energy dispersion, while wavelength dispersion requires high-powered x-ray tubes. Although the energy-dispersion technique records the intensities of all the elements simultaneously, the limited resolution compared to crystal spectrometers can make the quantitative interpretation somewhat more difficult. Automatic multichannel wavelength-dispersion x-ray analyzers can measure as many as 24 elements at the same time. The best-of-all-worlds of x-ray analysis would have an energy-dispersion system to survey the samples and crystal spectrometer(s) to perform the quantitative analysis.

5. WHAT ARE SURFACES FROM AN X-RAY PHOTON'S POINT OF VIEW?

It was said in Section 1 that x-ray fluorescence analysis was a surface technique in the sense that the x rays being measured originate in a surface layer. This statement is true but that fact does not mean that x-ray fluorescence gives the same information about the surface as does, for example, photoelectron spectroscopy. On the other hand, it is much more of a surface technique than neutron activation. The purpose of this section is to define what the "surface" represents when we talk about x-ray fluorescence as a surface technique.

5.1. The Absorption Criterion

The statement that x-ray fluorescence analysis is a surface technique would be disputed by many because the thickness measured is very large compared to several other methods. It is in fact the substance of that argument that x-ray fluorescence is used for bulk analysis. However because of the fact that x rays are attenuated in passing through rather thin layers of solid materials, it is necessary that the surface of a sample be representative of the bulk for the analytical result to characterize the whole specimen. The fraction of the generated photon absorbed in passing through a finite layer of some material is a very strong function of the elements present in the material and of the energy (or wavelength) of the emitted x ray. A rather simple method of determining the thickness of the surface layer which contributes the x rays being measured is a consideration of this "absorption criterion." Absorption is an exponential process and, therefore, it is impossible to say that x rays of a particular wavelength emerge from a layer of a certain thickness and not any deeper. It is possible, however, to say that 95 or 90% of the information comes from a certain thickness by calculating what that thickness is which absorbs 95 or 90%.

By rearranging Eq. (3), one can calculate that for 90% absorption (10% transmission)

$$(\mu/\rho)\rho t = 2.30 \tag{8}$$

As a function of atomic number, mass absorption coefficients may vary over about two orders of magnitude for any given wavelength. For long-wavelength radiation, the (μ/ρ) values are the highest; e.g., Na K_α (11.9 Å) has absorption coefficients varying from about 200 to about 13,000 cm^2/g, P K_α (6.2 Å) has absorption coefficients varying from about 50 to about 4500 cm^2/g. Shorter wavelengths have lower values; Cu K_α (1.54 Å) varies from about 10 to about 450 cm^2/g. Since from Eq. (7) we can see that a large value of (μ/ρ) gives a small value of t, the use of long-wavelength x rays will enable the analysis of a thinner surface layer. It is possible to show how in copper, the layer being analyzed using Cu radiation is

$$t = \frac{2.30}{9.0 \times 52.7} = 0.0048 \text{ cm or } 48\ \mu$$

if Cu K_α (1.54 Å) is used, but is only

$$t = \frac{2.30}{9.0 \times 1900} = 0.00013 \text{ cm or } 1.3\ \mu$$

if Cu L_α (13.3 Å) is used.

Referring back to Fig. 12-3 to see how the mass absorption coefficient varies with energy and remembering that an element's characteristic radiation is of slightly less energy than the absorption edge, it must be recognized that the above calculation represents a rather unique situation. That is, because the characteristic x rays occur at a position on the absorption-coefficient curve close to a minimum, an element is said to be particularly transparent to its own radiation. Speaking of surfaces and of the ability of x-ray fluorescence to analyze only a thin layer, the calculation above must be considered a most pessimistic case. At the other extreme it is possible to consider the situation where the matrix has a high absorption coefficient for the radiation being measured. Again, let us consider measuring the Cu x rays, but this time coming from the surface of an iron sheet. The 90% criterion shows that for Cu K_α

$$t = \frac{2.30}{7.9 \times 317} = 0.00092 \text{ cm or } 9.2\ \mu$$

and for Cu L_α

$$t = \frac{2.30}{7.9 \times 16{,}900} = 0.000017 \text{ cm or } 1700 \text{ Å}$$

5.2. Surface Thickness as a Function of Atomic Number and Wavelength

There is nothing sacrosanct about the 90%-absorption criterion. One could easily make the calculations for 99% [for which the (μt) product is 4.60], 50% [for which the (μt) product is 0.692], or any other absorbed fraction. There is an argument that the 50%-absorption criterion is the most appropriate because it includes the 50% absorption of the primary radiation penetrating into the sample and the 50% absorption of the generated x rays in emerging from the sample. It has also been suggested that $1/e$ is an appropriate fraction. For the sake of this discussion, however, we shall use the 90% criterion, recognizing that the thickness calculated can be converted to the 99% value simply by multiplying by 4.60/2.30, and to the 50% value using the factor 0.692/2.30.

As an example of the type of calculation which might be made, the analysis of a bulk stainless steel sample by x-ray fluorescence actually only measures the "surface," and if something has happened to make the surface unrepresentative of the bulk, the result will not reveal the bulk composition. In this case, a calculation of the thickness of the layer being analyzed might shed some light on the problem. Let us suppose that the steel has a nominal composition of 18% Cr, 8% Ni, and the balance Fe. We measure the characteristic x rays of all three elements. The density of the alloy is ~ 8 g/cm^3. For Cr K_α the (μ/ρ) of the alloy is

$$(\mu/\rho)^{\text{alloy}}_{\text{Cr } K_\alpha} = (W_{\text{Cr}})(\mu/\rho)^{\text{Cr}}_{\text{Cr } K_\alpha} + W_{\text{Fe}}(\mu/\rho)^{\text{Fe}}_{\text{Cr } K_\alpha} + W_{\text{Ni}}(\mu/\rho)^{\text{Ni}}_{\text{Cr } K_\alpha} \qquad (9)$$

where W is the weight fraction. A similar summation can be made for

TABLE 12-3. Stainless Steel Example

	Radiation	$(\mu/\rho)_{\text{alloy}}$	μ_{alloy}	Thickness for 90% absorption, μ
	Cr K_α	115	920	25
Using K_α lines	Fe K_α	156	1,248	18
	Ni K_α	354	2,832	8.1
	Cr L_α	4,000	32,000	0.72
Using L_α lines	Fe L_α	5,200	41,600	0.55
	Ni L_α	3,100	24,800	0.93

the other elements, giving the results shown in Table 12-3 for radiation using K_α lines. In the last column of this table is listed the thickness of the layer which satisfies the 90%-absorption criterion. In this same example it is possible to demonstrate the improvement in depth resolution possible by using the low-energy x rays, as shown in Table 12-3 for radiation using L_α lines. This type of calculation can be made for any material to determine if the layer analyzed by x-ray fluorescence satisfies the "surface" criterion.

There is a second consideration of importance when discussing surface analysis by x-ray fluorescence. If the surface layer of interest contains elements not present in the substrate, that layer can be analyzed directly or indirectly. The indirect method is applicable where the composition of the surface layer is known and its thickness is to be measured. A characteristic line of the substrate material is chosen and its attenuation by the overlaying material is a measure of its thickness. For example, in tin-plated steel, it is possible to measure the ratio of the Fe K_α from a coated and uncoated specimen in order to determine the thickness of the tin. A more direct method of measuring the tin thickness would be to measure the Sn L_α intensity which for thin coatings is proportional to the thickness. It has been demonstrated[(19)] that a fraction of a monolayer of material can be measured quantitatively if it is deposited on a low-mass (a few mg/cm^2) substrate and if sufficient further care is taken to ensure that the background is minimized. If the material of interest is present as a surface layer on a bulk substrate, the detectable limit will suffer significantly but should make possible the analysis of surfaces a few monolayers thick. This direct technique can be used theoretically for both thickness and composition measurements since the x-ray intensity is a direct function of the mass thickness of the element being determined, provided that enough additional information is available about the material on the surface (e.g., density). If a measurement indicates that a surface contains 6.7 μg Cu/cm^2 and 3.3 μg Zn/cm^2 it might be suspected that the coating is brass which, having a density of $\sim$8.5 g/cm^3, would be $(10 \times 10^{-6}$ g/cm^2/(8.5 g/cm^3) $= 1.2 \times 10^{-6}$ cm, or 120 Å thick. These are concentrations which give readily measurable x-ray intensities in today's instruments. This is a most useful application of surface analysis by x-ray fluorescence, but it does not seem to be practiced to a great extent. It has, however, been examined theoretically[20,21] and the calculations confirmed by experiment to demonstrate that empirical calibrations are not required. It must be emphasized that this technique is only applicable for surface layers on a substrate of different material.

6. APPLICATION OF X-RAY FLUORESCENCE TO THE ANALYSIS OF SURFACES

There is not an overabundance of references in the literature to the use of x-ray spectroscopy to characterize the surface of solids. What information is available includes both photon and electron excitation. Both types of excitation will be included here even though the case for electron excitation can be controlled more precisely by varying the primary electron energy.

6.1. Chemical Composition

The study of the surfaces of metallurgical materials using x-ray fluorescence as we know it today began about twenty years ago. Koh and Caugherty[22] conducted a theoretical examination of the minimum thickness of the surface which emitted x-ray intensity equal to a bulk sample. From these calculations, they concluded that it should be possible to analyze corrosion layers *in situ*, and demonstrated experimentally that the measurements could be made. Rhodin,[23] on the other hand, stripped off the surface layer and analyzed it as a thin film, demonstrating detection limits of 0.037 μg/cm^2 for nickel. Liebhafsky and Zemany[24] have defined the parameters involved in different types of x-ray thickness gauges, and in their book[25] they discuss the analysis of thin films in considerable detail. Schreiber *et al.*[26] determined the contamination on bulk steel disks caused by exposure at elevated temperatures to lubricating oils containing various additives. They also claim detection limits of a fraction of a microgram per square centimeter, but it is not easy to fathom the criteria by which this is established.

Even though the volume of the literature describing surface analysis by x rays is small, it is impossible to list all of the references because they are widely scattered and difficult to locate. It does seem appropriate, however, to mention a few, selected to illustrate several of the types of analysis about which this chapter is concerned.

The use of on-line x-ray coating-thickness gauges suffers from the problem associated with flutter in the fast-moving strip material. An interesting solution to this problem has been presented[27] in which the collimator is automatically adjusted to view the most intensely excited part of the sheet. A method has been proposed[28] for measuring the thickness of aluminum on silicon in the range of 0–4 μ, with a claimed accuracy of ± 6 Å of aluminum. In a paper[29] comparing various methods for determining the thickness of tin plate, the x-ray fluorescence technique gave the best results. Combination of a high-energy electron diffraction unit with an x-ray spectrometer provided a means for simultaneously obtaining electron diffraction data and elemental composition of surfaces and their reaction products.[30] The diffusion of uranium through fuel-element cladding alloys was investigated[31] by using x-ray emission to measure the uranium that migrates to the surface. And finally, Francis and Jutson[32] studied the oxidation of stainless steels by measuring the relative amounts of five elements in thin oxide films.

6.2. Other Information

The other information, relative to the characterization of materials, which can be obtained by x-ray spectroscopy is an elucidation of the electronic structure. A major advantage of x-ray fluorescence analysis in the usual sense is that the emitted photons are unaffected by the state of chemical combination of the elements of interest. However, if the electrons involved in the transitions leading to emission are in the valence band, this statement is not true. To the contrary, the x rays emitted as a result of electron transitions from the valence band exhibit intensity, wavelength, and line-shape changes as a function of the chemical combination. Because these "valence-band x rays" are always of low energy (*K* series for very low atomic number, *L* series for low–intermediate atomic number, and *M* or *N* series for higher atomic number), there exists a probe to examine the chemical state of the atoms (hence the electronic structure) on the surface.

A classic example of using x-ray measurements to examine the structure of a surface was reported by White and Roy,[33] who compared the Si K_β spectrum of SiO_2 and Si with that from a number of samples of material reputed to be thin films of silicon monoxide. The suspect samples showed a Si K_β line shape characteristic of a mixture of SiO_2 and Si rather than the single compound. In a similar fashion, White and

Gibbs[34] have shown how it is possible to determine the coordination number of Si atoms in complex silicates by measuring the wavelength shift of the Si K_β.

This particular aspect of x-ray emission spectroscopy has been studied in an empirical fashion for a number of years. More recently, however, several authors[35–37] have demonstrated an ability to correlate the molecular orbitals of certain types of materials with the fine structure in the soft x-ray emission spectra. This approach holds great promise for the total characterization of materials by x-ray techniques, since the determination of electronic structure by the soft x-ray emission bands complement the presently available determinations of elemental concentration by x-ray emission spectroscopy and of atomic structure by x-ray diffraction.

7. CONCLUSION

It is obviously impossible in a chapter of this sort to describe all there is to know about a subject. For further, or more detailed, information on x-ray fluorescence analysis, the reader is referred to several good books on the subject.[38–41]

Perhaps the one important thought which this author would like to convey is that soft x rays probe only a fairly thin surface layer on a sample. With modern instrumentation, it is practical to make measurements in the 10 to 100 Å wavelength region which ought to make x-ray fluorescence analysis a valuable tool for the surface scientist because of, and/or in spite of, the valence effects which trouble these soft x rays.

8. REFERENCES

1. H. G. J. Moseley, The high frequency spectra of the elements, *Phil. Mag.* **26**, 1024–1034 (1913).
2. D. Coster and G. Von Hevesy, On the missing element of atomic number 72, *Nature* **111**, 79 (1923).
3. H. Friedman and L. S. Birks, A geiger counter spectrometer for x-ray fluorescence analysis, *Rev. Sci. Instr.* **19**, 323–330 (1948).
4. A. H. Compton and S. K. Allison, *X-Rays in Theory and Experiment*, 2nd ed., pp. 104–115, Van Nostrand, Princeton, New Jersey (1935).
5. W. J. Veigele, E. Briggs, B. Bracewell, and M. Donaldson, *X-Ray Cross Section Compilation*, Kaman Nuclear Corp., Colorado Springs, Colorado, KN-798-69-2(R), (October 1969).
6. L. S. Birks, *Electron Probe Microanalysis*, 2nd ed., pp. 147–171, Wiley-Interscience, New York (1971).
7. E. J. Brooks and L. S. Birks, Compton scattering interference in fluorescent x-ray spectroscopy, *Anal Chem.* **29**, 1556 (1957).
8. B. D. Cullity, *Elements of X-Ray Diffraction*, pp. 78–85, Addison-Wesley, Reading, Massachusetts (1956).

9. J. V. Gilfrich and L. S. Birks, Spectral distribution of x-ray tubes for quantitative x-ray fluorescence analysis, *Anal. Chem.* **40**, 1070–1080 (1968); J. V. Gilfrich, P. G. Burkhalter, R. R. Whitlock, E. S. Warden, and L. S. Birks, Spectral distribution of a thin window rhodium target x-ray spectrographic tube, *Anal. Chem.* **43**, 934–936 (1971).
10. B. L. Henke, X-Ray Fluorescence Analysis for Sodium, Fluorine, Oxygen, Nitrogen, Carbon, and Boron, in *Advances in X-Ray Analysis* (W. M. Mueller, G. R. Mallet, and M. J. Fay, eds.) Vol. 7, pp. 460–488, Plenum Press, New York (1964).
11. J. W. Criss and L. S. Birks, Calculation methods for fluorescent x-ray spectrometry, *Anal. Chem.* **40**, 1080–1086 (1968).
12. L. S. Birks, *X-Ray Spectrochemical Analysis*, 2nd ed., pp. 85–89, Interscience, New York (1969).
13. S. D. Rasberry and K. F. J. Heinrich, Computations for quantitative x-ray fluorescence analysis in the presence of interelement effects, Colloquium Spectroscopicum Internationale XVI, Heidelberg, Germany, October 1971.
14. L. S. Birks, *X-Ray Spectrochemical Analysis*, 2nd ed., pp. 36–38, Interscience, New York (1969).
15. J. Holliday, Soft x-ray emission spectroscopy in the 13-Å to 44-Å region, *J. Appl. Phys.* **33**, 3259–3265 (1962).
16. L. S. Birks, Naval Research Laboratory (private communication).
17. H. R. Bowman, E. K. Hyde, S. G. Thompson, and R. C. Jared, Application of high-resolution semiconductor detectors in x-ray emission spectrography, *Science* **151**, 562–568 (1966).
18. L. S. Birks, R. J. Labrie, and J. W. Criss, Energy dispersion for quantitative x-ray spectrochemical analysis, *Anal. Chem.* **38**, 701–707 (1965).
19. J. V. Gilfrich, P. G. Burkhalter, and L. S. Birks, X-ray fluorescence analysis of air pollution particulates, Colloquium Spectroscopicum Internationale XVI, Heidelberg, Germany, October 1971.
20. T. Shiraiwa and N. Fujino, Theoretical formulas for film thickness measurement by means of fluorescence x-rays, in *Advances in X-Ray Analysis* (W. M. Mueller, G. R. Mallet, and M. J. Fay, eds.) Vol. 12, pp. 446–456, Plenum Press, New York (1969).
21. J. W. Criss, Naval Research Laboratory (private communication).
22. P. K. Koh and B. Caugherty, Metallurgical applications of x-ray fluorescent analysis, *J. Appl. Phys.* **23**, 427–433 (1952).
23. T. N. Rhodin, Jr., Chemical analysis of thin films by x-ray emission spectrography, *Anal. Chem.* **27**, 1857–1861 (1955).
24. H. A. Liebhafsky and P. D. Zemany, Film thickness by x-ray emission spectrography, *Anal. Chem.* **28**, 455–459 (1956).
25. H. A. Liebhafsky, H. G. Pfeiffer, E. H. Winslow, and P. D. Zemany, *X-Ray Absorption and Emission in Analytical Chemistry*, pp. 146–159, John Wiley & Sons, New York (1960).
26. T. P. Schreiber, A. C. Ottolini, and J. L. Johnson, X-ray emission analysis of thin films produced by lubricating oil additives, *Appl. Spectry.* **17**, 17–19 (1963).
27. J. A. Dunne, Continuous determination of Zn coating weights on steel by x-ray fluorescence, in *Advances in X-Ray Analysis* (W. M. Mueller and M. J. Fay, eds.) Vol. 6, pp. 345–351, Plenum Press, New York (1963).
28. J. E. Cline and S. Schwartz, Determination of the thickness of aluminum on silicon by x-ray fluorescence, *J. Electrochem. Soc.* **114**, 605–608 (1967).
29. J. Smuts, C. Plug, and J. Van Niekirk, Coating thickness determination of tin plate by x-ray methods, *J. S. African Inst. Mining Met.*, April 1967, 462–472.

30. P. B. Sewell, D. F. Mitchell, and M. Cohen, High-energy electron diffraction and x-ray emission analysis of surfaces and their reaction products, *Develop. Appl. Spectry.* **7A**, 61–79 (1969).
31. E. A. Schaefer, P. F. Elliot, and J. O. Hibbits, The x-ray fluorescence measurements of surface uranium on oxidized fuel elements, *Anal. Chem. Acta* **44**, 21–28 (1969).
32. J. M. Francis and J. A. Jutson, The application of thin layer x-ray fluorescence analysis to oxide composition studies on stainless steel, *J. Sci. Instr., Ser. E.* **1**, 772–774 (1968).
33. E. W. White and R. Roy, Silicon valence in SiO films studied by x-ray emission, *Solid State Commun.* **2**, 151–152 (1964).
34. E. W. White and G. V. Gibbs, Structural and chemical effects on Si K_β x-ray line for silicates, *Amer. Mineralogist* **52**, 985–993 (1967).
35. D. W. Fischer, X-ray band spectra and molecular-orbital structure of rutile TiO_2, *Phys. Rev.* **B5**, 4219–4226 (1972).
36. R. Manne, Molecular orbital interpretation of x-ray emission spectra: Simple hydrocarbons and carbon oxides, *J. Chem. Phys.* **52**, 5733–5739 (1970).
37. E. Gilberg, Das K-Röntgenemissionsspektrum des chlors in freien molekülen, *Z. Physik* **236**, 21–41 (1970).
38. H. A. Liebhafsky, H. G. Pfeiffer, E. H. Winslow, and P. D. Zemany, *X-Ray Absorption and Emission in Analytical Chemistry*, John Wiley & Sons, New York (1960).
39. R. Jenkins and J. L. DeVries, *Practical X-Ray Spectrometry*, Springer-Verlag, New York, (1967).
40. L. S. Birks, *X-Ray Spectrochemical Analysis*, 2nd ed., Interscience, New York (1969).
41. E. P. Bertin, *Principles and Practice of X-Ray Spectrometric Analysis*, Plenum Press, New York (1970).

13

*SURFACE CHARACTERIZATION BY ELECTRON SPECTROSCOPY FOR CHEMICAL ANALYSIS (ESCA)**

Shirley H. Hercules and David M. Hercules

Department of Chemistry
University of Georgia
Athens, Georgia

1. INTRODUCTION

Among the many new instrumental techniques to be added to the chemist's arsenal, electron spectroscopy for chemical analysis (ESCA) is one of the most ideally suited to the examination of surfaces. It is sensitive and quantitative, and signals can be obtained from nearly every element.

Although it is a surface technique, ESCA can also "look through" a layer of adsorbed material to see both the surface and the adsorbate. ESCA chemical shifts reflect changes in chemical bonding or oxidation state and thus may be used to follow such changes on surfaces.

The ESCA technique was originated by Professor Kai Siegbahn and his co-workers at Uppsala University in Sweden. A description of their early work and instrumentation can be found in their book describing many possible applications of ESCA.(1) A second book has also described the application of ESCA to free molecules.(2) A review

* This work was supported, in part, by the National Science Foundation under grant No. GP-32484X.

of ESCA literature through December 1971 has appeared as part of the Fundamental Annual Review Series published by *Analytical Chemistry*.[3] The first international conference on ESCA was held in September 1971, and the proceedings of that conference are now available.[4]

The present chapter gives an overview of ESCA with particular emphasis on those aspects related to surface analysis. We thought it appropriate to include sections on fundamentals and instrumentation, although because of space limitations they are abbreviated. A more detailed review may be consulted for additional information on these two topics.[5] Likewise we have minimized treatment of the effects of chemical structures on ESCA spectra. Because it will become increasingly important as ESCA is applied to catalyst studies, the reader is referred to a recent review on this topic as well.[6]

Because ESCA is a relatively new technique, the literature is undergoing a rapid expansion. Applications of ESCA to surface studies have only begun to appear, and it is safe to predict that this chapter will be dated before the manuscript is even typed. Thus the present contribution should be viewed as an illustration of the type of surface studies possible by ESCA, rather than a compendium of neatly completed results.

2. THEORETICAL ASPECTS

2.1. Electron Ejection

Several physical techniques for studying atomic and molecular structure utilize the ejection of electrons from the species studied. Depending upon the information sought, different mechanisms for ejection are employed. Three such mechanisms are illustrated in Table 13-1.

Electron bombardment may, depending upon the energy of the impinging electron beam, cause the ejection of either core or valence electrons. Because of the interactions between the electrons, the ejected electrons do not exhibit discrete energy values, in contrast to the electrons studied by ESCA, as discussed below. The excited ion produced undergoes relaxation via one of two available routes (refer to Section 2.2). The major use of the electron-bombardment technique is in the production of Auger electrons. If an inner electron is merely excited rather than ejected by the bombarding beam, the scattered electron can experience quantized changes in its kinetic energy which reflect the energy of excitation.

Electron ejection may be induced also by a photon beam. Ionization using vacuum uv photons involves the ejection of the less tightly bound

TABLE 13-1. Fundamental Processes in ESCA

Primary Processes	
1. $A + h\nu \rightarrow A^{*+} + e^-$	Photoionization
2. $A + e_1^- \rightarrow A^{*+} + e_1^- + e_2^-$	Electron-impact ionization
3. $A + e^- \rightarrow A^* + e^-$	Electron-impact excitation
Relaxation Processes	
1. $A^{*+} \rightarrow A^+ + h\nu$	X-ray fluorescence
2. $A^{*+} \rightarrow A^{++} + e^-$	Auger electron emission

valence electrons, which emerge at discrete kinetic energy values. The energy of the ejected electron may be measured, and knowing the energy of the impinging photon, the ionization potential of the valence electrons may be calculated.

The term ESCA is used to refer only to the ejection of electrons using x rays. Radiation in this range provides sufficient energy for the ejection of core electrons, which also emerge with discrete values of kinetic energy. Since the energy of the exciting radiation is known, analysis of the kinetic energies leads to calculation of the core-electron binding energies.

2.2. Relaxation Processes

The ejection of either a core or a valence electron by any of three previously discussed processes yields an excited ion that relaxes via one of the two pathways shown in Table 13-1. An outer electron may drop to fill the vacancy in the inner orbital, dissipating its energy by emitting a photon. This process is known as x-ray fluorescence.

In a competing process, known as Auger emission, a second electron is emitted as an outer electron drops in to fill an inner vacancy in the excited ion. The energy dissipated is dependent only upon the energies of the initial and final states of the species, as in x-ray fluorescence. The kinetic energy of an Auger electron is therefore not dependent upon the energy of the exciting radiation or electron beam, but is constant. Auger lines, as a consequence, may serve as a convenient standard in an ESCA spectrum.

Auger emission and x-ray fluorescence do not occur with equal probability in all elements. In the lighter atoms, i.e., below atomic number 11, x-ray fluorescence is virtually nonexistent whereas Auger emission occurs with a probability of about 1.0. Above atomic number 11, the probability of Auger emission decreases as that for x-ray fluorescence increases, the 50% probability point being at about atomic

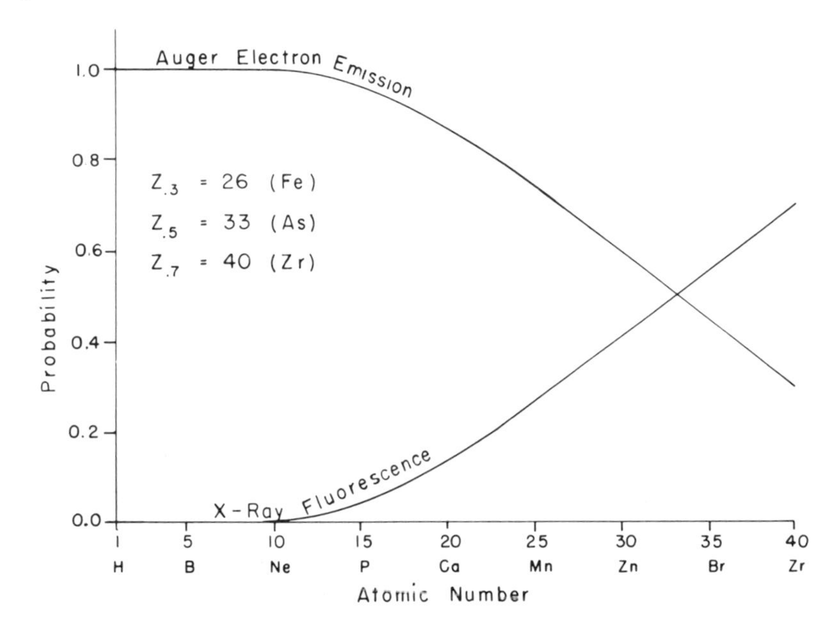

Fig. 13-1. Probability of Auger electron emission and x-ray fluorescence as a function of atomic number.[3]

number 33 (arsenic). X-ray fluorescence therefore is not a useful technique for the study of the lighter elements. The variation of x-ray fluorescence and *KLL* Auger electron emission with atomic number is shown in Fig. 13-1.

2.3. Relation to Existing Techniques

Because electron spectroscopy, x-ray fluorescence, and x-ray absorption involve basically the same phenomena, one might expect to obtain much the same information from the three techniques. The upper energy limit of an x-ray absorption band corresponds to the energy required to promote a given electron to the free electron level and thus should yield binding energies in much the same manner as ESCA. The spacings of the absorption bands should be identical to the spacings of the peaks on a photoelectron spectrum.

Also, since the emission bands in x-ray fluorescence reflect the energy given off as an outer electron drops in to fill an inner orbital vacancy, the chemical shifts, or the effects of the chemical environment on core-electron binding energies, should also be discernible from the fluorescence spectrum.

Although the information is theoretically available from all techniques, it is more easily obtainable from electron spectroscopy. In

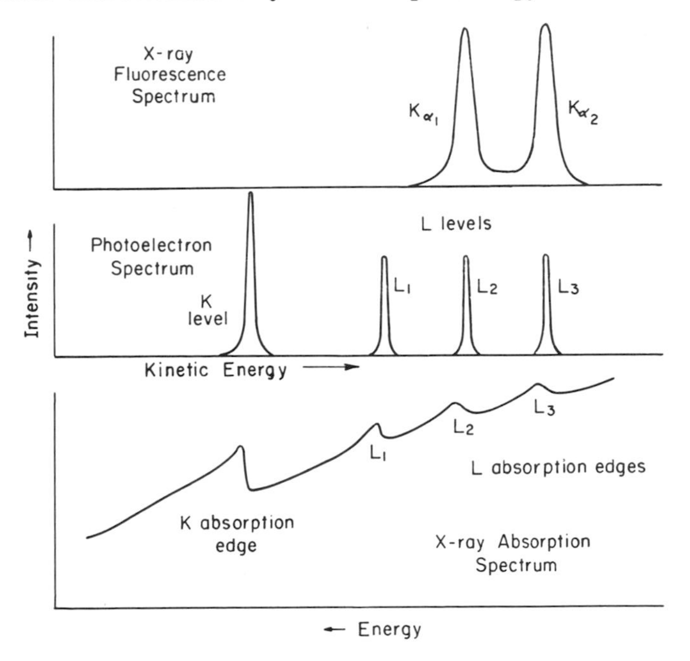

Fig. 13-2. Relationships between x-ray fluorescence, x-ray absorption, and ESCA spectra.[3]

general, line widths for a photoelectron line are narrower than those of an x-ray fluorescence peak. Chemical shift data also are more easily obtained from shifts in a photoelectron line than from shifts in an absorption edge. The relative energy spacings for the three techniques are shown in Fig. 13-2.

2.4. Energy Relationships

The relationships among the measured energy (kinetic energy of the electron), the calculated energy quantities (binding energies, chemical shifts) and other parameters are most easily explained by a diagram such as Fig. 13-3.

The energy of the exciting photon is depicted by the arrow to the far left on Fig. 13-3. This energy, when imparted to a core electron, may be viewed as being distributed among the four processes shown immediately to the right of the photon-energy arrow. A certain energy, the binding energy (E_b), is required to promote the electron from its ground state to the Fermi level (or to the free electron level in an isolated atom in the gaseous state).

In a condensed phase, an additional increment of energy known as the work function (ϕ_s) is required to dissociate the electron completely from the field of the matrix and bring it to the free electron level of the

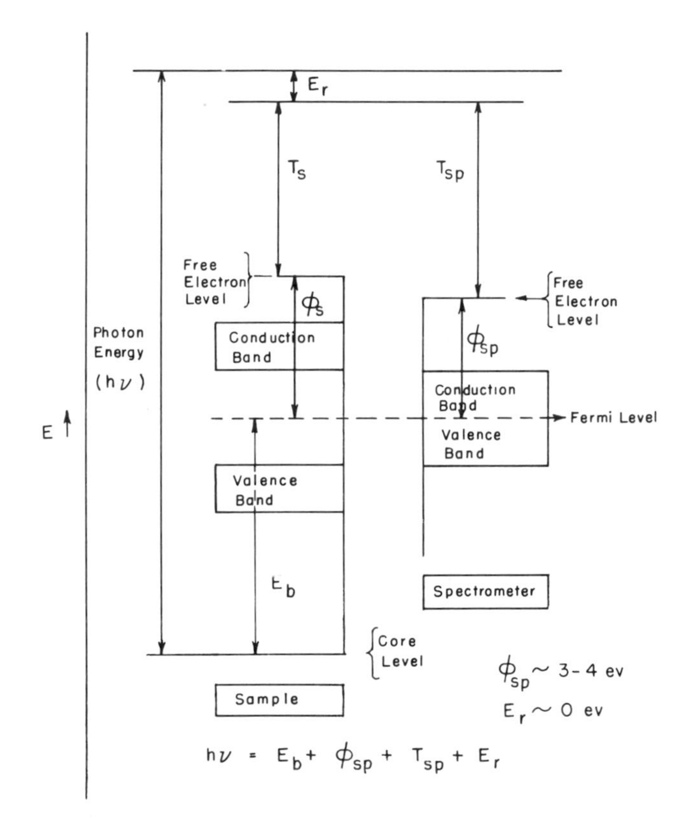

Fig. 13-3. Energetics of electron-binding measurements in solids,[3] where $h\nu$ is the exciting photon energy, E_b is the electron binding energy, ϕ_s is the sample work function, ϕ_{sp} is the spectrometer work function, T_s is the electron kinetic energy as ejected from the sample, T_{sp} is the electron kinetic energy measured by the spectrometer, and E_r is the recoil energy.

system. By virtue of the law of conservation of momentum, a small amount of the energy of the ejected electron is allocated to recoil (E_r), arising from proper distribution of energy as the electron is ejected. The recoil energy, a function of the atomic number Z and the energy of the exciting radiation, has been shown to be generally negligible even in the lighter elements. For example, with the aluminum K_α line as the exciting radiation, recoil energy may be neglected in atoms heavier than lithium. With copper K_α, it is negligible in atoms heavier than sodium.

The remainder of the energy provided by the exciting radiation appears as kinetic energy (T_s) of the ejected electron. Since the ejected electron comes under the influence of the electrical field of the spectrometer material before entering the monochromator, the mathematics of the interaction must be also considered. Whereas the sample is assumed to be an insulator with a Fermi level midway between the conduction and valence bands, the spectrometer is a conductor with a Fermi level, by definition, at the boundary between its conduction and valence bands. When two materials are in electrical contact, as is assumed to be the case with the sample and spectrometer, the Fermi levels are at a common level; the spectrometer and sample energy levels are therefore so placed relative to one another.

As the free electron with kinetic energy T_s travels the space between the sample and the monochromator slits, it encounters a potential gradient due to the difference between the work functions of the sample and the spectrometer. It will be accelerated or decelerated to kinetic energy T_{sp} as it reaches the free electron level of the spectrometer.

The energy distribution, expressed in terms of the measured quantity T_{sp}, is

$$h\nu = E_b + \phi_{sp} + T_{sp} + E_r \tag{1}$$

If absolute binding energies are to be calculated, the work function of the spectrometer (ϕ_{sp}) must be known accurately. However, in most chemical applications only relative binding energies or shifts in their values are relevant. For such calculations, it is necessary only that the spectrometer work function remain constant. If such is the case, assuming recoil energy to be negligible, shifts in binding energies are reflected as changes in the electron kinetic energies.

2.5. Shake-Up

In the simple case of photoejection, a core electron leaves the atom, having had the entire energy of the photon imparted to it to be distributed according to the discussion of the previous section. It is focused by the spectrometer and appears as a single spectral line at kinetic energy T_{sp} (as indicated by Fig. 13-2).

Upon close inspection of photoelectron spectra, however, one often finds on the low-energy side of a main spectral line, satellite lines of very low relative intensity, known as "shake-up" lines. Shake-up occurs when, instead of imparting its entire quantum of energy to the primary, photoejected electron, the x-ray photon gives up a discrete portion of its energy to the excitation of a second electron. The kinetic energy left for the ejected electron is T_{sp} less than expended in excitation of the electronic transition. Satellite lines occur with ΔE's corresponding for each allowed transition with intensities proportional to the probability of the transition (see Fig. 13-4). In general, the total intensity of all shake-up lines is not more than 10% of the intensity of the main line.

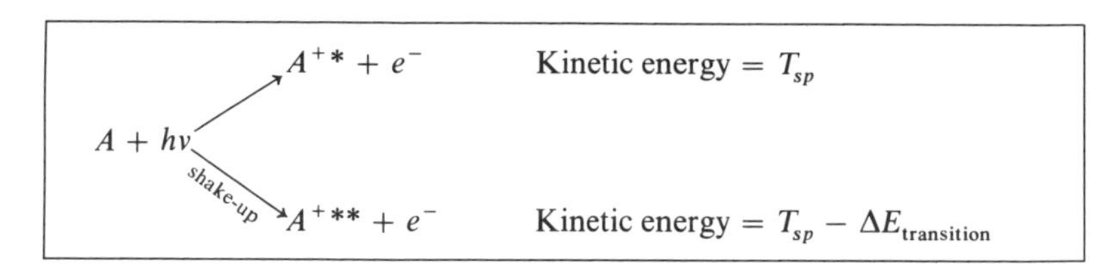

Fig. 13-4. A comparison of photoejection and shake-up.

3. INSTRUMENTATION

For any of the electron spectrometric techniques, the instrument is composed of five basic components: (1) source, (2) sample compartment, (3) electron energy analyzer, (4) detector, and (5) readout system. The source must produce an intense, monoenergetic, stable beam of photons, electrons, or x rays, of sufficient energy to excite the electrons of interest. The beam is directed into the sample compartment where it impinges upon the material to be studied, causing the ejection of electrons. The electron energy analyzer sorts the resulting electrons according to their kinetic energies and focuses them on a detector at the output. The detector produces an electrical signal proportional to the intensity of the electron beam and the readout system translates it into graphic form.

The specific choice of source, monochromator, or detector is dependent upon the purpose of the instrument, i.e., the type of electron spectroscopy to be employed.

3.1. Sources

Gaseous discharge lamps are the source of uv photons for the excitation of valence-shell electrons. Photoelectron spectroscopy therefore studies those electrons with a binding energy of about 0–40 eV. Typically used excitation lines are He^+ (40.812 eV), He (21.218), Kr (10.032), and Hg (4.8878).

For Auger and electron impact spectroscopy, beams from electron guns are accelerated, selected according to energy, and focused on the sample. In Auger spectroscopy a series of baffles and accelerator plates provides sufficient energy selection, but electron impact spectroscopy requires a more narrowly defined monoenergetic beam. A hemispherical electrostatic electron monochromator is used for this purpose.

For the photoejection of core electrons in ESCA soft x rays are used. The source is an x-ray tube of conventional design, consisting of a heated cathode and a cooled anode. The high power ($\sim$5 kW) required necessitates the dissipation of large quantities of heat. To avoid the effects of localized heating on the anode, it is water-cooled and sometimes rotated. The anode also should be located in such a way that it is not in a direct line with the cathode, thus minimizing the deposition of tungsten on the anode.

The precision of the final energy measurement is dependent upon using the most narrowly defined source of exciting radiation available. Since the linewidth of x-ray emission is proportional to atomic number, the lightest elements with well-defined intense radiation are the most desirable sources. Although Cu and Cr were used historically, the targets in most common use today are Al and Mg with K_α lines at 1486.6 and 1253.6 eV, respectively.

To minimize the stray radiation entering the sample chamber, a thin aluminum window is placed between the source and the sample. For further narrowing of the x-ray beam and elimination of stray radiation, an x-ray monochromator may be used. However, with some monochromators, the alignment of the sample becomes extremely critical, and a drastic loss in excitation energy must be expected from all such devices.

3.2. Samples

To ensure that most of the ejected electrons traverse the path to the monochromator without loss of energy, the sample chamber must be maintained at pressures below 10^{-6} Torr. Although differential pumping techniques make the study of gaseous samples possible, without such a facility the high vacuum necessary requires that samples be of low vapor pressure. The use of cryogenic probes permits the analysis of volatile materials. One must be sure, however, that freezing has no effect on the data so obtained. Liquid solutions are not suitable for study at present.

Another restriction on the sample is, of course, that it be stable to x-ray bombardment. Most materials have been found to be stable; however, some have been found to decompose during the course of a measurement. Such occurrences are readily recognized by the appearance of the spectrum or of the sample after analysis or by the time dependence of line intensities.

ESCA is a very sensitive technique, requiring only very small samples on the order of 10^{-6} g for typical measurement. Spectra have been obtained on samples as small as 10^{-8} g, however.

Because the ejected electrons do not emerge from depths of over about 100 Å, ESCA is extremely well suited to the study of surfaces. If, however, bulk properties are to be studied, assurance must be had that the surface properties are representative of the bulk.

3.3. Electron Energy Analyzers

The electrons ejected from the sample enter the electron energy analyzer, or monochromator, to be sorted according to kinetic energy. Various types of analyzers have been developed, with varying degrees of usefulness.

The simplest such device, known as the retarding-field analyzer, is perhaps the most limited in its utility. The electron beam produced from the sample is directed through a series of grids whose negative potential is varied as a function of time. When the potential, or retarding field, is low, the whole range of electron energies is passed through to the detector. As the potential is made more negative, electrons of increasingly high energy are repelled backward, thus decreasing the

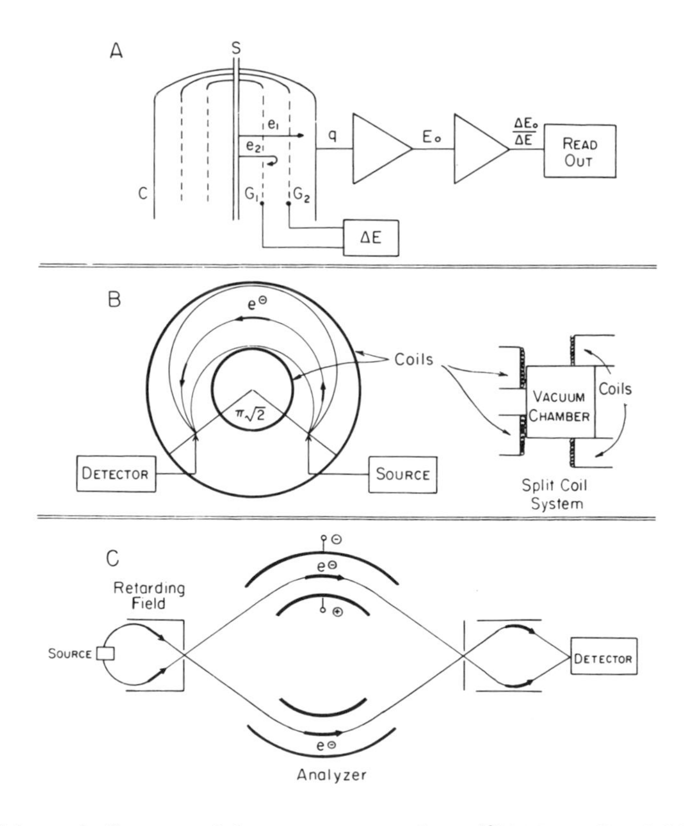

Fig. 13.5. Schematic diagrams of electron energy analyzers.[3] (a) Retarding field spectrometer. S = source; C = collector; G_1, G_2 = grids; ΔE = voltage source; e_1, e_2 = electrons having different kinetic energies; q = charge on cup; and E_0 = voltage from charge-sensitive amplifier. (b) Double-focusing magnetic spectrometer. (c) Spherical electrostatic spectrometer using a retarding field.

intensity of the beam arriving at the detector. A diagram of a retarding-field electron analyzer is shown in Fig. 13-5a.

The retarding-field electron analyzer presents two basic problems in the analysis of high-energy electron beams. For the resolution required in the sorting of such electron energies, an extremely uniform stable potential surface is necessary, a requirement that it is very difficult to satisfy. The nature of the resulting signal presents an additional problem. The initial signal represents the sum of the entire range of electron energies, and differentiating the small relative change in signal as the potential is increased is difficult at such a high signal level.

Retarding-field electron analyzers have been used to study low-energy electrons produced in photoelectron spectroscopy of gases. If a preretarding device is used, even high-energy electrons can be reduced to energies that may be sorted with this type of analyzer.

The first electron energy analyzer, used historically in Siegbahn's laboratory, was of the magnetic type.[1] This type of iron-free instrument is constructed of aluminum or brass and is built with a radius of 30 or 50 cm. A diagram of the magnetic instrument is shown in Fig. 13-5b. The circular vacuum chamber, or electron trajectory, is set within a set of cylindrical coils, which provide an inhomogeneous magnetic field. This design allows for double focusing of the electron beam, which travels through an angle of $\pi\sqrt{2}$ radians before reaching the detector.

Since a highly controlled field is used to direct the electron path, the magnetic field existing apart from the spectrometer, e.g., the earth's field, must be reduced to effectively zero. This represents the most serious disadvantage of the magnetic monochromator. Shielding to cancel the earth's field has been accomplished through a series of Helmholtz coils wound around the entire instrument. All perturbations on the field of the environment must be automatically compensated to better than 0.1 % to provide the resolution of electron energies required. Because of the difficulties so presented, commercial electron spectrometers are solely of the electrostatic type.

Although the electrostatic monochromator also requires that external fields and perturbations be reduced to effectively zero, paramagnetic materials such as μ-metal or netic and conetic shielding may be used to accomplish the purpose. These materials may not be used with the magnetic instrument since they shield by cutting magnetic force lines; thus they would perturb the spectrometer field also.

An electrostatic analyzer consists of a section of two concentric spherical surfaces with a potential difference between them. The ejected electrons are directed in a trajectory between these plates, and are deflected to varying extents by the electrostatic field. Prior to entering the electron analyzer, the electrons are also retarded using a lens system. An example is shown in Fig. 13-5c.

With such a system, two methods exist for scanning the electron spectrum. The retarding field of the lens system may be set at a constant value and the electrostatic field varied, linearly changing the energy of the electrons focused at the detector. This method is used in the McPherson instrument, a spherical sector of 30-cm radius. Alternately, the field between the plates may be set to pass a single electron energy and the retarding field varied. The Hewlett–Packard, AEI, and Varian instruments are so designed.

Some electrostatic instruments use, instead of concentric spherical sector plates, a segment on a pair of concentric cylindrical plates. Because the resolution of such an instrument is not good at high electron energies, this design has been used mainly for low-energy uv-excited electrons. A modification, however, has been made using the entire cylinder, that yields 0.08 % resolution at 1 % T.[7]

3.4. Detectors

A detector must be basically an electron-sensitive device that produces a response proportional to the intensity of an electron beam. Although several detectors have been used, the electron multiplier offers the maximum in convenience and sensitivity for chemical applications. Photographic detection has been used but is less convenient, although for specific applications it offers certain advantages. GM counters require a post-acceleration system for the electrons because of their lack of sensitivity to low-energy electrons. The charge-cup detector is necessary in the retarding-field electron energy analyzer.

Both single-channel and continuous-channel electron multipliers have been employed as detectors in electron spectrometers. The continuous-channel detector is sensitive and efficient at very low electron energies and is convenient to use. Recently ESCA workers have begun to use multichannel electron multipliers in conjunction with a multichannel scan system. The advantage of such a setup is that one increases the effective signal intensity in proportion to the number of channels used in the detector. Such a detector–scan system is particularly important when using an x-ray monochromator ahead of the sample, to narrow the energy widths of the exciting radiation. By so doing, it is possible to achieve signal levels comparable to those when a single-channel detector is used and an aluminum foil isolates the primary x-ray beam.

3.5. Readout Systems

The readout system is driven by the output to the detector and produces a graphic presentation of the electron energies. Three basic approaches are used; the linear scan, the incremental scan, and the multichannel analyzer.

In the first type, the field of the spectrometer is varied linearly as a function of time, scanning the energy spectrum once. The intensity of the electron signal is plotted as a function of the field strength or, alternatively, as a function of binding energy.

The incremental scan method divides the field range into a large number of finite steps or increments. The field is varied incrementally, either manually or by computer, and the signal counted at each increment.

The multichannel analyzer approach combines the first two methods. The field range is again divided into a large number of finite increments. Instead of scanning the spectrum only once, however, and counting each increment for a relatively long time, the field is scanned many times and the electron current summed at each pass for each increment. A visual display informs the operator of the progress of the analysis. This approach has the advantage that any fluctuations in the source intensity as a function of time will be felt equally by all channels rather than by just a few.

4. CHEMICAL SHIFTS

The quantity ultimately measured using electron spectroscopy is the binding energy E_b of the ejected electron. In their original work, Siegbahn and co-workers found that the binding energy for a given electron in an atom did not remain constant over a series of compounds containing that atom. Instead, it was markedly affected by the chemical environment and was roughly a function of the atomic charge.

4.1. Origin of Chemical Shifts

A core electron is subject to a combination of forces, the resultant of which is known as its binding energy. From the nucleus, it experiences a strong attractive force proportional to the magnitude of the nuclear charge or atomic number. The outer, or valence-shell electrons, exert a repulsive force which, in effect, screens the core electron from the nuclear charge, diminishing the nuclear attractive force. The resulting force by which the electron is bound to the atom is E_b.

If an electron is removed from the outer shell, the screening of the inner electrons is reduced by one electron charge, and the core electrons therefore feel an increased force of attraction from the nucleus. A negative change in oxidation state, i.e., the gain of an outer electron, has the opposite effect, effectively increasing the shielding and decreasing E_b.

Binding energies are therefore responsive to changes in the chemical environment and, as such, represent a new key to the study of chemical structure.

4.2. Koopman's Theorem

Some attempt has been made to predict binding energies *a priori* from theoretical calculations. The simplest method assumes that Koopman's theorem is valid, i.e., that the electron orbitals remain frozen during electron ejection. The binding energy would then be equal to the energy difference between the core orbital and the free electron level of the atom. If, however, the orbitals are assumed to relax adiabatically during electron ejection, the ejected electron would be imparted an additional increment of kinetic energy and the binding energy so measured would be that much smaller than calculated assuming Koopman's theorem. Allowing for orbital relaxation, the energy of the atom before ejection and of the ion after ejection, estimating a relaxation correction, must be calculated. The difference between the two is the binding energy. Comparison with experimental data seems to indicate that orbital relaxation does take place. However, for a given atom, the relaxation energy may be assumed to be the same in all compounds and, therefore, correlations between chemical shifts and one-electron ionization energies are valid.

4.3. Correlation with Atomic Charge

To a simple approximation, the binding energy of an electron may be expressed as

$$E_b = k_A q_A + \sum q_B / r_{ab} + E_R^0 \tag{2}$$

where $k_A q_A$ represents the contribution from the net formal charge on the atom considering that, in bonding, the electron is not removed to an infinite distance. The charge environment of an atom is not confined to its own electrons and those of atoms to which it is bonded. The second term takes into consideration the molecular and lattice potentials, i.e., the field contribution from nearby atoms within the molecule or within the lattice. E_R^0 is a reference level.

Several approaches have been used in an attempt to arrive at atomic charges that correlate with observed chemical shifts. The first and simplest of these uses Pauling electronegativity calculations:

$$q_i = Q_i + \sum_{i \neq j} n I_{ij} \tag{3}$$

where Q_i is the formal charge on atom i, n is the number of bonds in which it is involved, and I_{ij} is the partial ionic character of the bond between atoms i and j. I_{ij} may be expressed as

$$I_{ij} = i - e^{-0.25(\chi_i - \chi_j)^2} \tag{4}$$

where χ_i and χ_j are the Pauling electronegativity values of i and j, respectively. If chemical shifts are calculated only for molecules in the gas phase and with no major structural differences, e.g., the C(1s) binding energy in a series of halomethanes,[8] the molecular potential term and E_R^0 may be assumed to be constant. The calculations are simple and only k_A must be determined. For this type of example, good correlations are found between chemical shifts (ΔE_b) and calculated atomic charge. However, in molecules with significantly varied structure and in the solid phase, the approximation that the molecular potential term is constant is not necessarily valid.

Quantum mechanical methods for calculating atomic charge have also been attempted with some success. The CNDO approach appears to give a better correlation than does the EHMO, and the inclusion of the molecular-potential term in calculating chemical shifts improves the correlation. A summary of results of various calculations has been given.[9]

Yet another approach has been taken by Jolly *et al.*[10,11] to calculate shifts in binding energy directly using only thermodynamic data. The process of electron ejection, in this method, is divided into a series of hypothetical reactions for which thermodynamic data are available. Good correlations are obtained for those reactions that may be so broken down. However the examples to which this method may be applied are few, largely because of the lack of readily available thermochemical data.

4.4. Empirical Correlations

Fortunately, the utility of the ESCA technique to chemists is not dependent upon strict agreement of experimental observation with theoretical calculations. Much work has been done which indicates that an empirical approach or empirical correlations can yield significant information concerning oxidation states and molecular structure in both inorganic and organic compounds. Typical ESCA spectra are shown in Figs. 13-6 and 13-7.

Some examples of the kinds of chemical shifts that may be expected from a series of inorganic ions in various oxidation states are presented in Table 13-2.[12]

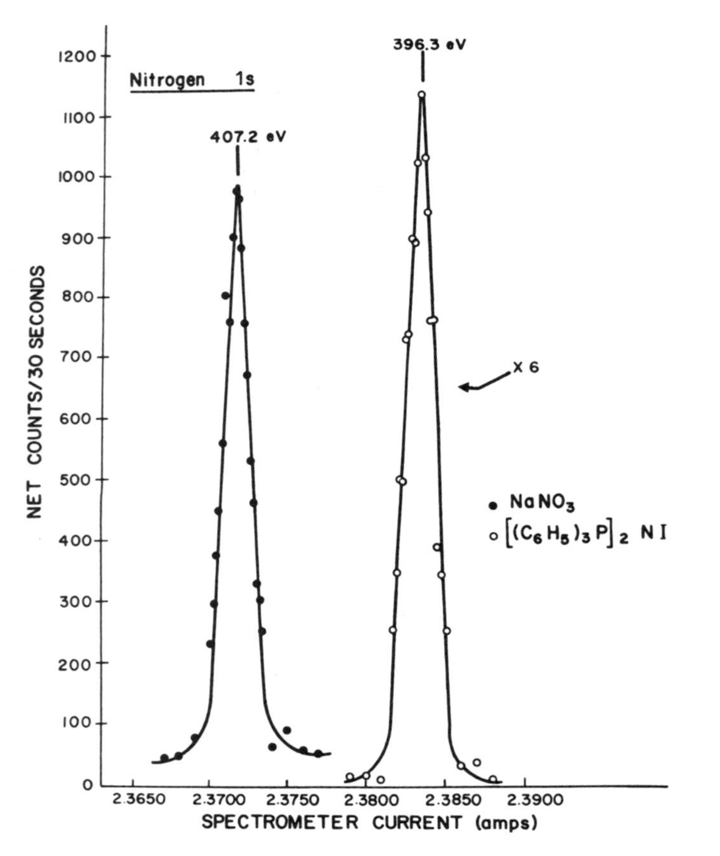

Fig. 13-6. Nitrogen 1s ESCA spectra of $NaNO_3$ and $[C_6H_5P]_2NI$. Note the magnitude of the shift between the two nitrogen atoms.

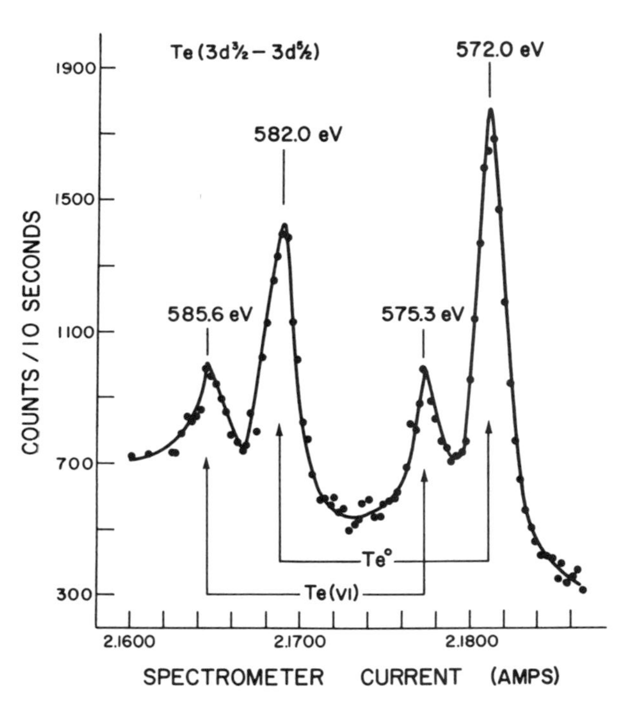

Fig. 13.7. Tellurium ($3d_{3/2}$–$3d_{5/2}$) electron spectrum for an elemental tellurium sample exposed to the atmosphere.[12] The 585.6-eV and 575.3-eV peaks are due to Te (+6). The 582.0-eV and 572.0-eV peaks are due to Te^0. This spectrum is indicative of the sensitivity of ESCA for detecting surface oxidation.

TABLE 13-2. Representative ESCA Chemical Shifts[a,b]

Element[d]	Oxidation state[c]										
	−3	−2	−1	0	+1	+2	+3	+4	+5	+6	+7
[5] Boron (1*s*)	—	—	—	0	—	—	+5.7	—	—	—	—
[7] Nitrogen (1*s*)	—	—	0[e]	—	+4.5[f]	—	+5.1	—	+8.0	—	—
[17] Silicon (2*p*)	—	—	—	0	—	—	—	+4	—	—	—
[15] Phosphorus (2*p*)	−1.3	—	—	0	—	—	+2.8	—	+3.1	—	—
[16] Sulfur (2*p*)	—	−2.0	—	0	—	—	—	+4.5	—	+5.8	—
[17] Chlorine (2*p*)	—	—	0	—	—	—	+3.8	—	+7.1	—	+9.5
[24] Chromium (3*p*)	—	—	—	0	—	—	+2.2	—	—	+5.5	—
[29] Copper (1*s*)	—	—	—	—	+0.7	+4.4	—	—	—	—	—
[33] Arsenic (3*d*)	0	—	—	—	—	—	−3.9	—	+5.3	—	—
[34] Selenium (3*p*)	—	−1.0	—	0	—	—	—	+3.6	—	+4.5	—
[35] Bromine (3*d*)	—	—	0	—	—	—	—	—	+609	—	+7.6
[42] Molybdenum (3*d*)	—	—	—	0	—	—	—	+4.3	—	+6.0	—
[52] Tellurium (3*d*)	—	−0.7	—	0	—	—	—	+2.4	—	+3.5	—
[53] Iodine (4*s*)	—	—	0	—	—	—	—	—	+5.3	—	+6.5
[63] Europium (4*p*)	—	—	—	0	—	—	+7.8	—	—	—	—

[a] From Hercules and Hercules.[12]
[b] All shifts in electron volts, measured relative to the oxidation state given as zero.
[c] These are formal oxidation states given for the elements indicated. They indicate the approximate magnitude of chemical shifts for an element in a given state.
[d] The atomic number is given in brackets [ ] to the left of the element; the electrons measured, in parentheses ( ) to the right.
[e] End nitrogen in NaN_3.
[f] Middle nitrogen in NaN_3.

Considerable work has been done to correlate chemical shifts with molecular structure of organic compounds. A recent review article[4] discusses in some detail a series of studies illustrating the use of ESCA in organic structural studies with respect to carbon, nitrogen, phosphorus, and sulfur.

The C(1*s*) binding energy was studied in a wide variety of configurations and in a broad range of compound types including such diverse species as insulin, nucleic acids, halogenated hydrocarbons, and acetone. Certain types of carbons showed narrow ranges of binding energy over a series of compounds, facilitating identification. However, others, e.g., esters and carboxylic acids, showed very broad ranges. In most cases, the intensities of the signals were proportional to the compositional ratio of the compound. A correlation chart for C(1*s*) binding energies and structures is shown in Fig. 13-8.

In one study[13] described, ESCA was used to identify the structure of a reaction product from a series of postulated compounds. The electron spectrum for each possible product was predicted using previously compiled data for carbons of a similar type. Comparison of the actual spectrum with the predicted spectra in combination with

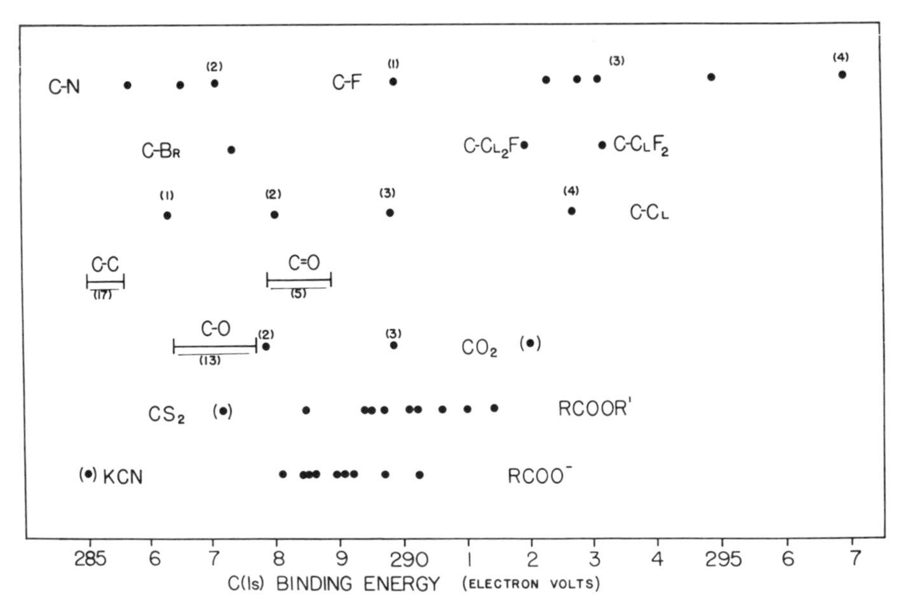

Fig. 13-8. Correlation chart for carbon 1*s* electron binding energies and functional groups.[6] ⊢⊣ = Range of observed energies for a given bond, ( ) = number of compounds used for the range above, (●) = energies observed for single compounds, ( ) = number of atoms attached to carbon, and ● = individual binding energies. The particular functional group compound or bond is shown beside the range or individual points.

NMR and mass spectrometric data clearly identified the structure of the product.

Correlation diagrams for various compounds of nitrogen, carbon, and sulfur have been made.[4,14] Some types of compounds in each of the elements show narrow ranges of binding energies, facilitating identification of them by spectral analysis. However, since several types, e.g., the primary carbon of esters and carboxylic acids, show a broad range of binding energies, it is clear that single-atom correlations, as in infrared analysis, will not be particularly useful for most compounds. Siegbahn, however, has suggested that a group approach might be useful[15] and has applied this to a variety of carbon compounds. Generally, the method has met with some success. However, for organic compounds containing a charged center[16] it is not certain that this type of correlation will be effective.

Similar studies are summarized in the same review article concerning nitrogen, phosphorus, and sulfur. A wide variety of compounds for each element was studied and attempts were made to correlate the observed binding energies with atomic charge. Correlations so made were somewhat more successful with sulfur and nitrogen than with phosphorus. It was concluded that the difficulty lay in the calculation of the charge on phosphorus.

4.5. Structure Determination

Some examples of the use of ESCA in structure determination should be mentioned. In the studies of metal–phosphine complexes,[17] the observed P(2*p*) binding energies were instrumental in indicating the high electron density on phosphorus, implicating the presence of back-bonding in the complexes.

In yet another example, the bonding characteristics of the bis(triphenylphosphine)iminium cation $[C_6H_5P]_2N^+$ (abbreviated PPN) were elucidated.[18] Nuclear magnetic resonance indicated the basic molecular structure and x-ray crystallography yielded a P–N–P bond angle of 137–142°. ESCA information indicated that the nitrogen had a negative charge similar to that in the azide ion and that the phosphorus atoms were partially positive, having $P(2_p)$ binding energies similar to the phosphorus in the phosphonium ion. The structure indicated by the accumulated data is

```
φ                 φ
  \             ⊕/
φ—P⊕           P—φ
  /  \  ⊖  /  \
φ       N        φ
```

In a study involving sulfur, ESCA was used to determine which of two possible structures represented the oxidation product of the disulfides.[19] Clearly the two structures would yield different spectra, and in fact, structure II was confirmed:

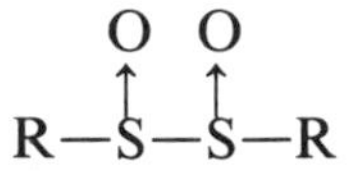

$$\begin{array}{c} \mathrm{O}\quad\mathrm{O} \\ \uparrow\quad\uparrow \\ \mathrm{R{-}S{-}S{-}R} \\ \\ \text{(I)} \end{array} \qquad\qquad \begin{array}{c} \mathrm{O} \\ \uparrow \\ \mathrm{R{-}S{-}S{-}R} \\ \downarrow \\ \mathrm{O} \\ \text{(II)} \end{array}$$

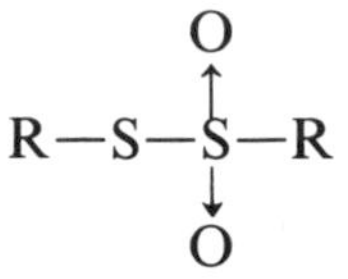

As more work is done and more data are gathered, ESCA will probably become a significant technique for structure determination. Its utility for determining structural changes for molecules interacting with a surface remains to be demonstrated; the possibilities here seem to be very great.

5. FACTORS RELATED TO USE OF ESCA FOR SURFACE ANALYSIS

5.1. Mean Escape Depth

As mentioned previously, electron spectroscopy by its very nature is ideally suited for surface analysis. Although the x rays penetrate deep into the sample, the ejected electrons emerge from within only a few molecular layers of the surface. The mean escape depth of an electron is dependent upon the kinetic energy of the electron and the atomic number of the material, low-Z materials being more permeable than high-Z materials. The chemical properties deduced, then, from ESCA data are actually those of surface molecules. Before extrapolations may be made to the bulk material, some assurance must be had that the surface properties are representative of the bulk.

Several studies have been done to determine the mean escape depth of a photoejected electron in a few materials. The first such work was done by Siegbahn on α-iodostearic acid.[1] He deposited on a metallic support a series of three α-iodostearic acid samples, containing 1, 3, and 10 double layers. Each double layer was about 40 Å thick. The relative intensities of the iodine $3d_{5/2}$ electron peaks obtained from the three samples were 1:2.1:3.5. From these data he calculated the mean escape depth ($I_{1/2}$) to be less than 100 Å.

In yet another experiment, Siegbahn deposited 200 molecular layers, amounting to about 8000 Å, of α-bromo-stearic acid on two chromium slides. On one, he also placed two molecular layers of stearic acid. No chromium signals were observed from either slide, indicating that the 200 molecular layers were sufficient to completely shield the chromium electrons. The bromine signals were observed in both samples, although the ratio of C(1*s*) signal to Br (3*d*) was less for the sample containing the extra layers of stearic acid.

Since then, other studies have pinpointed the mean escape depth at considerably lower values than 100 Å. Electrons from silver of kinetic energies 72 and 362 eV, respectively, were shown by Palmberg and Rhodin[20] to have ($I_{1/2}$) values of 4 and 8 Å through the silver matrix. Baer *et al.*[21] using gold samples have indicated a mean escape depth of 22 Å for electrons of 1200 eV. Finally, a study of the mean escape depths of electrons from a gold support through varying thickness of carbon film has shown the following:[22]

C(1*s*) energy, eV	$I_{1/2}$, Å
920	10
970	10
1169	13

These values correspond to a thickness of only a few molecular layers of carbon.

5.2. Surface Charging Effects

To minimize the buildup of charge on the surface of the sample, which would affect the measured binding energies of the ejected electrons, samples have been deposited in thin layers on metallic supports, making electrical contact with the spectrometer. However, the atoms or molecules involved in electron ejection as mentioned above, lie within a few molecular layers of the surface. In a nonconductor, surface layers cannot be assumed to be in good electrical contact with the spectrometer and an electrostatic charge can build up on the surface. The effect of the surface charges on measured binding energies would be proportional to the magnitude of the charge, which may be on the order of several volts. Thus, it will cause a shift of the ESCA lines, as well as possible line broadening.

To compensate for charging effects, considerable work has been done to find a suitable reference. Since carbon is present as a contaminant in virtually all measurements, the C(1*s*) line has been used by many researchers. However, the signal from such a contaminant is often so broad as to make precise estimation of its position difficult. Also the exact nature of the carbon giving rise to the signal is uncertain.

Physical mixtures of LiF (10%), MoO_3, and powdered gold have been successfully used.

Hnatowich *et al.*[23] have proposed the use of a thin film of vacuum-deposited gold or palladium on the sample surface as an excellent means of compensating for the effects of surface charging. In their experiments, they partially covered an aluminum support with $BaSO_4$ and then vacuum deposited Au "islands" on the surface of the $BaSO_4$ and Al. By running spectra with the aluminum support externally charged at +1.5, 0.0, and −1.6 V, they were able to estimate the amount of charging that existed on the $BaSO_4$ surface. The Au (N_6 and N_7) and the Ba(M_5) lines were monitored. A pair of gold N_6 and N_7 doublets was noted, one that shifted directly with the shift in the potential of the support and one that shifted by the same amount as the Ba line over the three spectra. A similar experiment using Pd deposition confirmed that the difference in the shifts noted for the gold on the $BaSO_4$ and the gold on the metal support was a good measure of the charge present on the $BaSO_4$ surface. The vacuum deposition of a noble metal such as Au or Pd is therefore a suitable method for compensating for charging effects.

5.3. Sensitivity

The absolute amount of material required to obtain an adequate ESCA spectrum is quite small—on the order of 10^{-6} g. Measurements however have been performed on samples of 10^{-8} g or less. The technique is so sensitive that 0.01 monolayer of many materials should produce a detectable signal.

Although the sensitivity quoted is quite good, one must remember that the figures quoted are samples in which the entire amount of material is deposited onto an area of about 0.5 cm^2. If the 1 μg of material to be studied is dispersed as a trace contaminant in a large amount of another material, the total amount of the contaminant that will be able to be placed into the sample holder will be much less than the minimum detectable quantity, unless some method exists for concentrating it. Thus ESCA, though sensitive to extremely small amounts of many species, is not yet a technique for trace analysis.

Studies or analyses may be performed on a given atomic species even if it is present as a minor constituent in the presence of many other atoms in a molecule. For example, Siegbahn *et al.*[1] observed the cobalt signal in a 100-Å layer of vitamin B_{12} in which the cobalt was present as 1 atom in 180 atoms of carbon, hydrogen, nitrogen, and oxygen. In yet another study, the disulfide bridges could be studied in insulin (mol. wt. 6000), though only three such bridges exist in the molecule.

Although better methods exist for the determination of trace amounts of metals in alloys, ESCA may represent a good technique for studying the segregation of components at phase boundaries.

5.4. Qualitative and Quantitative Nature of ESCA

Electron spectroscopy is sensitive to all elements but hydrogen and helium. Since the measured electron binding energies of most elements are sufficiently unique, a broad range scan of the electron energy spectrum provides a good means of qualitative analyses. Although accidental overlap of some peaks does occur, in general the electron signals from adjacent elements are quite well separated.

The integrated intensities of the electron signals are directly proportional to the number of similar atoms in the sample, representing a convenient quantitative tool. After correction for various effects, i.e., variation in photoelectric cross section, attenuation of electrons emerging from the sample, and shake-up effects, the relative peak intensities show good agreement with the stoichiometric formulae.

5.5. Surface Contamination

The high sensitivity of electron spectroscopy is both an asset and a liability. Mere exposure of a sample to the atmosphere for short periods of time permits surface adsorption of carbon and oxygen. Also, once the sample is placed in the spectrometer, sufficient back-diffusion of pump oil into the sample chamber through the vacuum system can deposit thin layers of carbon on the sample.

Several methods for the elimination of such surface contamination have been attempted. The time that the sample is exposed to air should, of course, be minimized. A method has also been developed for actually "scrubbing" the sample surface to clean off any deposited material. In the "argon-ion scrubbing technique," the sample is placed in an argon atmosphere and a high potential is placed between the sample and a positively charged electrode. In the resulting glow discharge, positively charged argon ions bombard the surface of the sample, cleaning it. If the sample is then transferred from the argon atmosphere into the spectrometer without contacting air, the amount of trace contamination is decreased significantly.

To eliminate the pump-oil deposition, a metallic screen may be placed close to the sample, between it and the vacuum port within the compartment. The screen is cooled to liquid-nitrogen temperature, effectively freezing out any contaminants, like pump oil, that diffuse into the chamber. This technique is useful only for samples having very low sublimation pressures.

6. APPLICATION OF ESCA TO SURFACE STUDIES

6.1. Surface Analyses

Environmental pollution is one of the major problems facing chemists today. In the development of abatement methods, for example, the specific oxidation state of a sulfur pollutant must be known. Because the amount of compound involved is small and it is often transported as a surface film on air-borne particles, sampling and analysis has been difficult. Research has indicated that ESCA may make a significant contribution in the effort to identify and monitor such compounds in the environment.

Hulett *et al.*[24] studied sulfur compounds adsorbed onto smoke particles and fly-ash with specific efforts at identifying the oxidation state of the sulfur. The interaction between the gaseous sulfur compounds and the solids onto which they adsorb was also investigated for application to pollution abatement.

Combustion analysis of the smoke and fly-ash indicated that the sulfur compounds under study were present mostly as an adsorbed layer on the particles. Spectral analysis showed the presence of two types of sulfur in the fly-ash, which were postulated to be the sulfate or adsorbed SO_3. The smoke particles contained three species of sulfur, postulated on the basis of the ESCA spectrum to be sulfite, sulfate, and, at much lower binding energy, H_2S or a mercaptan. A study was also performed which showed that many variables, including the nature of the particulate matter itself, affected the specific form of the sulfur species on the particle.

Another study showing the application of ESCA to a pollution analysis was conducted by Araktingi *et al.*[25] Particulate samples from the air were collected on fiberglass filter paper and submitted to electron spectrometric analysis. Broad range scans identified several elements, the most pertinent to air pollution being N, Pb, and S. In an attempt to identify the species giving rise to the broad Pb signal, synthetic mixtures of PbO_2 and $PbCl_2$ (suspected to be present in air from automobile exhaust) were prepared and their spectra compared with the air sample. Significant similarities were noted in the data; however, such a match does not represent positive identification. Sulfur in the sample was identified as the sulfate. The nitrogen compounds, however, were not identified.

ESCA has also been investigated as an analytical tool in the textile field. Knowledge of the surface characteristics of carbon fibers, i.e., the form of carbon and other elements present, is necessary in the development of synthetic fibers. In such a study, Barber *et al.*[26] have reported the surface analysis of five fiber types. Oxygen in two different forms was found to be present and its concentration estimated at 5 oxygens per 100 carbons in one sample. Some information was also gained about the type of carbon involved.

6.2. Analyses of Oxide Mixtures

An area of analysis that has been extremely difficult to date is the determination of one oxide of a metal in the presence of another. Swartz and Hercules[27] undertook to determine the applicability of ESCA for the determination of molybdenum oxides. The determination of mixed molybdenum oxides has been done until now only by time-consuming wet chemical analysis, since no instrumental technique could differentiate between MoO_2 and MoO_3.

The quantitative analysis of the mixed MoO_2/MoO_3 oxides was shown to be good to better than 2% in mixtures containing more than 15% MoO_2. The authors suggested that such differentiation between metal oxide pairs in bulk analysis and surface studies is possible for any two commercial oxides that conform to two boundary conditions—the peak separation must be sufficiently large, and the species with the lowest oxidation state must be stable to further oxidation. At least three other pairs have been found to fit these qualifications—PbO/PbO_2, Cr_2O_3/CrO_3, and AsO_3/As_2O_5.

6.3. Adsorption Studies

In studies of surface adsorption, provided the adsorbed species is bound strongly enough to resist the vacuum necessary for analysis, ESCA should be extremely valuable. Delgass *et al.* have done a number of preliminary studies that indicate the utility of ESCA in this area.[28] In one example, they studied the surface adsorption properties of a nitrogen-containing Y-zeolite. A zeolite wafer was exchanged with NH_4^+ and the N (1*s*) spectrum followed to determine the effects of various treatments. Following a spectral analysis of the exchanged wafer, it was subjected to heat at 150°C for 10 h *in vacuo*. Comparison of the pretreatment spectrum with the spectrum of the heated wafer indicated that over half of the nitrogen had been desorbed by the heat treatment.

Wafers of NH_4^+–Y-zeolite and of silica–alumina were evacuated, the zeolite at 550°C, the silica–alumina at room temperature and both were subjected to 1 atm of NH_3. Spectra of the two samples resembled that of the NH_4^+-exchanged zeolite wafer.

In yet another treatment, a heated NH_4^+-exchanged zeolite wafer was heated at low pressure of H_2 and then exposed to low-pressure pyridine. Spectra subject to this treatment showed two nitrogen peaks, one of which may be attributed to the pyridine. Suggestions have been offered to explain a small shift noted in the residual nitrogen line.

6.4. Surface Oxides

The exact nature of the oxides produced at the surface of a platinum electrode has long presented a problem. Identification of these oxides would be invaluable in the determination of the processes at work during electrolysis. However, because they are present in such small quantities, such information has been difficult to obtain. Since ESCA is by definition a surface technique and is sensitive to films less than a monolayer deep, it represents an ideal technique for such studies.

Kim *et al.*[29] have undertaken some studies of the platinum oxides produced at an electrode surface. They chemically prepared a series of platinum and platinum oxides as reference samples, ran electron spectra of each, and determined the Pt($4f_{7/2}$ $4f_{5/2}$) binding energies for each sample. Platinum oxides were then prepared electrochemically and analyzed by electron spectroscopy.

By comparison with the reference spectra, the oxide species produced in the electrochemical cell could be identified as a function of applied potential. The authors found that at +1.2 V, platinum and chemisorbed oxygen were the predominant species. PtO_2 began to be formed above +1.6 V and became predominant at +2.2 V.

Another important application of electron spectroscopy is the study of thin, submicron films of oxide in metal surfaces. (This is a very difficult area to study by techniques other than ESCA.) Two such investigations have been performed on stainless steel.[30,31] The shake-up satellite lines have been used in the analysis of oxide films on Ni and Ni-Cu alloys. Electron spectroscopy permits identification of the metals present as well as these oxidation states, even in very thin films.

6.5. Catalysts

The area of heterogeneous catalysis is yet another field for which electron spectroscopy is well suited. The nature of the catalyst surface may be studied as well as that of an adsorbed reactant, if present at the low pressures of the sample chamber. Several preliminary studies have been done showing the possible application of ESCA to this field.

Delgass *et al.*[28] have studied several active metals dispersed on supports, comparing them with the unsupported metals and their oxides. They studied platinum both supported on silica and unsupported and examined the Pt($4f_{5/2,7/2}$) peaks as an indication of the oxidation state of the metal. Since they did not make charge corrections for some spectra, this study serves mainly as an indication that the oxidation state can be qualitatively identified as a function of dispersion.

They also studied the characteristics of copper deposited on MgO. The samples were prepared by impregnating MgO with a solution of cupric acetate and subjecting the samples to various treatments. After being heated to 1000°C in air, the sample produced a broad peak, attributed to Cu^{+2} in or on the MgO surface. Merely drying the impregnated samples gave rise to a doublet that could only be explained as two copper species, one bonded on the surface and one present as cupric acetate. When samples were heated in the presence of H_2 at low pressure, narrow peaks appeared at lower binding energies, indicating the reduction of copper. However, the $2p_{3/2}$ binding energies of copper foil did not match that of the reduced sample. Although erroneous charging corrections may have caused the discrepancy, further investigation is needed to clarify the data.

In an ESCA investigation of nickel supported on silica–alumina, a preliminary spectrum of NiO showed a doublet separated by 6.1 eV in the region of the Ni($2p_{3/2}$) binding energy. No conclusive explanation could be offered, but the authors suggested that even without assignment of the two peaks, the observed spectrum could still function as a "fingerprint" for NiO. Spectra were obtained for pure Ni, NiO, and nickel foil which showed the characteristics of both metal and oxide.

A nickel nitrate-impregnated support wafer gave the following spectra: when calcined in air at 760°C, the spectrum of NiO was obtained; following treatment of the same wafer with H_2 at 480°, a spectrum indicated that most of the nickel was reduced to Ni metal; after exposure to air, the wafer returned to the NiO form. When a supported Ni sample was sulfided in an H_2–H_2S flow at 480°C, the Ni($2p_{3/2}$) line showed a split, unlike that of NiO.

The aging of FeV_2O_4 catalysts was also studied by Delgass *et al.*[28] Spectra were taken of the catalyst before and after use in the dehydrogenation of cyclohexane at 425°C. Although x-ray diffraction patterns of both samples were identical, indicating that no bulk change had taken place, significant differences in the ESCA spectra were noted. Vanadium lines were shifted to higher binding energies, oxygen lines were split and shifted to lower binding energies, and the iron was unchanged. Since the lines affected were shifted rather than broadened, the authors suggested that the entire surface layer was changed rather than just the surface itself.

In another study of the application of ESCA to catalysis, Wolberg *et al.*[32] investigated the dispersed surface phases of copper oxide supported on alumina. Having studied the same system with x-ray diffraction, ESR, and x-ray *K*-absorption edge spectroscopy, they compared results obtained with these techniques to those obtained with ESCA. The data were in complete agreement.

Two samples of Cu(10.37%) on a support surface area of 72 m^2/g were prepared and calcined at 500°C and 900°C, respectively. An additional sample of 8.8% Cu on an alumina with a surface area of 301 m^2/g was prepared and calcined at 500°C. Spectra were taken of the $Cu(2p_{3/2})$ line for both the reference and supported catalyst samples. Data indicated, as did previous studies, that the high-surface-area alumina with 10% cupric ions calcined at 500°C was in the copper aluminate structure and the low-surface-area alumina under the same conditions was in the CuO structure. On the low-surface-area sample, however, when heated to 900°C, the copper assumed the copper aluminate structure also.

In comparing the ESCA with x-ray diffraction, ESR, magnetic susceptibility, and x-ray absorption for the study of catalytic surfaces, the authors suggest that ESCA presents several advantages over all others. Whereas the ESR and magnetic susceptibility measurements are applicable only to paramagnetic materials, ESCA is applicable to all elements except hydrogen and helium. Also, it is more sensitive than x-ray diffraction.

Miller *et al.*[(33)] were interested in elucidating the surface structural characteristics of some Mo/Al_2O_3 systems. They prepared mixed powders of pseudoboehmite with molybdenum oxide and alumina with molybdenum oxide, and studied the electron spectra of the Mo(3*d*) levels. Prior to heating, the binding-energy spectra consisted of a well-resolved doublet for the $3d_{3/2}$ and $3d_{5/2}$ levels. However, after heating past 360°C, the peaks broadened. Well-resolved peaks were obtained for pure samples of molybdic oxide, and some molybdate salts. The authors concluded that an electron donor–acceptor process was occurring between the aluminum and the molybdenum in the calcined catalysts.

The surface of molybdenum hexacarbonyl in alumina was studied by Whan *et al.*[(34)] using electron spectroscopy. Catalysts of this type cause the disproportionation of olefins only after thermal activation. The samples were prepared and spectra were taken at various steps in the activation process. The binding-energy spectrum of the molybdenum (3*d*) region indicated that the thermal-activation treatment drove off the hexacarbonyl, leaving different molybdenum species on the alumina surface. Exposure to air produced yet a different species or mixture of Mo species.

7. REFERENCES

1. K. Siegbahn, C. Nordling, A. Fahlman, R. Nordberg, K. Hamrin, J. Hedman, G. Johansson, T. Bergmark, S. Karlsson, I. Lindgren, and B. J. Lindberg, *ESCA Atomic Molecular and Solid State Structure Studies by Means of Electron Spectroscopy*, Almquist and Wiksells, Uppsala (1967).
2. K. Siegbahn, C. Nordling, G. Johansson, J. Hedman, P. F. Heden, K. Hamrin, U. Gelius, T. Bergmark, L. Werme, R. Manne, and Y. Baer, *ESCA Applied to Free Molecules*, North Holland-American Elsevier, Amsterdam, New York (1969).
3. D. M. Hercules, *Anal. Chem.* **44**, 106R (1972).
4. D. A. Shirley (ed.), *Electron Spectroscopy*, North Holland, Amsterdam (1972).
5. D. M. Hercules, *Anal. Chem.* **42** (1), 20A (1970).
6. S. H. Hercules and D. M. Hercules, *Rect Chem. Progress* **32**, 183 (1971).
7. P. H. Citrin, R. W. Shaw, Jr., and T. D. Thomas, in *Electron Spectroscopy* (D. A. Shirley, ed.) pp. 105–120, North Holland, Amsterdam (1972).
8. T. D. Thomas, *J. Am. Chem. Soc.* **92**, 4184 (1970).
9. J. M. Hollander and D. A. Shirley, *Ann. Rev. Nucl. Sci.* **20**, 435 (1970).
10. W. Jolly and D. Hendrickson, *J. Am. Chem. Soc.* **92**, 1863 (1970).
11. J. M. Hollander and W. Jolly, *Accounts Chem. Res.* **3** (1970).
12. S. H. Hercules and D. M. Hercules, *Intern. J. Environ. Anal. Chem.* **1**, 169 (1972).
13. J. Hedman, P. F. Heden, R. Nordberg, C. Nordling, and B. Lindberg, *Spectrochim. Acta Part A*, **26**, 761 (1970).
14. B. Lindberg, K. Hamrin, G. Johansson, U. Gelius, A. Fahlman, C. Nordling, and K. Siegbahn, *Phys. Scripta* **1**, 286 (1970).
15. U. Gelius, P. F. Heden, J. Hedman, B. J. Lindberg, R. Manne, R. Nordberg, C. Nordling, and K. Siegbahn, *Phys. Scripta* **2**, 70 (1970).
16. J. J. Jack and D. M. Hercules, *Anal. Chem.* **43**, 729 (1971).
17. J. Blackburn, R. Nordberg, F. Stevie, R. G. Albridge, and M. M. Jones, *Inorg. Chem.* **9**, 2374 (1970).
18. W. E. Swartz, Jr., J. K. Ruff, and D. M. Hercules, *J. Am. Chem. Soc.* (in press).
19. G. Axelson, K. Hamrin, A. Fahlman, C. Nordling, and B. J. Lindberg, *Spectrochim. Acta* **23A**, 2015 (1967).
20. P. W. Palmberg and T. N. Rhodin, *J. Appl. Phys.* **39**, 2425 (1968).
21. Y. Baer, P. F. Heden, J. Hedman, M. Klasson, and C. Nordling, *Solid State Commun.* **8**, 1479 (1970).
22. R. G. Steinhardt, J. Hudes, and M. L. Perlman, *Phys. Rev. B.* **5**, 1016 (1972).
23. D. J. Hnatowich, J. Hudes, M. L. Perlman, and R. C. Ragaini, *J. Appl. Phys.* **42**, 4883 (1971).
24. L. D. Hulett, T. A. Carlson, B. R. Fish, and J. L. Durham, paper presented at 161st National American Chemical Society Meeting, Los Angeles, California (April 1971).
25. Y. E. Araktingi, N. S. Bhacca, W. G. Proctor, and J. W. Robinson, *Spectroscop. Letters* **4**, 365 (1971).
26. M. Barber, P. Swift, E. Evans, and J. Thomas, *Nature* 227, 1131 (1970).
27. W. E. Swartz and D. M. Hercules, *Anal. Chem.* **43**, 1774 (1971).
28. W. N. Delgass, T. R. Hughes, and C. S. Fadley, *Catal. Rev.* **4**, 179 (1970).
29. K. Kim, N. Winograd, and R. Davis, *J. Am. Chem. Soc.* **93**, 6296 (1971).
30. H. Fischmeister and I. Olejford, *Montash. Chem.* **102**, 1486 (1971).
31. I. Olejford, *6th Scandinavian Corrosion Congress Proceedings, 1971, II*, Swedish Corrosion Institute, Stockholm, Sweden.

32. A. Wolberg, J. Ogilvie, and J. Roth, *J. Catal.* **19**, 86 (1970).
33. A. Miller, W. Atkinson, M. Barber, and P. Swift, *J. Catal.* **22**, 140 (1971).
34. D. A. Whan, M. Barber, and P. Swift, *J. Chem. Soc.* (*London*) *Chem. Comms.*, 1972, 198.

14

RESONANCE METHODS

D. Haneman

School of Physics
University of New South Wales
Sydney, Australia

1. INTRODUCTION

In this article we will discuss the application to surfaces of two powerful techniques used widely for bulk studies: electron paramagnetic resonance (EPR) and nuclear magnetic resonance (NMR). EPR detects the presence of unpaired electrons through their magnetic moments; NMR detects nuclei with net (spin) magnetic moments. The detection sensitivity of EPR, however, is about 10^6 times that of NMR. This tends to make it a more useful tool in surface studies, where the number of entities is so limited by the available surface area that NMR is often insufficiently sensitive. Hence the major portion of this article will be devoted to EPR studies. There is also a technique known as cyclotron resonance which picks up effects due to surfaces, but it has not been exploited much for surfaces to date since these effects are due to relatively thick layers (50 Å and more) through which an electron describes substantial parts of orbits. There are few detailed specific surface results, and we shall not go into this technique here.

In a measurement of magnetic resonance, one must detect entities with net magnetic moments. The behavior of the moments can then be followed as various perturbations are applied, e.g., heating and cooling, exposure to various ambients, exposure to radiation, mechanical stress of the substrate material, and combinations of these. In this way it is often possible to obtain valuable clues concerning the entities and their surroundings, even if a proper theoretical analysis of the behavior is not available. However, considerable theoretical progress has been made, and frequently it is possible to make quite detailed deductions concerning electron wave functions in the regions of the magnetic entities. It is this aspect which makes these resonance methods one of the most powerful techniques for deriving fundamental detailed information about solids and surfaces.

Data that have been obtained include positive identification of adsorbed species and the particular adsorption sites, wave functions of electrons on adsorbed species and on the surface itself, and even electric field strengths at the surface. Most of these details are unobtainable by other methods.

2. THEORY OF RESONANCE METHODS

There are a number of detailed treatments of the theory, including necessary discussions of a large variety of special cases.[1–7] Hence only the principles will be mentioned here. For specificity we will refer to electrons and will later point out the differences where nuclei are concerned. The electrons are assumed to be unpaired so that their magnetic moments are not cancelled out. This includes electrons trapped at various defects and surfaces, and conduction electrons in solids, as well as free radicals.

An electron has energy levels that are split when a magnetic field **H** is applied (the Zeeman effect).[8–10] The amount of splitting depends essentially on the environment of the electron, i.e., its wave function and the potential it occupies, including effects due to the presence of nearby nuclear and electron moments. Hence by measuring the splitting, much information about the surroundings can be deduced. The splitting is measured by illuminating the specimen with radiation and detecting the absorption of those energies which induce transitions between the split levels. This is simple spectroscopy, but the term resonance is used because of the detection techniques used in the wavelength region in which absorption occurs. For an applied field of 0.3 T (3000 G) the splitting ΔE is $\sim 4 \times 10^{-5}$ eV so that, from $hf = \Delta E$, the frequency f is about 10^{10} Hz, corresponding to a wavelength of 3 cm, the microwave X-band region. Practical considerations make this the most used but not the only region. (In the case of NMR, the splitting is much less, and frequencies of about 50 MHz are used.) For lower magnetic fields, the splitting is proportionately reduced, yielding lower resolution and detection sensitivity. For higher applied magnetic fields, the smaller-wavelength radiation that is then necessary requires correspondingly

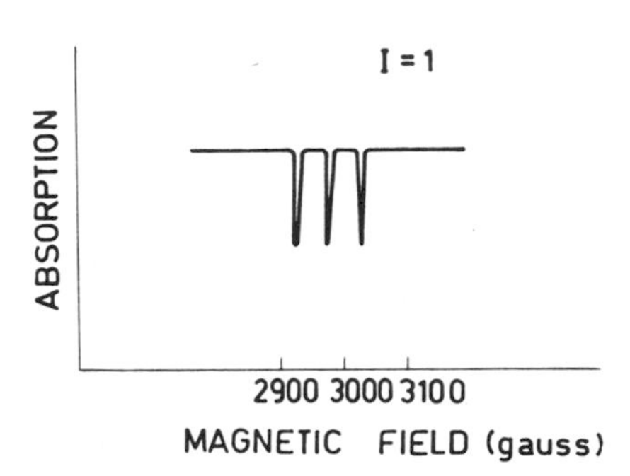

Fig. 14-1. Schematic graph of the absorption of microwave power by the sample as the magnetic field is varied. Three peaks are shown, corresponding to the three transitions possible when the magnetic field of a nucleus of spin operator I interacts with an unpaired electron. The number of lines is $(2I + 1)$.

smaller waveguides for the microwaves, and smaller effective sample sizes. Thus, most work is done in the X-band (3-cm wavelength) region to Ka-band (about 8 mm) region.

Usually one applies a fixed frequency f and varies the magnetic field **H**. The spectrum is then a plot of intensity of absorption of radiation *vs.* magnetic field, as shown schematically in Fig. 14-1. If the nucleus of an atom about which the electron is orbiting has a magnetic spin angular momentum $I\hbar$, then one would observe $(2I + 1)$ lines. This is basically because there are $(2I + 1)$ quantized energy states of the nucleus, each one of which affects the unpaired electron differently.

It is customary to define a tensor $\mathbf{g}_{ij}$ which contains the various terms due to the interaction of the electron spin and orbital magnetic moments with their surroundings. These terms vary in magnitude according to the directions of the components of the moments that are being considered relative to the crystal axes. Hence $\mathbf{g}$ is a tensor. One also defines a so-called hyperfine tensor $\mathbf{A}_{ij}$ which contains terms due to the interaction of nuclear and electron moments. The relationship between the microwave energy hf absorbed in inducing transitions between levels split by the magnetic field **H**, and the field can then be written

$$hf = \hbar\gamma \mathbf{g}_{zz} H_z + \mathbf{A}_{zz} m_{\mathrm{I}} \tag{1}$$

where γ is $(e/2mc)$, m is the electron mass, c is the velocity of light [$\gamma = (e/2m)$ in mks units], and the field is taken in the z direction. In this simple expression one has taken the system as possessing a principal axis in, say, the z direction for both $\mathbf{g}$ and $\mathbf{A}$ tensors, a situation which commonly occurs. The nuclear spin quantum number is m_{I}, having $(2I + 1)$ values, hence accounting for the $(2I + 1)$ absorption lines in Fig. 14-1.

2.1. Powders and Single Crystals

Many surface experiments are carried out on powders, where the surfaces present a variety of random orientations. It is clear that **H** will be parallel to each of the principal axes for a fraction of the surfaces. The spectrum would thus consist basically of three lines, each split into $(2I + 1)$ components, these lines being smeared or spread because of the variety of orientations. Fortunately, it is frequently possible to identify these lines in a powder spectrum (see article by Adrian[11] and references therein) and thus establish the principal values of **g** and thus also the principal values of **A** which determine the spacing of the component lines, as seen from Eq. (1).

For single crystals, analysis is best carried out by rotating the specimens until the applied field is parallel to principal axes. In the case of single-crystal surfaces, the orientation dependence of the EPR spectrum usually establishes the principal axes of the tensors involved. For powders, however, the smearing of the spectra by the multiple orientations makes identification more difficult. A number of theoretical spectra have been calculated for various special cases.[12–15] If the spectrum is not readily fitted by these, it is necessary to assume line shapes and **g** and **A** tensor angles and simulate the spectrum on a computer. The resulting match gives clues on how to modify the assumptions and generally it is possible to obtain reasonable fits between computed and observed spectra. Various techniques of computer simulation are available.[15] A typical powder spectrum and computer simulation are shown in Fig. 14-2.

We continue our summary further by mentioning two important subjects—line shape (and width) and temperature dependence.

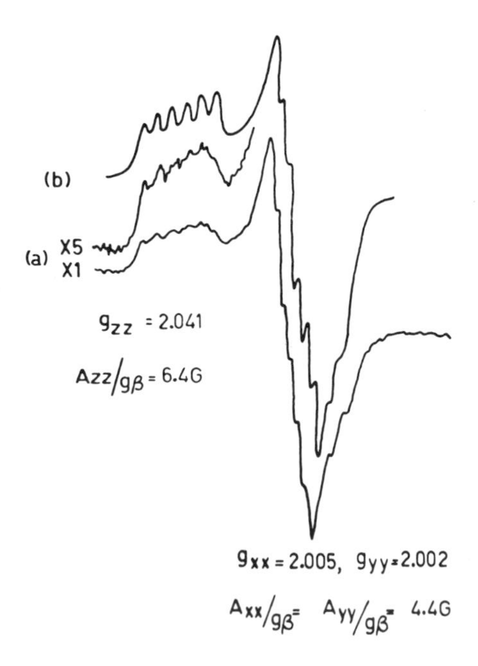

Fig. 14-2. (a) A typical EPR-derivative spectrum for a powder showing a characteristic broad hump at $g_{//}$ (g_{zz}) and a derivativelike shape at $g_{\perp}$ (g_{xx}, g_{yy}). The hyperfine lines are visible on both main parts of the spectrum, obtained at 77°K from O_2^- ions on vacuum-crushed AlSb.[66] The values of g_{xx} and g_{yy} are nearly equal. (b) The computed spectrum, based on the **g** values and hyperfine constants shown, and a Lorentzian single-crystal line width of 2.3 G.

2.2. Line Shape and Width

In the case of absorption occurring at optical wavelengths, electrons excited to upper levels decay to lower states in about 10^{-8} sec. This spontaneous emission process has an efficiency[8] proportional to the cube of the frequency, and is thus very inefficient at microwave (EPR) or radio (NMR) frequencies. Hence, in the absence of other effects, the excited levels will become occupied until they are populated equally with the ground levels. At this point, the probability of an incident photon being absorbed by a ground-state electron is equal to its probability of stimulating the emission of a like photon by causing an electron to decay from an excited state. Thus no net absorption can occur. This is a problem in EPR and NMR spectroscopy, but it is alleviated by the presence of the various interactions between an electron or nuclear spin and its surroundings which stimulate the set of excited electrons or nuclei to decay, giving their energy to the surroundings, eventually appearing as lattice vibrations, i.e., heat. This is best described by saying that there is a spin-lattice relaxation time, T_1, which characterizes the time for the spin system to exchange energy with its surroundings. The stronger the coupling between spin system and the "lattice," the shorter the time to come to equilibrium, and thus, the shorter T_1. This interaction with the surroundings means that the energy levels of the system are perturbed, hence leading to a range of field values at which resonance occurs [due to the spread in $\mathbf{g}$ values in Eq. (1)], i.e., the absorption line is broadened. The spread of the line is in fact proportional to $1/T_1$. This broadening may be swamped by another source of broadening, the direct interactions between electron spins on different sites which also perturb the energy levels. The phenomenon is described by referring to a spin–spin relaxation time, T_2, describing the time for the spin system to distribute absorbed energy uniformly within itself, as distinct from exchanging this with the lattice. The observed line spread is then proportional to $(T_1{}^{-1} + T_2{}^{-1})$, and if, as often happens, $T_2 < T_1$, the width is approximately proportional to $T_2{}^{-1}$.

The shape of the line can be calculated by assuming the major causes of broadening to be due to spin–spin interactions, known as dipole broadening. This results in a Gaussian shape (see e.g. Pake[1]). If so-called exchange effects are strong, the theoretical line shape is altered to one known as Lorentzian, and the line is said to be exchange-narrowed. Similar narrowing occurs if an entity bearing the unpaired electron, such as a molecule, is rapidly changing its orientation (tumbling) in which case the line is said to be motionally narrowed.

In practice, lines sometimes fit a Gaussian or Lorentzian shape quite closely, but often the shape is intermediate. In these cases one has

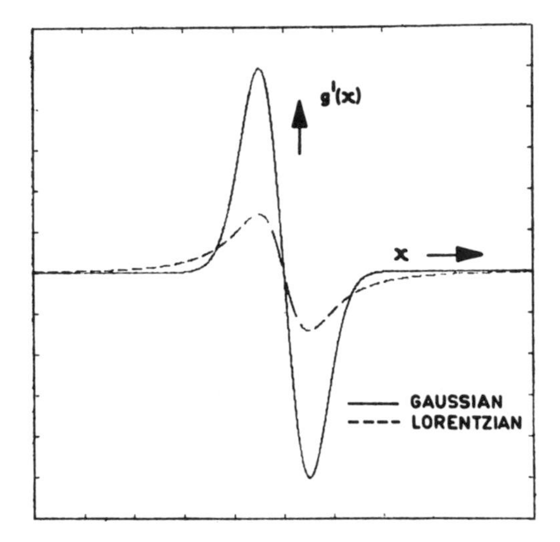

Fig. 14.3. The appearance of a derivative spectrum for a single absorption line having the Lorentzian shape and also the Gaussian shape. These lines both have the same intensity (integrated area under the absorption line, i.e., under the doubly integrated form of the curves shown). Note the smaller peak height of the Lorentzian curve, compensated by the greater spread in the "wings."

no simple theoretical formula for the shape and therefore computer simulation of the spectrum, much used for powder spectra, is made more difficult. As we shall see later, experimental data are often plotted as derivatives of absorption *vs.* magnetic field. The appearance of these is shown in Fig. 14-3. It is important to note that the Lorentzian curve has a very large "tail." This tail must not be neglected when obtaining the total number of spins which is related to the area under the absorption curve (see Section 6.3).

2.3. Temperature Dependence

Measurements of the total number of paramagnetic centers and of the relaxation times as a function of temperature give valuable information regarding the nature of the centers. If they are isolated from each other, one can show that the number of measured spins is inversely proportional to the temperature (Curie law). On the other hand, if they are in a band, which is broad with respect to kT, with Fermi level in the center region of the band, the measured number is independent of temperature. While the inverse temperature dependence is obeyed by

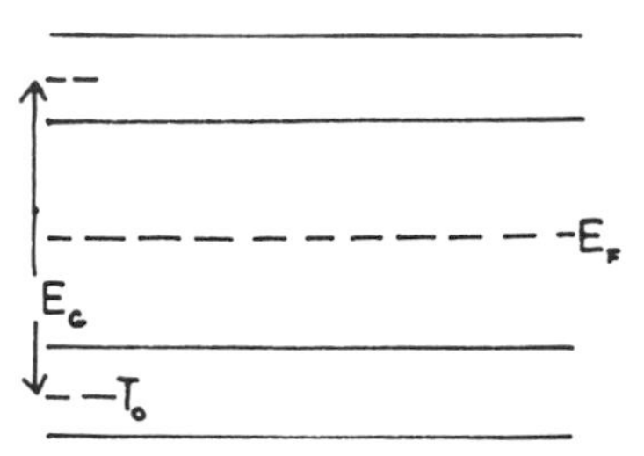

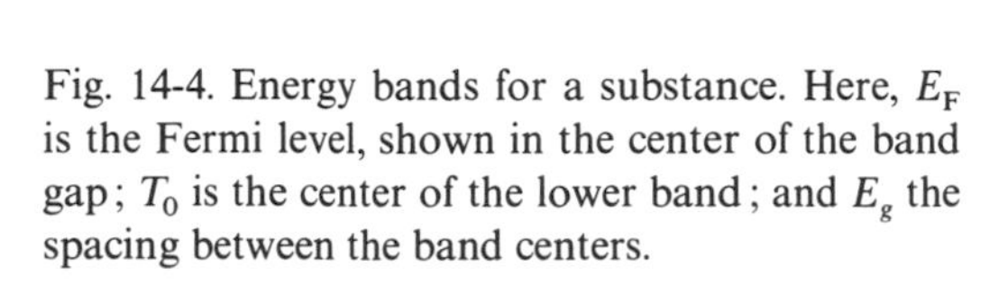

Fig. 14-4. Energy bands for a substance. Here, E_F is the Fermi level, shown in the center of the band gap; T_0 is the center of the lower band; and E_g the spacing between the band centers.

centers in many kinds of samples, and the temperature independence has been approximately verified in the case of conduction electrons in certain metals,[17] there are several cases of temperature variation which are intermediate between these types. These include surface centers in the presence of adsorbed gases.[18] Also temperature variation greater than T^{-1} has been observed.[19]

It is possible to derive an expression applicable to the intermediate cases.[19] We take a simple situation where the density of states is constant but split into two bands whose centers are separated by an energy gap E_g, as in Fig. 14-4. If Δ is the width of the upper and lower bands and T_0 is the center of the lower band, then the difference δn in numbers of electrons per atom in upper and lower Zeeman levels is

$$\delta n = (2b/\Delta)\sinh(\Delta/2kT)\{\cosh(\Delta/2kT) + \cosh(E_g/2kT)\}^{-1} \tag{2}$$

The EPR signal is proportional to δn and clearly can have a complex temperature dependence which is of particular interest for surface-state bands which may have widths neither narrow nor broad. Extra considerations for such bands are given below.

2.3.1. Corrections for Correlation Effects in Bands—Application to Surfaces

Recently, surface EPR results for Si were interpreted as due to electrons in a surface-state band.[20] Such electrons are confined to the surface and may have a lower density than bulk electrons. In such a case, the one-electron theory which is used instead of a proper many-electron theory may lead to greater errors than for bulk electrons since the neglect of correlation effects in this theory can be more serious when the electrons are more separated. In Fig. 14-5 we show an ideal (111) surface of a diamond-structure semiconductor like Si. If there is one unpaired electron per surface atom, and these electrons can be regarded as forming a separate group, the average nearest-neighbor spacing in the (111) surface plane is (for Si) 3.84 Å, as against 2.45 Å for the bulk electrons. At this large spacing, the conventional Hartree–Fock one-electron treatment fails to give a satisfactory account, since it does not reduce the

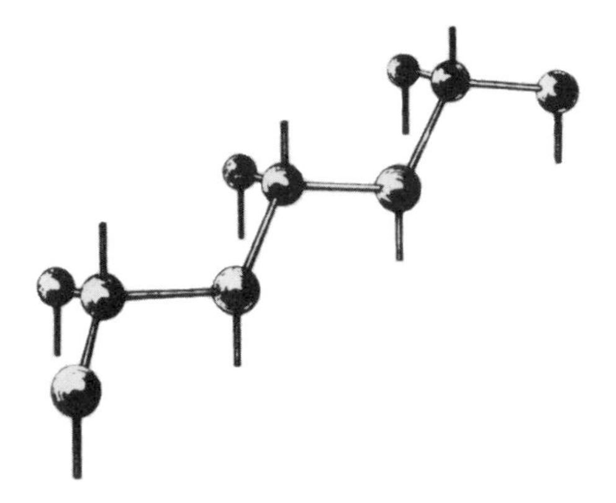

Fig. 14-5. Appearance of (111) surface of a diamond-structure material.

probability of finding two electrons on a given site, whereas it is obvious that the larger the spacing the less probable this becomes because of Coulomb repulsions. This matter has been discussed in some detail and a solution has been obtained by using the so-called Hubbard Hamiltonian, which is a somewhat better approximation in the case of separated centers and narrow bands than is the Hartree–Fock theory. We will not repeat the mathematics here. The theory predicts that a Hartree–Fock band of width 2Δ will split into two bands of reduced width if $2\Delta/I$ is less than about 1, I being a measure of electron repulsion and becoming equal to the separation of the two bands (E_g in Fig. 14-4). For such a system, with a constant density of states in the original Hartree–Fock band and assuming that the centers of the bands are many kT below the Fermi level (which is a case of practical interest for Si surfaces), one obtains[20]

$$\delta n_{\mathrm{Hub}} \simeq (2b/\Delta)\tanh(\Delta/2kT) \tag{3}$$

One may compare this with the non-Hubbard equation [Eq. (2)] by using mathematical approximations similar to those involved in deriving Eq. (3), namely, $E_g > \Delta$, $E_g > 2kT$, and obtain

$$\delta n \simeq (4b/\Delta)\sinh(\Delta/2kT)\exp(-E_g/2kT) \tag{4}$$

The major difference between the formulas is that the Hubbard term is free from the exponential dependence on $E_g/2kT$, and thus yields a much greater number of observed spins. This feature was essential in interpreting results from Si (see Section 5.2.2). The expressions are plotted in Fig. 14-6 for several values of Δ and I ($E_g = I$). Note that unless the bands are very narrow, the quantity δn actually decreases as the temperature is lowered, and even then only the Hubbard expression increases. In the latter case, the variation is approximately as T^{-1} for the range shown.

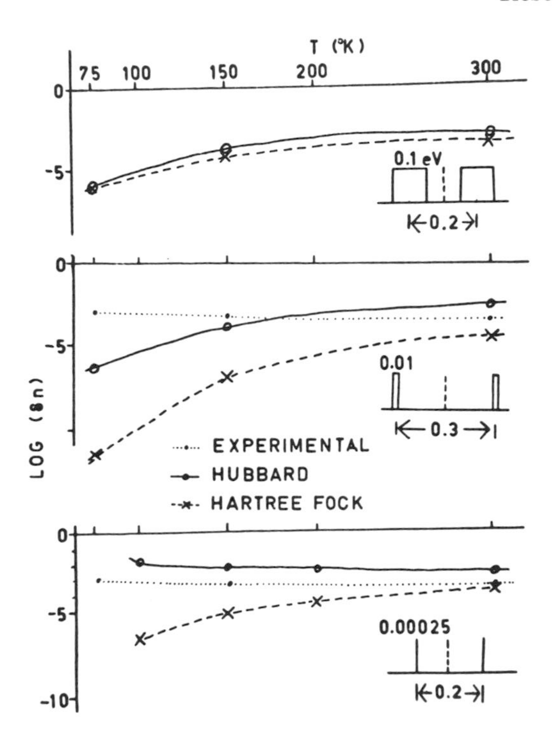

Fig. 14-6. Theoretical plots *vs.* temperature of the quantity δn, the difference in population, per atom, of up and down spin levels. The curves labeled "Experimental" show a Curie-law behavior. The curves labeled "Hartree–Fock" are a plot of δn in Eq. (4), and the curves labeled "Hubbard" are a plot of δn_{Hub} in Eq. (3), for three different kinds of bands.

3. EXPERIMENTAL METHODS

EPR and NMR spectrometers have been described in the literature in detail.[23,24] It is the purpose of this section merely to summarize some salient features and then concentrate on aspects of particular relevance to surface studies.

3.1. Electron Paramagnetic Resonance (EPR)

For electrons, as described carlier, the energy-level separation induced by a magnetic field is about 4×10^{-5} eV at 0.3 T (3000 G), corresponding to a photon frequency of 10^{10} Hz and wavelength of 3 cm. This X-band region is the most used. Larger splittings, using large magnetic fields, are desirable, but the shorter wavelengths are less convenient to use so that a practical upper-frequency limit is some

3.5 times higher, in the Ka band. This involves waveguide techniques since at these high frequencies the skin depth in metal conductors is very small, and therefore metal tubes which guide the electromagnetic radiation onto the sample are used. In practice, the waveguide terminates in a box, called the microwave cavity, which contains the sample. The incident electromagnetic waves, or photons, excite electrons between the split energy levels in the sample inside the cavity and then reflect back up the same waveguide. Instead of returning to the wave source, they are diverted to a detector (as indicated in Fig. 14-7), usually a silicon point-contact diode operating in the linear region, across which a voltage is developed. If power has been absorbed by the sample in the cavity, the voltage across the diode is reduced by ΔV, where[25]

$$\Delta V \propto \Delta Q_0 P^{1/2} Q_0^{-1} \tag{5}$$

and P_0 is the input power, Q_0 is the cavity Q off resonance (input power/power absorbed), and ΔQ_0 is the change in Q at resonance, i.e., when the frequency of the radiation is such that energy absorption by the spin system occurs.

Referring to Fig. 14-7, the source of radiation is usually a klystron tube, a vacuum-tube device in which an electron cloud emitted from a

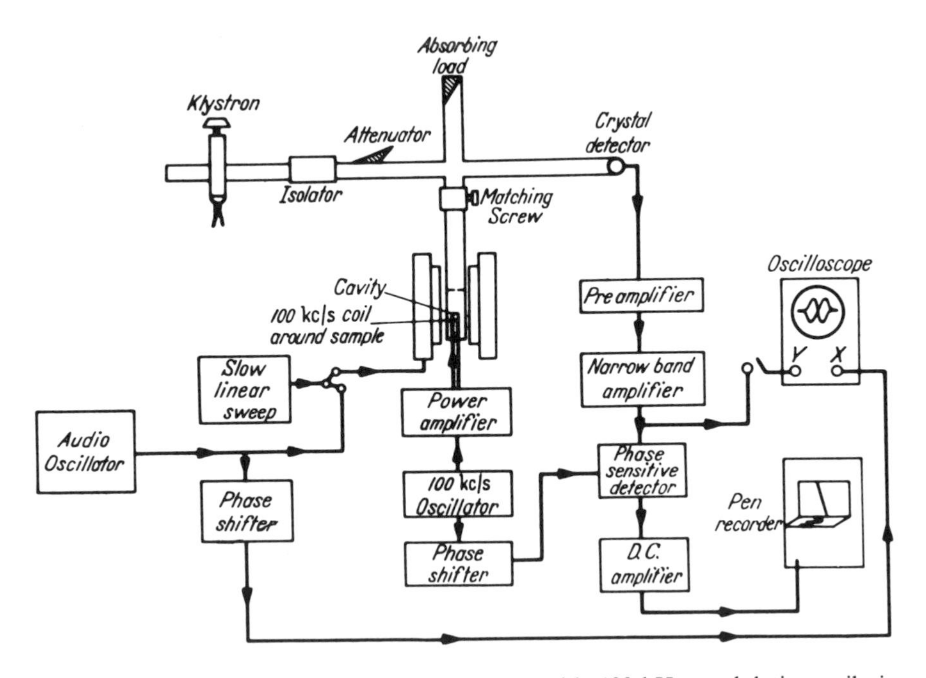

Fig. 14-7. Block diagram of EPR spectrometer with 100-kHz modulation coils in microwave cavity. After Ingram.[23]

filament is caused to oscillate by an appropriate oscillatory voltage applied to a repeller electrode. The electron cloud in turn emits electromagnetic radiation into the coupled waveguide leading to the resonant cavity. This cavity is designed for a high Q (minimum power loss), consistent with the incorporation of a hole through which the sample may be inserted. The cavity must be held in the magnetic field, conveniently provided by an electromagnet. The pole faces must be carefully designed for high homogeneity. If resonances separated by 10^{-2} G at X-band are to be resolved (i.e., spin systems whose energy splittings differ from each other by as little as 10^{-10} eV), such as occurs in some systems such as organic molecules, then clearly the magnetic field must be homogeneous to about 1 part per million. It is difficult to achieve this unless the pole faces are at least 15 cm in diameter and no more than about 6 cm apart. Likewise, the frequency of the klystron source, giving powers of a few hundred milliwatts, must be controlled to similar precision. This is achieved by a feedback control system.

The other portion of the electronic system which requires care is the detection system. The voltage change ΔV at the detector diode is usually small, and the sensitivity of detection depends on how small a ΔV can be measured. High signal-to-noise ratios can be achieved by modulation and phase-sensitive detection techniques (lock-in amplifier). These methods have come into increasing use in various branches of science in recent years, but resonance spectroscopy was one of the first fields where they were applied. A modulation is applied to the system giving the signal, and the amplifier is narrow-band tuned to the modulation frequency. In addition, the phase of the detector following the amplifier is set to the phase of the modulation signal. Therefore, only noise that happens to have the modulation phase as well as frequency passes through the system. In this way, a high noise rejection is achieved. The modulation is conveniently applied by modulating the magnetic field with amplitudes ranging from a small fraction to several gauss at frequencies up to 10^5 Hz. The dominant noise source(26) is the silicon-crystal detector diode. The noise power per unit of bandwidth varies inversely with frequency ($1/f$ noise), hence the higher the modulation frequency the better the signal-to-noise ratio, increasing as the square root of the modulation frequency. Practical considerations such as modulation field penetration and sample relaxation times set 10^5 Hz as about the useful upper limit. The klystron source also contributes noise, but this is less important provided incident power levels less than about 100 mW are used. Coils incorporated in the microwave cavity are mostly used to apply the modulation. For work at very low temperatures, a special cavity is used which can be immersed in liquid helium.

In this case the modulation coils are mounted on the Dewar, and lower frequencies are used to reduce eddy-current heating of the cavity walls. The output signal is passed through a filter whose time constant is matched to the time required to sweep the magnetic field through the region of interest (e.g., 3050–3100 G in 1 min). Due to the synchronous (phase-sensitive) detection, the signal comes out, not as a plot of ΔV *vs.* field H, which we call an absorption line, but approximately as the derivative. The lower the modulation amplitude ΔM, the more accurately the derivative is obtained. It is in general an acceptable derivative shape provided that ΔM is less than half the width of the absorption curve ΔH measured between the points of steepest slope. These points will be the peaks of the derivative curve, e.g., in Fig. 14-3. Hence the peak-to-peak width in derivative curves is the line width ΔH. One should keep $\Delta M < \frac{1}{2}\Delta H$, otherwise the curve will be distorted and the peak-to-peak width will no longer be the width of the original absorption curve.

Various kinds of cavities can be used. If they are rectangular, it is possible to incorporate slits or gaps in the side so that optical radiation may be shone onto the sample without spoiling the cavity Q seriously. The highest Q is obtained with a cylindrical cavity. Until recently, holes incorporated in the top and bottom of the X-band cavity, ending in short, open tubes, were up to 11 mm in diameter. However, recently[27] a design using holes of 2.5 cm in diameter has been shown to be successful. This design allows the insertion of larger tubes. However, to avoid serious loss of Q in the latter cavity, the tubes must be not only signal- and loss-free (hence quartz or Vycor), but of very even wall thickness, since the large-hole cavity is more sensitive to inhomogeneous loading than ones with smaller holes. As a practical point, it is difficult to obtain quartz or Vycor tubing of sufficiently even wall thickness, and grinding and etching of the tube usually has to be applied.

3.2. Detected Signal

We stated above that the EPR signal is usually detected as a change in voltage ΔV in a Si diode mounted in the path of the electromagnetic waves that pass through the sample in the cavity. The cavity acts as a kind of reflecting terminal to the waveguide, and a standing wave pattern of the electric and magnetic components of the microwave radiation is set up in it. The case of a rectangular and cylindrical cavity is illustrated in Fig. 14-8, which shows that the magnetic vector H_f is concentrated along the axis of the cavity, being at all times orthogonal to the dc magnetic field H. The electric vector is orthogonal to the magnetic vector. Clearly the size of the sample is important, since portions of it a few millimeters outside the axis receive less microwave field

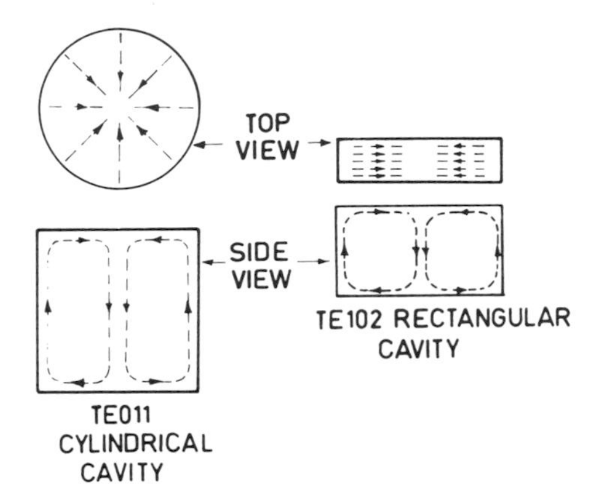

Fig. 14-8. Diagrams of two widely used microwave cavities, showing magnetic vector lines. The electric vector is orthogonal to the magnetic vector.

intensity. Furthermore, the presence of the sample and its supports can distort the field distribution in the cavity. These practical factors must be taken into account in considering EPR results.

It can be shown[25] that the detected signal as a function of field $S(H)$ is proportional to ΔV. This can be expressed in terms of the cavity Q (input power/power absorbed) off resonance, namely, Q_0, and P_0 the input power. It is also proportional to the filling factor η, which is the ratio of power intercepted by the sample to that in the whole cavity, and to δn, the spin population difference between upper and lower Zeeman levels [Eqs. (2), (3), and (4)], as well as to a term W, related to the microwave induced transition probability. The expression is

$$S(H) \propto \Delta V \propto \delta n \eta Q_0 P_0{}^{1/2} W \tag{6}$$

which justifies the statements in Section 2 that the signal is proportional to δn. The linear dependence on the square root of the power P_0 falls off when P_0 becomes so large that the upper levels become populated at a rate faster than that at which they can decay. This effect was assumed to be absent in the derivation of the formula. This is known as saturation, and it limits the useful power in many cases to only a few milliwatts, and even less as the temperature is lowered.

By taking samples which occupy all the central portion of the cavity where H_f is strongest, the filling factor η can be made close to unity. Sometimes quartz tubes inserted in the cavity act to concentrate H_f in the axial region, thus effectively increasing η. This effect is noticeable when quartz Dewars (and also quartz vacuum envelopes, for surface studies) are inserted in the cavity. Practical values of Q can be as high as about 20,000 for well-made unloaded cylindrical cavities. However this figure often falls off seriously for samples such as semiconductors, which have appreciable conductivity. Fortunately, a side benefit of crushing, which is used to obtain large surface areas, is that the cavity loading is reduced due to poor interparticle contact. Hence values of Q_0 of at least several thousand are possible.

3.3. Nuclear Magnetic Resonance (NMR)

The principal difference between EPR and NMR arises because the factor of γ in the Hamiltonian, Eq. (1), which determines the energy level splitting is $e/2mc$, and m for a nucleus is much greater than for an electron, by a factor of about 2000 in the case of a proton. (The value of g_e is also altered, being different for different nuclei and even for the same nucleus in an excited state. However, this effect is relatively minor.) Hence, the energy splitting is in the region of 10^{-7} eV or less for practical applied magnetic fields of several thousand gauss. This corresponds to absorbed radiation in the region of 10^8 Hz, which is in the range of ordinary radio frequencies. This causes a profound difference in technique since no waveguides or microwave cavities are required; instead, wire coils simply surround the specimen. We will now tabulate some points of difference.

1. The signal strength $S(H)$ is smaller for NMR. The physical reason is that the population difference, according to Boltzmann (or Fermi) statistics, reduces as the levels become closer together.
2. For microwave cavities Q can be made very high, e.g., 20,000, due to the very low resistive loss in the skin depth of the cavity walls, most of the energy being concentrated in the hollow space in the cavity. However, at the lower frequencies of about 10^8 Hz, the skin depth is large and the current passes through all the coil wire, thus contributing a greater resistive loss. Therefore, values of Q of less than 100 usually result, and this lowers the detected signal strength compared with EPR.
3. Since the nuclei interact with the lattice only indirectly via interaction with the electron clouds, the relaxation times T_1 tend to be much longer, often milliseconds up to seconds, and T_2 ranges from microseconds to seconds.
4. All nuclei of a given kind contribute to the spin magnetic moment, unlike the case of electrons where most pair off to give zero net spin.
5. The lower frequencies in NMR make possible the use of pulse techniques. These depend on applying a pulse of length τ, where τ

is of order $\pi/\gamma H_f$ and H_f is the microwave field amplitude. The quantity τ must be less than T_1 and T_2. For NMR this implies pulses of about 10^{-6}–10^{-3} sec, which are a practical possibility, whereas the corresponding EPR pulses would have to be about 10^{-9} sec, which are far less convenient. Pulse methods are a widely used part of NMR techniques.

6. Since the fine-structure effects in NMR (due to electron fields at the nucleus) can be small, lines may be separated by very small amounts, in the region of milligauss. Furthermore, the natural line width of liquid samples may be only a fraction of a milligauss due to the averaging out of broadening effects caused by the tumbling of the molecules (motional narrowing). This means that the homogeneity of the magnetic field, of strength about 10,000 G in the case of iron-core magnets, must be to within 1 part in 10^7 or better if such lines are to be resolved, or if narrow lines are not to be broadened by the instrument. Recently, superconducting magnets, of fields in the region of 50,000 G, have been used. Homogeneity is obtained by the use of extra trimming coils.

An NMR spectrometer therefore consists of a strong magnet, usually an electromagnet, with homogeneity requirements several decades better than for an EPR magnet if fine structure is to be studied. This means that pole faces of 30-cm diameter or more are needed. As with EPR, it is customary to apply a fixed frequency, say 40 MHz, and vary **H** until resonance is obtained. In the case of hydrogen nuclei, i.e., protons, resonance occurs at 42.577 MHz for 10^4 G, since the nuclear spin I is 1/2, and the equivalent of g_e is 2×2.7927. Most NMR experiments have in fact been done with proton resonances since hydrogen is present in water and most organic substances. The sample is held within a small coil of diameter about 1 cm or less which is fed from a high-stability rf oscillator. As with EPR, modulation and phase-sensitive detection techniques are used, the modulation usually being applied to the magnetic field by means of separate coils. It may also be applied by an af variation to the rf of the oscillator. These methods are called amplitude and frequency modulation, respectively.

The occurrence of absorption is detected in two ways. In the Pound–Purcell technique it is observed in the emitting sample coil as a small back emf. In the Bloch or cross-coil technique, a small coil is placed orthogonal to the sample coil. At resonance, a signal is picked up since scattering of the electromagnetic waves occurs in addition to absorption. The orthogonality minimizes pickup of the exciting radiation.

Most work on surface properties has been concerned with measuring relaxation times T_1 and T_2. These may be derived, as for EPR, from line-width analysis and by detecting the onset of saturation. Saturation occurs when the incident power excites entities into upper states faster than they can decay. At this stage, the signal height $S(H)$ is no longer proportional to the square root of the incident power [Eq. (6)] but depends on T_1. The above methods belong to so-called C.W. (continuous wave) techniques. In addition, however, one has available in NMR a technique involving pulses, as mentioned above, due to the relatively low frequencies involved. The approach to equilibrium after appropriate pulses have been applied has been analyzed in detail.[3] From the time behavior and the pickup of pulses (echos) from the system, it is possible to deduce the relaxation times from real-time measurements. These techniques are collectively called spin–echo methods. This theory takes into account time-dependent perturbations and will not be gone into here. We note that pulse methods have an inherently worse signal-to-noise ratio than C.W. methods since broad-band amplifiers (which pass a broad noise spectrum) are needed. Hence these methods are only suited to cases where there are relatively strong signals.

4. SPECIAL CONSIDERATION FOR SURFACE STUDIES

There are two principal approaches in EPR surface studies. One is to measure the resonances from paramagnetic molecules adsorbed on surfaces. It is sometimes possible to deduce properties of the substrate from the nature of the resonance of the adsorbed species. The properties of the latter are in any case of great interest, since much of the interest in surfaces is the reaction between them and gases or liquids. EPR in particular provides a unique method of characterizing the adsorbed species when it is paramagnetic or if it becomes so upon adsorption.

The second approach is useful when the surface itself shows detectable paramagnetism. Here one can obtain not only direct information about the surface from its own signal, but also study it as a function of the adsorption of both paramagnetic and nonparamagnetic species. In fact, adsorption studies are essential to demonstrate that the original observed resonance does in fact come from the surface, and not from the bulk. The change in signal after exposure to gases provides strong evidence that the paramagnetic centers are on or close to the surface. In the cases of porous substances, the centers are on or close to the surfaces of the pores. However, to conclude whether the centers are actually on or in the topmost surface layer itself requires careful examination of the evidence for the particular case being studied.

4.1. Vacuum

Obviously the use of interactions with gases requires that the ambient surrounding the sample be controlled. This is usually achieved by containing the sample in a vacuum envelope. This vacuum envelope must be inserted in the microwave cavity in EPR studies, and should itself contain no paramagnetic centers that interfere with observation of the surface resonance. Quartz and Vycor tubes are used, being bonded to the usually glass portion of the vacuum system by a quartz-to-glass seal. The reliability of the work is related to the degree to which the initial surface can be characterized and to the purity of the gases. Ideally, one should use an ultrahigh vacuum system (10^{-9}–10^{-10} Torr or better) into which gases of the highest possible purity are admitted. Such conditions have not always been used, due to the experimental problems involved. In the case of a paramagnetic gas, the presence of the gas on the surface can be positively established, particularly if its resonance remains after evacuation. However, other gases may also be present on the surface and affect, for example, the width of the signal. They may also occupy available sites and cause less coverage by the paramagnetic gas than might otherwise occur. There is the additional possibility of their causing the reverse effect, namely, activating the adsorption. The degree to which results are free of these effects depends on the experimental conditions used (not always fully described in the reports) and the nature of the results themselves. Methods that have been used to perform EPR measurements in ultrahigh vacuum will be described in Section 5.

4.2. Sensitivity

4.2.1. Electron Paramagnetic Resonance (EPR)

The principal problem in applying resonance techniques to surfaces is that of obtaining sufficient sensitivity. In Section 3.2 we quoted an expression for the signal height $S(H)$ in an EPR spectrometer, namely $S(H) \propto \delta n Q_o P_o^{1/2} W$. This however does not tell us the detection sensitivity unless we also know the noise level. As described earlier, the noise is kept down by modulating the magnetic field H at a high frequency (10^5 Hz) and detecting the signal with a narrow-band amplifier, phase-sensitive detector, and filter. Using a 1-second time constant for the latter, modern EPR spectrometers are capable of detecting in the region of 10^{11} spins/G. This means that a signal 1 G wide could be detected above the noise, from 10^{11} spin centers, provided there is no Q degradation of the cavity, optimum modulation is used, the filling factor is unity, and saturation does not occur at high power levels of order tens of milliwatts. If the width were 10 G, the minimum number detectable would be 10^{12}.

Theoretically, a Ka-band spectrometer operating at a higher frequency of 3.5×10^{10} Hz has higher sensitivity since the larger energy-level splitting causes a larger value of δn in the expression for $S(H)$. However, in practice better sensitivity is not achieved at Ka band due to electronic quality factors. An exception to this is the case where only small samples are available which are not sufficient to fill the sensitive axial portion of an X-band cavity but would automatically fill a larger fraction of the Ka-band cavity since its dimensions are smaller than X band, in proportion to the wavelength. An additional advantage of Ka-band work is that effects due to g anisotropy are spread out more as H is varied since a higher field is used. This aids identification but not sensitivity.

As a practical consideration, the modulation coils are usually incorporated in the cavity, and the modulation field falls off toward the cavity edges. Since the signal-to-noise ratio improves as the modulation field increases, it is desirable to have all of the sample experience the same modulation field. Hence a sample concentrated in the central portion of the cavity axis may give better signal-to-noise ratio than the same quantity of sample distributed along the cavity axis, despite the fact that the filling factor may be worse. Individual cases of cavity design and sample size and shape must be examined to optimize results where signals are very weak.

4.2.2. Surface Area

Assuming the maximum sensitivity of 1×10^{11} spins/gauss is available, we estimate the amount of surface area required to give a detectable EPR signal. Taking a center of 2 G width with a density of 1 per surface atom, and taking 10^{15} atoms per cm^2 of surface, the theoretical minimum area required is $4 \times 10^{11} \times 10^{-15}\,cm^2 = 4 \times 10^{-4}$ cm^2. This is quite small and easily obtainable. However in practice the situation is usually worse. Firstly, the Q value of the cavity may be degraded by the sample and its supports. If these are insulators of low dielectric loss, this effect is minor. However, semiconductors of resistivity less than about 100 ohm-cm have a noticeable effect. This effect can be minimized by powdering the sample.

More importantly, large values of power P_0, on whose square root the signal depends, may not be usable due to the onset of saturation. Furthermore, the filling factor is usually considerably less than one due to limitations of sample size and shape. If hyperfine structure is present, the intensity of resonance is distributed over more lines, so that the number of detectable spins is reduced. The most important factor in practice is that the number of resonance centers is often only in the region of 10^{13} cm^{-2} or less. Taking the above effects into account, one often needs of the order of 1 cm^2 of surface area to obtain a detectable resonance. However, in order to make precise measurements, the signal-to-noise ratio should be 100 : 1 and preferably 1000 : 1. For these reasons, most surface EPR work has been done on materials of area 10^2–10^6 cm^2, such as porous and powdered substances.

For application to single crystals which, because of the small volume available in the sensitive part of the EPR cavity, will have a small surface area, and for all cases where low signals are found, means of enhancing the sensitivity are required. The figure of about 1×10^{11} spins/G quoted above was for a time constant T_c of 1 second in the output filter following the phase-sensitive detector. This means that noise signals of frequency less than about $1/T_c$ will still be in the output. The noise can be reduced by increasing the time constant T_c, but this affects the time of scanning the magnetic field H through the signal. This time must appreciably exceed T_c. The relationship depends on the shape of the signal, but obviously the rate of scanning through a part which is changing must be sufficiently slow for the signal to pass through the filter without serious truncation due to the time constant. A rough guide is that $dS(H)/dt < S(H)/3T_c$. Thus if a smoothly varying signal is measured with a time constant of 30 minutes, a scanning time in excess of 30 minutes must be used. Such times, in which the instrument performance, sample temperature, etc., are held constant, are practical but inconvenient. It can be shown that the noise is reduced approximately as the increase in square root of the filter time constant. Thus an increase from 1 sec to 1 hr gives an improvement in signal-to-noise ratio of about 60, but from 1 to 9 hr gives only another factor of 3.

4.2.3. Signal Accumulation

There is another, related, method of obtaining sensitivity enhancement which uses a short time constant but with repeated scanning through the signal. The output is fed into a memory device, usually a multichannel analyzer. The accumulated signal increases in proportion to the number of scans but the noise, being random, only increases as the square root of the number of scans. Thus a signal-to-noise ratio improvement proportional to the square root of the number of scans, and therefore approximately to the root of the total accumulation time, is obtained. The number of channels, or memory locations if a computer is used, which are required depends on the detail in the signal. For a simple resonance 100 locations may be enough, whereas with hyperfine structure 1000 may be required.

Since such accumulation devices are expensive, it may be asked why they are used instead of simply an output time constant in the filter matched to a single scan of time length equal to the time if an accumulator had been used. The signal-to-noise ratio would apparently be the same. The answer is twofold. First, there is a small theoretical improvement in that noise of very low frequency that would penetrate the long-time-constant filter would be averaged in the accumulator. Second, the accumulation devices are much more convenient in that one does not have to choose the time constant and scan time in advance. The contents of the accumulator can be continuously inspected, either on a built-in oscilloscope or by displaying them on a recorder. Thus the accumulation can be stopped when sufficient sensitivity has been obtained, or if one sees that something has gone wrong, such as an instrument malfunction, incorrect placing of a marker, etc. Thus, in practice, considerable experimental time is saved. To prevent effects due to drift of magnetic field strength during the accumulation, it is possible to insert an extra sample which gives a sharp resonance at a field slightly different from that of the sample of interest. This sharp resonance signal is used electronically to trigger the accumulator, which thus accepts a scan through the signal starting at precisely the same value of H, regardless of any field drift. In practice, modern electromagnets with highly regulated power supplies are very stable and such precautions are usually unnecessary. Furthermore, one can determine if a drift has taken place by observing whether any increase in width has taken place in a marker signal.

The devices are marketed under various names, signal accumulator, signal averager, computer of average transients, digital memory oscilloscope, etc. Any computer with analog-to-digital converter will perform the same function if suitably programed. However, arrangements need to be incorporated for convenient display of the contents, ranging from alphanumeric print out on a teletype to analog display on a recorder or oscilloscope. (Overflow of a channel or memory location must also be checked.) Signal averagers of the so-called boxcar type (waveform eductor, etc.) have maximum scan times of several seconds which are not long enough for many classes of resonance experiment where the magnet must be swept over 50 G or more. (In addition, the magnet power supply needs a recovery time of a few seconds between sweeps due to the jump in magnet current when the sweep is restarted; this wasted period should be small with respect to the useful scan time.)

4.2.4. Nuclear Magnetic Resonance (NMR)

As described in Section 3.3, the sensitivity of NMR is much lower than EPR due to the reduced value of δn and the lower Q value of resonant circuits in the 50-to-100-MHz range. By optimizing power and modulation and using long filter time constants, it is possible to detect in the region of 10^{17} nuclear spins. This means that if each nucleus on a surface were magnetic, one would need 100 cm^2 of surface for detection, and much more than this for accurate measurements. Hence NMR is a much more limited technique than EPR. Nevertheless, a number of useful measurements have been made in the case of large-area samples covered with adsorbed layers. These will be described in Section 6.

4.3. Surface Spin Density

A very useful parameter is the number of spin centers per unit area. For example if this is about 10^{15} cm^{-2}, one deduces that there is one center per surface atom, and correspondingly for other values. This involves accurate measurement of the total number of centers and of the surface area.

The total number of centers is most readily obtained by comparing the signal with that of a sample of known number of spins N_S. If A and A_S are the areas under curves of signal height *vs.* field for the unknown and standard sample, one may show, for spin-1/2 centers

$$\frac{A}{A_S} = \frac{N\delta n\eta}{N_S\delta n_S\eta_S} \tag{7}$$

where δn is the excess per atom of spins in the lower Zeeman levels and η is the filling factor. This expression assumes off-resonance Q values are the same, the same incident power is used, and transition probabilities between the Zeeman levels are the same for both samples. For a standard sample containing an isolated nonoverlapping spin system, $\delta n_S = hf/2kT$. Thus δn_S, N_S, and η_S are known. Assuming η can be estimated, one then obtains a value from Eq. (7) of the product $N\delta n$. In the literature, workers have usually assumed that the centers of interest are isolated and obey the T^{-1} Curie law, in which case δn is $hf/2kT$ and thus the number of spins N is obtained. It is necessary to point out, however, that this assumption need not hold and the temperature dependence of the sample resonance must be checked. In the case of adsorbed gases, there are known cases[17] where the temperature behavior is slower than T^{-1}. Hence the applicability of the other expressions for δn discussed in Section 2.3 should be considered. Unless one knows the formula for δn, one cannot precisely deduce the number of spins N.

On an experimental note, the value of N_S is usually only quoted to within $\pm 25\%$. Furthermore, there are errors in obtaining the area A from the absorption curve. If it precisely obeys a known formula, e.g.,

Gaussian or Lorentzian, the integration may be performed mathematically. In intermediate cases, it must be done numerically and there are likely to be appreciable errors due to area hidden in the wings of the curve. For example, due to the large spread, the numerical integration of a Lorentzian curve from the visible area gives only about 50% of the area obtained from mathematical integration. This error is much less for Gaussian curves. Hence even if δn is known it is difficult to obtain N to an accuracy of even $\pm 50\%$.

The measurement of surface area is straightforward for a single crystal, but for powders and porous materials, on which almost all work has been done, indirect methods must be used. The most common technique is that devised by Brunauer, Emmett, and Teller,[28] known as the BET method. The material is placed in an evacuated tube. One then admits an inert gas, usually krypton, into the closed volume containing the sample and measures the change in pressure with a McLeod or calibrated Pirani gauge. This is performed with a sample both at room temperature and at liquid-nitrogen temperature, the latter by raising a Dewar containing liquid nitrogen around the tube containing the sample. Theoretically, up to a monolayer will physisorb at 77°K and, hence, from the drop in pressure and known volume, one can deduce the number of adsorbed krypton atoms and, hence, the surface area. The monolayer point is found by performing the experiment for several different initial pressures. If p_0 is the vapor pressure of krypton at 77°K, p_i the pressure when the sample is at 300°K, and p_f the pressure when it is at 77°K, then the area B of the sample is

$$B = V\sigma/kT(b + m) \tag{8}$$

where V is the closed volume, σ is the area covered by an adsorbed krypton atom, b is the intercept, and m is the slope in a plot of $p_f/(p_0 - p_f)(p_i - p_f)$ *vs.* p_f/p_0.

If the surface area exceeds about 50 cm^2, errors due to adsorption on the unavoidably cooled walls of the sample container are small, and can in any case be estimated from dummy experiments with an empty tube. Results obtained by this method are fairly reproducible, but errors can arise from factors such as fitting of a straight line to the experimental points and imperfect applicability of the basic theory (e.g., the value of effective area covered by krypton atom may vary). In practice, an error of up to 100% can be regarded as not improbable, although claims of higher confidence are made.

The final figure for the surface spin density is obtained from N/B. As noted from the individual errors, it is difficult to obtain this figure to better than a factor of 2. However, this is still very useful.

5. EPR RESULTS FOR SURFACES

The applications of EPR to surfaces are various. They include the properties of clean and real surfaces, the nature of adsorbed species and adsorption sites, effects on surfaces of irradiation by particles and photons, and effects of mechanical damage such as cleavage, abrasion, and polishing.

Measurements on clean surfaces are generally directed toward obtaining fundamental information, whereas measurements on real surfaces, i.e., produced by some reproducible but not necessarily contamination-free technique, are usually to clarify their role in various processes such as catalysis. We shall summarize some recent representative studies under the two headings, clean surfaces, namely those that are relatively well characterized, and real surfaces, those that are generally not well characterized, regarding surface atom composition and structure. This latter group is the larger one and we shall discuss it first.

5.1. Real Surfaces

The interest in much of this work arises from the practical importance of surfaces in industrial processes, particularly catalysis. Such surfaces are usually of large area per gram so that high sensitivity for resonance measurements is available. Applications of EPR to problems in chemisorption and catalysis started about 15 years ago. This method has been used on transition metal oxides, free radicals on surfaces of metals, semiconductors and insulators, and on nonconductors involving surface defects produced by irradiation and other methods. Many valuable results have been obtained which could not be found by other methods. The requirement that the system possess unpaired electrons for EPR study is not overly restrictive since many common adsorbed molecules exhibit paramagnetism. Species that have been identified include O^+, O_2^+, O_2^-, O_3^-, CO, CO_2, CO_2^-, NO, NO_2^{2-}, NO_3, Cl, and CH_3.

The details of adsorption depend on a knowledge of the details of the substrate atoms. This cannot necessarily be inferred by assuming that the bulk material simply terminates at a surface. Changes in relative concentration of atoms and in structure can occur during the process of producing the surface, and contamination is the rule rather than the exception. It may arise from remnants of chemicals such as zeolites and various oxides used in producing the material, from aggregation of impurities and defects at the surface during the heat treatments that are commonly employed, and also from surface interactions with background gases in the ambient surroundings, unless great precautions with the vacuum have been taken. Hence, in the class of work described in this section, the arrangements and types of atoms on the surface layer of the adsorbate are not well known. However, in some cases the nature of the hyperfine interaction between an adsorbed molecule and a substrate atom identifies the latter, usually from its nuclear spin I, so that one does know the surface atom on which the adsorbent molecule is sited.

Reviews of the early work[29] and of EPR in catalysis up to 1972 have appeared.[30] We will refer to some of this, but will mainly select examples from later work to illustrate the type of results obtainable.

5.1.1. Zeolites

Type-Y zeolites are an important catalyst, particularly for reactions of petroleum hydrocarbons such as isomerization, alkylation, and cracking. Linde Molecular Sieve type-Y zeolite[31] is a crystalline zeolite with a uniform pore size of about 10 Å and a Si/Al ratio ranging from 1.5 to about 3.0. When the univalent cations in the material have been replaced by base exchange with multivalent cations, hydrocarbons are cracked at considerably lower temperatures. The structure consists of a complex network of AlO_4 and SiO_4 tetrahedra between which at certain sites are located the cations.

EPR spectra have been obtained by Lunsford and co-workers[32,33] for a variety of zeolites. A decationated zeolite, NH_4Y, prepared by exchanging about 90% of the sodium cations with ammonium ions, was degassed at 600°C in 10^{-5} Torr for 12 hr, heated in 1 atm H_2 at 600°C, evacuated, and then γ-irradiated (413 r/min for 24 or 48 hr) in the presence of 1–400 Torr of oxygen. The resultant EPR spectrum observed at 77°K, after heating to 150–250°C for 30 min–24 hr, is shown in Fig. 14-9. This is an example of where hyperfine splitting identifies at least one of the substrate atoms. The spectrum is a powder shape, identified as due to adsorbed O_2^- ions, with 6 hyperfine lines due to interaction between the unpaired electron of O_2^- and the ^{27}Al nucleus with $I = 5/2$ [number of hyperfine lines equals $(2I + 1) = 6$]. A previous investigation using adsorbed NO gave similar results, but the Al hyperfine splitting was less sharply defined due to the presence of extra hyperfine lines caused by the spin of the ^{14}N nucleus ($I = 1$) of NO. The spectrum in Fig. 14-9 is remarkably similar to that of O_2^- on Al atoms of clean, crushed AlSb (see Fig. 14-2);[66] there is thus little doubt that its interpretation as being due to O_2^- ions sited on Al atoms of the zeolite surfaces is correct.

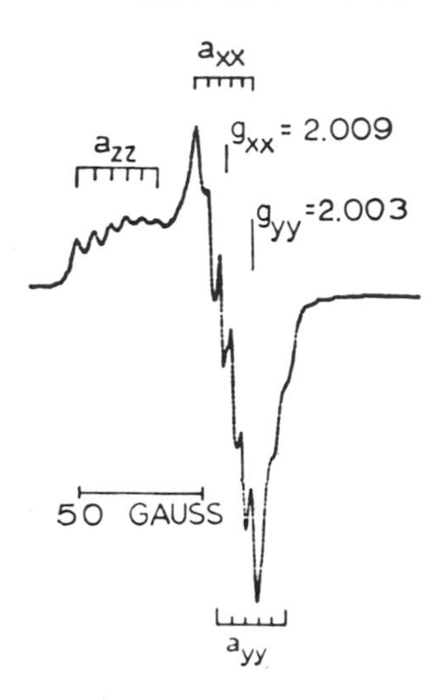

Fig. 14-9. EPR powder spectrum from O_2^- ions on decationated zeolite after γ irradiation and heating, recorded at 77°K.[32] Note the resemblance with Fig. 14-2.

Further information is gained by analyzing both the **A** and **g** tensors. The principal values of these can be obtained because the powder spectrum has a shape which is characteristic for a system with components having three principal axes in space (see Section 2.1).

The separations between the hyperfine lines at the positions of g_{xx}, g_{yy}, and g_{zz} in the powder spectrum give the principal values of the **A** tensor [from Eq. (1)]. Assuming that **A** is parallel to **g**, the fact that A_{xx}, A_{yy}, and A_{zz} are practically equal indicates that the anisotropic portion is small and hence that the wave function of the unpaired O_2^- electron around the Al atom is principally *s* type. The energy-level separations of the O_2^- molecule can also be obtained by analyzing **g**. For this one needs to know the surface symmetry, which is uncertain on surfaces such as these of doubtful composition and structure. We shall give an example of such analysis for clean AlSb surfaces with O_2^- ions, below.

A variety of zeolites were studied[33] where the cations were separately Mg, Ca, Sr, and Ba. These alkaline-earth Y-type zeolites were degassed by heating to 500°C in vacuum, followed in some cases by similar heating in H_2 and O_2. EPR spectra due to adsorption of NO at 77°K were obtained, showing a 3-line ^{14}N hyperfine splitting ($I = 1$) which showed less resolution on moving up the series from B_2Y to MgY. The spectra were affected by the degree of preceding dehydration and the degree of Ca exchange (for Na) in the case of CaY material. The data were interpreted to indicate that the lattice oxide ions are removed upon extensive dehydration of MgY, CaY, and SrY zeolites, leaving defects which adsorb NO. The presence of a second type of adsorption which did not involve Al ions was also established.

In the case of Linde Molecular Sieve type-X material[34] (Si/Al ratio about 1.2), adsorption of N_2O_4 and NO_2 gave a sharp NO_2 spectrum, detectable at room temperature, in which broadening effects were averaged away by the rapid tumbling motion of the molecule. This fact leads one to an inference concerning the pore sizes which are presumably such that only one NO_2 molecule is trapped per "cage" and cannot undergo broadening exchange interactions with other molecules, in this case N_2O_4. This is consistent with the small pore sizes of about 10 Å estimated by other methods.

EPR methods have also been applied to other sorption phenomena in synthetic zeolites. These include the sorption of diphenylamine,[35] univalent nickel ions, Na_4^{3+} and Na_6^{5+} complexes,[36] and ClO_2 and Cl_2^-.[37]

5.1.2. Chromium Oxide

The oxides of chromium of different valences have been the subject of many investigations. The early work centered around the EPR signals of Cr_2O_3 deposited on alumina, a combination which has long been known as an effective hydrogenation–dehydrogenation catalyst.[29,38] The results at low concentrations showed the Cr_2O_3 consisted mainly of isolated Cr^{3+} ions, while at higher concentrations a bulk phase occurred of clusters of Cr^{3+} ions with strong exchange interactions. A similar situation occurs in coprecipitated chromia–alumina, except that solid solutions of Cr_2O_3 in γ-alumina are formed at higher concentrations of Cr_2O_3.[39]

The application of EPR to the characterization of the higher-valent oxides of chromium became very important when it was discovered that these oxides, supported on silica and silica–alumina, polymerize ethylene to linear polyethylene of high molecular weight.[40] The signal at $g = 1.97$ and width about 60 G was identified as due to Cr^{5+} ions[38] being in tetrahedral coordination on silica, and pyramidal on silica–alumina and also on alumina. (The latter combination is inactive for polymerization.) The Cr^{5+} signal has been related to the polymerization activity of the catalysts.[41]

5.1.3. Magnesium Oxide

Irradiation of magnesium oxide leads to the formation of defect centers which catalyze the hydrogen–deuterium exchange reaction. A number of studies of this material, produced in large-area form of the order of order 100 m^2/gm, have been made. In the experiments of Lunsford and co-workers,[42–44] some of the MgO was prepared from reagent-grade powder, hydrolyzed to $Mg(OH)_2$ in hot water, extruded into pellets, dried in air at 100°C, and then vacuum degassed at up to 800°C. Ultraviolet or γ irradiation produces paramagnetic surface defects (S and S^1, respectively) which are believed to be electrons trapped at anion vacancies. Upon exposure to oxygen at about 20 μ, the room-temperature EPR spectrum of these defects rapidly disappears. However, cooling to 83°K in low oxygen pressure gives a new spectrum, identified as due to the presence of O_2^- ions. Unirradiated MgO shows[45] very little oxygen adsorption, presumably due to the absence of suitable sites.

Oxygen adsorption was also found in the experiments of Nelson, Tench, and co-workers.[46–48] Their material was prepared by various methods. These included dissolving Specpure Mg rods or crystals in nitric acid, precipitating as $MgCO_3 \cdot 3H_2O$ with ammonium carbonate, and igniting to the oxide in a silica tube in air at 820°C for 4 hr. This material was then dehydrated at up to 1000°C for 16 hr. EPR signals were obtained after γ and neutron irradiation, and the S center resonance disappeared upon oxygen adsorption. A detailed model for the surface center and formation of O_2^- was proposed. It should be noted that the particle sizes in these high-area materials are only a few hundred angstroms, and therefore all the material is within range of strong effects from surface charges.

Other experiments involved γ irradiation in the presence of hydrogen,[47] leading to a strong EPR signal (10 times that produced by vacuum irradiation) which was also destroyed by oxygen; the center was labeled S_H. Hyperfine splitting from ^{25}Mg nuclei of the lattice was observed. Reaction of N_2O with ultraviolet-irradiated MgO at room temperature[48] gave an EPR signal identified as O^-, and similar work,[44] but at 77°K, gave a somewhat different and less-complex signal which was also ascribed to O^-. This work is of interest in that the EPR signal from the O^- species on various substances has been controversial. The difference in the present signals also leaves some room for doubt. Exposure subsequently to O_2, H_2, CO, CO_2, or additional N_2O gave a signal that was similar in the case of both reports (only O_2 used in the former work) and was ascribed to O_3^-. Other surface species identified on MgO were chlorine radicals when chlorine-containing species were formed on surfaces of MgO and detected at 77°K by EPR.

Measurements have also been made on nonirradiated MgO. Adsorption of the paramagnetic gas NO_2 (in equilibrium with N_2O_4) gave a number of paramagnetic species depending on conditions of pressure and temperature, including ones interpreted as due to burial of NO_2 with N_2O_4 in the pore structure (sizes up to 60 Å) so that little interaction with a gas phase could occur. This was deduced from the failure of gas-phase $^{14}NO_2$ to exchange with sorbed $^{15}NO_2$ and from the fact that up to 760 Torr of O_2 did not broaden the NO_2 spectrum (by dipole–dipole interactions known to occur among coadsorbed species).

Other data were obtained using MgO doped with the paramagnetic impurity Mn.[50] At 600°C, hydrogen and oxygen were adsorbed to an extent dependent on the Mn concentration, showing that the latter provided adsorption sites. EPR spectra of MgO with 235 ppm Mn ($S = 5/2$, $I = 5/2$) gave a 6-line $Mn^{2+}(m_s = +\frac{1}{2} \leftrightarrow -\frac{1}{2})$ spectrum, with weak pairs of lines between, showing the so-called forbidden, i.e., low probability, transitions ($\Delta m_s = \pm 1, \Delta m_I = \pm 1$). The sets of 6 hyperfine lines due to other m_s transitions are smeared out in powder. However, the spectra differed depending on the method of preparation of the MgO. Low-specific-area (LSA) material with mean particle size of about 650 Å was prepared by calcination in air followed by evacuation at 1000°C, whereas high-specific-area (HSA) material with mean particle size of about 135 Å was prepared by vacuum decomposition of the carbonate with a final evacuation at 1000°C. The Mn^{2+} spectrum in LSA material was characteristic of Mn^{2+} in cubic-symmetry single crystals, whereas the spectrum in HSA material showed considerable line-shape distortion, interpreted as due to an axial component in the environment. The presence of this axial component in the normally octahedral environment of Mn was interpreted to show that the ions were very close to the surface. This is an example of how EPR spectra can give useful information even if accurate quantitative data are not obtainable.

5.1.4. Other Metal Oxides

A much-investigated system is ZnO, which when evacuated at temperatures above about 150°K gives a signal at $g = 1.96$ attributed to an O_2^- lattice vacancy. Other signals with higher g values also appear.[51,52] Irradiation at low temperature (wavelength 3650 Å) under evacuation also produces these signals.[53] On exposure to small pressures of oxygen the $g = 1.96$ signal is much reduced and a number of new signals in the range $g = 2.002$–2.030 appear, ascribed to chemisorbed peroxo radicals (OO^-)[52,53] and also O_2^-.[42] Adsorption of NO_2 gives NO species.[54] The ease of formation of O_2^- species depends on the manner of formation of the ZnO, indicating that the surfaces are affected in detailed structure and composition by the preparative process.

Similar measurements have been reported for CaO,[47] SrO,[46] ZrO_2,[56] TiO_2,[57] and also ZnS,[54] but the above details are sufficient to illustrate the EPR approach.

5.1.5. Miscellaneous Results

We mention here examples of other systems that have been studied. At low temperatures, radicals of aliphatic hydrocarbons and hydrogen atoms (H, CH_3, C_2H_5) can be formed on various alumina and other insulator surfaces and identified from their EPR spectra.[58] It was found by Rooney and Pink[59] that positively charged radicals form in the chemisorption of condensed polynuclear aromatic compounds on insulator surfaces such as alumina and silica–alumina. The EPR signal strength is much increased if oxygen is present.[60] EPR methods have also been fruitful in the study of the effects of radiation on catalysts and catalytic reactions, including hydrogen–deuterium exchange on alumina.[61] Experiments on chlorine adsorbed on vacuum-degassed ultraviolet-irradiated silica gel[62] gave a many-line spectrum which permitted detailed analysis of the trapped chlorine species, including an estimate of the electric field strength at the surface. This has been possible because the **g** tensor [Eq. (6) in Section 2] involves the energy-level separations of the unpaired electron states, and these separations can be related to the field strength[63] surrounding the unpaired electron using crystal field theory.

An interesting application of EPR techniques to surfaces by Tench[64] involved measurement of the temperature dependence of hyperfine splitting constants A_{ij} in the case of defects on MgO that was γ-irradiated in the presence of hydrogen. The temperature dependence of A_{ij} was nearly linear from 500°K to about 100°K, being much faster than for equivalent bulk centers. This behavior for surface defects was explained by assuming a larger mean square vibration displacement of surface ions than for bulk ones due to the weaker surface–neighbor interactions. The hyperfine constants were related theoretically to the ion motion with a few necessary assumptions, thus leading to the observed temperature dependence. Further results along these lines would be of interest to confirm this interpretation of the temperature behavior of the hyperfine tensor.

5.2. Clean Surfaces

Most of the work summarized under this heading has been done on nonconductors crushed in a vacuum. Hence one can assume that the surfaces produced are initially free of contamination, although they may be damaged by the fracture processes. Contamination will commence from the background gases to an extent depending on the degree and composition of the vacuum, sticking coefficients of the gases, and the number of available surface sites compared to the number of available gas molecules. Where hundreds of square centimeters and more of surface are produced in, say, 1 liter at 10^{-7} Torr, the ratio of surface atoms to gas molecules in the volume is about 10^4 so that, assuming slow replenishment of any gas lost by adsorption, contamination is small and, in general, less of a problem than in experiments with small-area crystals.

On simple reasoning, one expects a resonance when 2-electron bonds, as in many semiconductors, are broken. In a few cases (Si, Ge, Ge–Si alloys, graphite, SiO_2), EPR signals have been found from the freshly vacuum-crushed material and are subsequently affected by exposure to gases. In most cases, however, only negligible resonances occur on crushed surfaces. (Sometimes the crushing causes internal damage centers.[65]) Adsorption of gases, particularly oxygen at liquid-nitrogen temperatures, has led to EPR signals in several cases which could be interpreted to give information about the surfaces on which the gases absorbed. This has been a powerful technique in certain favorable cases.[66]

A number of experiments on crushed powders have been performed. We summarize some of the results for these and will give details for the case of AlSb which is a good example of how detailed surface information is obtainable by EPR.

5.2.1. Carbon

An EPR signal has been reported from graphite powdered at room temperature in air or medium vacuum.[67] In the latter case, the line was partially destroyed upon exposure to air or oxygen, and ascribed to broken σ bonds of carbon. However, no signal was observed by Demidovitch *et al.*[68] when graphite was crushed at room temperature, unless it was first bombarded by atomic hydrogen in a discharge. However, they did obtain a signal from ordinary graphite when it was crushed (for 20 hr) at 77°K. This signal was composed of lines of about equal intensity at $g = 2.027$, width 55 G, and $g = 2.0036$, width 12 G, corresponding to a total of about 2×10^{14} centers cm^2. The signal was unaffected on heating to 300°K, but it was irreversibly altered by exposure to oxygen, and also by atomic hydrogen.

Colorless diamond, on pulverizing in air, gave a line at $g = 2.0027$, width 5.5 G.[65] No effect was found, in another work,[69] by exposure to oxygen. The lack of detail in the above signals from graphite and diamond limits the possibilities for detailed interpretation. However, carbon signals have been found in other environments as well.[70]

Many carbonaceous materials exhibit a surface-type resonance after heating in vacuum in the range of 400–600°C. The line has

$g = 2.0027$ (cf., diamond) and width approximately 1 G, which is reversibly broadened, in some cases to below detection, upon admission of air or oxygen. Signals with such properties have been also reported for a variety of materials crushed and heated in high vacuum (Si, Ge, GaAs, InAs, GaSb, glass, quartz, ZnO, CdS, CdSe, ZnS).[71] An analysis of such cases suggests strongly that the signals are in all cases due to traces of carbon contamination.[71,72] Much of this comes from background hydrocarbons which make up some of the ambient even in oil-free ultrahigh vacuum systems. The EPR signal is not observed until the samples (including even an empty quartz tube) are heated to several hundred degrees Centigrade. Evidence was presented that this causes formation of small carbon aggregates, pyramids in the case of Si, and that the EPR signal originates from unpaired electrons stabilized in carbon rings on the surfaces of the materials.[71]

5.2.2. Silicon

It was found many years ago[73] and confirmed with more details since,[74] that abraded surfaces of Si gave an EPR signal. The same signal appears on crushing Si in air,[73,75,76] medium vacuum,[65] and in ultrahigh vacuum.[18,77] It has $g = 2.0055 \pm 0.0001$, irrespective of doping; but small random variations in the width, which is about 6.5 G, are found. The surface density is about 1×10^{14} spins/cm^2 to within a factor of 2. Exposure to oxygen alters the shape from off Lorentzian to near-perfect Lorentzian; but the total number of spins is hardly affected. Effects are also caused by hydrogen and water vapor.[18]

A similar resonance was found from single-crystal (111) cleavage surfaces produced in ultrahigh vacuum and aligned in the cavity.[78] The failure to observe hyperfine structure from the 4.7 % abundant ^{29}Si was analyzed to show that the unpaired electron wave functions were delocalized, with no more than 5 % of the electron density on any atom. The gas-adsorption effects were considered in detail to show that the centers were on the surface and not in an interior space-charge layer. A model of the surface was proposed,[78] consistent with the EPR data and also low-energy electron diffraction data from vacuum-cleaved surfaces. The same EPR signal was found from amorphous evaporated Si films, and interpreted as due to the surfaces of the voids in the film.[79]

The EPR data has recently been correlated with the surface-state distribution.[20] It was shown that it is necessary, in order to account for the observed density of spins, for correlation effects in the surface-state analysis to be taken into account (see Section 2.3.1).

5.2.3. Germanium

This material gives a broader and weaker signal than Si. When crushed in high vacuum,[80,81] it displays an EPR line (seen at 77°K, at $g = 2.02$, width about 30 G) corresponding to a surface spin density of about 2×10^{13} cm^{-2}. Exposure to oxygen at 77°K (10^{-3}–10^{-2} Torr) causes the signal to desaturate and also produces a small new signal, probably due to O_2^- ions.[81] However, exposure to oxygen at room temperature destroys all resonances. Bulk as well as surface EPR signals are broader in Ge than in Si due to a larger spin–orbit coupling constant and hence a greater spread in g values, greater unresolved hyperfine splittings, and also greater sensitivity to the distribution of microstrains in the solid. This signal also appears strongly on amorphous evaporated films[82] due to the large surface area in these porous structures.

5.2.4. Ge–Si Alloys

A full range of these alloys was crushed and studied in high vacuum.[19] The 92%Si alloy gave a signal related to that of Si but shifted slightly in g and broadened. All other alloys of 77%Si and less gave essentially the same signal as pure Ge, with similar effects upon oxygen adsorption.

The difference between the 92%Si alloy and pure Si resonances could be explained by the change in the average spin–orbit coupling constant caused by the replacement of 1 in 12 of the Si atoms by Ge atoms. However, for this to affect the Si resonance, the electrons contributing to it must sample a larger number of atoms, which is consistent with the delocalized nature of these electrons deduced from the pure-Si data.[78]

The fairly sudden change in signal that occurs at Si concentrations between 92% and 77% and persists for lower concentrations was correlated with a change in bulk band structure that is known to occur at about 90% Si in the alloys. The band structure affects the surface-state structure and thus the EPR signal.[20]

5.2.5. Silicon Oxide

Crushing of silica in high vacuum produced a resonance line[83] showing in addition two weak hyperfine lines due to ^{29}Si. Hence, the resonance is due to unpaired electrons associated with Si atoms. It disappears on exposure to CO_2, or O_2, indicating its surface origin. The surface density of centers is similar for silica or quartz of different origins, being about 1% of the total number of Si dangling bonds theoretically obtainable by cutting the solid. Adsorption of CO_2 yields CO_2^- ions, and of oxygen, O_2^- ions. The CO_2 is reversibly desorbed at up to 250°C, with recovery of the original resonance; but at up to 850°C, only 10% of the oxygen is desorbed.

The measured value of the hyperfine splitting, 465 G, leads to an estimate of the *s* character of the dangling bond of nearly 27%.[83] This is very close to the figure of 25% for a pure sp^3 bond, as expected on simple reasoning, for the SiO_2 structure, since it is built up of $Si(\frac{1}{2}O)_4$ tetrahedra, and the angles between sp^3 bonds are tetrahedral.

5.2.6. GaAs and AlSb

The III–V compound semiconductors, which cleave preferentially along (110), gave a very small resonance on vacuum crushing.[66] However they gave an easily detectable O^- signal when exposed to oxygen at 77°K.[66] This resonance had the powder shape (see Fig. 14-10) and showed hyperfine splitting in the case of AlSb (as seen in Fig. 14-2). As an example of EPR hyperfine structure analysis, we give a few steps used for AlSb.

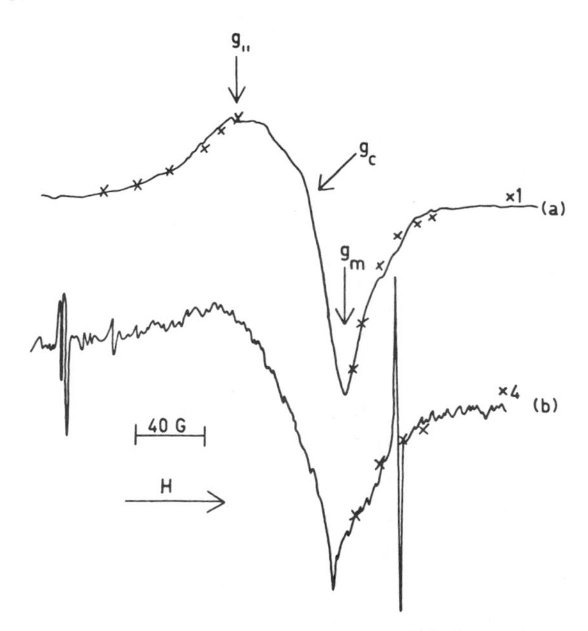

Fig. 14-10. EPR powder spectrum[66] from O_2^- ions on vacuum-crushed GaAs, recorded at 77°K. (a) Curve taken after evacuation, showing positions of **g** tensors. (b) Curve taken with 3.6×10^{-2} Torr oxygen in the system. The sharp lines are due to molecular oxygen. The crosses would fit a Lorentzian curve of width 30 G.

1. The hyperfine structure is due to interaction with a surface nucleus since oxygen does not have an abundant isotope with a nuclear spin. Both Al and Sb have a high nuclear spin, and significant interaction with both would yield more than 20 lines near the principal g values, whereas only 5 or 6 lines are seen. A computer simulation shows that it is the Al atom.

2. Assume the unpaired electron wave function $\psi(r)$ is composed of a fraction $C_2{}^2$ of Al orbital and $C_1{}^2$ of O_2 orbital. We assume the Al orbital is a hybrid of s function $\phi_s(r)$ and p function $\phi_p(r)$. Then

$$\psi(r) = C_1\phi_0(r) + C_2[\alpha\phi_s(r) + \beta\phi_p(r)] \tag{9}$$

where $\phi_0(r)$ is the wave function of the extra electron on $O_2{}^-$ in the absence of neighboring atoms, and α^2 and β^2 are the fractions of s and p Al functions, respectively. One has $(\alpha^2 + \beta^2) = 1$ and also, neglecting overlap between ϕ_0, ϕ_s, and ϕ_p, $(C_1{}^2 + C_2{}^2) = 1$. The isotropic and anisotropic parts of the hyperfine tensor $\mathbf{A}_{ij}$ can be written

$$\begin{aligned} a &= \tfrac{8}{3}\pi g g_N \mu_\beta \mu_N C_2{}^2 \alpha^2 |\phi_s(0)|^2 \\ b &= \tfrac{2}{5} g g_N \mu_\beta \mu_N C_2{}^2 \beta^2 \langle r^{-3} \rangle \end{aligned} \tag{10}$$

where $\mu_\beta = \hbar\gamma$ for an electron and $\mu_N = \hbar\gamma_N$ for a nucleus. If the system has axial symmetry, say about the z direction, then $A_{xx} = A_{yy}$. Furthermore, one can show that if the $\mathbf{A}$ and $\mathbf{g}$ tensors are parallel, then the expression for the $\mathbf{A}$ tensor becomes simply

$$A_{xx} = A_{yy} = a - b \qquad A_{zz} = a + 2b$$

The more complicated case where they are not parallel is treated in Miller and Haneman.[66] Hence one can solve for a and b from the experimental values of the hyperfine splitting A_{xx}, A_{yy}, and A_{zz} occurring at fields corresponding to g_{xx}, g_{yy}, and g_{zz}. Having a and b, one inserts values of $|\phi_s(0)|$ and $\langle r^{-3}\rangle$, which can be obtained from atomic wave function calculations like those of Watson and Freeman,[84] and thence deduces C_1, C_2, α, and β from Eq. (10). In the case of AlSb, it was shown in this way that the Al orbital charge density was over 90% p type ($\beta^2 > 0.90$). This is the first time that a semiconductor surface wave function has been obtained from experiment.

The $\mathbf{g}$ tensor with axial symmetry about the z direction has two principal components: g_{zz}, called $g_{\|}$, and g_{xx} $(=g_{yy})$ which is called $g_\perp$. These can be obtained from a powder spectrum using the expression[85]

$$(g_\perp)^{-1} = g_m{}^{-1} - 0.15 \times 2.28(g_m{}^{-1} - g_c{}^{-1}) \tag{11}$$

where g_m is the minimum of the curve and g_c is the crossover, as illustrated for the GaAs resonance in Fig. 14-10. The $\mathbf{g}$ tensor was analyzed to give the energy-level splittings of the adsorbed O_2^- ion. It was shown by Kanzig and Cohen[86] that the **g** tensor for O_2^- ions has the form, when higher-order terms are ignored,

$$g_{xx} = g_e + 2\lambda/E \qquad g_{yy} = g_e \qquad g_{xx} = g_e + 2\lambda/\Delta \tag{12}$$

where g_e is 2.0023, Δ is the separation of the Π_g levels of the extra electron of O_2^-, and E is the separation between the σ_g levels ($\sigma_g = p_z^1 - p_z^2$). With a value of 0.014 eV for λ, the spin–orbit coupling constant for O_2^-, the values of E and Δ could be deduced from the experimental **g** values. The terms g_{xx} and g_{yy} could not be separated from the experimental curve, but a small difference had to be assumed in order to obtain the best fit of the computer-simulated curve (see Fig. 14-2).

5.2.7. Other Compounds

EPR experiments have also been conducted on a variety of other semiconductors and insulators crushed and measured in high vacuum. These include InAs,[75] ZnO,[87] CdSe,[88] CdS,[89,90] and ZnS,[90] from which a signal at about $g = 2.0027$ was reported, and GaP, GaSb,[66] ZnSe, PbS, and PbTe,[90] from which negligible signals were found. The signal found at $g = 2.0027$ is almost certainly due to carbon contamination[71,72] as described in Section 5.2.1. It is concluded that none of the above materials gives an appreciable EPR signal. This may be because the number of localized unpaired electrons on the cleavage surfaces is small, as is not unexpected for substances with appreciable ionic content. Another possibility is that the surface electrons form a band (a surface-state band) of appreciable energy spread. The EPR signal from these would also be difficult to detect if it were an ordinary band. A stronger signal would be expected from conduction electrons if correlations play an important role (Section 2.3.1). However, the signal observed have been too weak to analyze in detail.

5.2.8. Indirect Effects

An interesting application of EPR techniques that has relevance to surfaces was reported by Lepine.[92] He measured the photoconductivity in a thin Si slice situated in an EPR spectrometer cavity under microwave irradiation. The conductivity peaked at a certain value of the scanning magnetic field, indicating a spin-dependent scattering or a recombination of the carriers. This was confirmed by using saturation microwave power to alter the strength of the effect. The paramagnetic recombination centers were tentatively identified with surface centers. This method of detecting such centers from their effects on conductivity was found to be considerably more sensitive than direct observation of their resonance.

6. NMR RESULTS FOR SURFACES

The type of phenomena that have been examined differ somewhat from EPR studies due to the much lower (by about 10^{-6}) detection ability. Very-large-area samples are needed. Furthermore, it is not possible to study a signal from an intrinsic surface since if the surface nuclei have a magnetic moment so will the bulk nuclei (for a homogeneous material), and the bulk signal dominates. Hence studies have been mostly applied to adsorbed layers, which are usually at least a substantial fraction of a monolayer.

Very detailed information has been difficult to obtain, although detailed models have been proposed. Much work has concentrated on measuring the spin relaxation times and relating them to the mobility and diffusion of the molecules in the adsorbed layers. The relaxation times of nuclei in an adsorbed molecule are affected because the relative motion of the nuclei are altered after adsorption. Also intermolecular interactions are altered and may affect the nuclei in the individual molecules, although this effect is usually the lesser of the two.

A number of reviews of this type of work have been published.[93–95] We will only briefly summarize some of the results as an illustration of the kind of information obtainable. Much work has been done with adsorbed water, studying spin relaxation times of the proton by C.W. and spin–echo techniques. Substrates have included silica, alumina, charcoal, magnesium oxide, cellulose, and keratin fibers (wool).

The silica studies are prompted by the importance of this material in many industrial processes. Various dehydration and rehydration sequences result in hydroxyl groups bonded to the surface. NMR positively identifies the presence of protons. From the resonance line shape, which reflects the nature of the interactions between the nuclear dipoles, and from the behavior with temperature, it was concluded[96] that surface hydroxyl groups existed, with nearest-neighbor protons at 5.2–5.4 Å, or at 2.3–2.6 Å. In the case of physically adsorbed water on microporous silica, the behavior of T_1 and T_2 with temperature indicated a two-phase behavior,[97] but details were difficult to establish. Similar studies were made for water adsorbed on alumina and thorium oxide. The temperature behavior of T_1 and T_2 was also measured for protons in water, and for benzene adsorbed on high-surface-area charcoal. The values obtained were consistent with a model[98] of benzene molecules rotating rapidly about hexad axes and having a distribution of surface sites. Benzene adsorbed on silica gel was studied by similar methods,[99] also leading to the conclusion that there were at least two kinds of surface sites of different energy. Diffusion rates were also deduced in the above work.

Studies on magnesium oxide dehydrated at 300°C revealed the presence of 0.08 molecules of water per Å^2.[100] The shape of the resonance line was analyzed, particularly the second moment, which was consistent with a model of a double layer of hydroxyl groups. The variation of the second moment with surface coverage led to the conclusion that desorption resulted in the formation of small clusters of hydroxyl groups rather than pairs or a random distribution.

In general, NMR studies alone do not give sufficiently detailed information for unambiguous interpretation and it is important to obtain concomitant results by as many other techniques as possible. These include EPR, infrared adsorption, electron–nuclear double resonance (this requires a high signal-to-noise ratio), heats of adsorption, and others. However, the positive identification of the presence of nuclei with spin (in sufficient numbers) is always a valuable feature.

7. ACKNOWLEDGMENTS

The author has had valuable discussions with D. J. Miller and J. Higinbotham on EPR studies, and with K. H. Marsden on NMR methods.

8. REFERENCES

1. G. E. Pake, *Paramagnetic Resonance*, W. A. Benjamin, New York, (1962).
2. D. J. E. Ingram, *Free Radicals as Studied by E.S.R.*, Butterworths, London (1958).
3. C. P. Slichter, *Principles of Magnetic Resonance*, Harper and Row, New York (1963).
4. B. Bleaney and K. W. H. Stevens, *Rept. Progr. Phys.* **16**, 108–180 (1953).
5. A. Abragam and M. H. L. Pryce, Theory of the nuclear hyperfine structure of paramagnetic resonance spectra in crystals, *Proc. Roy. Soc.* (*London*) **A205**, 135–153 (1951).
6. A. Abragam and B. Bleaney, *Electron Paramagnetic Resonance of Transition Ions*, Clarendon Press, Oxford (1970).
7. C. P. Poole, Jr. and H. A. Farach, *Relaxation in Magnetic Resonance*, Academic Press, New York (1971); *The Theory of Magnetic Resonance*, Interscience, New York (1972).
8. L. D. Landau and E. M. Lifshitz, *Quantum Mechanics*, Pergamon Press, London (1965).
9. P. A. M. Dirac, *The Principles of Quantum Mechanics*, Clarendon Press, Oxford (1958).
10. R. M. Golding, *Applied Wave Mechanics*, Van Nostrand, New York (1969).
11. F. J. Adrian, Guidelines for interpreting electron spin resonance spectra of paramagnetic species adsorbed on surfaces, *J. Colloid Interface Sci.* **26**, 317–360 (1968).
12. S. M. Blinder, *J. Chem. Phys.* **33**, 748 (1969).
13. F. J. Adrian, E. L. Cochran, and V. A. Bowers, *J. Chem. Phys.* **36**, 1661 (1962).
14. R. Lefebre and J. Maruani, *J. Chem. Phys.* **42**, 1480 (1965).
15. D. L. Griscom, P. C. Taylor, D. A. Ware, and P. J. Bray, *J. Chem. Phys.* **48**, 5158 (1968).
16. G. Feher and A. F. Kip, *Phys. Rev.* **98**, 337 (1955).

17. M. F. Chung and D. Haneman, Properties of clean Si surfaces by paramagnetic resonance, *J. Appl. Phys.* **37**, 1879–1889 (1966).
18. D. J. Miller and D. Haneman, EPR investigation of the surfaces of Si–Ge alloys, *Surface Sci.* **33**, 477–492 (1972).
19. D. J. Miller, Ph.D. Thesis, Univ. of New South Wales (1972).
20. D. J. Miller, D. L. Heron, and D. Haneman, Semiconductor surface states considered on the Hubbard model; correlation with EPR data, *J. Vac. Sci. Techn.* **9**, 906–914 (1972).
21. J. Hubbard, Electron correlations in narrow energy bands, *Proc. Roy. Soc. (London)* **A276**, 238–257 (1963).
22. J. Hubbard, Electron correlations in narrow energy bands. III. An improved solution, *Proc. Roy. Soc. (London)* **A281**, 401–418 (1964).
23. D. J. E. Ingram, Electron Spin Resonance, in *Handbuch der Physik*, Vol. 18/1, pp. 94–144, Springer-Verlag, New York (1968).
24. J. W. Emsley, J. Feeney, and L. H. Sutcliffe, *High Resolution Nuclear Magnetic Resonance Spectroscopy*, Vol. 1, Pergamon Press, London (1965).
25. G. Feher, *Bell System Tech. J.* **36**, 449 (1957).
26. J. S. Hyde, Principles of EPR instrumentation, in *Seventh Annual NMR-EPR Workshop*, Varian Associates, pp. 1–15 (1963).
27. Varian Associates, *Large Sample Access Cavity*, Data Sheet 18-A-017-17 (1967).
28. D. M. Young and A. D. Crowell, *Physical Adsorption of Gases*, Butterworths, London (1962).
29. D. E. O'Reilly, *Adv. Catalysis* **12**, 311 (1960).
30. J. H. Lunsford, Electron spin resonance in catalysis, in *Advances in Catalysis* **22**, 265 (1972).
31. P. E. Pickert, J. A. Rabo, E. Dempsey, and V. Schomaker, Zeolite cations with strong electrostatic fields as carboniogenic catalytic centers, in *Proceedings of the Third International Congress on Catalysis, Amsterdam*, pp. 714–726, North-Holland, Amsterdam (1965).
32. K. M. Wang and J. H. Lunsford, Electron paramagnetic resonance evidence for the presence of aluminum at adsorption sites on decationated zeolites, *J. Phys. Chem.* **73**, 2069–2071 (1969).
33. J. H. Lunsford, An electron paramagnetic resonance study of γ-type zeolites. II. Nitric oxide on alkaline earth zeolites, *J. Phys. Chem.* **74**, 1518–1522 (1970).
34. C. B. Colburn, R. Ettinger, and F. A. Johnson, *Inorg. Chem.* **2**, 1305 (1963).
35. J. Turkevich, F. Nozaki, and D. Stamires, Studies on nature of active centers and mechanism of heterogeneous catalysis, in *Proceedings of the Third International Congress on Catalysis, Amsterdam*, pp. 586–595, North-Holland, Amsterdam (1965).
36. J. A. Rabo, C. L. Angell, P. H. Kasai, and V. Schomaker, *Discussions Faraday Soc.* **41**, 328 (1966).
37. J. A. R. Cope, C. L. Gardner, C. A. McDowell, and A. J. Pelman, An ESR study of ClO and Cl adsorbed on zeolites, *Mol. Phys.* **21**, 1043 (1971).
38. D. E. O'Reilly and D. S. McIver, *J. Phys Chem.* **66**, 276 (1962).
39. C. P. Poole, W. L. Kehl, and D. S. McIver, *J. Catalysis* **1**, 407 (1962).
40. A. Clark, J. P. Hogan, R. L. Banks, and W. C. Lanning, *Ind. Eng. Chem.* **48**, 1152 (1956).
41. J. H. Lunsford, A study of irradiation induced active sites on magnesium oxide using electron paramagnetic resonance, *J. Phys. Chem.* **68**, 2312–2316 (1964).
42. J. H. Lunsford and J. P. Jayne, Electron paramagnetic resonance of oxygen on ZnO and ultraviolet-irradiated MgO, *J. Chem. Phys.* **44**, 1487–1492 (1966).

43. J. H. Lunsford, EPR spectra of radicals formed when NO_2 is adsorbed on magnesium oxide, *J. Colloid Interface Sci.* **26**, 355–360 (1968).
44. W. B. Williamson, J. H. Lunsford, and C. Naccache, The EPR spectrum of O^- on magnesium oxide, *Chem. Phys. Letters* **9**, 33–34 (1971).
45. H. B. Charman and R. M. Dell, *Trans. Faraday Soc.* **59**, 470 (1963).
46. R. L. Nelson, A. J. Tench, and B. J. Harmsworth, Chemisorption on some alkaline earth oxides, *Trans. Faraday Soc.* **63**, 1427–1446 (1967).
47. A. J. Tench and R. L. Nelson, Paramagnetic defects associated with hydrogen adsorbed on the surface of magnesium and calcium oxides, *J. Colloid Interface Sci.* **26**, 364–373 (1960).
48. A. J. Tench and T. Lawson, The formation of O^- and $O_3{}^-$ adsorbed on an oxide surface, *Chem. Phys. Letters* **7**, 459–460 (1970).
49. A. J. Tench and J. Kibblewhite, ESR study of chlorine radicals stabilized on an oxide surface, *J. Chem. Soc.* **A14**, 2282–2284 (1971).
50. D. Cordischi, R. L. Nelson, and A. J. Tench, Surface reactivity of magnesium oxide doped with manganese: An E.S.R. and chemisorption study, *Trans. Faraday Soc.* **65**, 2740–2757 (1969).
51. P. J. Kasai, *Phys. Rev.* **130**, 989 (1963).
52. R. J. Kokes, *J. Phys. Chem.* **66**, 99 (1962).
53. E. V. Baranov, V. E. Kholmogorov, and A. N. Terenin, *Dokl. Phys. Chem.* **146**, 125 (1962).
54. J. H. Lunsford, Surface interactions of zinc oxide and zinc sulphide with nitric oxide, *J. Phys. Chem.* **72**, 2141–2144 (1968).
55. R. D. Iyengar and V. V. Subba Rao, ESR studies on zinc and on zinc oxide obtained from a decomposition of zinc peroxide, *J. Phys. Chem.* **75**, 3089–3092 (1971).
56. T. Kwan, Photoadsorption and photodesorption of oxygen on inorganic semiconductors and related photocatalysis, in *Symposium on Electronic Phenomena in Chemisorption and Catalysis on Semiconductors*, pp. 184–195, Walter de Gruyter and Co., Berlin (1969).
57. R. D. Iyengar, M. Codell, J. S. Karra, and J. Turkevich, Electron spin resonance studies of the surface chemistry of rutile, *J. Am. Chem. Soc.* **88**, 5055 (1966).
58. V. B. Kasansky and G. B. Pariisky, in *Proceedings of the Third International Congress on Catalysis, Amsterdam*, p. 367, North-Holland, Amsterdam (1965).
59. J. J. Rooney and R. C. Pink, *Trans. Faraday Soc.* **58**, 1632 (1962).
60. H. P. Leftin, M. C. Hobson, and J. S. Leigh, *J. Phys. Chem.* **66**, 1214 (1962).
61. V. V. Voevodski, in *Proceedings of the Third International Congress on Catalysis, Amsterdam*, p. 88, North-Holland, Amsterdam (1965).
62. C. L. Gardner, Electron-spin-resonance study of chlorine atoms adsorbed on a silica-gel surface, *J. Chem. Phys.* **46**, 2991–2994 (1967).
63. C. L. Gardner and E. J. Casey, Tumbling of methyl radicals adsorbed on a silica gel surface studied by electron spin resonance, *Can. J. Chem.* **46**, 207–210 (1968).
64. A. J. Tench, Temperature effects on the hyperfine coupling of a surface center, *Surface Sci.* **25**, 625–632 (1971).
65. G. K. Walters and T. L. Estle, Paramagnetic resonance of defects introduced near the surface of solids by mechanical damage, *J. Appl. Phy.* **32**, 1854–1859 (1961).
66. D. J. Miller and D. Haneman, Electron-paramagnetic-resonance study of clean and oxygen exposed surfaces of GaAs, AlSb, and other III–V compounds, *Phy. Rev.* **B3**, 2918–2938 (1971).
67. S. Mrozowski and J. F. Andrew, in *Proceedings of the Fourth Conference on Carbon*, p. 207, Pergamon Press, New York (1960).

68. G. B. Demidovitch, V. F. Kiselev, N. N. Lejnev, and O. V. Nikitina, Nature of freshly crushed surfaces of graphite and mechanism of their interaction with oxygen and hydrogen, *J. Chim. Phys.* **65**, 1072–1078 (1968).
69. H. P. Boehm, *Angew. Chem.* **78**, 617 (1966).
70. L. S. Singer, A review of electron spin resonance in carbonaceous materials, in *Proceedings of the Fifth Conference on Carbon*, pp. 37–64, Pergamon Press, New York (1963).
71. D. J. Miller and D. Haneman, Carbon EPR signal from vacuum heated surfaces, *Surface Sci.* **24**, 639–642 (1971).
72. D. J. Miller and D. Haneman, Evidence for carbon contamination on vacuum heated surfaces by EPR, *Surface Sci.* **19**, 45–52 (1970).
73. G. Feher, *Phys. Rev.* **114**, 1219 (1959).
74. D. Haneman, M. F. Chung, and A. Taloni, Comparison of thermal behavior of vacuum crushed, air crushed and mechanically polished Si surfaces by EPR, *Phys. Rev.* **170**, 719–723 (1968).
75. P. Chan and A. Steinemann, EPR study on Si, Ge, and GaAs surfaces interacting with adsorbed oxygen, *Surface Sci.* **5**, 267–282 (1966).
76. T. Wada, T. Mizutani, M. Hirose, and T. Arizumi, Annealing effects of paramagnetic defects introduced near silicon surfaces, *J. Phys. Soc. Japan* **22**, 1060–1065 (1967).
77. M. F. Chung, The effects of bulk doping on the ESR signal of clean Si surfaces, *J. Phys. Chem. Solids* **32**, 475–485 (1961).
78. D. Haneman, EPR from clean single-crystal cleavage surfaces of silicon, *Phys. Rev.* **170**, 705–718 (1968).
79. R. S. Title, M. H. Brodsky, and B. L. Crowder, EPR studies in amorphous Si, in *Proceedings of the Tenth International Conference on the Physics of Semiconductors*, United States Atomic Energy Commission, Cambridge, Massachusetts (1971).
80. G. B. Demidovich and V. F. Kiselev, EPR from clean Ge and Si surfaces, *Phy. Stat. Sol.* (*b*) **50**, K33–K35 (1972).
81. J. Higinbotham and D. Haneman, Paramagnetic surface states of Ge, *Surface Sci.* **34**, 450–456 (1973).
82. M. H. Brodsky, R. S. Title, K. Weiser, and G. D. Pettit, *Phys. Rev.* **B1**, 2632 (1970).
83. G. Hochstrasser, J. F. Antonini, and I. Peyches, MS and ESR studies of dangling bonds and adsorbed ions on the pristine surface of silica, in *The Structure and Chemistry of Solid Surfaces* (G. Somorjai, ed.) pp. 36-1–36-11, Wiley, New York (1969).
84. R. E. Watson and A. J. Freeman, *Phys. Rev.* **123**, 521 (1961).
85. J. W. Searl, R. C. Smith, and S. J. Wyard, *Proc. Phys. Soc.* (*London*) **A78**, 1174 (1961).
86. W. Kanzig and M. H. Cohen, *Phys. Rev. Letters* **3**, 509 (1959).
87. R. J. Kokes, *Proceedings of the Third International Congress on Catalysis, Amsterdam*, p. 484, North-Holland, Amsterdam (1964).
88. Z. Z. Ditina and L. P. Strakhov, Investigation of the surface of CdSe by the EPR method, *Sov. Phys.—Solid State* **9**, 2000–2003 (1968).
89. Z. Z. Ditina, B. A. Kazennov, and L. P. Strakhov, Paramagnetic centers on CdS surface, *Sov. Phys.—Semicond.* **1**, 1434 (1968).
90. T. Arizumi, T. Mizutani, and K. Shimakawa, EPR study on surface properties of ZnS and CdS, *Japan. J. Appl. Phys.* **8**, 1411–1416 (1969).
91. J. Higinbotham and D. Haneman, EPR from II–VI and IV–VI semiconductor surfaces, *Surface Sci.* **32**, 466 (1972).

92. D. J. Lepine, Spin-dependent recombination on silicon surface, *Phys. Rev.* **B6**, 436 (1972).
93. H. Winkler, *Math.-Naturwiss.* **4**, 913 (1965).
94. K. J. Packer, Nuclear spin relaxation studies of molecules adsorbed on surfaces, in *Progress in NMR Spectroscopy*, Vol. 3, pp. 87–128, Pergamon Press, London (1967).
95. H. A. Resing, NMR relaxation of molecules adsorbed on surfaces, *Advan. Mol. Relax. Proc.* **1**, 109–154 (1967–1968).
96. V. I. Kvlividze, *Dokl. Adad. Nauk SSSR* **157**, 673 (1964).
97. D. E. Woessner, *J. Chem. Phys.* **39**, 2783 (1963).
98. J. K. Thompson, J. J. Krebs, and H. A. Resing, *J. Chem. Phys.* **43**, 3853 (1965).
99. D. E. Woessner, *J. Phys. Chem.* **70**, 1217 (1966).
100. R. K. Webster, T. L. Jones, and P. J. Anderson, *Proc. Brit. Ceram. Soc.* **5**, 153 (1965).

15

MÖSSBAUER SPECTROSCOPY

Melvin C. Hobson, Jr.

Virginia Institute for Scientific Research
Richmond, Virginia

1. INTRODUCTION

The discovery of the existence of nuclear γ-ray resonance in solids by R. L. Mössbauer[1] in 1958 led rapidly to the development of a new spectroscopic tool for the study of the structure and composition of matter. The technique quickly expanded across interdisciplinary lines from physics to metallurgy, chemistry, and biology. The applications have been diverse and often imaginative as witnessed by the observation of changes in the abdominal movements of large Transylvanian ants during their death throes.[2]

The theory and applications of Mössbauer spectroscopy have been presented in several monographs,[3–6] and a number of reviews on applications to surface studies[7–10] have been published. A brief review of the literature reveals that the number of elements easily available for Mössbauer effect studies is limited, but this disadvantage is offset by the wealth of detailed information obtainable from a Mössbauer spectrum. In principle, the valence state and type of bonding of an atom to the surface, the dynamics of motion of an atom about its equilibrium site, the symmetry of the electric field surrounding an atom, the presence and magnitude of a magnetic field, and diffusion of an atom over the surface can all be derived from Mössbauer measurements. In practice, many of these properties have been measured, some more convincingly than others.

2. THEORY

The energy of radiation emitted or absorbed by an atom or molecule is not strictly monochromatic but is distributed about a maximum as a consequence of the Heisenberg uncertainty principle. The shape of this distribution is given by the Lorentzian function

$$y(E) = y_0\left[1 + \left(\frac{E - E_0}{\frac{1}{2}\Gamma}\right)^2\right]^{-1} \quad (1)$$

where y_0 is the maximum of the distribution, E_0 is the energy at the maximum, and Γ is the width of the distribution envelope at $\frac{1}{2}y_0$. The theoretical width of the distribution may be calculated by

$$\Gamma\tau = \hbar \quad (2)$$

where τ is the mean life of the excited state. This relation fulfills the conditions imposed by the Heisenberg uncertainty principle.

In the emission and absorption of radiation some energy is always lost to recoil processes. The recoil energy in the optical region is much smaller than the width of the emission and absorption lines, and the two overlap producing the phenomenon of resonance absorption. But the recoil energy lost by a γ ray emitted from a free atom is much larger than the theoretical width of the line. The two maxima in the emission and absorption envelopes will be many line widths apart and resonance absorption cannot take place. Experimentally, it is very difficult to bring the two lines back into coincidence.[11]

2.1. Resonance Absorption in Solids

The essential feature of the discovery by Mössbauer was a new mechanism for the dissipation of the recoil energy from low-energy γ rays emitted by atoms in a solid lattice. The recoil energy may be dissipated by displacement of the atom from its lattice site, or by heating the lattice. However, if the average recoil energy is less than a quantum of energy needed to excite a low-lying energy level in the phonon spectrum of the solid, there is a finite probability that no energy will be lost by the γ ray and the event is free of recoil. Absorption of the emitted γ ray by nuclei of the same isotope may likewise be free of recoil. Since energy is not lost by the γ ray in either the emission or absorption processes, the emission and absorption lines can overlap and nuclear γ-ray resonance, the Mössbauer effect, occurs. If the emitter and absorber lines do not overlap because of shifts in the nuclear energy levels by interaction with the extranuclear environment, they may be brought back into coincidence by Doppler shifting. The heart of a Mössbauer spectrometer is a

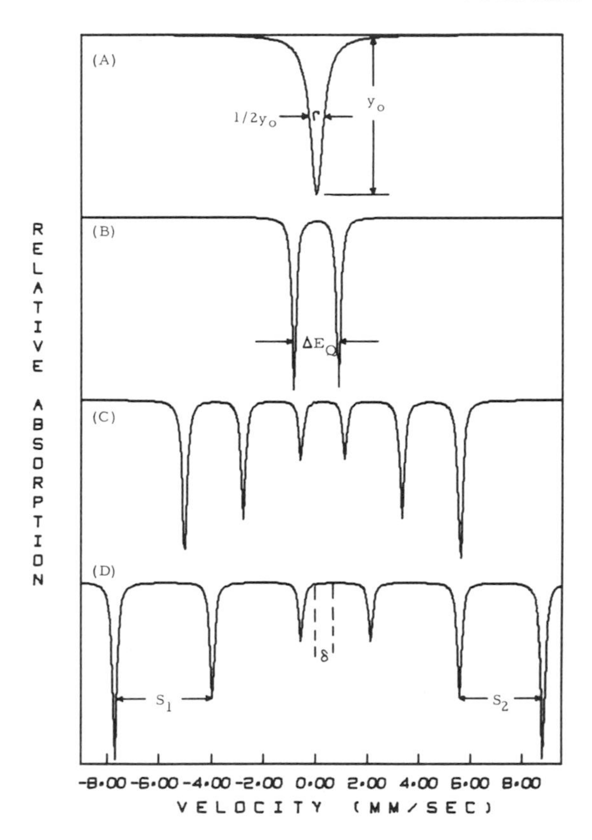

Fig. 15-1. Typical Mössbauer spectra showing the Lorentzian shape, half-width, isomer shift, quadrupole splitting, and magnetic hyperfine splitting. (a) Lorentzian shape, (b) sodium nitroprusside, (c) metallic iron, (d) α-Fe_2O_3.

mechanical or electromechanical device for moving either the emitter or absorber relative to the other to produce the resonance condition. Some typical Mössbauer spectra of ^{57}Fe are shown in Fig. 15-1.

The more strongly the atom is bound in the lattice the larger the percentage of recoil-free events observed. The recoil-free fraction f, often called the Debye–Waller factor, can be related to the vibration of the atom by

$$f = \exp\left(-\frac{\langle \mathbf{x}^2 \rangle 4\pi^2}{\lambda^2}\right) \tag{3}$$

where $\langle \mathbf{x}^2 \rangle$ is the component of the mean square displacement of the atom in the direction of the γ ray and λ is the wavelength of the γ ray. The f factor is measured by the area under the absorption curve.

2.2. Nuclear Interactions with the Environment

There are several ways in which a nucleus interacts with its surroundings. The s electrons of an atom have a finite density at the nucleus which affects the nuclear energy levels. Any changes in the s-electron wavefunction cause shifts in the nuclear energy levels and, consequently, a shift in the point of maximum resonance. If the nuclear spin quantum number $I \geqslant 1$, the nucleus will possess a quadrupole moment. The nuclear quadrupole moment will couple with an aspherical electric field and the energy levels will split into sublevels. Transitions between the sublevels will produce multiple absorption peaks in the Mössbauer spectrum. Finally, a magnetic field will interact with a nuclear magnetic dipole to split the nuclear energy levels into magnetic sublevels. Again multiple absorption peaks are observed in the Mössbauer spectrum.

2.2.1. Isomer Shift

The difference in the energy levels between an excited state and the ground state of a nucleus may be expressed as the sum of the difference in the energy levels for a theoretical point nucleus, δE_0, and a term expressing the effect of the finite size of the nucleus on the energy levels. Thus, the energy of a γ ray emitted from a Mössbauer radiation source may be written,

$$\delta E_s = \delta E_0 + \tfrac{2}{5}\pi Z e^2 |\psi_s(0)|^2 (R_e{}^2 - R_g{}^2) \tag{4}$$

where Z is the atomic number, e is the unit charge, R_e the radius of the nucleus in the excited state, R_g the radius in the ground state, and $|\psi_s(0)|^2$ the electronic charge density at the nucleus of the source. A similar expression may be written for the absorption of the Mössbauer γ ray by another nucleus of the same isotope. If the environment is the same for both the source and absorber nuclei, the centroid of the Doppler-modulated spectrum will be at zero velocity (Fig. 15-1a). On the other hand, any change, such as a change in valence, which causes a change in the electronic charge density at the nucleus of either the source, $|\psi_s(0)|^2$, or absorber, $|\psi_a(0)|^2$, will shift the position of the centroid of the spectrum. This shift, called the isomer shift, is given by

$$\delta = \delta E_a - \delta E_s = \tfrac{2}{5}\pi Z e^2 \{|\psi_a(0)|^2 - |\psi_s(0)|^2\}(R_e{}^2 - R_g{}^2) \tag{5}$$

The isomeric term, $(R_e{}^2 - R_g{}^2)$, is a physical property of the Mössbauer isotope. It is the "chemical term," $\{|\psi_a(0)|^2 - |\psi_s(0)|^2\}$, that is responsible for the variations in δ with changes in the structure and chemical composition. It is primarily, but not exclusively, changes in the s electrons that produce changes in δ. In iron for example, the spectrum of a ferrous species is shifted to large positive velocities relative to that of metallic iron. The loss of the two 4*s* electrons on the formation of the ferrous ion decreases the charge density at the nucleus. Since the isomeric term for iron has a negative value, the overall result is a shift to higher energies, i.e., positive velocities. On further oxidation to the ferric ion, the spectrum, instead of shifting to still higher velocities, shifts back toward the centroid of the metallic-iron spectrum. The removal of the 3*d* electron decreases the screening of the core electrons, particularly the 3*s* electrons. Their orbitals contract and the electronic charge density at the nucleus increases. Thus, the isomer shift may readily distinguish between valence states in some cases, but variations in the shift caused by π bonding, covalency effects, and, particularly, formation of strong ligand bonds may make the identification uncertain if the nature of the compound is completely unknown except for its Mössbauer spectrum.

2.2.2. Quadrupole Interaction

The energy-level diagram for quadrupole coupling of a ^{57}Fe nucleus to the electric field surrounding it is shown in Fig. 15-2a. The ground state does not split because the nucleus is spherically symmetrical in this state; the excited state, $I = 3/2$, does split, provided the electric field is not spherically symmetric. Often this is true and the principle components of the diagonallized electric field gradient, EFG, tensor are written by convention as $V_{zz} > V_{xx} > V_{yy}$. These components may be reduced to two terms, V_{zz} and an asymmetry parameter η. The asymmetry parameter is given by

$$\eta = \frac{V_{xx} - V_{yy}}{V_{zz}} \tag{6}$$

The other term, V_{zz}, is usually written eq when V_{zz} is given in units of electron charge e. The coupling of the nuclear quadrupole moment Q with the EFG splits the spectrum of ^{57}Fe compounds into a doublet as illustrated in Fig. 15-1b. The magnitude of the splitting is given by

$$\Delta E_Q = \tfrac{1}{2}e^2qQ[1 + (\eta^2/3)]^{1/2} \tag{7}$$

The asymmetry parameter vanishes in cases of axial symmetry. In surface studies V_{zz} is taken normal to the surface plane[12] and assumed to be large compared to V_{xx} and V_{yy}. In effect, this model approximates axial symmetry since the asymmetry parameter would be small if not zero. Although not too satisfying, this approximation for surface species is about the best that can be made, at least for measurements with ^{57}Fe and ^{119}Sn.

From the measurement of ΔE_Q the magnitude but not the sign of V_{zz} can be calculated from Eq. (7) for nuclides with an $I = 1/2$ ground

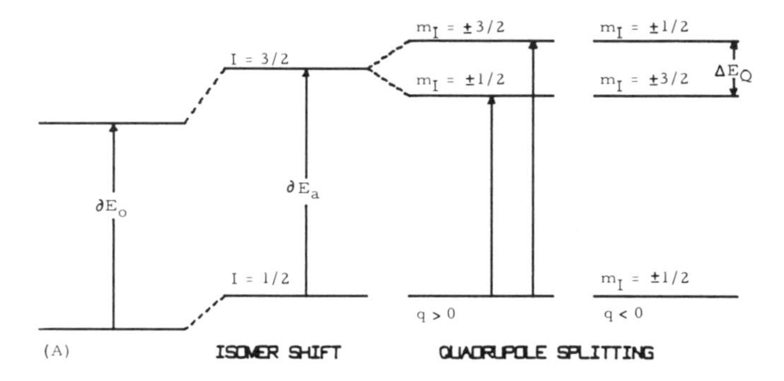

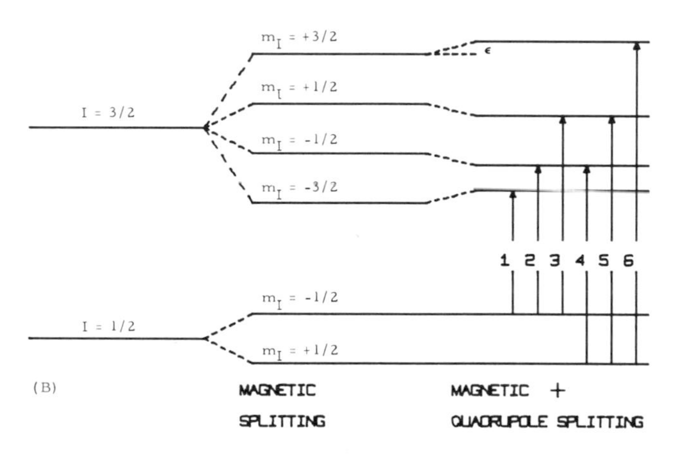

Fig. 15-2. Energy-level diagrams of nuclear transitions in ^{57}Fe. (a) Quadrupole splitting of the excited state, and (b) magnetic hyperfine splitting of the metallic state and combined magnetic hyperfine and quadrupole splitting as found in α-Fe_2O_3.

state and $I = 3/2$ excited state. The sign of V_{zz} may be determined from the intensities of the two peaks as a function of the orientation of a single-crystal sample in the γ-ray beam. The intensities vary with the angle, ϕ between the γ-ray beam and the principal crystallographic axis, i.e., axis of highest symmetry, according to the ratio

$$\frac{A(\pm\frac{3}{2})}{A(\pm\frac{1}{2})} = \frac{1 + \cos^2\theta}{\frac{5}{3} - \cos^2\theta} \tag{8}$$

where A is the area under the peak for the given transition. The sign cannot be obtained for a powdered sample with random orientation since this ratio will average to one, but in some cases the transition can be identified by lifting the magnetic degeneracy with an applied magnetic field.[13,14]

2.2.3. Magnetic Interaction

The degeneracy of the energy levels of the nuclear magnetic dipole will be lifted if the nucleus "sees" a magnetic field either from magnetic ordering in the solid itself or from an externally applied field. The splitting of the magnetic levels is illustrated for ^{57}Fe in Fig. 15-2b. The energy of each level is given by

$$E_m = -\frac{\mu H m_I}{I} \tag{9}$$

where μ is the nuclear magnetic dipole moment, m_I is the nuclear magnetic quantum number, I is the nuclear spin quantum number, and H is the magnetic field. Six transitions, governed by the selection rule $\Delta m_I = 0, \pm 1$, produce a six-line spectrum as shown for metallic iron in Fig. 15-1c.

If a quadrupole interaction is present also, the energy levels are perturbed and the spacings between transitions become unequal as shown in Fig. 15-2b. The spectrum of α-Fe_2O_3 (Fig. 15-1d) shows this effect. The splitting of the two higher-energy peaks S_2 less the splitting of the two lower-energy peaks S_1 is a measure of the quadrupole interaction.[15,16] If the EFG has axial symmetry, this difference directly measures the quadrupole coupling constant, e^2qQ. Otherwise, the interaction is too complex to resolve without the use of single-crystal samples.

The magnetic field that the nucleus "sees" in magnetically ordered materials, such as metallic iron and α-Fe_2O_3, is usually quite large, in the order of 300–500 kOe. The peaks are several half-widths apart and the quadrupole interaction represents only a small perturbation of the energy levels. The opposite situation is encountered when a small external field, say 20 kOe, is applied to a powdered sample to find the sign of V_{zz} for a quadrupole-split pair of lines.[14] The splitting by the applied magnetic field is approximately of the same magnitude as the quadrupole splitting, and the analysis of the resulting spectrum can become complex.

2.2.4. Line Shape

The absorption lines may broaden or deviate from a Lorentzian shape for a variety of reasons. These include instrumental broadening, sample broadening, diffusion broadening, relaxation effects, and the Goldanskii–Karyagin effect. Instrumental broadening can be decreased by careful collimation of the γ-ray beam, and sample broadening can be minimized by using "thin" absorbers. The latter three can be identified by their temperature dependence.

Two relaxation effects can result in the collapse of a magnetic pattern into a distorted singlet or doublet.[17,18] Direct spin–spin relaxation is temperature independent, but produces a complicated magnetic splitting as the separation of spin species increases. The spin–lattice relaxation mechanism is independent of concentration, but is strongly temperature dependent. A spectrum magnetically split by an effective internal field will collapse into a distorted singlet as the temperature is raised through the compound's Néel point. In the presence of quadrupole interaction an asymmetric doublet appears in which the $(\pm\frac{3}{2}, \pm\frac{1}{2})$ transition exhibits a peak with a smaller height and larger half-width than the $(\pm\frac{1}{2}, \pm\frac{1}{2})$ transition. Different precessional frequencies for the transitions producing the two peaks cause this effect.[17] The peaks become more symmetrical as the temperature is raised further. The asymmetric doublet has two distinctive features: (1) although the heights of the two peaks are not the same the areas under the two peaks are equal, and (2) the peaks become more symmetrical as the temperature increases.

Asymmetry in the two peaks may also arise from anisotropy of the Debye-Waller factor. The intensities of the two transitions are dependent on the angle θ between the direction of the γ-ray beam and the principal crystallographic axis of the absorber. The intensity ratio of the two transitions, Eq. (8), is 1 when averaged over a sphere for randomly oriented powder samples. However, if the mean square displacement of the Mössbauer atom is different along the principal crystallographic axis compared to the others, the Debye-Waller factor will be anisotropic and then the intensity ratio is given by

$$\frac{A(\pm\frac{3}{2})}{A(\pm\frac{1}{2})} = \frac{\int_0^{\pi} f(\theta)(1 + \cos^2\theta)\sin\theta\, d\theta}{\int_0^{\pi} f(\theta)(\frac{5}{3} - \cos^2\theta)\sin\theta\, d\theta} \tag{10}$$

and $f\ (\theta)$ is

$$f(\theta) = \exp\left[-\left(\frac{2\pi}{\lambda}\right)^2 \langle \mathbf{x}^2 \rangle\right] \exp\left[-\frac{4\pi^2\cos^2\theta}{\lambda^2}(\langle \mathbf{z}^2 \rangle - \langle \mathbf{x}^2 \rangle)\right] \tag{11}$$

where $\mathbf{z}$ is the projection of the mean square displacement on the z axis and $\mathbf{x}$ is the projection on the x and y axis. Only when $\mathbf{z} = \mathbf{x}$ will the intensity ratio be 1. The phenomenon is known as the Goldanskii–Karyagin effect[19,20,21] and its presence may be anticipated in surface studies since it seems unlikely that the vibrational amplitudes of a surface atom would be the same in all directions. The distinguishing features of this effect are just the opposite to those for spin-lattice relaxation. The areas under the two peaks are not equivalent and the asymmetry increases with temperature.

Finally, the absorption line may be broadened by diffusion of the Mössbauer atom.[22,23] If diffusion is considered to follow a jump mechanism, the change in line width $\Delta\Gamma$ is given by

$$\Delta\Gamma = 2\hbar/\tau_0 = 12\hbar D/l^2 \quad (12)$$

where τ_0 is the mean resting time between jumps, D is the diffusion coefficient, and l is the distance between jump sites. At temperatures where τ_0 is about 280 nsec an iron line would be broadened by a theoretical linewidth.

3. EXPERIMENTAL

Mössbauer spectrometers are operated in either a constant-velocity or constant-acceleration mode.[24] The latter mode is the more popular and is available in all the commercial spectrometers. It has the advantage of being able to monitor the accumulation of a spectrum and of averaging out baseline drifts. On the other hand, the constant-velocity mode, in principle, can obtain higher precision in the data. Auxiliary equipment, for example, furnaces and cryostats, has often been obtained by modifications or adaptations from similar equipment used for other types of spectroscopy. Replacement of quartz by beryllium windows in an optical Dewar can convert it to use for a Mössbauer spectrometer. Care must be taken to avoid contamination of the window material by the nuclide under study. Iron is particularly difficult to study since it is difficult to obtain beryllium free of traces of iron. High-purity aluminum may be favored in some cases, but it should be remembered that iron is a common impurity in aluminum as well.

Reference standards have not been adopted for Mössbauer nuclides except for iron. Calibrated single crystals of sodium nitroprusside and standard iron foils are available from the National Bureau of Standards as standards for iron measurements. High-purity iron foils are recommended for velocity calibration, and, as in this text, isomer-shift data are referenced with respect to sodium nitroprusside.

3.1. Dynamics of Motion of Surface Atoms

Of the many theoretical treatments concerned with the peculiarities of the vibrational properties of surface atoms a few have been directed specifically at the effect of the surface properties on the Mössbauer spectrum.[12,25,26] Theory dictates that surface atoms should have larger mean square displacements than the corresponding bulk atoms, consequently the f factor will be smaller than the corresponding bulk value. In fact, Burton and Godwin[12] suggest a model for the surface in which the ratio of the mean square displacements of surface to bulk atoms will vary over a wide range depending on the position and number of nearest neighbors of the surface atom.

Ideally, surface studies should be carried out on single crystals. Experimentally, surface studies on single crystals are very difficult and only a few Mössbauer studies have been reported.[12,27] To achieve the necessary sensitivity for Mössbauer measurements of a monolayer or less surface coverage, Burton and Godwin[12] vacuum-evaporated ^{57}Co, the radioactive parent of iron, onto single crystals of silver and tungsten. Spectra of these sources were obtained using a standard stainless steel absorber. All of the spectra were broad, somewhat distorted, single lines which could be resolved into two or more peaks by computer fitting. Uncertainties in the resolution of the components of the observed lines made interpretation difficult, but, in general, the results supported the concept that the anisotropy in the components in the mean square displacement of the surface atom may vary markedly depending on the location of the atom either above or in the plane of the surface.

3.1.1. Polycrystalline Surfaces

The dynamics of motion of atoms on polycrystalline surfaces can be analyzed by combining measurements on the absolute value of the f factor and the Goldanskii–Karyagin effect.[28,29] Although not in closed form, Eq. (10) can be solved by numerical integration to give values of $(\langle \mathbf{z}^2 \rangle - \langle \mathbf{x}^2 \rangle)$ as a function of the ratio R. By writing Eq. (3) in the form

$$f = \exp\left[-\left(\frac{2\pi}{\lambda}\right)^2 (\langle \mathbf{z}^2 \rangle + 2\langle \mathbf{x}^2 \rangle)\right] \tag{13}$$

values for $(\langle \mathbf{z}^2 \rangle + 2\langle \mathbf{x}^2 \rangle)$ can be calculated from the measurement of the Debye–Waller factor. Then these two expressions can be solved for the mean square displacement in the z direction taken as normal to the surface and the x direction taken as parallel to the surface.

The vibrational anisotropy of tin ions on a surface has been investigated by Suzdalev *et al.*[28] by ion-exchanging Sn^{2+} onto silica gel. They identified an asymmetric doublet as a SnO species on the surface formed by the ion-exchange process. The measured ratio R was found to be 0.80. On the assumption that the z component of the EFG was positive and in a direction normal to the surface, the value for $(\langle \mathbf{z}^2 \rangle - \langle \mathbf{x}^2 \rangle)$ was found to be 1.3×10^{-18} cm^2. The amount of the SnO species was determined analytically and the Debye–Waller factor measured. From this measurement, $(\langle \mathbf{z}^2 \rangle + 2\langle \mathbf{x}^2 \rangle)$ was found to be 2.67×10^{-18} cm^2. Using these two values they found $\langle \mathbf{z}^2 \rangle = 1.76 \times 10^{-18}$ cm^2 and $\langle \mathbf{x}^2 \rangle = 0.46 \times$

10^{-18} cm^2. Thus, the rms displacement normal to the surface is about twice that parallel to the surface.

The analysis of the vibrational motion of surface atoms is based on several assumptions that do not rest on as firm a foundation as one would wish. The use of the Goldanskii–Karyagin effect rests on an assumed model of the surface site that is not established by experimental observation. The integral asymmetry may be caused also by the superposition of several similar, but not identical, doublets resulting from structural heterogeneity of the surface sites. If the temperature of the asymmetry is not large and the half-widths of the peaks are large and unequal, then the validity of the application of the Goldanskii–Karyagin effect is suspect. Furthermore, heterogeneity of the surface sites probably leads to variations in the f factor for each type of site. The measured f factor is an average value, f_{avg}, but f_{avg} does not measure necessarily an average mean square displacement for the atoms in a distribution of sites.[30] We may conclude that determination of the vibrational properties of surface atoms is not very precise unless there is good reason to believe all of the sites are structurally the same.

3.2. Surface Structure

The decreased site symmetry expected for surface atoms implies that the Mössbauer spectra of surface atoms will exhibit a number of distinctive features. In addition to the Goldanskii–Karyagin effect and a decrease in the Debye–Waller factor, changes in bond lengths and a decrease in nearest neighbors at the surface produce negative shifts in the spectra,[31] and the discontinuity of the surface causes unusually large quadrupole splittings.

A number of Mössbauer studies have been carried out on the structure of the cation site in ion-exchange materials, both organic resins[32–39] and zeolites.[40–45] Typical are the results of Delgass *et al.*[43] for the ferrous ion in Y-zeolite. The dehydrated and outgassed zeolite had a spectrum consisting of two overlapping and only partially resolved doublets. For the inner doublet, $\delta = +1.14$ mm/sec and for the outer doublet, $\delta = +1.53$ mm/sec with respect to sodium nitroprusside. The corresponding ΔE_Q values were 0.62 and 2.37 mm/sec, respectively. Based on the smaller isomer shift, the inner doublet was assumed to represent Fe^{2+} in a site with fewer nearest neighbors (distorted tetrahedral) than the outer doublet (distorted octahedral). The ion in the tetrahedral site was assigned to a position in the sodalite cage of the zeolite structure while the ion in the octahedral site was assigned to the hexagonal prism. This assignment was supported by the observation that the spectrum gradually disappears on the addition of small polar molecules, e.g., water and ammonia, to the sample. However, larger molecules, e.g., ethanol and piperidine, affect only the inner peaks, which represent ions in the more accessible position, the large sodalite cage. This "solvation" effect on ferrous ions, in which the spectrum vanishes on addition of an adsorbate, has been generally observed in all the Mössbauer studies of ion-exchange materials.

3.2.1. Chemisorption on Supported Iron Oxides

High dispersions of iron on inert materials offer another type of sample for surface investigations. The function of the inert material is to prevent sintering of the deposited catalyst and maintain a high surface area of active material. High-surface-area silica gel and alumina have been found not to be as inert as previously supposed for iron deposits.[46] If 1–5 wt % iron-on-silica gel samples are made by impregnating the silica gel with a ferric nitrate solution, the deposit, after drying and calcining, consists of small clumps, or microcrystallites, of a ferric oxide or silicate.[47] The typical Mössbauer spectrum of such samples after outgassing consists of a doublet with line widths ten times the natural line width, an isomer shift of about +0.6 mm/sec and a quadrupole splitting between 1.8 and 2.2 mm/sec. The unusually large line widths suggest a heterogeneity of ferric-ion sites which produce similar, but not identical, unresolved doublets. The isomer shift is characteristic of ferric ions in an oxide-ion environment. The quadrupole splitting is quite large for ferric ions and is indicative of a highly distorted electric field surrounding a surface ion. The variation in the quadrupole splitting from sample to sample is a qualitative measure of variations in the degree of dispersion of the deposits. A high degree of dispersion, and consequently a large fraction of ferric ions on the surface of the deposits, has been confirmed by ammonia chemisorption. The addition of sufficient NH_3 to form a monolayer coverage on the ferric-ion sites reduces the quadrupole splitting to 1.2–1.4 mm/sec. The NH_3, an additional nearest-neighbor, reduces the distortion of the EFG caused by the discontinuity of the surface. The process is completely reversible. The NH_3 can be pumped off at elevated temperatures and the original quadrupole splitting restored.

The structure of the ferric oxide deposit remains obscure. It cannot be simply microcrystalline particles of α-Fe_2O_3 sitting on the silica-gel surface since reduction by hydrogen produces only ferrous species and no metallic iron. Even after treatment in flowing hydrogen at 600°C for 72 hr, the Mössbauer spectrum of the final state consists entirely of two partially resolved doublets[48] as shown in Spectrum (a) of Fig. 15-3. Peak 1 is the sum of the two low-energy peaks of the two doublets.

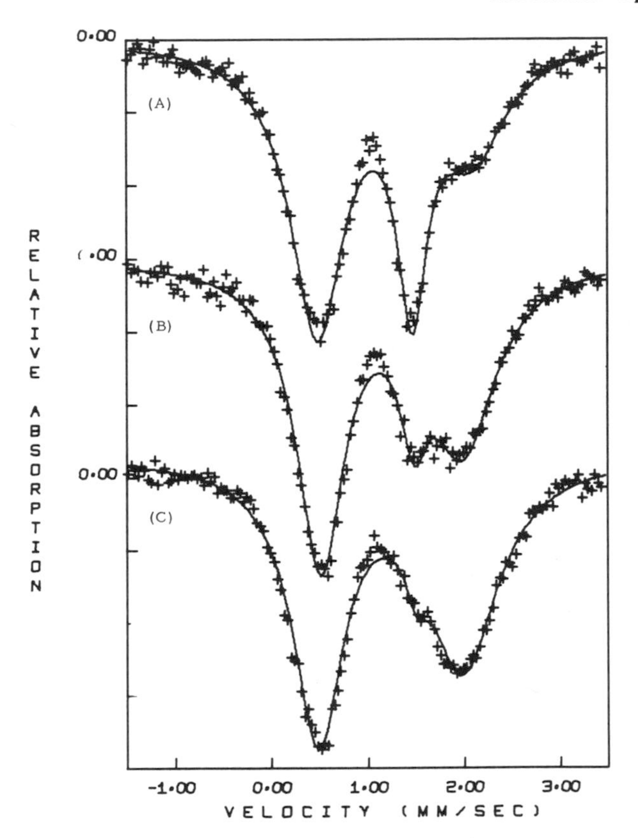

Fig. 15-3. Effect of ammonia chemisorption on the Mössbauer spectrum of a highly dispersed iron-on-silica gel in its ferrous state. (a) 0.5×10^{-2} mmole NH_3; (b) 2.99×10^{-2} mmole NH_3; (c) 4.31×10^{-2} mmole NH_3 [from Hobson and Gager[48]].

Peaks 1 and 2 make up one doublet with $\delta = +1.02$ mm/sec and $\Delta E_Q = 0.98$ mm/sec; peaks 1 and 3 make up the other doublet with $\delta = +1.39$ mm/sec and $\Delta E_Q = 1.65$ mm/sec. The two doublets may be assigned to high-spin ferrous ions with two different site symmetries. The doublet with the smaller isomer shift and quadrupole splitting presumably is a surface ion while the other doublet represents ferrous ions in the interior of the deposit. Since the iron is not reduced any further than the ferrous state, even under rather drastic conditions, reaction with the silica-gel surface must have taken place. Regardless of the exact nature of the deposit, a large fraction, 50 % or more, of the iron atoms are exposed ions on the surface of the deposit, since the area under the surface doublet is 50 % or more of the total area under the absorption curve.

The addition of ammonia to the reduced and outgassed sample has a marked effect on peak 2 of the spectrum as shown in Fig. 15-3. Peak 2 decreases in relative area as small doses of NH_3 are added. Since peak 2 has been assigned as half of a doublet produced by surface ferrous ions, the relative area of peak 2 is used as a measure of surface sites available for chemisorption. Surface coverage θ is given by $[1 - S(NH_3)/S(0)]$ where $S(0)$ is the initial relative area under peak 2 and $S(NH_3)$ the area after the addition of a given amount of ammonia. The results are plotted in Fig. 15-4. There is no noticeable change in the spectrum with the addition of the first small amounts of adsorbate. However, the relative area under peak 2 begins to decrease with the further addition of ammonia, starting at approximately 2.0×10^{-2} mmole of added adsorbate. The decrease in the area is rapid at first and then slowly levels off to a minimum value of 0.02 at about 5.0×10^{-2} mmole of added adsorbate.

Desorption of NH_3 at room temperature does not cause any changes in the spectrum. On raising the temperature of the sample to 100°C, NH_3 begins to desorb and peak 2 begins to increase in size. Increments of NH_3 continue to desorb as the temperature is raised in steps until the original spectrum is recovered at a desorption temperature

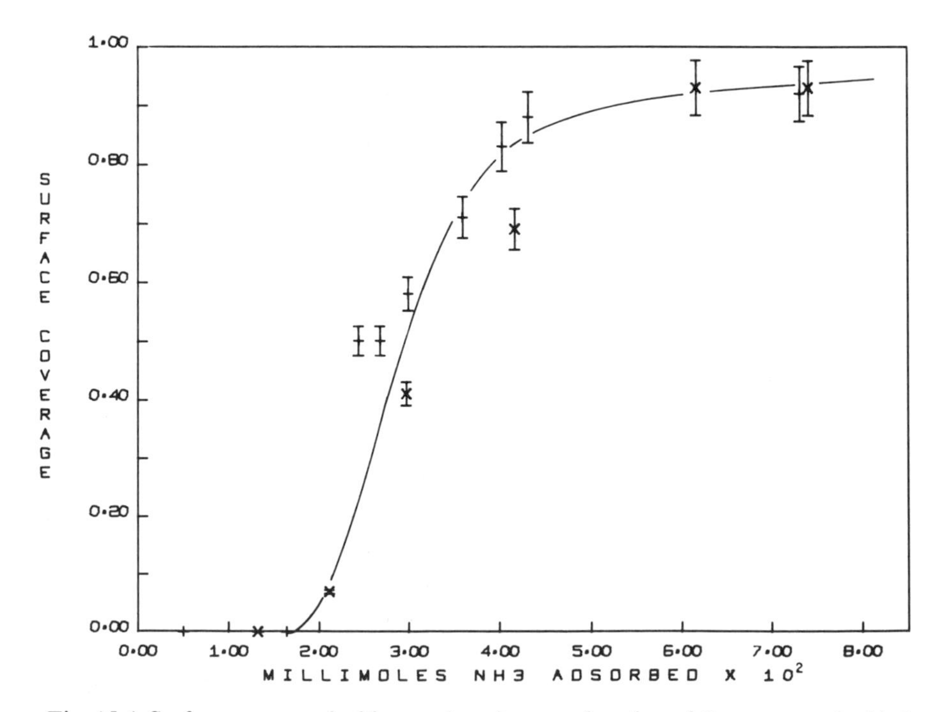

Fig. 15-4. Surface coverage θ of ferrous-ion sites as a function of the amount of added ammonia.

of 300°C. At this temperature of desorption, the ammonia mass balance is not complete. Some of the NH_3 remains strongly bound to surface sites on the silica gel. From Fig. 15-4 it is seen that the amount of NH_3 that forms a monolayer coverage of the ferrous-ion surface sites is approximately 3×10^{-2} mmole, or just enough to form a 1:1 surface complex with the ferrous species.

Since the area under peak 2 does not begin to decrease until a small amount of NH_3 has been added, and fully recovers before all of the NH_3 has desorbed, the initial increment of NH_3 must be strongly adsorbed on silica sites. Similar results have been obtained in infrared studies of ammonia adsorption on porous glass.[(49)] The addition of NH_3 had no effect on the infrared band of surface silanol groups until enough NH_3 had been added to cover approximately 1 % of the surface.

A difference in behavior may be noted between the ferric state and the ferrous state on chemisorption of polar molecules. The quadrupole splitting of the outgassed Fe^{3+}–SiO_2 state decreases on addition of NH_3 while the quadrupole splitting of the Fe^{2+}–SiO_2 surface species increases on addition of NH_3. This difference may be explained by the factors that produce the EFG in these two species. The five electrons in the 3*d* level of a high-spin ferric ion are symmetrically distributed about the nucleus, and the EFG is produced primarily by the ligands surrounding the ion. The addition of NH_3 to a Fe^{3+} site increases the symmetry of the EFG, and the quadrupole splitting decreases. On the other hand, the EFG at the ferrous-ion site consists of contributions from the ligands and from the additional 3*d* electron. According to Ingalls,[(50)] the latter is usually quite large compared to the ligand, or lattice, contribution and of opposite sign. In his analysis, q in Eq. (7) may be written

$$q = (1 - R)q_{\text{val}} - (1 - \gamma_{\infty})q_{\text{lat}}$$

where q_{val} is the z component of the EFG produced by the aspherical 3*d* electron, q_{lat} is the z component produced by the ligands, and the coefficients are the appropriate Sternheimer antishielding factors. With a very large lattice contribution from the surface site, the two contributions sum to a relatively small quadrupole splitting for the surface Fe^{2+} ion. Again, the addition of NH_3 increases the symmetry at the surface site, but this only decreases the contribution to the EFG from the ligands. The sum of the 3*d*-electron and ligand contributions increases and, consequently, the quadrupole splitting increases.

The Mössbauer spectra of the ferrous Y-zeolite and the reduced silica-gel samples are surprisingly similar in several ways. The inner doublet of the Y-zeolite and the surface doublet of the silica gel have the smaller isomer shifts and quadrupole splittings. Both are the first to be affected by adsorption of polar molecules. But there the similarity ends. The Mössbauer spectrum gradually vanishes with the addition of excess amounts of water or ammonia to the Y-zeolite. The ferrous ions are apparently "solvated" and the binding to the crystalline lattice becomes weak. This "solvation" effect is not observed for the ferrous species on silica gel.

3.3. Microcrystalline and Thin-Film Properties

The properties of microcrystalline catalysts, corrosion products on surfaces of materials of construction, and other thin films are dependent to some degree on surface properties and surface reactions. Mössbauer studies of thin films have concentrated mostly on magnetic properties as a function of film thickness.[51–53] The identification of corrosion products has been the main concern of Mössbauer studies in this area.[54–57] However, nucleation and growth of microcrystalline material are of great interest both in corrosion and catalyst formulation. The specific rate and selectivity of some catalytic reactions are highly dependent on the size of the microcrystalline catalyst. This dependence appears to be related to the concentration of certain surface structures that vary with crystallite size, e.g., the density of surface sites with five nearest neighbors has been shown to go through a maximum as the size of the crystallite increases.[58]

3.3.1. Measurement of Microcrystalline Size

Of the several methods investigated for the determination of crystallite size from Mössbauer parameters,[59–61] the only reliable one at present is based on the observation of the transition of a magnetically ordered structure to its superparamagnetic state as a function of temperature. According to theory,[62,63] the relaxation of the magnetic moment of ferromagnetic and antiferromagnetic particles the size of a single magnetic domain is given by

$$\tau = \frac{1}{f} \exp\left(\frac{Kv}{kT}\right) \tag{14}$$

where T is the temperature, k is Boltzmann's constant, and f is usually set equal to the gyromagnetic precessional frequency of the magnetization vector about the effective field. The relaxation time τ is the average time between spontaneous changes in the direction of the magnetic moment in a single domain crystallite. The energy barrier to this change, Kv, contains the uniaxial, magnetocrystalline anisotropy constant K and the crystallite volume v. If the structure of the crystallite has less than uniaxial symmetry, the expression for K becomes more complex. With decreasing crystallite size at a given temperature, the value of the energy barrier approaches that of thermal energy, and τ becomes small. Kündig *et al.*[64–67] have applied this theory to the measurement of crystallite size of α-Fe_2O_3 and other material.

In a single-domain particle of α-Fe_2O_3, the magnetization vector is held in the c plane perpendicular to the c axis by the magnetocrystalline field. In Mössbauer effect studies the measuring device or "observer" is the ^{57}Fe nucleus. Therefore, the six-line Mössbauer spectrum, Fig. 15-1d, collapses into a doublet when the relaxation time for the spontaneous change in the magnetization vector is shorter than the period for the precession of the nuclear spin about the direction of the effective magnetic field. Kündig *et al.*[64,65] calculated a Larmor frequency for the ^{57}Fe nucleus in α-Fe_2O_3 of 4×10^7 sec^{-1}, or an "observer" relaxation time of 2.5×10^{-8} sec. On substituting this and an expression for the frequency factor f proportional to the specific volume and anisotropy constant of the oxide into Eq. (14), they obtained

$$\ln(2 \times 10^{-4}K) = \frac{Kv}{kT} \tag{15}$$

If the crystallites were all the same size, the six-line pattern would collapse completely to a doublet at the appropriate transition temperature. The slow decrease in the superparamagnetic doublet with decreasing temperature (Fig.15-5) simply means that there is a distribution of crystallite sizes. For a random distribution, an average crystallite size is calculated at the temperature where 50% of the crystallites are in the superparamagnetic state. The anisotropy constant K may be evaluated by obtaining spectra at constant temperature of a series of samples having known average particle sizes, or from spectra as a function of temperature of a sample having a known average particle size.

Once the value for K has been established, the average crystallite size for an unknown sample can be determined by plotting the superparamagnetic fraction *vs.* the temperature as shown by curve a in Fig. 15-6. The data for this plot were calculated from the spectra in Fig. 15-5, and the temperature at which 50% of the sample is superparamagnetic is 304 ± 10°K. On substituting in Eq. (15) the value 7.6×10^4 ergs/cm^3 for the anisotropy constant of α-Fe_2O_3 obtained by Kündig *et al.*,[65] the average diameter of the α-Fe_2O_3 crystallites in this sample is found to be 140 ± 10 Å. An independent determination by x-ray line broadening gave a comparable figure of 158 ± 10 Å. This method complements the x-ray line-broadening technique and should be useful for the determination of oxide crystallite sizes below the useful range of x-ray line broadening.

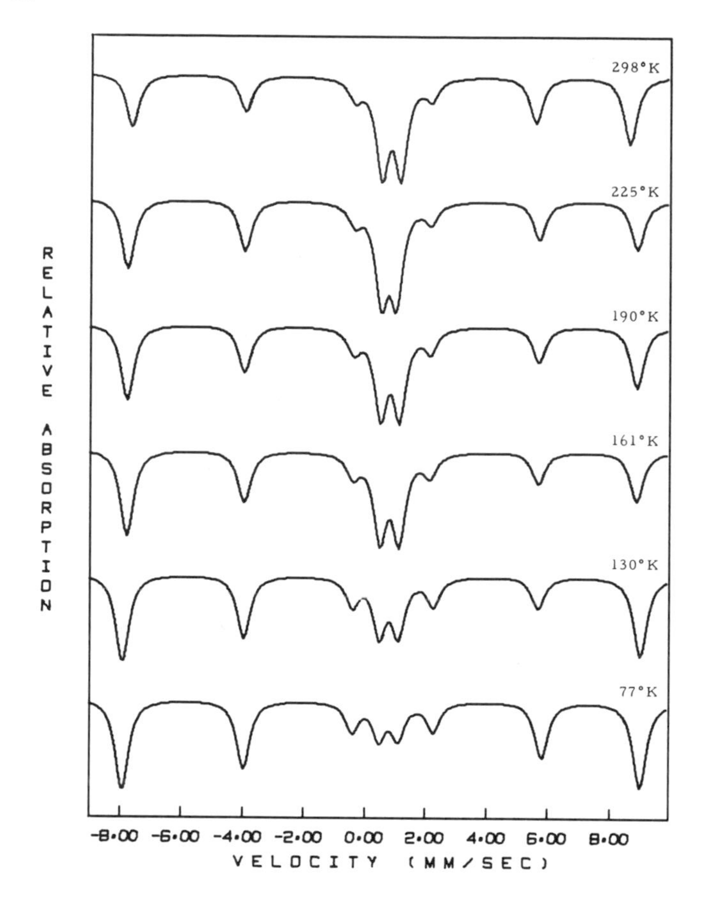

Fig. 15-5. Mössbauer spectra of microcrystalline α-Fe_2O_3 supported on silica gel as a function of sample temperature. (From Hobson and Gager.[46])

This method of crystallite size measurements must be applied with caution in studies on catalysts. In an investigation of the effect of typical pretreatments and reaction conditions on microcrystalline α-Fe_2O_3 supported by silica gel and alumina, Hobson and Gager[46] found that the α-Fe_2O_3 reacted to some extent with the supposedly "inert" support materials. The sample used for Fig. 15-5 was prepared by thermally decomposing a ferric nitrate-impregnated silica gel and calcining it at 500°C for 2 hr. Since curve a in Fig. 15-6 extrapolates to 0% superparamagnetic fraction at or above 0°K, almost all of the deposit must be sitting on the silica surface as α-Fe_2O_3 of varying crystallite size. However, calcining for longer periods of time, say 18 hr, increases the relative area under the doublet from about 45% initially to 70% in spectra recorded at room temperature. On decreasing the temperature to 77°K, the super-

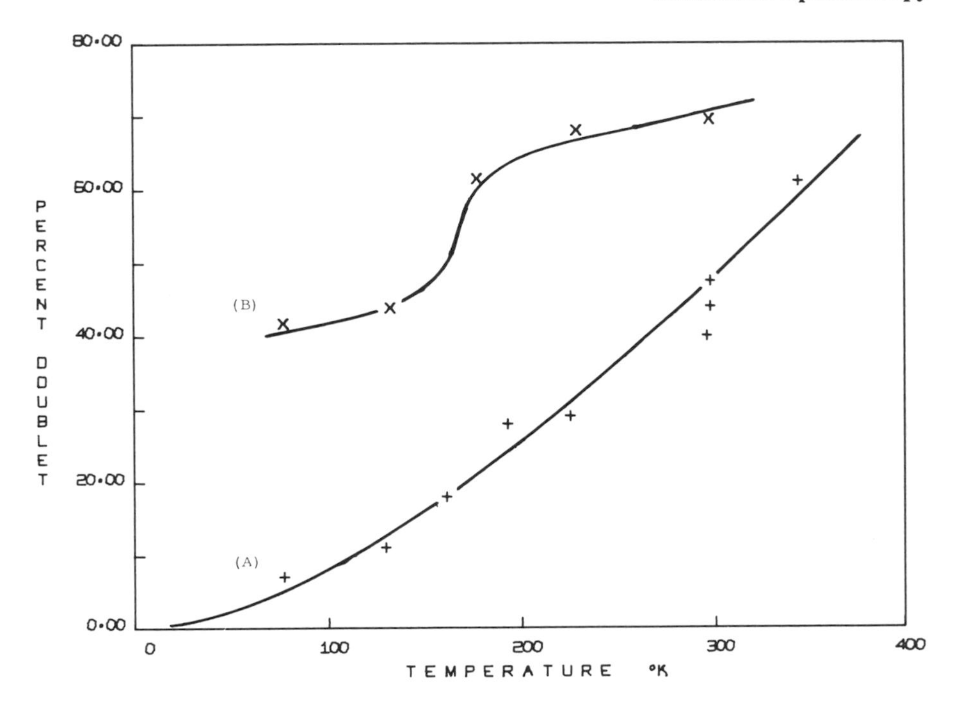

Fig. 15-6. The percentage of α-Fe_2O_3 in the superparamagnetic state as a function of temperature. (a) 12% α-Fe_2O_3 on silica gel calcined 2 hr at 500°C, (b) 17% α-Fe_2O_3 on silica gel calcined 18 hr at 500°C.

paramagnetic fraction does not parallel curve a in Fig. 15-6, but follows curve b and tends to level out at the low temperatures. Part of the contribution to the doublet must be from some new species, probably a product from the reaction of the ferric oxide with the support. The deposit now consists of antiferromagnetic crystallites of α-Fe_2O_3, superparamagnetic crystallites of α-Fe_2O_3, and an unidentified ferric species. When alumina is used as the support, similar results are obtained on prolonged calcinations. Irreversible increases in the area under the doublet are also observed following reduction and reoxidation of the samples at 500–600°C.

X-ray diffraction patterns of such samples do not indicate the presence of any compound other than α-Fe_2O_3. These observations suggest an explanation for the reason catalyst preparations are so often difficult to reproduce. The physical methods used to characterize the catalyst simply do not measure all of the important species on the surface.

4. SUMMARY

The application of the Mössbauer effect to surface studies has produced several significant contributions to the field. A new technique for the measurement of crystallite size and size distribution has been developed around the observation of the superparamagnetic transition in magnetically ordered materials. Strong evidence for the interaction of iron oxide deposits with "inert" support materials has been obtained from studies of the superparamagnetic transition of α-Fe_2O_3 and the resistance to reduction of iron oxides by hydrogen as a function of typical pretreatment conditions of catalyst preparations. The observations of interaction with support materials have also shown the presence of species not previously identified by other experimental techniques. Chemisorption studies have revealed the formation of surface complexes with certain similarities to transition metal complexes in solution. Finally, investigations of the dynamics of motion of surface atoms have shown that cations on the surface of silica gel are located in sites in which their displacement normal to the surface is greater than parallel to the surface.

Future prospects for surface studies look very encouraging despite the limitation of the technique to only a few of the elements. Applications may be anticipated in the growing area of ferrite catalysts and in corrosion studies where some of the newest innovations in instrumentation should find extensive use.

5. ACKNOWLEDGMENT

The author gratefully acknowledges the many contributions by Dr. H. M. Gager to surface studies in this laboratory and her successful negotiations with a computer for the drawings contained herein.

6. REFERENCES

1. R. L. Mössbauer, Kernresonanzfluoreszenz von gammastrahlung in Ir^{191}, *Z. Phys.* **151**, 124–143 (1958).
2. T. Bonchev, I. Vassilev, T. Sapundzhiev, and M. Evtimov, Possibility of investigating movement in a group of ants by the Mössbauer effect, *Nature* **217**, 96–98 (1968).
3. J. Danon, *Lectures on the Mössbauer Effect*, Gordon and Breach, Science Publishers, Inc., New York (1968).
4. H. Frauenfelder, *The Mössbauer Effect*, W. A. Benjamin, New York (1962).
5. V. I. Goldanskii and R. H. Herber (eds.), *Chemical Applications of Mössbauer Spectroscopy*, Academic Press, New York (1968).
6. G. K. Wertheim, *Mössbauer Effect: Principles and Applications*, Academic Press, New York (1964).
7. M. J. D. Low, in *The Gas–Solid Interface* (E. A. Flood, ed.) Vol. 2, pp. 947–974, Marcel Dekker, Inc., New York (1967).
8. W. N. Delgass and M. Boudart, Application of Mössbauer spectroscopy to the study of adsorption and catalysis, *Catal. Rev.* **2**, 129–160 (1968).
9. M. C. Hobson, Jr., Surface studies by Mössbauer spectroscopy, *Adv. Colloid Interface Sci.* **3**, 1–43 (1971).

10. I. V. Goldanskii and I. P. Suzdalev, in *Proceedings of the Conference on Applications of the Mössbauer Effect*, pp. 269–305, Tihany (1969), Akadémai Kiado, Budapest, 1971.
11. K. G. Malmfors, in *Beta and Gamma Ray Spectroscopy* (K. Siegbahn, ed.) Vol. 2, pp. 1281–1292, North-Holland, Amsterdam (1965).
12. J. W. Burton and R. P. Godwin, Mössbauer effect in surface studies: Fe-57 on W and Ag, *Phys. Rev.* **158**, 218–224 (1967).
13. R. L. Collins, Mössbauer studies of iron organometallic complexes. IV. Sign of the electric field gradient in ferrocene, *J. Chem. Phys.* **42**, 1072–1080 (1965).
14. R. L. Collins and J. C. Travis, in *Mössbauer Effect Methodology* (I. J. Gruverman, ed.) Vol. 3, pp. 123–161, Plenum Press, New York (1967).
15. Y. Bando, M. Kiyama, N. Yamamoto, T. Takada, T. Shinjo, and H. Takaki, The magnetic properties of α-Fe_2O_3 fine particles, *J. Phys. Soc. Japan* **20**, 2086 (1965).
16. T. Zemcik, Mössbauer six-line spectra positions analysis for Fe^{57} in metallic iron, *Czech. J. Phys. B* **18**, 551–566 (1968).
17. M. Blume, Magnetic relaxation and asymmetric quadrupole doublets in the Mössbauer effect, *Phys. Rev. Letters* **14**, 96–98 (1965).
18. H. H. Wickman, in *Mössbauer Effect Methodology* (I. J. Gruverman, ed.) Vol. 2, pp. 39–67, Plenum Press, New York (1966).
19. S. V. Karyagin, A possible cause for the doublet component asymmetry in the Mössbauer absorption spectrum of some powdered tin compounds, *Dokl. Akad. Nauk SSSR* **148**, 1102–1105 (1963).
20. V. I. Goldanskii, E. F. Makarov, and V. V. Khrapov, On the difference in two peaks of quadrupole splitting in Mössbauer spectra, *Phys. Letters* **3**, 344–346 (1963).
21. V. I. Goldanskii, G. M. Gorodinskii, S. V. Karyagin, L. A. Korytkeo, L. M. Krizhanskii, E. F. Makarov, I. P. Suzdalev, and V. V. Khrapov, The Mössbauer effect in tin compounds, *Dokl. Akad. Nauk SSSR* **147**, 127–130 (1962).
22. K. S. Singwi and A. Sjolander, Resonance absorption of nuclear γ-rays and the dynamics of atomic motion, *Phys. Rev.* **120**, 1093–1102 (1960).
23. N. N. Greenwood, Applications of Mössbauer spectroscopy to problems in solid-state chemistry, *Angew. Chem.* (*Intl. ed.*) **10**, 716–724 (1971).
24. J. J. Spijkerman, in *An Introduction to Mössbauer Spectroscopy* (L. May, ed.) pp. 23–43, Plenum Press, New York (1971).
25. A. A. Maradudin and J. Melngailis, Some dynamical properties of surface atoms, *Phys. Rev.* **133A**, 1188–1193 (1964).
26. A. A. Maradudin, in *Solid State Physics* (F. Seitz and D. Turnbull, eds.) Vol. 19, pp. 1–134, Academic Press, New York (1966).
27. F. G. Allen, Mössbauer effect from ^{57}Co on a clean silicon surface, *Bull. Am. Phys. Soc.* **9**, 296 (1964).
28. I. P. Suzdalev, V. I. Goldanskii, E. F. Makarov, A. S. Plachinda, and L. A. Korytko, An investigation of the dynamics of motion of tin atoms on a silica gel surface by means of the Mössbauer effect, *Sov. Phys.—JETP* **22**, 979–983 (1966).
29. P. A. Flinn, S. L. Ruby, and W. L. Kehl, Mössbauer effect for surface atoms: Iron–57 at the surface of η-Al_2O_3, *Science* **143**, 1434–1436 (1964).
30. R. M. Housley and F. Hess, Analysis of Debye–Waller factor and Mössbauer-thermal-shift measurements. I. General theory, *Phys. Rev.* **146**, 517–526 (1966).
31. N. E. Erickson, in *Advances in Chemistry Series* (R. F. Gould, ed.) Vol. 68, pp. 86–104, American Chemical Society, Washington (1967).
32. J. L. Mackey and R. L. Collins, The Mössbauer effect of iron in ion exchange resins, *J. Inorg. Nucl. Chem.* **29**, 655–660 (1967).

33. I. P. Suzdalev, A. S. Plachinda, E. F. Makarov, and V. A. Dolgopolov, Study of ion-exchange resins by gamma-resonance spectroscopy (Mössbauer effect), *Russ. J. Phys. Chem.* **41**, 1522–1526 (1967).
34. V. I. Goldanskii, I. P. Suzdalev, A. S. Plachinda, and V. P. Korneev, Hyperfine structure of iron (3+) Mössbauer spectra in an ion exchange sulfo resin with varying degrees of dehydration, *Dokl. Akad. Nauk SSSR* **185**, 203–205 (1969).
35. A. Johansson, Mössbauer spectra of ^{57}Fe in ion-exchange resins, *J. Inorg. Nucl. Chem.* **31**, 3273–3285 (1969).
36. I. P. Suzdalev, A. M. Afanasiev, A. S. Plachinda, V. I. Goldanskii, and E. F. Makarov, Spin lattice relaxation studied from the hyperfine structure of iron (3+) ion Mössbauer spectra, *Sov. Phys.—JETP* **28**, 923–930 (1969).
37. I. P. Suzdalev, V. P. Korneev, and Yu. F. Krupiansky, in *Proceedings of the Conference on Applications of the Mössbauer Effect*, pp. 148–151, Tihany (1969), Akadémai Kiado, Budapest, 1971.
38. A. S. Plachinda, I. P. Suzdalev, V. I. Goldanskii, and I. E. Neimark, Mechanism of the interaction of water molecules with iron ions in ion-exchange resins studied by Mössbauer spectroscopy, *Teor. Eksp. Khim.* **6**, 347–352 (1970).
39. G. Pfrepper, K. Hennig, and S. Usmanowa, Untersuchungen zur bindung von Fe (III)—Ionen in starksauren kationenaustauschern mit hilfe der Mössbauerspectroskopie, *Z. Phys. Chem.* **244**, 113–116 (1970).
40. V. I. Goldanskii, I. P. Suzdalev, A. S. Plachinda, and L. G. Shtyrkov, Investigation of the structure and adsorption properties of zeolites by the nuclear γ-resonance method, *Dokl. Akad. Nauk SSSR* **169**, 511–514 (1966).
41. J. Morice and L. V. C. Rees, Mössbauer studies of Fe-57 in zeolites, *Trans. Faraday Soc.* **64**, 1388–1395 (1968).
42. W. N. Delgass, R. L. Garten, and M. Boudart, Mössbauer effect of exchangeable ferrous ions in Y-zeolite and Dowex 50 resin, *J. Chem. Phys.* **50**, 4603–4606 (1969).
43. W. N. Delgass, R. L. Garten, and M. Boudart, Dehydration and Adsorbate interactions of Fe-Y zeolite by Mössbauer spectroscopy, *J. Phys. Chem.* **73**, 2970–2979 (1969).
44. R. W. J. Wedd, B. V. Liengme, J. C. Scott, and J. R. Sams, Mössbauer investigation of iron species in a zeolite, *Solid State Commun.* **7**, 1091–1093 (1969).
45. R. L. Garten, W. N. Delgass, and M. Boudart, A. Mössbauer spectroscopic study of the reversible oxidation of ferrous ions in Y-zeolite, *J. Catal.* **18**, 90–107 (1970).
46. M. C. Hobson, Jr. and H. M. Gager, A Mössbauer effect study on crystallites of supported ferric oxide, *J. Catal.* **16**, 254–263 (1970).
47. M. C. Hobson, Jr. and A. D. Campbell, Mössbauer effect spectra of a supported iron catalyst, *J. Catal.* **8**, 294–298 (1967).
48. M. C. Hobson, Jr. and H. M. Gager, Mössbauer effect studies of chemisorption. Titration of surfaces of iron catalysts with polar molecules, *J. Colloid Interface Sci.* **34**, 357–364 (1970).
49. M. Folman and D. J. C. Yates, Infrared studies of physically adsorbed polar molecules and of the surface of a silica adsorbent containing hydroxyl groups, *J. Phys. Chem.* **63**, 183–187 (1959).
50. R. Ingalls, Electric field gradient tensor in ferrous compounds, *Phys. Rev.* **133A**, 787–795 (1964).
51. K. Rabinovitch and H. Shechter, Mössbauer measurements on magnetic anisotropy in thin films of iron, *J. Appl. Phys.* **39**, 2464–2466 (1968).
52. A. C. Zuppero and R. W. Hoffman, Mössbauer spectra of monolayer iron films, *J. Vac. Sci. Technol.* **7**, 118–121 (1970).

53. W. Zinn, Mössbauer effect studies on magnetic thin films, *Czech. J. Phys.* **B21**, 391–406 (1971).
54. D. D. Joye and R. C. Axtmann, Quantitative analysis for corrosion studies by the Mössbauer effect, *Anal. Chem.* **40**, 876–878 (1968).
55. J. H. Terrell and J. J. Spijkerman, Determination of surface compound formation by backscatter Mössbauer spectroscopy, *Appl. Phys. Letters* **13**, 11–13 (1968).
56. A. M. Pritchard and C. M. Dobson, Mössbauer effect and iron corrosion kinetics, *Nature* **224**, 1295 (1969).
57. A. M. Pritchard, J. R. Haddon, and G. N. Walton, Study of some products of the corrosion of iron under hydrothermal conditions using the Mössbauer effect, *Corros. Sci.* **11**, 11–23 (1971).
58. R. Van Hardeveld and A. Van Montfoort, The influence of crystallite size on the adsorption of molecular nitrogen on nickel, palladium and platinum. Infrared and electron-microscopic study, *Surface Sci.* **4**, 396–430 (1966).
59. N. Yamamoto, The shift of the spin flip temperature of α-Fe_2O_3 fine particles, *J. Phys. Soc. Japan* **24**, 23–28 (1968).
60. D. Schroeer and R. C. Ninninger, Jr., Morin transition in α-Fe_2O_3 microcrystals, *Phys. Rev. Letters* **19**, 632–634 (1967).
61. D. Schroeer, in *Mössbauer Effect Methodology* (I. J. Gruverman, ed.) Vol. 5, pp. 141–162, Plenum Press, New York (1970).
62. L. Néel, Influence des fluctuations thermiques sur l'aimantation des grains ferromagnetique trés fins, *Compt. Rendu.* **228**, 664–666 (1949).
63. C. P. Bean and J. D. Livingston, Superparamagnetism, *J. Appl. Phys.* **30**, 120S–129S (1959).
64. W. Kündig, H. Bömmel, G. Constabaris, and R. H. Lindquist, Some properties of supported small α-Fe_2O_3 particles determined with the Mössbauer effect, *Phys. Rev.* **142**, 327–333 (1966).
65. W. Kündig, K. J. Ando, R. H. Lindquist, and G. Constabaris, Mössbauer studies of ultrafine particles of NiO and α-Fe_2O_3, *Czech. J. Phys.* **B17**, 467–473 (1967).
66. K. J. Ando, W. Kündig, G. Constabaris, and R. H. Lindquist, Mössbauer effect of Fe-57 in ultrafine particles and bulk NiO, *J. Phys. Chem. Solids* **28**, 2291–2295 (1967).
67. W. Kündig, M. Kobelt, H. Appel, G. Constabaris, and R. H. Lindquist, Mössbauer studies of Co_3O_4; Bulk material and ultrafine particles, *J. Phys. Chem. Solids* **30**, 819–826 (1969).

16

RUTHERFORD SCATTERING

W. D. Mackintosh

Atomic Energy of Canada, Limited
Chalk River Nuclear Laboratories
Chalk River, Ontario, Canada

1. INTRODUCTION

By 1911, Rutherford and his co-workers had described the elastic scattering of energetic ions, yet it was not until 1957 that Rubin, Passell, and Bailey[1] exploited this nuclear interaction as an analytical tool. Until 1966, only a few applications were found for this method since it is not generally useful in identifying and determining impurities scattered throughout the bulk of a sample. Since that time, however, three factors have promoted its use: (1) an increasing need, particularly in the field of solid-state devices for the determination of foreign atoms concentrated on or within a few microns of the surface of an otherwise pure material, (2) a realization of the potential for obtaining concentration changes both with depth and across surfaces, and (3) the facility with which elastic scattering can be combined with channeling (see Section 2.5) to reveal the location of foreign atoms in the unit cell of single crystals.

2. PRINCIPLES

When a beam of mono-energetic ions strikes a surface, some will be elastically scattered, losing energy in an amount related to the mass of the atom struck. Furthermore, the number of particles so scattered is a function of the number of atoms present. Hence by making a sample the

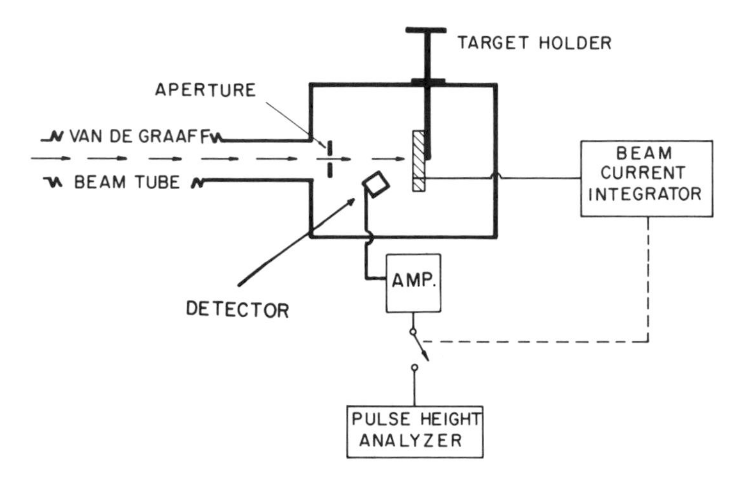

Fig. 16-1. Typical target-chamber arrangements for Rutherford-scattering analysis.

target of the particle beam in an accelerator and by providing means for measuring the energy and counting the number of scattered particles, the identity and number of atoms in the solid can be determined. An example of such an arrangement is illustrated in Fig. 16-1, where energy and number of counts are considered to be obtained and recorded by a solid-state detector and associated electronic circuitry.

2.1. The Energy of Backscattered Ions

The energy E_1 of a particle scattered from a surface atom of the target is given by the equation

$$E_1 = k^2 E_0 \tag{1}$$

where

$$k = \frac{M_1 \cos \theta_s}{M_1 + M_2} + \left[\left(\frac{M_1 \cos \theta_s}{M_1 + M_2}\right)^2 + \frac{M_2 - M_1}{M_1 + M_2}\right]^{1/2}$$

E_0 is the incident energy, M_1 is the mass of the incident ion, M_2 is the mass of the struck atom, and θ_s is the laboratory scattering angle, i.e., the angle between the beam and detector measured in the forward direction.

If an incident particle is not scattered by the surface atoms, it will lose energy by ionization and excitation processes as it penetrates to lower layers. Thus just before being scattered, its energy will have been reduced to E_0'. The scattered particle will also be slowed down on its way back to the surface and will emerge with the further-reduced energy E_1'. If the rate of energy loss by the slowing-down process is given by the stopping power S, then the energy of a particle emerging from a depth x below the surface is expressed by

$$\left(E_0 - \int_0^{l_1 = x \sec\theta_1} S(E)\, dl\right) k^2 - \int_{l_2 = x \sec\theta_2}^{0} S(E)\, dl \qquad (2)$$

where the first and second integrals are the energy losses along the incoming and outgoing trajectories l_1 and l_2. The values of the stopping power depend on the energy of the particle E and the composition of target material. The angles θ_1 and θ_2 are between the normal to the surface and the ingoing and outgoing trajectories. These geometric relationships are illustrated in Fig. 16-2.

Some features of an energy spectrum of elastically scattered particles can be understood by reference to these equations. Figure 16-3 contains a typical spectrum. The sample, silicon covered with 0.5 μg/cm^2 of gold, was bombarded with He^+ at an energy E_0 of 1000 keV. The peak to the right was derived from the gold. Its position (at peak-height maximum) on the energy scale E_1 is given by Eq. (1), where M_1 and M_2 are the He and Au masses, respectively, and θ_s is 150°. The width of the peak, in this instance where the surface film is thin, is entirely due to the finite resolution of the detector. The thick silicon gives rise to the sloping

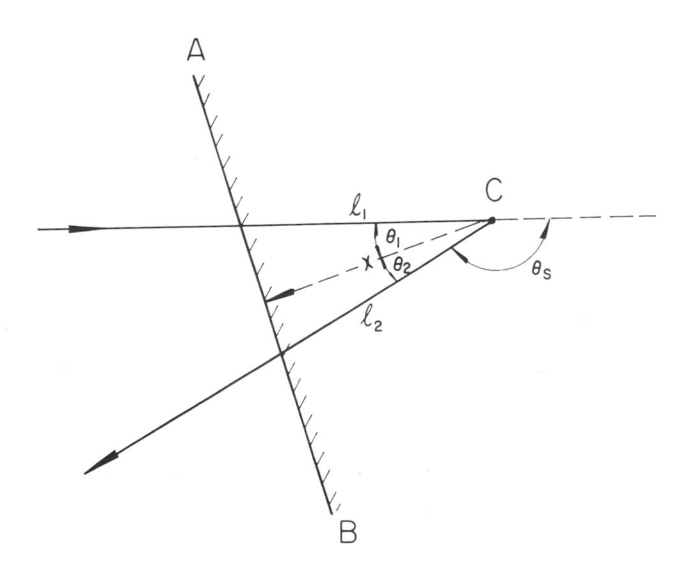

Fig. 16-2. The geometrical relationship of trajectories, angles, and depth, where AB is the surface plane, C is the point of impact at depth x, l_1 and l_2 are the ingoing and outgoing trajectories which make angles θ_1 and θ_2, respectively, with the normal to the surface plane, and θ_s is the scattering angle.

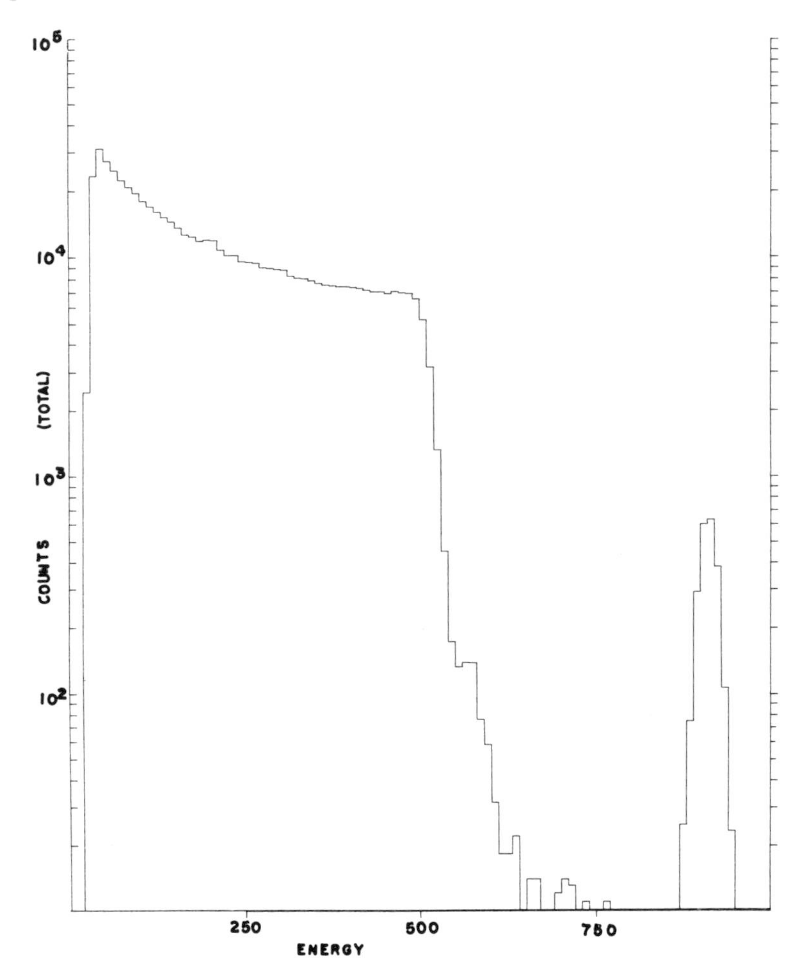

Fig. 16-3. The Rutherford-scattering spectrum of a thin gold film on the surface of a silicon substrate.

plateau to the left. The sharp edge terminating the plateau represents scattering from the surface layers; its slope depends on detector resolution. Its position, taken to be the point at half-height of the plateau, is given by Eq. (1), where M_2 is now the mass of the silicon. Counts recorded at lower energies are derived from particles scattered from progressively greater depths. As x increases, the energy E_1' decreases according to Eq. (2).

Some facts relevant to the use of Rutherford scattering for analysis can be inferred from Eqs. (1) and (2).

1. A surface contaminant can only be identified and determined if its atomic mass is greater than all components of the substrate.

If the mass is less, the peak is superimposed on the plateau and can rarely be properly resolved. As will be seen later, this does not necessarily apply to samples which are single crystals where channeling techniques may be employed.

2. There is a limit to the thickness of the surface film that can be analyzed. Theoretically this is obtained when the ion scattered from the substrate surface is not sufficiently energetic to traverse and to emerge from the film surface. The maximum thickness can be calculated for any substrate–impurity combination. In practice the thickness permissible is somewhat less due to energy straggling in the film.
3. Foreign atoms concentrated into a thin layer but buried in the substrate cannot be identified since the backscattered energy is dependent on location as well as the mass of the impurity. However, if the identity of the impurity is known, the number of atoms can be determined. Furthermore, their depth can be found by solving Eq. (2) for x using the experimentally determined value of E_1'.

2.2. The Yield of Backscattered Ions

The number of particles detected is related to the number of atoms present in the target through the scattering cross-section. The Rutherford-scattering cross section σ into the solid angle $d\Omega$ is given by

$$\frac{d\sigma}{d\Omega} = \frac{(Z_1 Z_2 e^2)^2}{16(E_{cm})^2} \frac{1}{\sin^4(\theta_s'/2)} \frac{(1 + R^2 + 2R\cos\theta_s')^{3/2}}{1 + R\cos\theta_s'} \tag{3}$$

where $Z_1 Z_2$ are the atomic numbers of incident and target atoms, respectively, e is the electronic charge, E_{cm} is the incident ion energy, θ is the scattering angle, and $R = (M_1/M_2)$, where M_1 and M_2 are the masses of the incident ions and target atoms respectively. The incident ion energy E_{cm} and the scattering angle θ_s' are center of mass variables. The former is related to the laboratory energy E_0 by $E_{cm} = [(M_2/M_1 + M_2)]E_0$. The latter is related to the laboratory angle θ_s by $\tan\theta_s = \sin\theta_s'/(\cos\theta_s' + R)$.

This law is not always applicable. No significant errors are introduced when the incident particles are He^+ or heavier ions at energies between 500 and 2000 keV. At higher energies there may be nuclear excitation, i.e., the scattering is not elastic. At lower energies the scattering is not by the nuclear charge; it is screened field scattering and the scattering potential is different.

The number of target atoms can be calculated using the cross section, the detected count, the geometry of the detector, and the integrated beam current. However, this is not usually done in practice. Since particles backscattered from substrate and from impurity are detected and counted simultaneously, the ratio of the two yields may be used instead. If C_i is the number of counts observed in an impurity peak and C_s the number of counts in one channel (i.e., a known energy interval) of the substrate spectrum, then

$$\frac{C_i}{C_s} = \left[\frac{d\sigma}{d\Omega}\right]_i \left[\frac{d\Omega}{d\sigma}\right]_s \frac{N_i}{N_s} \tag{4}$$

where N_i is the number of impurity atoms and N_s is the number of substrate atoms contributing to the selected channel. This number can be calculated provided the stopping power of the substrate is known. For example, in Fig. 16-3 one channel represents an energy interval of 10 keV; the energy loss of the He ion in silicon ($E_0 = 1000$ keV) along the ingoing and outgoing trajectories is 115.2×10^{-15} eV/atom.[2] Therefore,

$$N_s = \frac{10 \times 10^3}{115.2 \times 10^{-15}} = 8.68 \times 10^{16} \text{ atoms}$$

Only a small error ($\sim 2\%$) is introduced if the ratio of the cross sections is simply taken to be Z_i^2/Z_s^2 since (1) E_{cm} and θ' are very similar for a surface impurity and for the surface layer of the substrate, and (2) the differences in R for the two cases are insignificant as both are much less than one. However, if the impurity is buried, the incident ions striking the impurity atoms will have suffered energy losses in traversing the overlying layers of the substrate. Since the cross section is inversely proportional to E_0^2 [Eq. (3)] then

$$\frac{C_i}{C_s} = \frac{Z_i^2}{Z_s^2}\frac{N_i}{N_s}\frac{E_0^2}{(E_0 - \int_0^{l = x\sec\theta_1} S(E)\, dl)^2} \tag{5}$$

As noted earlier, when the identity of the impurity and the stopping power of the substrate (S) are known, the depth x of the impurity can be calculated from Eq. (2).

The increase of cross section with increasing Z again favors the analysis of heavy elements on or in light matrices. Conversely, the statistics are worsened for light elements in a heavier matrix.

2.3. Stopping Power

Equations (2) and (5) contain terms dependent on the stopping powers $S(E)$ of substrates. There are a number of tabulations from which these may be obtained for incident ions and energies most often used.[3–5] In order to obtain values for materials which are not single elements, the stoichiometry of compounds or the composition of mixtures must be known and the Bragg additivity rule[6] applied. Since stopping powers are a function of energy and since the energy changes with depth, accurate solutions of Eqs. (2) and (5) require computer programs designed to adjust E continuously as a function of depth. However, where the depth being considered is less than a few thousand angstroms, there is no significant error introduced by taking E to equal E_0 and E_0k in the first and second integrals of Eq. (2), respectively. Similarly, E may be taken to be E_0 in Eq. (5).

2.4. Dependence of Yield on Crystal Orientation

When a sample is a single crystal, the yield of particles backscattered from the substrate is strongly dependent on the orientation of the crystal with respect to the beam. An accelerated ion entering a crystal lattice within a small predictable angle of a crystal axis or plane becomes channeled, i.e., steered, by successive gentle collisions and is prevented from having violent collisions with individual lattice atoms. As long as it remains channeled it will not undergo elastic scattering and the yield will be less than that observed when the beam enters in an unaligned (random) direction. A theoretical description of the actual steering mechanism has been provided by Lindhard.[7]

The yield of particles backscattered from surface contaminants on a crystal aligned with the incident beam is the same as that for the random direction and there is no difference between one aligned direction and another (corrections for changing trajectory lengths are necessary). Buried foreign atoms, on the other hand, may give yields which differ for random and aligned directions, as well as for one aligned direction and another. The exact behavior depends on the location of the foreign atoms in the unit cell of the crystal lattice. These facts may be exploited by the analyst: (1) If scattering from the substrate is suppressed by channeling, the sensitivity for determining lower-mass surface contaminants is enhanced because the impurity peak (not suppressed) is then clearly resolved. (2) When foreign atoms are embedded in the host crystal, their location in the lattice of the unit cell can be established as follows: A sample is bombarded in a random direction and in two or more aligned directions. If the foreign-atom yield for all the aligned directions is less than that for the random direction, then at least some of the foreign atoms are located on substitutional sites and the proportion can be determined. If there is no decrease in any direction, this indicates that the atoms are in some nonregular position such as a dislocation or other lattice defect, or in a precipitate. When there is suppression in some directions and not in others, a quantitative measure of the atoms occupying regular interstitial holes along any particular axis can be obtained.

3. EQUIPMENT

3.1. Accelerators

The characteristics required of an accelerator for Rutherford scattering analysis are not particularly exacting. Preferably it should be capable of working in a low energy range where interfering nuclear reactions are not promoted, i.e., 500–2500 keV, and it should have a small energy spread at the target ($\pm 0.1\%$). Only modest beam currents are required; for most experiments ~ 10 nA is sufficient. Some choice of accelerated ions is desirable. He^+ is the most generally useful as there is little chance of interference from nuclear reactions; but H^+ may be wanted to obtain deeper penetration, or heavier ions such as C^+, N^+, or O^+ may be wanted to give greater mass separation. Van de Graaff machines, which meet these requirements nicely, have been most frequently used.

The forms of target chambers and target holders may vary greatly, but a few guidelines can be set down. It is useful to have one part of the target chamber designed to take several different forms of interchangeable target holders. If the setup is to be used for charged-particle activation as well, another part capable of accepting interchangeable detectors is good practice. The solid-state detector for elastic-scattering analysis should be mounted so that the scattering angle can be altered.

It is convenient to have different types of target holders available: (1) One capable of holding a number of samples and so designed that each may be brought into the beam successively without breaking the vacuum. (2) One with provision for moving a sample across the beam so that concentration gradients in a surface plane may be located and determined. (3) A goniometer having two independent axes of rotation so that any desired crystallographic axis or plane of a single crystal may be oriented with respect to the incoming beam.

All these devices should have the target electrically insulated from the rest of the chamber so that beam currents can be measured and integrated.

3.2. Counting Systems

Two quite different systems for detection and energy measurement of the scattered particles have been used.

1. Magnetic or electrostatic spectrometers have the best energy resolution. However, these instruments have a low solid accep-

tance angle and do not give high counting rates, and therefore require high beam currents or long bombardment in order to accumulate sufficient counts for good statistics. They are not as readily available and are more expensive than the alternative system.

2. Solid-state detectors combined with multichannel pulse-height analyzers are readily available commercially and are much less expensive. They make it possible to accumulate good counting statistics at low beam currents. This minimizes sputtering, excessive heating of the target, and radiation damage, all of considerable importance in channeling studies.

The finite resolution of the counting system sets limits to the minimum difference in the masses of surface contaminant and substrate acceptable for analysis. Spectrometers with high resolution can distinguish nuclides differing by two or three atomic mass units. The best solid-state detectors, which typically have a resolution of ~ 15 keV (fwhm) at 1 MeV, require that the masses differ by $> 20\,\%$. In both cases it is assumed that the surface layer is thin. When the film thickness is such that

$$\int_0^{l = x\sec\theta_1} S(E)\, dl > 20 \text{ keV}$$

the peaks become proportionately broader and the mass differences must be greater to be resolvable. Similarly, detector resolution determines the precision of depth measurements. For thin distributions this is calculated by equating the integral

$$\int_0^{l = x\sec\theta_1} S(E)\, dl$$

to the detector resolution and solving for x.

4. APPLICATIONS

4.1. Applicability

The foregoing discussion clearly implies that Rutherford scattering cannot profitably be used as a general analytical tool because of the limitations imposed by mass difference requirements and by the necessity of having the impurity atoms concentrated in rather thin layers. However, where it is applicable, its essential simplicity and capability for depth perception makes it attractive. Except for investigation of dopants in single-crystal semiconductor materials, the published literature is not extensive, but the method is more widely used in accelerator laboratories than the volume of published work would indicate.

No indication of the sensitivity was given earlier as it depends on many factors, e.g., masses of impurity and substrate, depth, distribution, detector geometry, etc., and consequently differs with circumstance. The sections which follow describe different situations which lend themselves to the application of Rutherford-scattering techniques; in each example the sensitivity obtained is given.

4.2. Determination of Surface Contaminants

The determination of a single, identified contaminant on the surface of a lower-mass high-purity substrate is simple, rapid, sensitive, and nondestructive. This kind of analysis is required with surprising frequency, e.g., in the investigation of residues left on solid surfaces after immersion in etching, anodizing, chemical-polishing, etc., solutions. Thompson, Barber, and Mackintosh[8] studied the retention of gold on silicon after etching in HF solutions containing small amounts of gold. Many samples were analyzed; each required only about 10 minutes. The sensitivity was 3 ng/cm^2. Another investigation revealed the amount of chromium found on the surface of aluminum metal following dissolution of an anodic oxide surface layer in phosphate–chromate solutions. In these analyses the incident particles were 2-MeV He ions.

This technique is also used for the measurement of films evaporated on a substrate surface. Anders[9] used 2-MeV protons to determine silver on glass in the range of 10–49 μg/cm^2 and chromium on magnesium in the range of 20–150 μg/cm^2. Peisach,[10] bombarding with He, found 1–10 μg/cm^2 of gold on aluminum and confirmed the validity of the method by comparison with neutron-activation analysis. Rubin,[1] using protons and performing the energy analysis with a magnetic spectrometer, was able to identify and determine silver/antimony and lead/mercury (finer mass resolution was not possible) on steel surfaces exposed to an explosion.

Ball, Buck, and Wheatley[11] established the sensitivity for films of gold, iron, nickel, copper, aluminum, silicon, and oxygen on graphite and silicon, employing both protons and He ions accelerated with a relatively inexpensive low-voltage (100 kV) machine primarily intended for ion implantation. The low energy of the incident particle introduced complications in calculating results, but their accuracies were verified by comparison with values obtained by x-ray fluorescence and neutron activation. Good energy resolution was obtained with an electrostatic analyzer. Sensitivities of 3 ng/cm^2 of oxygen on graphite were observed.

The requirement that elements to be determined be concentrated in a thin layer has lead to the development of techniques for depositing materials to be analyzed on a suitable substrate. Rubin[1] electrostatically precipitated aerosols on aluminum foils and obtained 5, 0.13, and 0.3 $\mu g/cm^2$ of oxygen, sulfur, and silicon, respectively. A magnetic spectrometer had made possible the analysis of elements which are close in mass to the supporting substrate. Nicolet, Mayer, and Mitchell[12] have carried out some preliminary studies of films deposited by drawing "smog" through cellulose ester filters. Lead was dominant in the spectra obtained but no quantities were given.

4.3. Thickness and Composition of Chemically Altered Surface Layers

When a surface of a sample has been altered chemically, information on the thickness and composition of the layer may be extracted from Rutherford-scattering spectra. The most commonly encountered cases are oxide and nitride films on elemental substrates. As seen in Fig. 16-4, the spectrum for such a sample exhibits the usual sharp edge at an energy representing scattering from the higher-mass component at the surface. The sloping plateau is interrupted by a second sharp step which represents scatter from the interface of substrate and compound. The count at this point is greater because the number of atoms contributing to the yield per unit of depth sharply increases from compound to substrate. Superimposed on the plateau derived from scattering deeper and deeper in the substrate, there is a peak derived from the lower-mass component of the surface compound. As noted earlier, the area of this peak is difficult to obtain with sufficient precision to calculate the thickness by Eq. (5). However, when the stoichiometry of the compound is known so that the rate of energy loss can be calculated from tabulations, the energy scale of the spectrum can be converted to a depth scale using Eq. (2) and the position of the interface step will indicate the thickness of the compound. The thickness of film that can be measured has a lower limit set by the minimum energy differences detectable by the counting system and an upper limit set by the requirement that the ion backscattered from the substrate must have sufficient residual energy to emerge from the film surface. Thicknesses varying from 10 to 100 $\mu g/cm^2$ of anodically formed films on aluminum (the stoichiometry is well established) have been measured by Peisach[10] and by Brown and Mackintosh[13] with precisions of $\pm 3\%$. Both found good agreement with other methods.

Gyulai, Meyer, and Mayer[14] have shown how the width of the compound plateau and the width of the superimposed peak of the lower-mass component may be used to determine the ratios of nitrogen

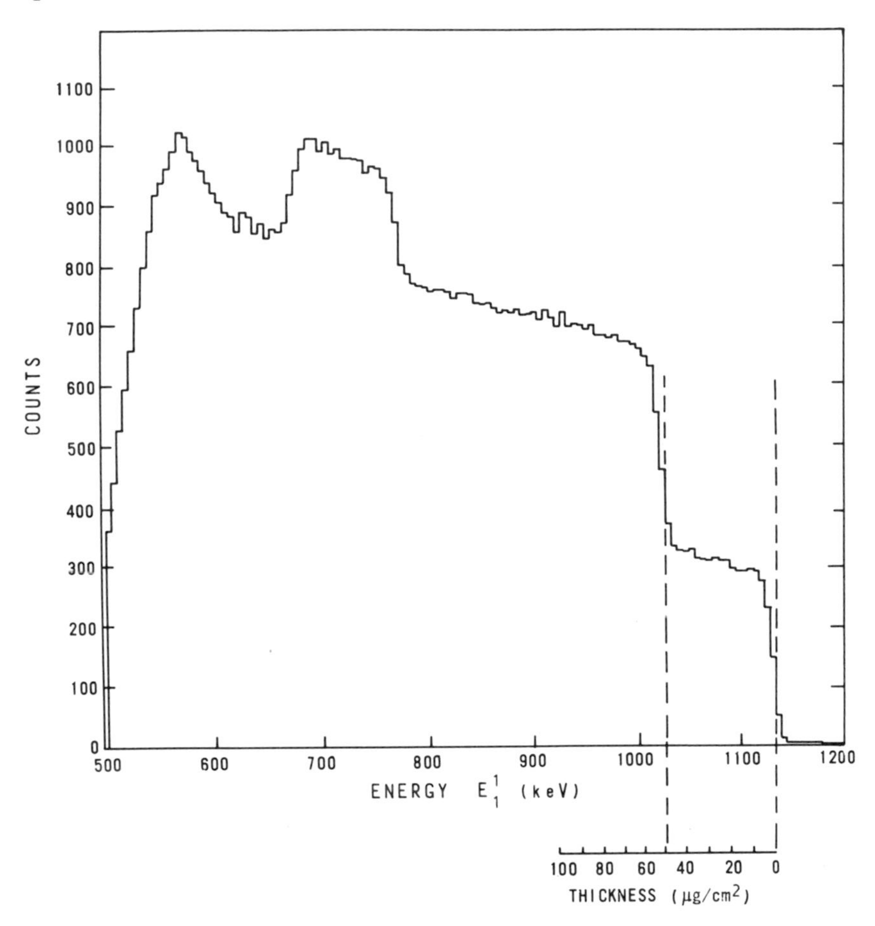

Fig. 16-4. The Rutherford-scattering spectrum of aluminum with a surface film of oxide which is 50 μg/cm^2 thick.

and oxygen to silicon in nonstoichiometric surface layers. Absolute concentrations can be obtained if the layer thickness is known. In their work, 1-to-2-MeV He ions were adequate to study surface layers of 300–3000 Å.

4.4. Determination of Concentration Gradients

A number of published papers describe the determination of change of concentration with depth of solutes that have been introduced by diffusion into the surface layers of a substrate. In principle this is accomplished by observing in the spectrum the shape of the peak arising from particles backscattered from the solute atoms. When the concentration is decreasing with depth, the low-energy side slopes gently downward reflecting the decreasing number of atoms giving rise to backscattering events. Relating the spectral curve to absolute concentration from point to point involves complex calculations. Sippel[15] and Ziegler and Baglin[16] described approaches to this problem.

Sippel[15] studied the diffusion of gold in copper to depths of $< 10^{-5}$ cm and then showed that the Rutherford-scattering method was suitable for obtaining diffusion constants of the order of 10^{-18} cm^2/sec. Ziegler and Baglin[16] compared the scattering method with that of neutron-activation and Hall-resistivity measurements for determining the profile of arsenic diffused in silicon. These agreed within 10%, at concentration of the order of 10^{21} atoms/cm^3. Gyulai *et al.*[17] investigated the diffusion of gold and gold–germanium to form ohmic contacts in GaAs. Bower and Mayer[18] carried out similar experiments to observe the formation of palladiun, titanium, chromium, and molybdenum silicides on silicon wafers of the type used in the solid-state devices.

4.5. Depth Location of "Markers" in Oxidation Studies

The movement of deliberately added foreign atoms during oxidation of surface layers may be followed by Rutherford scattering as an alternative to radiotracer techniques. As well as the obvious advantage of not requiring radioactive materials, the method reveals the depths directly, eliminating the necessity for sectioning. It is also nondestructive. By implanting some 10^{14}–10^{16} atoms/cm^2 of inert gases as markers of the surface of aluminum and then oxidizing anodically, Brown and Mackintosh[13] demonstrated that both aluminum and oxygen ions must have crossed this marked surface layer during oxidation because the final position of the markers was roughly halfway between the oxide surface and the metal–oxide interface. Since aluminum is a low-mass material, the final position of many other foreign atoms could be determined as well. The studies included chlorine, bromine, iodine, potassium, rubidium, and cesium. The oxide films were as thick as 100 μg/cm^2 and it was concluded that the positions could be found within $\pm 4\%$.

4.6. Extension of Applicability by Employing Channeling

In Section 2.5 it was stated that elements lighter in mass than their substrate could be analyzed in the case of single crystals because the yield from the substrate could be suppressed by channeling. In analyses of this kind, the crystal is mounted on a goniometer and a principal crystal axis or plane is aligned with the impinging beam. Fortunately this orientation can be done quite easily by means of the channeling effect and the elastic-scattering process themselves.[19] The crystal is tilted at some small angle to the beam and then rotated through 360°.

During rotation the position of the sharp attenuations in yield which occur as the various close-packed planes become successively aligned are noted and plotted on a stereogram to obtain the aligned azimuthal and tilt angles. The process requires only 15–30 minutes. The amount of attenuation of yield is dependent on the crystal perfection—30- to 100-fold suppressions have been obtained in crystals of silicon, germanium, and tungsten.

Mackintosh and Davies[19] report the determination of 0.2 $\mu g/cm^2$ of oxygen on the surface of silicon when this technique was used. Gyulai *et al.*[14] used channeling, where applicable, to enhance the precision for determining the thickness of the silicon nitride and oxide films.

4.7. Foreign-Atom Location in Single Crystals

In recent years, ion implantation,[21] a method of introducing atoms into the surface layer of a solid, has become an important technique for doping semiconductor materials in the production of electronic devices. The electrical properties of the doped materials depend on the dopant, its distribution and location in the unit cell lattice. The number of foreign atoms introduced by ion implantation, in contrast to diffusion processes, is not controlled by solubility considerations but may greatly exceed the solubility limit. Thus a sufficient number of foreign atoms may be introduced to make detection by elastic scattering practical–i.e., 10^{15}–10^{17} atoms/cm^2. This means that not only can the amount be assessed, but distribution and location of the atoms may be obtained simultaneously by using the channeling technique as discussed in Section 2.5. It is also worth noting here that this method can also reveal the extent of disorder produced in the lattice by the implanting process.

Lattice location studies of implanted semiconductor materials have been very extensive and have given rise to voluminous literature. Details of the technique are to be found in several references,[20,21] while results of studies involving backscattering are to be found in Mayer, Eriksson, and Davies[21] (and references therein) and in the proceedings of a number of conferences.[22,23]

5. SUMMARY

The principles of Rutherford scattering described in the first part of this chapter indicate that the method is applicable to the analysis of solid surfaces where (1) the impurities to be determined are concentrated in a thin layer either on or buried within a few thousand angstroms of the surface, (2) the impurities are greater in atomic mass than the substrate (except in single crystals), and (3) few impurities are present. The method has two distinct advantages: (1) its characteristics can be exploited to yield depth information without employing ancillary physical or chemical sectioning techniques, and (2) it can be readily combined with channeling methods to find positions of foreign atoms with respect to a crystal lattice.

These features have made it a strikingly useful tool in the study of changes in surface properties effected by adventitious impurities or foreign atoms deliberately introduced into surface layers. It is equally useful for the converse problem—investigation of the effect of surface processes in redistributing impurities.

6. REFERENCES

1. S. Rubin, T. O. Passell, and L. E. Bailey, Chemical analysis of surfaces by nuclear methods, *Anal. Chem.* **29**, 736 (1957).
2. D. A. Thompson and W. D. Mackintosh, Stopping cross sections for 0.3 to 1.7 MeV helium ions in silicon and silicon dioxide, *J. Appl. Phys.* **42**, 3969 (1971).
3. C. F. Williamson, J. P. Boujot, and J. Picard, Tables of range and stopping power of chemical elements for charged particles of energy 0.05 to 500 MeV, Centre D'Etudes Nucleaires de Saclay, Report CEA-R3042 (1966).
4. J. F. Janni, Calculations of energy loss, range, pathlength, straggling, multiple scattering, and the probability of inelastic nuclear collisions for 0.1 to 1000 MeV protons, Air Force Weapons Laboratory, Report AFWL-TR-65-150 (1966).
5. L. C. Northcliffe and R. F. Schilling, Range and stopping-power tables for heavy ions, *Nuclear Data Tables* **7**, 233 (1970).
6. W. H. Bragg and R. Kleeman, On α particles of radium, and their loss of range in passing through various atoms and molecules, *Phil. Mag.* **10**, 318 (1905).
7. J. Lindhard, Influence of crystal lattice on motion of energetic charged particles, *Mat. Fys. Medd: Dansk. Vid. Selsk.* **34** (1965).
8. D. A. Thompson, H. D. Barber, and W. D. Mackintosh, The determination of surface contamination on silicon by large angle ion scattering, *Appl. Phys. Letters* **14**, 102 (1969).
9. O. U. Anders, Use of charged particles from a 2-megavolt Van de Graaff accelerator for elemental surface analysis, *Anal. Chem.* **38**, 1442 (1966).
10. M. Peisach and D. O. Poole, Analysis of surfaces by scattering of accelerated alpha particles, *Anal. Chem.* **38**, 1345 (1966).
11. D. J. Ball, T. M. Buck, and G. H. Wheatley, Studies of solid surfaces with 100 keV He^+ and H^+ ion beams, *Surface Science* **30**, 69 (1972).
12. M. A. Nicolet, J. W. Mayer, and I. V. Mitchell, Microanalysis of materials by backscattering spectrometry, *Science* **117**, 841 (1972).
13. F. Brown and W. D. Mackintosh, The use of Rutherford backscattering methods to study the behavior of ion implanted marker atoms during anodic oxidation of aluminum; Ar, Kr, Xe, K, Rb, Cr, Cl, Br, and I, *J. Electrochem. Soc.* **120**, 1096 (1973).
14. J. Gyulai, O. Meyer, and J. W. Mayer, Analysis of silicon nitride layers on silicon by backscattering and channeling effect measurements, *Appl. Phys. Letters* **16**, 232 (1970).
15. R. F. Sippel, Diffusion measurements in the system Cu–Au by elastic scattering, *Phys. Rev.* **115**, 1441 (1959).

16. J. F. Ziegler and J. E. E. Baglin, Determination of surface impurity concentration profiles by nuclear backscattering, *J. Appl. Phys.* **42**, 2031 (1971).
17. J. Gyulai, J. W. Mayer, V. Rodriguex, A. Y. C. Yu, and H. J. Gopen, Alloying behavior of Au and Au–Ge on GaAs, *J. Appl. Phys.* **42**, 3578 (1971).
18. R. W. Bower and J. W. Mayer, Growth kinetics observed in the formation of metal silicides on silicon, *Appl. Phys. Letters* **20**, 359 (1972).
19. W. D. Mackintosh and J. A. Davies, Rutherford scattering and channeling—A useful combination for chemical analysis of surfaces, *Anal. Chem.* **41**, 26A (1969).
20. J. A. Davies, J. Denhartog, L. Eriksson, and J. W. Mayer, Ion implantation of silicon, I. Atom location and lattice disorder by means of 1.0 MeV helium ion scattering, *Can. J. Phys.* **45**, 4053 (1967).
21. J. W. Mayer, L. Eriksson, and J. A. Davies, *Ion Implantation in Semiconductors*, Academic Press, New York (1970).
22. *Atomic Collision and Penetration Studies, Proceedings of the International Conference on Atomic Collision and Penetration Studies with Energetic (keV) Ion Beams*, University of Toronto Press (1968).
23. D. W. Palmer, M. W. Thompson, and P. D. Townsend (ed.), *Atomic Collision Phenomena in Solids*, American Elsevier, New York (1970).

17

ACCELERATOR MICROBEAM TECHNIQUES

T. B. Pierce

Analytical Sciences Division
U.K. Atomic Energy Authority
Atomic Energy Research Establishment
Harwell, near Didcot, Berkshire, England

1. INTRODUCTION

The analytical potential of nuclear interactions induced by accelerated charged particles was demonstrated during the earliest investigations into nuclear structure, when yields of some emitted radiations were interpreted in terms of the number of nuclei available for interaction in the target material. Subsequently, shortly after Hevesy[1] demonstrated that neutron irradiation could provide the basis of analytical measurement, Seaborg and Livingood[2] measured the gallium content of iron after irradiation with accelerated deuterons. Analytical techniques based upon charged-particle irradiation were initially overshadowed by methods exploiting neutron irradiation, since simple, sensitive neutron-activation procedures were available and the nuclear reactor offered an intense radiation source capable of irradiating many samples simultaneously. Application of charged-particle techniques was thus largely restricted to the determination of those elements which could not be conveniently measured after neutron activation. These were mainly light elements since (1) high specific activity could be induced in light elements by charged-particle bombardment, and (2) small quantities of characteristic activity from the element to be determined could be isolated by chemistry from large quantities of other activities induced in the sample, for example from a major constituent.

In parallel with the development of "conventional" methods of charged-particle-activation analysis based upon the measurement of induced activity, limited studies were also in progress to assess the potential of so-called "prompt" radiations emitted rapidly after nuclear interaction to provide both qualitative and quantitative information from the sample under irradiation. Since prompt radiations must be measured while the irradiation is in progress, and the different sources of radiation in the sample cannot be isolated by chemistry between irradiation and counting, instrumental methods are essential for distinguishing between the several components of the radiation emitted from the sample. The capability of prompt techniques is therefore closely tied to the ability of available instrumentation and data processing methods to identify the individual components of complex spectra and to provide detailed information about the specific components present in realistic analytical samples. Over recent years, a growing appreciation of the analytical potential of such methods, together with the improvement in the performance of the instrumentation available for radiation detection, signal processing, and data interpretation, has stimulated interest in the techniques and increased the range of measurements to which they can be applied. Although the number of papers published in the analytical use of prompt-radiation techniques is at present small, a new family of methods has been demonstrated which promises to add additional useful techniques to the armory of the analytical scientist. The low incident-particle energies of a few MeV or less most often used mean that work can be carried out with small accelerators, and since determination is not based upon the measurement of the induced radioactivity, methods are no longer limited to those interactions which yield radioisotopes. Several different types of radiation can result from charged-particle irradiation so that alternative analytical methods are available. Prompt γ radiation, particle groups emitted as a result of the nuclear reaction, elastically scattered particles, and more recently x radiation, which is also emitted during sample irradiation with charged particles, have all provided the basis for analytical determination; each radiation, having its own specific characteristic which defines the type of measurement to which it can be employed, provides particular information about sample composition.

Usual methods of charged-particle-activation analysis based upon measurement of induced radioactivity have not been devised with the aim of providing information from restricted areas of sample, although the distribution of carbon and oxygen in steel samples has been followed by autoradiography of induced radioactivity after ^{3}He irradiation.[3] Nevertheless the composition of small samples and positional variation in elemental content of larger samples is important in characterizing the nature and behavior of many materials. The measurement of prompt radiation can provide localized information about sample composition if it is produced by small-diameter charged-particle beams. Since, in contrast to induced radioactivity, radiation emission ceases when the incident particle beam is switched off, elemental variations across samples can be followed by sequential irradiation of different locations of the sample surface associated with the measurement of some appropriate emitted radiation. Work has shown that when appropriate modifications have been made to the beam-handling system, small-diameter ion beams of the order of 3 μ can be produced in a conventional high-voltage accelerator at particle energies and beam currents which are adequate to provide measurable yields of radiation for a variety of elements present over a range of concentrations. Thus, a genuine nuclear positive-ion microprobe is possible.[4] Production and use of accelerator microbeams for analysis is a relatively new innovation and few groups are as yet working on the development and application of microprobe techniques. Therefore, any discussion in this chapter must be limited by the very small amount of published experience so far available. In the absence of published material from many sources, the author has had to draw heavily for this article on experience built up in the Analytical Sciences Division at Harwell, where the analytical use of small-diameter positive-ion beams with energies varying from a few hundred keV to a few MeV has been in progress for a number of years.

2. ANALYTICAL METHODS

A number of alternative modes of decay may follow nuclear reaction. These are summarized in general form in Fig. 17-1 where it is assumed that a compound nucleus is formed between the incident particle and the reacting target nucleus. The excited state of the compound nucleus may decay by emission of γ-radiation ($\gamma_1 - \gamma_3$) thus giving rise to radiative capture reactions, or by particle emission ($P_0 - P_2$) which may subsequently feed one or more states of the residual nucleus. The residual nucleus may be in an excited or ground state, and the excited state may again lose its excitation energy by γ-ray emission ($\gamma_4 - \gamma_5$). The ground state of the residual nucleus may itself be unstable, for example, decaying by β- or α-particle emission, forming the final nucleus in ground or excited states. These decay mechanisms are not included in Fig. 17-1; only those decay routes are given which are normally classified as prompt.

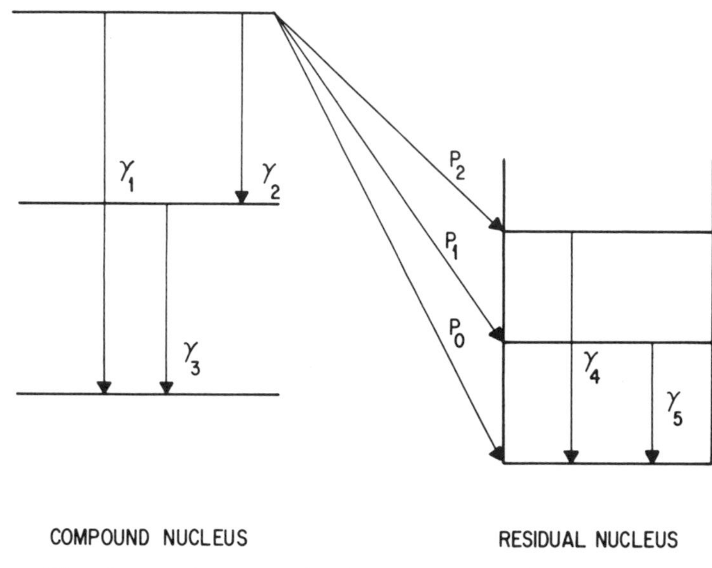

Fig. 17-1. Nuclear decays leading to prompt-radiation emission.

Nonnuclear interactions leading to the emission of measurable radiation include elastic scattering, caused by electrostatic repulsion of an incident charged particle by a positively charged nucleus, and x radiation. The major characteristics of analytical techniques based upon the various radiations will now be summarized briefly.

2.1. Prompt Gamma Radiation

Emission of characteristic γ radiation produced by the de-excitation of excited nuclear states offers the basis for both qualitative and quantitative analysis since γ lines of specific energies produced during the de-excitation process can be identified with the emitting nuclei, and the total radiation yield is a function of the number of nuclei reacting. Since γ lines of characteristic energies are emitted during stable-to-stable transitions, such as $^{12}C(d, p)\,^{13}C$ and $^{16}O(d, p)\,^{17}O$, reaction products may be entirely different from those produced in conventional activation procedures, and γ-ray spectroscopy can be used to identify the individual radiation yields from light elements. This is not generally possible for the positron-emitting radionuclides produced from many of the light elements. The hard γ radiation produced from lighter elements contrasts with the low-energy x radiation emitted by these elements, and complicating contributions from heavy elements to the accumulated γ-ray spectra can often be avoided by limiting the incident particle energy to less than that required to penetrate the Coulomb barrier. The prompt γ-ray spectrum from a geological sample irradiated with 4.0 MeV protons and detected with a germanium counter is shown in Fig. 17-2 and illustrates the type of spectrum that can be obtained by prompt-γ-ray techniques. Gamma-lines from a number of light elements are clearly visible, and the characteristic energies of the emitted γ rays enable the reacting nuclides to be identified. The measurement of γ-line intensity permits the analysis of the samples to be placed on a quantitative basis.

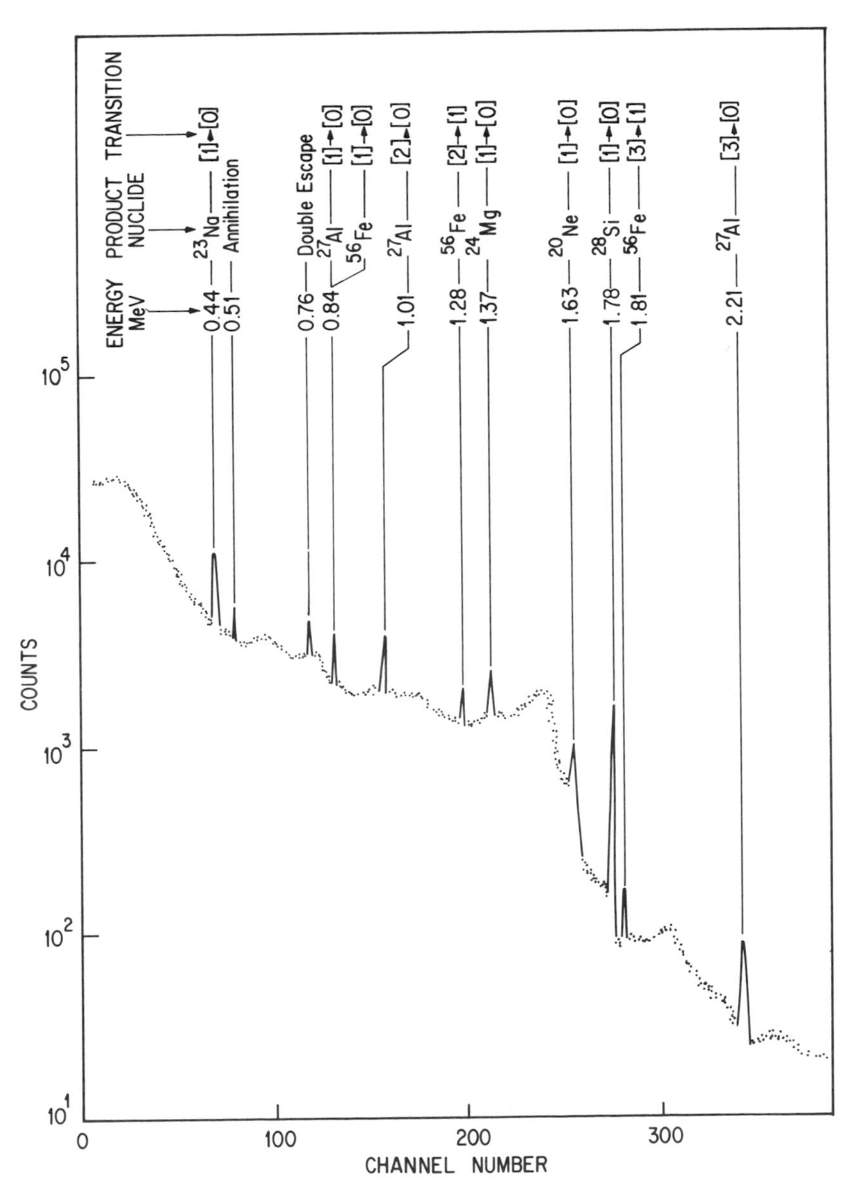

Fig. 17-2. Prompt γ-ray spectrum of a geological specimen.

TABLE 17-1. Some Analytical Determinations Based on Measurement of Prompt γ Radiation

Element	Reaction	Reference
Lithium	$^{7}Li(p, \gamma)\,^{8}Be$	5
Beryllium	$^{9}Be(p, \gamma)\,^{10}B$	5
	$^{9}Be(\alpha, n)\,^{12}C$	6
Boron	$^{11}B(p, \gamma)\,^{12}C$	5
Carbon	$^{12}C(d, p)\,^{13}C$	7
	$^{12}C(p, \gamma)\,^{13}N$	8
Nitrogen	$^{14}N(d, p)\,^{15}N$	9
	$^{15}N(p, \alpha)\,^{12}C$	10
Oxygen	$^{16}O(d, p)\,^{17}O$	9
	$^{18}O(p, \gamma)\,^{19}F$	11
	$^{16}O(t, n)\,^{18}F$	15
Fluorine	$^{19}F(p, \alpha)\,^{16}O$	5, 6, 12–14
Magnesium	$^{24}Mg(p, p')\,^{24}Mg$	16
Aluminum	$^{27}Al(p, \gamma)\,^{28}Si$	17
	$^{27}Al(p, p')\,^{27}Al$	16
Silicon	$^{28}Si(p, p')\,^{28}Si$	18
Sulfur	$^{32}S(p, p')\,^{32}S$	20

A number, although by no means all, of the applications found for prompt-γ-ray measurement are summarized in Table 17-1 and are intended to illustrate the extent of the uses so far investigated. Clearly, interest has extended over most of the low-Z region of the periodic table, with particular elements being determined by more than one reaction. Some control of the γ-ray spectrum emitted from the sample is possible by selecting particle and energy to excite the most suitable line for measurement and interference from overlapping lines can be avoided in particular cases. The success of the approach is limited, however, by the relative sensitivities available for the reactions contributing to the radiation yield.

Since γ lines must be measured in the presence of machine background and all other γ radiation produced from the sample, restrictions to the use of prompt-γ-ray measurement are imposed by the usual limitations of γ-ray spectroscopy when counting one component of a complex spectrum. Thus the prompt-γ-ray technique has in general found application to the determination of one or a group of elements present at major and minor levels (albeit sometimes in very small samples), and relatively few examples have been reported of elemental determinations at concentrations of parts per million or less. Measurements of fluorine at low levels have been most consistently pursued, exploiting both the high radiation yield of the $^{19}F(p, \alpha)\,^{16}O$ reaction and the high energy of the emitted γ rays, which usually exceed that of major components of both the emitted and background radiation. Even for this determination, measurement has been generally at levels of greater than a few parts per million.

The resonance character of γ-ray emission from some reactions, particularly those based upon a proton irradiation showing sharp increases in yield at a particular energy, has been exploited with effect to restrict reaction to limited depths in the sample and has consequently provided a means for examining variations in elemental concentrations with depth.[12] While such depth profiles require each point examined on the sample surface to be irradiated a number of times (and so would be tedious, if applied to a very large number of sample locations) valuable information can be obtained about surface composition from the intact sample without any need for chemical or mechanical processing of the sample surface. Irradiation at resonance energies can also provide an enhancement of specific, wanted reactions at the expense of others which yield interfering radiation from the sample.

2.2. Particle Groups

Both neutrons and charged particles resulting from nuclear reactions offer a further basis for analytical determination but, although neutrons from, for example, the (d, n) reaction have been counted for analytical purposes,[26] the complexity of the instrumentation necessary for energy analysis of emitted neutrons has resulted in charged-particle measurement receiving greater attention. Since the emitted particles are characterized by energies which are determined by the transitions occurring, multielement analysis is possible and again the yields can be interpreted in terms of the number of reacting atoms. The high positive Q values of many reactions can lead to the production of energetic particles with relatively low-energy incident particle beams, and a careful choice of irradiating particle and energy can sometimes simplify spectral interpretation. Table 17-2 summarizes some of the determinations which have been carried out by particle group measurement and illustrates the rather more restricted range of elements that have so far been investigated by the technique. Particular attention has been paid to carbon and oxygen measurement and a number of applications have been reported which involve the determination of these two elements. A simple spectrum of protons from a carbon- and oxygen-rich film is shown in Fig. 17-3 and illustrates the type of information on which measurement is often based. When spectra are as uncomplex as the one shown, spectral interpretation is straightforward; but at lower

TABLE 17-2. Some Analytical Methods Based on Charged-Particle Detection

Element	Reaction	Reference
Lithium	$^{6}Li(d, \alpha)\,^{4}He$	19
	$^{7}Li(d, \alpha)\,^{5}He$	
Boron	$^{10}B(d, p)\,^{11}B$	21
Carbon	$^{12}C(d, p)\,^{13}C$	22, 23
Nitrogen	$^{14}N(d, p)\,^{15}N$	23
Oxygen	$^{16}O(d, p)\,^{17}O$	22–24
Fluorine	$^{19}F(p, \alpha)\,^{16}O$	19
Sulfur	$^{32}S(d, p)\,^{33}S$	25

carbon and oxygen levels reactions such as $^{2}H(d, p)\,^{3}H$ provide additional particles which complicate the accumulated spectrum. The number of charged particles produced by nuclear reaction is likely to be small compared with the total yield of particles scattered elastically from the sample. Consequently, it is often desirable to prevent these scattered particles from reaching the detector since they may not only contribute to instrumental dead time, but in some cases also overlap and swamp the peaks. A foil of suitable thickness placed in front of the detector can sometimes absorb the interfering elastically scattered component while still allowing the required particles to pass through into the detector. The sensitivity of analytical techniques based upon particle group

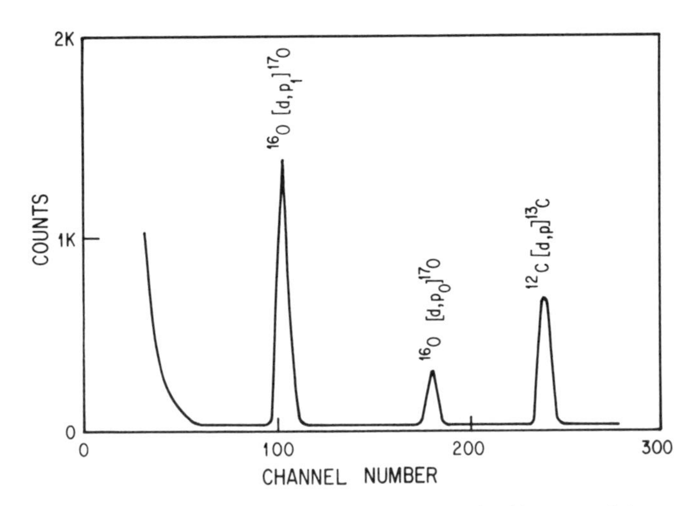

Fig. 17-3. Spectrum of protons from a thin film containing carbon and oxygen.

measurement can be high, particularly since measurement is often carried out on a relatively low background.

2.3. Elastic Scattering

The use of elastic scattering for analysis is described in detail in Chapter 16 and it would be inappropriate in this chapter to do more than outline those general characteristics of the technique which are of particular value to microprobe methods. Elastic scattering has primarily found application to the determination of thin films of heavy elements present on lighter substrates, since the energies of particles scattered from the thin films are greater than the energies of those scattered from the substrate. Provided that the mass difference is adequate, a peak is obtained for the heavy element clear of the Rutherford plateau in the particle energy *vs.* intensity curve. This peak is therefore present on a relatively low background. Figure 17-4 shows a spectrum of α particles scattered from a thin film of gold present on an aluminum substrate and covered with copper. The peaks due to particles scattered from the thin films of gold and copper are clearly visible, as is the edge resulting from scattering from the aluminum substrate. Clearly in this case, the mass difference of the two elements which comprise the thin films is adequate to provide particles of sufficiently different energies for them to be distinguished by the detector and to permit simple quantitative

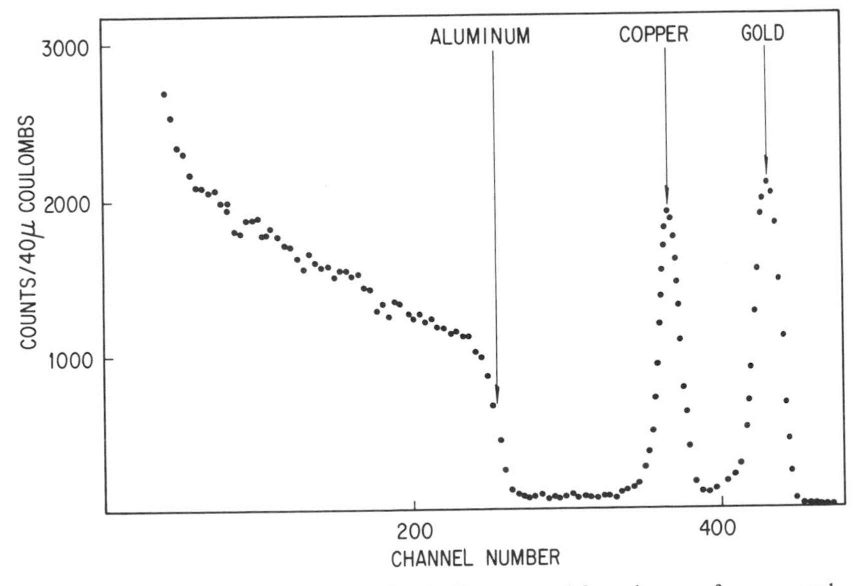

Fig. 17-4. Spectrum of α particles elastically scattered from layers of copper and gold on aluminum.

measurement to be made. When semiconductor detectors are used in conjunction with pulse-height analysis, an elastically scattered particle spectrum can be accumulated, displayed, and processed in much the same manner as for conventional γ-ray spectrometry, thus providing a convenient method of measurement which for many applications is likely to outweigh the higher resolution attainable by magnetic spectrometers.

Preliminary examinations of sample surfaces prior to scanning can also be carried out by elastic scattering, providing an application peculiar to microbeam techniques.[27] When sample areas are large compared to the diameter of the ion beam, detailed examination of the whole surface area may not be practicable, and some means of preliminary examination is desirable to define the areas of particular interest for more detailed investigation, provided, of course, that these cannot be identified by some other simple direct means such as visual observation of specific phases. Any survey technique should permit rapid examination of areas which are many times greater than areas irradiated by the stationary beam, should not demand any special changes in machine configuration prior to more detailed examination, and should be applicable to the range of particles likely to be used for sample irradiation. Since the gross yield of particles scattered from any location on the target will, among other factors, be dependent on the charge of the scattering nucleus, regions of high particle yield are likely to correspond to areas containing nuclei of high atomic number. High counting rates of scattered particles are achieved in practice since good energy resolution is not required and high counting geometries are acceptable. In this manner, the ion beam can be moved rapidly to permit examination of a substantial number of positions on the sample surface. Since survey examinations by elastic particle scattering can usually be achieved with the particle subsequently used to provide specific information about sample composition, the need for changing either particle or machine settings before the actual analysis commences is thereby avoided. This not only provides for economy in operator effort, but simplifies re-alignment problems which might be necessary in finding the areas of interest for detailed examination after identification by gross scattering methods.

2.4. X Radiation

The potential of charged-particle bombardment for the production of x rays has been known since Chadwick[28] observed and subsequently identified the characteristic x radiation emitted from a number of elements during irradiation with α particles, but subsequent analytical interest in charged-particle excitation of x radiation has been intermittent and concerned mainly with particles of energies of a few hundred kilovolts or less.[29] The particular attractions of charged-particle-excited x rays include the production of lines associated with low bremsstrahlung and of relatively high x-ray yields, particularly from low-Z elements. The penalty paid for the better spectral shapes of the emitted x radiation over those obtained by, for example, electron excitation, is that much higher particle energies are required to achieve the same x-ray output. Thus, protons of about 1.5 MeV are roughly equivalent to 20-keV electrons for the production of Cu K_α x rays.[30] Particle energies of the order of millions of electron volts are needed to induce many of the nuclear reactions leading to the formation of prompt-reaction products; at these energies x-ray yields may be high, with some K-shell ionization cross sections being of the order of thousands of barns,[31] so that high sensitivity may be possible. Where x-ray production with MeV particles has been investigated for analytical purposes, signal-to-background ratios of 40,000:1 have been measured for the silicon K_α line in a prototype system, and substantially better ratios approaching 250,000:1 have been predicted with improvements to the experimental assembly.[30] Detection of 10^{-12} g or less of particular elements seems possible,[32] although the actual levels at which reliable information can be obtained will be completely dependent upon the sample composition and the quality of the experimental system used for radiation detection.

Measurement of characteristic x radiation excited by charged-particle irradiation is clearly complementary to the prompt-γ and particle-group methods discussed previously, which have been primarily applied to light-element determination, since it offers a means of extending determinations to elements of high Z, and indeed offers an alternative method of determining light elements provided that the long wavelengths of x radiation emitted from these elements is acceptable. For certain applications the very limited depths of penetration of these low-energy x radiations can be a positive advantage when, for example, information is only required from thin layers of the sample close to the surface. Since x radiation will be emitted during the production of other prompt radiations, x-ray counting can often be combined with particle or with γ-ray measurement to provide more detailed information about sample composition, providing that adequate radiation detection and signal processing equipment is available. Nondispersive detectors again offer a convenient means of examining the structure of x-ray spectra over a range of energies, but the relatively poor resolution of these detectors does limit their ability to distinguish between x rays of similar energies, and dispersive systems will undoubtedly be necessary for examining samples containing certain elemental combinations. Although the measurement of x radiation emitted as a result of charged-particle irradiation has not so far been extensively used for analysis, conventional x-ray fluorescence using either x-ray or electron excitation has an established place in analytical laboratories and much of the experience gained in that work is relevant to charged-particle x-ray techniques.

3. PRODUCTION OF MICROBEAMS

A reduction in the diameter of accelerator ion beams to sizes suitable for microbeam work can be achieved either by beam collimation or by focusing conventional accelerator ion beams down to small diameters. Collimation offers a means of obtaining small-diameter ion beams which usually requires a minimum of investment in new beam-handling equipment and can provide ion beams for microbeam work provided that the smallest beam diameters are not required.[33] An important requirement of any ion beam used for microbeam work is that the beam should be well defined, with no halo around the central core, so multiple collimators are generally necessary to remove edge-scattered components. A simple collimation system has been used for a considerable time for producing beams down to about 20 μ in diameter in the Analytical Sciences Division at Harwell, and a 3-stop system has been found to effectively remove edge-scattered particles and give a sharply defined particle beam suitable for microbeam scanning. The difficulty which naturally accrues from the use of a number of small-diameter stops in the beam line is the problem of achieving correct alignment to provide useable beam currents on target. Stops can be mounted on adjustable holders and positioned by micrometer movements, or they can be formed by two offset pairs of adjustable knife edges at 90°; positioning of the stops can be checked by several methods, including measurement of transmitted particle count or optical techniques. However, these processes can be tedious, and complete rigidity of the beam line is absolutely essential to avoid loss of alignment during use of the target chamber or by movement near the beam line. A further difficulty in the use of collimation techniques is the inefficient use of the available particle beam, resulting in low ion counts on target. Nevertheless, collimation has been used to provide relatively small-diameter ion beams with success over a long period of time for analytical work.

Beam focusing offers an alternative means of producing small-diameter ion beams provided that a suitable beam-handling system is available, and it has the major advantage of not reducing the beam intensity on target to the same extent as collimation. Relatively little work has so far been carried out aimed at focusing beams of ions with MeV energies down to small diameters, but where microbeam production has been investigated, well-focused beams have been achieved. One detailed study has demonstrated that beams of 3-MeV protons can be obtained only 3 μ in diameter by employing quadrupole lenses which are carefully designed to avoid aberration and to give a clean beam spot.[34] The focusing system used was a Russian quadruplet[35] consisting of four quadrupoles which alternately focused and defocused the beam in a particular plane to provide an image of the object shape. Symmetrical magnetic fields were obtained by careful construction of the lenses, and the distances between the opposite poles differed by ± 0.013 mm while the average variation in the shortest distance between the neighboring poles on a single magnet was ± 0.045 mm in 14.25 mm. The quadrupoles were placed on a common geometrical axis with an accuracy of about $\pm 50\ \mu$ by optical methods using a telescope and graticule. The magnetic centers of the four quadrupoles coincided with the geometrical centers to within about $\pm 130\ \mu$, which was within the limit of the technique of measurement used. The objective focused by the quadrupole assembly was produced by one of five holes drilled in tantalum disks present in the beam-handling system. The focal length of the overall beam optics was 60.5 cm and the magnification was 0.178. Careful alignment of beam-line components on their support was initially necessary to provide optimum performance, and rigidity was again essential to maintain the quality of the focused beam. While small beam diameters of the order of 3 μ were achieved by focusing, when beam deflection was employed to move the ion beam relative to the sample for scanning, the beam diameter varied with the deflection of the beam, the aberration being roughly proportional to the extent of the deflection. In the experimental system described above, when voltage was applied to a single deflector plate, the additional aberration was found to be about 5.5% of the lateral displacement in the x direction and 6.7% in the y direction; but when equal but opposite potentials were applied to both y plates, the aberration was reduced to about 2.7% of the displacement. Figure 17-5 shows a typical beam profile obtained by measuring the change in yield of protons scattered from a gold film, which was vacuum-evaporated on to the center of an aluminum plate, as the beam was scanned over a gold–aluminum edge. The plot is in the form of dc/dx *vs.* the distance, where c is the particle count and x is the distance moved. The beam width given by the full width at half maximum (fwhm) in this case was 3.8 μ, although smaller diameters of less than 2 μ have been reported for beam diameters measured in a single dimension. Clearly, deviations from circularity of the ion beam

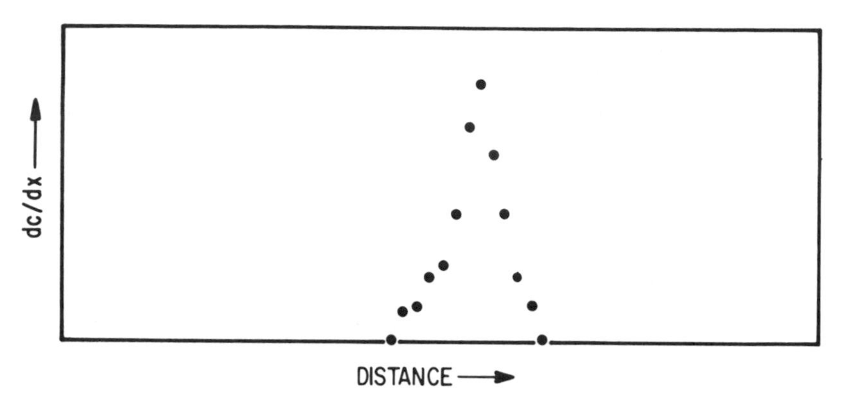

Fig. 17-5. Example of microbeam shape.

may result in variations in distance across the beam along various diameters.

4. EXPERIMENTAL TECHNIQUE

Analytical utilization of small-diameter ion beams requires that they be used in conjunction with experimental techniques and equipment which permit the small diameters to be exploited. Thus sample preparation, beam-line equipment, including target-chamber assembly, and techniques of data processing and interpretation must be able to match the needs of microbeam usage. The levels of sophistication necessary will, of course, be entirely dependent upon the nature of analytical information demanded, so when ion beams of the smallest diameters are not required some simplification to the experimental method is possible. At the present state of the art, accelerators used for microbeam work are also often employed for nuclear physics and other purposes, with the result that equipment must be assembled for microbeam work after the line has been in use for an entirely different purpose. If this is so, any simplification that can be permitted to the specification of analytical measurement can often be exploited in simpler beam assemblies and shorter alignment procedures. The type of radiation measured must govern the detector chosen (i.e., charged-particle, γ-ray, x-ray) and will also dictate the most appropriate type of system for data accumulation and processing. Thus where the variation in intensity with position of a clear component of a particular spectrum is being examined, a single-channel analyzer feeding a rate meter and chart recorder may be adequate, whereas for more complex spectral interpretation, computer data accumulation and processing may be necessary both to limit the analytical effort involved in interpretation of raw data and to permit a more detailed investigation of spectral composition.

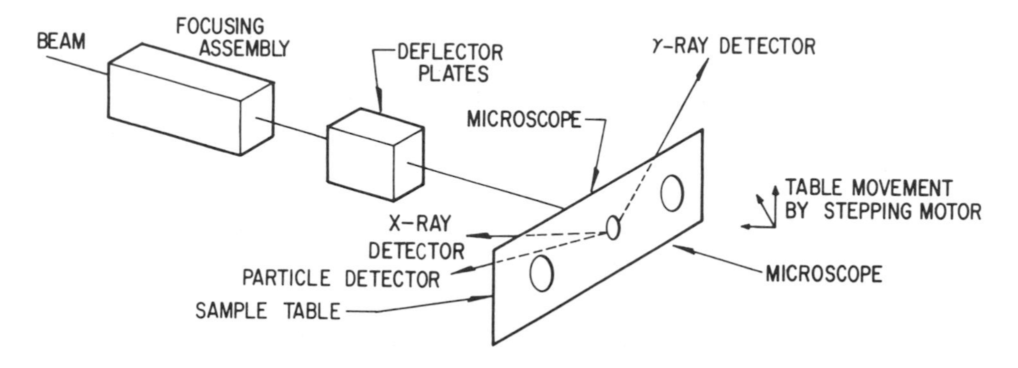

Fig. 17-6. Outline of target-chamber functions.

A general-purpose target chamber used for microbeam studies on the accelerator IBIS at A.E.R.E. Harwell is shown in Fig. 17-6 as a line drawing to simplify the various functions available. The chamber itself is mounted on movements which permit it to be adjusted either horizontally or vertically for initial beam alignment and is electrically isolated from the rest of the beam line to permit beam intensity to be monitored by charge collection. Samples are mounted on a table which can be moved in three dimensions by stepping motors remotely controlled from outside the machine vacuum and are used during initial positioning of the target to ensure that the region of interest of the sample is in the ion beam. The table has positions for the samples, one on either side of a quartz disk which is used for examining beam shape and position. Two microscopes are available to assist in beam focusing and sample viewing, and graticules either in the microscope optics or in the quartz disk can assist in gauging the size of the focused beam and of the extent of beam deflection with applied voltage. Before using the microbeam for sample examination, the sample table is moved until the ion beam falls on to the quartz disk, enabling the beam shape to be examined, and adjustments are made if necessary to the beam-handling system to obtain a suitable microbeam. The sample table is then moved until the region of the target of interest is in the ion beam, and the scanning is started, if positional variations are to be followed. Prompt γ-radiation is measured either with large thallium-activated sodium iodide scintillators or germanium counters positioned outside the main body of the target chamber, or alternatively, with a high resolution 2×1.5-inch thallium-activated sodium iodide scintillator in an insert in the chamber wall close to the point of interaction of the ion beam with the sample. Elastically scattered particles or charged-particle groups are measured by means of a silicon surface-barrier detector placed at a high angle with respect to the incident ion beam to provide reasonable resolution of backscattered particles and is electrically isolated from the chamber; a stop of variable size is placed in front of the detector to limit the sensitive solid angle subtended. Provision is also made for positioning an absorber between the target and detector aperture to prevent low-energy particles, such as those scattered elastically, from entering the detector during particle-group measurements, while permitting the more penetrating particles resulting from nuclear reaction to be detected. X rays are counted with a silicon detector, and the output is fed to either single or multichannel analyzers for data accumulation. Figure 17-6 shows the number of facilities which must be grouped around the sample-irradiation position in a versatile chamber assembly, emphasizing the difficulties in design and practical construction.

The movement of the ion beam relative to the sample for scanning is usually achieved by electrostatic deflection, moving the beam discontinuously and counting only while the beam is stationary. The beam deflection control is therefore used to inhibit the counting circuit so that no counts are accumulated during beam movement, and so that counts can be ascribed to specific positions on the sample surface. Where the high beam intensities produced by microbeams are liable to damage the sample surface, the ion beam is repeatedly scanned rapidly over the sample surface to a predetermined raster, thus reducing the overall localized heating, and reducing the consequent heat damage. Information from a number of high-speed scans is integrated on an oscilloscope or accumulated digitally by collecting counts from particular regions of the sample in specific sectors of an addressable memory.

One of the major problems associated with the determination of carbon and certain other elements by accelerator techniques is the high background which can result from the machine itself. In general, accelerators are not specifically designed for analytical work and the cleanliness available does not approach that attained with specially constructed analytical instrumentation such as, for example, a mass spectrometer. Care must consequently be taken to reduce the overall machine background by careful usage, particularly when the machine is shared with other groups for entirely different purposes, and to avoid memory effects due to appreciable quantities of elements finding their way into the beam-handling system from previous experiments and thus affecting the quality of the analytical results. The presence of carbon is extremely troublesome since carbon determinations are very likely to be part of the analytical load of a microbeam facility. Good housekeeping in maintenance of target assemblies and those parts of the beam lines which are open to the target is essential, and some benefit has been found by having a cold finger cooled by liquid nitrogen very close to the point of interaction between beam and target.

Sample preparation for positive-ion-microprobe work bears many resemblances to long established electron-microprobe techniques. Samples must of course be able to tolerate irradiation *in vacuo* and heat generation by the incident ion beam, while an adequate means of sample mounting must be available both to ensure reliable positioning in the target chamber and to assist in the sample-preparation stage. Many of the standard metallographic techniques of sample preparation are appropriate to the preparation of samples for irradiation, but the more common sample-mounting materials are made up of light elements which may not be appropriate if the purpose of the analytical method is to determine the same light elements present in the sample. Heavy-element mounting materials are therefore more suitable for light-element determination provided that the x-ray yield produced does not interfere with measurement. Similarly, small specimens must be mounted in such a way as to ensure that the radiation yield from the sample does not contain any contribution from the mounting material which would interfere with the analytical measurement and thus invalidate the result.

The extent of data processing which is required will depend upon the nature of the contribution of the measured radiation to the total yield emitted from the sample under charged-particle irradiation. Where the particular radiation upon which analytical measurement is to be based can be uniquely identified, then single-channel analysis is often adequate and provides good analytical results. However, complex spectra require more sophisticated methods of data interpretation, and established techniques of processing γ-ray, particle and x-ray spectra can be applied. Data accumulation can be a problem, particularly when multichannel spectra are obtained from irradiation of a large number of points, and on-line computers, or at least data-accumulation systems permitting rapid output of information in a form convenient for computer processing, can save substantial amounts of experimental time.

5. EXAMPLES OF THE APPLICATION OF CHARGED-PARTICLE-MICROPROBE TECHNIQUES TO ANALYTICAL INVESTIGATIONS

The analytical outlets for the positive-ion microprobe are primarily governed by the measurement techniques described in Section 2, combined with the particular characteristics imposed by the use of small-diameter ion beams. The versatility of the methods available offers a choice which may well be modified by practical considerations. Thus the measurement of prompt γ radiation, charged-particles, or characteristic x rays for the determination of the light elements will depend not only upon the ease with which the radiation from the wanted element can be distinguished from the other radiations likely to be emitted by the sample at the existing levels under investigation, but also upon the nature of the emitting region compared with the region of the sample to be investigated. The depth of the yield from charged-particle reactions will depend entirely upon the nature of the reaction and upon the variation of radiation yield with incident particle energy. Thus, for example, major yields of γ radiation resulting from reactions with sharp resonances enable 0.1-μ resolution in subsurface profiles to be obtained, while it has been calculated for x rays that 90% of the Cu K_α line produced from the solid copper sample by irradiation with microbeams of 2.5-MeV protons will be emitted from a sample depth of 10 μ of the sample surface. In view of the many types of measurement possible with positive-ion microbeams, the objective of the analysis must be clearly defined, if this is possible, before sample irradiation commences, both to permit the best analytical technique to be selected and to restrict the amount of experimental work and accelerator time to that necessary to provide useful analytical information, avoiding the accumulation of unnecessarily detailed information about elemental or spatial compositions. The examples given in this section are chosen solely to illustrate the potential analytical capabilities of accelerator microbeams, and have been chosen to demonstrate one application of each of the radiations available for measurement. Figure 17-7 shows the variation in levels of fluorine across a coated metal surface. The 6.1-MeV γ rays emitted during the de-excitation of ^{16}O produced by the reaction

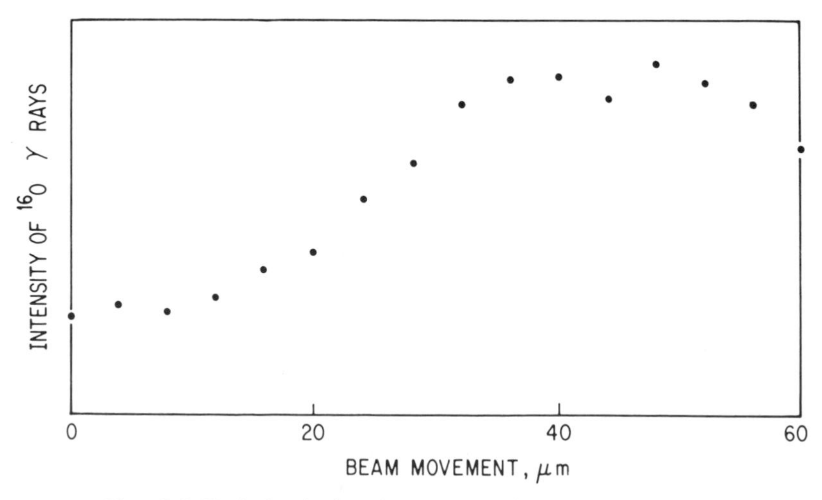

Fig. 17-7. Variation in fluorine content of coated metal surface.

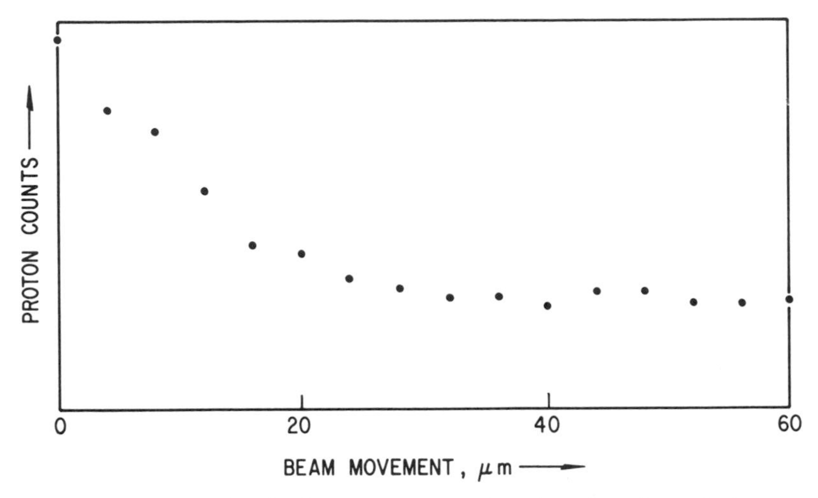

Fig. 17-8. Carbon distribution across a steel surface found by deuteron-microbeam scanning.

$^{19}F(p, \alpha)\,^{16}O$ provide a measure of fluorine content, and the yield is followed with position of the incident proton beam. The variation in carbon content across a thin, sectioned metal plate is given in Fig. 17-8. Protons from the reaction $^{12}C(d, p)\,^{13}C$ are counted to provide a measure of carbon content. The α particles emitted as a result of the reaction $^{18}O(p, \alpha)\,^{15}N$ have been measured to give the distribution of oxygen across welds.[36] Figure 17-9 also illustrates the use of microbeam

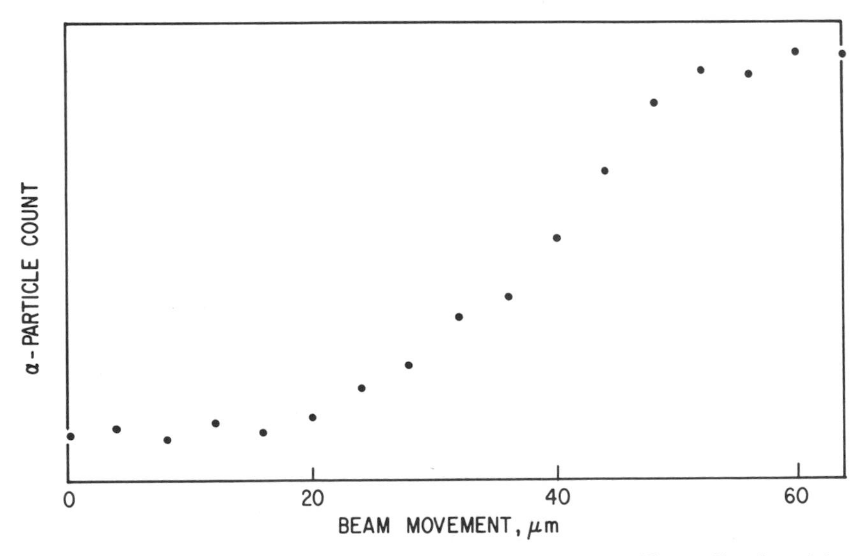

Fig. 17-9. Variation in thickness of gold layer present on silicon slice found by elastic α-particle scattering.

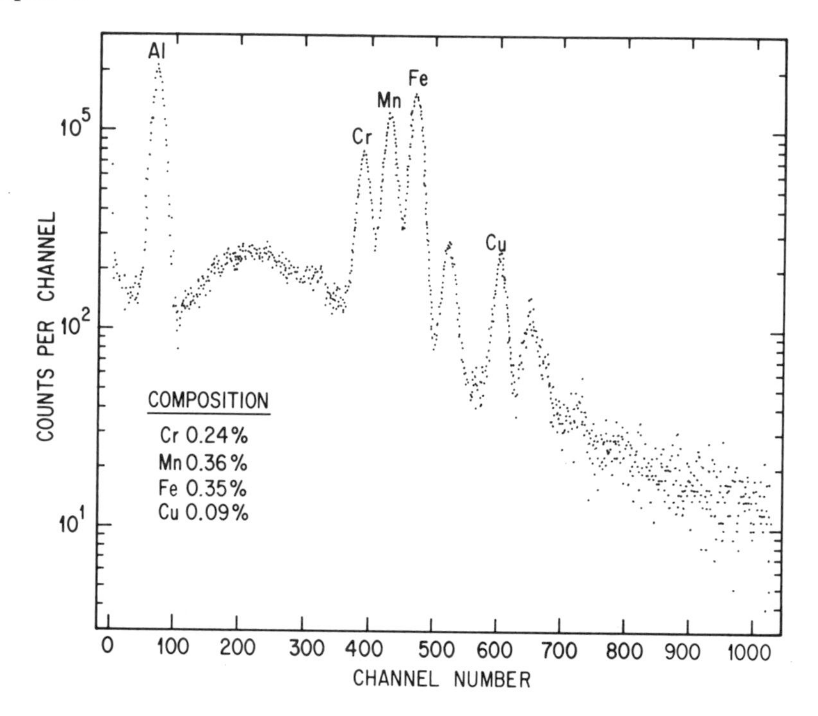

Fig. 17-10. X-ray spectrum for Al alloy. Alloy composition: 0.24% Cr, 0.36% Mn, 0.35% Fe, 0.09% Cu.

techniques for examining elemental distributions. In this case, the distribution of gold in a thin film present on a silicon surface is found by elastic α-particle scattering. The x-ray spectrum in Fig. 17-10 permits identification of an aluminum alloy and was obtained from a fragment approximately 100 μ across.

A chapter such as this ought perhaps to conclude with an assessment of the potential analytical value of accelerator microbeams. As stated in an earlier section, so little work with real microbeams has so far been reported that a meaningful judgment cannot yet be made. However, even with target-chamber and beam-handling systems that are far from fully developed, the technique has found valuable application in these laboratories, thus suggesting that in spite of the high capital costs of instrumental installations, charged-particle-microbeam techniques warrant further attention from the analytical scientist.

6. REFERENCES

1. G. Hevesy and H. Levi, *Det. Kgl. Dansk. Viden. Selsk. Math.-Fys. Medd.* **14**, 3 (1936).
2. G. T. Seaborg and J. J. Livingood, *J. Am. Chem. Soc.* **60**, 1784 (1938).
3. D. M. Holm, W. M. Sanders, W. L. Briscoe, and J. L. Parker, in *Nucleonics in Aerospace* (P. Polishuk, ed.) p. 306, Plenum Press, New York (1968).

4. T. B. Pierce, *Proc. Soc. Anal. Chem.* **7**, 59 (1970).
5. Y. A. Dzemard'Yan, G. I. Mikhailov, and L. P. Starchik, *Ind. Lab.* **37**, 708 (1971).
6. R. F. Sippel and E. D. Glover, *Nucl. Instr. Meth.* **9**, 37 (1960).
7. T. B. Pierce, P. F. Peck, and W. M. Henry, *Analyst* **90**, 339 (1965).
8. T. B. Pierce, P. F. Peck, and W. M. Henry, *Nature* **204**, 571 (1964).
9. D. J. Macey and W. B. Gilboy, *Nucl. Instr. Meth.* **92**, 501 (1971).
10. E. Ricci (private communication).
11. V. F. Zelenskii, O. N. Khar'Kov, V. S. Kulakov, and N. A. Skakum, *Protect. Met.* **6**, 235 (1970).
12. E. Moller and N. Starfelt, Aktiebelaget Atomenergi, Sweden, **237** (1966).
13. G. M. Padawer, *Nuclear Appl. Technol.* **9**, 856 (1970).
14. W. D. Mackintosh, Nuclear Technology 13 (1972), 65.
15. M. Peisach, *J. Radioanal. Chem.* **12**, 257 (1972).
16. T. B. Pierce, in *Proceedings of the 2nd Conference on Practical Aspects of Activation Analysis with Charged Particles, Liege, 1967* (H. G. Enert, ed.) Report EUR3896, d-f-e, p. 389, Brussels (1968).
17. G. Deconninck and G. Demortier, in *Colloquium on the Application of Nuclear Methods in the Basic Metal Industries, Helsinki, 1972*, I.A.E.A., Vienna (1973).
18. T. B. Pierce, P. F. Peck, and D. R. A. Cuff, *Anal. Chim. Acta* **39**, 433 (1967).
19. R. Pretorius and P. Coetzee, *J. Radioanal. Chem.* **12**, 301 (1972).
20. J. F. Chemin, J. Roturier, B. Saboya, and G. Y. Petit, *J. Radioanal. Chem.* **12**, 221 (1972).
21. C. Olivier and M. Peisach, *J. Radioanal. Chem.* **12**, 313 (1972).
22. G. Amsel, J. P. Nadai, E. D'Artemare, D. David, E. Girard, and J. Moulin, *Nucl. Inst. Meth.* **92**, 481 (1971).
23. G. Weber and L. Quaglia, *J. Radioanal. Chem.* **12**, 323 (1972).
24. M. Cuypers, L. Quaglia, G. Robaye, P. Dumont, and J. N. Barrandon, in *Proceedings of the 2nd Conference on Practical Aspects of Activation Analysis with Charged Particles, Liege*, 1967 (H. G. Ebert, ed.) Report EUR3896 d-f-e, p. 371, Brussels (1968).
25. E. A. Wolicki and A. R. Knudson, *Int. J. Appl. Rad. Isotop.* **18**, 429 (1967).
26. E. Moller, L. Nilsson, and N. Starfelt, *Nucl. Inst. Meth.* **50**, 270 (1967).
27. T. B. Pierce, P. F. Peck, and D. R. A. Cuff, *Analyst* **97**, 171 (1972).
28. J. Chadwick, *Phil. Mag.* **24**, 594 (1912).
29. P. B. Needham and B. D. Sartwell, in *Advances in X-ray Analysis* (C. S. Barret, J. B. Newkirk and C. D. Rund, eds.) Vol. 14, p. 184, Plenum Press, New York (1971).
30. D. M. Poole and J. L. Shaw, U.K. Atomic Energy Authority Report AERE-R 5918 (1968).
31. G. A. Bissinger, J. M. Joyce, E. J. Ludwig, W. S. McEver, and S. M. Shafroth, *Phys. Rev.* **A1**, 841 (1970).
32. T. B. Johansson, R. Akselsson, and S. A. E. Johansson, *Nucl. Inst. Meth.* **84**, 141 (1970).
33. T. B. Pierce, P. F. Peck, and D. R. A. Cuff, *Nucl. Instrum. Meth.* **67**, 1 (1968).
34. J. A. Cookson and F. D. Pilling, U.K. Atomic Energy Authority Report AERE-R-6300 (1970).
35. A. D. Dymnikov, T. Fishkova, and S. Yavor, *Sov. Phys.—Tech. Phys.* **10**, 340 (1965).
36. P. B. Price and J. R. Bird, *Nucl. Instr. Meth.* **69**, 277 (1969).

18

ELECTRON PROBE MICROANALYSIS

Gudrun A. Hutchins

Sprague Electric Company
Research and Development Laboratories
North Adams, Massachusetts

1. INTRODUCTION

Electron probe microanalysis is an analytical technique that may be used to determine the chemical composition of a solid specimen weighing as little as 10^{-11} gram and having a volume as small as one cubic micron. The primary advantage of electron probe microanalysis over other analytical methods is the possibility of obtaining a quantitative analysis of a specimen of very small size.

During the analysis, the selected area of the specimen is bombarded with a beam of electrons. The accelerating voltage of the electrons (typically 10–30 kV) determines the depth of penetration into the specimen. The degree of beam focusing determines the diameter of the analyzed volume. The electron bombardment of the specimen causes the emission of an x-ray spectrum that consists of characteristic x-ray lines of elements present in the bombarded volume. The chemical analysis is accomplished by the dispersion of this x-ray spectrum and the quantitative measurement of the wavelength and intensity of each characteristic line. The wavelengths present identify the emitting elements, and the line intensities are related to the concentration of the corresponding elements.

This chapter describes the variety of specimens suitable for electron probe microanalysis and the advantages and limitations of the analytical technique. A brief description of the instrument, measurement procedures, x-ray generation, and interelement effects is followed by the discussion of specific types of analyses and examples. Less-common applications, such as thin-film analysis, are stressed because typical applications in metallurgy and mineralogy have been treated previously in excellent general papers.[1,2]

2. FUNDAMENTALS OF ELECTRON PROBE MICROANALYSIS

2.1. The Instrument

A number of good electron probe microanalyzers are commercially available and are described in detail in the literature.[3,4] For the types of analyses to be discussed in this chapter, the following four major-instrument subsystems are required.

1. An electron optical system of high stability is needed to produce a focused beam of electrons on the specimen. The electron energy should be variable in steps from 5 to 30 keV. A magnetic beam-sweeping system for two-dimensional x-ray or current scans should be included.
2. A specimen airlock, a stage with *xyz* motion, and an optical microscope must be incorporated into the instrument so that the desired area of the specimen can be positioned under the electron beam. A stage drive for one-dimensional continuous or step scanning is a useful accessory.
3. An energy or wavelength spectrometer is required to disperse the x rays so that the characteristic lines can be assigned to specific elements. Measurements with a wavelength spectrometer are more time consuming, but have a much higher resolution and are, therefore, more accurate for quantitative and trace analyses. An x-ray detector and associated electronics are necessary to measure x-ray line intensities.
4. Readout and recording electronics are needed to display and record the characteristic x-ray intensities as a function of energy, wavelength, and/or specimen position.

Subsystems 1–3 cannot be individually optimized and incorporated into the same instrument because of space and mechanical limitations. The compromises made depend upon the specific design and account for most of the differences between commercial instruments. One important instrument parameter is the x-ray take-off angle ψ. It is the angle between the measurement direction of the spectrometer and the tangent to the specimen surface. To minimize absorption and surface roughness effects, it is desirable to make ψ as large as possible. This requires an objective lens of special design (minilens, inverted lens, lens with an x-ray window) or a tilted specimen with nonnormal electron beam incidence.

2.2. Types of Analyses

There are two basic types of analyses, and both may be either qualitative or quantitative. A spot analysis consists of an analysis for all detectable elements on one spot of a much larger specimen. This analysis may be representative of the entire specimen or it may be an analysis of an unusual region. A distribution analysis determines the distribution of one or more elements as a function of position on the specimen. A distribution analysis is used to detect compositional gradients on a specimen surface; the average composition of the specimen is very often known from a bulk analysis performed by other methods.

A qualitative spot analysis can be completed relatively quickly by scanning the spectrometer(s) through the portion of the x-ray spectrum detectable with the instrument. A strip-chart recording of x-ray intensity *vs.* wavelength or an oscilloscope trace of x-ray intensity *vs.* energy is obtained. Peaks are assigned to emitting elements with the aid of tables.[5,6] An experienced operator can often estimate the weight concentration of an element from the recorded x-ray intensity with an uncertainty of less than a factor of two.

A quantitative analysis requires considerably more analysis time. Relative x-ray intensities of specimen to standard are measured to cancel out uncertain terms, such as the absolute spectrometer efficiency. The standards may be either pure elements or well-characterized compounds or alloys. The measured x-ray intensities are corrected for background, detector nonlinearity, and possible instrumental drift. The corrected intensity ratio for each element is then adjusted by the corrections for quantitative analysis (see Section 2.5) to obtain the specimen weight concentration for that element. The magnitude of these corrections depends upon the composition of the specimen and calculations are done by successive approximations.

Distribution analysis has become increasingly more important in material and surface characterization. Many physical and electrical properties of materials depend upon homogeneity or upon the size and distribution of a second phase to a greater degree than they do upon average composition. To determine the variation in a specific element, a spectrometer is peaked for an x-ray line of that element and the beam is moved across the specimen while the x-ray intensity *vs.* position is recorded.

A one-dimensional line scan may be produced by sweeping the beam across a stationary specimen with the magnetic deflection coils, or by driving the specimen stage under a stationary beam. The stage-drive method yields more accurate x-ray intensities for long scans with the curved-crystal focusing spectrometers in common use, because it avoids the problem of spectrometer-defocusing as the beam is swept away from the focusing circle. Two-dimensional scans may be of considerable help in the analysis of samples containing several phases. They are produced by sweeping the beam over an area on the specimen in synchronization with the sweeping spot on an oscilloscope. The intensity of x-ray, current, or other signals is used to modulate the brightness of the oscilloscope spot. The resulting images are qualitative compositional maps.[7–9] Two-dimensional scans are best suited for large compositional variations in very complex samples. Small variations in an x-ray signal of high intensity and large compositional variations in a very small area tend to be indistinguishable from noise.

For specimens containing more than two elements, quantitative distribution analyses require point-by-point analyses of all elements present. If these analyses are done routinely, they are very time-consuming. The capacity for data accumulation can be greatly increased by computer-operated instruments with on-line data reduction that can be operated 24 hours a day.

2.3. Electron Scattering

As the high-energy electrons penetrate into the specimen, they are scattered elastically by the screened atomic nuclei and inelastically by the atomic electrons. The proportion of elastic and inelastic collisions depends upon the atomic number of the specimen. *Elastic scattering is more pronounced in an element with a high atomic number; energy loss is more rapid in an element with a low atomic number.* The following approximate numbers for gold and carbon will give an indication of the degree of variation with atomic number. For an electron energy of 30 keV, the mean free path between scattering events in Au is ~ 30 Å, approximately 70% of all scattering events are elastic, and the average scattering angle per event is near 2°. The mean free path in C is ~ 80 Å, less than 20% of all scattering events are elastic, and the average scattering angle per event is approximately 0.35°. The energy lost per unit mass distance $\Delta \rho x$ along the path of a 30-keV electron is at least three times as great in C as in Au.

The scattering and energy loss of electrons in a specimen containing a mixture of Au and C is intermediate and considerably different from both pure elements. This difference in electron propagation affects the production of characteristic x rays and causes interelement effects. A thorough understanding of electron scattering is, therefore, a requirement for quantitative electron probe analysis. A detailed discussion of electron scattering applicable to electron probe analysis can be found in the thesis of Bishop;[10] a brief description of the most important features is given in the following paragraphs.

2.3.1. Energy Losses

The loss of energy at each inelastic collision is difficult to calculate. In practice, it is necessary to replace discrete energy losses by a continuous mean loss. In the treatment of Bethe and others, energy loss is described by the stopping power S defined as

$$S = -(1/\rho)\, dE/dx \tag{1}$$

where dE is the *average* increment of energy lost by an electron along the path length dx in a material of density ρ. A suitable expression for S is

$$S = \text{const}(Z/A)(1/E)\ln(1.166\, E/J) \tag{2}$$

where Z and A are the atomic number and weight, respectively, E is the electron energy in electron volts, and J is the mean ionization potential of the target atoms (roughly equal to $11.5Z$ in electron volts). Both Z/A and the logarithmic term decrease with increasing Z.

2.3.2. Elastic Scattering

The degree of elastic scattering determines the spatial profile of electrons below the specimen surface, and also the number and energy of electrons scattered back out of the surface. In a high-Z element, the electrons lose their initial direction very quickly and the electron profile within the specimen approaches the shape of a hemisphere. In a low-Z element, elastic scattering is less pronounced and the electron profile approaches the shape of a sphere truncated by the specimen surface.

The high-energy electrons that are scattered back out of the specimen while still retaining much of their energy are called backscattered electrons. Their number and energy distribution as a function of Z can be measured experimentally with a good degree of accuracy. The backscattered electron fraction η is defined as

$$\eta = \frac{i_b - i_s}{i_b} \tag{3}$$

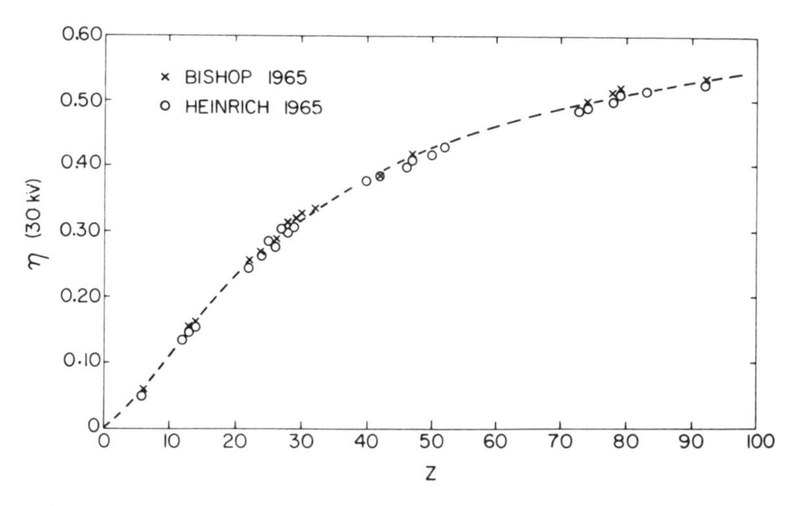

Fig. 18.1. Experimentally determined backscattered electron fraction η as a function of atomic number Z of the specimen.

where i_b is the incident beam current and i_s is the current measured through the specimen. Measurements of η by Bishop[11] and Heinrich[12] are shown in Fig. 18-1. Electrons backscattered from a high-Z element have a greater average energy than those backscattered from a low-Z element.[11] Thus, much more of the incident beam energy is unavailable for x-ray production in an element with a high atomic number; more electrons are backscattered and the average backscattered electron has a greater energy.

2.4. X-Ray Emission

The generation of a characteristic x-ray photon first requires the ejection of an inner-shell electron from a specimen atom by a high-energy incident beam electron. The second step is the transition of an outer-shell electron to the vacant site. The final step is the emission of a photon with an energy equal to the energy difference between the excited and final atomic states. The energy (or wavelength) of the x-ray photon is thus characteristic of the emitting atom. The production of an x-ray photon is a relatively rare event. A 30-keV electron has 3.3 times the energy required for copper K-shell ionization, yet its probability of generating a Cu K_α photon before it comes to rest is only approximately 3×10^{-3}.[13]

2.4.1. Ionization Cross Section

The probability of c-shell ionization is given by the ionization cross section $Q_c(E)$. (The letter c will be used as a general subshell designation;

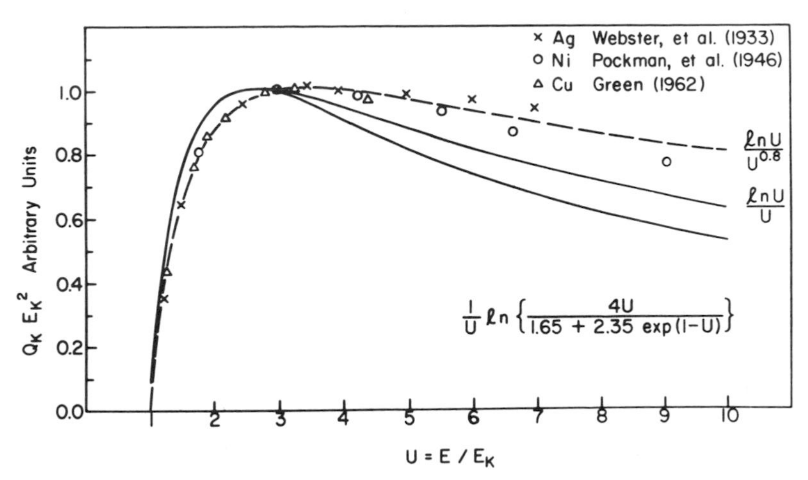

Fig. 18.2. Relative values of the ionization cross section as a function of U. All curves and experimental data have been fitted to the same value for $U = 3$. Experimental data are from Webster *et al.*,[16] Pockman *et al.*,[17] and Green.[55] The curves show three different energy dependencies.[15]

$c = K, L_{\mathrm{I}}, L_{\mathrm{II}}$, etc.) The ionization cross section is defined as the number of c-shell ionizations per unit length along the path of an electron of energy E, divided by the number of atoms per unit volume. The magnitude of Q_K is in the range of 10^{-22}–10^{-20} for typical electron-probe beam energies. Absolute values of Q are extremely difficult to measure.[14]

The energy dependence of Q is most easily expressed as a function of the overvoltage $U = E/E_c$, the ratio of the electron energy E to the ionization energy E_c required to remove an electron from the c shell. The precise form of the energy dependence is open to debate and depends upon the boundary conditions chosen in the derivation of Q.[15] Figure 18-2 gives relative values of Q as a function of U obtained experimentally and calculated from several energy dependencies. The writer favors an expression of the form

$$Q = \mathrm{const}(1/E_c^{\,2})(\ln U)(1/U^m) \qquad (4)$$

with m between 0.7 and 1. More complex equations offer no apparent advantage.

2.4.2. Fluorescence Yield

The probability of an electron vacancy in the c subshell resulting in the emission of an x-ray photon is given by the atomic fluorescence yield ω_c. Values of the fluorescence yield vary from <0.01 to 0.60 for x-ray lines used in electron probe analysis. The fluorescence yield for any particular subshell increases with E_c. For low energy values, a competing process, Auger electron emission, is much more efficient than characteristic x-ray emission. Auger electron emission also requires the creation of an inner-shell vacancy and the transition of an outer-shell electron to the vacant site. However, the difference in binding energy is transferred to another bound electron which is ejected from the atom. Thus, the atomic fluorescence yield is really the probability that the difference in binding energy is released as an x-ray photon rather than as an Auger electron. Extensive theoretical calculations and experimental measurements of atomic fluorescence yields have been made and are reviewed in a comprehensive paper by Fink *et al.*[18]

2.4.3. Calculation of Emitted X-Ray Intensity

It would be ideal to be able to calculate accurately the absolute number of x-ray photons emitted by a specimen under certain analytical conditions. Inaccuracies are reduced when the calculation is limited to relative numbers of x-ray photons or relative x-ray intensity because the constants in the equations for Q and S need not be known and ω_c cancels out of much of the calculation. This is one of the reasons why relative x-ray intensities are measured in quantitative electron probe analysis. The primary limitation in the calculations is the statistical nature of electron scattering and energy losses. It is necessary to devise a model of the physical situation which can be treated mathematically; these models by necessity involve a high degree of simplification. At this time, there is no fundamentally derived mathematical model that will accurately predict the experimentally measured values for the backscattered electron fraction, the energy distribution of backscattered electrons, the depth distribution of generated x rays, and the various interelement effects. Two approaches which have come very close are a Boltzmann transport equation adapted to electrons by Brown *et al.*[19,20] and a Monte Carlo procedure used by Bishop[10,21] and others to make a random sampling of synthesized individual electron trajectories. Both of these calculations use a continuous-energy-loss approximation because detailed information about discrete energy losses for inelastic collisions is not available. In practice, most electron probe analysts calculate the corrections necessary for quantitative electron probe analysis with a system of individual models for different aspects of the problem. Used within their range of applicability, these models are fairly successful.

2.5. Corrections for Quantitative Analysis

The measurement of Cu K_α in a 40Cu–60Au specimen and a pure-Cu standard is used as an example throughout this section. As a first approximation, the measured intensity ratio of Cu K_α in the alloy and standard is equal to the weight fraction of Cu in the alloy. To obtain a more accurate analysis, the effect of the presence of the Au on the measurement of Cu in the specimen has to be calculated. The accuracy of this calculation rather than the precision of the x-ray intensity measurement will limit the absolute accuracy of the analysis.

The interelement effects can be separated into the following four categories:

1. The atomic number effect is due to the differences in electron scattering and energy losses in sample and standard. For the Cu–Au alloy, the presence of Au increases the elastic scattering and reduces the stopping power. The net effect is an enhancement of the Cu K_α signal in the alloy relative to Cu K_α in Cu.
2. The x-ray absorption effect is due to the difference in the degree of absorption of characteristic x-ray photons in the specimen and in the standard. Absorption of generated Cu K_α photons is greater in the Cu–Au alloy than in pure Cu. This causes a net decrease in the Cu K_α signal measured in the alloy.
3. The characteristic-line fluorescence effect is due to the production of characteristic x rays by the characteristic x-ray lines of other elements present in the specimen. In the Cu–Au alloy, some Cu K_α photons are produced as a result of the absorption of Au L_α and Au L_β photons. This causes a net enhancement of the Cu K_α signal measured in the alloy.
4. The continuous-spectrum fluorescence effect is due to the production of characteristic x-rays by the continuous x-ray spectrum. In pure Cu, some Cu K_α photons are produced by the Cu continuous spectrum. Au has a stronger continuous spectrum and, therefore, more Cu K_α photons are produced by the continuous spectrum in the alloy.

If a good mathematical model were available, corrections for the various effects could be calculated in one step. At this time, it is more accurate to treat them individually, relying on experimental data and simple physical models. The current status of the corrections is discussed in recent reviews by Heinrich[22] and by Beaman and Isasi.[4] A brief explanation of the most common correction procedure is given in the rest of this section.

The corrections are generally expressed as multiplying factors and their order is not important. The weight concentration C_{Cu} of Cu in the

Cu–Au specimen is given by

$$C_{\mathrm{Cu}} = k\begin{bmatrix}\text{Atomic number}\\ \text{correction}\\ \text{factor}\end{bmatrix}\begin{bmatrix}\text{Absorption}\\ \text{correction}\\ \text{factor}\end{bmatrix}\begin{bmatrix}\text{Fluorescence}\\ \text{correction}\\ \text{factors}\end{bmatrix} \quad (5)$$

where k is the experimentally measured Cu K_α x-ray intensity ratio. Each of the correction factors depends upon the composition of the specimen and standard. Thus, several rounds of iteration through all applicable corrections are required for a specimen of unknown composition. The current trend is to feed all quantitative data into a time-shared computer for calculation; a number of good programs have been written for this purpose.[22,23]

2.5.1. Atomic Number Correction

The variation of electron scattering with the atomic number of the specimen has the following three separate consequences:

1. A decrease in stopping power for high Z.
2. An increase in the backscattered electron fraction for high Z.
3. A shallower distribution of generated x-ray photons within the specimen for high Z.

The first two items directly affect the *number* of x-ray photons produced and are calculated as the two parts of the so-called atomic number correction. The third item only affects the *distribution* of the generated x-ray photons within the specimen and is treated with the absorption correction.

For the 40Cu–60Au alloy, a decrease in the stopping power S due to the presence of Au means that the average beam electron will travel a greater mass distance in the alloy than in pure Cu. During the longer trajectory in the alloy, the average electron has a greater probability of producing a Cu K_α photon. If backscattering of electrons and fluorescence effects are neglected temporarily, the ratio of the number of Cu K_α photons generated in the alloy and in pure Cu can be expressed as

$$C_{\mathrm{Cu}}\frac{\int_{E_{\mathrm{Cu}\,K}}^{E_0}(Q_{\mathrm{Cu}\,K}(E)/S_{\mathrm{CuAu}})\,dE}{\int_{E_{\mathrm{Cu}\,K}}^{E_0}(Q_{\mathrm{Cu}\,K}(E)/S_{\mathrm{Cu}})\,dE} \quad (6)$$

where E_0 is the incident electron energy and $E_{\mathrm{Cu}\,K}$ is the ionization energy below which Cu K_α cannot be excited. Constants and the atomic fluorescence yield common to sample and standard have been dropped.

After substitution of Eqs. (2) and (4) in Eq. (6), the integrals are of the form

$$\int_{E_{\mathrm{Cu}\,K}}^{E_0} \frac{(1/E^m)\ln(E/E_{\mathrm{Cu}\,K})}{(Z/A)(1/E)\ln(1.166\,E/J)}\,dE \tag{7}$$

The values of Z, A, and J depend upon the specimen composition and S for a multielement specimen is given by

$$S = \sum_i C_i S_i \tag{8}$$

where C_i and S_i are the weight concentration and stopping power of the ith element. The integrals may be evaluated as terms of a logarithmic integral function.[24] Alternatively, the values of S in Eq. (6) may be evaluated at a mean energy and taken out of the integrals which then cancel. The stopping power S varies slowly with E, and this approximation introduces an error of less than 1 % in most cases.[25] Equation (6) then reduces to

$$\text{Stopping-power effect} = \frac{C_{\mathrm{Cu}}(1/\bar{S}_{\mathrm{CuAu}})}{(1/\bar{S}_{\mathrm{Cu}})} \tag{9}$$

where $\bar{S}$ is evaluated from Eq. (2) at a mean energy $\bar{E} = \frac{1}{2}(E_{\mathrm{Cu}\,K} + E_0)$. The value of Eq. (9) calculated for Cu K_α in the 40Cu–60Au alloy with $E_0 = 20$ keV is 1.279 C_{Cu}. The largest source of error in the calculation of $\bar{S}$ is the uncertainty in the values of the mean ionization potential J.[56]

The second part of the atomic number correction is the calculation of the x-ray intensity lost because some of the beam electrons are backscattered while they still retain much of their initial energy. More energy is lost due to backscattering in the Cu–Au alloy than in Cu, and this causes a decrease in the number of Cu K_α photons generated in the alloy. The backscatter effect is always in the opposite direction as the stopping-power effect. The stopping-power correction is generally dominant and causes an enhancement of a low-Z element in a multielement sample.

The correction is made by the introduction of a coefficient R defined as the ratio of the intensity actually generated to that which would be generated if all of the incident electrons remained within the specimen. The coefficient R is always less than unity, and $(1 - R)$ is the intensity lost due to backscattering. If all backscattered electrons retained their original energy E_0, then $(1 - R)$ would be equal to η. Since they possess a range of energies from E_0 down, $(1 - R)$ is less than η by an amount which depends upon the energy distribution of the backscattered electrons.

Both the values of η and the shape of the energy distribution curves are very nearly independent of E_0 in the 10 to 30-keV range. It is thus most convenient to plot the energy distribution as a function of $W = E/E_0$ rather than E. The $d\eta/dW$ curve for a particular element can then be used for all values of E_0. From experimental data for η and $d\eta/dW$ as a function of Z, it is possible to calculate $(1 - R)$ by numerical integration of the equation

$$1 - R = \frac{\int_{W_c}^{1} d\eta/dW \int_{E_c}^{WE_0} Q/S \, dE \, dW}{\int_{E_c}^{E_0} Q/S \, dE} \tag{10}$$

This computation has been carried out by Duncumb and Reed, who have published a table of R *vs.* Z and $U_0 = E_0/E_c$ that may be interpolated to calculate all values of R.[25] This table is used almost exclusively by electron probe analysts for the calculation of R.

For the 40Cu–60Au example, the decrease in the Cu K_α signal due to backscattering is given by

$$\text{Backscattering effect} = C_{\text{Cu}} R_{\text{CuAu}}/R_{\text{Cu}} \tag{11}$$

Calculated values for $E_0 = 20$ keV are: $R_{\text{CuAu}} = 0.789$, $R_{\text{Cu}} = 0.878$, R ratio $= 0.898$. To determine the atomic-number correction factor in Eq. (5), Eqs. (9) and (11) have to be inverted and combined. Thus, for the Cu–Au specimen, the

$$\text{Atomic number correction factor} = \frac{\bar{S}_{\text{CuAu}}}{\bar{S}_{\text{Cu}}} \frac{R_{\text{Cu}}}{R_{\text{CuAu}}} \tag{12}$$

and has a value of 0.870. The probable error in this value due to uncertainties in the experimental data and the physical model is estimated to be ± 0.015.

2.5.2. Absorption Correction

Some x-ray photons generated below the specimen surface will be absorbed by overlying layers of the specimen. The degree of absorption varies as an exponential function of $\mu\rho z \csc \psi$. The parameter μ is the mass absorption coefficient in cm^2/g for a particular wavelength in a particular absorbing element; values of μ have been tabulated by Heinrich[26] and others. The absorption path length is $\rho z \csc \psi$, where ρz is the perpendicular mass depth and ψ is the instrumental x-ray take-off angle defined in Section 2.1. If the absorber is a compound or an alloy, a weight-fraction average of μ is used. The absorption parameter $\chi = \mu \csc \psi$ is substituted for ease of notation in much of the probe literature and in the remainder of this discussion.

The absorption function $f(\chi)$ is defined as the ratio of the intensity actually measured to the intensity that would be measured in the absence of absorption. The evaluation of $f(\chi)$ requires integration over the x-ray ionization function $\phi(\rho z)$, which gives the distribution of excited x-ray intensity as a function of mass thickness ρz.

$$f(\chi) = \frac{\int_0^\infty \phi(\rho z) \exp(-\chi \rho z)\, d(\rho z)}{\int_0^\infty \phi(\rho z)\, d(\rho z)} \tag{13}$$

If $\phi(\rho z)$ is known from experimental measurements, $f(\chi)$ can be evaluated by numerical integration of Eq. (13). The function $\phi(\rho z)$ varies rapidly with E_0 and more slowly with Z. It has been determined experimentally by the following trace-layer method: A thin trace layer of element X is evaporated onto a polished solid piece of element Y. The trace layer is then covered by different thicknesses of evaporated layers of Y. The x-ray intensity from the trace layer of X buried under different thicknesses of Y is measured and corrected for absorption by the overlying layers of Y. The data are normalized by the intensity measurement of trace layer X without a substrate (i.e., supported on a fine grid); this intensity measurement is given the value of $\phi = 1$.

The original trace-layer data of Castaing and Descamps[27,28] at 29 keV are replotted on a linear scale in Fig. 18-3. Each curve has a maximum at some depth below the surface because the incident electrons lose their original direction while still retaining much of their energy. Thus, at some depth, the layer $\Delta\rho z$ is traversed at various angles by the scattered electrons and is of greater effective thickness than $\Delta\rho z$ at the surface. The maximum for gold occurs at a smaller mass depth than that for aluminum because the incident electrons are scattered more rapidly in gold. The value of ϕ at $\rho z = 0$ (often referred to as ϕ_0) is greater than one due to the presence of the bulk substrate. Some of the incident electrons are backscattered and may excite x rays during both traverses through the trace layer on the surface.

Any $\phi(\rho z)$ function determined experimentally by the trace-layer method is strictly valid only for the particular electron energy, trace layer, and primary element chosen. The three variables are only loosely interconnected and their effects can be separated to a large degree by varying them one at a time.

The first $f(\chi)$ curves were calculated by numerical integration over the $\phi(\rho z)$ curves shown in Fig. 18-3. Philibert derived an analytical expression for $f(\chi)$ from a simplified scattering model and fitted it to the available experimental data.[29] The Philibert model was improved by Duncumb and Shields through the introduction of a dependence on

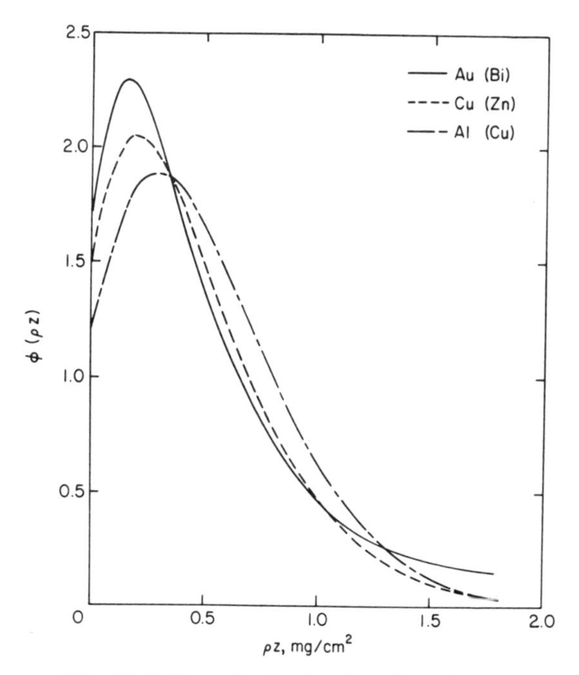

Fig. 18-3. X-ray ionization functions $\phi(\rho z)$ experimentally determined by Castaing and Descamps for a beam energy of 29 keV. The curves are for gold with a bismuth trace layer, copper with a zinc trace layer, and aluminum with a copper trace layer.

E_c, and new constants were suggested by Heinrich.[30-32] The latest expression, which is fairly accurate for $f(\chi) > 0.5$, is

$$f(\chi) = \frac{1 + h}{(1 + \chi/\sigma_c)[1 + h(1 + \chi/\sigma_c)]} \tag{14}$$

with $h = 1.2\ A/Z^2$ and $\sigma_c = 4.5 \times 10^5/(E_0^{1.65} - E_c^{1.65})$ (E in keV). The relatively small Z dependence is contained in the h term. The primary dependence of $f(\chi)$ is on χ/σ_c. Since $\chi = \mu \csc \psi$ is generally fixed by the specimen and the available instrument, the only means of minimizing absorption is to use a relatively low beam energy E_0. Equation (14) is not applicable for very soft x rays when χ/σ_c becomes very large and $f(\chi)$ becomes very small. Alternate expressions are difficult to evaluate because of the very large errors in μ for soft x rays.

For the 40Cu–60Au example, the absorption correction factor in Eq. (5) is given by

$$\text{Absorption correction factor} = \frac{f(\chi)_{\text{Cu}}}{f(\chi)_{\text{CuAu}}} \tag{15}$$

Calculated values for $E_0 = 20$ keV are 1.027 ($\psi = 52.5°$) and 1.079 ($\psi = 15.5°$). Probable errors in these values are estimated to be less than 10% of the difference from unity.

2.5.3. Characteristic-Line Fluorescence Correction

Fluorescence by characteristic x-ray lines is possible only if spectral lines of higher energy than the line being measured are emitted by another element present in the specimen. The effect is strongest when the exciting lines have an energy between approximately 1.1 and 1.5 times the energy of the line that is being enhanced. This situation exists for the analysis of Cu in Cu–Au, and all of the Au L lines can excite Cu K_α.

To correct for fluorescence by characteristic lines, it is necessary to calculate the ratio

$$\gamma_f = \frac{I_f}{I_p} \tag{16}$$

where I_f is the measured intensity due to excitation by characteristic-line fluorescence and I_p is the measured intensity due to primary excitation by electrons. For the fluorescence of Cu by Au in the Cu–Au specimen, the following factors must be included in an expression for γ_f: (1) The ratio of the absolute x-ray intensities generated by electrons in Cu and Au, (2) the probability of Au L photon absorption in Cu, (3) the probability of an absorbed Au L photon resulting in the generation of a Cu K_α photon, and (4) the distribution of primary radiation with depth and the absorption of primary and fluorescent radiation. These factors yield a very lengthy equation that will not be given here, but is derived in a generally applicable form by Reed.[33]

The value of γ_f is directly proportional to the atomic fluorescence yield ω_c of the exciting element; the uncertainty in the values for ω_c is one of the primary sources of error in γ_f. Characteristic-line fluorescence becomes negligible for x-ray lines with energies below 3 keV because the values of ω_c are very small. For a particular binary system, the magnitude of γ_f will increase with increasing E_0, ψ, and weight fraction of the exciting element. The correction factor in Eq. (5) is of the form

$$\text{Characteristic fluorescence correction factor} = \frac{1}{1 + \gamma_f} \tag{17}$$

Values for Cu K_α in 40Cu–60Au calculated from the Reed equation for $E_0 = 20$ keV are 0.966 ($\psi = 15.5°$) and 0.954 ($\psi = 52.5°$). The probable error is estimated to be $\pm 20\%$ of the deviation from unity.

2.5.4. Continuous-Spectrum Fluorescence Correction

Every element emits a continuous x-ray spectrum of relatively low intensity in addition to the sharp characteristic x-ray lines. The continuous spectrum results from the deceleration of the incident electrons through the interaction with the strong fields surrounding the atomic nuclei; the intensity in a particular energy range increases with Z. In the appropriate energy range, photons of the continuous spectrum can excite characteristic x-ray photons in any pure element. The correction deals with the difference in the excitation of characteristic photons due to the variation of the intensity and energy distribution of the continuous spectrum with specimen composition.

The continuous spectrum fluorescence correction is treated in detail by Hénoc.[34] The computation is mathematically complex and typically requires more than half of the computation time in a computer correction program. There are large uncertainties in both the model and the constants used. Fortunately, this correction is small in many cases of practical interest. The omission is likely to result in significant error ($>1\%$) only if the measured radiation is of high energy and a large spread of atomic numbers is present in the specimen. For Cu K_α in 40Cu–60Au at 20 keV, the correction factors are approximately 0.995 ($\psi = 15.5°$) and 0.987 ($\psi = 52.5°$).

2.6. Limitations

Before the discussion of a number of successful analyses in the remaining portion of the chapter, it is appropriate to consider the limitations of the method. The electron probe is by no means a magical instrument to solve all problems. Every electron probe analyst has had at least some dismal failures; these are generally carefully omitted in the published literature on electron probe microanalysis. As materials and problems become more complex, it occurs more and more frequently that samples "defy analysis" in the sense that the problem which prompted the analysis is never satisfactorily solved. These analytical failures are typically due to one or more of the following reasons:

1. The specimen was not stable under the analytical conditions.
2. The sensitivity for the problem element(s) was insufficient under the analytical conditions.
3. The problem is related to localized trace impurities; the "right" elements and/or sample areas were missed.
4. The sample geometry makes specimen preparation the limiting factor.

5. The problem is unrelated to composition.

These five topics will now be discussed in somewhat greater detail.

2.6.1. Specimen Stability

The specimen must be stable in a vacuum. Nonconductors should be coated with an evaporated conducting film to prevent electrical charging of the specimen surface and resulting beam instability. The resistivity of the conducting layer should be 10^7 Ω/square or less. Most instability problems are due to local heating which causes decomposition or melting of the specimen. The problem can be severe with glasses, organic materials, and easily decomposed compounds. Local heating of poor thermal conductors can be minimized by placing a relatively thin layer of the specimen on a metal backing and coating the surface with a conducting film. A defocused electron beam will lower the current density on the specimen and may improve stability.

2.6.2. Sensitivity

Under optimum conditions, the detectability limit will not be lower than approximately 50 ppm, and it may be much higher. This means, for example, that typical semiconductor doping levels are not detectable. The poorest sensitivity is for very soft x rays in a specimen which is highly absorbing for the measured radiation. The low-atomic-number elements (below magnesium) typically are detectable only if present in concentrations $>0.1\%$. The lowest detectability limits are achieved with focusing wavelength spectrometers; energy spectrometers are poorer by an order of magnitude.

To achieve the detectability limit of 50 ppm, the peak and background of the analytical line must be counted for a specific time interval (typically 100 sec) so that N, the number of counts collected, is statistically significant. Unless a fairly elaborate instrument is available, the counter, spectrometer, and specimen position must be manually reset; this type of analysis can consume a considerable amount of instrument and operator time. Spectral scans or distribution scans are less time-consuming for the operator, but are also less sensitive to traces.

There are some cases of spectral interference, although they are relatively rare. The most severe interference is for the L spectra of the elements ncar silver in the periodic table. Several lines of adjacent elements in this region are superimposed and a small quantity of element Z is difficult to detect in the presence of a large quantity of $(Z - 1)$. Other interferences involve the elements below atomic number 18, which have only one strong x-ray line. If the K_α line is masked by a strong line of another element, the sensitivity may be poor.

2.6.3. Localized Trace Impurities

The usefulness of electron probe analysis for the detection and identification of trace impurities that cover a small fraction of the analyzed surface area is generally overrated by scientists who are only slightly familiar with the technique. If the impurity sites are not visible and the identity of the impurity element is not known, the statistical odds for a successful analysis are very unfavorable. Thus, the analysis may be abandoned in favor of other work after a "reasonable attempt." The "right" sample areas and/or elements may have been missed completely. (Approximately 50 areas can be analyzed for 10 elements with a typical three-spectrometer instrument in two working days.) If the problem areas can be "decorated" to make them visible, or can be "tagged" with a known element by electrochemical means, the probability of a successful analysis is good. It is, therefore, advantageous to spend considerable effort on the development of a technique to localize the areas to be analyzed on specimens of this type.

2.6.4. Specimen-Geometry Limitations

In some samples, the material of interest may be part of a complex geometry. An example of this is an impurity in a deep hole or crevice. The x rays generated within the deep hole will be very highly attenuated by the surrounding material before they can be detected by the spectrometer. If the specimen surface is polished, some of the material of interest may be removed or the projecting material may be smeared over it.

In some multilayer specimens, problems occur primarily at interfaces. The success of these analyses depends upon the possibility of evenly removing the upper layers chemically or mechanically to expose the material to be analyzed.

2.6.5. Problem Unrelated to Composition

Some specimen problems are unrelated to chemical composition, but are rather due to allotropic transformations, line- or point-defect structure or density, and other crystallographic variations. The probe has no sensitivity for long-range order or structural variations in its normal mode of operation. Accessories for electron diffraction or divergent-beam x-ray diffraction (Kossel technique) are available for some instruments and might be used for problems of this type.

3. SURFACE ANALYSIS

3.1. Introduction

The specimens analyzed with the electron probe may be separated into two basic groups. The first type of sample has been cut from a larger specimen and the surface has been carefully polished. The aim of the analysis is a characterization of the bulk material, even though the analysis is of a 1-μ surface layer. Many of the applications in metallurgy and mineralogy require the analysis of samples of this type, and they comprise a large fraction of the quantitative analyses completed in the average electron probe laboratory. The specimens are generally large compared to the volume analyzed by the electron probe, and their average composition may be known. The small beam size is required to look for compositional variation within the specimen or to limit the analysis to a secondary phase. Characterization of bulk materials is beyond the scope of this book, but a few of the applications will be summarized in Section 3.2 to indicate the analytical advantages and difficulties. For a more comprehensive treatment, the reader is referred to two general papers[1,2] and many individual application papers in the literature.

The second type of sample is the as-reacted or as-deposited surface obtained after the specimen has undergone a specific chemical or mechanical process. In this case, the surface is known (or suspected) to be chemically different from the bulk material and the aim of the analysis is a qualitative or quantitative characterization of the surface. These samples will be discussed in Sections 3.3–3.5.

3.2. Bulk Characterization

3.2.1. Multiple Phases and Phase Equilibria

Some of the earliest electron probe studies dealt with the analysis of secondary phases and the determination or confirmation of equilibrium phase diagrams. Both are still important and widely published applications. The electron probe can be used to measure the interface composition of two coexisting phases. Since equilibrium conditions must exist at phase interfaces, the solubility limits of the coexisting phases can be determined even if the various phases are not homogeneous. This makes electron probe analysis a powerful tool for basic studies in systems in which homogenization is extremely slow. Specimens for phase-diagram studies may be prepared by a diffusion-couple technique or a quench-and-anneal technique. A sharp discontinuity in specimen composition is indicative of a phase boundary in both types of samples. The study of the Fe–Ni system by Goldstein and Ogilvie is a good example of the method.[35]

In many industrial materials, conditions are nonideal for the compositional determination of a second phase. The primary limitation is normally the small size or dendritic nature of the second phase. This is

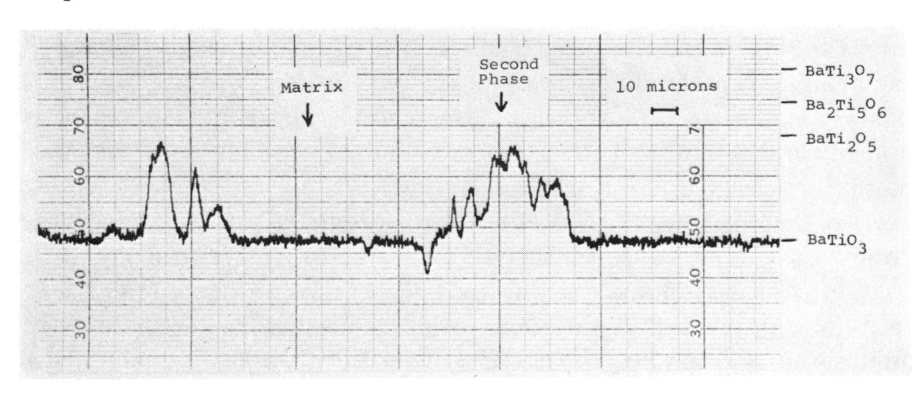

Fig. 18-4. Line scan for Ti K_α on TiO_2-rich $BaTiO_3$. The composition of the second phase cannot be determined accurately because it is mixed with the primary phase.

illustrated in Fig. 18-4, which shows a line scan for titanium on a sample of TiO_2-rich $BaTiO_3$. The secondary phase is titanium-rich and grows in widely separated snowflakelike structures that contain many fine branches. The primary and secondary phases are intermixed on a smaller scale than can be completely resolved. As a result, the composition of the secondary phase cannot be determined accurately since some of the primary phase is always mixed with it.

3.2.2. Diffusion Studies

Diffusion gradients can be measured directly on a sectioned specimen with the electron probe analyzer. The accuracy depends primarily on the relative length of the compositional gradient and the beam diameter, but uncertainties in the corrections for quantitative analysis may also be limiting. Very impressive results have been obtained for diffusion studies in bicrystals. Isoconcentration lines determined for Ni diffused into Cu bicrystals and Zn diffused into Ag bicrystals clearly demonstrate the effect of grain-boundary diffusion.[36,37]

3.2.3. Homogeneity

Homogeneity testing on a micron scale is a frequent application of electron probe analysis. The detection of small differences in concentration ($<1\%$ of the amount present) is a statistical problem and requires a stable instrument and an accumulation of many counts on a number of specimen areas. A variation of $\pm 5\%$ in a major constituent can be detected with a line scan and occurs surprisingly often in single-phase materials. An example is shown in Fig. 18-5. The sample is an alloy with the nominal composition 42Cr–58Ni. This alloy is in a single-phase region of the Cr–Ni phase diagram above 1000°C, and in a two-phase region below 1000°C. The left side of Fig. 18-5 shows a Cr K_α line scan of the alloy after solidification and quenching to room temperature. An expanded scale has been used; the actual intensity is twice full scale plus what is shown. The scan shows a deviation up to $\pm 5.7\%$ from the average Cr intensity. This corresponds to a compositional range of 42 ± 2.4 wt% Cr. The right side of Fig. 18-5 shows a Cr K_α scan of the same specimen after a long high-temperature anneal and quench. Any remaining inhomogeneity is less than $\pm 1.7\%$ of the average Cr intensity and is difficult to distinguish from statistical fluctuations.

3.3. Surface Contamination, Localized Surface Impurities

The electron probe is a very good tool for the identification of surface contamination that might be introduced in a manufacturing process. The qualitative composition of the contaminant will very often lead to the contamination source and a solution to the problem. An extreme example is given in Fig. 18-6, showing a portion of a silicon-based integrated circuit that has failed high-humidity testing. The horizontal aluminum interconnection strip in the center of the photograph has been almost totally etched away. The staining in the area (more vivid when seen in color) corresponds to the reaction product and was found to contain large amounts of Al and lesser concentrations of S, Cl, K, and Ca. This particular failure could be traced to the presence of moisture and a photoresist residue left on the circuit. The elements S, Cl, K, and Ca are frequent trace contaminants in photoresist materials used for masking and in many other organic materials.

A gross failure or contamination of the type shown in Fig. 18-6 can be seen with a microscope because the area of interest is discolored. A spectral scan of the contaminated area and a comparison scan of an adjacent "clean" area are all that is required to complete the analysis.

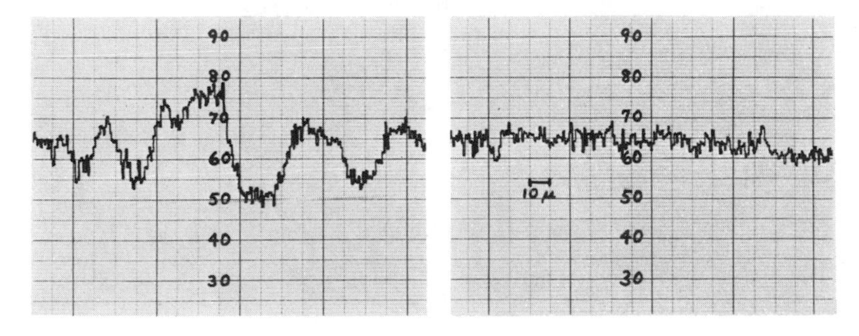

Fig. 18-5. Line scan for Cr K_α in 42Cr–58Ni alloy before and after long anneal for homogenization. The actual intensity is twice full scale plus what is shown.

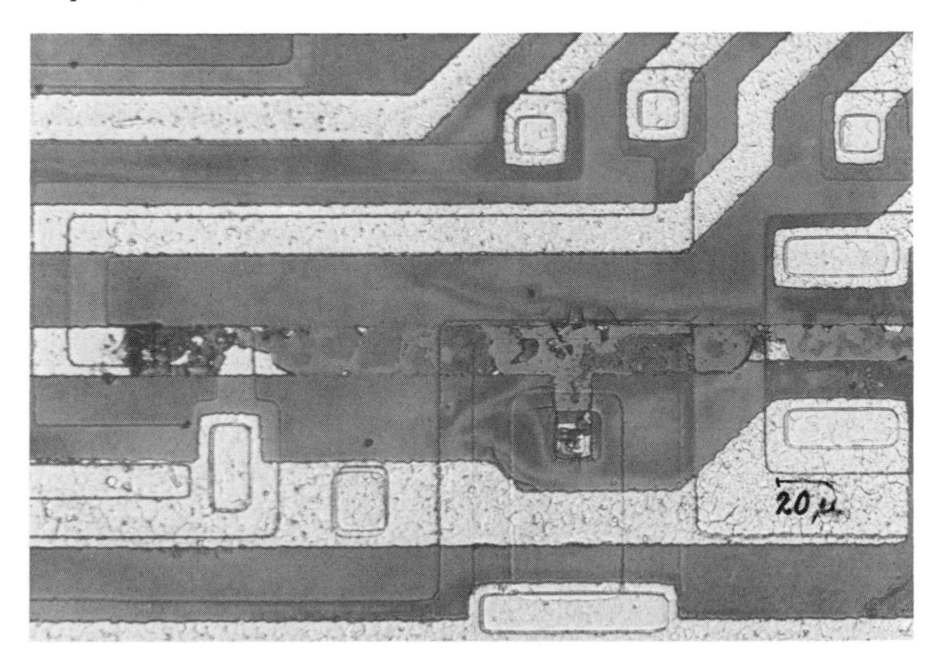

Fig. 18-6. Photomicrograph of silicon-based integrated circuit with etched aluminum interconnection.

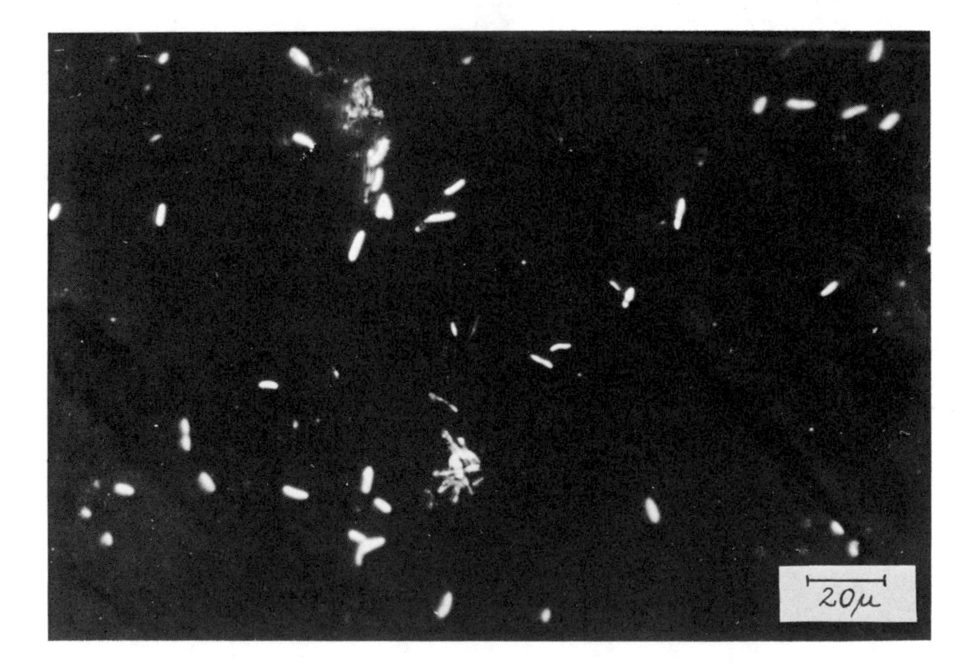

Fig. 18-7. Photomicrograph of tantalum foil surface after anodic oxidation at 20 V. The white rod-shaped areas showed a high surface concentration of iron.

The example shown in Fig. 18-7 is a little more subtle. In this case, the surface impurity was not visible on the original specimen, but had to be "decorated." Also, the impurity concentrations were small enough so that they could not be detected on a scan, but had to be counted manually. The specimen is high-purity tantalum foil. The decoration procedure consisted of anodic oxidation at 20 V in dilute phosphoric acid. Most of the sample is covered by an oxide approximately 300 Å thick, which gives the sample a deep-purple interference color. The rod-shaped areas in the photomicrograph are bare or covered by a much thinner oxide and are lighter in color. The thin spots were all found to contain Fe; for this reason, the oxide did not grow properly. In a capacitor, these regions would contribute to a very high leakage current. The Fe concentrations measured in the rod-shaped spots with a beam energy of 20 keV correspond to a bulk concentration of 500–1000 ppm. Analyses with a 14-keV beam give considerably higher Fe levels and indicate that the Fe impurity is concentrated in a shallow surface layer. The average bulk Fe content of this specimen, as measured by emission spectroscopy, is 8 ppm; this concentration is not detectable with the electron probe.

It should be emphasized that the analysis on the tantalum specimen was successful primarily because the spots could be seen on the anodized tantalum. The spots could be checked for a number of likely impurity elements, and iron was found to be the culprit. Without the decoration technique, random searching for several elements would probably have been unsuccessful, and the analysis would have been a failure of type 3 discussed in Section 2.6.

3.4. Surface Distribution Analyses

The distribution of a major or minor constituent on a specimen surface may be determined quickly by a line scan or a two-dimensional scan. An example is given in Fig. 18-8, which shows a line scan for zinc on α brass. The specimen was the surface of a polished piece of large-grained brass that was part of a fixture in a vacuum system and had been heated on numerous occasions. A fair amount of Zn was lost from the surface. The interesting aspect is that the Zn loss was more pronounced for certain grain orientations and near large-angle grain boundaries. The unevenness is apparent in the scan. At a depth of approximately 10 μ below the surface, the material retained its original composition and was very uniform.

On surfaces that have been exposed to various treatments or environments, intensity variations in a line scan may be due to topography or the presence of an additional, unsuspected element. These possibilities must be carefully checked before any conclusions can be

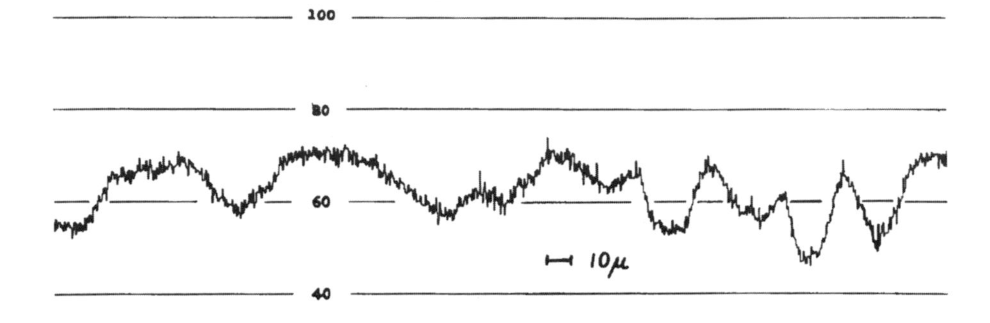

Fig. 18-8. Line scan for zinc on an α-brass surface that shows zinc loss due to heating in a vacuum. The actual measured intensity is twice full scale, plus what is shown.

drawn. Thus, a "mirror-image" line scan for Cu was required to confirm that the Zn scan in Fig. 18-8 was not due to topography or the deposition of some foreign material.

Topography can be a severe problem on rough surfaces, especially if a quantitative analysis is required. As the irregular surface is traversed, a number of negative and positive tilt angles relative to the horizontal plane of a flat specimen are encountered. This affects the x-ray measurement in two ways: (1) the electron beam incidence is no longer normal and (2) the absorption path length becomes longer or shorter because of a change in the effective value of ψ. (For nonnormal beam incidence and/or tilted specimen instruments, the tilt becomes greater or smaller, and similar relationships hold.) The effect of the nonnormal beam incidence is small for $\pm 10°$. The larger or smaller effective ψ primarily affects the absorption correction. The sample $f(\chi)$ must now be calculated for $\chi = \mu \csc \psi$ with ψ above and below the nominal instrumental value. The magnitude of the absorption change depends strongly upon the nominal value of ψ and upon the value of μ for the specimen, as can be seen from typical values listed in Table 18-1. The desirability of a high-ψ instrument to minimize topography difficulties is obvious from the table.

If all of the elements present can be measured in a single spot with the same linear spectrometer, it is possible to complete a quantitative analysis on an uneven surface. The value of ψ will be unknown, but it will be the same for all elements on the one spot. The matrix corrections are calculated for a series of ψ values above and below the instrumental value. The ψ value for which the calculated concentrations add up to 1 is then chosen from this series of evaluations. Armstrong *et al.*[38] have determined the effective values of ψ by using spectrometers on opposite

TABLE 18-1. Typical Changes in $f(\chi)$ Due to a Change in ψ of $+5°$ and $-5°$[a]

μ	Nominal ψ 15.5°	30°	52.5°
100	+2.6% −4.7%	+0.8% −1.1%	+0.2% −0.3%
2000	+24.0% −29.5%	+8.7% −10.4%	+2.9% −3.5%

[a] Evaluated for 20 keV, $\sigma_c = 3500$, $h = 0.1$.

sides of the specimen and measuring intensity ratios for a number of elements. It is then necessary to solve the matrix-correction equations simultaneously for several pairs of element intensity ratios in order to obtain a unique solution for the ψ values. Both of these procedures are tedious and have the flaw that a large error in the measurement of one element is propagated through the entire analysis. However, they are the only possibilities of obtaining quantitative data from moderately rough surfaces (e.g., rolled sheet, small particles) with a low-ψ instrument. An example of a quantitative analysis on the surface of a rolled sheet is shown in Fig. 18-9. The nominal composition of the alloy was 1Al–99Ti,

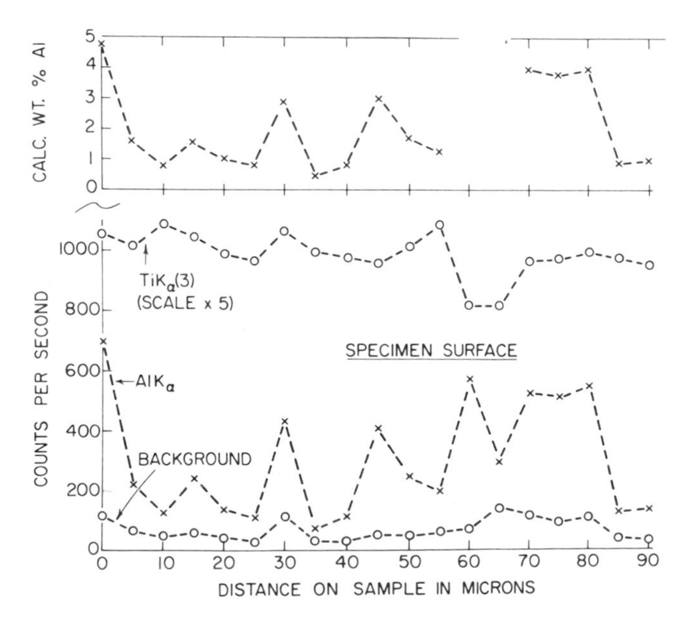

Fig. 18-9. Determination of Al concentration on the surface of a rolled sheet.

but it showed an unusual surface enrichment of Al that was to be characterized by the analysis. By treating ψ as a variable, the contribution to the Al K_α intensity of topography and Al concentration could be separated.

3.5. Shallow Depth Gradients

In the last two sections, the lateral distribution of impurities and major or minor constituents was considered. The distribution in depth may also be important, and it may be necessary to differentiate between surface and bulk impurities. A long-range depth gradient is treated by cross-sectioning the specimen and treating it like a lateral gradient. Shallow depth gradients may be characterized by surface analyses at several beam voltages or by analysis of an angle-lapped specimen.

3.5.1. Analysis at Several Beam Voltages

This method is based on the fact that the depth of the analyzed volume is a function of the electron beam energy. At a low beam voltage, a surface layer will be a much greater fraction of the total analyzed volume than at a high beam voltage. The primary advantage of the method is that it allows the characterization of the depth distribution of an element present on a very small surface area, such as the iron on the tantalum foil shown in Fig. 18-7. Lapping or sectioning of such a sample may be extremely difficult. There are also applications in forensic studies or failure analysis in which the evidence may not be destroyed or distorted by lapping.

If the element of interest has been measured on the sample and on a homogeneous standard at several beam voltages, the depth distribution can be characterized as follows: (1) For a uniform bulk distribution, the calculated concentration (after quantitative corrections) should be constant with increasing voltage. (2) A surface layer should show a rapid decrease in calculated bulk concentration with increasing voltage, but a thin-film calculation should give a constant film thickness. (3) An intermediate case, showing a gradual decrease of the calculated concentration with increasing accelerating voltage, indicates a compositional gradient close to the surface. Examples of such gradients are a bulk constituent concentrated at the surface, or a surface impurity extending several microns into the bulk at reduced concentration. A semi-quantitative concentration profile with depth can be calculated for such cases from measurements at four or more voltages and experimental $\phi(\rho z)$ curves. The practical depth limit of this method is 2×10^{-3} g/cm^2.

3.5.2. Analysis of Angle-Lapped Specimens

Compositional gradients that are several microns deep are easily characterized by the analysis of an angle-lapped sample. This technique is most effective for samples having a depth gradient, but little or no lateral gradient. The purpose of the angle lap is to magnify the depth direction; on a 3° lap, approximately 20 μ laterally correspond to 1 μ in depth.

The depth resolution of an analysis on an angle-lapped sample is limited by the depth of the layer in which x rays are excited. It is desirable to make this effective range as small as possible by using low accelerating voltages and soft x rays for the analysis. For quantitative analyses on angle-lapped specimens, it is necessary to take into account a possible change in the effective ψ, which may increase or decrease from the nominal value depending on the relative orientation of the lap and the spectrometer crystal.

The Pb concentration on a 3° lap of a flux-grown $Pb_xBa_{1-x}TiO_3$ crystal is shown in Fig. 18-10. The effective depth of Pb M_α excitation was 0.3 μ; thus, a discontinuous boundary would be broadened to approximately 6 μ on the scan. At the lap edge there is a very slight peak

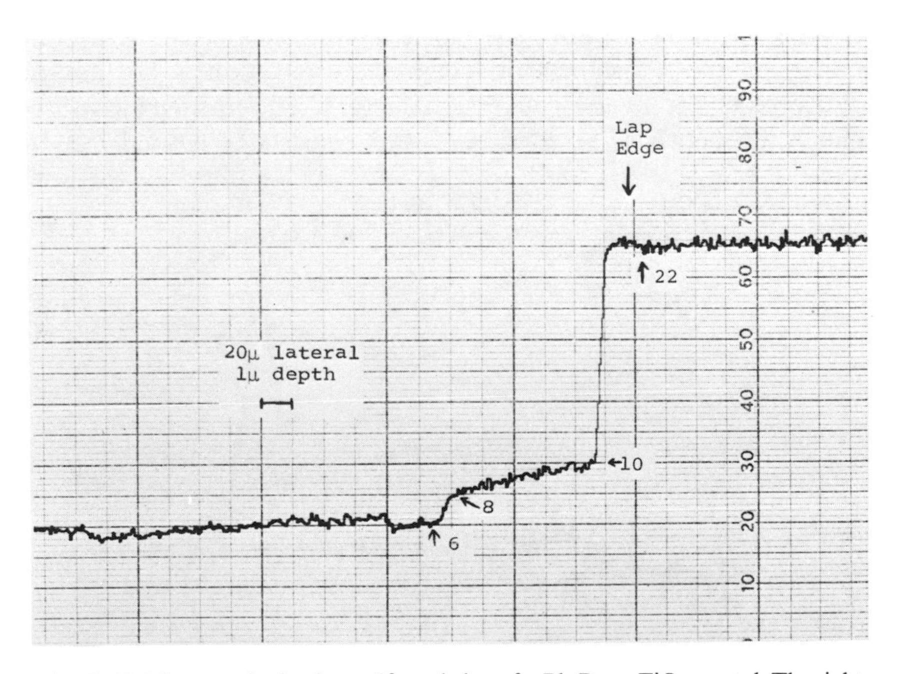

Fig. 18-10. Line scan for lead on a 3° angle lap of a $Pb_xBa_{1-x}TiO_3$ crystal. The right side of the scan corresponds to the crystal surface. The numbers give the value of x in mole % Pb at the indicated positions in the scan.

in the Pb signal due to a 3° increase in the effective ψ. After approximately 12 μ (0.6 μ in depth), the Pb content drops rapidly from 22 to 10 mole %. The drop is very sharp and the total breadth on the scan can be attributed to effective range broadening. After a gradual decrease to 8 mole %, the Pb concentration again drops suddenly to 6 mole %. The difference between the sharp and gradual drops is rather striking, and the 20 × magnification achieved by angle lapping makes quantitative determinations possible. The regions bounded by the sharp concentration gradients correspond to different stages of crystal growth.[39]

4. THIN-FILM ANALYSIS

4.1. Introduction

The electron probe has very good sensitivity for the thickness measurement and compositional analysis of thin films. The primary advantages are that very little material is required for an analysis and that the film need not be removed from the substrate or chemically altered in any way. An example of the sensitivity is shown in Fig. 18-11. The sample consists of high-purity aluminum foil covered by a very thin layer of chemically deposited nickel and copper.

The measurement precision of a quantitative thin-film analysis is typically $\pm 0.05\ \mu g/cm^2$, or $\pm 1\%$ of the film thickness. The absolute accuracy in mass thickness or weight concentration is not as good because of errors in the conversion factors. The absolute error will typically be between ± 5 and 10% and consists of a small random

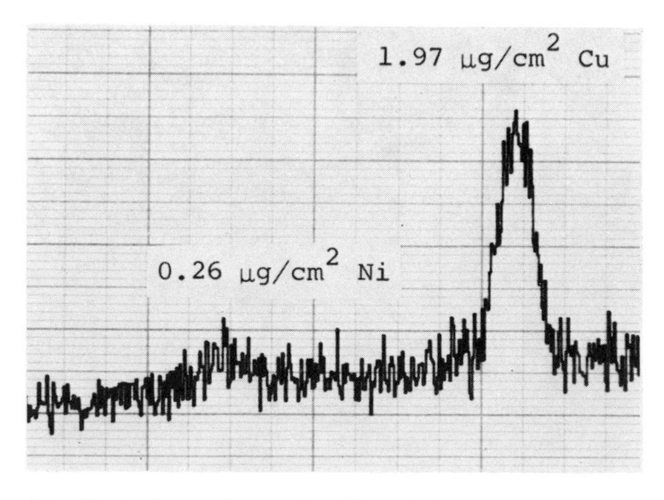

Fig. 18-11. Spectral scan for Ni and Cu on Al foil. The corresponding linear thicknesses at bulk density are 3 Å and 22 Å.

measurement error and a larger systematic normalization error. The comparison of several films or of several areas on one film can, therefore, be done more accurately than an absolute determination.

In electron probe analysis, mass thickness ρz in g/cm^2 is the basic thickness unit. Because of the great popularity of interferometry, many readers probably have a much better "feel" for linear thickness in angstroms. For this reason, z values are occasionally included in the text; the bulk density ρ is used in the conversion.

The films to be discussed are divided somewhat arbitrarily into thick, thin, and very thin films. These designations are not absolute, but depend on the relative size of the film thickness and the depth of x-ray excitation. The excitation depth is approximately proportional to $E_0^n - E_c^n$ with $n = 1.65$ for typical electron probe conditions.

The thick-, thin-, and very-thin-film classifications are most easily understood with reference to the x-ray ionization functions $\phi(\rho z)$ shown in Fig. 18-3. For 29-keV electrons, any film with a thickness greater than 1500 μg/cm^2 is considered thick, and is treated as a bulk sample. (The tails of the ionization curves contribute very little to the total x-ray intensity and can be corrected for by a normalization multiplier.) The thick-film conditions are only useful for the determination of composition. At 29 keV, any film significantly thinner than 1500 μg/cm^2 may be treated by a general thin-film model that requires numerical integration of $\phi(\rho z)$ curves. A mathematically much simpler very-thin-film model applies for thicknesses less than approximately 150 μg/cm^2 for 29-keV electrons. The cutoff for the very-thin-film model is a ρz value somewhat before the peak of the $\phi(\rho z)$ curve. Both thickness and composition may be determined by the thin-film and very-thin-film models.

As the beam voltage is lowered, the depth of x-ray excitation decreases and the ρz values corresponding to the cutoff for the various models are proportionately smaller. The sensitivity for extremely thin layers is much greater at a low beam energy. It is important, therefore, that the analytical conditions be optimized for a particular thickness range.

4.2. Substrate Effects

A thin-film analysis can be affected by the substrate in several ways. First of all, some of the elements present in the substrate may emit x-ray lines that will interfere with the measurement of x-ray lines emitted by the film. An extreme example is a specimen in which some of the same elements are present in the film and in the substrate; this makes the analysis of the film for these elements nearly impossible. The substrate spectra may also cause excitation of the film elements through secondary fluorescence by spectral lines or the continuum under the conditions detailed in Section 2.5. Both of these effects can normally be eliminated or minimized by careful choice of the substrate, and will not be discussed further.

The favorite substrates of the writer are high-purity silicon, pyrolytic graphite, and beryllium. All have a weak continuous spectrum and few spectral lines, can be sliced and polished readily, are electrical and thermal conductors, and are reasonably stable under most deposition and analysis conditions. Microscope slides are very poor substrates, because they tend to soften during analysis and often contain many poorly characterized impurity elements.

For very thin films, the substrate is a much larger portion of the analyzed sample volume than the film. The substrate, therefore, largely determines the electron penetration and scattering within the specimen. The backscattering of high-energy electrons is of particular importance for thin-film analysis, because each backscattered electron passes through the film a second time in the opposite direction. In the limiting case of an extremely thin film, the electron scattering will depend only upon the substrate.

For an extremely thin film, the enhancement of the film intensity due to backscattering from the substrate is given by ϕ_0, the value of the x-ray ionization function at $\rho z = 0$. The value of ϕ_0 is always greater than unity and is a function of the atomic number of the substrate and of the overvoltage ratio $U_0 = E_0/E_c$ for the film x-ray line measured. The dependence of ϕ_0 on the Z value of the substrate actually involves three separate variables: the backscattered electron fraction η, the energy distribution of backscattered electrons $d\eta/dW$, and the angular distribution of backscattered electrons as they pass through the surface film. The dependence on U_0 is due to the form of the ionization cross section of the measured x-ray line

$$Q = \text{const}\,\frac{\ln U}{U^m} = \text{const}\,\frac{\ln WU_0}{(WU_0)^m} \tag{18}$$

where the exponent m is between 0.7 and 1.0, and $W = E/E_0$. This variable is introduced for convenience, so that the same backscattered electron energy distribution $d\eta/dW$ can be used for all beam energies E_0.

If an effective mean emergence angle α of the backscattered electrons can be chosen, the value of ϕ_0 may be calculated by numerical integration of the following equation:

$$\phi_0 = 1 + \sec\alpha \int_{W_c}^{1} \frac{d\eta}{dW}\,\frac{Q(WU_0)}{Q(U_0)}\,dW \tag{19}$$

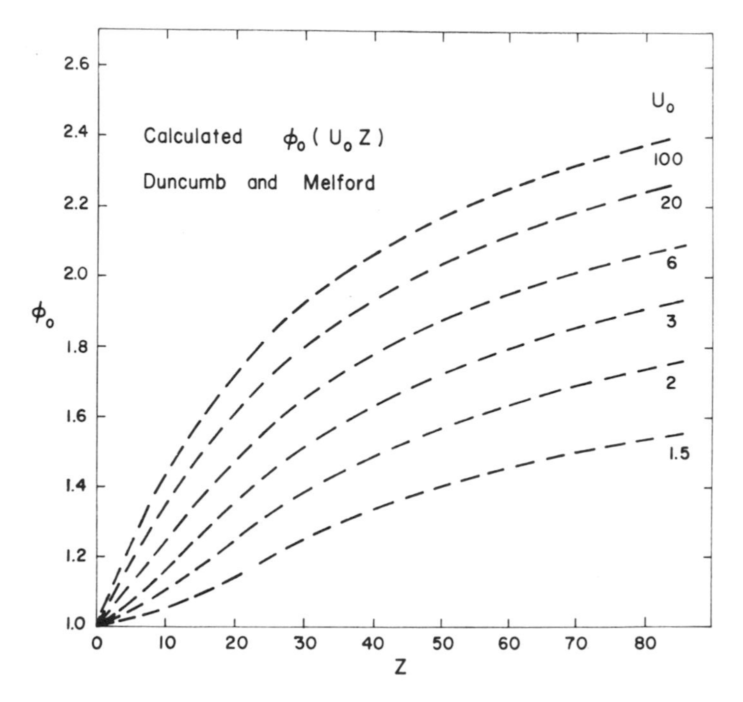

Fig. 18-12. Values of ϕ_0 calculated by Duncumb and Melford by numerical integration of Eq. (19).

This calculation was done by Duncumb and Melford,[40] who used the $d\eta/dW$ data of Bishop, a value of 2 for the sec α term, and $m = 1$ in the expression for Q. The family of calculated curves is shown in Fig. 18-12.

Experimental measurements of ϕ_0 may be made by two different techniques. The first method is the one originally used by Castaing and Descamps and consists of measuring the x-ray intensity of film F on substrate S and dividing the result by the x-ray intensity of an identical film F supported on a fine grid.[27,28] Although simple in principle, these measurements are difficult to perform accurately because the grid-supported films tend to buckle and tear. A second method, originally suggested by Hutchins,[41] avoids the difficulties inherent in the measurement of grid-supported films. A film is evaporated onto a number of pure-element substrates that have been mounted and polished together. The film x-ray intensity is measured on each substrate, plotted as a function of substrate Z, and extrapolated to $Z = 0$. All measured intensities are then divided by this extrapolated value to give ϕ_0 directly.

The variable substrate method has two pitfalls which resulted in considerable error in early measurements:

1. Several low-Z substrates must be included to get a good extrapolated value for $Z = 0$. This is especially true for curves with low U_0 which have a definite inflection below $Z = 10$.
2. The films must be very thin to make scattering by the film negligible. The effect of film thickness is shown in Fig. 18-13. A measurement error of $\pm 2\%$ is probably the best that can be achieved for experimental ϕ_0 measurements.

Reuter[42,43] has measured ϕ_0 by the multiple-substrate technique and has suggested the expression

$$\phi_0 = 1 + 2.8(1 - 0.9/U_0)\eta \qquad (20)$$

as a good fit to his experimental data. The value of η *vs.* Z may be read from Fig. 18-1 or may be calculated from Reuter's polynomial fit to Heinrich's 20-keV data.

$$\eta = -0.0254 + 0.016Z - 0.000186Z^2 + 8.3 \times 10^{-7} Z^3 \qquad (21)$$

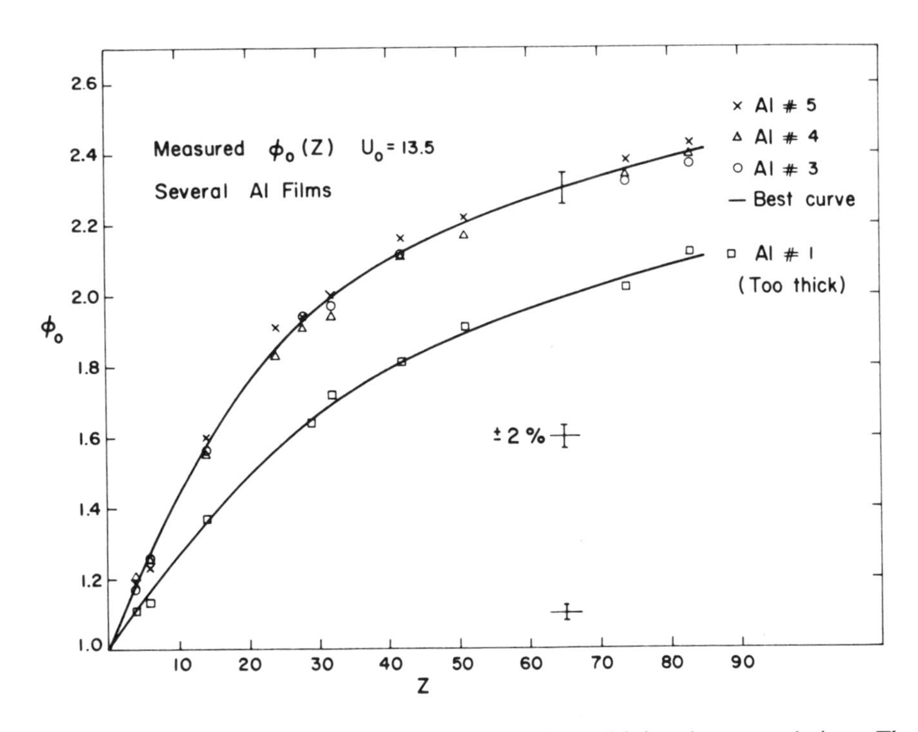

Fig. 18-13. Values of ϕ_0 obtained for Al films by the multiple-substrate technique. The upper curve is an average of measurements on three very thin Al films. Due to the error in extrapolating to $Z = 0$, the points for any one film tend to lie above or below the curve. The data for film 1 were treated in the same manner and show the effect of scattering by the thick film. The effective ϕ_0 for the thick film can be obtained by moving up the curve so that it coincides with the thin-film curve for $Z = 13$ (G. A. Hutchins and R. D. Wantman, unpublished).

The ϕ_0 values calculated from Reuter's equation [Eq. (20)] are always higher than the Duncumb and Melford calculations shown in Fig. 18-12. For $U_0 = 20$, the difference between the curves is within experimental error (<2%), but for $U = 1.5$, the difference is as large as 10%. The discrepancy exists because Eq. (20) contains no separate dependency upon the energy distribution $d\eta/dW$. This is a valid approximation for $U_0 = 4$, but introduces large errors for $U_0 < 2$. There is an additional problem with Eq. (20) for low U_0, because the U_0 dependence becomes too strong and ϕ_0 approaches $1 + 0.28\eta$ instead of unity as $U_0 \to 1$.

The writer suggests that ϕ_0 be calculated from Eq. (20) for all Z values when $U_0 > 4$; for $Z > 30$, Eq. (20) may be used down to $U_0 = 2$. For $U_0 < 2$ and the combination of $U_0 < 4$ and $Z < 30$, the Duncumb and Melford curves are likely to be more accurate.

The value of ϕ_0^* for a moderately thick film (e.g., aluminum film 1 in Fig. 18-13) should be calculated for an effective value η^* that is between the value of η for the substrate and the film element. The value of η^* can be determined easily if consecutive sample-current measurements of the sample and a standard of known η are made at the same time that the x-ray data are taken. From the definition of η in Eq. (3) and the fact that the beam current i_b is the same for the two measurements, it follows that

$$\frac{i_s^*}{i_s} = \frac{1 - \eta^*}{1 - \eta} \tag{22}$$

With reasonably stable equipment, η^* can be measured routinely to a precision of ± 0.01.

4.3. Thickness Determination

The x-ray intensity measured for a film F of thickness ρz on a substrate S and normalized to the emitted intensity from a bulk standard of the film element can be expressed as:

$$k_F = \frac{I_{FS}}{I_{\text{bulk}}} = \frac{\int_0^{\rho z} \phi^*(\rho z) \exp(-\chi \rho z)\, d(\rho z)}{\int_0^{\infty} \phi(\rho z) \exp(-\chi \rho z)\, d(\rho z)} \tag{23}$$

The asterisk is used to indicate the required modification of the ionization function due to the differences in the backscatter properties of the substrate and the film material. The denominator of Eq. (23) is a function of E_c, E_0, Z, and χ, and will be constant for a particular element measured under specific conditions. The value of the integral in the numerator will increase systematically with ρz. The intensity ratio k_F can typically be measured to a precision of 1%. What is needed is an accurate conversion from k_F to ρz. From the form of Eq. (23), it is apparent that an accurate conversion is not to be achieved easily. Three approaches will be considered that increase in complexity, but also in general applicability.

4.3.1. Empirical Methods

If only a few film elements are to be measured routinely on one or two substrates, it may be easiest to construct empirical calibration curves with films of known thickness. The conversion accuracy will then depend upon the accuracy with which the absolute thickness of the standard films can be determined by other techniques. For this standard thickness measurement, it is desirable to determine ρz by chemical analysis on a separate control sample prepared simultaneously with the microprobe sample. Marshall and Hall claim a 2% accuracy for colorimetric chemical weighing of thin films.[44] Interferometric measurements are not as reliable since they only determine z, and thin films are not necessarily of bulk density.

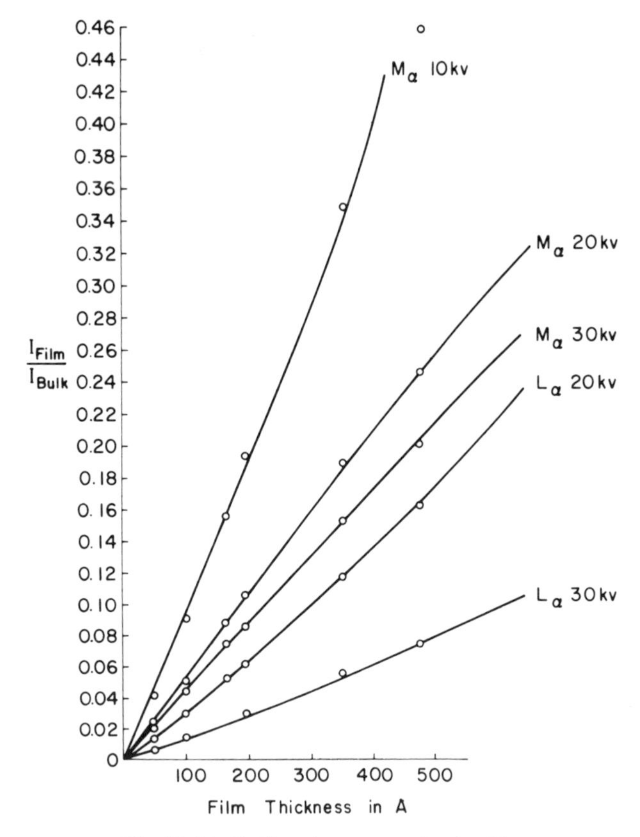

Fig. 18-14. Calibration curves for Au films on Si measured with a low-ψ instrument.

After the standard films have been prepared and chemically weighed, k_F may be measured with the electron probe for a number of different characteristic lines and beam energies. Calibration curves similar to Fig. 18-14 can then be drawn for the appropriate analytical conditions. These curves will be nearly linear for small ρz with the slope dependent upon E_0, E_c, and χ. For somewhat thicker films, the curves will show a positive or negative deviation from linearity.

Films of known thickness are also required to test the accuracy and validity of the more complex calculations to be discussed subsequently. Thus, the accuracy to which the absolute thickness of standard films can be determined by other methods is a limitation to the ultimate accuracy obtainable by electron probe analysis even with the more complex models.

4.3.2. The Very-Thin-Film Model

This model is valid for ρz somewhat less than the peak of the $\phi(\rho z)$ curve or approximately one-tenth of the effective electron range. The model requires two approximations.

The first approximation is that the substrate and film dependence can be separated:

$$I_{FS} = \phi_0{}^* I_F \tag{24}$$

The value of $\phi_0{}^*$ is calculated from Eq. (20) and the measured η^* as detailed in the previous section. The intensity I_F corresponds to the intensity measured on an *unsupported* thin film of thickness ρz and

$$I_F \propto \int_0^{\rho z} \phi_F(\rho z) \exp(-\chi \rho z)\, d(\rho z) \tag{25}$$

The value of $\phi_F \to 1$ as $\rho z \to 0$, by definition (see Section 2.5).

The second approximation is that ϕ_F can be given by

$$\phi_F(\rho z) = 1 + B\rho z \qquad \rho z \ll \text{total range} \tag{26}$$

and that B can be treated as a constant for any specific combination of E_0, E_c, and film Z. The parameter B is a measure of electron scattering within the film material. Its value increases rapidly with decreasing E_0 and more slowly with increasing film Z. Typical values for B are given in Table 18-2. These values may be computed from the initial slope of $\phi(\rho z)$ curves after a substrate correction or from experimental electron-scattering data for thin films. A $\pm 10\%$ accuracy in B is sufficient since $B\rho z < 1$.

After substitution of Eq. (26) in Eq. (25), the integral may be evaluated directly. Since the value of $\chi\rho z$ is less than one in the valid

TABLE 18-2. Values of the Electron Scattering Parameter B

	Low Z	High Z
10 keV	9,000	12,000
20 keV	4,000	5,000
30 keV	2,000	3,000

thickness range for the model, the exponential factor may be replaced by the first few terms of a rapidly converging series in ρz. This gives

$$I_F \propto \rho z + (\rho z)^2\left(\frac{B}{2} - \frac{\chi}{2}\right) - (\rho z)^3\left(\frac{B\chi}{3} - \frac{\chi^2}{6}\right) + \cdots + (-1)^n(\rho z)^n\left(\frac{B\chi^{n-2}}{(n-1)!} - \frac{B\chi^{n-2}}{n!} - \frac{\chi^{n-1}}{n!}\right) \tag{27}$$

Equation (27) predicts a positive or negative deviation from linearity of I_F with ρz as the film thickness increases. If $B \simeq \chi$, I_F will increase nearly linearly with ρz. For $B \neq \chi$, the $(\rho z)^2$ term may be as much as 20% of ρz and should always be included. The $(\rho z)^3$ term will be less than 1% of ρz under all conditions for which the approximations are justified; higher-order terms are never required.

In order to make absolute measurements, the bulk normalization has to be included by substitution of Eqs. (24) and (27) in Eq. (23).

$$k_F = \frac{I_{FS}}{I_{\text{bulk}}} = \frac{\phi_0{}^*[\rho z + (\rho z)^2(B/2 - \chi/2) - (\rho z)^3(B\chi/3 - \chi^2/6)]}{\int_0^\infty \phi(\rho z)\exp(-\chi\rho z)\,d(\rho z)} \tag{28}$$

The factor in the denominator can be evaluated most accurately from standard films of known ρz. The error in the bulk normalization will then depend primarily on the ρz error of the standard film.

Since the integral

$$\int_0^\infty \phi(\rho z)\exp(-\chi\rho z)\,d(\rho z) = f(\chi)\int_0^\infty \phi(\rho z)\,d(\rho z) \tag{29}$$

by definition, the bulk normalization may also be determined by calculation of $f(\chi)$ from Eq. (14) and graphical or numerical integration of experimental $\phi(\rho z)$ curves. This approach requires no standard films, but generally results in larger errors. Voltage series of $\phi(\rho z)$ curves have been determined experimentally for Mg K_α in Al,[45] V K_α in Ti,[46] Zn K_α in Cu,[47] and Bi L_α in Pb.[46] These curves may be interpolated for other beam voltages and may also be used for adjacent elements. However, there are large gaps in film elements and x-ray lines that cannot be interpolated. Examples are all M spectra and the soft L spectra in the atomic-number range 38–53. The bulk normalization for these elements can only be done with standard films. Another problem is that $\phi(\rho z)$ curves determined for one instrument geometry are not necessarily valid for other instrument geometries. Brown and Parobek have recently measured the same trace-layer samples with several instruments and have demonstrated a dependence on the angle of electron beam incidence.[48]

The form of Eq. (28) used for a thickness determination is

$$\rho z = \frac{k_F \int_0^\infty \phi(\rho z) \exp(-\chi \rho z)\, d(\rho z)}{\phi_0^*[1 + \rho z(B/2 - \chi/2) - (\rho z)^2(B\chi/3 - \chi^2/6)]} \tag{30}$$

Successive approximations are used to calculate the unknown ρz; three iterations are generally sufficient. The modification of Eq. (30) for multielement films is discussed in Section 4.4.2.

4.3.3. A General Thin Film Model

Let us take another look at Eq. (23):

$$k_F = \frac{I_{FS}}{I_{\text{bulk}}} = \frac{\int_0^{\rho z} \phi^*(\rho z) \exp(-\chi \rho z)\, d(\rho z)}{\int_0^\infty \phi(\rho z) \exp(-\chi \rho z)\, d(\rho z)} \tag{23}$$

If it were possible to develop an expression for ϕ and ϕ^*, k_F *vs.* ρz could be calculated by numerical integration. This approach would be valid for all film thicknesses and would be completely general.

An expression for $\phi(\rho z)$ was derived by Philibert[29] in connection with his treatment of the absorption correction more than ten years ago. Since that time, considerable experimental data have been generated concerning the propagation of relatively slow, 5- to 30-keV electrons through solid films.[49–52] These data have given a much better insight into the scattering of electrons in the electron probe energy range. They have also proven many of the expressions extrapolated from high-energy-electron measurements to be very inaccurate. Reuter[42,43] has recently utilized some of the new data in an expression for $\phi(\rho z)$. He has also applied numerical integration of $\phi(\rho z)$ to the analysis of thin films and has tested this approach with a number of standard films. The overall absolute accuracy of approximately 10% found by Reuter is somewhat disappointing. However, his measurements included extremes in E_c, E_0, and substrate Z that would not generally be encountered in practice.

4.4. Determination of Composition

The composition of thin films can be calculated with relative ease only for thick-film and very-thin-film conditions. X-ray intensities can be measured equally well for intermediate films, but the conversion to absolute weight concentration becomes very difficult.

4.4.1. Thick Films

These films are treated as bulk samples and measured at relatively low electron-beam energies (typically 5–10 keV). The film thickness should be 100 μg/cm^2 or greater for this approach to be successful. Only very soft x-ray lines can be excited efficiently at these low beam energies; the uncertainties in soft x-ray mass absorption coefficients may be a limitation in some cases.

4.4.2. Very Thin Films

The composition of very thin films can be determined by a procedure that is an extension of the thickness determination. As a first approximation, $(\rho z)_X$ and $(\rho z)_Y$ can be determined separately from Eq. (30) as detailed in Section 4.3. For an XY alloy, the weight concentration of element X is given by

$$C_X = \frac{(\rho z)_X}{(\rho z)_X + (\rho z)_Y} \tag{31}$$

This approximation ignores interelement effects, but is a good approximation for very thin films.

For a more exact solution, electron scattering and x-ray absorption in the total film have to be considered. For a one-element film, the integral leading to the series in Eq. (27) is:

$$I_F \propto \int_0^{\rho z} (1 + B\rho z) \exp(-\chi\rho z)\, d(\rho z) \tag{32}$$

The corresponding expression for element X in a two-element film XY is

$$I_X \propto C_X \int_0^{(\rho z)_T} (1 + B_{XY}\rho z) \exp(-\chi_{X \text{ in } XY}\, \rho z)\, d(\rho z) \tag{33}$$

where $(\rho z)_T = (\rho z)_X + (\rho z)_Y$, and C_X is given by Eq. (31). The expression corresponding to Eq. (30) for element X in XY then becomes:

$$(\rho z)_X = \frac{k_X \int_0^\infty \phi_X(\rho z) \exp(-\chi\rho z)\, d(\rho z)}{\phi_{0X}^*\left[1 + (\rho z)_X\left(\dfrac{B_X}{2} - \dfrac{\chi_{X \text{ in } X}}{2}\right) + \dfrac{(\rho z)_Y{}^2}{(\rho z)_X}\left(\dfrac{B_Y}{2} - \dfrac{\chi_{X \text{ in } Y}}{2}\right)\right]} \tag{34}$$

The third-order terms have been dropped in Eq. (34). The only new term for the two-element film is the $(\rho z)_Y^2/(\rho z)_X$ term in the denominator. Initial values for $(\rho z)_X$ and $(\rho z)_Y$ are calculated without the mixed terms; final results are obtained by successive approximations for both X and Y. The presence of additional elements in the film leads to additional mixed terms. Both ϕ_0 and the bulk normalization integral are generally different for different elements in the same film.

4.5. Examples of Thin-Film Studies

In this section, a few of the thin-film analyses performed in the writer's laboratory are summarized to give an indication of the great variety of film studies that are suitable for electron probe microanalysis.

4.5.1. Very Thin Nickel–Chromium Films

Sputtered nickel–chromium films were prepared in a wide compositional range by sputtering from alloy targets and also by sputtering simultaneously from two pure-element targets. The composition of the alloy sputtering targets was determined by bulk electron probe microanalysis. Films sputtered from these targets onto polished silicon slices were found to be of the same composition within experimental error.[53] For films with ρz of approximately 6 μg/cm^2 ($\sim$75 Å), the measurement error in both mass thickness and composition was less than 2% in all cases. The absolute errors were somewhat higher because of errors in the conversion factors.

The composition of nickel–chromium films evaporated from a large volume of melt was studied as a function of melt composition and temperature. The evaporated films all had a much higher chromium concentration than the melts. This chromium enrichment was found to vary systematically with evaporation temperature.

4.5.2. Thick Films of Barium Titanate

Barium titanate films approximately 5000 Å thick were rf-sputtered from a stoichiometric target onto polished silicon slices. Some of the deposition variables were the substrate temperature, silicon resistivity, the sputtering gas mixture, and the sputtering power. Film stoichiometry as a function of these variables was measured with the electron probe; the Ti:Ba atomic ratio was found to vary from approximately 0.99 to 1.22. X-ray diffraction, electron microscopy, and reflection electron diffraction were required in addition to electron probe microanalysis to sort out the effects of all the deposition variables. It was eventually concluded that stoichiometric $BaTiO_3$ could be deposited only if the sputtered film was crystalline as deposited. Amorphous films were always titanium-rich and some dissociated into two phases upon annealing.[54]

4.5.3. Reaction- or Etch-Rate Studies

Mass thickness measurements may be used to monitor various chemical reaction rates. Examples are the reaction of two films, the reaction of a film with its substrate, oxidation, corrosion, and preferential etching. Very often it is only necessary to determine the shape of the rate curve (linear, exponential, asymptotic, etc.) and this can be done with relative data.

An example is the testing and optimization of a preferential etchant that will rapidly remove a surface layer of material X, but will attack the underlying layer Y only at a much slower rate or not at all. To be useful, a preferential etchant has to have an asymptotic rate curve with a relatively long period during which all X, but no Y, has been removed. The analysis can be done most easily by cutting the initial specimen into a number of small pieces (the writer typically uses $\frac{1}{4} \times \frac{1}{8}$ inch rectangles). These pieces can then be etched or reacted individually, mounted together, and analyzed successively.

The growth of an oxide or other compound film on a metal can be monitored by measuring a film element not present in the base metal. An alternate, but less sensitive, method is to measure the decrease of the base-metal intensity as the film thickness increases. Reaction rates between two different films or a film and its substrate can be determined if suitable differential etchants can be found to remove the unreacted material.

Electron probe analysis is often used in conjunction with other techniques for basic thin-film studies. If the substrate and film contain some of the same elements, the film composition must be determined by another technique (e.g., reflection electron diffraction) unless the films can be grown thick enough for a low-keV bulk analysis. The composition in turn must be known to calculate the mixed terms in Eq. (34) and obtain the highest accuracy for ρz with the electron probe.

5. SYMBOLS

This list contains only those symbols that occur several times in the text and are not standard units.

A Atomic weight
B Electron-scattering parameter for thin-film analysis
C_X Weight concentration of element X
E Electron energy
E_0 Incident beam electron energy

E_c Critical electron energy necessary to excite measured x-ray line
F Subscript referring to film
I Measured intensity, used with descriptive subscript
k_X Measured relative intensity ratio (sample-to-standard) for element X
Q X-ray ionization cross section
R Backscatter coefficient
S Subscript referring to substrate, or Stopping power
U Overvoltage, $U = E/E_c$, $U_0 = E_0/E_c$
W Fractional incident-electron energy $W = E/E_0$, $W_c = E_c/E_0$
X Analyzed element
Y Second analyzed element, or element affecting the analysis of element X
Z Atomic number
z Linear film thickness
η Backscattered electron fraction
μ Mass absorption coefficient for x rays, or Micron $= 10^{-6}$ m
ρ Density
ρz Mass thickness of film
$\phi(\rho z)$ X-ray ionization function
ϕ_0 Value of $\phi(\rho z)$ at surface; substrate correction for thin films
χ X-ray absorption parameter, $\chi = \mu \csc \psi$
$f(\chi)$ X-ray absorption correction function
ψ X-ray take-off angle
ω_c Atomic fluorescence yield for c subshell

6. REFERENCES*

1. J. I. Goldstein, Electron probe analysis in metallurgy, in *Electron Probe Microanalysis* (A. J. Tousimis and L. Marton, eds.) pp. 245–290, Academic Press, New York (1969).
2. C. W. Mead, Electron probe microanalysis in mineralogy, in *Electron Probe Microanalysis* (A. J. Tousimis and L. Marton, eds.) pp. 227–244, Academic Press, New York (1969).
3. P. Duncumb, Recent advances in electron probe microanalysis, *J. Phys. E* **2**, 553–560 (1969).
4. D. R. Beaman and J. A. Isasi, *Electron Beam Microanalysis*, ASTM STP 506, American Society for Testing and Materials, Philadelphia (1972).
5. E. W. White and G. G. Johnson, *X-Ray Emission and Absorption Wavelengths and Two-Theta Tables*, ASTM DS 37A, American Society for Testing and Materials, Philadelphia (1970).

* This bibliography contains only the key references on the topics discussed. A more extensive general bibliography is included in the work by Beaman and Isasi,[4] and in a more detailed treatment of the subject by the author.[57]

6. G. G. Johnson and E. W. White, *X-Ray Emission Wavelength and keV Tables for Nondiffractive Analysis*, ASTM DS 46, American Society for Testing and Materials, Philadelphia (1970).
7. K. F. J. Heinrich, Instrumental developments for electron microprobe readout, in *Advances in X-Ray Analysis* (W. M. Mueller, G. Mallett, and M. Fay, eds.) Vol. 7, pp. 382–394, Plenum Press, New York (1964).
8. K. F. J. Heinrich, Oscilloscope readout of electron microprobe data, in *Advances in X-Ray Analysis* (W. M. Mueller and M. Fay, eds.) Vol. 6, pp. 291–330, Plenum Press, New York (1963).
9. K. F. J. Heinrich, *Scanning Electron Probe Microanalysis*, Technical Note 278, National Bureau of Standards, Washington (1967).
10. H. E. Bishop, Electron Scattering and X-Ray Production, Ph.D. Thesis, Univ. of Cambridge (1966).
11. H. E. Bishop, Some electron backscattering measurements for solid targets, in *Optique des Rayons X et Microanalyse* (R. Castaing, P. Deschamps, and J. Philibert, eds.) pp. 153–158, Hermann, Paris (1966).
12. K. F. J. Heinrich, Electron probe microanalysis by specimen current measurement, in *Optique des Rayons X et Microanalyse* (R. Castaing, P. Deschamps, and J. Philibert, eds.) pp. 159–167, Hermann, Paris (1966).
13. M. Green and V. E. Cosslett, Measurements of *K*, *L*, and *M* shell x-ray production efficiencies, *J. Phys. D* **1**, 425–436 (1968).
14. J. C. Clark, A measurement of the absolute probability of *K*-electron ionization of silver by cathode rays, *Phys. Rev.* **48**, 30–42 (1935).
15. C. R. Worthington and S. G. Tomlin, The intensity of emission of characteristic x-radiation, *Proc. Phys. Soc. (London)* **69**, 401–412 (1956).
16. D. L. Webster, W. W. Hansen, and F. B. Duveneck, Probabilities of *K*-electron ionization of silver by cathode rays, *Phys. Rev.* **43**, 839–858 (1933).
17. L. T. Pockman, D. L. Webster, P. Kirkpatrick, and K. Harworth, The probability of *K*-ionization of nickel by electrons as a function of their energy, *Phys. Rev.* **71**, 330–338 (1947).
18. R. W. Fink, R. C. Jopson, H. Mark, and C. D. Swift, Atomic fluorescence yields, *Rev. Mod. Phys.* **38**, 513–540 (1966).
19. D. B. Brown and R. E. Ogilvie, An electron transport model for the prediction of x-ray production and electron backscattering in electron microanalysis, *J. Appl. Phys.* **37**, 4429–4433 (1966).
20. D. B. Brown, D. B. Wittry, and D. F. Kyser, Prediction of x-ray production and electron scattering in electron probe analysis using a transport equation, *J. Appl. Phys.* **40**, 1627–1636 (1969).
21. H. E. Bishop, Electron scattering in thick targets, *Brit. J. Appl. Phys.* **18**, 703–715 (1967).
22. K. F. J. Heinrich, Errors in theoretical correction systems in quantitative electron probe microanalysis—A synopsis, *Anal. Chem.* **44**, 350–354 (1972).
23. D. R. Beaman and J. A. Isasi, A critical examination of computer programs used in quantitative electron microprobe analysis, *Anal. Chem.* **42**, 1540–1568 (1970).
24. J. Philibert and R. Tixier, Electron penetration and the atomic number correction in electron probe microanalysis, *J. Phys. D* **1**, 685–694 (1968).
25. P. Duncumb and S. J. B. Reed, The calculation of stopping power and backscatter effects in electron probe microanalysis, in *Quantitative Electron Probe Microanalysis* (K. F. J. Heinrich, ed.) Special Publication 298, pp. 133–154, National Bureau of Standards, Washington (1968).

26. K. F. J. Heinrich, X-ray absorption uncertainty, in *The Electron Microprobe* (T. D. McKinley, K. F. J. Heinrich, and D. B. Wittry, eds.) pp. 296–377, John Wiley & Sons, New York (1966).
27. R. Castaing and J. Descamps, Sur les bases physiques de l'analyse ponctuelle par spectrographie X, *J. Phys. Rad.* **16**, 304–317 (1955).
28. R. Castaing, Electron probe microanalysis, in *Advances in Electronics and Electron Physics* (L. Marton, ed.) Vol. 13, pp. 317–386, Academic Press, New York (1960).
29. J. Philibert, A method for calculating the absorption correction in electron probe microanalysis, in *X-Ray Optics and X-Ray Microanalysis* (H. H. Pattee, V. E. Cosslett, and A: Engström, eds.) pp. 379–392, Academic Press, New York (1963).
30. P. Duncumb and P. K. Shields, Effect of critical excitation potential on the absorption correction, in *The Electron Microprobe* (T. D. McKinley, K. F. J. Heinrich, and D. B. Wittry, eds.) pp. 284–295, John Wiley & Sons, New York (1966).
31. K. F. J. Heinrich, The absorption correction model for microprobe analysis, *Proceedings of the Second National Conference on Electron Probe Analysis*, paper 7 (1967).
32. P. Duncumb, P. K. Shields-Mason, and C. daCasa, Accuracy of atomic number and absorption corrections in electron probe microanalysis, in *Fifth International Congress on X-Ray Optics and Microanalysis* (G. Möllenstedt and K. H. Gaukler, eds.) pp. 146–150, Springer-Verlag, Berlin (1969).
33. S. J. B. Reed, Characteristic fluorescence correction in electron probe microanalysis, *Brit. J. Appl. Phys.* **16**, 913–926 (1965).
34. J. Hénoc, Fluorescence excited by the continuum, in *Quantitative Electron Probe Microanalysis* (K. F. J. Heinrich, ed.) Special Publication 298, pp. 197–214, National Bureau of Standards, Washington (1968).
35. J. I. Goldstein and R. E. Ogilvie, A re-evaluation of the iron-rich portion of the Fe–Ni system, *Trans. Met. Soc. AIME* **233**, 2083–2087 (1965).
36. A. E. Austin and N. A. Richard, Grain boundary diffusion, *J. Appl. Phys.* **32**, 1462–1471 (1961).
37. D. M. Koffman, Sc.D. Thesis, Massachusetts Institute of Technology (1964).
38. J. T. Armstrong, P. R. Busek, and E. F. Holdsworth, The effect of take-off angle on particle analysis with the electron microprobe, paper 36, *Proceedings of the Seventh National Conference on Electron Probe Analysis* (1972).
39. F. W. Perry, G. A. Hutchins, and L. E. Cross, Compositional inhomogeneity of (Ba, Pb) TiO_3 crystals, *Mater. Res. Bull.* **2**, 409–418 (1967).
40. P. Duncumb and D. A. Melford, A simple correction procedure for ultra-soft x-ray microanalysis, paper 12, *Proceedings of the First National Conference on Electron Probe Analysis* (1966).
41. G. A. Hutchins, Thickness determination of thin films by electron probe microanalysis, in *The Electron Microprobe* (T. D. McKinley, K. F. J. Heinrich, and D. B. Wittry, eds.) pp. 390–404, John Wiley & Sons, New York (1966).
42. W. Reuter, The ionization function and its application to the electron probe analysis of thin films, paper 34, *Proceedings of the Seventh National Conference on Electron Probe Analysis* (1972).
43. W. Reuter, The ionization function and its application to the electron probe analysis of thin films, IBM Research Report RC 3590, IBM Watson Research Center, Yorktown Heights, New York (1971).
44. D. J. Marshall and T. A. Hall, Electron probe x-ray microanalysis of thin films, *J. Phys. D* **1**, 1651–1656 (1968).

45. R. Castaing and J. Hénoc, Repartition en profondeur du rayonnement caracteristique, in *Optique des Rayons X et Microanalyse* (R.Castaing, P. Deschamps, and J. Philibert, eds.) pp. 120–126, Hermann, Paris (1966).
46. A. Vignes and G. Dez, Distribution in depth of the primary x-ray emission in anticathodes of titanium and lead, *J. Phys. D* **1**, 1309–1322 (1968).
47. J. D. Brown, The sandwich sample technique applied to quantitative microprobe analysis, in *Electron Probe Microanalysis* (A. J. Tousimis and L. Marton, eds.) pp. 45–71, Academic Press, New York (1969).
48. J. D. Brown and L. Parobek, Comparison of $\phi(\rho z)$ curves measured on instruments of different geometries, paper 5, *Proceedings of the Seventh National Conference on Electron Probe Analysis* (1972).
49. V. E. Cosslett and R. N. Thomas, The plural scattering of 20 keV electrons, *Brit. J. Appl. Phys.* **15**, 235–248 (1964).
50. V. E. Cosslett and R. N. Thomas, Multiple scattering of 5–30 keV electrons in evaporated metal films I. Total transmission and angular distribution, *Brit. J. Appl. Phys.* **15**, 883–907 (1964).
51. V. E. Cosslett and R. N. Thomas, Multiple scattering of 5–30 keV electrons in evaporated metal films II. Range-energy relations, *Brit. J. Appl. Phys.* **15**, 1283–1300 (1964).
52. V. E. Cosslett and R. N. Thomas, Multiple scattering of 5–30 keV electrons in evaporated metal films III. Backscattering and absorption, *Brit. J. Appl. Phys.* **16**, 779–796 (1965).
53. W. L. Patterson and G. A. Shirn, The sputtering of nickel–chromium alloys, *J. Vac. Sci. Technol.* **4**, 343–346 (1967).
54. G. H. Maher, Physical and Electrical Properties of Thin Film Barium Titanate Prepared by rf Sputtering on Silicon Substrates, Ph.D. Thesis, Rensselaer Polytechnic Institute (1971).
55. M. Green, Ph.D. Thesis, Cambridge University (1962).
56. K. F. J. Heinrich and H. Yakowitz, Propagation of errors in correction models for quantitative electron probe microanalysis, in *Fifth International Congress on X-Ray Optics and Microanalysis* (G. Möllenstedt and K. H. Gaukler, eds.), pp. 151–159, Springer-Verlag, Berlin (1969).
57. G. A. Hutchins, Electron probe microanalysis, in *Techniques of Surface and Colloid Chemistry and Physics* (R. J. Good, R. R. Stromberg, and R. L. Patrick, eds.), Vol. 2, Marcel Dekker, New York (in press).

19

X-RAY EMISSION FINE FEATURES

William L. Baun

Mechanics and Surface Interactions Branch
Air Force Materials Laboratory
Wright Patterson AFB, Ohio

1. INTRODUCTION

1.1. History and Reviews

For a time following the discovery of x rays it was believed that characteristic x rays were simple functions of the emitting atoms and that combining the atoms with dissimilar atoms did not change the characteristic spectrum. However, it was predicted by Swinne[1] in 1916 that under certain conditions characteristic x rays would be affected by chemical combination. The first changes observed were not in emission lines but in *K* absorption edges by Bergengren[2] in 1920. Following this discovery, Lindh and Lundquist[3] and several others found changes due to chemical combination in emission line energies. Shortly afterward, there were many reports of changes in emission spectra, especially in third-period elements. The early literature was reviewed in Siegbahn's book in 1931.[4] A complete review and reference list through 1951 was prepared by Herglotz.[5] Faessler's chapter in the Landolt–Bornstein tables[6] emphasizes experimental absorption and emission data for many pure elements and compounds. The Encyclopedia of Physics, Vol. XXX (X Rays) includes a section by Sandstrom on experimental methods of x-ray spectroscopy[7] and by Tomboulian on soft x-ray spectroscopy and the valence-band spectra of light elements.[8] Blokhin's book,[9] translated from Russian in 1961 and available from the Department of Commerce, contains much useful data and comment on effects of chemical combination, especially Chapters 8 and 9, which deal with fine structures in emission and absorption spectra. Shaw's excellent review[10] on the x-ray spectroscopy of solids contains a bibliography of 355 references as well as some unpublished data by Shaw and his co-workers. An annotated bibliography by Yakowitz and Cuthill[11] contains over 500 references and emphasizes work done during the ten-year period of 1950–1960. Reviews by Baun[12] and Nagel[13] and a cooperative book chapter by these two authors[14] have brought the subject up to date. There have been several meetings or sessions at meetings devoted to the effect of chemical combination on x-ray spectra. A conference was held on the applications of x-ray spectroscopy to solid-state problems in Madison, Wisconsin in 1950,[15] and on the physics of x-ray spectra at Cornell University in 1965.[16] Also in 1965, a complete program was presented on the chemical effect at the Karl Marx University in Leipzig.[17] The published proceedings of the 1967 Strathclyde meeting[18] present many recent results of work in soft x-ray spectroscopy and the use of fine features on x-ray spectra.

1.2. What Else is X-Ray Spectroscopy Good For?

Naturally, x-ray spectra are useful for the purpose to which they are usually directed as outlined in the chapters in this volume by Hutchins (Electron Probe Microanalysis) and Gilfrich (X-Ray Fluorescence Analysis). However, in addition to these uses for qualitative and quantitative analysis, one may use x-ray spectra in many cases to determine just how each element is chemically combined in the sample. The purpose of this review is to outline the kinds of spectral effects which are observed, review special problems and measuring techniques, and show some practical examples of the use of fine features of x-ray spectra. Emphasis will be placed on techniques for surface or near-surface characterization.

2. CHANGES IN FINE FEATURES WITH CHEMICAL COMBINATION

2.1. Shifts

Wavelength shifts to both longer and shorter wavelengths have been observed in x-ray spectra for various elements upon chemical combination. These shifts result from energy-level changes due to electrical shielding or screening of the electrons when the valence electrons are drawn into a bond. The greatest shifts are seen in spectra elements with low atomic numbers. Generally, the so-called last or highest-energy member of a given series is most affected by chemical combination. For instance, K bands (K_β) of oxides of Mg, Al, and Si shift over 4 eV compared to the pure metal while the K_α doublet shifts less than 1 eV (in the opposite direction). Satellite lines shift more than the parent line. X-ray shifts are usually not as great as the actual energy-level shifts as measured by electron spectroscopy, because in x-ray spectroscopy we measure only energy-level differences. Where each level, say the K and L, shift about the same amount and in the same direction, as is often the case, the resulting K_α line will show little shift with chemical combination. Although the shifts are sometimes small, V. H. Sanner concluded in his dissertation (Uppsala, 1941) that the wavelength shift of the K_α doublet can be followed to $_{25}$Mn, that of K_{β_1} to $_{27}$Co, and the quadrupole line K_{β_5} to $_{28}$Ni. Similar measurements by other authors have also established practical shift-detection limits for L and M series.[7]

2.2. Shape Changes

The shape of a band gives some indication of the energy distribution of the electrons occupying positions in or near the valence shell. It was once thought that the electron density of states could be deduced directly from band spectra, but it is now obvious that many factors govern band shapes and that corrections are necessary to determine the actual electron population. Recent interpretations by numerous authors based on molecular orbital theory account nicely for nearly all absorption- and emission-spectra fine features for individual compounds. However, no "broad brush" theoretical treatment has been advanced to cover each spectral series for all materials.

2.3. Intensity Changes

Large intensity changes are seen in certain lines and bands due to changes in excitation probabilities of the electrons undergoing transitions. Several phenomena combine to affect transition probabilities. One possibility is that with chemical combination, there is some change in electron character which allows either an increased or decreased number of electrons to undergo a given transition. Another possibility is the electron depopulation of a given level or state and the subsequent formation of exciton or excitation states. A third contributing factor is the occurrence of nonradiative transitions. These nonradiative transitions increase with decreasing atomic number and undoubtedly contribute to absolute intensity changes observed in low-Z elements with chemical combination. Relative changes in intensity of various line and band components have been attributed to shifts within a band contour which give the appearance of true intensity changes. These shifts have been explained as noted above using molecular orbital theory. Certain lines or bands appear or disappear with chemical combinations. The origin of these features, such as $K_{\beta'}$ were, until only recently, very unclear. Now that molecular orbital theory has been applied to x-ray spectra and the occurrence of satellites has been better explained, we have a far more lucid picture of the origin of x-ray spectral features.

3. X-RAY EMISSION SPECTRA

In the following section, some of these changes due to chemical combination are summarized for a number of elements with emphasis on low-Z elements. The wavelength range to be discussed is rather arbitrary, based only on the current instrumental state of the art. The long-wavelength cutoff is placed at beryllium K (116 Å) since wavelengths to 120–130 Å appear to be a practical limit for microprobes employing curved-crystal optics and flow-proportional counter detection. Naturally, when microbeam-probe equipment using grating spectrometers and windowless detectors becomes more prominent, this long-wavelength limit will be extended to perhaps 500 Å, which will allow measurements of Li K and Mg L spectra, for instance. This development probably will only be applied to combination SEM–microprobe instruments which have ultrahigh vacuum systems since the current low-vacuum ($\sim 10^{-5}$ Torr) instruments create carbon contamination on the sample which attenuates ultrasoft x rays.

3.1. *K* Series

It was mentioned earlier that the most noticeable effects are seen in spectra from low-Z elements. Figure 19-1 shows some of the effects of chemical combination on K spectra of very light elements beginning with beryllium. The Be K spectrum in the metal consists of just one asymmetrical band and illustrates an effect of chemical bonding even in the pure element. The $K_{\alpha_{1,2}}$ line originates from the transition $L_{\text{II,III}} \rightarrow K$. In beryllium the $L_{\text{II,III}}$ level is empty, having only two electrons in the L_{I} level. Therefore, K_α appears sooner than the first electron appears in the $L_{\text{II,III}}$ level of the free atom. This indicates the role of chemical bonding in the solid; the external electrons of the atom are excited to the next optical level where these electrons then can complete the transition to the vacancy in the K level. With chemical combination to the oxide, the K_α line shifts significantly to longer wavelengths and becomes quite symmetrical, and a new line is observed on the short-wavelength side of the K_α line. Similar effects are noted on the boron K spectra also shown in this figure. Carbon K and nitrogen K spectra do

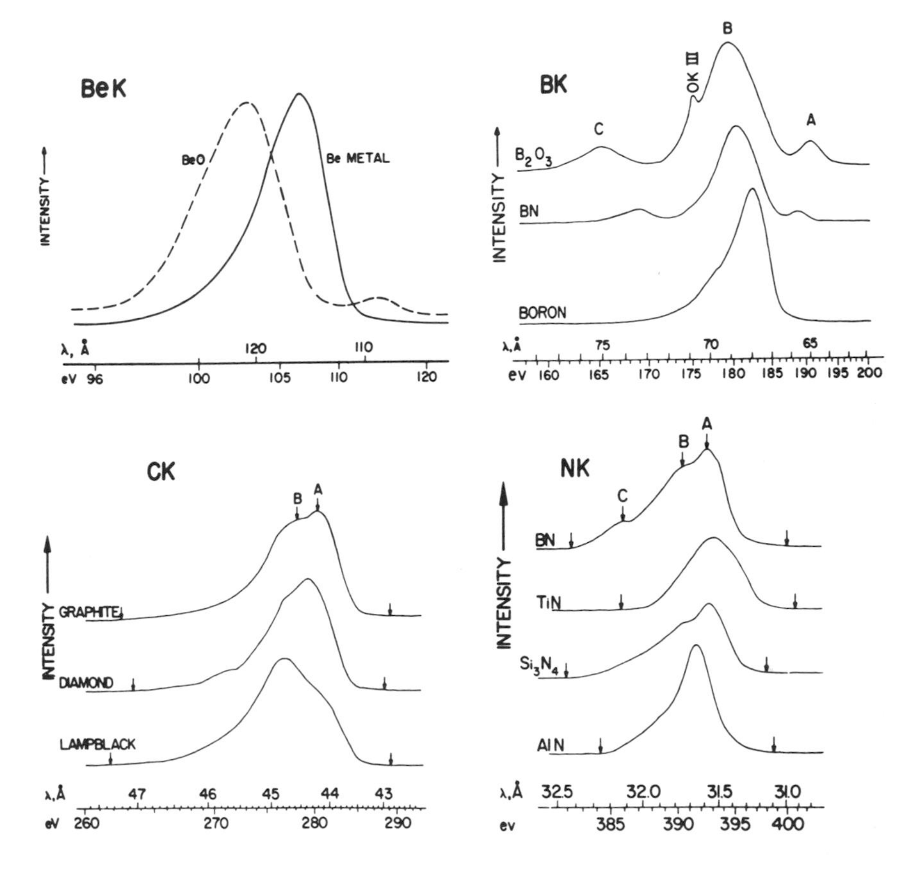

Fig. 19-1. The K emission band spectra from some low-Z elements.

not change as drastically perhaps, but sufficiently to allow determination of chemical combination or even structure in the case of carbon polymorphs.

The K spectra of Mg, Al, and Si are quite similar and are greatly affected by chemical combination. Baun and Fischer[19] have studied these elements in the form of pure elements, oxides, and various compounds and have tabulated wavelengths, energies, intensities, and energy differences for diagram and satellite lines. Table 19-1 is an example of some of the data tabulated. It is included to show lines and bands existing in the Al K series and the changes which occur in going from the metal to the oxide (anodized coating). From this table it is seen that the largest changes due to chemical combination are in K_α satellite lines, the K emission band, and in satellites of the K band (it has not been positively established that all extra lines near the K band are true satellites). The three satellite lines observed closest to the K_α line,

TABLE 19-1. Aluminum *K* Lines and Bands from Aluminum and Al_2O_3[a]

Line	λ, Å	E, eV	Intensity[b]	ΔE, eV[c]
Aluminum metal				
$\alpha_1\alpha_2$	8.3393	1486.3	1000	—
α'	8.3080	1492.0	13	—
α_3	8.2854	1496.0	78	—
α_4	8.2744	1498.0	39	—
α_5	8.2284	1506.3	5.0	—
α_6	8.2098	1509.8	3.9	—
β	7.9590	1557.3	6.5	—
β''	7.8330	1582.4	<0.5	—
β'''	7.8048	1588.1	<0.3	—
Aluminum oxide, Al_2O_3				
$\alpha_1\alpha_2$	8.3380	1486.6	1000	+0.3
α'	8.3037	1492.7	18	+0.7
α_3	8.2820	1496.6	64	+0.6
α_4	8.2718	1498.6	60	+0.6
α_5	8.2240	1507.2	4.6	+0.9
α_6	8.2050	1510.7	3.5	+0.9
β'	8.0618	1537.5	1.3	—
β	7.9819	1552.9	7.0	−4.4
β''	7.8522	1578.6	<0.3	−3.8
β'''	7.8204	1585.0	<0.2	−3.1

[a] See Baun and Fischer.[19]
[b] Peak intensity.
[c] Shift between metal and oxide.

$K_{\alpha'}$, K_{α_3}, and K_{α_4}, are shown in Fig. 19-3 for the pure element and the oxide for Mg, Al, and Si. The spectrum shown for Al_2O_3 is for anodized Al. The spectrum is different for α-Al_2O_3. With oxidation, and in general with any form of chemical combination, the satellites shift to shorter wavelengths and the intensity relationships change significantly. The K band (K_β) shows similar gross changes as seen in Fig. 19-2. In all three elements an asymmetrical band is obtained from the pure metal much the same as shown earlier for pure Be and B. With oxidation, the band shifts over 4 eV and becomes symmetrical. Also the molecular orbital band $K_{\beta'}$ appears in the oxide; $K_{\beta'}$ is very useful for characterizing thin oxide films on several materials.

Alloying has a significant effect on x-ray spectra of low-Z elements. In Al–Cu, for instance, and in the spectra of nearly all polyvalent metals alloyed with monovalent elements, there are significant changes,

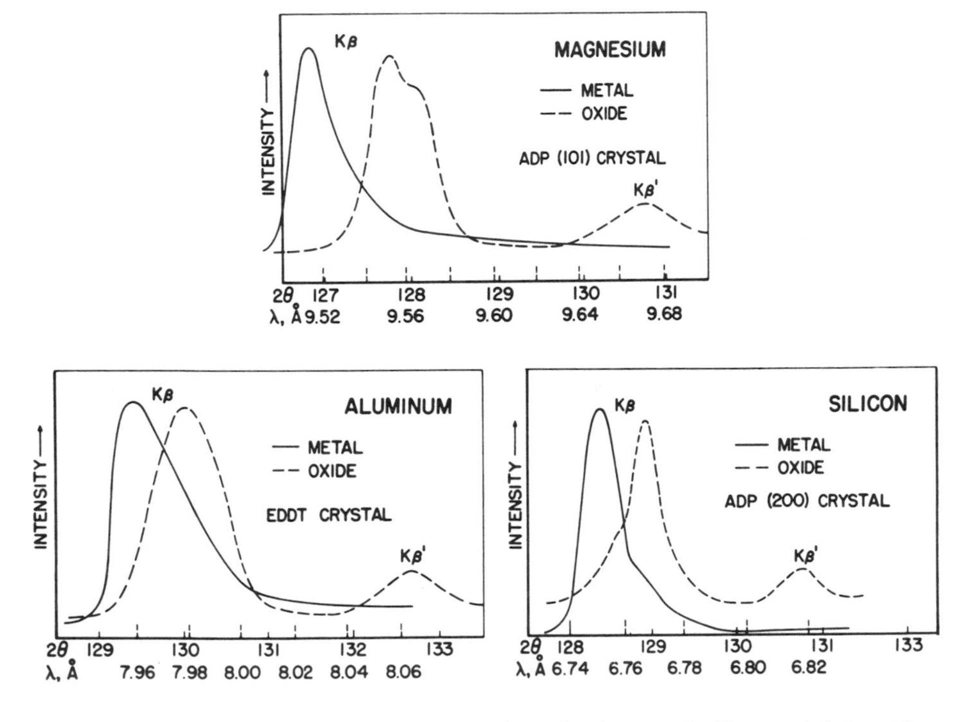

Fig. 19-2. The K emission bands from magnesium, aluminum, and silicon and their oxides.

especially in the K band. In the Al–Cu system the Al K band splits into two components, as will be shown later in Section 6. In this system, the short-wavelength component retains the sharp short-wavelength limit and shifts only slightly with alloying, while the long-wavelength component shifts linearly with composition toward longer wavelengths. Other alloys behave differently. A summary of the systematic nature of Al K spectral changes has been prepared.[20]

The K spectra of the elements phosphorus, sulfur, and chlorine exhibit many of the same changes observed for the previous elements, but sometimes K_α shifts are equal to or greater than K_β shifts. A very complete report on the relationship between chemical bonding and the x-ray spectrum of sulfur was published by Wilbur.[21] In this work, K_α and K_β energies were determined for a number of compounds, and the intensity dependence of K_β on chemical binding was investigated. Wilbur was able to find good correlation of K_β intensity and energies for higher sulfur valences, but within the 0 and -2 states there was no particular correlation. The K_β energies and widths exhibited large differences, but there was no correlation with bonding. A representative set of K_β profiles from Wilbur's report is shown in Fig. 19-4.

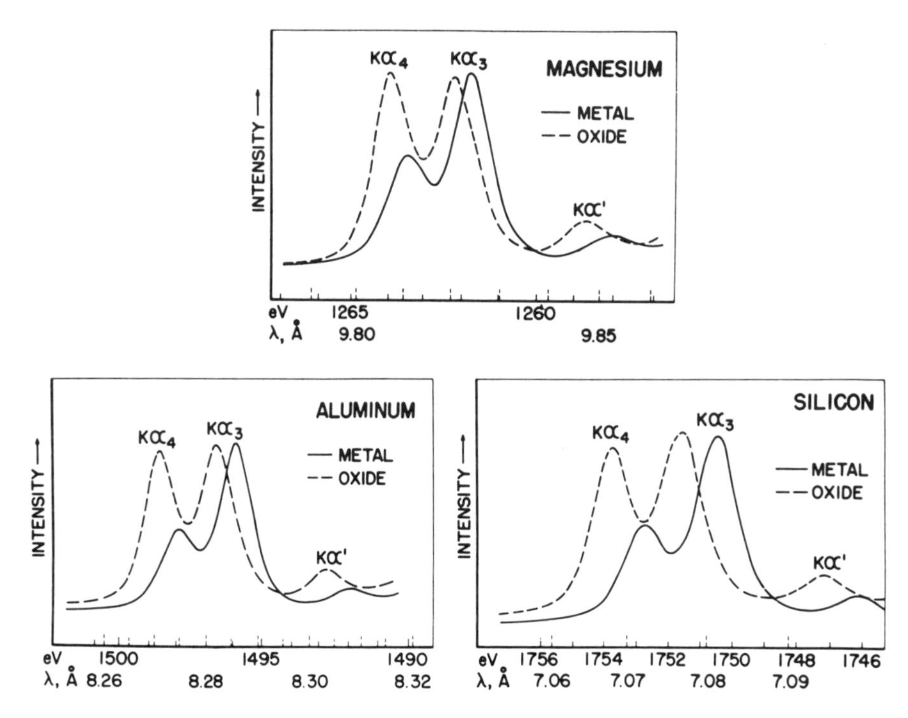

Fig. 19-3. The K emission satellite lines from magnesium, aluminum, and silicon and their oxides.

3.2. *L* Series

There are L spectra that show as large or even larger changes with chemical combination of the elements than K spectra. Examples of L spectra from compounds containing chlorine, sulfur, and potassium are shown in Fig. 19-5.[22] The chlorine L spectrum consists of one main band having structure on the low-energy side. The main band shift may be correlated with the electronegativity of the cation, as was shown by Fischer and Baun,[22] or with an estimation of the percent ionic character of a single bond, using Pauling's electronegativity values. This relationship shows that the greater the ionic character of the bond, the shorter the wavelength of the main L band in simple chlorides.

The sulfur L spectrum undergoes extremely large changes with chemical combination of sulfur. Elemental sulfur gives one main broad band ($3s$) with a strong component ($3p$) which is almost 4 Å toward the shorter wavelengths. Sulfides, on the other hand, show the main band ($3s$) and a weaker short-wavelength component ($3d$) that averages a nearly 5-Å separation from the main band. Care must be taken in these materials to ensure·that specimen decomposition has not taken place if direct electron-excitation methods are used.

Changes with chemical combination take place in L spectra of elements of the first transition series and of elements nearby in the periodic table. Fischer[23] has shown that the L_{III} band from oxides of elements from ${}_{23}Cr$ to ${}_{30}Zn$ shift predictably with respect to the pure element. Shifts are to higher energy in oxides of the transition elements ($Z = 24$–28) and to lower energy in Cu and Zn oxides. Intensities have been shown to change significantly between the pure elements and oxides.

There are some changes with alloying in L spectra of the first transition elements, but these changes are not as large as in K spectra of low-Z elements described earlier.

3.3. *M* Series

A complete review of changes in M spectra due to chemical combination will not be attempted. Very little recent M spectral data exist, especially where the emphasis of the work was on effects of chemical combination. The M spectra in the long-wavelength region are in general

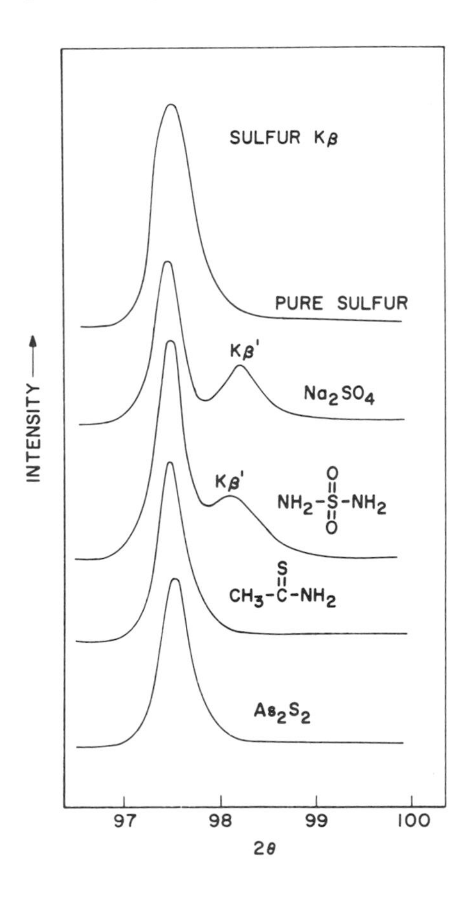

Fig. 19-4. The sulfur K band from a group of sulfur-containing compounds.

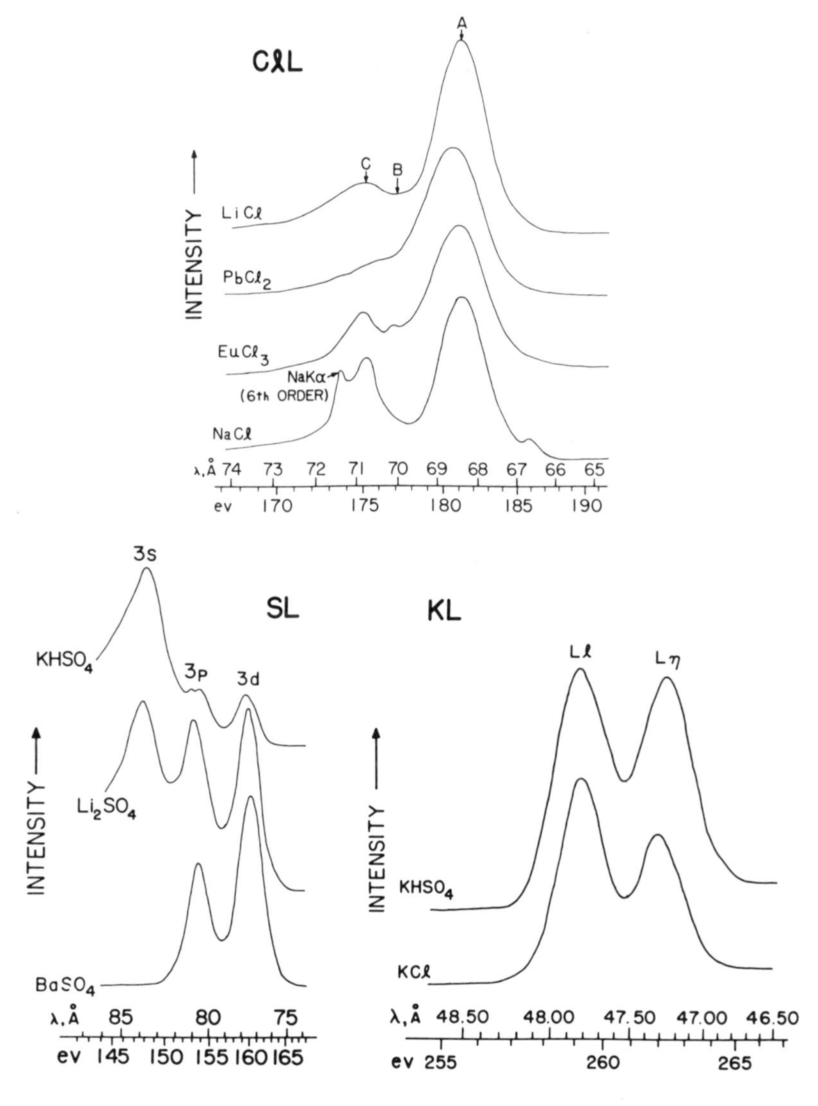

Fig. 19-5. The L emission spectra from chlorine, sulfur, and potassium in various compounds.

quite weak, which probably accounts for the paucity of data on M bands. Furthermore, the majority of M spectra are beyond the wavelength capability of most microbeam probe instruments. It seems unlikely that the microbeam probe will be used for M spectra of elements below about ${}_{39}Y$; for very-high-Z elements there is little effect on M lines due to chemical combination. As an example of the kind of changes induced in M spectra by chemical combination, Fig. 19-6 is shown from work of

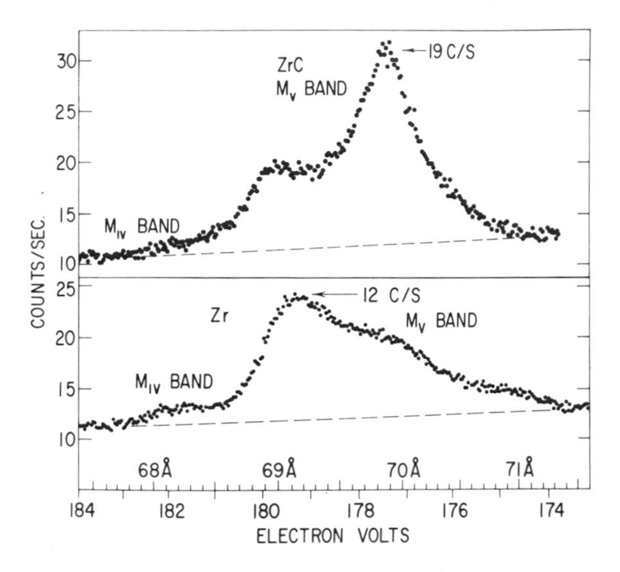

Fig. 19-6. The zirconium M spectra from Zr and ZrC.

Holliday.[24] The shape of the Zr M_V band is significantly different in ZrC compared to Zr metal.

4. PROBLEMS AND REPRODUCIBILITY

4.1. Carbon Contamination

The problem of carbon contamination is well known to all microprobe users. Carbon contamination results from the decomposition on the specimen of hydrocarbons from vacuum pump oil, cleaning solvents, and from within the specimen itself. It is not always considered a problem, of course, because the carbon spot shows the microprobe operator where the beam is and the approximate size of the beam. It can generally be ignored when using hard x rays, but not when using x rays softer than about 10 Å. There are several techniques that one can use to decrease carbon contamination. If one is not doing microanalysis on a small spot, one can defocus the beam so that the specific loading is reduced and the carbon is deposited over a larger area and/or the specimen may be scanned under the beam to present constantly a new specimen surface. Carbon contamination may also be minimized by heating the sample and by cooling the area surrounding the specimen, thus condensing hydrocarbons on the cold plate rather than on the sample. Corrections for carbon buildup may also be applied by measuring intensities over the period of the experiment and then extrapolating back to zero time. Such a set of measurements at 10 kV for Fe L_α from pure ion at 17.6 Å

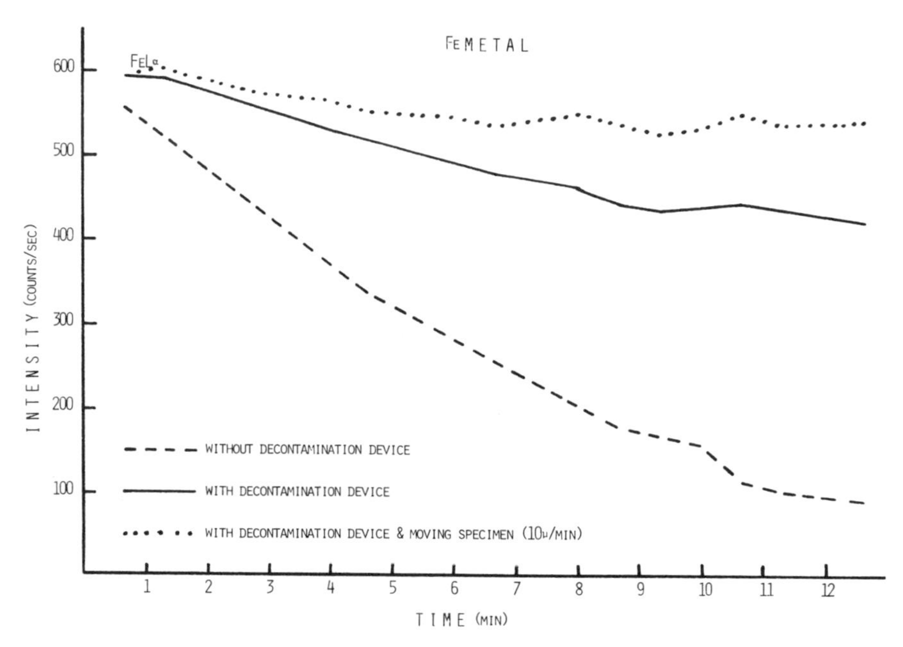

Fig. 19-7. Decrease in Fe L intensity with time due to carbon contamination.

is shown in Fig. 19-7. The dashed line shows the very rapid decrease in intensity with a fixed sample, small beam, and no decontamination device (cold plate). When the decontamination device is cooled with liquid nitrogen, the results improve and the solid curve is obtained. Finally, when the decontamination device is used and the sample is moved under the beam, the dotted curve is obtained. It is quite obvious that even for very short experiments, one must correct for carbon contamination when conditions are not minimized for carbon production. Table 19-2 summarizes the intensity decrease of Fe L_α from Fe

TABLE 19-2. Intensity Decrease in Fe L due to Carbon Contamination

	Intensity decrease after 10 min, %		
Line	Without decontamination device	With decontamination device	With moving specimen and decontamination device
Fe L_a (metal)	77.1	25.4	11.1
Fe L_α (FeO)	29.4	9.5	—
Fe L_{β_1} (FeO)	25.4	10.2	—

metal and Fe L_α and L_β from FeO after the beam has been impinging on the sample for ten minutes. As can be seen, the problem is not nearly as severe for the poorer conductor, FeO, as for the good conductor, Fe metal.

4.2. Sample Self-Absorption

Another absorption phenomenon, self-absorption in the sample, may cause serious problems in the interpretation of spectra. Work by Fischer[23] on L spectra of elements of the first transition series showed large variations in relative intensities of L_α and L_β with changes in voltage. These changes can be correlated with the depth of electron penetration. For instance, the ratio L_β/L_α for copper is shown to vary linearly with electron penetration. However, it does not vary as one might expect. Generally, it would be expected that as electron penetration increases and the effective x-ray production depth increases, the relative intensity of softer radiation compared to harder radiation would decrease. This is what happens for K_α and K_β where the K absorption edge lies to the short-wavelength side of both lines. Cu L_α and L_β behave in an opposite manner, however, with the harder radiation, L_β, being selectively absorbed in the sample with increasing electron penetration. The strong L_{III} absorption edge between L_β and L_α emission bands accounts for this behavior.

The penetration of electrons is appreciable in the 1- to 10-keV energy region, and therefore x-ray spectroscopy is primarily a bulk technique, not capable of analyses of the first few monolayers. In order to make it a near-surface technique, beam voltages must be kept small. Feldman[25] has measured the range of energies of electrons in solids (1–10 keV) and has formulated the following expression

$$R = 250\frac{A}{\rho Z n/2}E^n \quad (1)$$

where $n = 1.2\,(1 - 0.29 \log_{10} Z)$, E is the energy in keV, R is the range in angstroms, A is the atomic weight, ρ is the density, and Z is the atomic number.

Of course, this range of electron energies does not tell us from what depth x rays are actually being produced or from what depth the spectrometer sees the x rays, but it does give us an idea of how the penetration changes with energy from element to element.

Similar absorption changes and subsequent spectral changes are caused by changes in take-off angle (the angle at which the x rays leave the sample surface). Therefore, the exact conditions, including take-off angle and exciting voltage, should be specified for each spectrum.

The effects due to changes in excitation conditions can be quite dramatic. Self-absorption and, to some extent, changes in satellite emission, account for many changes in spectra. Spectra taken at ten or more times the excitation potential will be very similar because each is greatly affected by self-absorption. For instance, rate-meter traces from Gadolinium M at 10 kV and at 30 kV are nearly the same. When the spectra are normalized and one spectrum is divided into the other, however, significant differences are seen, and a self-absorption curve can be generated as shown in Fig. 19-8. Here the self-absorption curve is plotted with the emission spectrum obtained at 10 kV to show the correspondence of emission and self-absorption features. This self-absorption curve is plotted by the computer just as in the case of the emission spectra. Therefore, although the microbeam probe is not normally used for absorption measurement, it is possible to deduce absorption features from emission spectra in this way. Such absorption

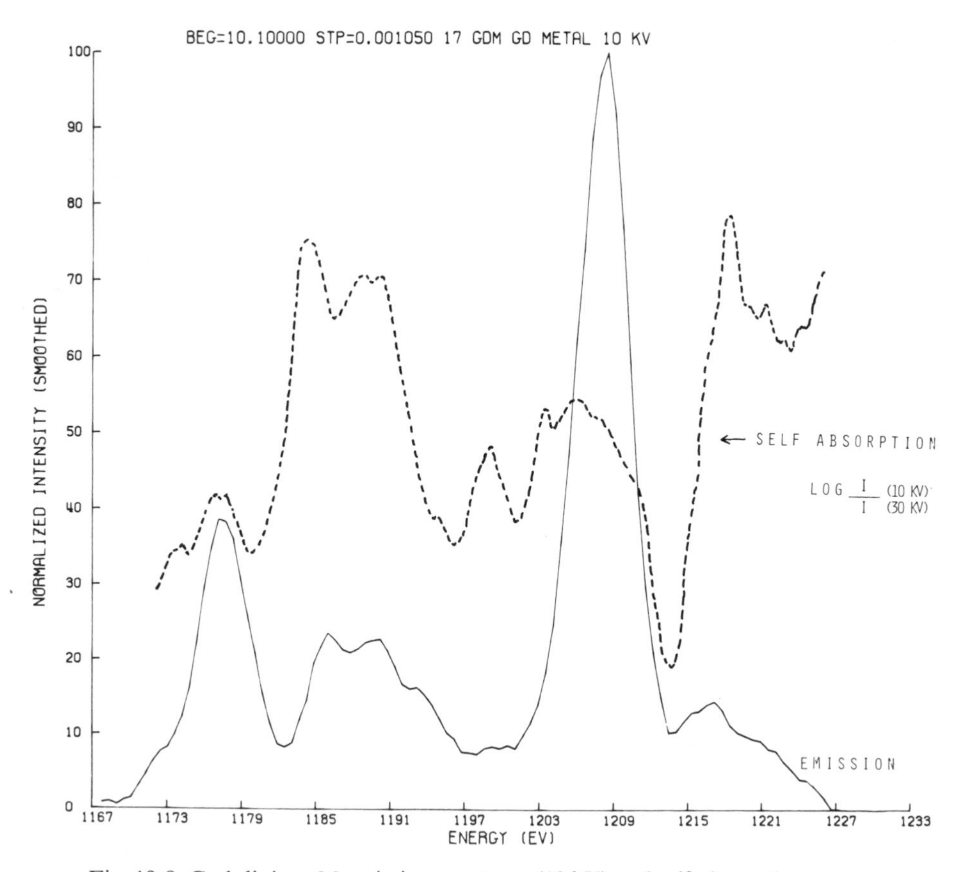

Fig. 19-8. Gadolinium M emission spectrum (10 kV) and self-absorption curve.

replicas allow absorption features to be obtained on materials and in spectral regions which would be impossible using conventional transmission methods.

5. DATA ACQUISITION AND HANDLING

Generally the fine-feature changes in x-ray spectra with changes in chemical combination are quite subtle and require careful experimental procedures. Dr. E. W. White of the Pennsylvania State University outlined the general procedures which must be followed for recording spectra from the microprobe for use of fine features in chemical-combination studies.[26] His comments are summarized as follows:

The following general procedure for recording soft x-ray shift information applies particularly to strip-chart-recording data. The analogy to step-scanned data should be obvious:

1. Select a mechanically sound spectrometer and crystal that yields good resolution, dispersion, and intensity for the emission line of interest. Align the spectrometer for the particular spectral range to be covered by using a reference line (typically in high order) not subject to the chemical effect. The reference specimen must be well polished and positioned exactly on the fixed focal plane of microscope during the alignment process.
2. Be certain that any stress in the spectrometer drives, resulting from the alignment procedure, has been relieved by repeated high-speed scanning over a range in the vicinity of the peak.
3. Carefully synchronize the starting of the spectrometer motor drive to coincide with a fiducial mark on the strip-chart recorder.
4. The rate-meter time constant must be short enough not to artificially "displace" the peak.
5. Select a scale factor that gives as near full-scale deflection as possible without going off scale.
6. Select one sample as a reference "standard" and rerun a peak on if after every two or three unknowns in order to detect any instrumental drift.
7. Be sure to select a fresh area of sample for each repeat scan and "refocus" the specimen each time. Do not change anything (oculars, etc.) about the microscope throughout a given set of scans.
8. Work only with an instrument that has come to thermal equilibrium in a constant-temperature room.
9. Always scan from the same direction when recording peaks.
10. Always use as large a spot size as feasible to reduce effects of beam damage.

Dr. White feels that two factors in particular limit the precision with which x-ray spectra may be measured: (1) Chemical and structural changes in the sample induced by the electron beam and the high vacuum contribute the most troublesome source of error. Obviously, the lower the specimen current loading, the less will be the beam damage. (2) Inconsistent positioning of the specimen surface on the focal plane of the microscope is the chief source of random error.

Colby[28a] described a digital technique utilizing step scanning and computer plotting of data to enhance resolution of electron microprobe spectra. Figure 19-9 summarizes some results Colby has achieved in allowing the computer to analyze curves and pinpoint spectral features. In this figure the raw data for silicon $K_{\beta,\beta'}$ from SiO_2 is compared with

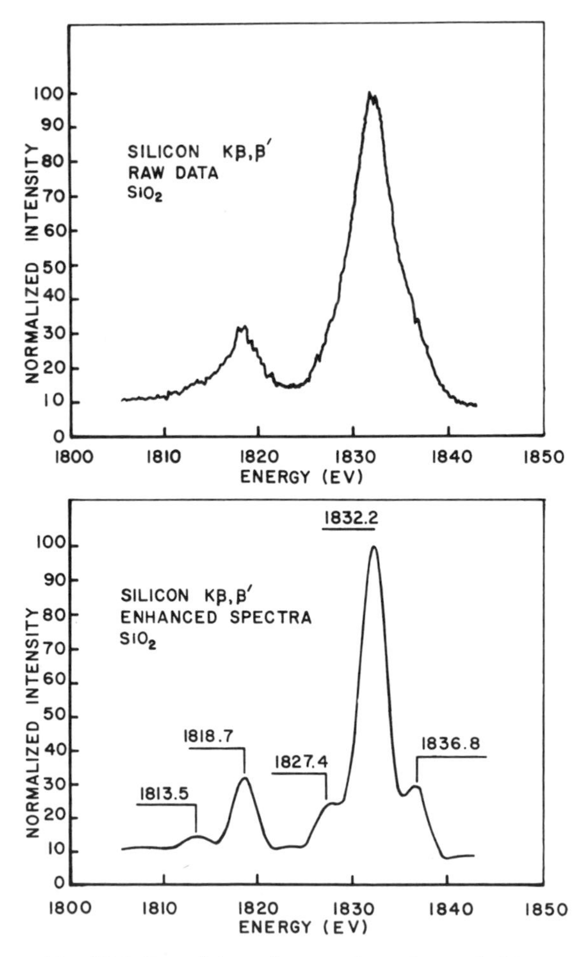

Fig. 19-9. Raw data and computer-enhanced silicon K emission bands for SiO_2.

the computer-resolved and computer-plotted spectra. This work offers convincing evidence that digital techniques and objective analysis by computer is far superior to rate-meter spectral traces.

6. EXAMPLES OF PRACTICAL USE OF X-RAY FINE FEATURES

Fine features in x-ray emission spectra may be used to partially characterize materials which do not lend themselves to conventional techniques. For instance, poorly crystalline materials and very thin films often pose problems to the individual faced with the task of characterizing them. One such problem encountered in the author's laboratory concerned vacuum-deposition characteristics of binary alloys. Particularly, it was desired to determine for the Al–Cu system[27] if vapor deposition resulted in thin films having the same stoichiometry and state of chemical combination as in the original bulk material. It was felt that x-ray spectroscopy would be useful in this problem since Baun and Fischer[20] had shown that fine features of the Al K band and Al K satellites are considerably different in Al–Cu alloys than in pure aluminum. In particular, they found that the Al K band (Al K_β) exhibited a splitting effect in Al–Cu alloys, and the energy difference between high- and low-energy components of the band varied linearly with composition, as seen in Fig. 19-10a. A typical K band observed from Al–Cu alloys is plotted below the ΔE curve. Above the line, the asymmetrical K band from pure aluminum is plotted. In both cases wavelength increases to the right.

The vacuum-evaporated alloy that was of greatest interest was the η_2 phase (49Al–51Cu). Bulk material used for evaporation was obtained from the stock of alloys prepared for the original work on Al–Cu alloys. Vacuum evaporation was carried out from a tungsten boat at temperatures just above the melting point in a vacuum of 5×10^{-6} Torr. The evaporated material was collected on a polished copper plate which fitted into the anode or sample holder in the spectrometer. Samples of the same thickness were also collected on glass microscope slides. No x-ray diffraction peaks were observed from this film on glass when subjected to a routine diffractometer trace using copper radiation. Apparently, a softer radiation or special film techniques would have been necessary to obtain a diffraction pattern.

The Al K band obtained from vacuum-evaporated 49Al–51Cu is shown in Fig. 19-10b. This same curve is obtained even when the evaporation rate is varied over a wide range. In the insert of this figure, a K-band trace from a bulk 49Al–51Cu alloy is shown. The intensity relationships and the energy difference between high- and low-energy

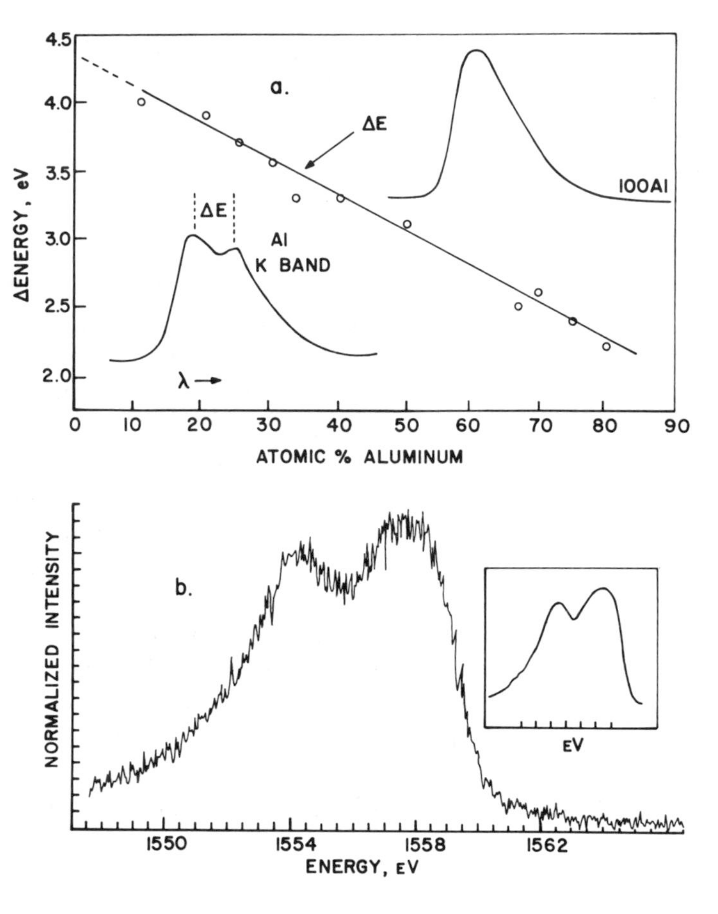

Fig. 19-10. (a) Energy difference between high- and low-energy components of the Al K band with change in composition in the Al–Cu system. Typical Al K band shown from Al (top) and from Al–Cu alloy (bottom). Al K band from vacuum-evaporated 49Al–51Cu alloy. Insert shows same band from bulk alloy.

components (3.1 $\pm$ 0.1 eV) are identical in the two curves. Even without running the bulk material as a standard, the composition of this film can be determined by referring to the ΔE curve in Fig. 19-10a. The conclusion that is reached is that the 49Al–51Cu alloy is a "constant-evaporation alloy" and is condensed in a form having the same chemical combination as in the bulk material. Other alloys on the copper-rich end of the Al–Cu system also were evaporated, but were two-phase alloys. In these alloys the evaporation temperature had a significant effect on the K-band features. In these two-phase alloys the K-band high- and low-energy-component differences did not fall on the same point as in the bulk materials. It appears that the copper or copper-rich phase evaporates more rapidly than the aluminum-rich alloy.

The change in intensity and energy of the K_α satellites α_3 and α_4 may also be used to determine the composition of Al–Cu and many alloy thin films. These lines have the advantage that they are stronger than the K band, but they show relatively small changes with changes in composition. The Al $K_{\alpha 4}/K_{\alpha 3}$ ratio in pure aluminum is 0.48 and changes in a near-linear fashion to 0.70 in 10Al–90Cu.[20] Energy changes in $K_{\alpha 3}$ and $K_{\alpha 4}$ are small, but are about twice that observed for the parent K_α line.

X-ray emission spectra provide convincing evidence that SiO in a condensed state consists of an intimate mixture of elemental silicon and silicon dioxide. White and Roy[28] used the K-band splitting in SiO films and concluded that the films were variable mixtures of Si and SiO_2. Baun and Solomon[29] used other fine features in Si K and O K spectra to corroborate the results of White and Roy. Figure 19-11a shows a typical Si K-band contour from SiO along with the spectra obtained from pure Si and SiO_2. As can be seen, the two components in SiO match the positions for the two pure materials. Further evidence that SiO is a mixture may be found in the position and intensity of $K_{\beta'}$, which is the feature to the left of the main K band in Fig. 19-11a.

Another feature of the Si K spectrum which may be used to characterize SiO films is the K_α satellite pair $K_{\alpha 3,4}$ which involve transitions from multiply ionized levels. These lines are several times as strong as K_β and an order of magnitude stronger than $K_{\beta'}$, this making measurements easier and statistics better. The satellite lines are very sensitive to chemical combination with $K_{\alpha 4}/K_{\alpha 3}$, about 0.60 in pure Si and nearly unity in SiO_2. Many different line profiles in the $K_{\alpha 3,4}$ region are obtained for SiO films, depending on vaporization conditions. Nevertheless, it has been possible in this work to match each contour by synthesizing curves for SiO_2 and Si and then adding these curves on the curve resolver. It is possible to obtain a calibration curve relating the composition of films with the area or peak heights in the individual curves. Analyzed thick films and bulk specimens may be used as secondary standards. Figure 19-11b shows a typical $K_{\alpha 3,4}$ spectrum from an SiO film along with a curve (III) obtained by adding curves from Si (I) and SiO_2 (II). Curve III was synthesized on the DuPont Curve Resolver by (1) matching curves from pure Si and SiO_2, (2) placing the curve from SiO_2 (II) on the screen, and (3) gradually adding the curve from Si (I) until a match was achieved with the experimental SiO curve which had been placed on the resolver table. The area under each curve may be read directly from the instrument and related to composition of chemically analyzed samples; a calibration curve may then be prepared.

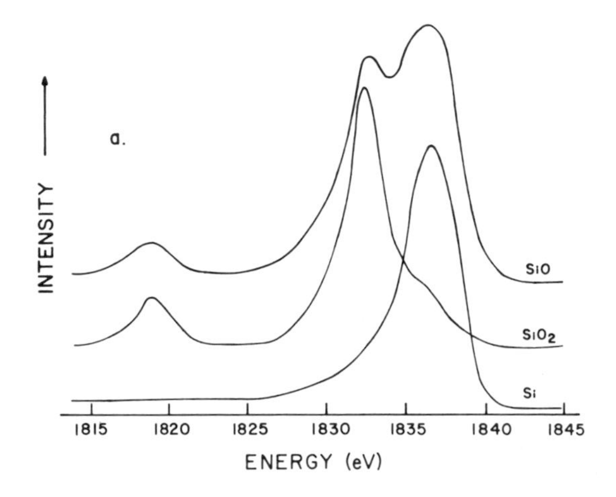

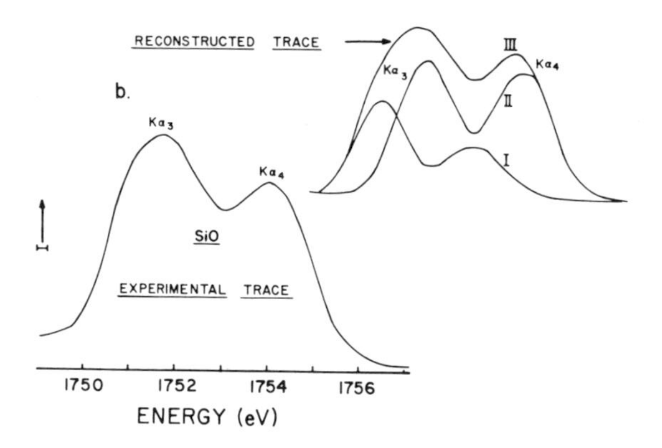

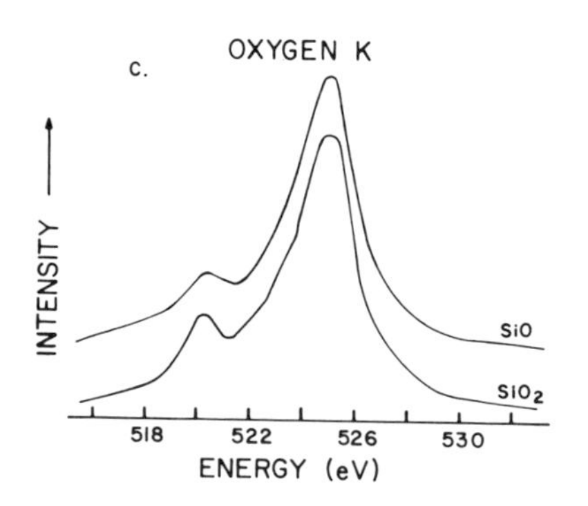

Fig. 19-11. (a) Si K emission bands from Si, SiO_2, and SiO. (b) Experimental K satellite lines from SiO along with a reconstructed synthetic curve obtained by adding spectra obtained from Si and SiO_2. (c) Oxygen K emission spectra from SiO and SiO_2.

The oxygen K spectrum from SiO films is virtually identical to the spectrum obtained from SiO_2 (α-quartz), as seen in Fig. 19-11c. The oxygen K band requires at least four main Gaussian components to match the curve shown in Fig. 19-11c. A small change in intensity of one of these components accounts for the slight difference noted on the low-energy portion of the SiO curve. This may be interpreted as a change in population of a molecular orbital, and could be an indication that the SiO_2 phase in SiO (1) is one of the high temperature phases, (2) contains stresses which have distorted the typical SiO_4 tetrahedral structure, or (3) is slightly changed in Si–O stoichiometry. In addition, it is possible that the addition of impurities has disturbed one or more of the molecular orbitals compared to α-quartz.

White and his group at the Pennsylvania State University have published widely in the area of the use of fine features of x-ray spectra for characterizing materials. Some of their work on surface characterization has included characterization of thin films of aluminum oxides and hydroxides,[30] thin-film characterization by electron microprobe and ellipsometry of SiO_2 on Si,[31] and characterization of corrosion layers on aluminum.[32]

Many other examples could be cited, but space limitations preclude further review of practical work which has been accomplished in the area of surface characterization. Hopefully, as time passes, more people who use x rays for qualitative and quantitative analysis will realize that the fine features on x-ray spectra are a valuable tool for deduction of chemical bonding.

7. REFERENCES

1. R. Swinne, *Phys. Z.* **17**, 481 (1916).
2. J. Bergengren, On the x-ray absorption of phosphorus, *Z. Physik* **3**, 247 (1920).
3. A. E. Lindh and O. Lundquist, Structure of the K_{β_1} line of sulfur, *Ark. Mat. Astr. Fys.* **18**, (14), 3 (1924).
4. M. Siegbahn, *Spektroskopie der Röntgenstrahlen*, Springer-Verlag, Berlin (1931).
5. H. Herglotz, Einflüsse der Bindung auf das Röntgenspektrum, *Trans. Doc. Center Technol. Econ.* (*Vienna*) **13** (1955).
6. A. Faessler, 1508—Röntgenspektrum und bindungszustand, in *Landolt-Bornstein Tables, Zahlenwerte und Funktionen*, 6th ed., Vol. I, part 4, 769–808 (1955).
7. A. Sandström, Experimental methods of x-ray spectroscopy: Ordinary wavelengths, in *Encyclopedia of Physics*, Vol. XXX (X-Rays), pp. 78–245, Springer-Verlag, Berlin (1957).
8. D. H. Tomboulian, The experimental methods of soft x-ray spectroscopy and the valence band spectra of the light elements, in *Encyclopedia of X-Rays*, Vol. XXX, pp. 246–304, Springer-Verlag, Berlin (1957).
9. M. A. Blokhin, *The Physics of X-Rays*, 2nd ed., State Publishing House of Technical-Theoretical Literature, Moscow (1957).

10. C. H. Shaw, The x-ray spectroscopy of solids, in *Theory of Alloy Phases*, pp. 13–62, American Society for Metals, Cleveland (1956).
11. H. Yakowitz and J. R. Cuthill, Annotated Bibliography on Soft X-Ray Spectroscopy, National Bureau of Standards Monograph 52 (1962).
12. W. L. Baun, Instrumentation, spectral characteristics and applications of soft x-ray spectroscopy, *Appl. Spectry. Rev.* **1**, 379–433 (1968).
13. D. J. Nagel, Interpretation of valence band x-ray spectra, in *Advances in X-Ray Analysis* (B. L. Henke, J. B. Newkirk, and G. R. Mallett, eds.) Plenum Press, New York (1963).
14. D. J. Nagel and W. L. Baun, *X-Ray Spectroscopy* (L. Azaroff, ed) Chapter 8, McGraw-Hill, New York (1973).
15. Conference on the Applications of X-Ray Spectroscopy to Solid State Problems, Office of Naval Research NP-4287 NAVEXOS, p. 1033, AT1204728 (1950).
16. Abstracts of papers presented at the International Conference on the Physics of X-Ray Spectra, Cornell University, Ithaca, N.Y. 1965, *Bull. Am. Phys. Soc.* **10** (1965).
17. *Proceedings of the Conference on Röntgenspectren und Chemische Bindung, Karl Marx University, Leipzig, 1965*, Leipzig (1966).
18. D. J. Fabian (ed), *Soft X-Ray Band Structure and Electronic Structure of Metals and Materials*, Academic Press, London (1968).
19. W. L. Baun and D. W. Fischer, The Effect of Chemical Combination on *K* X-Ray Emission Spectra from Magnesium, Aluminum, and Silicon, Air Force Materials Laboratory Technical Report 64-350 (December 1964).
20. D. W. Fischer and W. L. Baun, The effects of electronic structure and interatomic bonding on the soft x-ray Al *K* emission spectrum from aluminum binary systems, *Adv. x-ray Anal.* **10**, 374–389 (1967).
21. D. W. Wilbur, Relationship Between Chemical Bonding and the X-Ray Spectrum: Studies with the Sulfur Atom UCRL-14379, TID-4500 AEC Contract No. W-7405-eng.-48.
22. D. W. Fischer and W. L. Baun, Effect of chemical combination on the soft x-ray *L* emission spectra of potassium, chlorine, and sulfur using a stearate soap film crystal, *Anal. Chem.* **37**, 902 (1965).
23. D. W. Fischer, Changes in the soft x-ray *L* emission spectra with oxidation of the first series transition metals, *J. Appl. Phys.* **36**, 2048 (1965).
24. J. Holliday, *Handbook of X-Rays* (E. Kaeble, ed.) Chapter 38, McGraw-Hill, New York (1967).
25. C. F. Feldman, Range of 1–10 keV electrons in solids, *Phys. Rev.* **117**, 455 (1960).
26. E. W. White, Tutorial Session: 7th National Conference, Electron Probe Analysis Society of America, San Francisco, July 17, 1972.
27. W. L. Baun, Al *K* x-ray emission fine features for characterizing Al–Cu films, *J. Appl. Phys.* **40**, 4210 (1969).
28. E. W. White and R. Roy, Silicon valence in SiO films studied by x-ray emission, *Solid State Commun.* **2**, 151 (1964).
28*a*. J. Colby, Bell Laboratories, private communication.
29. W. L. Baun and J. S. Solomon, Characterization of SiO using fine features of x-ray *K* emission spectra, *Vacuum* **21**, 165 (1971).
30. E. W. White and R. Roy, Use of x-ray emission spectroscopy in the characterization of thin films of aluminum oxides and hydroxides, *Mater. Res. Bull.* **2**, 395 (1967).
31. W. H. Knausenberger, K. Vedam, E. W. White, and W. Zeigler, Thin film characterization by electron microprobe and ellipsometry: SiO_2 films on silicon, *App. Phys. Letters* **14**, 43 (1969).

32. P. D. Gigl, G. A. Savanick, and E. W. White, Characterization of corrosion layers on aluminum by shifts in the aluminum and oxygen x-ray emission bands, *J. Electrochem. Soc.* **117**, 15 (1970).

20

ANALYTICAL AUGER ELECTRON SPECTROSCOPY

Chuan C. Chang

Bell Laboratories
Murray Hill, New Jersey

1. INTRODUCTION

Auger spectroscopy explores the electronic energy levels in atoms and solids. As an analytical technique, it has been applied to investigations of the first few atom layers of surfaces. The term "Auger process" has come to denote any electron de-excitation in which the de-excitation energy is transferred to a second electron, the "Auger electron." Because of the discrete nature of most electronic energy levels, Auger electrons are best characterized by their energies. Therefore, the Auger process can be analyzed by measuring the energy distribution $N(E)$ of Auger electrons, where E is the electron energy. Experimental $N(E)$ curves contain much more information than just those concerning the Auger process, and this extra information adds greatly to the usefulness of the data. Low-energy Auger electrons ($\lesssim 1$ keV) can escape from only the first several atom layers of a surface because they are strongly absorbed by even a monolayer of atoms. This property gives Auger spectroscopy its high surface sensitivity.

The purpose of this article is to review the basic physical concepts, experimental techniques, and capabilities and limitations of Auger spectroscopy, particularly in surface analysis. Methods of data interpretation, quantitative analysis, and recent developments that are of interest to those active in this field are discussed, and a bibliography has been provided. Numerous reviews on the Auger effect have already appeared.[1] Most of the fundamentals presented in those publications are covered in Sections 1–4. Later sections contain some new material that are of particular importance to analytical work. Several topics, such as the Auger spectra from carbon and silicon, the properties of the cylindrical electron analyzer, and some aspects of quantitative surface analysis, have been considered in detail, because true insight is often best gained from specific examples and numerical calculations.

2. THE AUGER PROCESS

2.1. Mechanism of Auger Ejection

Auger electrons were first discovered in 1925 by Pierre Auger[013]* who observed their tracks in a Wilson cloud chamber and correctly explained their origin. These electrons are produced by the process depicted in Fig. 20-1, shown for silicon. The electronic energy levels of the singly ionized atom are listed on the left, taking the Fermi level E_F as the zero of energy; these are the x-ray levels, $K, L, M, \ldots,$ as labeled on the right. Methods of ionization are discussed in Sections 4.3 and 5.3.1. When atoms are brought together to form a solid, the valence level V broadens into an effectively continuous band[2] and the calculated[3] density of states (shaded area) has been drawn in this band. Figure 20-1 shows the atom immediately after K-shell ionization by a primary electron; the Auger process initiates when an outer electron fills this "hole," as shown for an L_1 electron. The energy released by this transition is either emitted as a photon or given to another electron. If this energy is sufficient, an electron can be ejected from the atom, as illustrated for another electron at $L_{2,3}$. The process just described is called the $KL_1L_{2,3}$ Auger transition. Unfortunately, after Auger ejection, the atom is doubly ionized and the energy level diagram in general changes so that Fig. 20-1 cannot be used for estimating the energy of the Auger electron. We return to this point in Section 2.3, but will first describe a very useful semiempirical method for estimating the energy of the Auger electron.

2.2. Energies of Auger Electrons: Semiempirical Method

The energy E of the ejected electron can be estimated by assuming the diagram of Fig. 20-1 to be approximately valid. The energy released is $E_K - E_{L_1}$ (exactly), but the ejected electron must expend the energy $(E'_{L_{2,3}} + \phi)$ to escape the atom, where ϕ is the work function and $E'_{L_{2,3}} \neq E_{L_{2,3}}$ due to the extra positive charge of the atom. Because of

* All three-digit numbers denote works listed in the bibliography of Section 11.

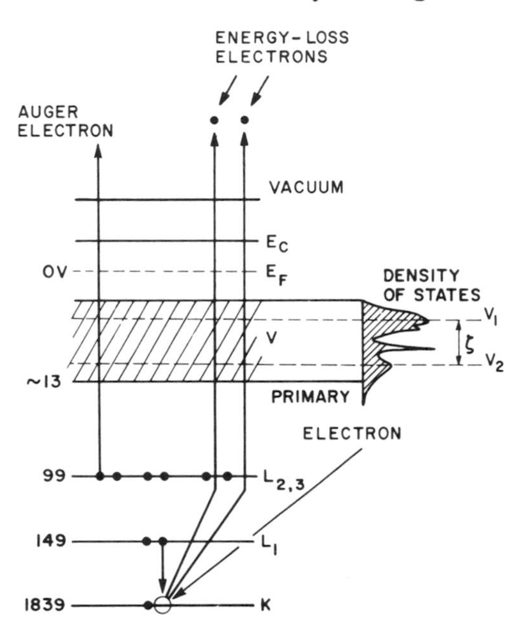

Fig. 20-1. The singly ionized Si atom. The electronic energy levels are listed on the left (in eV) with the zero of energy at the Fermi level E_F ; E_C is the bottom of the conduction band. The x-ray nomenclature is given on the right, and the density of states has been drawn into the valence band. A $KL_1L_{2,3}$ Auger process is depicted, after primary electron ionization.

this positive charge, $E'_{L_{2,3}}$ should be approximately equal to the $L_{2,3}$ ionization energy of the next heavier element, or $E'_{L_{2,3}}(Z) = E_{L_{2,3}}(Z + \Delta)$, where Z is the atomic number and $\Delta \approx 1$ to account for the extra charge. We therefore have

$$E(Z) = E_K(Z) - E_{L_1}(Z) - E_{L_{2,3}}(Z + \Delta) - \phi \tag{1}$$

Equation (1) is useful because E is expressed in terms of single ionization energy levels which can be found in the x-ray[(4)] and photoelectron energy tables.[(1d)] Equation (1) can be generalized to any Auger event involving the levels $W_oX_pY_q$; also, the actual measured energy will have an additional term $-(\phi_A - \phi)$, which is the difference between the work functions of the energy analyzer ϕ_A and the material under investigation. This gives

$$E_{WXY}(Z) = E_W(Z) - E_X(Z) - E_Y(Z + \Delta) - \phi_A \tag{2}$$

Note that ϕ drops out, and therefore cannot be measured using Auger energies. Experimental values of Δ are generally between $\frac{1}{2}$ and $\frac{3}{2}$.[(1a,j,5)]

Equation (2) is not self-consistent if experimental values from x-ray tables are inserted, since it does not give $E_{WXY} = E_{WYX}$ although we

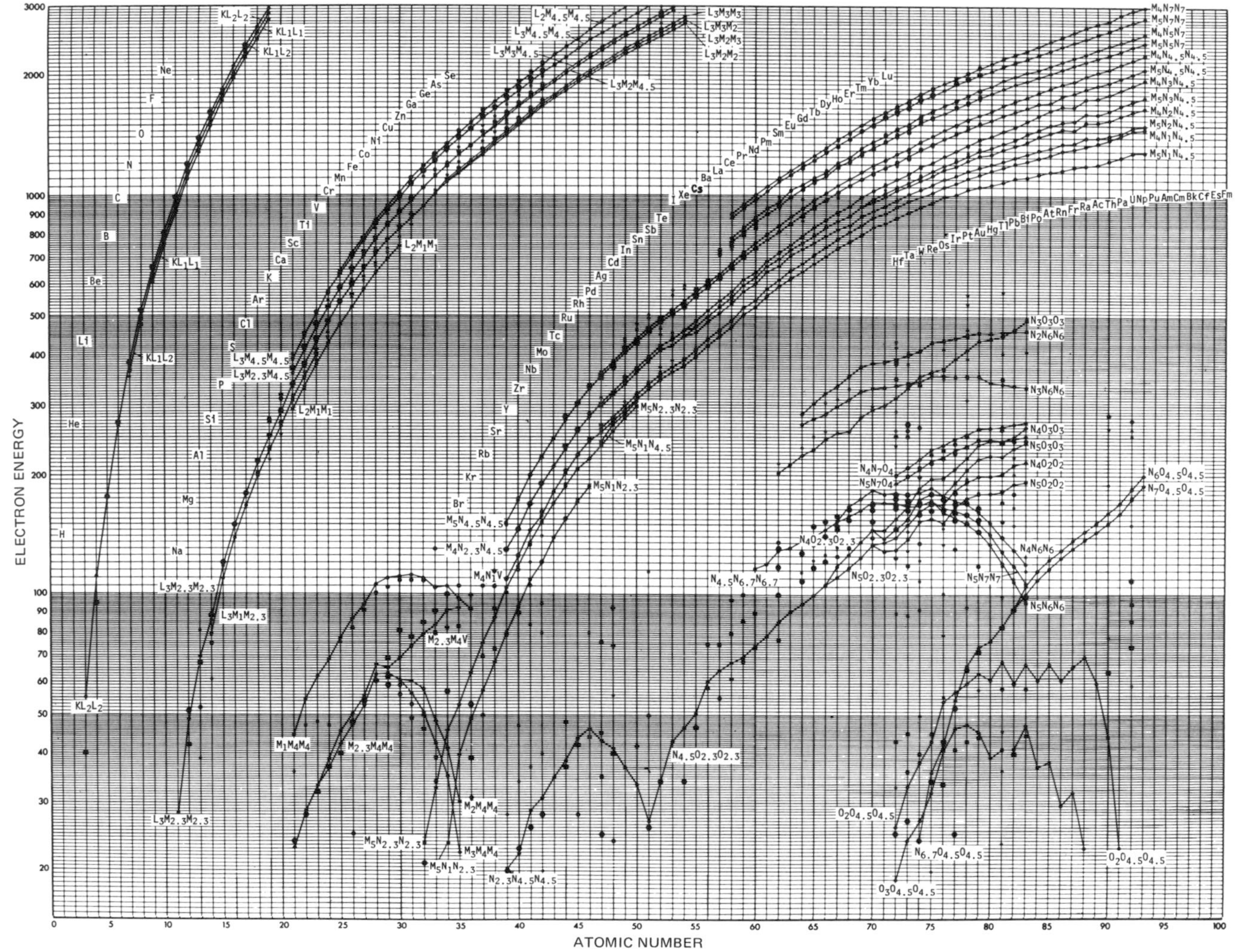

Fig. 20-2. Auger energies E_{WXY} *vs.* atomic number, Z. Lines connect calculated points computed from Eq. (2), and fall fairly close to the experimental points. The largest points represent the strongest observed transitions. Some of the experimental points were obtained from compounds and may not correspond to transitions in the pure element, especially below ~100 eV. (Courtesy of Y. Strausser and J. J. Uebbing, Varian Associates.)

know this must be so because the initial and final states for both transitions are identical. This discrepancy arises because Eq. (2) neglects quantum mechanical exchange effects. Methods of accounting for such effects are well known, and have led[048,049] to refinements of Eq. (2) which give values of E_{WXY} that agree with experiment within $\lesssim 1\,\%$ for the simpler families of Auger transitions. Figure 20-2 is a plot of measured and calculated Auger energies, and shows their remarkable agreement, especially for the KLL, LMM, and MNN families of transitions. The strongest transitions observed experimentally are of the type WXY with $X, Y = W$ or $(W + 1)$, where $(W + 1)$ is the next higher level; e.g., $K + 1 = L$. This "selection rule" follows from the fact that electron–electron interactions are strongest between electrons whose orbitals are closest together.

At this point, it is instructive to determine the lightest element that can be detected with Auger spectroscopy. Since three electrons are needed, lithium appears to be the first detectable element. However, a Li atom has only one electron in the L shell and cannot undergo an Auger transition. Therefore, for isolated atoms, Be is the lightest element that gives Auger electrons. In a solid, the valence electrons are shared, so that lithium gives Auger electrons of the type KVV, where V is the valence level of the solid. Similarly, He embedded in a solid should also produce Auger electrons.

2.3. Theoretical Calculations

The WXY notation is almost universally used for designations of Auger energies in analytical work. However, Eq. (2) was not derived from first principles and therefore exhibits numerous inadequacies in detailed analyses. For example, the WXY method does not even give the correct *number* of Auger transitions. To illustrate this, consider the simplest family of transitions, the KLL group. There are four x-ray levels: K, L_1, L_2, L_3, giving a maximum of six distinct transitions:

$$\begin{array}{ll} KL_1L_1 & KL_2L_2 \\ KL_1L_2 & KL_2L_3 \\ KL_1L_3 & KL_3L_3 \end{array}$$

The actual number of transitions is shown in Fig. 20-3. There are six transitions in the limits of high and low atomic numbers; however, there are *nine* in between, and this is confirmed by experiment.[6,7] Clearly, from the point of view of basic physics, the WXY notation is inadequate and we must examine the theoretical approach.

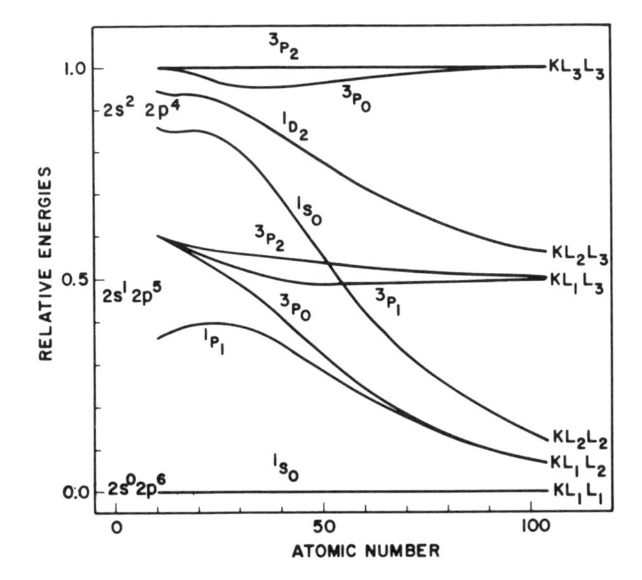

Fig. 20-3. Relative positions of *KLL* Auger energies, normalized to the $KL_1L_1(^1S_0)$ – $KL_3L_3(^3P_2)$ interval. The doubly ionized atomic configurations ($2s^02p^6$, etc.) and the spectroscopic terms ($^1S_0$, etc.) are indicated. There are six transitions in the limits of low and high Z, but nine in between. The *KLL* designation is valid only for heavy atoms ($Z > 80$), when *j–j* coupling dominates. (From K. Siegbahn *et al.*[1d])

The basic theoretical Auger problems are the evaluation of the type and number of transitions, their energies, and their intensities (transition rates). To determine the number of possible transitions, it is necessary to know the number of initial and final electron configurations; that is, what happens to the electron configuration as an atom is singly, and then doubly, ionized. It is instructive to start by first considering the neutral atom. Energy levels of neutral atoms cannot be measured without perturbing them, as by ionization. However, the electronic configurations are well known,[8] and are listed on the left-hand side of Table 20-1. The unperturbed atom has filled shells *s*, *s*, *p*, *s*, *p*, *d*, etc.

The simplest measurable energy levels are the x-ray levels, obtained from singly ionized atoms, as shown on the right side of Table 20-1. Note that the only added complication upon single ionization is the splitting of the $l \neq 0$ levels. This splitting is caused by the spin–orbit interactions (*j–j* coupling, see below). For example, a filled *p* shell is symmetric and has no net angular momentum; however, a singly ionized *p* shell can have a net angular momentum of $j = \frac{1}{2}$ or $\frac{3}{2}$, corresponding to the angular momentum of a single hole with $\mathbf{j} = \mathbf{l} + \mathbf{s} = 1 \pm \frac{1}{2}$. Each higher level (*d*, *f*, etc.) also splits into two.

TABLE 20-1. Atomic *vs*. Ionic Energy Levels

Atomic levels				Singly ionized atom
n	l	Electron shell	Number of electrons	X-ray symbol
1	0	1s	2	K
2	0	2s	2	L_1
	1	2p	6	L_2
				L_3
3	0	3s	2	M_1
	1	3p	6	M_2
				M_3
	2	3d	10	M_4
				M_5
4	0	4s	2	N_1
	1	4p	6	N_2
				N_3
	2	4d	10	N_4
				N_5
	3	4f	14	N_6
				N_7
5	0	5s	2	O_1
				etc.

The number of configurations after double ionization is shown for LL ionization in Fig. 20-4. Each step leading to the final configuration shown in the figure is best understood by reviewing the fundamental basis of atomic physics.[8,9a,10] Calculations of electronic energy levels are usually carried out using the central-field approximation which is based on the quantum mechanical properties of the (approximate) Hamiltonian[10]

$$H = \sum_{i=1}^{N} \left[\frac{1}{2m} \mathbf{p}_i^2 - \frac{Ze^2}{r_i} + \xi(r_i) \mathbf{L}_i \cdot \mathbf{S}_i \right] + \sum_{i>j=1}^{N} \frac{e^2}{r_{ij}} \tag{3a}$$

The solutions are obtained using perturbation methods in which the last three terms of H are replaced by an artificial potential $U(r)$ which is chosen to approximate the true potential, but to be sufficiently simple so as to allow an initial computation using the simplified Hamiltonian

$$E = \sum_{i} \left[\frac{1}{2m} \mathbf{p}_i^2 + U(r_i) \right] \tag{3b}$$

The perturbation potential is then given by $V = H - E$. The necessity

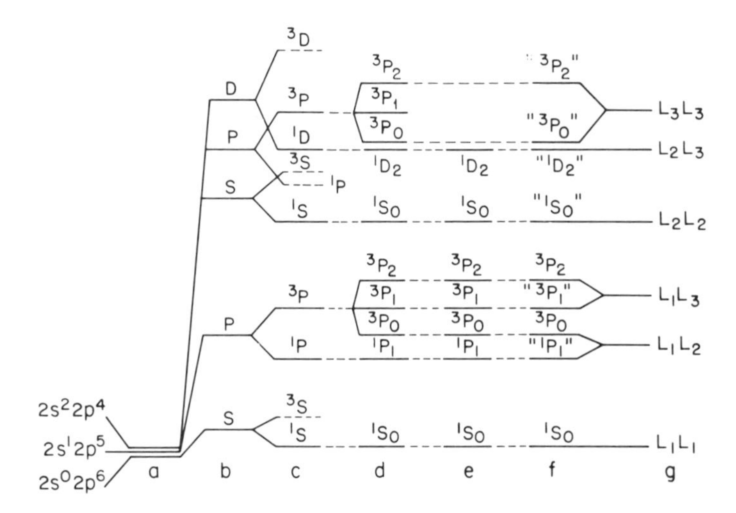

Fig. 20-4. Final state of Auger transitions in the *KLL* group under various assumptions for the electron interaction and the type of coupling (level positions not to scale): (a) No electron interaction, central potential of $1/r$ shape, completely degenerate energies; (b) screening and Coulomb interaction between electrons generate levels of different orbital angular momentum; (c) exchange interaction causes a splitting in singlet and triplet terms. Those excluded by the Pauli principle are indicated by broken lines; (d) Spin-orbit interaction decomposes terms into individual levels of definite angular momentum; (e) allowed final states conservation of angular momentum and parity), pure *L–S* coupling; (f) allowed final states, intermediate coupling, no configuration interaction. Quotation marks indicate that the states are mixed; (g) allowed final states, pure *j–j* coupling. (From I. Bergstrom and C. Nordling.[1a])

for introducing $U(r)$ arises because of the complexity of the electron–electron interactions; these interactions are given by

$$\sum_{i,j(i \neq j)} \frac{e^2}{r_{ij}} + \sum_i \xi(r_i)\mathbf{L}_i \cdot \mathbf{S}_i \tag{4a}$$

$$\xi(r) = \frac{e}{2m^2c^2} \frac{1}{r} \frac{\partial V}{\partial r} \tag{4b}$$

where the e^2 term gives the electrostatic coulomb repulsion between pairs of electrons, and the $\mathbf{L} \cdot \mathbf{S}$ term is the magnetic interaction between spin and orbital motion of electrons. The perturbation potential is further simplified by assuming that one of the above two interactions is much smaller than the other. At low energies, the electrostatic contribution is larger and the $\mathbf{L} \cdot \mathbf{S}$ term can be temporarily neglected. The electrons are then in states of fixed L and S, by conservation of angular momentum, since electrostatic interactions cannot affect the total spin or orbital angular momentum. The electronic configurations resulting from fixed L and S follow the L–S coupling scheme.[8,10]

At high energies (core electrons of heavy elements), the magnetic interactions become strong as seen by the increase in ξ of Eq. (4). The magnetic interaction couples the spin to the orbital angular momentum of each electron, and the conservation of angular momentum applies to the total angular momentum j. The electronic configurations then follow the j–j coupling scheme[8,10] in which each electron is in a state of definite l, j. For intermediate energies, the coupling scheme is mixed.[10]

In calculations of Auger electron energies, we are interested in the case of one or two "holes" in a system of otherwise filled electronic shells.[9a,10] Coupling to the valence electrons is neglected. In the energy range of interest ($\leq$2.5 keV), L–S coupling dominates regardless of atomic number; but unfortunately, the WXY notation borrowed from x-ray language is derived from j–j coupling. Therefore confusions and difficulties in line assignments do arise, but the WXY + (Spectroscopic) notation,[9a] e.g., $KL_1L_1(^1S_0)$, etc. has been adopted by most workers; this notation "works" for data taken at low resolution ($\gtrsim$1 eV).

First-principles calculations of Auger energies have been made only with wave functions of the free atom.[9] These computations have not yielded sufficiently accurate results to be competitive with the semi-empirical procedures. Nevertheless, the method will be examined here, applied to the KLL spectrum, because the same formalism is used for transition-rate calculations.

The problem of one or two missing electrons is equivalent to that of one or two holes,[10] with appropriate modifications in Eqs. (3) and (4). The quantization of angular momenta is included by use of the $(SLJM)$ representation appropriate to L–S coupling. Deviations from strict L–S coupling are then calculated in terms of energies $E(J)$, using the mixing coefficients for intermediate coupling. The initial- and final-state energies are then calculated in terms of the matrix elements of the electrostatic interaction between two electrons:[9a,e]

$$F^{\nu}(nl, n'l') = e^2 \int\int_{r_1, r_2 = 0}^{\infty} \gamma_\nu R_{nl}^2(r_1) R_{n'l'}^2(r_2) r_1^{\,2} r_2^{\,2}\, dr_1\, dr_2 \tag{5}$$

$$G^{\nu}(nl, n'l') = e^2 \int\int_{r_1, r_2 = 0}^{\infty} \gamma_\nu R_{nl}(r_1) R_{n'l'}(r_1) R_{nl}(r_2) R_{n'l'}(r_2) r_1^{\,2} r_2^{\,2}\, dr_1\, dr_2 \tag{6}$$

$$\gamma_\nu = \begin{cases} r_1^{\,\nu}/r_2^{\,\nu+1} & (r_1 < r_2) \\ r_2^{\,\nu}/r_1^{\,\nu+1} & (r_2 < r_1) \end{cases} \tag{7}$$

where 1, 2 refer to the two electrons, $R_{n,l}$ and $R_{n',l'}$, are radial parts of the wave functions of electrons in the $nl, n'l'$ states, and v is an index related to the angular momentum (see E.U. Condon and G.H. Shortley,[10] where it is labeled k).

For the KLL spectrum, the energies are given by:

$$
\begin{aligned}
&(2s)^0(2p)^6: 2I(2s) - F^0(20,20) - 6[2F^0(21,20) - \tfrac{1}{3}G^1(21,20)] \\
&(2s)(2p)^5: I(2s) + I(2p) - F^0(20,20) - 4[2F^0(21,20) - \tfrac{1}{3}G^1(21,20)] \\
&\qquad - 3[\tfrac{5}{3}F^0(21,21) - \tfrac{2}{15}F^2(21,21)] + E_i^{(2s)(2p)}(J) \qquad (8) \\
&(2s)^2(2p)^4: 2I(2p) - 2[2F^0(21,20) - \tfrac{1}{3}G^1(21,20)] \\
&\qquad - 6[\tfrac{5}{8}F^0(21,21) - \tfrac{2}{15}F^2(21,21)] + E_i^{(2p)^2}(J)
\end{aligned}
$$

where $I(2s)$, $I(2p)$ are the energies of $2s$ and $2p$ electrons, respectively, in the field of the nucleus and electrons other than those in the L shell. The energies calculated from Eq. (8) are in poor agreement with experiment, but provide the basis for alternative methods using similar equations with adjustable parameters.[7,9a,d] Such calculations have yielded energies within $<1\%$ of measured values.[7,9a,b] For the LXY and MXY transitions, the computations become very long because of the large number of terms that must be computed for each final state, even if coupling to the valence shell is neglected. Relativistic corrections are also important.[9b]

Very few first-principles calculations of Auger energies have been attempted because of the difficulty of calculating energies with accuracies comparable to those attainable using the semiempirical procedures. In contrast, the number of publications on transition probabilities is very large (see Refs. 11–18 and 029, 044, 049, 061, 150, 175–178, 290–292). One reason why so many transition-rate calculations have been made is the lack of a simple formula, like Eq. (2) for the energies, which can be used to predict intensities.

The formalism developed above for calculating the energies also applies to transition rates. Again for the KLL series, the transition probabilities can be calculated in terms of the amplitudes:[9]

$$
\begin{aligned}
\{nl, n'l', v, k\} = \frac{4\pi^2 e^2}{h^2} \int\int_{r_1, r_2 = 0}^{\infty} \gamma_v R_{10}(r_1) R_{nl}(r_1) R_{n'l'}(r_2) R_{\infty}{}^k(r_2) \\
\times r_1{}^2 r_2{}^2 \, dr_1 \, dr_2 \qquad (9)
\end{aligned}
$$

where R_{10} and $R_{\infty}{}^k$ are radial parts of the wave functions of an electron in the $(1s)$ state and a positive energy state of orbital angular momentum

k, respectively. For L–S coupling, the transition probabilities are given by

$$
\begin{aligned}
&2s^0 2p^6(^1S_0): |\{2s, 2s, 0, 0\}|^2 \\
&2s^1 2p^5(^1P_1): \tfrac{3}{2}|\{2s, 2p, 0, 1\} + \tfrac{1}{3}\{2p, 2s, 1, 1\}|^2 \\
&\quad (^3P_{0,1,2}): (J + \tfrac{1}{2})|\{2s, 2p, 0, 1\} - \tfrac{1}{3}\{2p, 2s, 1, 1\}|^2 \\
&2s^2 2p^4(^1S_0): \tfrac{1}{3}|\{2p, 2p, 1, 0\}|^2 \\
&\quad (^3P_{0,1,2}): 0 \\
&\quad\ (^1D_2): \tfrac{2}{3}|\{2p, 2p, 1, 2\}|^2
\end{aligned}
\tag{10}
$$

For intermediate coupling, the rates can be obtained by use of the mixing coefficients mentioned earlier. The calculated relative rates agree with experiment usually within a factor of 2, but there are occasional outstanding exceptions. Where significant disagreement is found, relativistic corrections[9b,17,18] can account for much of the discrepancy.

2.4. Coster–Kronig Transitions

Coster–Kronig transitions (WWX type) have particular significance because their transition rates are much higher than the others.[19] This can be seen qualitatively in Eq. (9) if the appropriate wave functions are substituted: two of the wavefunctions can be fairly similar, giving a large transition rate.[029] This high transition rate produces several important consequences:

1. Line broadening. Energy levels with short lifetimes Δt are broadened by an amount ΔE given by the uncertainty equation $\Delta E \Delta t = \hbar$. Since line broadening of 2–10 eV is commonly observed,[12,044,175,178] the lifetimes can be $\lesssim 10^{-16}$ sec, which is about the time it takes the electron to make a single orbit.
2. Changes the relative intensities of Auger lines. For example, if an L_1 shell is ionized, the L_1L_2Y transitions will be rapid (Coster–Kronig). Therefore, transitions of the type L_1XY ($X, Y \neq L$) are reduced and those of type L_2XY will be increased due to the enhanced ionization caused by L_1L_2Y transitions.[20]

3. THE SECONDARY-ELECTRON ENERGY DISTRIBUTION

3.1. Gross Features

To observe Auger electrons, it is necessary to measure the electron energy distribution $N(E)$. The initial ionization required for the Auger process is accomplished by bombardment with some energetic primaries

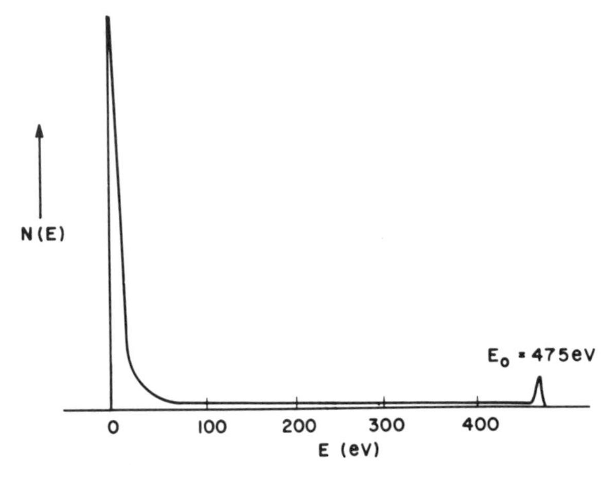

Fig. 20-5. Gross structure in the secondary-electron energy distribution $N(E)$ displaying the large true secondary peak near $E = 0$, a region with relatively few electrons at higher energies, and the sharp peak at the primary energy E_0.

(see Section 4.3) and the Auger electrons are found among the various secondary electrons ejected. A typical $N(E)$ curve obtained using primary electron bombardment is shown in Fig. 20-5. Three major features are visible: (1) a sharp peak at the incidence energy E_0 due to elastically reflected primaries, (2) a large peak with a half-width of $\lesssim 10$ eV near $E = 0$, and (3) a region with relatively few electrons between features (1) and (2).

The elastic peak at E_0 represents the electrons used in electron diffraction work, such as LEED[21] and HEED.[21] In Auger work, the elastic peak is useful for calibrating the energy scale, for aligning the electron analyzer, and for measuring the analyzer resolution. It also serves as the zero-point of energy for the "loss peaks" described below.

The large number of slow electrons near $E = 0$ is created in the following way.[22,008,009] A primary electron will produce secondaries by collisions with atoms; these secondaries in turn create more secondaries, initiating a cascade of ever more electrons with ever decreasing energy. The peak near $E = 0$ would rise to infinity (strictly valid only for metals) if it were not for the fact that internal electrons must overcome the work function to escape the surface. The cutoff point of this peak is not at $E = 0$, but at a value equal to the difference between the work functions of the surface emitting the secondaries and of the surface of the electron-energy analyzer. Therefore, if the work function of the analyzer surface remains constant, changes in the work function of the target surface will be reflected in changes in the cutoff energy.

3.2. Energy-Loss Spectra

When the vertical scale of the $N(E)$ curve of Fig. 20-5 is magnified, it is found to contain a number of peaks. There are two major classes: Auger peaks and loss peaks (see Refs. 23, 24, 181, 188, 235, 237, 239, 243, 249, 270). For the type of secondary electron spectra obtained in surface-analysis work, most of the observed loss peaks are either plasmon or ionization losses. Plasmon *gain* also appears to have been detected.[25,141,262]

Plasmons are collective oscillations of the valence electrons against the positive atom cores.[23] The plasmon energy can be simply calculated by considering the motion of a homogeneous mixture of electrons and positive cores. If the "electron gas" is momentarily expanded radially outward from some origin 0, the number of electrons leaving a sphere of radius r is $4\pi nr^2\xi(r)$, where n is the electron density and ξ is the radial displacement of expansion. This creates a radial electric field of magnitude $4\pi ne\xi(r)$ at r and a force $-4\pi ne^2\xi(r)$ on the electrons, where e is the electronic charge. Classically, this force is equal to $m\ddot{\xi}$ or

$$\ddot{\xi} + \frac{4\pi e^2 n}{m}\xi = 0 \tag{11}$$

where m is the electron mass. This is an equation for harmonic motion with oscillation frequency

$$\omega_p = \sqrt{\frac{4\pi ne^2}{m}} \tag{12}$$

and energy $\hbar\omega_p$. In metals, ω_p is of the order of 10^{16} rad/sec and $\hbar\omega_p \sim 15$ eV. In a crystal, m is replaced by the effective mass m^* and the calculations must be carried out in terms of interband transitions so that the band structure of the material enters directly into the calculation of ω_p.[11,23]

At a surface, the electric field is weaker and the ideal "surface plasmon" has a smaller energy[23]

$$\hbar\omega_s = \frac{1}{\sqrt{2}}\hbar\omega_p \tag{13}$$

If the surface is covered by a layer with dielectric constant ε, the "modified surface" plasmon energy becomes[23]

$$\hbar\omega_{ms} = \frac{1}{\sqrt{1 + \varepsilon}}\hbar\omega_p \tag{14}$$

In addition to plasmon losses, incident electrons can lose energy in ionization collisions, and suffer "ionization losses." The term ionization loss can be misleading because it appears to associate the process with the ionization energy E_i as in photoelectron emission. In the latter case, a photon of energy $h\nu$ is absorbed by a bound electron which escapes with energy $(h\nu - E_i)$, leaving an ion with energy E_i. In the electron case, ionization also produces an ion of energy E_i, but the process creates *two* electrons (Fig. 20-1) the sum of whose energies equals $(E_p - E_i)$ where E_p is the primary energy. The maximum energy of an ionization-loss electron is $(E_p - E_i - E_c)$, where E_c is the lowest unoccupied level. One consequence of the creation of two energy-loss electrons is that ionization losses do not produce peaks on $N(E)$ curves, but instead produce steplike structures with the step at the high-energy limit, $(E_p - E_i - E_c)$.

The remaining prominent features on the $N(E)$ curves are the Auger peaks. Before discussing details of Auger spectra, we shall first review the experimental methods for obtaining them. This is necessary because the spectra, and the amount of information contained in them, depend strongly on how the data were taken.

4. EXPERIMENTAL APPARATUS

4.1. General Requirements

The housing for an electron spectroscopy apparatus is the vacuum chamber. Vacuum levels $<1 \times 10^{-8}$ Torr are generally desirable; for example, if electron bombardment is used for excitation, this bombardment can cause surface reactions with adsorbed hydrocarbons, water, CO, etc. Vacuums $<1 \times 10^{-8}$ Torr can now be attained in less than several hours without bakeout (see Ref. 26 for details of vacuum technology), so that attainment of vacuum need not be a major time-consuming operation.

Gas molecules collide with surfaces at rates that would deposit roughly one atomic monolayer ($\sim 10^{15}$ atoms/cm^2) per second at 1×10^{-6} Torr, if all the molecules stick to the surface. However, sticking probabilities are usually $\ll 1$, especially after the first monolayer of adsorption. Therefore at pressures $<1 \times 10^{-8}$ Torr there is usually sufficient time for obtaining the data before reactions with adsorbates can alter the surface appreciably.

The vacuum system should have an attached gas-handling network for admitting controlled amounts of gases, such as Ar for ion milling (see Section 7.1) and other gases for surface cleaning and adsorption work. A quadrupole gas analyzer is useful for measuring gas compositions and for a host of auxiliary experiments. An ion gun giving intense ion beams ($>10\ \mu A/cm^2$) is indispensable for surface cleaning and obtaining depth profiles (see Section 7.1).

The heart of the electron spectroscopy apparatus is, of course, the electron-energy analyzer, which we now consider.

4.2. Electron-Energy Analyzers

4.2.1. The Ideal Analyzer

In order to judge the relative merits of different analyzers, we first consider the properties of the ideal analyzer: one that will obtain data with maximum signal-to-noise (S/N) ratio and highest resolution in the shortest possible time. To attain these goals, the analyzer should:

1. Allow only electrons in the desired energy range to filter through (window devices). If other electrons are admitted, the resolution could be degraded and the S/N ratio will be decreased. The latter follows because the shot noise I_N is given by[26,27]

$$I_N = \sqrt{2eIB} \quad (15)$$

   where e and I are the electronic charge and collected current, respectively, and $B = 1/\tau$ is the bandwidth where τ is the time constant. Shot noise is a consequence of the fact that electrons arrive at the collector as discrete particles and there is no feedback scheme, etc., that can be used to eliminate this noise.
2. Simultaneously analyze a range of electron energies so as to allow multichannel operation. Otherwise, the transmission will decrease as the resolution is increased. When multichannel operation is possible, the number of channels can be increased to compensate for the decrease in signal per channel.
3. Have high transmission. A value $>1\%$ is sufficient for most applications.
4. Have high resolution. A value $<1\%$ is adequate for most Auger spectra used in surface analysis (see Sections 5 and 7.3).
5. Possess a maximum of "convenience" features such as simplicity of construction and operation, accessibility to entrance and exit apertures, compatibility with other analytical instruments, etc.

There are two basic physical forces that can be used to analyze the electron energy: magnetic and electrostatic.

4.2.2. Magnetic Analyzers

Magnetic analyzers[1d,28] will only be briefly discussed here because the complexity of their construction makes them worthwhile only for high-resolution work ($\leq 0.01\%$), which in turn imposes stringent requirements on magnetic shielding. However, at high resolutions there appears to be no analyzer clearly superior to the magnetic type, which can be constructed to satisfy all of the requirements discussed above fairly well.[28] High-resolution work necessarily involves low transmission and long data acquisition times. However, the low transmission can be largely offset by multichannel operation, which can increase the overall transmission by over a factor of 100 compared to single channel operation.[28]

4.2.3. Electrostatic Analyzers

The simplest energy analyzers, in principle, are the electrostatic retarding-potential analyzers (RPA's).[29] There are many ways of constructing RPA's; two are shown in Fig. 20-6. The RPA's are generally high-pass or low-pass filters; that is, when the analyzer energy is set at E, they either pass only electrons with $E' > E$ (high pass) or $E' < E$ (low pass). The major advantages are: (1) simplicity, (2) high transmission[135] and, because the resolution is proportional to the energy,

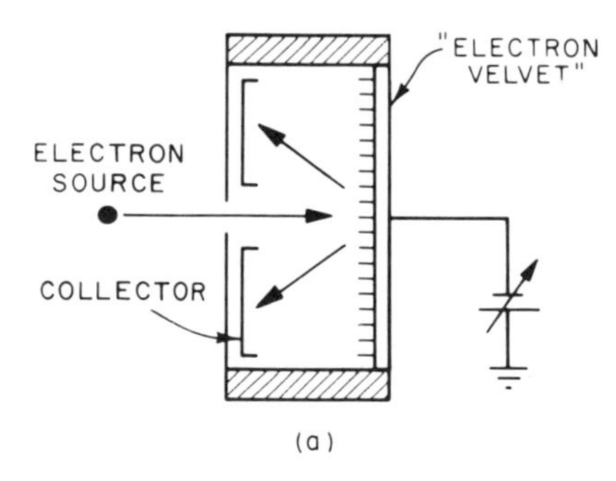

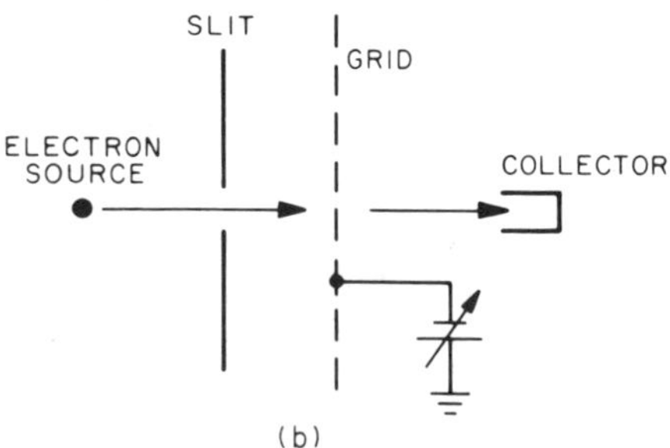

Fig. 20-6. Examples of retarding-potential analyzers: (a) low-pass and (b) high-pass devices.

(3) high resolution at low energies. A high-pass RPA can be combined with a low-pass RPA to make a window device (see for example Refs. 30, 31, 135).

A retarding-potential analyzer used in Auger work is shown in Fig. 20-7. This is the "LEED-Auger" device;[038,203,264,297] it has the unique feature that it can perform both LEED and Auger experiments. In Fig. 20-7, $G1$ and $G4$ are shielding grids, and $G2$ and $G3$ are the grids used for energy analysis; they are tied together to improve the resolution over the use of a single grid.[203,204,264] The fluorescent screen is used as the electron collector. Note that this is a high-pass device; the collected current I, when the analyzing grids are set at E, is

$$I(E) = \int_{E}^{\infty} N(E')\, dE' \tag{16}$$

According to Eq. (15), shot noise will be high because a large number of electrons, in addition to those at E, are collected. The most useful Auger

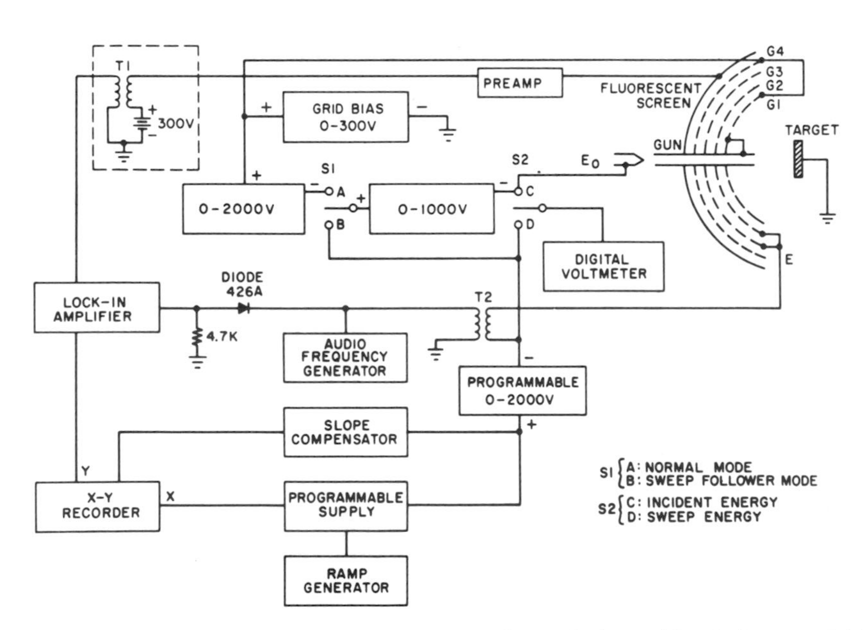

Fig. 20-7. The LEED–Auger device. $G2$, $G3$ are the analyzing grids; their potential is controlled by the ramp generator and programable supplies. The audio-frequency generator provides the ac modulation and the diode arrangement allows for tuning the lock-in amplifier to 2ω for electronic differentiation. $G1$, $G4$ are shielding grids; the secondary electrons created at these grids can be "biased out" using the grid bias supply. The electrons are collected by the fluorescent screen and detected with the lock-in amplifier. The slope compensator is useful for subtracting background slope from the data, and the sweep-follower mode produces spectra with the primary loss peaks eliminated.

data are acquired using electronic differentiation to obtain $dN(E)/dE \equiv N'(E)$, by applying a time-varying voltage $k \sin \omega t$ to the retarding grids.[235,270,297] If the dependence on resolution is neglected, $N'(E)$ is given by the second harmonic ac term[038]

$$I(2\omega) = \tfrac{1}{4}Ak^2N'(E) \tag{17}$$

where A is a constant. Useful values of k range from $\sim$0.1 V (below $\sim$50 eV where $N(E)$ is very large) up to $\gtrsim$20 V for data above 500 eV where the Auger peaks are small.

Another type of electrostatic analyzer is the single-focusing parallel-plate configuration[32] shown in Fig. 20-8a. Figure 20-8b illustrates the 127°-sector analyzer,[33,34] which is also single-focusing and has a first-order focus at 127°. Single-focusing analyzers focus along one direction only. First-order focus occurs when $\partial L/\partial\theta = 0$, where L is

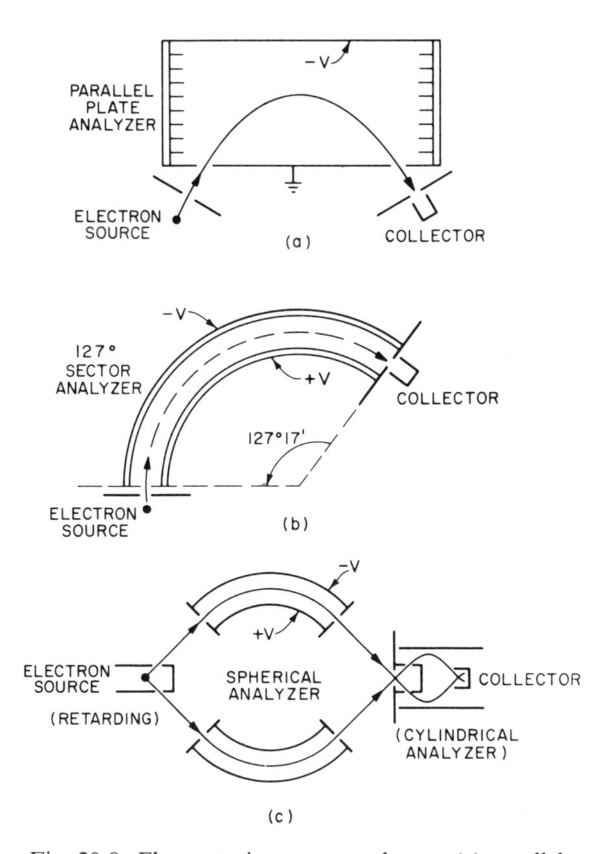

Fig. 20-8. Electrostatic energy analyzers: (a) parallel-plate, (b) 127° sector, and (c) spherical (center only) analyzers.

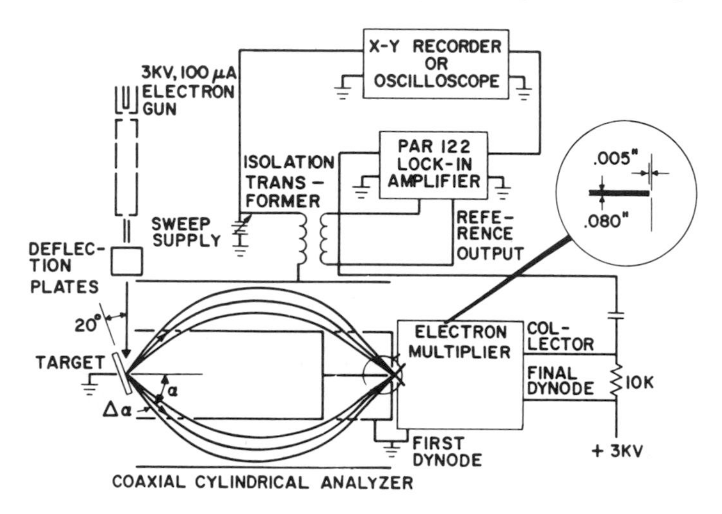

Fig. 20-9. The cylindrical-mirror analyzer (CMA). The electron gun can be mounted on the side as shown, or coaxially, inside the CMA. (After Ref. 207.)

the focal distance and θ is the electron entrance angle. Figure 20-8c shows the spherical analyzer,[35–37] which is double-focusing (focuses along two axes) and has first-order focus. Double-focusing spectrometers can attain higher transmission than the single-focusing type.

The cylindrical-mirror analyzer[38–41,207] shown in Fig. 20-9 is double-focusing and has second-order focus: $\partial^2 L/\partial\theta^2 = 0$ at $\theta = 42.3°$. The latter allows a larger aperture for a given resolution and therefore a higher transmission. The cylindrical-mirror analyzer (CMA) would have been unquestionably the "best" device available today were it not for the fact that it is a single-channel device. The current collected by the CMA is

$$I(E) \propto \Delta E N(E) = REN(E) \tag{18}$$

where ΔE is the energy window and R is the resolution. For reasons discussed in Section 5.1, the $N'(E)$ curve is most useful for surface-analysis work.[119] The current I can be electronically differentiated by applying a voltage modulation $k \sin \omega t$ to the outer cylinder and detecting the first harmonic of the output current:

$$I(\omega) = RkA[N(E) + EN'(E)] \tag{19}$$

This simple modulation scheme gives a result which contains $N(E)$ in addition to $N'(E)$; but the $N(E)$ term is relatively small except at very low energies ($\lesssim 50$ eV). There are other modulation schemes[41] which eliminate the $N(E)$ term.

It is instructive to compare the outputs of the LEED–Auger (LA) device and the CMA, to illustrate the advantages of the desirable characteristics described at the beginning of this section.

4.2.3.1. Noise

The LA analyzer is a high-pass filter, whereas the CMA is a window device and typically collects a current $\sim 10^{-4}$ times that of the LA device. Therefore, shot noise is lower in the CMA by a factor of ~ 100. The CMA is double-focusing: the collected electrons are brought to a small spot, allowing the use of an electron multiplier so that pickup and amplifier noises are minimized. The CMA enjoys the unique advantage that its performance is often not limited by S/N considerations.

4.2.3.2. Transmission

The desirable effect of higher transmission is the larger S/N ratio. Using Eqs. (17) and (19), it is seen that the LA and CMA devices collect the same output current at $E \sim 50$ eV, assuming $k = 1$ eV, $R = 0.5\%$ and the same transmission (typically 5–10%). Above 50 eV, the CMA output is larger by the factor $E/50$. On the other hand, the effects of magnetic fields and other distortions that create the major weaknesses of the CMA become large below ~ 50 eV so that the LA device would be superior to the CMA below ~ 50 eV except for grid effects.[42,264]

4.2.3.3. Resolution

The resolution of the LA device is limited either by grid resolution,[42,264] or by the modulation voltage k. The CMA resolution is determined essentially by the apertures and k. Because of the low S/N ratio of the LA device, relatively large values of k are needed; this degrades the resolution and interferes with the approximations leading to Eq. (17). The CMA resolution is constant with energy, because the resolution is essentially a geometrical factor which scales with energy.

4.2.3.4. Speed

The rate of data acquisition is limited by the bandwidth $B = 1/\tau$ of Eq. (15). Therefore, the speed of operation is limited by noise considerations. According to Eq. (15), for a given noise level, the required time constant τ is proportional to the total collected current I. The CMA is then $\sim 10^4$ times faster than the LA device, all other conditions being equal.

4.3. Methods of Primary Excitation

To observe Auger electrons, it is necessary to ionize atoms in some core level. This is most efficiently accomplished using electron bombardment. Ionization by electron bombardment will be considered in Section 5.3; we discuss here some other methods of excitation. It should be noted, however, that electron guns are relatively easy to construct,[43] and the high beam-intensity capability ($>100\ \mu$A) and ease of focusing and deflection contribute to the usefulness of electron-beam excitation.

Another important method of ionization is by photon bombardment.[1d–i,44,215] For core ionization, x rays must be used. [Methods of x-ray generation are discussed by V. A. Tsukerman *et al.*[46]] The ionization of atoms by x rays is closely related to x-ray absorption,[8,47,48] and is complicated by diffraction for single crystals.[47,48] When x-ray bombardment is used, however, photoelectron spectroscopy[1d–i] becomes a more fruitful technique than Auger analysis.

Ion bombardment has also been used to excite Auger electrons.[49,50] The Auger yield using 350-keV protons is comparable to that using 3-keV electrons.[50] Ion bombardment creates its own characteristic secondary electrons and these can be analyzed to study chemical bonding.[51]

5. DATA INTERPRETATION AND SURFACE ANALYSIS

The use of Auger spectroscopy for surface analysis was first suggested by Lander[155] in 1953 and useful data were soon obtained by Harrower,[128] who also made use of the loss peaks, and by Siegbahn and co-workers (see Refs. 1*d*, 52, 53). Harris demonstrated the relative simplicity and remarkable sensitivity of Auger spectroscopy[119,121] ($\sim$1968), and Palmberg, Bohn, and Tracy[207] (1969) used the cylindrical mirror to obtain high-quality Auger spectra with subsecond data acquisition times. These developments established Auger spectroscopy as an important analytical technique. We shall now consider the methods of data interpretation for surface analysis.

5.1. Characteristic Auger Spectra

The most generally useful Auger spectra for surface analysis have been acquired with the CMA, applying electronic differentiation to obtain $I(\omega) \approx EN'(E)$, according to Eq. (19). The need for differentiation arises because $N(E)$ is a positive function, and the number of background secondary electrons is usually larger than the number of Auger electrons. This means that under high magnification, $N(E)$ becomes unmanageable owing to the need for unreasonably large zero-point suppression.

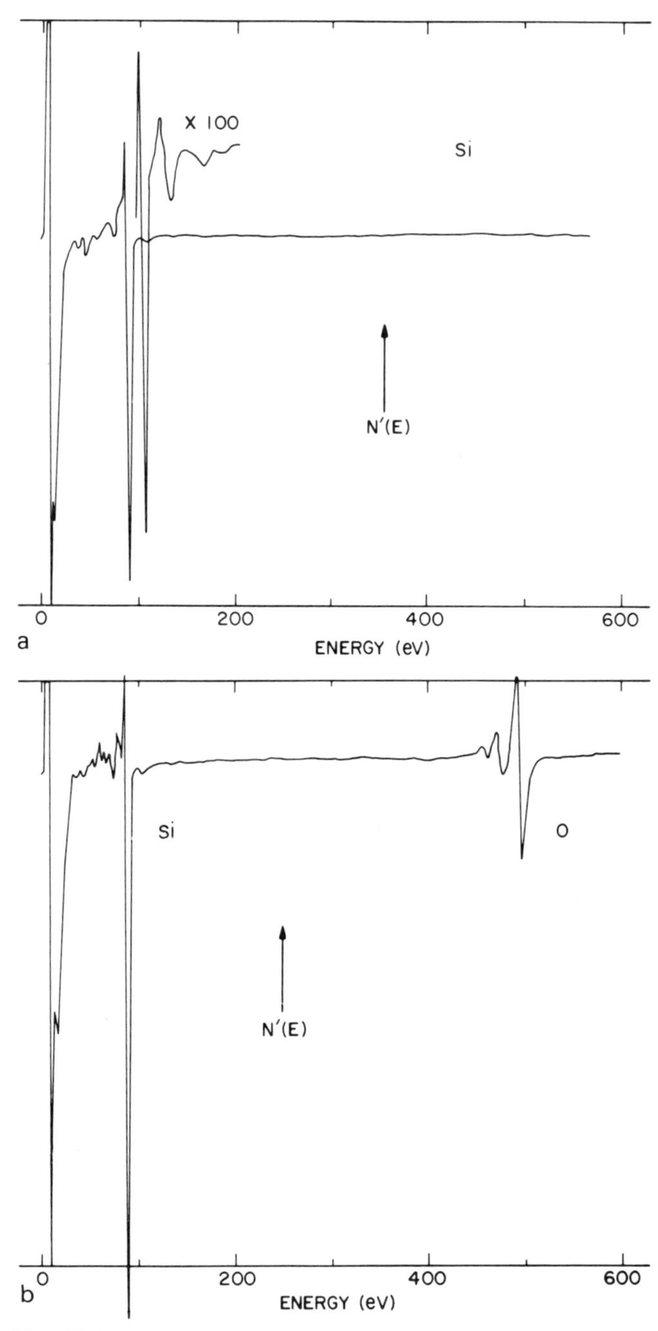

Fig. 20-10. Auger spectra from (a) single-crystal Si(111) after sputter-cleaning with 1-keV Ne ions, and (b) similar surface after annealing and adsorption of ~0.5 monolayer of oxygen atoms. Note the changes in the Si spectrum below ~90 eV compared to (a). CMA, 3.1-keV primaries.

Auger peaks are relatively sharp features appearing on a slowly varying background, and $N'(E)$ measures changes in $N(E)$, so that the Auger peaks appear more prominently on $N'(E)$; therefore, Auger spectra are usually displayed in differentiated form. $N'(E)$ spectra of every common element have been published in book form (see Ref. 54).

Auger spectra from Si and SiO_2 are shown in Figs. 20-10 and 20-11. These were taken with the CMA using a primary-electron current of 40 μA at 3.1 keV, $k = 1$ eV, and a 30-msec time constant, 12 db roll-off (except for the curve labeled $\times 100$). The Si surface was cleaned by Ne ion bombardment and was probably somewhat amorphous. Note the small size of the zero-volt peak in these spectra. If the LEED–Auger device is used, the zero-volt peak is seen to be almost 10^5 times larger than the oxygen peak (as shown in Fig. 20-12, which was taken from a chemically etched Si surface[036]). The zero-volt peak appears smaller in the CMA data due to the use of an electron multiplier for detection, without postacceleration. Also, $N'(E)$ is negative at low energy and subtracts from $N(E)$ according to Eq. (19).

Figure 20-10b was obtained from the surface of Fig. 20-10a after annealing and adsorption of about $\frac{1}{2}$ monolayer of oxygen. Note the

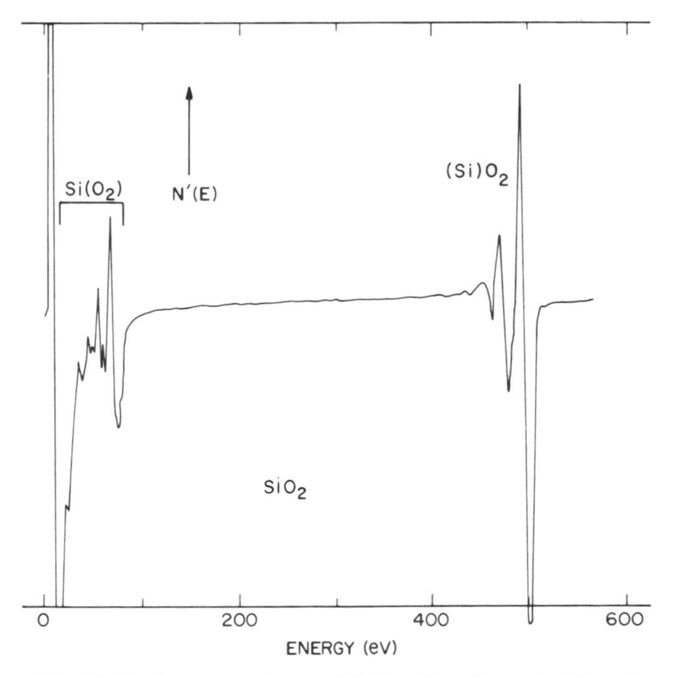

Fig. 20-11. Auger spectrum of SiO_2. Note the main Si peak at 78 eV and the complete disappearance of the 92- and 107-eV peaks which are seen in Fig. 20-10a. CMA, 3.1-keV primaries.

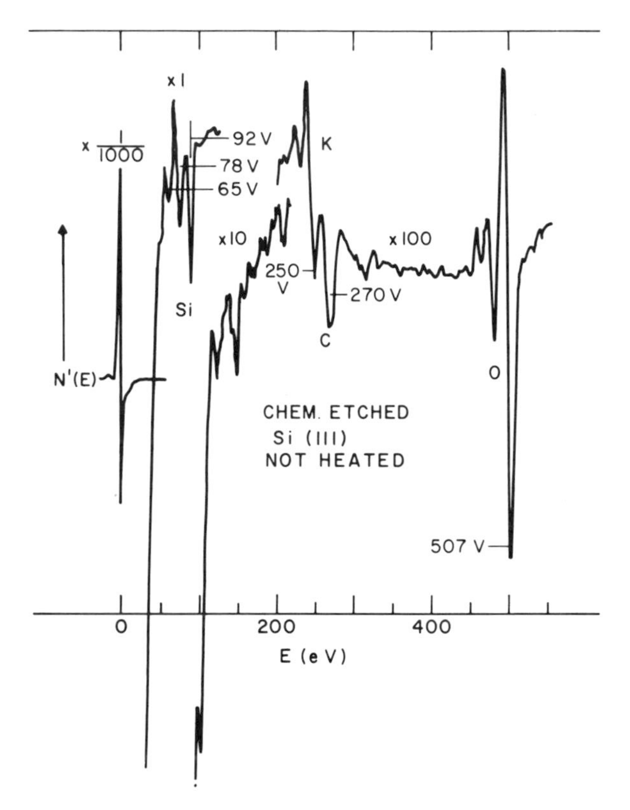

Fig. 20-12. Auger spectra from Si etched in hydrofluoric, nitric, acetic acid, and iodine solution, taken with the LEED–Auger apparatus. The surface contains O, C, Si(92 eV), SiO_2(78, 65 eV), and K(250 eV). The *K* peak is sharper than the others because the K is adsorbed in a thin surface layer. The zero-volt peak is almost five orders of magnitude larger than the oxygen peak.

changes in the Si spectrum below ~90 eV. The background noise was not faithfully reproduced in Fig. 20-10, but the S/N ratio for the oxygen peak of Fig. 20-10b was about 40.

The shape of the oxygen peak of Fig. 20-10b is fairly representative of an ideal Auger spectrum. The major peak, from the KL_2L_2 transition, falls near 500 eV, with two smaller ones from the KL_1L_2 and KL_1L_1 transitions. The largest peak is fairly symmetric; that is, the positive and negative portions corresponding to the respective slopes on the $N(E)$ curve are of nearly equal magnitude. Many Auger peaks are not symmetric, as illustrated by the Si peak at 92 eV in Fig. 20-10—only the negative portion is prominent. As we shall see, the negative part is almost always expected to be sharp and prominent, and for this reason, most workers have adopted the convention of assigning the energy of the minimum of the negative portion to that Auger transition.

One mechanism which diminishes the positive portion is energy loss. This is most clearly illustrated in Fig. 20-13 by the *KLL* transitions of Si taken from Si and SiO_2 using the CMA. Since these are core transitions, the differences between the two top traces are almost

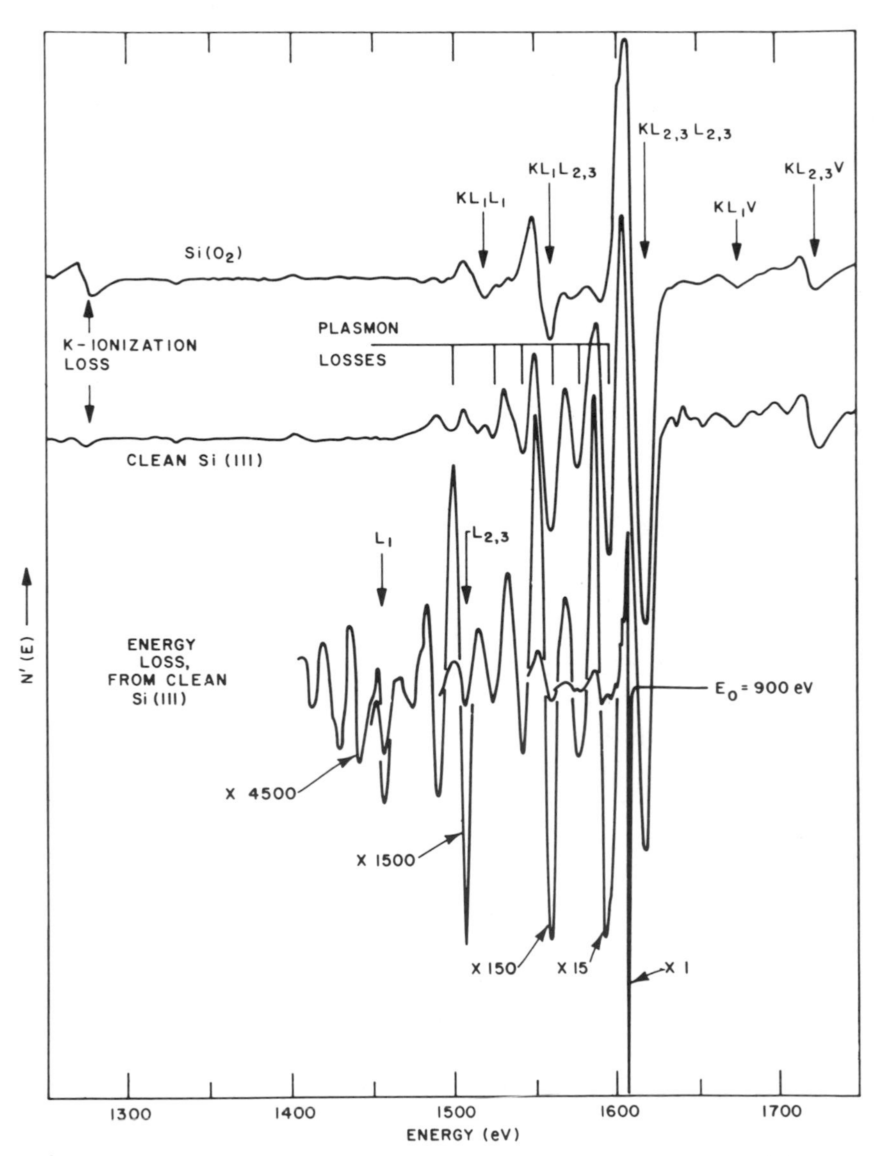

Fig. 20-13. The *KLL* spectra from SiO_2 and Si (CMA, 3.1-keV primaries) and the energy-loss spectra from clean Si (LEED–Auger apparatus, 900-eV primaries). All three spectra have been displaced in energy to coincide at 1619 eV. The spectrum from SiO_2 is relatively simple. The spectrum from Si is more complex because of the loss peaks, which are shown in the bottom trace.

entirely due to differences in loss mechanism. The valence electrons in SiO_2 are tightly bound and allow few loss transitions; the spectrum is therefore relatively simple. In the spectrum from Si, plasmon excitation is important, and the plasmon loss peaks are indicated in the figure. The positions of the loss peaks can be seen in the bottom curve, showing the reflected primaries and energy loss peaks, taken using the LA device in Fig. 20-7. The monotonic decrease in the amplitudes of the plasmon loss peaks is interrupted at the $L_1, L_{2,3}$ ionization energies. Note the sharp interband-transition loss peaks near E_0 and the comparative broadness of the plasmon loss peaks ($\times 1$ curve). These prominent plasmon loss peaks are not present in similar data from SiO_2.

The widths of Auger peaks are partly determined by loss mechanisms. This is why high-resolution spectrometers are not needed for obtaining spectra from solid surfaces. We also saw (see Section 2.4) that broadening over 2 eV is possible, for Coster–Kronig transitions.

5.2. Valence Spectra and Chemical Shifts

Valence spectra of type WXV have widths at least as wide as the valence band, and WVV transitions, at least twice as wide; the latter can be seen from Fig. 20-1. An $L_{2,3}V_1V_1$ transition would have the energy $E_1 = E_{L_{2,3}} - 2E_{V_1}$, and an $L_{2,3}V_2V_2$ transition would have the energy $E_2 = E_{L_{2,3}} - 2E_{V_2} = E_1 - 2\zeta$. Any energy $E_1 \geq E \geq E_2$ is also allowed (e.g., $L_{2,3}V_1V_2$), so that a broad peak of width 2ζ is created. In Fig. 20-10a, $\zeta \approx 12$ eV, and the valence spectrum is mixed with loss spectra down to ~25 V below the 92-eV peak. The WVV-type spectra are said to be "folded" because the valence-band density of states is used twice in determining the total transition density. Therefore, to obtain the density of states, the spectrum must be "unfolded" mathematically.[55,56,007,010]

The valence band is sensitive to the environment of an atom (see Refs. 052, 067, 082, 104, 111, 112, 113, 234). The changes in the Si spectra of Figs. 20-10 and 20-11 are such effects. Another example is shown in Figs. 20-14–20-16. Figure 20-14 is a spectrum from freshly cleaved single-crystal graphite which gave a LEED pattern in a separate experiment. Figure 20-15 is from the same area (vertical scale expanded 2.5 ×) after sufficient electron bombardment so that the surface is amorphous and does not give a LEED pattern. Note the disappearance, in Fig. 20.15, of the fine structure seen in Fig. 20-14. The carbon spectra from various carbides are shown in Fig. 20-16; these are very different from the spectra of Figs. 20-14 and 20-15. Spectra from metallic carbides are characterized by the two large low-energy peaks that are almost totally absent in data from pure carbon. The carbide spectra have been investigated using photoelectron spectroscopy[57a] and x-ray emission

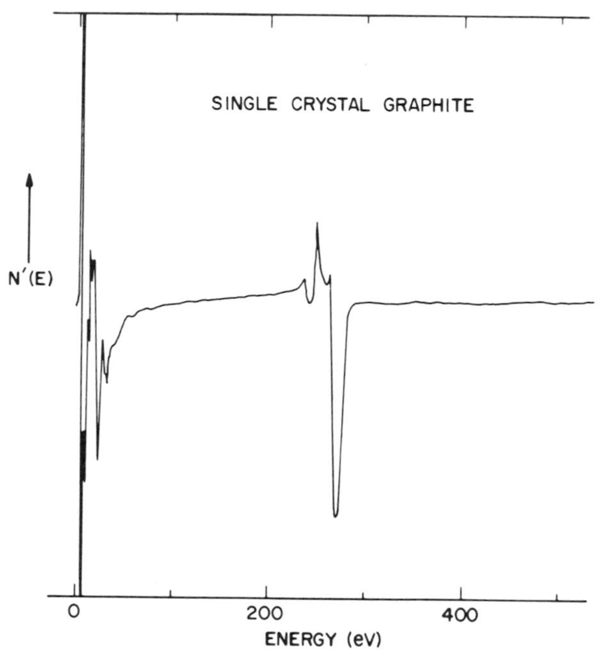

Fig. 20-14. Auger spectrum from single-crystal graphite (CMA, 3.1-keV primaries).

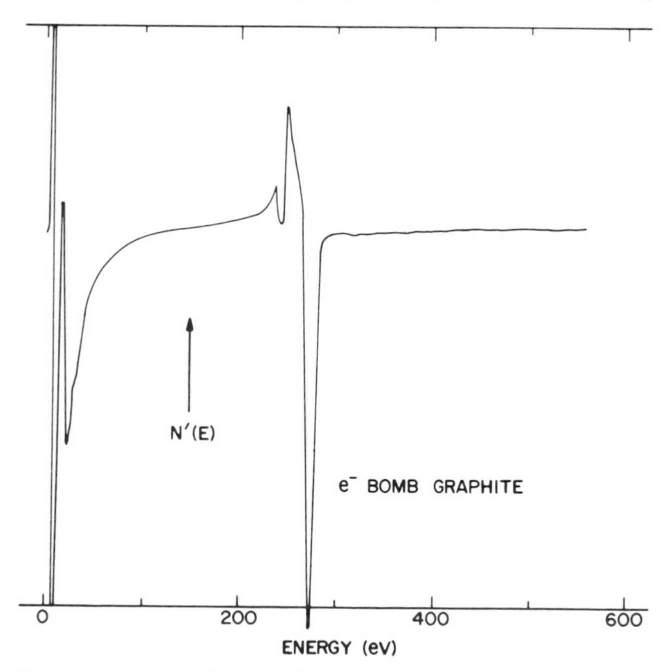

Fig. 20-15. The same area as Fig. 20-14, after 15 minutes of electron bombardment at 40 μA, 3.1 KeV (vertical scale expanded 2.5 $\times$). Note that most of the sharp peaks seen in Fig. 20-14 have disappeared.

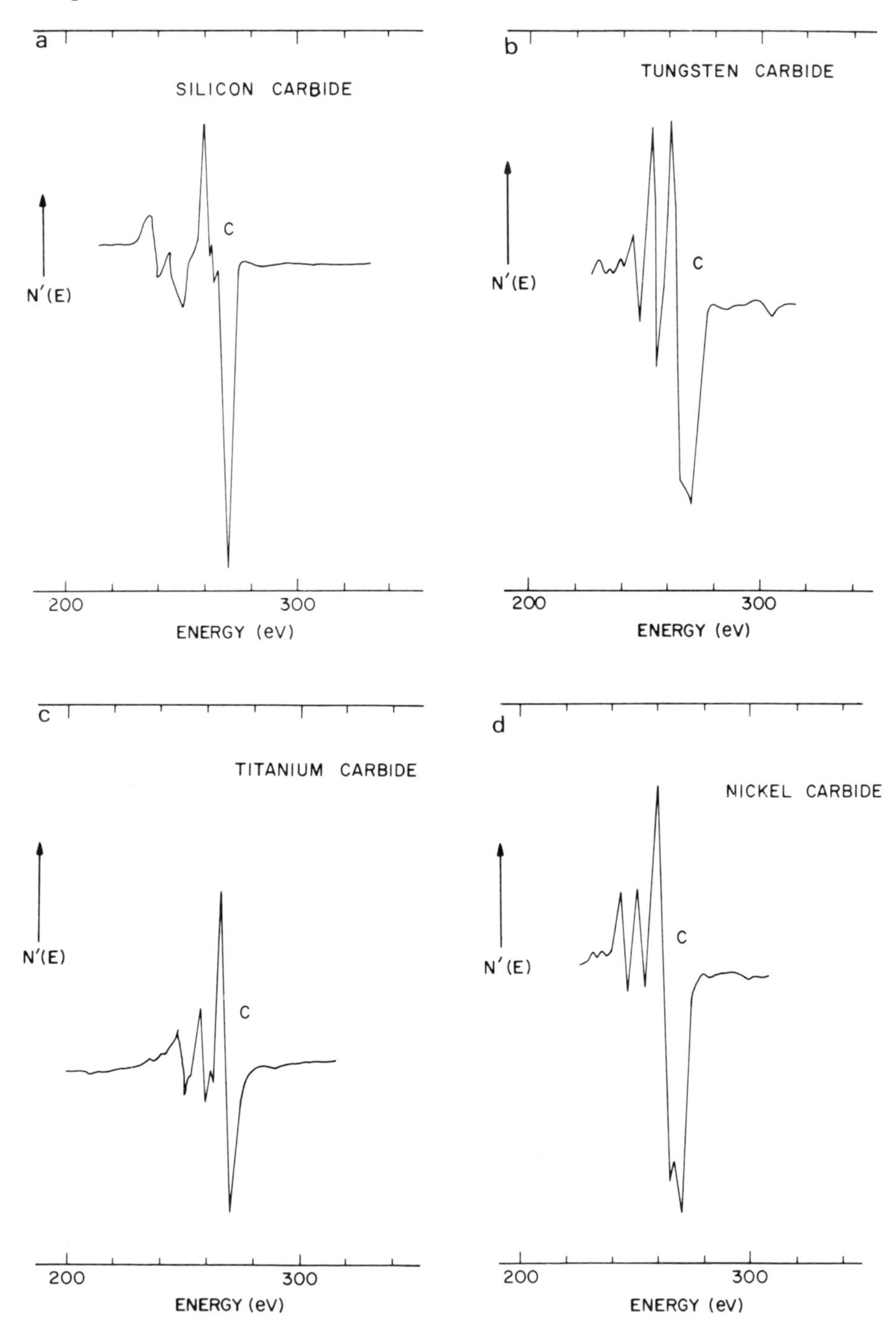

Fig. 20-16. Carbon spectra from various carbides: (a) silicon carbide, (b) tungsten carbide, (c) titanium carbide, and (d) nickel carbide. The traces have been displaced in energy to coincide at 270 eV. The metal carbides are characterized by two relatively large peaks on the low-energy side of the main peak (CMA, 3.1-keV primaries).

spectroscopy.[57b] Large shifts in spectra from sulfur compounds have also been observed and attributed to different valence states.[58]

Core levels also shift with changes in chemical environment.[1d–i] Such chemical shifts are due mainly to a net exchange of electrons among atoms in a solid and are difficult to measure using Auger spectra. Firstly, chemical shifts are often small (≤ 1 eV) and cannot be accurately measured using the relatively broad Auger peaks. Secondly, if Δ_W, Δ_X, Δ'_Y, (where Δ'_Y is the shift in the doubly ionized state) are the chemical shifts of the W, X, and Y levels, respectively, then the Auger peak will be shifted by

$$\Delta E = -\Delta_W + \Delta_X + \Delta'_Y \tag{20}$$

Since only ΔE is measured, the individual Δ's are not known. A more straightforward method is to measure the positions of the ionization loss peaks,[088,091,092] or use photoelectron spectroscopy. Because of the extra ionization, chemical shifts of Auger peaks are often larger than those in photoelectron spectra.[59]

5.3. Quantitative Analysis

5.3.1. The Auger Yield

Several reports on different aspects of quantitative analysis have appeared in the literature (see Refs. 1*k*, 60, 61, 022, 036, 041, 083, 194, 210, 214, 264, 279, 297). The factors that determine the size of Auger peaks in typical spectra will be considered in this section. The current I_i from an Auger transition i, collected by an analyzer, is given symbolically by

$$I_i = A I_p X_i \rho D B \phi \psi R T \tag{21}$$

where the product notation is only symbolic because each factor is not independent of the others. Each symbol represents a complicated function of many variables: A is the area which is irradiated by the primary beam and contributes electrons to the analyzer (cm^2), I_p is the primary excitation beam intensity (A/cm^2), ρ is the atom density (atoms/cm^3), X_i is the concentration of element i ($0 \leq X_i \leq 1 = 100\%$), D is the escape depth (cm), B is the backscattering factor (>1), ϕ is the ionization cross section (cm^2/atom), ψ_i is the Auger transition probability, R is the surface-roughness factor, and T is the instrument transmission.

The quantity which we wish to evaluate is X_i. In the following section we develop an empirical calibration procedure which eliminates the necessity of calculating most of the above factors, and enables us to calculate an approximate value for X_i. The accuracy of this approximate answer can then be improved to varying degrees depending on how well some of the above factors are known. Therefore, we must first understand the significance of each factor.

The factors A, ρ, and T are easily understood and require no further explanation. Note that several other minor factors, such as the sensitivity of the electron collector, the dependence of the signal on resolution, etc., have been neglected in Eq. (21).

The value of I_p is attenuated with depth and is in general difficult to calculate, especially for nonhomogeneous surfaces; it is influenced by diffraction effects and depends strongly on incidence angle. The value of D_i also varies with depth and angle. (Note that I_p applies to electrons with the incidence energy, whereas D_i applies to the escaping Auger electrons with energy E_{WXY}.) Therefore, the angle of incidence and angle of escape produce large effects on the Auger yield;[62,63,118,203] such effects can be used for investigating the depth distribution of various surface components.

Perhaps the most important factor for quantitative analysis is D_i.[60] Although D_i is a complex function,[55] for the purposes of this section we shall simplify it into a single quantity, the ejection depth, which we shall also call D_i. It specifies the "detected volume" from which Auger electrons can escape. The value of D_i depends on the material and the energy E_{WXY}, and is independent of the primary energy. To define D_i mathematically, we consider N electrons passing through a material of thickness dz and attenuated by an amount dN. Since the attenuation is proportional to N and to dz,

$$-dN = DN\,dz$$

or (22)

$$N = N_0\, e^{-z/D}$$

Moreover, the total number of escaping electrons is

$$N_0 \int_0^\infty e^{-z/D}\, dz = DN_0 \tag{23}$$

where D is the mean free path.[24]

The mean free path D is less than the electron range, $l(E)$; typically, if we choose $\frac{1}{3}E$ as the cutoff energy, then $l(E) \sim 5D$ (see H. Seiler[64]). Therefore, the detected volume is always smaller than the total volume in which Auger electrons are excited by the primary beam. Some experimental values of D are shown in Fig. 20-17. Note that D is a minimum near 75 eV and increases approximately as $\sqrt{E}$ from 100–2000 eV, but rises more rapidly thereafter; below ~75 eV, D again increases. These features are in general agreement with theoretical predictions.[24,65] The experimental points of Fig. 20-17 show a surprising lack of dependence on atomic number.

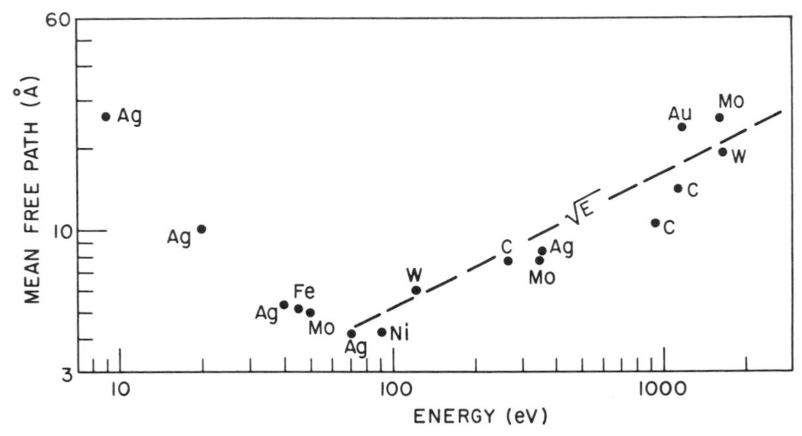

Fig. 20-17. Mean free path D of electrons in various materials (after J. C. Tracy.[1c]) Note that D is smallest near 75 eV, and increases approximately as $\sqrt{E}$ between 100 and 2000 eV.

The three terms B, ϕ, and ψ_i are closely related. The general behavior of the ionization cross section[66–71,022] ϕ is shown in Fig. 20-18 for electron bombardment. The relative ionization probability is shown plotted *vs.* the reduced energy E_0/E_W, where E_0 is the primary energy and E_W is the critical energy for ionization. Maximum ionization should occur near $E_0/E_W \sim 3$. However, primary electrons create secondaries, some of which may have energies $> E_W$ and therefore cause further ionization.[133] These secondaries increase, up to a point, with increased E_0, and the extra ionization gives rise to the backscattering factor B. This factor increases the ionization probability and the energy for maximum ionization increases to $E_0/E_W \sim 6$.[70,71]

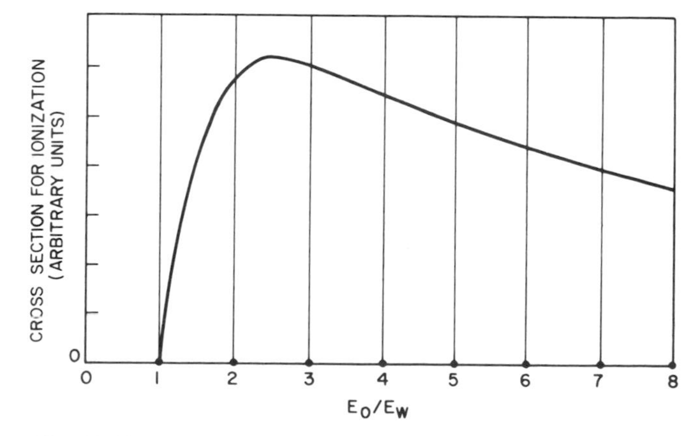

Fig. 20-18. Ionization cross section *vs.* reduced energy E_0/E_W, where E_0 is the primary electron energy and E_W is the critical energy of ionization (from Ref. 022).

Once a shell is ionized, the atom is ready to decay by the Auger process, but in general, it has more than one choice of decay mode.[72,029] The probability of decay by a particular Auger transition i is given by ψ_i, which is usually calculated by assuming that there are only two decay processes: (1) photon emission and (2) Auger transition. In addition, usually only dipole radiation is considered in (1), and only the basic two-electron Auger process is considered in (2). Thus, other radiative processes (e.g., quadrupole) and multiple-decay processes[73] (e.g., multielectron, electron + plasmon, etc.) are ignored. These assumptions have remained unchallenged because few such other processes have been identified experimentally.

The above assumptions lead to the following result: $\psi_T = \sum_i \psi_i = (1 - \omega)$, where ω is the probability of photon emission,[1a,022,029] or fluorescence yield. Therefore, the calculations of ψ described in Section 2.3 are also calculations of the fluorescence yields, and *vice versa*. The points of importance to Auger spectroscopy are:

1. The fluorescence yield decreases with decreasing Z. For transitions with Auger energies <2.5 keV, $\omega \lesssim 0.1$ for K-shell fluorescence yield,[1a,022,029] $\lesssim 0.05$ for the L shell,[178] and even smaller for the M shell.[12]
2. The Auger yield is essentially determined by the ionization $B\phi$ because ψ_T is almost unity for $E_{WXY} \leq 2.5$ keV.
3. The above two points apply to ψ_T, but for each ionization (e.g., K shell, L_1 shell, etc.) there are usually several transitions (e.g., KL_1L_1, KL_1L_2, etc.), and each has its own transition rate ψ_i. These transition rates are given in the references cited in Section 2.3.

The only term in Eq. (21) that has not been discussed is the roughness factor R. The escape probability of electrons from rough surfaces is smaller than for smooth surfaces because electrons leaving a smooth surface into field-free space never return, whereas those leaving a rough surface can be recaptured. This effect is shown in Fig. 20-19, which shows data from the LEED-Auger device of Fig. 20-7. The size of the Auger peaks in the spectrum of a Au film deposited on sintered alumina (rough surface) is ~0.6 times that from a similar Au film deposited on glazed alumina (smooth surface); therefore, $R \sim 0.6$ for the unglazed alumina, assuming $R = 1$ for the glazed surface. Results of roughness calculations depend on the exact surface geometry,[74] so that a generalized formula cannot be presented. However, R has some basic properties: (1) The escape probability depends on the electron escape angle. Therefore, R depends on the shape of a surface but not on the absolute scale. For

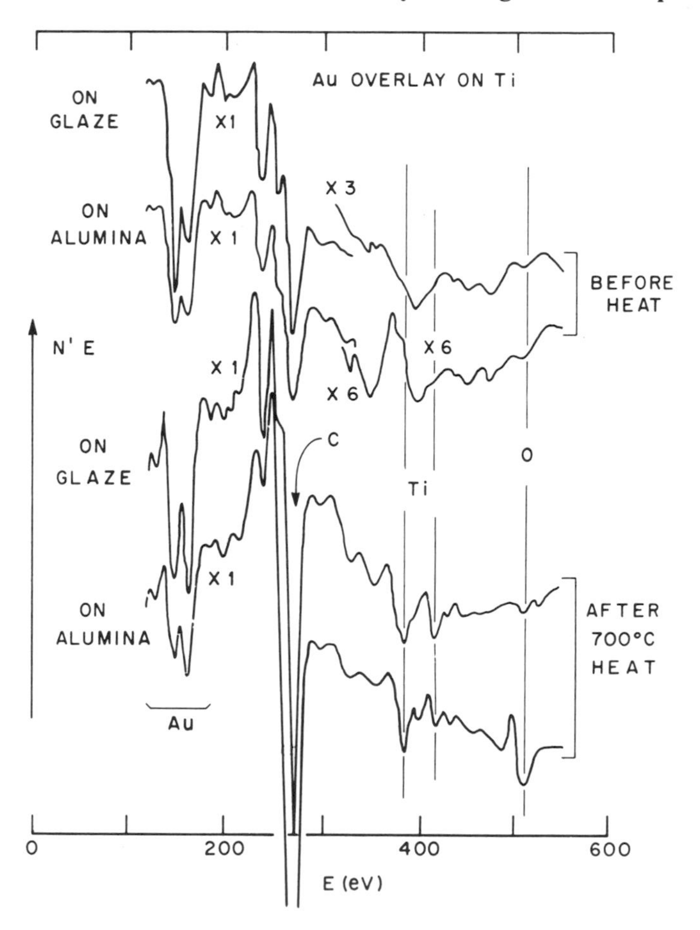

Fig. 20-19. Auger spectra from 2-μ thick Au films on 500 Å of Ti deposited on glazed and nonglazed alumina. The signal from the film on rough alumina is 0.6 times that from the smooth glazed substrate. Large amounts of C, and smaller quantities of Ti and O, diffused to the surface after heating to 700°C for 10 min in high vacuum; however, the film roughness does not appear to have changed significantly. The C contamination was probably deposited during, or prior to, Au deposition (LEED–Auger apparatus, 900-eV primaries).

example, a sinusoidal surface of amplitude A (normal to the surface) and period b (along the surface) will have the same R as another sinusoidal surface with amplitude A' and period b' if $A'/b' = A/b$ (within reasonable limits). (2) If the roughness has a characteristic amplitude A and a characteristic period b, R is smaller for larger A/b. (3) For glancing incidence, $R \to 1$ as the glancing angle $\theta \to 0$. This last fact is useful not only for quantitative analysis but also for preventing "charge-up" when investigating insulators (Section 6.2).

5.3.2. Calculation of Surface Composition

We will now present the equations for calculating X_i from the Auger data I_i. The significance of these equations will be best understood if they are derived from the basic Eq. (21) applied to some generalized surface. Therefore, consider a surface with a homogeneous distribution of elements i and concentrations X_i. For the pure element, $X_i = 1$ and Eq. (21) can be rewritten:

$$b_i I_i^0 = D_i(E_i) \tag{24}$$

where ^{0} denotes the pure element, b_i accounts for all the remaining factors, and E_i is the Auger transition energy, E_{WXY}. For a mixed (homogeneous) surface, we have

$$b_i' I_i = X_i D_a(E_i) \tag{25}$$

where D_a is some average escape depth. To calculate D_a, we define the electron stopping power

$$S_i(E_i) = 1/D_i(E_i) \tag{26}$$

Then, the total stopping power of all the elements is

$$\begin{aligned} S_a(E_i) &\approx \sum_j X_j S_j(E_i) \approx \sum_j X_j / D_j(E_i) \\ &= 1/D_a(E_i) \end{aligned} \tag{27}$$

where $\approx$ has been used because S_j is not independent of chemical environment. Note that all the D_j are calculated at the single energy E_i. The solution for X_i is now

$$X_i = \frac{b_i' I_i D_i(E_i)}{b_i I_i^0 D_a(E_i)} \tag{28}$$

Referring back to Eq. (21), we see that to a first approximation, we can temporarily set $b_i' \approx b_i$; also I_i^0 can be eliminated by introducing an "inverse Auger sensitivity factor" α defined by

$$\alpha_i I_i^0 = I_s^0 \tag{29}$$

where s is some standard element. Then

$$X_i = \frac{\alpha_i I_i D_i(E_i)}{I_s^0 D_a(E_i)} \tag{30}$$

Finally, I_s^0 is eliminated by using the equation

$$\sum_j X_j = 1 \tag{31}$$

so that

$$X_i = \frac{\alpha_i I_i D_i(E_i)}{I_s{}^0 D_a(E_i)} \bigg/ \sum_j \frac{\alpha_j I_j D_j(E_j)}{I_s{}^0 D_a(E_j)}$$

$$= \frac{\alpha_i I_i D_i(E_i)/D_a(E_i)}{\sum_j \alpha_j I_j D_j(E_j)/D_a(E_j)} \tag{32}$$

Evaluation of D_a is somewhat laborious, and it is desirable to replace it with a simpler quantity. To this end, we choose a standard element s with ejection depth D_s, and write

$$D_j(E_i) = \eta_{sj} D_s(E_i) \tag{33}$$

where $\eta_{ss} = 1$ by definition, and η_{sj} will in general depend not only on the element j but also on the energy E_i. Then,

$$D_a(E_i) = D_s(E_i) \bigg/ \sum_k \frac{X_k}{\eta_{sk}} \tag{34}$$

If η is independent of energy, Eq. (32) becomes

$$X_i = \frac{\alpha_i I_i D_i(E_i)/D_s(E_i)}{\sum_j \alpha_j I_j D_j(E_j)/D_s(E_j)} \tag{35a}$$

$$= \frac{\alpha_i I_i \eta_{si}}{\sum_j \alpha_j I_j \eta_{sj}} \tag{35b}$$

The effect of η on the Auger spectra can be seen by examining the simple example of a binary compound with $X_1 = X_2$ and $\alpha_1 = \alpha_2$; taking $s = 1$, we have $I_1 = \eta_{12} I_2$. The material with the shorter ejection depth gives a larger Auger peak, at the expense of the material with the longer ejection depth. When $\eta = 1$, Eq. (35) takes a particularly simple form:

$$X_i = \frac{\alpha_i I_i}{\sum_j \alpha_j I_j} \tag{36}$$

The absolute amount of material detected on a surface is then given by

$$C_i = X_i \rho D_a(E_i) \tag{37}$$

After rough values of X_i and C_i are thus obtained, they can be refined by correcting for the approximation $b_i' \approx b_i$ made earlier.

The most useful equation for quantitative analysis, Eq. (36), has now been derived rigorously. The assumptions and approximations needed for its derivation make it clear that Eq. (36) cannot be expected to provide answers with a high degree of accuracy, especially for real surfaces that are nonideal. More accurate quantitative analyses are possible if calibrations from known standards are used. Corrections to Eq. (36) are provided by introducing "matrix-effect" parameters M_j and writing:

$$X_i = \frac{M_i \alpha_i I_i}{\sum_j M_j \alpha_j I_j} \tag{36a}$$

$$\equiv \frac{\beta_i I_i}{\sum_j \beta_j I_j} \tag{36b}$$

where the β_j are evaluated empirically from known standards. Note that most matrix corrections affect all elements in similar ways (i.e., $M_i \approx M_j$ for all j). Therefore, Eq. (36) can be expected to approximate Eq. (36b) even if $\alpha_j \neq \beta_j$.

The Auger current I_i should ideally be measured as the area under the $N(E)$ curve, but if the Auger peak shape does not change with composition, then the peak-to-peak height on the $N'(E)$ curve should be an equally good measure.[264] Unfortunately, changes in the peak shape with composition are often encountered. Two classes of phenomena are most commonly observed: (1) the shapes of WXV and WVV transitions depend on chemical composition (already described in Sections 5.1 and 5.2), and (2) line broadening by energy loss depends on depth away from the surface and on the band structure of the material. An example of changes in peak shape with changes in loss mechanisms was described in Section 5.1 for the Si KLL transitions (Fig. 20-13). Also, atoms closer to the surface produce sharper Auger peaks because of the lower probability of suffering energy loss[036] (Fig. 20-12). Therefore, care must be exercised when using the $N'(E)$ instead of the $N(E)$ curve; for instance, removal of a single monolayer of contaminants will cause all the peaks from the substrate to become sharper and therefore appear much larger on the $N'(E)$ curve.

In summary, an approximate value of X_i can be calculated, for the homogeneous case, using Eq. (36). If η_{sj} is independent of energy, Eq. (35b) gives a more accurate value of X_i. The absolute amount of detected material is given by Eq. (37). When elements of short and long $D_j(E_i)$ are mixed, the element with the shorter D_j gives a larger Auger signal (and *vice versa* for the longer D_j element) than when all D_j are the same.

5.3.3. Measurement of Spatial Distribution

Some idea of the depth distribution can be obtained by using different angles of incidence[118,203] of the primary beam. Near glancing incidence, the surface atoms are preferentially excited, compared to normal incidence. A second method is direct depth profiling using ion milling (see Section 7.1).

A third method is to use Auger peaks at different energies. These also contain information on spatial distribution because of the dependence of D_i on energy. As an example of the use of different D_i, and as an illustration of how the methods described above are applied, consider the data of Fig. 20-20, which was taken from a ~3000-Å film of WSi_2 on a Si substrate. Data were taken with the CMA and a primary-electron energy of 3.1 keV. For data analysis, two Auger peaks from both Si and W were used: $E_{l,Si} = 92$ eV, $E_{h,Si} = 1619$ eV, $E_{l,W} = 169$ eV, and $E_{h,W} = 1736$ eV (the subscripts, l and h, denote low and high energy, respectively). We chose Si as the standard element, and to simplify matters, performed the low- and high-energy analyses separately. This separation allowed us to let both $\alpha_{l,Si} = \alpha_{h,Si} = 1$. By comparing data from pure Si and pure W, we found $\alpha_{l,W} \approx 3.5$ and $\alpha_{h,W} \approx 0.7$. The atomic percent of Si is then given by

$$X_{Si} = \frac{I_{Si}}{I_{Si} + \alpha_W I_W} \times 100 \tag{38}$$

where Eq. (36) has been used. According to Fig. 20-17, $D_l \sim 8$ Å and $D_h \sim 25$ Å; we shall call the low- and high-energy data "surface" and "bulk" concentrations, respectively. In Fig. 20-20, the Si concentration

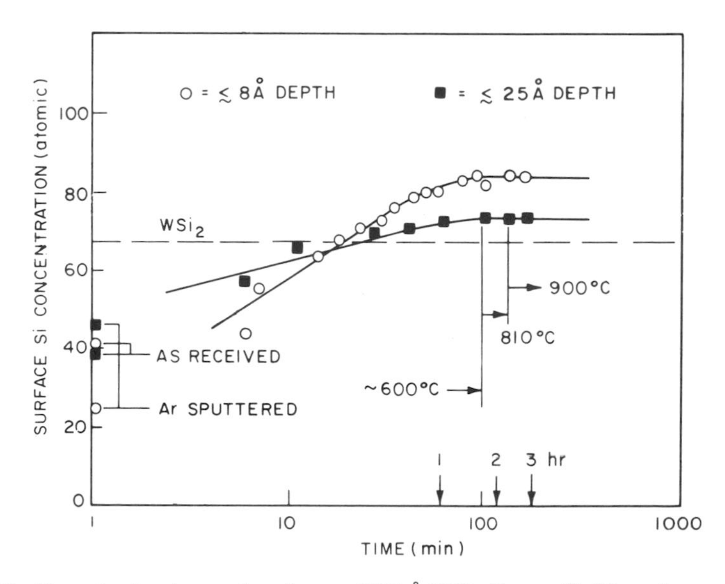

Fig. 20-20. Quantitative Auger data from a 3000-Å WSi_2 film on Si. The existence of WSi_2 was verified by x-ray diffraction. The surface was deficient in Si after Ar ion sputtering, but became covered with two monolayers of elemental Si (see text) upon annealing.

is seen to have increased upon annealing at 600°C due to Si out-diffusion from the substrate. The surface Si concentration increased faster than the bulk concentration, suggesting preferential surface segregation of Si. Eventually both concentrations increased *beyond* WSi_2 stoichiometry and saturated at a constant value. The surface Si saturated at 84%, which means that the W concentration was 16%, or only 32% of the Si can be combined as WSi_2. This leaves 52% of excess surface Si in the first 8 Å; this excess is therefore ~4 Å (about two atom layers or $\sim 2 \times 10^{15}$ atoms/cm^2). The same result should be obtainable using the bulk data. The bulk data saturated at ~74%, leaving 26% W or 52% (W)Si_2; this gives an excess surface Si of 22% in the first ~25 Å, or 5.5 Å, in fairly good agreement. It appears that when WSi_2 is covered with two atom layers of Si, the surface energy becomes almost the same as that of pure Si, and further rapid out-diffusion does not occur.

Many elements have both low- and high-energy Auger transitions, and a depth analysis technique utilizing only Auger data is often useful. To illustrate how such an analysis might be accomplished, consider the model of Fig. 20-21, where islands of material A with thickness d cover a fractional area s of the substrate B. Such a model might approximate heterogeneous nucleation, or a surface created by facture at the interface of a composite material. Consider the case when A and B have low- and high-energy Auger peaks. For simplicity the following assumptions are made: the ejection depths $D_l(A) = D_l(B)$ and $D_h(A) = D_h(B)$, and the Auger yield is constant for $d < D$, but zero for $d > D$. Then we can evaluate all the C_i:

$$\begin{aligned} C_{l,A} &\propto sd \\ C_{l,B} &\propto D_l - sd, \qquad d < D_l \end{aligned} \tag{39}$$

etc., for larger values of d. We now compute the ratios

$$\begin{aligned} r_l(A/B) &= C_{l,A}/C_{l,B} \\ r_h(A/B) &= C_{h,A}/C_{h,B} \end{aligned} \tag{40}$$

Then we can evaluate the ratio of ratios:

$$R(A/B) = \frac{r_l(A/B)}{r_h(A/B)} \tag{41}$$

Values of C, r, and R are plotted in Fig. 20-22 as functions of d, for $D_h = 3D_l$ and for $s = 0.6$. Note some useful properties of R:

1. As $d \to 0$, $R \to D_h/D_l$ independently of s.
2. When $R \geq D_h/D_l$, d is double-valued, but d can still be uniquely determined since, for $d < D_l$, $C_{l,A} = C_{h,A}$.

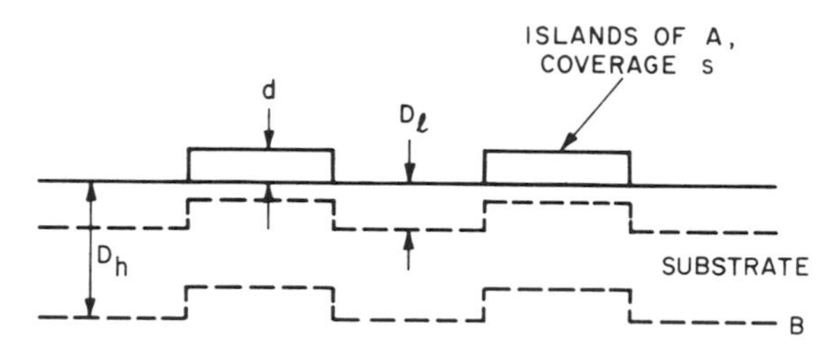

Fig. 20-21. Model of surface with overlayers of thickness d and coverage s. The low- and high-energy electron escape depths are denoted by D_l, D_h, respectively.

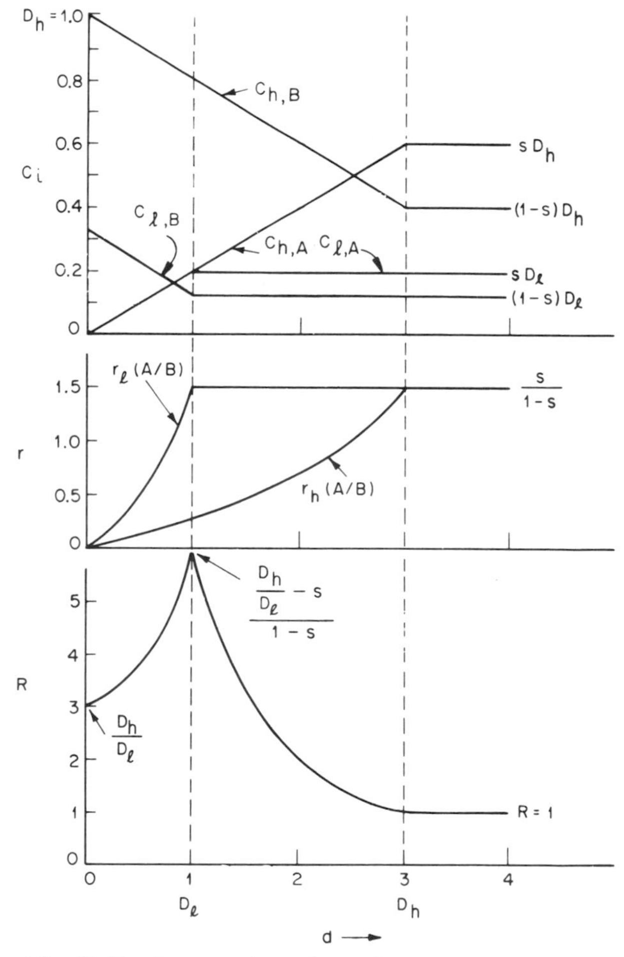

Fig. 20-22. Computation of R using C_A, C_B, and r (see text). These curves can be utilized for performing depth analysis using Auger data alone.

3. When $d = D_l$, R is a maximum and becomes very large as $s \to 1$.
4. When $D_l < d < D_h$, R decreases with increasing d and $s = C_{l,A}/D_l$. Therefore, if d is increased or decreased (as by deposition of A or by ion milling, respectively), the change in R will tell whether $d <$ or $> D_l$.
5. When $d > D_h$, $R = 1$, and

$$s = C_{l,A}/D_l = C_{h,A}/D_h \tag{42}$$

Therefore, d can always be uniquely determined, if $d < D_h$, and in many cases, s can also be found.

5.4. Sensitivity Limit and Trace Analysis

Auger spectroscopy is often cited for its high sensitivity, but it is sensitive only in the sense that $C = X\rho D$ is small because D is typically ~ 10 Å. However, the minimum detectable X is typically larger than for many other analytical techniques.

To calculate numerical sensitivity limits, consider the performance of the CMA. The largest measured Auger currents are $I_i \sim 10^{-4} I_p$; this is most easily seen from the size of the Auger peak compared to the zero-volt peak of Fig. 20-12, if we assume that the secondary-electron emission coefficient $\delta \approx 1$ (see Section 6.2). This value of I_i also agrees with calculated estimates.(022) For typical primary currents of $I_p \sim 5 \times 10^{-5}$ A and bandwidth of $B = 1$, Eq. (15) gives a shot-noise current of $\sim 5 \times 10^{-14}$ A, or a maximum S/N ratio of $\sim 10^5$. Even if the primary current were raised and data-acquisition time lengthened, it appears that S/N ratios above $\sim 10^6$ are not practically attainable. This imposes a limit of ~ 1 ppm for trace analysis, but this limit is academic in most instances because other factors impose larger sensitivity limits.

The most common limit encountered is the existence of background peaks in the Auger spectra. If elements with $Z \gtrsim 30$ are present in appreciable quantities, the probability of spectral overlap becomes high. This happens because as the sensitivity is increased, the number of detectable Auger peaks also increases. Each Auger peak creates a large number of loss peaks (see Fig. 20-13), so that the number of detectable peaks increases almost indefinitely with higher sensitivity. These peaks generally impose a detection limit of $X_i = 10^{-4}$–10^{-5}. In addition, it is usually not possible to create or maintain a surface in a vacuum which is free of contaminants in the ppm range.

In summary, the sensitivity limit is $< 10^{-2}$ (1 at %) for practically every element except hydrogen. When spectral overlap is not severe,

$X_i \leq 10^{-4}$ is usually possible, using data acquisition times of minutes and primary electron currents of $\sim 50\ \mu A$. The major sensitivity limit is imposed by the appearance of background peaks; the limit due to shot noise is ~ 1 ppm for reasonable data acquisition times.

6. SPECIAL PROBLEMS

6.1. Electron–Beam Artifacts

The electron beam can cause desorption,(193) decomposition(039, 212,274) out-diffusion,(231) and create other artifacts.(054) Figure 20-23 shows the electron-bombardment decomposition of SiO_2. The rate of dissociation was reduced when the SiO_2 was covered with $\gtrsim 1$ atom layer of C. Figure 20-24 shows the out-diffusion of phosphorus, due again to electron bombardment, from a deposited SiO_2 film doped with ~ 0.5 at % P. The minimum time required to scan the P peak at a sensitivity limit of 0.05 % was $\lesssim 6$ seconds. Extrapolation of the curve in the figure to zero time showed that the bulk concentration of ~ 0.5 at % could be measured with better than 10 % accuracy if the data are taken in < 12 seconds.

6.2. Studies of Insulators

Insulators present problems because they allow electrons or ions to be captured and cause the material to "charge-up." Electrons are captured in two general classes of traps: (1) isolated conduction bands, and (2) defect states. An example of (1) is a piece of metal on an insulator;

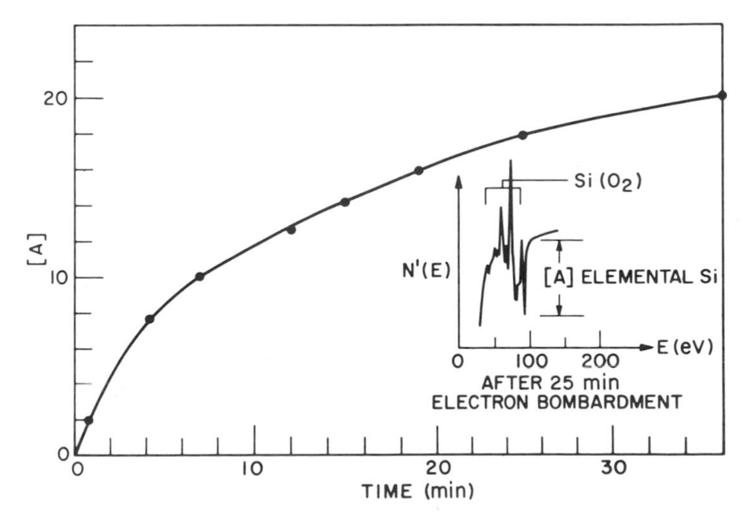

Fig. 20-23. Decomposition of SiO_2 to "elemental Si" by primary electron bombardment at 3.1 keV, 0.05 A/cm^2 (CMA data).

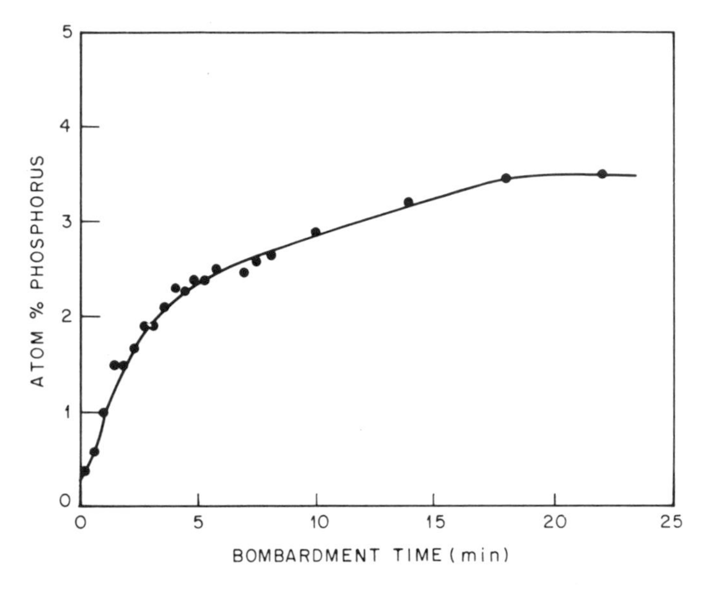

Fig. 20-24. Out-diffusion of phosphorus from deposited SiO_2 doped with 0.5 at % P, caused by primary electron bombardment at 3.1 keV, 0.05 A/cm^2.

the most important example of (2) is the surface state. It can be shown that all electron traps associated with defect states have energies in the forbidden gap.[75] For perfect single-crystal insulators, the only possible traps are surface states,[75] and many of them can be annealed out. Therefore, single-crystal insulators usually present no charge-up problems unless their surfaces are extensively damaged.[039] On the other hand, even well-annealed insulators are problematic if particles of narrower-band-gap materials reside on their surfaces, e.g., epitaxial Si islands on single-crystal sapphire.[039]

The electron traps can be filled when the secondary-electron emission coefficient, δ, is <1. This occurs at low primary energies (<100 eV) because many secondaries have insufficient energy to overcome the work function and escape the material, and at high energies ($>$ several keV) because most of the secondaries are created deep inside the material.[26,76] When $\delta < 1$, the charge-up is often unstable: after the first few electrons are collected, the material assumes a negative potential, thus repelling low-energy secondaries (created at the walls of the apparatus, etc.) and increasing δ somewhat. However, this effect is usually small and the charge-up continues until the surface potential attains that of the incident electron beam. When $\delta > 1$, the surface becomes positive and re-attracts sufficient secondaries so as to make

$\delta = 1$. Therefore, positive charging is usually self-arresting at several volts.

There are some general solutions that can be applied when charge-up problems occur:

1. Choose the incident energy range in which $\delta > 1$; δ is usually largest near 1 keV.[26]
2. Use glancing incidence so that the electron escape probability R of Section 5.3 is as large as possible; this, of course, increases δ.
3. Bias the surface with a negative potential so as to repel the low-energy electrons traveling toward the surface; this also increases the effective δ.
4. Place a grounded conductor (such as a wire mesh) close to, or on, the insulator surface. Although the mesh may interfere with experiment, this is sometimes the only method of preventing charge-up.[77]

Another problem encountered with insulators is that of ion drift.[78] Ions with energy levels in the forbidden gap are long-lived because there are no electrons above the bandgap to neutralize positive ions and there are no empty states below the bandgap to accept electrons from negative ions. In the presence of an electric field, these ions can migrate through the insulator, especially for the positive ions since they can be relatively small. Ion drift alters the original distribution of elements and can lead to erroneous analyses. The ions can be created either during electron bombardment or when ion milling.

7. SOME RECENT DEVELOPMENTS

7.1. Depth Profiles: Ion Milling

Depth profiles of chemical compositions can be obtained by use of ion milling.[79,254,260,263] This is accomplished by bombarding the surface with energetic ions (200–5000 eV) and sputtering off surface atoms.[26] Ions of noble gases, such as Ar^+, are generally used because they do not react with the bombardment surface and with hot filaments in the vacuum system. In Auger work, it is desirable to have a choice of two sputtering ions because these ions are embedded into the surface during sputtering and contribute their own Auger spectra which may interfere with the data. For Ar, sputtering yields are >1 at 1 keV for many materials,[26] and sputtering rates are typically $\gtrsim 1$ Å/μA-min/cm^2. Since ion beams $> 50\ \mu$A/cm^2 are readily attainable, films up to $\sim 1\ \mu$ can be sputtered in reasonable amounts of time.

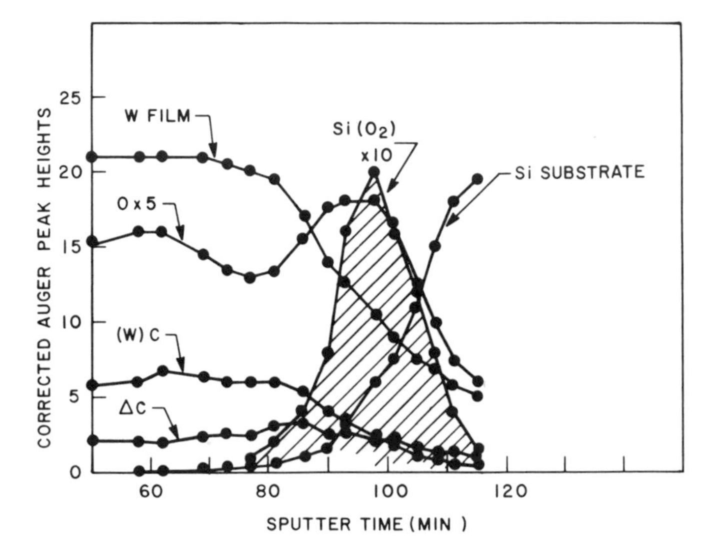

Fig. 20-25. Chemical depth profile obtained using simultaneous Auger spectroscopy and ion milling. A 9-Å SiO_2 layer between a 100-Å W film and the Si substrate is easily detected. The vertical scale is proportional to C_i (see text). The carbon signal is divided into that from tungsten carbide, $(W)C$, and the remainder, ΔC.

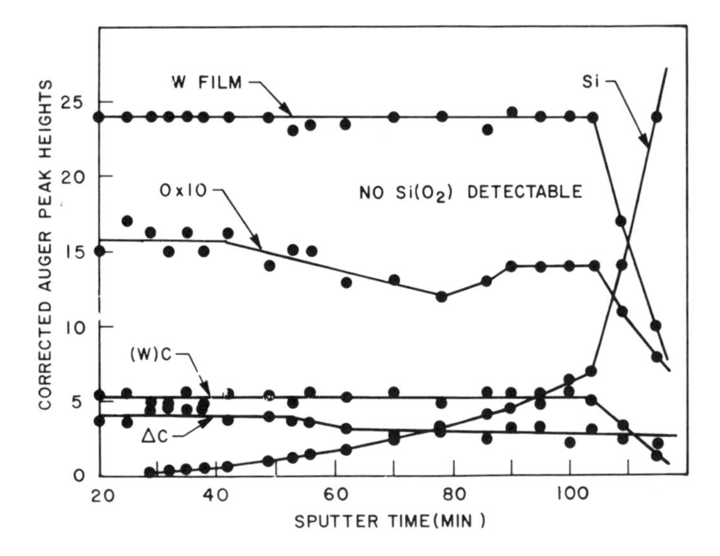

Fig. 20-26. Chemical depth profile obtained by the same means as Fig. 20-25, from a similar W film, deposited after removal of the native oxide on the Si substrate by back-sputtering. The detection of Si long before the interface is reached is probably due to premature punch through at weak spots in the W film; some Si diffusion may also have occurred during W deposition.

An example of depth profiling is shown in Fig. 20-25, taken from a 100-Å tungsten film sputter-deposited on a Si substrate which had ~9-Å native oxide (measured with ellipsometry) before W deposition. The Auger spectra were taken while the surface was being continuously ion milled. The figure shows that the interface oxide was easily detected, but the thickness of this oxide is difficult to estimate because the sputtering rate at the interface cannot be calibrated accurately. Figure 20-26 was obtained from a similar W film sputter-deposited after the native oxide was removed by backsputtering;[80] no SiO_2 could be detected. This latter result demonstrates that the SiO_2 seen in Fig. 20-25 was not an experimental artifact.

Depth resolution is a difficult parameter to determine in an ion-milling experiment. When ion milling and Auger spectroscopy are combined, the depth profile of a sharp interface between materials A and B will show that B is detected before the interface is reached, and A will still be detected after the nominal sputtered surface has passed the interface because an ion-milled surface is not perfectly smooth. In addition, atoms of A are driven into the surface as well as sputtered sideways along the surface so that small amounts of A will be detected at the surface long after the interface is sputtered away.

It is important to know the degree of surface roughening caused by ion milling. Electron micrographs of typical ion-milled surfaces are shown in Fig. 20-27, taken from a 7000-Å SiO_2 film deposited on Si. Figure 20-27a shows the SiO_2 surface after ion milling almost 7000 Å, and Fig. 20-27b shows the Si surface just after removal of all SiO_2. The roughness is difficult to estimate using such micrographs, and is often more simply measured using the Auger data. For example, in Figs. 20-25 and 20-26, the surface roughness is given by the apparent widths of the sharp interfaces, if we assume zero electron ejection depth and that the ion-milled surface was perfectly parallel to the interface. Using the Si curve of Fig. 20-25, the mean roughness of the ion-milled surface is estimated to be $\lesssim 20$ Å, but the local depth resolution is much smaller than the mean roughness, as attested by the fact that the 9-Å SiO_2 film was easily detected. However, surface compositions measured after ion milling can be quite erroneous near an interface because (1) the ion bombardment tends to "homogenize" the surface, (2) the roughness of the surface causes both sides of the interface to be simultaneously exposed, and (3) different materials may have different sputtering yields.

Finally, when ion milling insulators with positive ions, it is usually necessary to prevent charging by use of an auxiliary electron beam. It is also necessary to make sure that the depth profiles have not been affected by ion drift (see Section 6.2).

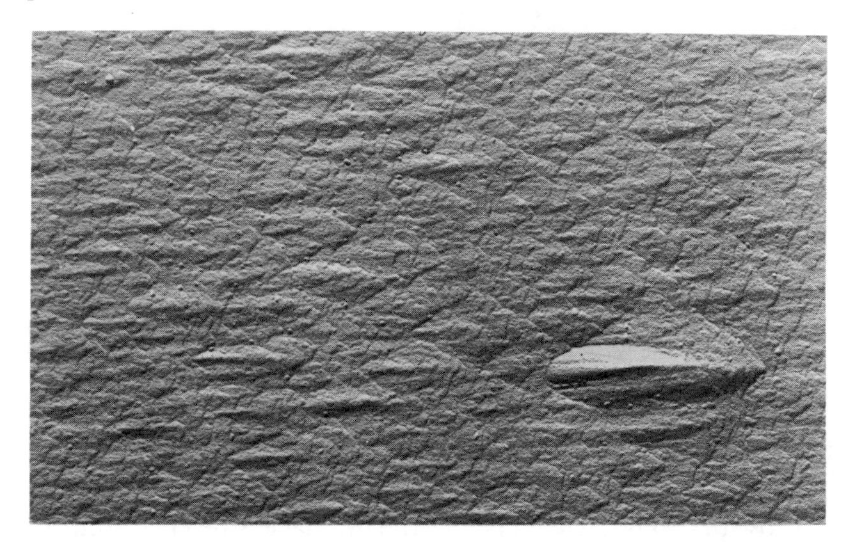

A

B

Fig. 20-27. Electron micrographs of Pt-shadowed C replicas of (a) a deposited 7000-Å SiO_2 film on Si after sputter-removal of most of the SiO_2, and (b) the Si substrate just after complete removal of SiO_2, illustrating surface roughening due to ion milling. The origin of the small circular defects visible in both micrographs, and the thin platelets in (b), are not known. The ion beam was incident at 30° in a horizontal plane. The latex ball in (b) has a diameter of 2640 Å.

7.2. Lateral Resolution: Surface Imaging

If the primary electron beam is rastered, the Auger instrument can be operated as a scanning electron microscope. Such a mode of operation is useful when small areas of samples with complex geometry are to be investigated, or when pictorial records of a surface are desired. Photographs of such scanned pictures are displayed in Fig. 20-28, showing an integrated-circuit chip, taken with the CMA.

Scanning surface photographs of each *element* can also be obtained. Figure 20-29 shows pictures of 20-mil dots of Cs ($\sim$0.5 monolayer coverage) deposited on GaAs, taken using a LEED–Auger apparatus.[81] High spatial resolution has also been attained by attaching a CMA to a scanning electron microscope, and using a computerized data acquisition system.[161,162]

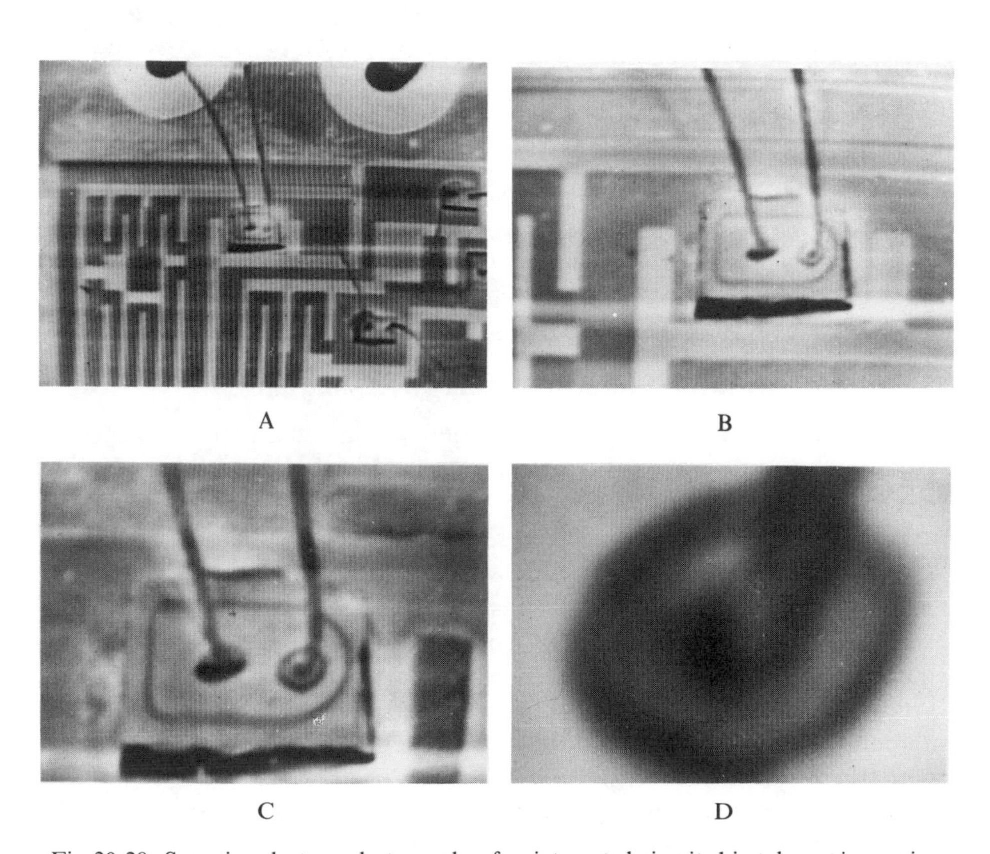

Fig. 20-28. Scanning electron photographs of an integrated-circuit chip taken at increasing magnification, (a–d). The diameters of the leads (Au) seen magnified in (d) are 20 μ. The imaging capability is useful for locating the precise position of the primary beam and obtaining pictorial records. (Courtesy of Vacuum Division, Varian Associates.)

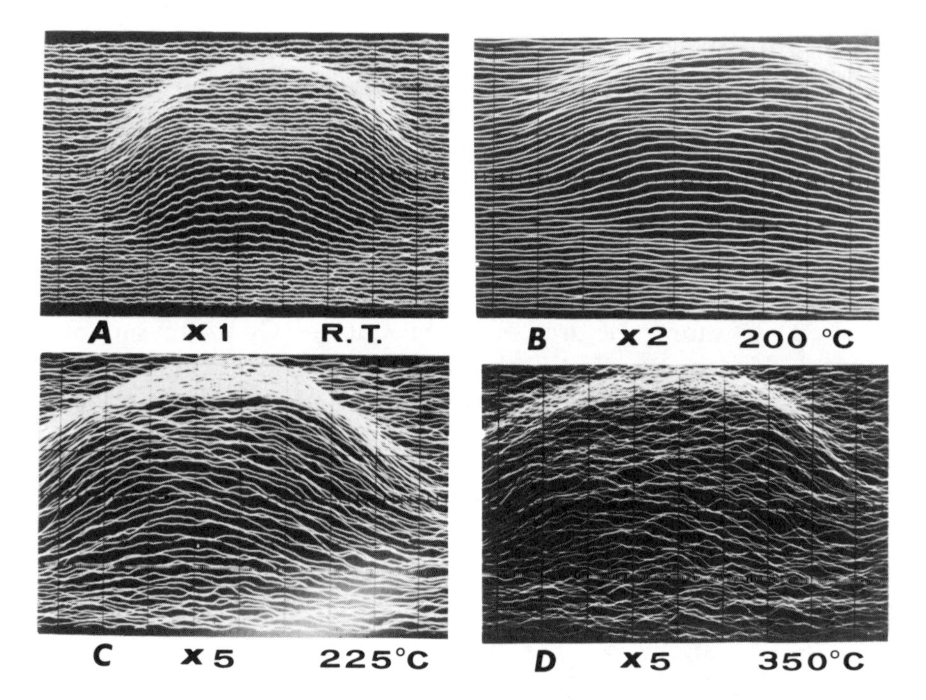

Fig. 20-29. Scanning Auger photographs of 20-mil Cs dots [~0.5 monolayer coverage in (A)] on GaAs. The gradual evaporation of the Cs upon heating is shown. These photographs were taken with the LEED–Auger apparatus, using the 47-eV Cs Auger peak; the vertical sensitivity scale (× 1– × 5) is indicated under each photograph. (From J. R. Arthur.[81])

7.3. Instrumentation: Improved Energy Resolution

It has been amply demonstrated in photoelectron spectroscopy[1d–i] and low-energy scattering work[82,83] that high-resolution electron spectroscopy is extremely fruitful. The CMA can be used to attain high resolution by first retarding the electrons before they enter the CMA.[84,85] Retardation improves the resolution because the CMA resolution is proportional to the pass energy; however, the transmission is also proportional to the pass energy, and therefore decreases with increased resolution. There is in principle no limit to the resolution attainable using retardation. In practice, retardation is accompanied by defocusing effects[85] and added requirements for magnetic shielding, which impose a practical limit. Figure 20-30 shows a double-pass CMA with retardation capability. The double analyzer configuration has two useful features: (1) the energy scale is accurate [in the single-pass analyzer, the energy scale can be in error by $\pm 0.5\%$ (caused by misalignment) without noticeable degradation of the spectra] and (2) the target can be bombarded by a large incident beam without degrading

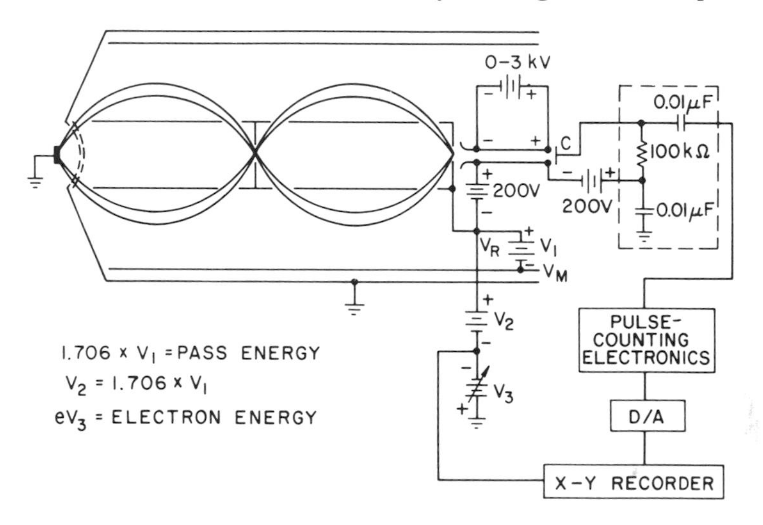

Fig. 20-30. The double-pass cylindrical analyzer with retarding capability. The electronics for pulse counting mode of operation is shown. (Courtesy of Physical Electronics Industries, Inc.)

the resolution, because only electrons originating at the "correct" area on the surface can traverse both sections of the analyzer. The increased resolution of this analyzer is illustrated in Fig. 20-31, where Auger spectra from Cu, taken with the single-pass CMA and with the double-pass CMA of Fig. 20-30, are shown.

8. SOME RELATED SPECTROSCOPIC TECHNIQUES

There are several spectroscopic techniques that are related to Auger spectroscopy and possess some unique advantages, but are not described in the other chapters of this book. We briefly describe some of these below:

1. Ion neutralization spectroscopy:[55,56,86] The surface is bombarded by low-energy ions, which undergo Auger neutralization, ejecting Auger electrons from the surface. These electrons are energy-analyzed and the data are used to infer the density of states of valence electrons of surface atoms.

2. EXAFS—x-ray absorption fine structure analysis: The x-ray absorption fine structure is measured up to ~2000 eV above the absorption edge. The fine structure is due primarily to diffraction of the photoelectron and can be analyzed to solve for the position of other atoms around a particular atom under investigation.[47,48]

3. X-ray appearance potential spectroscopy:[87,88] The surface is bombarded by primary electrons. As the primary electron energy is

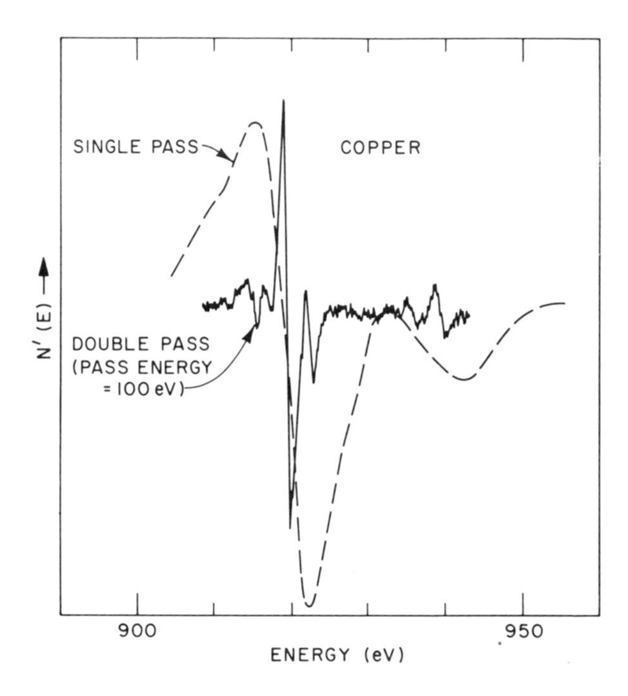

Fig. 20-31. Comparison of spectra from copper, obtained with the double-pass analyzer of Fig. 20-30 at a pass energy of 100 eV, and a single-pass analyzer. (Courtesy of Physical Electronics Industries, Inc.)

varied across an ionization threshold, the atom begins to emit the corresponding characteristic x rays. The x-ray emission threshold is measured to identify the element. One major advantage of this technique is simplicity; however, the method is relatively insensitive to many elements.[275]

4. Ionization spectroscopy:[89,088–092] Attention is focused on the loss peaks in the $N(E)$ curve. It is possible to obtain $N(E)$ curves which contain only primary loss peaks. These loss peaks can be used to identify elements much as in photoelectron spectroscopy.

9. ACKNOWLEDGMENTS

The author is grateful to many members of the Bell Laboratories staff who carefully read the manuscript and contributed useful comments, to J. C. Tracy and D. T. Hawkins for permission to use their bibliography, to G. A. Barber and his drafting staff for their excellent work, and to Miss S. Strond, our departmental secretary, for a most efficient job of manuscript preparation.

10. REFERENCES

1. Reviews and Books on Auger Spectroscopy and related topics.
   a. I. Bergstrom and C. Nordling, in *Alpha-, Beta-, and Gamma-Ray Spectroscopy* (K. Siegbahn, ed.) Vol. II, p. 1523, North-Holland, Amsterdam (1965).
   b. P. W. Palmberg, in *Electron Spectroscopy* (D. A. Shirley, ed.) p. 835, North-Holland, Amsterdam (1972).
   c. J. C. Tracy, NATO Summer School Lectures, Ghent (1972).
   d. K. Siegbahn, C. Nordling, A. Fahlman, *et al.*, in *Atomic, Molecular and Solid State Structure Studied by Means of Electron Spectroscopy, ESCA (Electron Spectroscopy for Chemical Analysis)*, Almquist and Wiksell Boktryckeri AB., Uppsala (1967).
   e. D. A. Shirley (ed.), *Electron Spectroscopy*, North-Holland, Amsterdam (1972).
   f. C. R. Brundle, *Appl. Spectry.* **25**, 8 (1971).
   g. D. M. Hercules, *Anal. Chem.* **42**, 20A (1970).
   h. D. Betteridge and A. D. Baker, *Anal. Chem.* **42**, 43A (1970).
   i. K. Siegbahn, D. Hammond, H. Fellner-Feldegg, and E. F. Barnet, *Science*, **176**, 245 (1972).
   j. K. D. Sevier, *Low Energy Electron Spectroscopy*, Interscience, New York (1972).
   k. T. E. Gallon and J. A. D. Matthew, *Rev. Phys. Technol., III* **1**, 31 (1972).
   l. In the Bibliography: 028, 029, 038, 060, 064, 119, 121, 122, 123, 126, 127, 179, 199, 202, 225, 230, 232, 267.
2. C. Kittel, *Introduction to Solid State Physics*, John Wiley & Sons, New York (1968).
3. E. O. Kane, *Phys. Rev.* **146**, 558 (1966).
4. a. J. A. Bearden and A. F. Burr, *Rev. Mod. Phys.* **39**, 125 (1967).
   b. G. Wiech and E. Zopf, *Z. Physik* **244**, 94 (1971).
5. M. A. Listengarten, *Bull. Acad. Sci. USSR Phys. Sci.* **26**, 182 (1962).
6. O. Hörnfeld, A. Fahlman, and C. Nordling, *Arkiv Fysik* **23**, 155 (1962).
7. O. Hörnfeld, *Arkiv Fysik* **23** 235 (1962).
8. F. K. Richtmeyer, E. H. Kennard, and T. Lauritsen, *Introduction to Modern Physics*, McGraw-Hill, New York (1955).
9. a. W. N. Asaad and E. H. S. Burhop, *Proc. Phys. Soc. (London)* **71**, 369 (1958).
   b. W. N. Asaad, *Proc. Roy. Soc. (London)* **A249**, 555 (1959).
   c. W. N. Asaad, *Nucl. Phys.* **44**, 399 (1963).
   d. W. N. Asaad, *Nucl. Phys.* **44**, 415 (1963).
   e. W. N. Asaad, *Nucl. Phys.* **63**, 337 (1965).
   f. W. N. Asaad, *Nucl. Phys.* **66**, 494 (1965).
10. E. U. Condon and G. H. Shortley, *The Theory of Atomic Spectra*, Cambridge Univ. Press, London (1970).
11. W. N. Asaad and W. Mehlhorn, *Z. Physik* **217**, 304 (1968).
12. a. E. J. McGuire, *Phys. Rev.* **A 5**, 1043 (1972).
    b. E. J. McGuire, *Phys. Rev.* **A 5**, 1052 (1972).
13. S. N. El Ibyari, W. N. Asaad, and E. J. McGuire, *Phys. Rev.* **A 5**, 1048 (1972).
14. E. J. Callan, *Phys. Rev.* **124**, 793 (1961).
15. E. J. Callan, *Rev. Mod. Phys.* **35**, 524 (1963).
16. D. L. Walters and C. P. Bhalla, *Phys. Rev.* **A 4**, 141 (1971).
17. B. Talukdar and D. Chattarji, *Phys. Rev.* **A 1**, 33 (1970).
18. D. Chattarji and B. Talukdar, *Phys. Rev.* **174**, 44 (1968).
19. D. Coster and R. de L. Kronig, *Physica* **2**, 13 (1935).
20. J. C. Tracy, *Surface Sci.* **38**, 265 (1973).

21. E. Bauer, in *Techniques of Metals Research* (R. F. Bunshah, ed.) Vol. II, Pt. 2, John Wiley & Sons, New York (1968).
22. G. F. Amelio, *J. Vac. Sci. Technol.* **1**, 593 (1970).
23. D. Pines, *Elementary Excitations in Solids*, W. A. Benjamin, New York (1964).
24. O. Klemperer and J. P. G. Shepherd, *Adv. Phys.* **12**, 355 (1963).
25. C. V. von Koch, *Phys. Rev. Letters* **25**, 792 (1970).
26. P. A. Redhead, J. P. Hobson, and E. V. Kornelsen, *The Physical Basis of Ultra-high Vacuum*, Chapman & Hall, London (1968).
27. K. K. Spangenberg, *Vacuum Tubes*, McGraw-Hill, New York (1948).
28. C. S. Fadley, R. N. Healey, J. M. Hollander, and C. E. Miner, *J. Appl. Phys.* **43**, 1085 (1972).
29. J. A. Simpson, *Rev. Sci. Instr.* **32**, 1283 (1961).
30. D. E. Golden and A. Zecca, *Rev. Sci. Instr.* **42**, 210 (1971).
31. P. Staib, *J. Phys. E* **5**, 484 (1972).
32. J. D. H. Eland and C. J. Danby, *J. Sci. Instr.* **1**, 406 (1968).
33. A. L. Hughes and V. Rojansky, *Phys. Rev.* **34**, 284 (1929).
34. A. L. Hughes and H. McMillan, *Phys. Rev.* **34**, 293 (1929).
35. J. A. Simpson, *Rev. Sci. Instr.* **35**, 1968 (1964).
36. E. M. Purcell, *Phys. Rev.* **54**, 818 (1938).
37. M. H. Weichert and J. C. Helmer, *Advan. X-Ray Anal.* **13**, 406 (1970).
38. H. Z. Sar-el, *Rev. Sci. Instr.* **38**, 1210 (1967).
39. H. Hafner, J. A. Simpson, and C. E. Kuyatt, *Rev. Sci. Instr.* **39**, 33 (1968).
40. S. Aksela, M. Karras, M. Pessa, and E. Suoninen, *Rev. Sci. Instr.* **41**, 351 (1970).
41. R. L. Gerlach, *J. Vac. Sci. Technol.* **9**, 1043 (1972).
42. T. H. DiStefano and D. T. Pierce, *Rev. Sci. Instr.* **41**, 180 (1970).
43. M. Pessa, *Acta Polytech. Scand. PH* **85**, 16 (1971).
44. D. L. Smith and D. A. Huchital, *J. Appl. Phys.* **43**, 2624 (1972).
45. M. Pessa, *J. Appl. Phys.* **42**, 5831 (1971).
46. V. A. Tsukerman, L. V. Tarasova, and S. J. Lobov, *Sov. Phys. Usp.* **14**, 61 (1971).
47. D. E. Sayers, Technical Report D180-14436-1, Boeing Research Center, Seattle, Washington (1972).
48. L. G. Parratt, *Rev. Mod. Phys.* **31**, 616 (1969).
49. L. Viel, C. Benazeth, B. Fagot, F. Louchet, and N. Colombie, *Compt. Rendu. B* **273**, 30 (1971).
50. R. G. Musket and E. Bauer, *Appl. Phys. Letters* **20**, 455 (1972).
51. G. N. Ogurtson, *Rev. Mod. Phys.* **44**, 1 (1972).
52. E. Sokolowski, C. Nordling, and K. Siegbahn, *Arkiv Fysik* **12**, 301 (1957).
53. K. Siegbahn, C. Nordling, and E. Sokolowski, *Proceedings of the Rehovoth Conference on Nuclear Structure*, North-Holland, Amsterdam, 291 (1957).
54. P. W. Palmberg, G. E. Riach, R. E. Weber, and N. C. MacDonald, *Handbook of Auger Electron Spectroscopy*, Physical Electronics Industries, Inc., Edina, Minnesota (1972).
55. H. D. Hagstrum, *Phys. Rev.* **150**, 495 (1966).
56. H. D. Hagstrum and G. E. Becker, *Phys. Rev. Letters* **22**, 1054 (1969).
57. a. L. Ramquist, *J. Appl. Phys.* **42**, 2113 (1971).
    b. L. Ramquist, B. Ekstig, E. Kallne, E. Noreland, and R. Manne, *J. Phys. Chem. Solids* **32**, 149 (1971).
58. A. Fahlman, K. Hamrin, R. Nordling, C. Nordling, and K. Siegbahn, *Phys. Letters* **20**, 159 (1966).

59. G. Schön, *J. Electron Spectroscopy*, **2**, 75 (1973).
60. M. P. Seah, *Surface Sci.* **32**, 703 (1972).
61. L. L. Levenson, L. E. Davis, C. E. Bryson, III, J. J. Melles, and W. H. Kou, *J. Vac. Sci. Technol.* **9**, 608 (1972).
62. J. H. Neave, C. T. Fox, and B. A. Joyce, *Surface Sci.* **29**, 411 (1972).
63. J. Kadlec and L. Eckertova, *Z. Angew. Phys.* **30**, 141 (1970).
64. H. Seiler, *Z. Angew. Phys.* **22**, 249 (1967).
65. a. J. J. Quinn, *Phys. Rev.* **126**, 1435 (1962).
 b. B. I. Lundqvist, *Phys. Stat. Sol.* **32**, 273 (1969).
66. M. Pessa and W. R. Newell, *Physica Scripta* **3**, 165 (1971).
67. H. S. W. Massey and E. H. S. Burhop, *Electron and Ionic Impact Phenomena*, Vol. I, Oxford Univ. Press, London (1969).
68. G. Glupe and W. Mehlhorn, *Phys. Letters* **25A**, 274 (1967).
69. G. Peach, *J. Phys. B* **4**, 1670 (1971).
70. a. R. L. Gerlach and A. R. DuCharme, *Surface Sci.* **32**, 317 (1972).
 b. R. L. Gerlach and A. R. DuCharme, *Surface Sci.* **32**, 329 (1972).
71. T. E. Gallon, *J. Phys. D* **5**, 822 (1972).
72. W. E. Maddeman, T. A. Carlson, M. O. Krause, B. P. Pullen, W. E. Bull, and G. K. Schweitzer, *J. Chem. Phys.* **55**, 2317 (1971).
73. A. Fahlman, K. Hamrin, G. Axelson, C. Nordling, and K. Siegbahn, *Z. Physik* **192**, 484 (1966).
74. D. H. Hensler, *J. Appl. Opt.* **11**, 2522 (1972).
75. P. Mark, *Surface Sci.* **25**, 192 (1971).
76. A. C. M. Chen, J. F. Norton, and J. M. Wang, *Appl. Phys. Letters* **18**, 443 (1971).
77. K. Muller and C. C. Chang, *Surface Sci.* **14**, 39 (1969).
78. D. V. McCaughan and V. T. Murphy, *IEEE Trans. Nucl. Sci. N.S.* **19**, 749 (1972). See also this volume Chapter 22.
79. N. Nakayama, M. Ono, and H. Shimizu, *J. Vac. Sci. Technol.* **9**, 749 (1972).
80. C. C. Chang, P. Petroff, G. Quintana, and J. Sosniak, *Surface Science* **38**, 341 (1973).
81. J. R. Arthur, *J. Vac. Sci. Technol.* **10**, 136 (1972).
82. F. M. Propst and T. C. Piper, *J. Vac. Sci. Technol.* **4**, 53 (1967).
83. H. Ibach, *Phys. Rev. Letters* **27**, 253 (1971).
84. K. Maeda and T. Ihara, *Rev. Sci. Instr.* **42**, 1480 (1971).
85. R. L. Gerlach, *J. Vac. Sci. Technol.* **10**, 122 (1973).
86. H. D. Hagstrum, *Science* **178**, 275 (1972).
87. R. L. Park, J. E. Houston, and D. G. Schreiner, *Rev. Sci. Instr.* **41**, 1810 (1970).
88. J. E. Houston and R. L. Park, *Solid State Commun.* **10**, 91 (1972).
89. R. L. Gerlach, in *Electron Spectroscopy* (D. A. Shirley, ed.) p. 885, North-Holland, Amsterdam (1972).

11. SELECTED BIBLIOGRAPHY ON AUGER ELECTRON SPECTROSCOPY

This bibliography includes mostly publications on Auger spectroscopy applied to surface analysis. The entries are arranged by author, and cover the time period up to early 1972 (prepared by D. T. Hawkins of the Libraries and Information Systems Center of Bell Laboratories, Murray Hill, New Jersey).

001 EVIDENCE FOR A RADIATIVE AUGER EFFECT IN X-RAY PHOTON EMISSION.
ABERG T
PHYS REV LETT 22: 1346-8 (1969)

002 AUGER EFFECT IN THE MULTIPLE IONIZATION OF MANGANESE AND CADMIUM BY ELECTRON IMPACT.
ABOUAF R
J PHYS (PARIS) 31: 277-83 (1970) (IN FRENCH)

003 MULTIPLE IONIZATION AND AUGER TRANSITIONS IN INDIUM AND SILVER BY ELECTRON IMPACT.
ABOUAF R
J PHYS (PARIS) 32: 603-8 (1971) (IN FRENCH)

004 HIGH RESOLUTION MNN AUGER SPECTRA OF SILVER, CADMIUM, INDIUM, ANTIMONY, TELLURIUM AND IODINE.
AKSELA S
Z PHYS 244: 268-74 (1971)

005 HIGH RESOLUTION LMM AUGER SPECTRA OF LOW ENERGY FROM SOLID SURFACES.
AKSELA S + PESSA M + KARRAS M
Z PHYS 237: 381-7 (1970)

006 KLL AUGER SPECTRUM OF FLUORINE. (IN FLUORIDE SALTS)
ALBRIDGE RG + HAMRIN K + JOHANSSON G + FAHLMAN A
Z PHYS 209: 419-27 (1968) (IN GERMAN)

007 BAND STRUCTURE OF SILICON BY CHARACTERISTIC AUGER SPECTRUM ANALYSIS.
AMELIO GF
SURFACE SCI 22: 301-18 (1970)

008 AUGER ELECTRON SPECTROSCOPY AND SECONDARY EMISSION IN SEMICONDUCTORS. (SILICON, GERMANIUM, GRAPHITE, ELECTRON DIFFUSION THEORY)
AMELIO GF
GEORGIA INST TECHNOL, THESIS 1968

009 TRUE SECONDARY ELECTRON ENERGY DISTRIBUTIONS.
AMELIO GF + SCHEIBNER EJ
P11-1 TO 11-15 OF STRUCTURE AND CHEMISTRY OF SOLID SURFACES, SOMORJAI GA (ED). NY WILEY 1969. (541.375/B51) 140258

010 AUGER SPECTROSCOPY OF GRAPHITE SINGLE CRYSTALS WITH LOW-ENERGY ELECTRONS.
AMELIO GF + SCHEIBNER EJ
SURFACE SCI 11: 242-54 (1968)

011 CLEAN TELLURIUM SURFACES STUDIED BY LEED.
ANDERSSON S + ANDERSSON D + MARKLUND I
SURFACE SCI 12: 284-98 (1968)

012 DEVELOPMENT AND APPLICATION OF AN "AUGER ELECTRON SPECTROMETER" FOR HIGH TEMPERATURE ADSORPTION STUDIES.
ARAMATI VS
MASS INST TECHNOL, MS THESIS 1971

013 ON THE COMPOUND PHOTOELECTRIC EFFECT.
AUGER P
J PHYS RADIUM 6: 205-8 (1925)

014 KLL AUGER ELECTRON SPECTRA OF THULIUM AND LUTETIUM.
BABENKOV MI + BOBYKIN BV
BULL ACAD SCI USSR, PHYS SER 32: 1840-3 (1968) (IN ENGLISH)

015 KLL AUGER ELECTRONS OF CADMIUM AND ANTIMONY.
BABENKOV MI + BOBYKIN BV + NORGOPODOV AF + TASKARIN BT
BULL ACAD SCI USSR, PHYS SER 33: 1205-9 (1969) (IN ENGLISH)

016 NATURE OF ANNEALED SEMICONDUCTOR SURFACES. (AUGER)
BAUER E
PHYS LETT A26: 530-1 (1968)

017 COMMENTS ON "ANGULAR DEPENDENCIES IN ELECTRON-EXCITED AUGER EMISSION".
BISHOP HE + RIVIERE JC
SURFACE SCI 17: 446-7 (1969)

018 CHARACTERISTIC IONIZATION LOSSES OBSERVED IN AUGER EMISSION SPECTROSCOPY.
BISHOP HE + RIVIERE JC
APPL PHYS LETT 16: 21-3 (1970)

019 AUGER SPECTROSCOPY OF TITANIUM.
BISHOP HE + RIVIERE JC
SURFACE SCI 24: 1-17 (1971)

020 SURFACE ABSORPTION BY POLYPEPTIDE FILMS EXAMINED BY AUGER EMISSION SPECTROSCOPY.
BISHOP HE + RIVIERE JC
J CCLLOID INTERFACE SCI 33: 272-7 (1970)

021 SEGREGATION OF GOLD TO SILICON (111) SURFACE OBSERVED BY AUGER EMISSION SPECTROSCOPY AND BY LEED.
BISHOP HE + RIVIERE JC
J PHYS D 2: 1635-42 (1969)

022 ESTIMATES OF EFFICIENCIES OF PRODUCTION AND DETECTION OF ELECTRON-EXCITED AUGER EMISSION.
BISHOP HE + RIVIERE JC
J APPL PHYS 40: 1740-4 (1969)

023 SURFACE SEGREGATION IN BORON DOPED IRON OBSERVED BY AUGER EMISSION SPECTROSCOPY.
BISHOP HE + RIVIERE JC
ACTA MET 18: 813-17 (1970)

024 AUGER SPECTROSCOPY OF SILICON.
BISHOP HE + RIVIERE JC + TAYLOR NJ
SURFACE SCI 17: 462-5 (1969)

025 SEGREGATION OF CARBON TO (100) SURFACE OF NICKEL.
BLAKELY JM + KIM JS + POTTER HC
J APPL PHYS 41: 2693-7 (1970)

026 AUGER ELECTRON SPECTROSCOPY OF A SULFUR- OXYGEN SURFACE REACTION ON A COPPER (110) CRYSTAL.
BONZEL HP
SURFACE SCI 27(3): 387-410 (1971)

027 DETECTION OF IMPURITIES ON COPPER SURFACES BY LEED AUGER ELECTRON SPECTROSCOPY AND THEIR INFLUENCE ON SURFACE SELF DIFFUSION OF COPPER.
BONZEL HP + GJOSTEIN NA
J METALS 21(3): A88 (1969)

028 ELECTRON SPECTROSCOPY OF SURFACES.
BRUNDLE CR
SURFACE SCI 27(3): 681-5 (1971)

029 THE AUGER EFFECT AND OTHER RADIATIONLESS TRANSITIONS.
BURHOP EHS
CAMBRIDGE UNIVERSITY PRESS, 1952. (539.76/B95) 110732 188PP

030 AUGER ELECTRON STUDIES OF SURFACES: URANIUM DIOXIDE, URANIUM, GRAPHITE, 300 SERIES STAINLESS STEEL, AND NIOBIUM.
CAMPBELL BD
LOS ALAMOS SCI LABS 1969. REPT NO LA-4010

031 AUGER ELECTRON STUDIES OF URANIUM DIOXIDE SURFACES.
CAMPBELL BD + ELLIS WP
J CHEM PHYS 52: 3303-4 (1970)

032 IDENTIFICATION OF HIGH ENERGY LINES IN THE KLL AUGER SPECTRUM OF NITROGEN.
CARLSON TA + MODDEMAN WE + PULLEN BP + KRAUSE MO
CHEM PHYS LETT 5: 390-2 (1970)

033 HIGH RESOLUTION ELECTRON SPECTROMETER FOR PHOTOELECTRON AND AUGER ELECTRON STUDIES.
CARLSON TA + PULLEN BP + MODDEMAN WE + KRAUSE MO + WARD FW
ACS ANN MEET 1969, 158TH PROC ABSTR PAPERS NO PHYS-179

034 STUDY OF MUSCOVITE AND SILICON BY AUGER ELECTRON SPECTROSCOPY.
CARRIERE B + DEVILLE JP + GOLDSZTAUB S
COMPT REND 271: 796-8 (1970) (IN FRENCH)

035 AUGER ELECTRON SPECTROSCOPY STUDY OF SOME OXYGENATED SILICON COMPOUNDS.
CARRIERE B + DEVILLE JP + GOLDSZTAUB S
COMPT REND B272: 951-4 (1971) (IN FRENCH)

036 CONTAMINANTS ON CHEMICALLY ETCHED SILICON SURFACES LEED- AUGER METHOD.
CHANG CC
SURFACE SCI 23: 283-98 (1970)

037 AUGER ELECTRON SPECTROSCOPY FOR CHEMICAL ANALYSIS.
CHANG CC
ACS ANN MEET 1970, 160TH PROC ABSTR PAPERS NO PHYS-7.

038 AUGER ELECTRON SPECTROSCOPY.
(REVIEW ARTICLE)
CHANG CC
SURFACE SCI 25: 53-79 (1971)

039 SILICON ON SAPPHIRE EPITAXY BY VACUUM SUBLIMATION - LEED- AUGER STUDIES AND ELECTRONIC PROPERTIES OF FILMS.
CHANG CC
J VACUUM SCI TECHNOL 8: 500-11 (1971)

040 CARBON CONTAMINATION OF SILICON (111) SURFACES.
CHARIG JM + SKINNER DK
SURFACE SCI 15: 277-85 (1969)

041 AUGER ELECTRON SPECTROSCOPY OF NICKEL DEPOSITS ON THE SILICON (111) SURFACE.
CHARIG JM + SKINNER DK
SURFACE SCI 19: 283-90 (1970)

042 LEED, AUGER, AND WORK FUNCTION STUDIES OF CLEAN AND SODIUM- COVERED SURFACES OF GALLIUM ARSENIDE.
CHEN JM
SURFACE SCI 25: 305-14 (1971)

043 DITUNGSTEN CARBIDE OVERLAYER ON TUNGSTEN (112).
CHEN JM + PAPGEORGOPOULOS CA
SURFACE SCI 20: 195-200 (1970)

044 THEORETICAL L(2)- AND L(3)- SUBSHELL FLUORESCENCE YIELDS AND L(2)- L(3)X COSTER-KRONIG TRANSITION PROBABILITIES.
CHEN MH + CRASEMANN B + KOSTROUN VO
PHYS REV A 4: 1-7 (1971)

045 LEED AND SURFACE POTENTIAL STUDY OF CARBON MONOXIDE AND XENON ADSORBED ON COPPER (100).
CHESTERS MA + PRITCHARD J
SURFACE SCI 28: 460-8 (1971)

046 CHARACTERISTIC ENERGIES IN SECONDARY ELECTRON SPECTRA FROM SILICON (111) SURFACES.
CHUNG MF + JENKINS LH
SURFACE SCI 26: 649-63 (1971)

047 LLM VERSUS LMM AUGER TRANSITIONS FOR THE LIGHT ELEMENTS.
CHUNG MF + JENKINS LH
SURFACE SCI 28: 637-44 (1971)

048 AUGER ELECTRON SPECTROSCOPY OF THE OUTER SHELL ELECTRONS.
CHUNG MF + JENKINS LH
SURFACE SCI 22: 479-85 (1970)

049 AUGER ELECTRON EMISSION SPECTRA: A SIMPLE TREATMENT OF ENERGIES AND INTENSITIES.
CLARKE TA + MASON R + RANDACCIO L + THOMAS JM
J CHEM SOC LONDON, PART A, INORG, PHYS, THEOR 9: 1156-60 (1971)

050 APPLICATION OF AUGER ELECTRON SPECTROSCOPY TO REACTOR MATERIALS RESEARCH.
CLAUSING RE
J VACUUM SCI TECHNOL 7: 5124 (1970)

051 KLL AUGER SPECTRUM OF CHLORINE. (IN CARBON CHLORINE(4))
CLEFF B + MEHLHORN W
Z PHYS 219: 311-24 (1969) (IN GERMAN)

052 CHEMICAL SHIFTS IN THE AUGER SPECTRA FROM OXIDIZED CHROMIUM AND VANADIUM.
COAD JP
PHYS LETT A35: 185-6 (1971)

053 LMM AUGER SPECTRA OF SOME TRANSITION METALS OF THE FIRST SERIES.
COAD JP
Z PHYS 244: 19-30 (1971)

054 ELECTRON-BEAM ASSISTED ADSORPTION ON SILICON (111) SURFACE.
COAD JP + BISHOP HE + RIVIERE JC
SURFACE SCI 21: 253-64 (1970)

055 COMBINATION OF AUGER SPECTROSCOPY AND CHARACTERISTIC LOSS SPECTROSCOPY FOR THE ELEMENTS VANADIUM TO COBALT.
COAD JP + RIVIERE JC
PHYS STATUS SOLIDI A 7: 571-5 (1971)

056 AUGER SPECTROSCOPY OF CARBON ON NICKEL.
COAD JP + RIVIERE JC
SURFACE SCI 25: 609-24 (1971)

057 STUDY OF MULBERRY SURFACES BY AUGER AND IEE SPECTROSCOPY.
COLMENARES CA
UNIV CALIF, LAWRENCE RADIATION LAB 1970. REPT UCID-15599

058 AUGER ELECTRON SPECTROSCOPY OF GRAPHITE FIBER SURFACES.
CONNELL GL
NATURE 230: 377 (1971)

059 AUGER ELECTRON SPECTROSCOPY OF LUNAR MATERIAL.
CONNELL GL + GUPTA YP
2ND NASA LUNAR SCIENCE CONF (1971) PROC

060 AUGER ELECTRON SPECTROSCOPY.
CONNELL GL + GUPTA YP
MATER RES STAND 11(1): 8-13 (1971)

061 AUGER AND COSTER-KRONIG TRANSITION PROBABILITIES TO THE ATOMIC 2S STATE AND THEORETICAL L(1) FLUORESCENCE YIELDS.
CRASEMANN B + CHEN MH + KOSTROUN VO
PHYS REV A 4: 2161-3 (1971)

062 COMMENTS ON "AUGER SPECTROSCOPY OF CARBON ON NICKEL" BY JP COAD AND JC RIVIERE.
DALMAI-IMELIK G + BERTOLINI JC + ROUSSEAU J
SURFACE SCI 27: 379 (1971)

063 ADSORBATE EFFECTS IN ELECTRON EJECTION BY RARE GAS METASTABLE ATOMS.
DELCHAR TA + MACLENNAN DA + LANDERS AM
J CHEM PHYS 50: 1779-87 (1969)

064 SPECTROSCOPY OF AUGER ELECTRONS.
REVIEW, 24 REFERENCES
DEVILLE JP
REV PHYS APPL, SUPPL J PHYS 3: 351-5 (1968) (IN FRENCH)

065 MUSCOVITE CLEAVAGE STUDIED BY AUGER ELECTRON SPECTROSCOPY.
DEVILLE JP + GOLDSZTAUB S
COMPT REND B268: 629-30 (1969) (IN FRENCH)

066 AUGER ELECTRON SPECTROSCOPY STUDIES OF REFRACTORY METAL SURFACES.
DOOLEY GJ + GRANT JT + HAAS TW
BULL AMER PHYS SOC 14: 948 (1969)

067 CHEMICAL EFFECTS ON KLL AUGER ELECTRON SPECTRUM FROM OXYGEN.
DOOLEY GJ + GRANT JT + HAAS TW
BULL AMER PHYS SOC 15: 1507 (1970)

068 BEHAVIOR OF REFRACTORY METAL SURFACES IN ULTRAHIGH VACUUM AS OBSERVED BY LEED AND AUGER ELECTRON SPECTROSCOPY.
DOOLEY GJ + HAAS TW
J VACUUM SCI TECHNCL 7: S90-100 (1970)

069 AUGER ELECTRON SPECTROSCOPY: METALLURGICAL APPLICATIONS.
DOOLEY GJ + HAAS TW
J METALS 22(11): 17-24 (1970)

070 LEED STUDY OF MOLYBDENUM (100) SURFACE.
DOOLEY GJ + HAAS TW
BULL AMER PHYS SOC 14: 270 (1969)

071 COMPARATIVE STUDY OF LEED INTENSITY DATA.
DOOLEY GJ + HAAS TW
BULL AMER PHYS SOC 15: 632 (1970)

072 CHEMISORPTION ON SINGLE CRYSTAL MOLYBDENUM (112) SURFACES.
DOOLEY GJ + HAAS TW
J VACUUM SCI TECHNCL 7: 49-52 (1970)

073 SOME PROPERTIES OF THE RHENIUM (0001) SURFACE.
DOOLEY GJ + HAAS TW
SURFACE SCI 19: 1-8 (1970)

074 FURTHER STUDIES OF GAS ADSORPTION ON MOLYBDENUM (100) SURFACE.
DOOLEY GJ + HAAS TW
J CHEM PHYS 52: 461-2 (1970)

075 AUGER ELECTRON ANALYSIS OF ELECTROPOLISHED HIGH-PURITY ALUMINUM.
DUNN CG + HARRIS LA
J ELECTROCHEM SOC 117: 81-2 (1970)

076 SECONDARY- ELECTRON ENERGY DISTRIBUTION STUDIES OF URANIUM DIOXIDE SURFACES.
ELLIS WP + CAMPBELL BD
J APPL PHYS 41: 1858-61 (1970)

077 ADSORPTION ON SINGLE CRYSTAL SURFACES OF COPPER- NICKEL ALLOYS. I.
ERTL G + KUPPERS J
SURFACE SCI 24: 104-24 (1971) (IN GERMAN)

078 THE INTERACTION BETWEEN CHLORINE AND THE (100) SURFACE OF GOLD.
FEDAK DG + FLORIO JV + ROBERTSON WD
PP 74-1 TO 74-18 OF STRUCTURE AND CHEMISTRY OF SOLID SURFACES,
SOMORJAI GA (ED). NY, WILEY, 1969. (541.375/B51) 140258

079 AUGER SPECTROSCOPY AND LEED STUDY OF EQUILIBRIUM SURFACE SEGREGATION IN COPPER- ALUMINUM ALLOYS.
FERRANTE J
ACTA MET 19: 743-8 (1971)

080 PHASE TRANSFORMATIONS OF SILICON (111) SURFACE.
FLORIO JV + ROBERTSON WD
SURFACE SCI 22: 459-64 (1970)

081 CHLORINE REACTIONS ON THE SILICON (111) SURFACE.
FLORIO JV + ROBERTSON WD
SURFACE SCI 18: 398-427 (1970)

082 CHEMICAL EFFECTS ON THE AUGER ELECTRON SPECTRA OF BERYLLIUM.
FORTNER RJ + MUSKET RG
SURFACE SCI 28: 339-43 (1971)

083 SIMPLE MODEL FOR DEPENDENCE OF AUGER INTENSITIES ON SPECIMEN THICKNESS.
GALLON TE
SURFACE SCI 17: 486-9 (1969)

084 (100) SURFACES OF ALKALI HALIDES PART-1: AIR AND VACUUM CLEANED SURFACES.
GALLON TE + HIGGINBOTHAM IG + PRUTTON M + TOKUTAKA H
SURFACE SCI 21: 224-32 (1970)

085 GROWTH OF SILVER ON POTASSIUM CHLORIDE OBSERVED BY LEED AND AUGER EMISSION SPECTROSCOPY.
GALLON TE + HIGGINBOTHAM IG + PRUTTON M + TOKUTAKA H
THIN SOLID FILMS 2: 369-73 (1968)

086 IMPROVED APPARATUS FOR MEASUREMENT OF AUGER ELECTRON SPECTRA.
GALLON TE + HIGGINBOTHAM IG + PRUTTON M
J PHYS E 2: 894-6 (1969)

087 LOW ENERGY AUGER EMISSION FROM LITHIUM FLUORIDE.
GALLON TE + MATTHEW JAD
PHYS STATUS SOLIDI 41: 343-51 (1970)

088 IONIZATION SPECTROSCOPY OF CONTAMINATED METAL SURFACES.
GERLACH RL
J VACUUM SCI TECHNCL 8: 599-604 (1971)

089 ELECTRON BINDING ENERGIES OF BARIUM FROM THE SECONDARY ELECTRON YIELD SPECTRUM.
GERLACH RL
SURFACE SCI 28: 648-50 (1971)

090 DIFFERENTIAL CROSS SECTIONS FOR K- SHELL IONIZATION OF SURFACE ATOMS BY ELECTRON IMPACT.
GERLACH RL + DUCHARME AR
PHYS REV LETT 27: 290-2 (1971)

091 IONIZATION SPECTROSCOPY OF SURFACES.
GERLACH RL + HOUSTON JE + PARK RL
APPL PHYS LETT 16: 179-81 (1970)

092 IONIZATION SPECTROMETER FOR ELEMENTAL ANALYSIS OF SURFACES.
GERLACH RL + TIPPING DW
REV SCI INSTRUM 42: 151-4 (1971)

093 MODERN TECHNIQUES FOR SURFACE STUDIES.
GJOSTEIN NA + BONZEL HP + CHAUKA NG
RES DEVELOP 21(10): 24-30 (1970)

094 DETECTION OF IMPURITIES ON A SILICON (111) SURFACE BY AUGER- LEED ANALYSIS.
GOFF RF + JACOBSON RL
BULL AMER PHYS SOC 13: 944 (1968)

095 SENSITIVITY VARIATIONS IN BAYARD- ALPERT GAUGES CAUSED BY AUGER EMISSION AT COLLECTOR.
GOPALARAMAN CP + ARMSTRONG RA + REDHEAD PA
J VACUUM SCI TECHNCL 6: 910 (1969)

096 LEED STUDY OF IRIDIUM (100) SURFACE.
GRANT JT
SURFACE SCI 18: 228-38 (1969)

097 STUDY OF IRIDIUM (100) SURFACE USING LEED AND AUGER ELECTRON SPECTROSCOPY.
GRANT JT
BULL AMER PHYS SOC 14: 794 (1969)

098 STUDIES ON THE IRIDIUM (111) SURFACE USING LEED AND AUGER ELECTRON SPECTROSCOPY.
GRANT JT
SURFACE SCI 25: 451-6 (1971)

099 STRUCTURE OF THE PLATINUM (100) SURFACE.
GRANT JT + HAAS TW
SURFACE SCI 18: 457-61 (1969)

100 AUGER STUDIES OF (111) SILICON SURFACES.
(ABSTRACT)
GRANT JT + HAAS TW
J VACUUM SCI TECHNCL 6: 903 (1969)

101 AUGER ELECTRON SPECTROSCOPY STUDIES OF CARBON OVERLAYERS ON METAL SURFACES.
GRANT JT + HAAS TW
SURFACE SCI 24: 332-4 (1971)

102 AUGER ELECTRON SPECTROSCOPY OF SILICON.
GRANT JT + HAAS TW
SURFACE SCI 23: 347-62 (1970)

103 LEED STUDY OF PLATINUM (100) SURFACE.
GRANT JT + HAAS TW
BULL AMER PHYS SCC 14: 948 (1969)

104 IDENTIFICATION OF THE FORM OF CARBON AT A SILICON (100) SURFACE USING AUGER ELECTRON SPECTROSCOPY.
GRANT JT + HAAS TW
PHYS LETT A33: 386-7 (1970)

105 COMBINED LEED AND AUGER ELECTRON SPECTROSCOPY STUDIES OF SILICON, GERMANIUM, GALLIUM ARSENIDE, AND INDIUM ANTIMONIDE SURFACES.
GRANT JT + HAAS TW
J VACUUM SCI TECHNOL 8(1): 94-7 (1971)

106 STUDY OF RUTHENIUM (0001) AND RHODIUM (111) SURFACES USING LEED AND AUGER ELECTRON SPECTROSCOPY.
GRANT JT + HAAS TW
SURFACE SCI 21: 76-85 (1970)

107 STUDY OF INDIUM ARSENIDE (111) AND (-1-1-1) SURFACES USING LEED AND AUGER ELECTRON SPECTROSCOPY.
GRANT JT + HAAS TW
SURFACE SCI 26(2): 669-76 (1971)

108 AUGER STUDIES OF CLEAVED SILICON (111) SURFACES.
GRANT JT + HAAS TW
J VACUUM SCI TECHNOL 7: 77-9 (1970)

109 NATURE OF SILICON (111) SURFACES.
GRANT JT + HAAS TW
APPL PHYS LETT 15: 140-1 (1969)

110 A BIBLIOGRAPHY OF LOW ENERGY ELECTRON DIFFRACTION AND AUGER ELECTRON SPECTROSCOPY.
(766 REFERENCES)
HAAS TW + DOOLEY GJ + GRANT JT + JACKSON AG + HOOKER MP
PROGRESS IN SURFACE SCIENCE, VOL. 1(2). PERGAMON, 1971.

111 CHEMICAL SHIFTS IN AUGER ELECTRON SPECTROSCOPY FROM INITIAL OXIDATION OF TANTALUM (110).
HAAS TW + GRANT JT
PHYS LETT A30: 272 (1969)

112 CHEMICAL EFFECTS ON THE KLL AUGER ELECTRON SPECTRUM FROM SURFACE CARBON.
HAAS TW + GRANT JT
APPL PHYS LETT 16: 172-3 (1970)

113 CHEMICAL SHIFTS IN AUGER ELECTRON SPECTROSCOPY.
HAAS TW + GRANT JT + DOOLEY GJ
BULL AMER PHYS SOC 14: 948 (1969)

114 AUGER ELECTRON SPECTROSCOPY OF TRANSITION METALS.
HAAS TW + GRANT JT + DOOLEY GJ
PHYS REV B 1: 1449-59 (1970)

115 AUGER ELECTRON SPECTROSCOPY OF SOME REFRACTORY METALS.
HAAS TW + GRANT JT + DOOLEY GJ
J VACUUM SCI TECHNOL 6: 903 (1969)

116 SOME PROBLEMS IN ANALYSIS OF AUGER ELECTRON SPECTRA.
HAAS TW + GRANT JT + DOOLEY GJ
J VACUUM SCI TECHNOL 7: 43-5 (1970)

117 SOFT X-RAY APPEARANCE POTENTIAL SPECTROSCOPY IN A DISPLAY LEED SYSTEM.
HAAS TW + THOMAS S + DOOLEY GJ
SURFACE SCI 28: 645-7 (1971)

118 ANGULAR DEPENDENCIES OF ELECTRON-EXCITED AUGER EMISSION.
HARRIS LA
SURFACE SCI 15: 77-93 (1969)

119 ANALYSIS OF MATERIALS BY ELECTRON-EXCITED AUGER ELECTRONS.
HARRIS LA
J APPL PHYS 39: 1419-27 (1968)

120 CARBON EVAPORATION FROM A THORIUM DISPENSER CATHODE OBSERVED BY AUGER ELECTRON EMISSION.
HARRIS LA
J APPL PHYS 39: 4862 (1968)

121 AUGER ELECTRON SPECTROSCOPY FOR SURFACE ANALYSIS.
HARRIS LA
J METALS 20(12): A16 (1968)

122 AUGER ELECTRON EMISSION ANALYSIS.
(GOOD "LAYMAN'S" INTRODUCTORY ARTICLE.)
HARRIS LA
ANAL CHEM A40(14): 24-34 (1968)

123 SECONDARY ELECTRON SPECTROSCOPY.
HARRIS LA
IND RES 10: 52 (1968)

124 REPLY TO COMMENTS OF HE BISHOP AND JC RIVIERE ON "ANGULAR DEPENDENCIES IN ELECTRON-EXCITED AUGER EMISSION."
HARRIS LA
SURFACE SCI 17: 448-9 (1969)

125 SOME OBSERVATIONS OF SURFACE SEGREGATION BY AUGER ELECTRON EMISSION.
HARRIS LA
J APPL PHYS 39: 1428-31 (1968)

126 SURFACE ANALYSIS BY AUGER ELECTRON SPECTROSCOPY.
HARRIS LA
J ELECTROCHEM SOC 115: C250 (1968)

127 SECONDARY ELECTRON SPECTROSCOPY. (SURFACE ANALYSIS OF MATERIALS BY ELECTRON-EXCITED AUGER ELECTRON EMISSION)
HARRIS LA
APPL SPECTROSC 22: 372 (1968)

128 AUGER ELECTRON EMISSION IN THE ENERGY SPECTRA OF SECONDARY ELECTRONS FROM MOLYBDENUM AND TUNGSTEN.
HARROWER GA
PHYS REV 102: 340-7 (1956)

129 IN-PROCESS CONTROL TECHNIQUES FOR COMPLEX SEMICONDUCTOR STRUCTURES. TASK-II. APPLICATION OF SECONDARY ELECTRON SPECTROSCOPY (AUGER ELECTRON ANALYSIS) TO SURFACE CONTROL PROGRAM IN THE MANUFACTURING OF SILICON DEVICES.
HARTMANN DK + HARRIS LA + AFFLECK JH
GENERAL ELECTRIC CO. · SEMICONDUCTOR PRODUCT DEPT.
PART I REPT NO AD 845596 (1968)
PART II REPT NO AD 845597 (1968)
PART III REPT NO AD 847775 (1969)
PART IV REPT NO AD 852179 (1969)
PART V REPT NO AD 856920 (1969)
PART VI REPT NO AD 861921 (1969)

130 METHOD FOR PACKET OF WAVES IN AUGER ELECTRON-ELECTRON PROCESSES.
HAYMANN P
COMPT REND B272: 1029-32 (1971)

131 EMISSION OF AUGER ELECTRONS BY ATOMS IN A METALLIC TARGET SUBJECTED TO AN IONIC BOMBARDMENT.
HENNEQUIN JF
J PHYS (PARIS) 29: 1053-65 (1968) (IN FRENCH)

132 LOW ENERGY PHOTO- AUGER ELECTRON SPECTROSCOPY.
HENKE BL
APPL SPECTROSC 22: 372 (1968)

133 AUGER EXCITATION BY INTERNAL SECONDARY ELECTRONS.
HOUSTON JE + PARK RL
APPL PHYS LETT 14: 358-60 (1969)

134 CROSS CORRELATION TECHNIQUES IN AUGER SPECTROSCOPY.
HOUSTON JE + PARK RL
BULL AMER PHYS SOC 14: 793 (1969)

135 HIGH SENSITIVITY ELECTRON SPECTROMETER.
HUCHITAL DA + RIGDEN JD
APPL PHYS LETT 16: 348-51 (1970)

136 EPITAXY OF ULTRATHIN METAL FILMS ON BODY CENTERED CUBIC SUBSTRATES USING LEED- AUGER TECHNIQUES.
(ABSTRACT)
JACKSON AG
J VACUUM SCI TECHNOL 8: 23 (1971)

137 AUGER -LEED INVESTIGATION OF THE DEPOSITION OF ALUMINUM ONTO THE MOLYBDENUM (110) SURFACE.
JACKSON AG + HOOKER MP
SURFACE SCI 28: 373-94 (1971)

138 AUGER -LEED INVESTIGATION OF TIN ON MOLYBDENUM (100).
JACKSON AG + HOOKER MP
SURFACE SCI 27: 197-210 (1971)

139 AUGER ELECTRON EMISSION OF THIN CARBON FOILS IN REFLECTION AND TRANSMISSION.
JACOBI K
SURFACE SCI 26: 54 (1971) (IN GERMAN)

140 AUGER AND OTHER CHARACTERISTIC ENERGIES IN SECONDARY ELECTRON SPECTRA FROM ALUMINUM SURFACES.
JENKINS LH + CHUNG MF
SURFACE SCI 28: 409-22 (1971)

141 ENERGY SPECTRUM OF BACK-SCATTERED ELECTRONS AND CHARACTERISTIC LOSS AND GAIN PHENOMENA OF COPPER (111).
JENKINS LH + CHUNG MF
SURFACE SCI 26: 151-64 (1971)

142 LEED AND AUGER INVESTIGATIONS OF COPPER (111) SURFACE.
JENKINS LH + CHUNG MF
SURFACE SCI 24: 125-39 (1971)

143 LOW-ENERGY ELECTRON DIFFRACTION, CIRCA 1968.
(REVIEW OF LEED AND AUGER.)
JONA F
HELV PHYS ACTA 41: 960-4 (1968)

144 AUGER SPECTROSCOPIC ANALYSIS OF BISMUTH SEGREGATED TO GRAIN BOUNDARIES IN COPPER.
JOSHI A + STEIN DF
J INST METALS 99: 178-81 (1971)

145 INTERANGULAR BRITTLENESS STUDIES IN TUNGSTEN USING AUGER SPECTROSCOPY.
JOSHI A + STEIN DF
MET TRANS 1: 2543-6 (1970)

146 SILICON- OXYGEN INTERACTIONS USING AUGER ELECTRON SPECTROSCOPY.
JOYCE BA + NEAVE JH
SURFACE SCI 27(3): 499-515 (1971)

147 THE INFLUENCE OF SUBSTRATE CONDITIONS ON THE NUCLEATION AND GROWTH OF EPITAXIAL SILICON FILMS.
JOYCE BA + NEAVE JH + WATTS BE
SURFACE SCI 15: 1-13 (1969)

148 THE ADSORPTION OF CARBON MONOXIDE ON COPPER (001) LEED AND AUGER EMISSION STUDIES.
JOYNER RW + MCKEE CS + ROBERTS MW
SURFACE SCI 26: 303-9 (1971)

149 THE INTERACTION OF HYDROGEN SULFIDE WITH COPPER (001)
JOYNER RW + MCKEE CS + ROBERTS MW
SURFACE SCI 27: 279-85 (1971)

150 ATOMIC RADIATION TRANSITION PROBABILITIES TO THE 1S STATE AND THEORETICAL K- SHELL FLUORESCENCE YIELDS.
KOSTROUN VO + CHEN MH + CRASEMANN B
PHYS REV A 3: 533-45 (1971)

151 OBSERVATIONS OF BETA- SILICON CARBIDE FORMATION ON RECONSTRUCTED SILICON SURFACES.
KRAUSE MO
PHYS STATUS SOLIDI A3: 899-906 (1970)

152 SECONDARY EMISSION AND CONTAMINATION OF METAL SURFACES.
KULOV SK + SHERTNEV LG
INSTRUM EXP TECH 4: 917-8 (1967)

153 ADSORPTION OF OXYGEN ON MOLYBDENUM (111): EFFECT OF TRACE IMPURITIES.
LAMBERT RM + LINNETT JW + SCHWARZ JA
SURFACE SCI 26: 572-86 (1971)

154 LEED- AUGER INVESTIGATION OF A STABLE CARBIDE OVERLAYER ON A PLATINUM (111) SURFACE.
LAMBERT RM + WEINBERG WH + COMRIE CM + LINNETT JW
SURFACE SCI 27: 653-8 (1971)

155 AUGER PEAKS IN THE ENERGY SPECTRA OF SECONDARY ELECTRONS FROM VARIOUS MATERIALS.
LANDER JJ
PHYS REV 91: 1382-7 (1953)

156 LEED- AUGER ANALYSIS OF THE BERYLLIUM (0001) SURFACE.
LE JEUNE EJ
J VACUUM SCI TECHNCL 8(1): 9 (1971)

157 CORRELATION OF ELECTRONIC, LEED, AND AUGER DIAGNOSTICS ON ZINC OXIDE SURFACES.
LEVINE JD + WILLIS A + BOTTOMS WR + MARK P
SURFACE SCI 29: 144-64 (1972)

158 KLL AUGER SPECTRUM OF MANGANESE.
LIU YY + ALBRIDGE RG
NUCL PHYS A92: 139-144 (1967)

159 IMPURITIES, INTERFACES, AND BRITTLE FRACTURE.
(GIVES SHORT REVIEW OF AUGER ELECTRON SPECTROSCOPY AS APPLIED TO INTERFACES.)
LOW JR
TRANS MET SOC AIME 245: 2481-94 (1969)

160 POTENTIAL MAPPING USING AUGER ELECTRON SPECTROSCOPY.
MACDONALD NC
ANN SCANNING ELECTRON MICROSCOPE SYMP 1970, 3RD PROC

161 AUGER ELECTRON SPECTROSCOPY IN SCANNING ELECTRON MICROSCOPY: POTENTIAL MEASUREMENTS.
MACDONALD NC
APPL PHYS LETT 16: 76-80 (1970)

162 AUGER ELECTRON SPECTROSCOPY FOR SCANNING ELECTRON MICROSCOPY.
MACDONALD NC
P89-96 OF ANN SCANNING ELECTRON MICROSCOPE SYMP 1971, 4TH PROC
PHYS ABSTR 74: 52003 (1971)

163 MICROSCOPIC AUGER ELECTRON ANALYSIS OF FRACTURE SURFACES.
MACDONALD NC + MARCUS HL + PALMBERG PW
ANN SCANNING ELECTRON MICROSCOPE SYMP 1970, 3RD PROC

164 AUGER EJECTION OF ELECTRONS FROM TUNGSTEN BY OXYGEN CHEMISORPTION.
MACLENNAN DA
BULL AMER PHYS SOC 13: 197 (1968)

165 ROLE OF WORK FUNCTION IN ELECTRON EJECTION BY METASTABLE ATOMS: HELIUM AND ARGON ON (111) AND (110) TUNGSTEN. (AUGER)
MACLENNAN DA + DELCHAR TA
J CHEM PHYS 50: 1772-8 (1969)

166 WORK FUNCTION EFFECTS ON AUGER ELECTRON EJECTION BY NOBLE GAS METASTABLE ATOMS.
MACLENNAN DA + DELCHAR TA
BULL AMER PHYS SOC 13: 197 (1968)

167 LOW ENERGY ELECTRON DIFFRACTION STUDY OF THE POLAR (111) SURFACES OF GALLIUM ARSENIDE AND GALLIUM ANTIMONIDE.
MACRAE AU
SURFACE SCI 4: 247-64 (1966)

168 AN ELECTRON DIFFRACTION STUDY OF CESIUM ADSORPTION ON TUNGSTEN.
MACRAE AU + MULLER K + LANDER JJ + MORRISON J
SURFACE SCI 15: 483-97 (1969)

169 ELECTRONIC AND LATTICE STRUCTURE OF CESIUM FILMS ADSORBED ON TUNGSTEN.
MACRAE AU + MULLER K + LANDER JJ + MORRISON J + PHILLIPS JC
PHYS REV LETT 22: 1048-51 (1969)

170 SCATTERING OF LOW-ENERGY ELECTRONS FROM A COPPER (111) SURFACE.
MARKLUND I + ANDERSSON S + MARTINSON J
ARKIV FYS 37: 127-39 (1967)

171 FRACTURE SURFACE ANALYSIS OF TEMPER EMBRITTLED STEEL BY AUGER ELECTRON SPECTROSCOPY.
MARCUS HL + PALMBERG PW
J METALS 211(3): A96 (1969)

172 AUGER FRACTURE SURFACE ANALYSIS OF A TEMPER EMBRITTLED 3340- STAINLESS STEEL.
MARCUS HL + PALMBERG PW
TRANS MET SOC AIME 245: 1664-6 (1969)

173 AUGER ELECTRON FINE PROFILES IN IONIC CRYSTALS.
MATTHEW JAD
PHYS LETT 32A: 261-2 (1970)

174 A TEMPERATURE DEPENDENT CONTRIBUTION TO AUGER ELECTRON ENERGY DISTRIBUTIONS.
MATTHEW JAD
SURFACE SCI 20: 183-6 (1970)

175 L- SHELL AUGER AND COSTER-KRONIG ELECTRON SPECTRA.
MCGUIRE EJ
PHYS REV A 3: 1801-10 (1971)

176 K- SHELL AUGER TRANSITION RATES AND FLUORESCENT YIELDS FOR ELEMENTS ARGON TO XENON.
MCGUIRE EJ
PHYS REV A 2: 273-8 (1970)

177 K- SHELL AUGER TRANSITION RATES AND FLUORESCENT YIELDS FOR ELEMENTS BERYLLIUM TO ARGON.
MCGUIRE EJ
PHYS REV 185: 1-6 (1969)

178 ATOMIC L- SHELL COSTER-KRONIG, AUGER, AND RADIATIVE RATES AND FLUORESCENCE YIELDS FOR SODIUM- THORIUM.
MCGUIRE EJ
PHYS REV A 3: 587-94 (1971)

179 LEED AND AUGER ELECTRON SPECTROSCOPY.
(REVIEW ARTICLE.)
MCKEE CS + ROBERTS MW
CHEM BRIT 6: 106-10 (1970)

180 ENERGY WIDTHS OF ROENTGEN LEVELS BY AUGER ELECTRON SPECTROSCOPY.
MEHLHORN W + STAHLHERM D + VERBEEK H
Z NATURFORSCH A23:287-94 (1968)

181 INELASTIC INTERACTIONS OF SLOW ELECTRONS WITH ABSORBED PARTICLES.
(SHORT SECTION ON AUGER SPECTROSCOPY)
MENZEL D
ANGEW CHEM INTERNAT ED 9: 255-66 (1970)

182 DETERMINATION OF KLL AUGER SPECTRA OF NITROGEN, OXYGEN, CARBON DIOXIDE, NITRIC OXIDE, WATER, AND CARBON MONOXIDE.
MODDEMAN WE + CARLSON TA + KRAUSE MO + PULLEN BP + BULL WE + SCHWEITZ G
J CHEM PHYS 55: 2317 (1971)

183 AUGER SPECTRA OF SIMPLE MOLECULES.
MODDEMAN WE + CARLSON TA + PULLEN BP + KRAUSE MO
ACS ANN MEET 1969, 158TH PROC ABSTR PAPER NO PHYS-178

184 APPLICATION OF TRIPLE GRID LEED SYSTEM TO AUGER SPECTRUM ANALYSES.
MORRISON J + LANDER JJ
J VACUUM SCI TECHNCL 6: 338-42 (1969)

185 THE ADSORPTION OF IONIC SALTS ON A TUNGSTEN (100) SURFACE.
MORRISON J + LANDER JJ
SURFACE SCI 18: 420-30 (1969)

186 EPITAXIAL GROWTH OF COPPER ON (110) SURFACE OF A TUNGSTEN SINGLE CRYSTAL STUDIED BY LEED, AUGER ELECTRON AND WORK FUNCTION TECHNIQUES.
MOSS ARL + BLOTT BH
SURFACE SCI 17: 240-61 (1969)

187 DECONVOLUTION TECHNIQUES IN AUGER ELECTRON SPECTROSCOPY.
MULARIE WM + PERIA WT
J VACUUM SCI TECHNOL 8: 90 (1971)
SURFACE SCI 26: 125 (1971)

188 INELASTIC EFFECTS IN AUGER ELECTRON SPECTROSCOPY.
MULARIE WM + RUSCH TW
SURFACE SCI 19: 469-74 (1970)

189 CESIUM ADSORPTION ON TUNGSTEN STUDIED BY LEED AND SECONDARY ELECTRON SPECTROSCOPY.
MULLER K
P1-5 OF SURFACE PHENOMENA OF THERMIONIC EMITTERS, 1969

190 AUGER ELECTRONS, INDICATORS IN ANALYSIS OF SOLID BODY SURFACES.
MULLER K
MIKROCHIM ACTA SUPPL 4: 1-9 (1969)

191 ROOM TEMPERATURE ADSORPTION OF OXYGEN ON TUNGSTEN SURFACES.
(REVIEW OF METHODS OF MEASUREMENT)
MUSKET RG
J LESS COMMON METALS 22: 175-91 (1970)

192 OBSERVATION AND INTERPRETATION OF THE AUGER ELECTRON SPECTRUM FROM CLEAN BERYLLIUM.
MUSKET RG + FARTNER RJ
PHYS REV LETT 26: 80-2 (1971)

193 AUGER ELECTRON SPECTROSCOPY STUDY OF ELECTRON IMPACT DESORPTION.
MUSKET RG + FERRANTE J
SURFACE SCI 21: 440-2 (1970)

194 AUGER ELECTRON SPECTROSCOPY STUDY OF OXYGEN ADSORPTION ON TUNGSTEN (110).
MUSKET RG + FERRANTE J
J VACUUM SCI TECHNCL 7: 14-7 (1970)

195 AUGER ELECTRON EMISSION FROM GOLD DEPOSITED ON SILICON (111) SURFACE.
NARUSAWA T
JAP J APPL PHYS 10: 280-1 (1971)

196 SINGULARITIES IN AUGER EMISSION SPECTRUM OF METALS.
NATTA M + JOYES P
J PHYS CHEM SOLIDS 31: 447-52 (1970)

197 EFFECT OF THE POLARITY OF INDIUM ANTIMONIDE ON THE EXTERNAL PHOTOEFFECT IN THE X-RAY SPECTRAL REGION.
(EFFECT OF IMPURITIES ON THE AUGER SPECTRUM.)
NIKOLAENYA AZ + NEKRASHEVICH IG + SEMERENKO VV
IZV VYSSH UCHEB ZAVED, FIZ 12(3):68-73 (1969)
(FOR TRANSLATION, SEE SOV PHYS J.)

198 AUGER SPECTROSCOPY ON COPPER- NICKEL ALLOY SURFACES RELATED TO CATALYSIS.
ONO M + TAKASU Y + NAKAYAMA K + YAMASHINA T
SURFACE SCI 26: 313-6 (1971)

199 AUGER SPECTROSCOPY FOR THE CHEMICAL ANALYSIS OF SURFACES.
OSWALD RC
UNIV OF MINNESOTA, ELECTRICAL ENG. DEPT. MS THESIS (1969)

200 PHYSICAL ADSORPTION OF XENON ON PALLADIUM (100).
PALMBERG PW
SURFACE SCI 25: 598-608 (1971)

201 CHEMICAL ANALYSIS OF PLATINUM (100) AND GOLD (100) SURFACES BY AUGER ELECTRON SPECTROSCOPY.
PALMBERG PW
P29-1 TO 29-18 OF STRUCTURE AND CHEMISTRY OF SOLID SURFACES,
SOMORJAI GA (ED) NY, JOHN WILEY 1969. (541.375/B51) 140258

202 TECHNIQUE AND APPLICATIONS OF AUGER ELECTRON SPECTROSCOPY.
PALMBERG PW
J VACUUM SCI TECHNCL 7: 76 (1970); 6: 903 (1969)

203 OPTIMIZATION OF AUGER ELECTRON SPECTROSCOPY IN LEED SYSTEMS.
PALMBERG PW
APPL PHYS LETT 13: 183-5 (1968)

204 AUGER ELECTRON SPECTROSCOPY IN LEEC SYSTEMS.
PALMBERG PW
PP 29-1 TO 29-18 OF STRUCTURE AND CHEMISTRY OF SOLID SURFACES,
SOMORJAI GA (ED), NY WILEY, 1969. (5419375/B51) 140258

205 STRUCTURE TRANSFORMATIONS ON CLEAVEC AND ANNEALED GERMANIUM (111) SURFACES.
PALMBERG PW
SURFACE SCI 11: 153-8 (1958)

206 SECCNDARY EMISSION STUDIES ON GERMANIUM AND SODIUM- COVERED GERMANIUM.
PALMBERG PW
J APPL PHYS 38: 2137-47 (1967)

207 HIGH SENSITIVITY AUGER ELECTRON SPECTROMETER.
PALMBERG PW + BOHN GK + TRACY JC
APPL PHYS LETT 15: 254-5 (1969)

208 AUGER SPECTROSCOPIC ANALYSIS OF GRAIN BOUNDARY SEGREGATION.
PALMBERG PW + MARCUS HL
TRANS AMER SOC METALS 62: 1016-8 (1969)

209 ATCMIC ARRANGEMENT OF GOLD (100) AND RELATED METAL OVERLAYER SURFACE STRUCTURES.
PALMBERG PW + RHODIN TN
J CHEM PHYS 49: 134-46 (1968)

210 AUGER ELECTRON SPECTROSCOPY OF FACE CENTERED CUBIC METAL SURFACES. (GOLD, SILVER, PALLACIUM, COPPER, NICKEL)
PALMBERG PW + RHODIN TN
J APPL PHYS 39: 2425-32 (1968)

211 ATOMIC ARRANGEMENT CF GOLD (100), GOLD-COVERED (100), AND SILVER-COVERED COPPER (100) SURFACES.
PALMBERG PW + RHODIN TN
BULL AMER PHYS SOC 13: 944 (1968)

212 SURFACE DISSOCIATION OF POTASSIUM CHLORIDE BY LOW ENERGY ELECTRON BOMBARDMENT.
PALMBERG PW + RHODIN TN
J PHYS CHEM SOLIDS 29: 1917-24 (1968)

213 CHARACTERIZATION OF CHEMISORPTION BY LEED.
PARK RL + HOUSTON JE
J METALS 21(3): A87 (1969)

214 QUANTITATIVE USE OF AUGER SPECTROSCOPY: STANDARDIZATION OF THE METHOD.
PERCEREAU M
SURFACE SCI 24: 239-47 (1971)

215 STUDY OF INTENSITIES OF AUGER LINES EXCITED BY ELECTRON BOMBARDMENT AND ALUMINUM K- ALPHA X-RAY IRRADIATION.
PESSA M
J APPL PHYS 42: 5831-6 (1971)

216 NEW FINE STRUCTURE IN ELECTRON-EXCITED AUGER SPECTRA FROM SOLID SURFACES.
PESSA M + AKSELA S + KARRAS M
PHYS LETT A31: 382-3 (1970)

217 INTERACTION OF LOW-ENERGY ATMOSPHERIC IONS WITH CONTROLLED SURFACES. (AUGER NEUTRALIZATION AT (100) FACE OF TUNGSTEN, POLYCRYSTALLINE MOLYBDENUM).
PIERCE RH + FRENCH JB
UNIV TORONTO REPT. AD682373 (1968) (CHEM ABS 71: 16846G (1969));
REPT. AD675206 (1969) (CHEM ABS 70: 81182 (1969))

218 A CCRRELATION OF AUGER SPECTROSCOPY, LEED AND WORK FUNCTION MEASUREMENTS FOR EPITAXIAL GROWTH OF THORIUM ON A TUNGSTEN (100) SUBSTRATE.
POLLARD JH
SURFACE SCI 20: 269-84 (1970)

219 SURFACE COMPOSITION CF MICA SUBSTRATES.
POPPA H + ELLIOT AG
SURFACE SCI 24: 149-63 (1971)

220 CHARACTERIZATION OF COPPER- GOLD ALLOY SINGLE SURFACES.
POTTER HC
CORNELL UNIV. PH.D THESIS (1970)

221 SPECTROSCOPY OF A METAL- GAS INTERFACE.
PRITCHARD J
ANN REP PROGR CHEM SECT A66: 65 (1969)

222 UHV EVAPORATOR FOR LEED AND AUGER EMISSION STUDIES.
PRUTTON M + TOKUTAKA H
THIN SOLID FILMS 3: 411-16 (1969)

223 IDENTIFICATION OF AUGER SPECTRA FROM ALUMINUM.
QUINTO DT + ROBERTSON WD
SURFACE SCI 27(3): 645-8 (1971)

224 AUGER SPECTRA OF COPPER- NICKEL ALLOYS.
QUINTO DT + SUNDARAN VS + ROBERTSON WD
SURFACE SCI 28: 504-16 (1971)

225 SECONDARY (AUGER) ELECTRON SPECTROSCOPY.
RAMSEY JA
VACUUM 21: 115-19 (1971)

226 AUGER ELECTRON SPECTRUM OF OSMIUM AT ENERGIES UP TO 300 EV.
REDKIN VS + ZASHKVARA VV + KORSUNSKII MI + TSVEIMAN EV
SOV PHYS SOLID STATE 13: 1269-70 (1971)

227 THE EFFECT OF TELLURIUM ON INTERGRANULAR COHESION OF IRON.
RELLICK JR + MCMAHON CJ + MARCUS HL + PALMBERG PW
MET TRANS 2: 1492-4 (1971)

228 AUGER SPECTRA AND LEED PATTERNS FROM NICKEL DEPOSITS ON CLEAVED SILICON.
RIDGWAY JWT + HANEMAN D
SURFACE SCI 26: 683-7 (1971)

229 AUGER SPECTRA AND LEED PATTERNS FROM VACUUM CLEANED SILICON CRYSTALS WITH CALIBRATED DEPOSITS OF IRON.
RIDGWAY JWT + HANEMAN D
SURFACE SCI 24: 451-8 (1971)

230 CHARACTERISTIC AUGER ELECTRON EMISSION AS A TOOL FOR THE ANALYSIS OF SURFACE COMPOSITION.
RIVIERE JC
PHYS BULL 20: 85- (1969)

231 DIFFUSION OF SULFUR TOWARD THE (110) FACE OF NICKEL.
RIWAN R
SURFACE SCI 27: 267-72 (1971)

232 CHARACTERISTIC ENERGY LOSS AND AUGER ELECTRON SPECTROSCOPY APPLIED TO THE STUDY OF ADSORPTION PHENOMENA AT METAL SURFACES.
ROUSSEAU J + PRALIAND H
J CHIM PHYS 67: 1493-505 (1970) (IN FRENCH)

233 THEORETICAL STUDY OF THE AUGER EFFECT IN THE LIGHT TO MEDIUM RANGE OF ATOMIC NUMBER.
RUBENSTEIN RA
UNIV OF ILLINOIS PHD THESIS (1955)
DISSERTATION ABST 15: 851 (1955)

234 EXPERIMENTAL OBSERVATION OF CHEMICAL SHIFTS IN AUGER SPECTRUM FROM SURFACE LAYERS OF SILICON DIOXIDE DURING ELECTRON BOMBARDMENT.
SALMERON M + BARO AM
SURFACE SCI 29: 300-2 (1972)

235 INELASTIC SCATTERING OF LOW ENERGY ELECTRONS FROM SURFACES.
SCHEIBNER EJ + THARP LN
SURFACE SCI 8: 247-65 (1967)

236 THEORY OF THE AUGER EFFECT IN III-V SEMICONDUCTORS.
SCHRENE D
Z NATURFORSCH A 24: 1752-9 (1969)

237 SLOW ELECTRON SCATTERING FROM METALS. PART-2: INELASTICALLY SCATTERED PRIMARY ELECTRONS.
SEAH MP
SURFACE SCI 17: 161-80 (1969)

238 SLOW ELECTRON SCATTERING FROM METALS. PART-1: EMISSION OF TRUE SECONDARY ELECTRONS.
SEAH MP
SURFACE SCI 17: 132-60 (1969)

239 FARADAY CUP LEED APPARATUS WITH FACILITY FOR INVESTIGATING ENERGY AND ANGULAR DISTRIBUTIONS OF INELASTICALLY SCATTERED OR PHOTOEMITTED ELECTRONS.
SEAH MP + FORTY AJ
J PHYS E3: 833-41 (1970)

240 SECONDARY EMISSION AND ELASTIC REFLECTION OF ELECTRONS FROM MONOCRYSTALS.
SHULMAN AR + KORABLEV VV + MOROZOV YA
IZV AKAD NAUK SSSR SER FIZ 35: 218 (1971) (IN RUSSIAN)

241 SECONDARY ELECTRON EMISSION OF SILICON DIOXIDE SINGLE CRYSTALS.
SHULMAN AR + KORABLEV VV + MOROZOV YA
SOV PHYS SOLID STATE 12: 519-20 (1970)

242 SECONDARY ELECTRON EMISSION OF MOLYBDENUM CRYSTALS.
SHULMAN AR + KORABLEV VV + MOROZOV YA
SOV PHYS SOLID STATE 12: 586-9 (1970)

243 ENERGY SPECTRA OF INELASTICALLY SCATTERED AND AUGER ELECTRONS FROM SINGLE CRYSTALS.
SHULMAN AR + KORABLEV VV + MOROZOV YA
SOV PHYS SOLID STATE 12: 1487-8 (1970)

242 SECONDARY ELECTRON EMISSION OF MOLYBDENUM CRYSTALS.
SHULMAN AR + KORABLEV VV + MOROZOV YA
SOV PHYS SOLID STATE 12: 586-9 (1970)

243 ENERGY SPECTRA OF INELASTICALLY SCATTERED AND AUGER ELECTRONS FROM SINGLE CRYSTALS.
SHULMAN AR + KORABLEV VV + MOROZOV YA
SOV PHYS SOLID STATE 12: 1487-8 (1970)

244 AUGER ELECTRON SPECTROSCOPY APPLIED TO SURFACE COMPOSITION PROBLEMS.
SICKAFUS EN
J METALS 21(3): A72 (1969)

245 SULFUR AND CARBON ON (110) SURFACE OF NICKEL.
SICKAFUS EN
SURFACE SCI 19: 181-97 (1970)

246 SECONDARY EMISSION ANALOG FOR IMPROVED AUGER SPECTROSCOPY WITH RETARDING POTENTIAL ANALYZERS.
SICKAFUS EN
REV SCI INSTRUM 42: 933-41 (1971)

247 LEED AND AUGER ELECTRON SPECTROSCOPY STUDY OF NICKEL (110) SURFACE EFFECTS DUE TO CARBON AND SULFUR.
SICKAFUS EN
BULL AMER PHYS SOC 14: 793 (1969)

248 A MULTICHANNEL MONITOR FOR REPETITIVE AUGER ELECTRON SPECTROSCOPY WITH APPLICATION TO SURFACE COMPOSITION CHANGES.
SICKAFUS EN + COLVIN AD
REV SCI INSTRUM 41: 1349-54 (1970)

249 AUGER ELECTRON SPECTROSCOPY AND INELASTIC ELECTRON SCATTERING IN THE STUDIES OF METAL SURFACES.
SIMMONS GW
J COLLOID INTERFACE SCI 34: 343-56 (1970)

250 ORDER- DISORDER PHENOMENA AT THE SURFACE OF ALPHA TITANIUM OXYGEN SOLID SOLUTIONS.
SIMMONS GW + SCHEIBNER EJ
J MATER 5: 933-49 (1970)

251 SURFACE MADELUNG POTENTIALS IN ELECTRON SPECTROSCOPY.
SLATER RR
SURFACE SCI 23: 403-8 (1970)

252 SURFACE ANALYSIS BY LOW-ENERGY ION REFLECTION.
SMITH DP
APPL SPECTROSC 25: 147 (1971)

253 ANALYSIS OF SURFACE COMPOSITION WITH LOW-ENERGY BACK-SCATTERED IONS.
SMITH DP
SURFACE SCI 25: 171-91 (1971)

254 SPUTTER CLEANING AND ETCHING OF CRYSTAL SURFACES (TITANIUM, TUNGSTEN, SILICON) MONITORED BY AUGER SPECTROSCOPY, ELLIPSOMETRY AND WORK FUNCTION CHANGE.
SMITH T
SURFACE SCI 27: 45-59 (1971)

255 INTENSITIES OF SPECTRA- AUGER AND PHOTOELECTRON.
STADNIKOV CG + NIKOLSHII AP
IZV AKAD NAUK SSSR, FIZ ESKA 35: 330 (1971) (IN RUSSIAN)

256 ENERGIES OF EXCITED STATES OF DOUBLY IONIZED MOLECULES BY MEANS OF AUGER ELECTRON SPECTROSCOPY. PART-1: ELECTRONIC STATES OF NITROGEN(2).
STAHLHERM D + CLEFF B + HILLIG H + MEHLHORN W
Z NATURFORSCH A24: 1728-33 (1969)

257 STUDY OF GRAIN BOUNDARY SEGREGATION USING AUGER ELECTRON EMISSION SPECTROSCOPY.
STEIN DF
J METALS 21(3): A88 (1969)

258 STUDIES USING AUGER ELECTRON EMISSION SPECTROSCOPY ON TEMPER EMBRITTLEMENT IN LOW ALLOY STEELS.
STEIN DF + JOSHI A + LAFORCE RP
TRANS AMER SOC METALS 62: 776-83 (1969)

259 STUDY OF GRAIN BOUNDARY SEGREGATION USING AUGER EMISSION SPECTRA.
STEIN DF + RAMASUBRAMANIAN PV
UNIV OF MINNESOTA ANN TECH PROG REPT I. N69-26102 (1968)

260 AUGER ELECTRON SPECTROSCOPY OF METAL SURFACES.
STEIN DF + WEBER RE + PALMBERG PW
J METALS 23(2): 39-44 (1971)

261 AUGER SPECTROSCOPY OF BERYLLIUM.
SULEMAN M + PATTINSON EB
J PHYS F 1: 124-7 (1971)

262 OBSERVATION OF A PLASMON GAIN IN THE FINE STRUCTURE OF THE ALUMINUM AUGER SPECTRUM.
SULEMAN M + PATTINSON EB
J PHYS F 1: L21-4 (1971)

263 ALLOY SPUTTERING STUDIES WITH IN SITU AUGER ELECTRON SPECTROSCOPY.
TARNG ML + WEHNER GK
J VACUUM SCI TECHNOL 8: 23 (1971)

264 RESOLUTION AND SENSITIVITY CONSIDERATIONS OF AN AUGER ELECTRON SPECTROMETER BASED ON LEED DISPLAY OPTICS.
TAYLOR NJ
REV SCI INSTRUM 40: 792-804 (1969)

265 AUGER ELECTRON SPECTROMETER AS TOOL FOR SURFACE ANALYSIS (CONTAMINATION MONITOR)
TAYLOR NJ
J VACUUM SCI TECHNOL 6: 241-5 (1969)

266 THIN REACTION LAYERS AND SURFACE STRUCTURE OF SILICON (111).
TAYLOR NJ
SURFACE SCI 15: 169-74 (1969)

267 TECHNIQUE OF AUGER ELECTRON SPECTROSCOPY IN SURFACE ANALYSIS.
TAYLOR NJ
IN TECHNIQUES OF METALS RESEARCH VOL-7, RF BUNSHAH (ED). NY
INTERSCIENCE 1971. (669.028/B94)

268 ROLE OF AUGER ELECTRON SPECTROSCOPY IN SURFACE ELEMENTAL ANALYSIS.
TAYLOR NJ
VACUUM 19: 575-8 (1969)

269 REPLY TO COMMENTS OF HE BISHOP AND JC RIVIERE ON "AUGER SPECTROSCOPY OF SILICON".
TAYLOR NJ
SURFACE SCI 17: 466-8 (1969)

270 ENERGY SPECTRA OF INELASTICALLY SCATTERED ELECTRONS AND LEED STUDIES OF TUNGSTEN.
THARP LN + SCHEIBNER EJ
J APPL PHYS 38: 3320-30 (1967)

271 AUGER SPECTROSCOPY STUDY OF THE ADSORPTION OF RUBIDIUM ON MOLYBDENUM (100).
THOMAS S + HAAS TW
SURFACE SCI 28: 632-6 (1971)

272 ELECTRON SPECTROSCOPY OF METAL BLACKS.
THOMAS S + SULEMAN M + PATTINSON EB
J PHYS D 3: L77-80 (1970)

273 K-, L-, AND M- AUGER AND L- COSTER-KRONIG SPECTRA OF PLATINUM.
TOBUREN LH + ALBRIDGE RG
NUCL PHYS A90: 529-44 (1967)

274 (100) SURFACES OF ALKALI HALIDES. II. ELECTRON STIMULATED DISSOCIATION.
TOKUTAKA H + PRUTTON M + HIGGINBOTHAM IG + GALLON TE
SURFACE SCI 21: 233-40 (1970)

275 SURFACE CHEMICAL ANALYSIS BY AUGER ELECTRON SPECTROSCOPY AND APPEARANCE POTENTIAL SPECTROSCOPY: A COMPARISON.
TRACY JC
APPL PHYS LETT 19: 353-6 (1971)

276 AUGER ELECTRON SPECTROMETER PREAMPLIFIER.
TRACY JC + BOHN GK
REV SCI INSTRUM 41: 591-2 (1970)

277 THE KINETICS OF OXYGEN ADSORPTION ON THE (112) AND (110) PLANES OF TUNGSTEN.
TRACY JC + BLAKELY JM
SURFACE SCI 15: 257-76 (1969)

278 STRUCTURAL INFLUENCES ON ADSORBATE BINDING ENERGY. I. CARBON MONOXIDE ON (100) PALLADIUM.
TRACY JC + PALMBERG PW
J CHEM PHYS 51: 4852-62 (1969)

279 USE OF AUGER ELECTRON SPECTROSCOPY IN DETERMINING THE EFFECT OF CARBON AND OTHER SURFACE CONTAMINANTS ON GALLIUM ARSENIDE- CESIUM- OXYGEN PHOTOCATHODES.
UEBBING JJ
J APPL PHYS 41: 802-4 (1970)

280 AUGER ELECTRON SPECTROSCOPY OF CONTAMINATED GALLIUM ARSENIDE SURFACES.
UEBBING JJ + JAMES LW
J VACUUM SCI TECHNOL 7: 81-3 (1970)

281 AUGER ELECTRON SPECTROSCOPY OF CLEAN GALLIUM ARSENIDE.
UEBBING JJ + TAYLOR NJ
J APPL PHYS 41: 804-8 (1970)

282 AUGER ELECTRON SPECTROSCOPY OF GALLIUM ARSENIDE PHOTOSURFACES.
UEBBING JJ + TAYLOR NJ
BULL AMER PHYS SOC 14: 792 (1969)

283 ANGLE OF INCIDENCE EFFECTS IN AUGER ELECTRON EMISSION FROM CLEAN MOLYBDENUM.
VANCE DW
BULL AMER PHYS SOC 13: 947 (1968)

284 AUGER ELECTRON EMISSION FROM CLEAN AND CARBON- CONTAMINATED MOLYBDENUM BOMBARDED BY POSITIVE IONS.
VANCE DW
PHYS REV 164: 372-80 (1967)

285 AUGER ELECTRON EMISSION FROM CLEAN AND CARBON- CONTAMINATED MOLYBDENUM BOMBARDED BY POSITIVE IONS. PART-2: EFFECT OF ANGLE OF INCIDENCE. PART-3: EFFECT OF ELECTRONICALLY EXCITED IONS.
VANCE DW
PHYS REV 169: 252-72 (1968)

286 AUGER TRANSITIONS AND SHAPE OF X-RAY SPECTRUM.
VEDRINSKII RV + KOLESNIKOV VV
BULL ACAD SCI USSR, PHYS SER 31: 904-10 (1967)

287 OBSERVATION OF AUGER ELECTRONS IN THE ENERGY SPECTRUM OF SECONDARY ELECTRONS EMITTED BY COPPER UNDER BOMBARDMENT BY RARE GAS IONS.
VIEL L + FAGOT B + COLOMBIE N
COMPT REND B272:623-8 (1971)

288 AUGER EMISSION SPECTROSCOPY VANADIUM(2) OXYGEN(5) (010) AND VANADIUM (100) SURFACES.
VIERMANS L + VENNIK J
SURFACE SCI 24: 541-54 (1971)

289 AUGER ELECTRON SPECTROSCOPY MADE QUANTITATIVE BY ELLIPSOMETRIC CALIBRATION.
VRAKKING JJ + MEYER F
APPL PHYS LETT 18: 226-8 (1971)

290 NONRELATIVISTIC AUGER RATES, X-RAY RATES, AND FLUORESCENCE YIELDS FOR THE 2P SHELL.
WALTERS DL + BHALLA CP
PHYS REV A 4: 2164-70 (1971)

291 NONRELATIVISTIC AUGER RATES, X-RAY RATES, AND FLUORESCENCE YIELDS FOR THE K SHELL.
WALTERS DL + BALLA CP
PHYS REV A 3: 1919-27 (1971)

292 Z DEPENDANCE OF THE KLL AUGER RATES.
WALTERS DL + BALLA CP
PHYS REV A 3: 519-20 (1971)

293 THIN FILM ANALYSIS BY AUGER ELECTRON SPECTROSCOPY.
(SHORT REVIEW)
WEBER RE
SOLID STATE TECH 13(12): 49-53 (1970)

294 DETERMINATION OF SURFACE STRUCTURES USING LEED AND ENERGY ANALYSIS OF SCATTERED ELECTRONS.
WEBER RE + JOHNSON AL
J APPL PHYS 40: 314-8 (1969)

295 DETERMINATION OF SURFACE STRUCTURES BY LEED AND AUGER ELECTRON SPECTROSCOPY.
WEBER RE + JOHNSON AL
BULL AMER PHYS SOC 13: 945 (1968)

296 WORK FUNCTION AND STRUCTURAL STUDIES OF ALKALI- COVERED SEMICONDUCTORS.
WEBER RE + PERIA WT
SURFACE SCI 14: 13-38 (1969)

297 USE OF LEED APPARATUS FOR THE DETECTION AND IDENTIFICATION OF SURFACE CONTAMINANTS.
WEBER RE + PERIA WT
J APPL PHYS 38: 4355-8 (1967)

298 AUGER RECOMBINATION IN GALLIUM ARSENIDE.
WEISBERG LR
J APPL PHYS 39: 6096-8 (1968)

299 THEORY OF AUGER EJECTION OF ELECTRONS FROM METALS BY IONS.
WENAAS EP + HOWSMAN AJ
P13-1 TO 13-22 OF STRUCTURE AND CHEMISTRY OF SOLID SURFACES,
SOMORJAI GA (ED). NY WILEY, 1969. (541.375/B51) 140258

300 SIZE EFFECT IN IONIC CHARGE RELAXATION FOLLOWING AUGER EFFECT.
WERTHEIM GK + GUGGENHEIM HJ + BUCHANAN DN
J CHEM PHYS 51: 1931-4 (1969)

301 AUGER ELECTRON ENERGIES (0- 2000 EV) FOR ELEMENTS OF ATOMIC NUMBER 5- 103.
YASKO RN + WHITMOYER RD
J VACUUM SCI TECH 8: 733-7 (1971)

302 SOME PARAMETERS AFFECTING AUGER AND PHOTOELECTRON SPECTROSCOPY AS AN ANALYTICAL TECHNIQUE.
YIN LI + ADLER I + LAMOTHE R
APPL SPECTROSC 23: 41-50 (1969)

303 X-RAY EXCITED LMM AUGER SPECTRA OF COPPER, NICKEL AND IRON.
YIN LI + YELLIN E + ADLER I
J APPL PHYS 42: 3595-600 (1971)

304 AUGER SPECTRA OF URANIUM.
ZENDER MJ
Z PHYS 218: 245-59 (1969)

305 A CARBON STRUCTURE ON THE RHENIUM (0001) SURFACE.
ZIMMER RS + ROBERTSON WD
SURFACE SCI 29: 230-6 (1972)

21

MASS SPECTROMETRY

J. M. McCrea*

P.O. Box 172, Monroeville, Pennsylvania
and
Department of Chemistry, Indiana University of Pennsylvania
Indiana, Pennsylvania

1. INTRODUCTION

1.1. Mass Spectrometers and Mass Spectrometry

The research and development work leading to the present-day wealth of instrumentation for mass spectrometry started around 1905, and applications of the technique grew in number and diversity along with the instrumentation. Until 1940, mass spectrometers were seldom found in laboratories other than those at a few centers of expertise, but substantial contributions by these laboratories to the identification and cataloging of stable isotopes and their nuclear masses helped build the data reservoir necessary for the birth of the era of nuclear energy, 1942–1945. Between 1945 and 1955, mass spectrometry became established as an accepted technique for research in chemical kinetics; for chemical analysis of gases, volatile and moderately volatile organic liquids, and petroleum fractions; and for thermodynamic and structural studies of solids. Since 1955, growth in the areas of instrumentation and application has been phenomenal, and many of the new developments have been of particular significance to the characterization of solid surfaces.

Basically a mass spectrometer may be considered as an apparatus that produces a supply of gaseous ions from a sample, separates the ions

* All correspondence should be directed to P.O. Box 172, Monroeville, Pennsylvania 15146.

in either space or time according to their mass-to-charge ratios, and provides an output record or display indicating the intensity of the separated ions. The specification just given for a mass spectrometer is deliberately broadened to include some of the newer apparatus, such as ion microprobes and ion microscopes, under the generic term mass spectrometer.

1.2. Terminology and Definitions

A few of the definitions related to mass spectrometry have been studied by a task group of mass spectrometrists and reduced to consensus form accepted by an overwhelming vote of the members of the American Society for Testing and Materials Committee E-14 on mass spectrometry. These definitions are published yearly in the *Book of ASTM Standards*, in methods designated E-137 and E-304.[1,2] This author participated in that work, and feels that, although the definitions are perhaps not ideal, they do represent accepted meanings for the terms defined.

1.2.1. Mass Spectrometer

"The term *mass spectrometer* or the adjectival form, mass spectrometric, shall apply to all apparatus in which an analysis of matter is effected by means of ionization of the matter followed by separation of the ions according to mass-to-charge ratio and recording of a measure of the numbers of the various ions."

1.2.2. Mass Spectrograph

"The terms *mass spectrograph* and *mass spectrographic* shall apply to that class of mass spectrometric apparatus in which the ionized particles are recorded by means of an ion-sensitive plate. Many instruments are designed in such a way that they may be converted from ion-sensitive plate recording to electrical recording, and *vice-versa*, or both types of recording may be used simultaneously. Such instruments are best described by the generic term *mass spectrometer* but data obtained by the use of ion-sensitive plates should be described as *mass spectrographic*."

1.2.3. Background

"The *background* at a specific mass refers to the response at that position resulting from influence extraneous to the components of a specimen giving rise to that specific mass. Background in mass spectrometers is caused by various processes, such as ionization of residual materials in the spectrometer, action of ions or electrons scattered from their normal trajectories by residual gas in the instrument, action of secondary particles generated under positive ion bombardment, and fogging of ion-sensitive plates by light or chemicals during processing and handling."

1.2.4. Sensitivity

"The *sensitivity* of a mass spectrometer for a particular component is a measure of the instrument's response to ions of the component at an arbitrary specific mass and under specified operating conditions. For a given sample, sensitivity may be measured alternatively, but not equivalently, in terms of (1) response per unit weight of material consumed during a run, (2) response per unit time of ion-source operation, or (3) response per unit charge of ions leaving the source. The response measured in the expressions for sensitivity may be measured at the focal position for the resolved ion beam either by the blackening of an ion-sensitive plate or by the charge received at a measuring device.

"The greater the sensitivity of an instrument, the smaller the amount of sample or the shorter the time required to detect a given level of a component in a sample. If neither amount of sample nor the time required is a limiting factor, the minimum level of a component for which a response can be observed may be of prime importance in a method of analysis. In such cases a limit of detection criterion may replace a sensitivity criterion and response-to-background or signal-to-noise factors govern the situation."

1.2.5. Resolution

"The *resolution* of a mass spectrometer pertains to the ability of a mass spectrometer to discriminate between ions of slightly different specific masses. The specific mass (or mass/charge ratio) of an ion is its mass expressed in atomic mass units divided by the magnitude of its charge in electron charge units. Resolution is often defined in terms of two slightly separated ion beams of equal magnitude. It is the ratio of the average specific mass to the specific mass interval between the beams for some arbitrary level of the ratio of the minimum response near the average specific mass to the maximum response of either of the ion beams. Any numerical value for resolution is ambiguous unless the arbitrary level of the response ratio is clearly stated; common choices of this ratio are 0.81 (corresponding to the Rayleigh criterion for optics), 0.5, 0.1 or 0.01."

Because the design and construction of mass spectrometers to achieve both high sensitivity and resolution at moderate cost is not easy to attain, resolution has become a selling feature of commercial instruments and, to avoid low-resolution figures, use of the 1% valley or "0.01" ratio level has fallen into disuse. The 10% valley, or "0.1" ratio, is now most commonly used for most mass spectrometers, but there is little consensus on the matter of mass spectrographic data recorded on ion-sensitive plates. An operational definition of resolution that is technically equivalent to the 10% valley definition for a wide range of situations is the 5% peak-width definition, as follows:

"The resolution R achieved at any isolated peak in a mass spectrum may be calculated from the equation $R = m/\Delta m$, where m is the mass corresponding to the peak and Δm is the width (in mass units) at 5 percent of the peak height."

This definition can be applied to a single peak in a mass spectrum. The corresponding 10%-valley definition requires two adjacent peaks of equal magnitude and consequently considerable experimental adjustments before it can be applied to a pair of peaks in many mass spectra.

1.2.6. Standard Sample

In the characterization of solids by mass spectrometry, an important basis for establishing validity of results is the *standard sample*.

> "This term refers to material that has been analyzed carefully by mass spectrometer and by at least one independent method, and which is considered to be of known composition and of a composition and fine-scale homogeneity suitable for analysis by the mass spectrometric method employed. Standard samples with certified analyses result from extensive evaluation of the materials in many laboratories."

Materials that meet the criteria for standard samples are not plentiful because of the large amount of cooperative laboratory work involved, and most research on the characterization of solid surfaces by mass spectrometry can only be supported by data on the best-characterized materials of the same general nature that are currently available to the investigator.

1.2.7. Other Terms

There are many terms used in mass spectrometry that have not been standardized into consensus definitions of the type quoted in Sections 1.2.1–1.2.6, inclusive. Information on such terms frequently appears in the "Instructions for Contributors," published by the editors of technical journals in the field, and in textbooks on mass spectrometry. Definitions and descriptions of a few of the more important terms are given here for the sake of balance *vis-a-vis* consensus definitions.

Ions are charged atoms, groups of atoms, or molecules formed by the process of ionization in which an electrically neutral atom, group of atoms, or molecule acquires a charge through gain or loss of an electron or electrons. Ion–solvent interactions are normally of no consequence in mass spectrometry because the ions are studied in the gas phase and usually at pressures below 10^{-6} Torr.*

The *specific mass* ("mass" in many articles), or m/e of an ion as normally used in mass spectrometry refers to the total mass, on the ^{12}C nuclear mass scale, of all the nuclei in the ion divided by the ionic charge expressed as a multiple of the charge on the electron. If the decimal portions of the nuclear masses are neglected and only the integral or mass-number portions of the component nuclear masses are used, a specific mass that is the total mass number divided by the number of net electronic charges results. The exact intentions of a writer can be inferred from the context and from the use of either integral or decimal notation in relation to the specific mass. When the quantity m/e appears

* In chemical-ionization mass spectrometry, pressures up to the region of 10^{-4} Torr are commonly encountered in the ion source. The technique has not been applied to studies of solids or surfaces however.

in equations related to ion trajectories, *m* and *e* will usually have to be expressed in physical units of mass and charge in order to keep a consistent system of units.

The *mass spectrum* is a listing of the observed specific masses and their relative intensities as obtained from a mass spectrometer, a chart recording or bar graph of the same data, or a spectral record of ion-position and ion-intensity information obtained on an ion-sensitive plate. Mass spectra are sometimes referred to as "cracking patterns" or "fragmentation patterns" in organic chemical applications; this historical terminology derives from the early application of mass spectrometry to petroleum technology and the conceptual similarity between thermal and catalytic cracking of molecules in petroleum processing and the various decompositions of an organic molecule during ionization in a mass spectrometer ion source.

The *mass range* of a mass spectrometer is a statement of the lowest and highest specific masses accessible experimentally. The lowest specific mass is 1, and corresponds to the hydrogen ion; the current high end of the range may be as large as 3000. In surface characterization where elemental composition is emphasized, the high end is frequently adjusted to lie near $m/e = 250$ so that ions of all naturally occurring elements lie within the instrumental recording range. When data are taken on an ion-sensitive plate, all m/e between 12 and 250 are generally recorded simultaneously. Electrical measurements of relative ion intensity are generally taken through a single ion-collector system, and it is necessary to *scan* in order to obtain data for the full range of m/e values. The *scan rate* gives the range of m/e values brought over the ion collector per unit time. It is obviously of importance in determining the effective resolution and sensitivity for a given selection of instrumental operation conditions.

1.3. General Applications of Mass Spectrometers

While the balance of this chapter will be restricted to the applications of mass spectrometry in the characterization of solid surfaces, these surface applications form only a small portion of the total application picture. The chart of stable nuclear species was compiled largely on the basis of mass spectrometric data, and the energy relationship among these nuclear species were worked out in part by precise determination of mass differences by mass spectrometric techniques. Mass spectrometry is used for a tremendous volume of characterization studies in organic chemistry and medicinal chemistry, and for quantitative analysis in the petroleum- and chemical-processing fields. Precise measurements of isotopic composition have been used for a variety of stable-isotope tracer studies, for geological and ecological studies and, once sample material became available, for selenology (lunar geology) research. Mass spectrometry is applied to things as prosaic as leak detection and as sophisticated as the elucidation of catalytic mechanisms and high-temperature thermodynamics. Its application to surface studies and solids are relatively new, but the application to bulk-solids analysis of high-purity materials is well established. Its applications to the study of solid surfaces, adsorption, surface coatings, surface impurities, and surface changes are being included here.

2. SAMPLING CONSIDERATIONS

2.1. Ionization and Mass Spectra

In mass spectrometry, the ions produced from the sample material are always examined in the gas phase and generally in ultrahigh vacuum, with instrument pressures under 10^{-6} Torr. A matter of prime importance in the characterization of solids and surfaces is the experimental method of converting sample material to ions in the gas phase, and the physical and chemical mechanisms involved in the process. The most widely used ion source in mass spectrometry is the electron-bombardment source, in which a beam of electrons is used to eject electrons from neutral gas molecules and produce positive ions. For a solid to be amenable to study by this ionization technique, it must have a vapor pressure great enough to produce the necessary gas for ionization. Furthermore, the vaporization process must be under the experimenter's control if data meaningfully related to the solid surface being evaporated are to be obtained. Thus, naphthalene, a crystalline, white, organic compound with a substantial vapor pressure over the solid form at room temperature, can be studied easily in a mass spectrometer with an electron-bombardment source. The same instrumentation is not effective for studying degassed graphite unless one resorts to an auxiliary step of selectively converting the graphite to a gaseous product, possibly by treatment with hydrogen or with oxygen. Thus, a small furnace or reaction chamber is frequently designed as a mechanically integral part of an electron-bombardment ion source on an instrument intended for studies of solids. The mass spectra produced in such an instrument are those typical of the gaseous species actually ionized. There are large catalogs of reference spectra for the gaseous species, and a large body of literature on the correlation of chemical structure before ionization with the proportions of the ions observed in the mass spectra, which have been published. These spectra and concepts will in many cases aid in analysis of the material desorbed from a solid surface as a function of time and surface temperature.

The thermal-ionization source for a mass spectrometer is used almost exclusively for analysis of solids. It depends on the fact that numerous substances, particularly salts of the alkali and alkaline-earth elements, will give off ions, in addition to neutral molecules, when evaporated at high temperature. The ions produced are separated to obtain a mass spectrum. The thermal-ionization source is most useful for isotopic analysis of a number of elements, but it presents difficulties in securing consistent operation necessary for ordinary quantitative analysis.

The vacuum-discharge class of ion source is used generally for solid samples and surfaces, and is occasionally used for liquid samples. The radio-frequency spark source is a common version in which a train of short rf pulses with voltages of 100 kV and over is applied between two solid electrodes in the ion source. The mass spectra of ions produced in the source are characteristic, and consist almost exclusively of elemental ions with one to as many as ten or more electronic charge units. Additional spectral lines with low intensities may be observed for dimer and trimer ions, ion clusters of two or more elements; lines for ions that have altered charge during transit through the mass spectrometer may also be observed. The triggered dc spark is another type of vacuum-discharge source frequently employed for studies of solids. The mass spectra produced by this source again are mainly elemental spectra, but the doubly and triply charged species are favored, and the charge levels greater than five are extremely weak. The triggered dc spark is reported to operate more stably than the rf spark, and as a result may be better suited to analysis of solids in bulk.

The laser-discharge source, similar in principle to both the vacuum-discharge and thermal ionization sources, has received some exploratory attention as a means for providing ions for mass spectrometric studies of surfaces. A field ion microscope with a mass analyzer for identifying the ions extracted from the specimen at extremely high electric field is really a mass spectrometer with a field ionization source.

Ion bombardment of a solid results in the ejection of material from the solid; some of this material is ionized and subject to mass spectrometric analysis. The sputtering or ion-bombardment emission process is complex, and there are many experimental parameters to consider, such as identity of the ion or ions used in the bombardment, ion energy, angle of incidence and take-off angle for the secondary ions, and the role of un-ionized, sputtered materials. Instruments that provide close control of the type and energy of the primary ion beam resemble two mass spectrometers in one, a primary-ion-control mass spectrometer in which the detector is actually the sample material in the ion-generation region of the second instrument used to characterize the secondary ions according to type, relative abundance, and energy distribution. By use of carefully designed ion optics, it is possible to image a surface region of the sample by use of the secondary ions, resulting in an ion microscope. It is also experimentally possible to confine the area to the sample bombarded by the primary ion beam to a very small surface area, giving an experimental setup called an ion microprobe or ion microanalyzer.

In addition to the types of ion sources mentioned as applicable to solid surfaces, there are numerous other types such as photoionization, chemical ionization, gas discharge, and arc discharge that are not applied to the study of solid surfaces. There have also been reports of mass spectrometric analysis of recoil ions from nuclear reactions; in principle this process might be used for solids.

The ion-bombardment source is similar in some ways to the ion-scattering process in which ions incident on the sample are scattered and then analyzed for any change in energy occurring during the scattering process. The distinction between the two is that the same ions are present in the incident and scattered beams in ion scattering, whereas the ions in the secondary beam from ion bombardment are ions derived from the solid surface.

2.2. Ionization from and above Surfaces

When a solid is studied by evaporation and ionization of the evolved vapor in an electron-bombardment source, the two processes of evaporation and ionization are distinct steps toward obtaining a spectrum. Known information about the stability of the components on evaporation and the possibility of catalytic or detrimental reactions with the materials exposed to the vapors can be clearly used to assess the first step. The ionization step is simply the ionization of a low-pressure gas by an electron beam as in normal analysis of gases and vapors by mass spectrometry.

The functioning of the thermal ionization is also reasonably well understood,[3] but work functions of the surface play a part in the ionization through the Langmuir-Saha relations

$$\begin{aligned} I^+/I^n &= \exp[11{,}600(\phi - IP)/T] \\ I^-/I^n &= \exp[11{,}600(EA - \phi)/T] \end{aligned} \tag{1}$$

where I^+ is the positive ion current, I^- is the negative ion current, I^n is the neutral flux, T is the temperature in °K, ϕ is the work function of surface in eV, IP is the ionization potential of neutral species in eV, and EA is the electron affinity of neutral species in eV.

The three types of ion sources most useful for the characterization of surfaces are the vacuum-discharge, the ion-bombardment, and the laser-discharge sources; all function through processes that produce localized but intense disruption of the surface as it is examined experimentally. Material is torn from the surface layers with a large amount of energy per atom involved, and in the end product ions are extracted from the scene of action by electric fields and measured according to specific mass and intensity by the analyzing and detecting system of the mass spectrometer. How much of the observed ionization takes place as the material leaves the surface and how much takes place subsequently in primary or secondary processes in the small zone of vapor or plasma adjacent to the sampling point is not easy to establish. Information deduced from the mass spectral data obviously can be related to surface composition and structure only if assumptions about the ion-production steps do not critically affect the nature of the derived information, or if the details of ion production are reasonably well worked out for the system under study. In other words, a conservative attitude in interpretation of the mass spectral data will pay off by yielding more firmly based conclusions about the surface composition and structure.

2.3. Mass Spectrometers as Destructive Analyzers with Small Sample Requirements

The scientist should entertain no illusions about mass spectrometric experiments not being destructive of the material under study. The name of the game is to convert surface material to ions that are physically transported through the ion optics of the spectrometer to a detector where they are collected and measured. Any part of a solid surface that is actually studied by mass spectrometry is destroyed in the course of the study. It is only the small mass requirements that make mass spectrometry a desirable technique for the purpose at all. Efficient ion optics and detector systems serve to keep the mass requirements so small that in many situations only a negligible amount of the available sample is consumed during the investigation. This low sample consumption has lead to erroneous statements that mass spectrometer methods are nondestructive. In fact, they must be clearly recognized as destructive, as must those studies of surfaces and solids involving irradiation with electrons or electromagnetic radiation (particularly hard x rays and γ rays) in which any type of surface change, no matter how slight, results. Experimental convenience usually dictates that the specimen mounted in the mass spectrometer has a surface much greater than the part of it actually used for sampling purposes. The unused portions of the specimen are acceptable for use in any further studies of the specimen. The recovery of unchanged specimen material from the mass spectrometer after the investigation often approaches 100 % in routine bulk analysis.

The mass of material actually consumed in a mass spectrometric analysis or study depends on at least five factors.

1. Efficiency of conversion of specimen material to ions. This is usually low and possibly in the 0.001–0.01 range for many source arrangements.
2. Efficiency of the ion optics in gathering the ions produced and delivering them to the detector system. This is often taken to be about 0.01, but can be raised to nearly 1 with some of the highly efficient systems now in use for studying secondary ion emission.
3. Efficiency of the ion-detection system. This can range from unity for efficient pulse counters down to 0.0001 or less for current-measuring devices.
4. The desired relative precision of detection. This is inversely proportional to the square root of the number of ions of a given type actually measured. (This characteristic is closely associated with 3.)
5. The number of types of ions to be studied and the capability of the mass spectrometer for making two or more of these measurements simultaneously.

The overall design of the mass spectrometer is one of compromise because efforts to excel in one area may involve design restrictions that necessarily yield a lowered efficiency in another part of the system. The low efficiency of current-measuring detectors is, in theory, compensated for by an automatic improvement in the relative precision of detection simply because a considerable number of ions must be collected to make the current device work. If 10,000 ions must be measured, the relative precision of detection can then approach a theoretical limit of 1 in 100. The various efficiencies of the individual parts of the mass spectrometer system combine multiplicatively to give the efficiency of the system as a whole, and $0.01 \times 0.01 \times 0.0001$ gives the disturbingly low result of 1×10^{-8} as perhaps a typical efficiency of a "mass spectrometer" that might be available for surface studies, with 1×10^{-4} as a theoretical limit for an ideal instrument assigned to a project requiring a 1 % precision. The redeeming feature in the whole situation is the large size of Avogadro's number, 6.02×10^{23} g-mole; it is simply the vast number of atoms in what the scientist might call a microsample that makes surface studies of any type at all possible with any reasonable precision.

The following example is instructive. Take a 0.001-cm (10-μ) cube of material with a density of 3.0 g/cm and an average atomic mass of 50 amu. Simple computation will show that such a cubic speck of matter will contain about 3.6×10^{13} atoms, and removal of a 0.1-μ surface slice from one face of the cube will involve the separation of 3.6×10^{11} atoms from the bulk material. Because the experimenter would like to have 10^8 atoms or more available for consumption in his experiment, his requirements for the surface slice are more than adequately met in the case of major components, and adequately met by minor components down to the 0.03% level on an atom basis. The situation is strained at the trace levels; either a compromise on precision or the use of special high-efficiency techniques is indicated. The fifth factor mentioned above becomes very important in a case such as this, when data on more than one component is to be obtained. If data for ten components have to be obtained by sequential runs, the sample requirement may be increased by as much as a factor of ten. If a system of multiple ion collectors and detectors can be utilized so that several sets of data may be obtained simultaneously, the increase in sample requirements for additional components may be held down substantially. The ion-sensitive-plate detector and the Mattauch–Herzog mass spectrometer geometry to be discussed later are of importance for the analysis of several elements on a surface because data on all can be gathered concurrently. Sequential runs involving several different surface areas on the same specimen introduce also a very serious assumption into the interpretation of results; namely the assumption that adjacent areas have the same composition and that this composition is unaltered during the time that the particular segment of surface is awaiting study in the instrument. This assumption requires that there be no loss of adsorbed material prior to the run and no deposit of sputtered debris from adjacent surface areas subjected to prior investigation. Obviously, repeated experiments are necessary to establish a sound basis for interpretation of experimental results when instrumental design dictates that several parts of the same analysis be carried out in sequence. The essence of the situation is, however, that a mass spectrometric approach is possible if 10^8 atoms of the element to be studied are present, and it may even be possible in special instances with as few as 10^4 or 10^5 atoms. Below these levels, a loss in precision must occur since mass spectrometry is essentially an ion-counting process. If an experimenter is simply interested in the detection of a specified ion arising from a specified point area on a surface, this may be done very sensitively with a combination of a field ion microscope and a mass spectrometer, where the detection of a single atom in principle can be achieved.

However, such a study by itself gives only one datum for a very particular situation, and the experiment has to be repeated many times before the result can be generalized to a conclusion that on the average, say, 1.33 ions of that type can be detected on a point of that type.

2.4. Mass Spectrometric Analysis of Samples Derived from Surfaces

Conventional analysis of samples derived from various sources has long been the stock in trade of the mass spectrometrist. A variety of instruments has been used for this purpose, and the most successful types used for this sort of experiment have typically been magnetic-sector instruments and cycloidal instruments. Since 1960, the time-of-flight and quadrupole designs have been effectively applied to such analyses, as have the more advanced double-focusing types with both electric and magnetic sectors for sharper focusing of ions of various energies. In these applications, the mass spectrometer is used simply as an analytical device, although for convenience it may be directly coupled to a vacuum-extraction or differential-pyrolysis unit to avoid the difficulties incurred in sample transfer between the preparative apparatus and the spectrometer. Space does not permit detailed descriptions of the many instruments that have been used, but they are amply described in many of the books now available on mass spectrometry, such as Beynon's treatise[4] or the chapters on mass spectrometry in books on instrumental analysis. In many cases, similarly prepared samples can and have been analyzed by other instrumental techniques.

Mass spectrometrists, in their search to provide accurate and precise analytical data on samples, have had to investigate surface properties of materials used for construction of the spectrometer inlet systems. Much of this work was done with the 180° and 60° magnetic-sector instruments equipped with electron-bombardment sources for gas and vapor analysis. Figure 21-1 shows a 180° instrument schematically. The sample reservoir necessary to hold a supply of gas or vapor to supply a stream of very-low-pressure gas to the ionization chamber has been the center of these studies. If the gas flow from the reservoir to the ionization chamber is controlled by a gold-foil or other pin-hole leak, the reservoir pressure is maintained typically at a steady value in the range of 10^{-3}–10^{-2} Torr to achieve molecular flow through the leak.[5] Instruments with a capillary leak designed for viscous flow require a reservoir pressure in the range of 3–10 Torr, but a smaller reservoir volume.[6] The molecular leak has been generally used for chemical analysis, and the viscous leak for isotopic-abundance measurements on stable gases such as N_2, CO_2, and CO. To extend the application of the

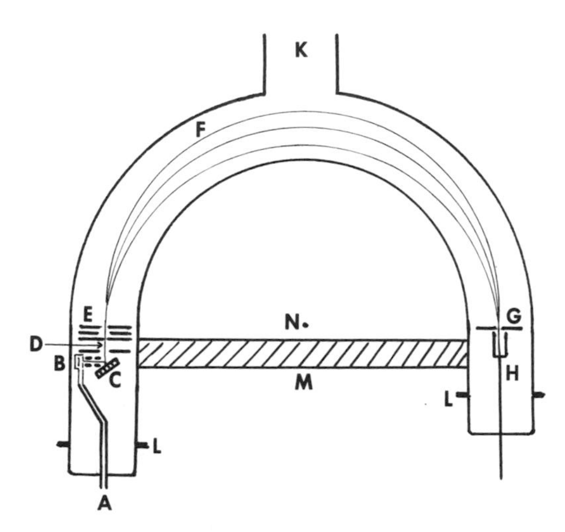

Fig. 21-1. A 180° mass spectrometer equipped with sputtering source and Faraday-cup ion collector. A = gas inlet, B = gas-ionization source, C = specimen, D = region of ionizing electron beam for ionization of neutrals, E = focusing plates of secondary-ion source, F = ion trajectories illustrating direction-focusing principle, G = collector slit, H = collector lead to amplifier, K = pump out tube, L = demountable flanges, M = mechanical brace for rigidity, N = center for central orbit. (Adapted from Honig.[40])

mass spectrometer to systems of low volatility, the reservoir volume associated with the molecular leak is usually maintained at an elevated temperature, and problems of surface adsorption and decomposition resulting from surface and homogeneous catalysis have been encountered. Considerable information about surfaces has been developed in the course of analytical mass spectrometry. An attempt will be made to summarize this information here because it is pertinent to the study of surfaces, *per se*.

The traditional materials of construction for the reservoir volume have been borosilicate glass (Pyrex, Kimax, or similar brands) and nonmagnetic stainless steel, along with small amounts of gold, halogenated polymers (Teflon, KEL-F, and Viton), and in some cases molten indium. Polished and gold-plated stainless steel may be used as a deluxe version of a stainless steel system. Basically the reservoir and associated transfer lines are either mostly glass or mostly stainless steel. With reservoirs operated at room temperature for analysis of light hydrocarbon gases, it was well known that the admission of a water-saturated gas would reveal the analytical history of the system through a preferential desorption of higher-molecular-mass components adsorbed on the inner surfaces of the reservoir. Meyerson devised a method of using this memory effect to increase the sensitivity of a mass spectrometer for the determination of low-volatility components.[7] After heated-reservoir systems were in use, Lumpkin and Aczel made a systematic study of inlet-system performance.[8] They used effusion measurements on the molecular leak made with the mass spectrometer itself to select an optimum temperature for operation of the reservoir surface. At too low a temperature, adsorption of some components was significant; at too high a temperature, thermal decomposition, possibly surface-catalyzed, became too great for accurate analysis. In the oxygenated aromatic system studied they were able to avoid both extensive adsorption and extensive decomposition with a reservoir temperature of 250°C. The author has suggested that specific studies of surface properties could be made by equipping a mass spectrometer reservoir volume with a sample chamber for solid samples of high surface area, but to date is unaware of any research that has made use of the suggested technique.[9]

As examples of a mass spectrometer used in conjunction with basically separate preparative facilities and applicable to some degree to the study of surfaces, the hot-extraction and vacuum-fusion apparatus of Roboz and Wallace and the thermal-analysis gas-release apparatus of Gibson may be mentioned.[10–12] In the first, a high-vacuum gas transfer system for heating the specimen to or near melting was combined with a cycloidal mass spectrometer. The latter consisted of a thermo-analyzer interfaced to a quadrupole mass spectrometer controllable either by computer or experimenter. The adjustable temperature or temperature program in these facilities makes it possible to obtain information on the distribution of gases between bulk and subsurface regions. Data on the bulk material—metal, mineral, pigment, or polymer—are necessary to aid in interpretation of data on partial evolution of gases from solids as it is related to surface composition. Some care is required in the selection of experimental parameters, and each solid system tends to be a small but special research problem.

3. MASS SPECTROMETERS IN SURFACE STUDIES

3.1. Some Economic Considerations

When a mass spectrometer is used for gas sample analysis, the cost of analysis may be fairly stated as about $25 per sample, and thus the cost is in line with outside laboratories performing a similar service on a commercial basis. If the spectrometer is directly interfaced to the sample-preparation apparatus and thus dedicated to a particular service, the cost is obviously greater, perhaps double the cost when the work is done on a service basis. The situation is drastically different when the sole function of the mass spectrometer is surface and solids studies and the equipment is quite sophisticated. Stewart has estimated that an $80,000 surface analyzer will cost about $25 per hour to operate, including depreciation or leasing cost, electricity, servicing, liquid and dry nitrogen, and the time of a single operator, but exclusive of any overhead costs.[13] A spark-source mass spectrometer with appropriate accessories including a plate reader is more likely to wind up costing over $125,000, and if computer interfacing and a small dedicated computer are added, $150,000. For an ion microprobe or ion microscope, the inclusive instrument cost is more like $200,000. Thus a realistic cost of in-house studies using these techniques is $40 per hour with a single operator. In the author's experience, any new and sophisticated installation will not be effectively utilized with less than three persons assigned to it, and thus the cost goes to $55–60 per hour. Any reasonable assignment of overhead immediately raises the cost to around $100 per hour, or perhaps as high as $150 per hour in a high-overhead establishment. The fact that solids mass spectrometers and ion microprobes and ion microscopes are very expensive research facilities is abundantly obvious. Their acquisition for in-house research programs can only be justified if a viable program and need are clearly demonstrated, and the economic balance between the need and the cost of an alternative purchased service is favorable. These instruments must be worked effectively to have a chance of paying off their initial cost, and attempts to stretch the equipment budget at the expense of auxiliary facilities and personnel will almost certainly lead to some agonizing retrospection when the showpiece fails to produce as expected.

3.2. The Spark-Source Mass Spectrometer

The spark-source mass spectrometer is a versatile instrument initially intended for bulk analysis of high-purity materials, but in many instances it is capable of some surface investigations without any modification. With minor modifications, it becomes a very flexible device for the study of solid surfaces and subsurface layers. The instrument, its variations and component parts, its applications, and its limitations are thoroughly treated in two books edited by A. J. Ahearn.[14,15] There are several commercial models of the instrument available, variously manufactured in the United States by the Nuclide Corporation and the DuPont Instrument Division (formerly the

Analytical and Control Division of Consolidated Electrodynamics), in Great Britain by Associated Electrical Industries, in Germany by Varian-MAT, and in Japan by Japan Electrooptical Corporation. A few laboratories have constructed custom-made instruments; these include Bell Laboratories, the Westinghouse Research Laboratories, and some university or research institute laboratories in the U.S.A., Germany, and the U.S.S.R. Instruments may be found in Canada, Norway, Sweden, France, and South Africa as well as in the countries already mentioned. Overall, between 80 and 100 instruments have been constructed. Some of the commercial instruments have been heavily modified to suit the user's specific needs. The instruments are generally large and will not fit conveniently into a single laboratory module of 10 × 20-ft in size. Air conditioning and humidity control are recommended, as is a vibration-free, load-bearing floor. Since photographic processing is necessary for the ion-sensitive-plate detectors used with these instruments, darkroom facilities are required. J. W. Guthrie at the Sandia Laboratories in Albuquerque, New Mexico, and the author when he was with the U.S. Steel Research Center in Monroeville, Pennsylvania, opted for the darkenable instrument laboratory rather than the more conventional darkroom. Modest sample-preparation facilities are desirable in an adjacent laboratory, and cleanroom conditions are preferred to conventional air conditioning for the instrument laboratory.

3.2.1. Instrument Geometry

Spark-source mass spectrometers show little of the variety in basic design that is characteristic of mass spectrometers for gas analysis. The operation and characteristics of spark sources provide restrictive conditions that can be met only with a limited number of ion optical systems. The rf-spark source was first developed by Dempster and, although variations of the basic electrical-discharge principle can be utilized, the Dempster-type source is the most widely used for elemental analysis of solids.(16) The spark between two self-electrodes to be studied is excited by a train of megahertz-frequency high-voltage pulses with an adjustable repetition rate of 1–3000 pulses per second. The duration of the pulses can be adjusted in many instances, but the typical pulse length lies in the range of 10–100 μsec. Because the discharge is established in a very high vacuum, the spark gap is small, perhaps 0.05–0.2 mm, and the voltage required to initiate and maintain the spark is 500,000 volts and up. Practically all material involved in the spark is atomized, and much of it is ionized. The ions produced have a wide spread of kinetic energy and a wide spread of ion charge level. The spark intensity tends to vary considerably over the short term with many sample materials, and the resulting ion production also varies. The spread of ion energy has forced the selection of double-focusing ion optics (those that focus ions according to the mass-to-charge ratio irrespective of energy) on instrument designers, and fluctuations in intensity have similarly placed an emphasis on simultaneous measurement for all ion beams. The simultaneous-collection feature dictates a planar focal locus for the ion optics because the ion-sensitive plate that can be used for simultaneous measurements is made on flat glass. The Mattauch–Herzog ion optics are double focusing to a planar locus, and are used for almost all spark-source mass spectrometers.[17] More complicated ion optical systems with the necessary properties are known, but are used infrequently. Figure 21-2 shows a schematic of the mass spectrometer geometry. The ions accelerated away from the spark by a dc accelerating voltage of 12–30 kV pass through a slit to a field-free region and then through another slit into an electrostatic lens. This lens may be either cylindrical or spherical; the spherical version provides a geometrical focusing in the direction of the magnetic field, while the cylindrical version merely permits the beam to diverge steadily in this direction. The exit slit of the electric lens defines the energy range of the ions passing into the second field-free region. After a second field-free region, the ions enter a magnetic field where they are separated according to mass-to-charge ratio. After a 90° deflection in the magnetic field, the separated ion beams emerge

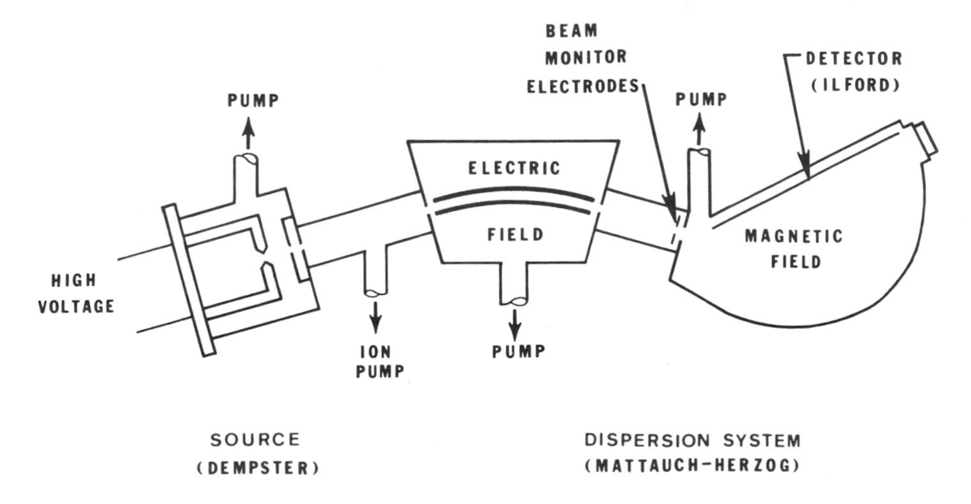

Fig. 21-2. A solids mass spectrometer equipped with Dempster-type spark source, Mattauch–Herzog ion optics, and ion-sensitive plate detector. (From McCrea,[31] by permission from *Applied Spectroscopy*.)

and impinge at 45° on the ion-sensitive plate detector located in the focal plane at the edge of the magnetic field. The angles and lengths of the ion optics are calculated according to theoretical formulas and must be precisely and accurately maintained to achieve an efficiently operating mass spectrometer with high resolution.[18]

A number of spark-source instruments are equipped with facilities for electrical- or multiplier-type detectors as an adjunct to the ion-sensitive plate. The relative importance of the electrical and plate detectors depends largely on the particular type of problem under study, but only the plate provides truly simultaneous data on all ion beams. The ion-sensitive plate in most common use is the Ilford Q-2 low-gelatin silver bromide plate.[19] This plate was developed especially for mass spectrometry at the suggestion of Aston, and is also used to some degree in visible and ultraviolet optical spectrography. The term photoplate, which the author considers to be jargon, derives from the spectrographic uses. Ion-sensitive plate conveys the meaning of a class of special products particularly sensitive to ions, although not necessarily particularly sensitive to photons. Indeed, detectors with an evaporated-silver bromide plate have a rather low photon sensitivity, and the exposure processes by ions and photons follow different mechanisms. A recording microphotometer is almost essential to effective quantitative data reduction for the position and blackness of the lines recorded on an ion-sensitive plate, although some individuals have done remarkably well with visual inspection and estimation techniques. Rapid data-reduction from plates with a large number of lines present can only be accomplished if the microphotometer is computer-interfaced and properly programed.

3.2.2. Source Adaptations

Because specimens used as self-electrodes in spark-source mass spectrometers have to be carefully positioned to maintain a suitable spark, all instruments are equipped with at least a passable set of electrode manipulators adjustable from outside the high-vacuum system and with a viewing port for watching the adjustments made. Adaptations for surface studies differ from those useful for bulk analysis of solids. For example, automatic spark-gap controllers useful for improving quantitative precision of bulk analysis will be a hindrance in surface studies. On the other hand, special surface scanners can be used to obtain averaged bulk analyses over a considerable volume of a specimen.

3.2.2.1. Point-to-Surface Probing

The existing facilities in spark-source mass spectrometers can be used directly for surface studies, including the identification of surface layers, stains, inclusions, and composition gradients. In place of the two self-electrodes frequently used for bulk analysis, the specimen is used as one electrode and a finely pointed wire or probe of a high-purity metal with only one or two naturally occurring isotopes is used as a counter electrode. Gold and silver are two useful elements for this purpose. The counter-electrode material must be selected with regard to the composition of the surface studied; for example a gold counter electrode would be a foolish choice if the nature of a partially gold-plated surface were to be investigated. If the surface is nonconducting, it may be necessary to apply a very thin layer of a conductive overcoat by painting, evaporation, or vacuum deposition in order to obtain a spark.

The first adaptation that is desirable for surface studies with a spark source is a long-focal-length microscope mounted on a track or stand near the viewing ports in the source chamber. The unaided eye is not adequate for observing specimens under the rather poor conditions provided by standard viewing ports. An additional viewing port to provide for inspection of the specimen from a different angle would be quite helpful. These aids of course are only effective for convex, plane, and modestly concave specimens. Although difficult to cope with, hollow specimens are not completely out of reach. Malm has reported an instance where a hollow object was successfully surveyed by manual control of the electrode manipulators using the ion monitor current and the spark pulse waveform on an oscilloscope as indicators of spark behavior.[20] Visual observations in the source region are also subject to clutter resulting from incidental reflections from the polished metal interior of the chamber and from the multiple-layer construction employed on some viewing ports. A port may have three glasses, a sealing glass to form the vacuum wall, a lead-glass window outside, and a disposable antisputter glass inside the vacuum system. The lead glass is needed to protect the viewer from x-rays generated by the high-voltage discharge of the spark, and the disposable glass is useful for intercepting material traveling by line-of-sight from the spark to the port. The vapor-deposited material on the inner face of this glass would otherwise build up on the main vacuum window and gradually reduce its light-transmitting capability.

3.2.2.2. Point-to-Rotating-Cylinder Probing

A special source for sparking between rotating cylinders described by Aulinger has the capability of sparking from point to rotating cylinder.[21] The required rotary motions are transmitted into the vacuum system by magnetic feed-throughs. Such a source is ideally suited to studying the surfaces of specimens of cylindrical form, but has been used mainly for improving the accuracy of bulk analyses of material that is somewhat inhomogeneous.

3.2.2.3. Point-to-Rotating-Disk Probing

This technique is useful for study of specimens that come as plane sheets, can be prepared on plane disks, or can be sectioned to give plane disks for investigation. The mechanical design of a successful source for this application is not simple, because the spark must remain geometrically in front of the entrance aperture of the ion optics. The axis of rotation for the disk must be moved in such a manner that fresh surface is progressively exposed in front of the aperture as rotation proceeds, and as this should be done at a constant rate, the angular velocity should be inversely proportional to the distance of the counter electrode from the center of the disk. The axis of rotation must also be aligned almost perfectly perpendicular to the surface of the disk, and the bearings of the shaft must be precise enough to allow rotation without appreciable wobble. These considerations restrict the permissible orientations of the disk to the one perpendicular to the ion-optical axis and those lying at any angle in the axis. The gap between the counter electrode and disk must be adjustable, either by motion of the counter electrode or of the entire rotating-disk assembly.

Hickam and Sandler reported the modification of the source housing of a spark-source mass spectrometer to accommodate a small battery-driven motor with a specimen disk mounted in the optical axis of the instrument.[22] The counter electrode and motor position could be adjusted manually for spark control. A motor with the high rotation speed of 1750 $\min^{-1}$, was selected to give a linear velocity of 2 μ (μsec)$^{-1}$, and each cycle of radio-frequency voltage was applied to a different, fresh surface. The motor and battery assembly operated successfully at both high vacuum and full dc accelerating voltage, and numerous results were obtained with this apparatus. The measured crater depth was about 0.2 μ and the area was about 0.2 μ^2, with an estimated material consumption of 1×10^9 atoms per crater on silicon. Malm has arranged a similar rotating-disk source in which the motor is variable in speed and external to the vacuum system.[23] Motion is transmitted magnetically through a viewing port. The axis of rotation may be translated along the optical axis. More recently, Clegg, Millett, and Roberts[24] and Meyer[25] have described similar source arrangements that achieve axis translation through gearing arrangements coupled to the shaft for the disk (see Fig. 21-3). The spark track over the specimen surface is an arithmetic spiral path when these devices are used. The Clegg device incorporates a ball drive that both translates the axis of

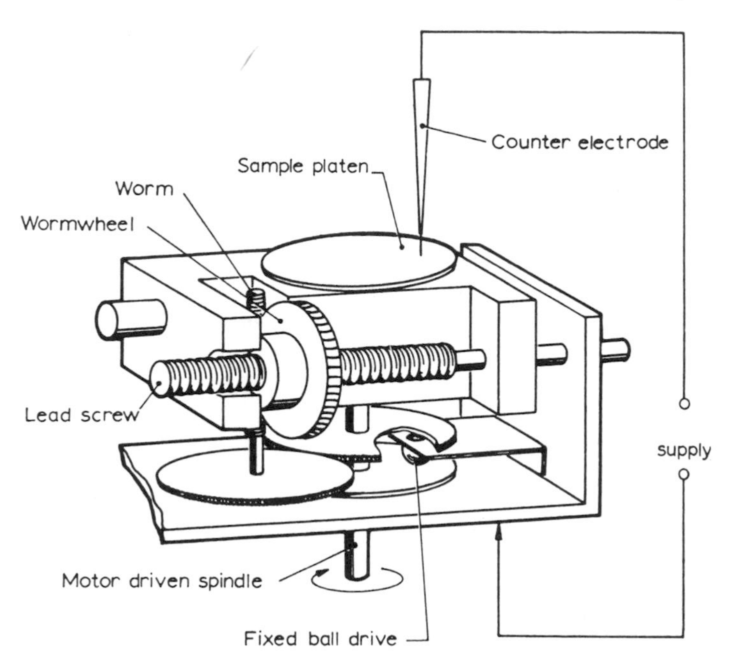

Fig. 21-3. Mechanical design of an arithmetic-spiral-scanning spark source. (From Clegg, Millett, and Roberts,[24] © 1970, American Chemical Society, by permission of the copyright owner.)

rotation and alters the angular velocity for a constant exposure rate of fresh surface area. It is powered by batteries mounted on the unit inside the source vacuum housing of a commercially designed spark-source mass spectrometer. The linear velocity of the area sweep is $6 \times 10^{-3}\ \mu$ $(\mu\text{sec})^{-1}$. This rate is much slower than that given by the high-speed motor of Hickam and Sandler. It was chosen so that the craters of spark trains repeated at a repetition rate of up to 300 sec^{-1} would yield separated clusters of craters resulting from individual radio-frequency sparks of a single rf pulse. The decision to go for separation of spark crater clusters due to different pulses rather than for separation of individual craters due to each individual radio-frequency cycle may well have been prompted by mechanical difficulty in transmitting high-speed motion through a ball drive. Crater dimensions appear to be dependent on the applied rf voltage, and are about 0.4–0.9 μ deep and 80–150 μ in diameter. The disk is apparently mounted in the optical axis, as far as can be told from the figures and data reported.

Meyer, working independently of and more or less concurrently with the Clegg group, tested a rotating-disk source assembly in which the disk is mounted perpendicular to the optical axis.[25] This arrangement can accept large-diameter disks, and specimens of 5.08 cm

(2.0 inch) in diameter are mounted in the source. Two external motors are used, one to drive a lead screw and the other to rotate the disk. The disk rotation can be preselected at any rate up to 3000 min^{-1} by choice of motor speed and drive gearing. An extendible Buna-N O-ring is used as a drive belt; the change in distance between the axis of the rotary feed-through and the axis of rotation of the disk is accommodated by stretch in the belt. The lead screw that translates the disk mount is equipped with a cam-and-stepping relay system that changes capacitors in the pulse repetition-rate circuit, and the repetition rate is decreased in many steps as the distance between the counter electrode and the axis of rotation of the disk is decreased. This programed repetition rate results in a nearly constant number of sparks per unit area over the entire disk surface, and provides for an automatic stop of the scan after a preselectable amount of translation. The rate of translation can be varied between 0.04 and 9.4 cm-min^{-1}. Meyer has found a slow rate of 0.08 cm-min^{-1} and a fast disk rotation of 2000 min^{-1} useful for covering the entire specimen surface with a random overlap of spark craters that are about 0.7 μ deep on silicon surfaces and 2 μ deep on silicon oxide surfaces. Attempts to cover the disks with a continuous spiral of craters resulted in a number of disk fractures ascribed to poor heat transfer, and the more moderate sparking was adopted to avoid fractures.

The author participated in the design of a rotating-disk source with automatic axis translation by means of a worm gear and chain drive operating against a spring-loaded carriage. The motor was mounted externally to the source chamber, and the motion transmitted into the vacuum system by a tumbler-drive rotary feed-through. A flexible shaft inside the vacuum system allowed for the relative motion between the axis of the rotating disk and the fixed axis of the rotary feed-through. A motor with a gearbox was used for adjusting the rotation rate to the pulse repetition rate, and to compensate for the variation of the area-sweep rate with distance of the counter electrode from the axis of rotation. With an external motor mounting, any sort of variable-speed motor could be substituted if desired. Because the system was matched to the pulse repetition rate as Clegg and his co-workers did, the rotation rate was slow. Partly due to the end-flange construction of the mass spectrometer housing and partly because efficient ion extraction was desired, the disk in this source was positioned normal to the optical axis. The only *in vacuo* adjustment provided for the counter electrode was translation parallel to the optical axis. Little difficulty was encountered in shaping the counter electrode externally so that a satisfactory spark could be obtained with the single spark-gap adjustment. Craters about 4 μ deep were produced.

3.2.3. Sample-Preparation Techniques

In bulk analysis it is conventional to cut specimens from the sample in the form of cylindrical or square rods, polish them metallographically in some way, possibly follow with an electrochemical bright-polishing step, and finally wash the specimen with deionized water and a high-purity volatile solvent such as acetone or ethanol. The goal of such specimen preparation is to reduce surface roughness and remove contamination from depressions at the same time. Materials of different properties characteristically require different treatments, and one of the hazards of specimen preparation is the embedding of polishing-compound particles in the surface. Unnecessary steps should be avoided, because each step involves some risk of surface contamination. As final steps, the surfaces are cleaned by outgassing in the mass spectrometer source chamber, possibly ion-cleaned by a very-low-pressure inert-gas discharge, and presparked to dispose of the remaining surface contaminants. If the ions from the first spark are separately detected on an ion-sensitive plate, they are almost always found to be mostly carbon ions, regardless of the nature of the specimen. This of course is a surface-layer analysis.

Cleaning and outgassing for surface studies must be tailored to the problem, or the very information wanted may be cleaned away. The experimenter might very well want to collect and analyze the outgassing material in some way, or attach a residual gas analyzer to the source chamber of the solids mass spectrometer to achieve the same end. The practice of vapor-depositing aluminum as a conducting layer over nonconducting samples has been employed to facilitate bulk analysis, and Orzhonikidze has mentioned that a controlled-thickness overcoat of aluminum on a thin surface layer can be useful in preventing the penetration of the spark through the layer to the base material.[26] Obviously, thickness and composition of the overcoat must be selected carefully if any real simplification of the analytical problem is to be achieved; indiscriminate use of overcoats would simply confuse the situation with ambiguous data.

Special sample-preparation techniques have been developed for use with spark-source mass spectrometers when the surface layer to be analyzed results simply from a deposition of material obtained by other means from surface layers or bulk samples. Selective electroplating of metals on an electrode also suitable for use as an electrode in the mass spectrometer is one example. The continuous-feed evaporator and electrode designed by Ahearn for concentration of impurities in deionized water is another.[27] The final analysis in the spark-source mass spectrometer is an analysis of a surface deposit, but the overall purpose of the analysis is bulk analysis in this case. The spark-source mass spectrometer may also be used for analysis of material stripped mechanically, chemically, or electrochemically in stages from a surface. The service-type analysis of the material from successive layers is a phase of this incremental method for characterizing surfaces.

3.2.4. Analytical Results

The results of investigations on surfaces by the spark-source method may perhaps be characterized as delighting to the elemental analyst on a qualitative basis, interesting to the device scientist and metallurgist, sobering to the scientist interested in high precision and accuracy, and generally discouraging to the structural chemist and surface scientist interested in bonding and atomic and molecular clustering on the surface. These varied impressions are intrinsically related to the ion-formation process in the spark source. There is little likelihood of any possible new apparatus or technique to change anyone's point of view.

3.2.4.1. Spectral Characteristics

With the exception of carbon, and to a lesser degree silicon, molecular ions are secondary features of spark-source mass spectra. Although molecular ions and fragment ions are obtained from high-molecular-mass aromatic hydrocarbons when these organic compounds are packed in metal tubes, a well-mixed system of these materials with high-purity metal powder yields little but ions of the metal and carbon. A typical spark-source spectrum consists mainly of singly, doubly, triply, and multiply charged ions of the elements present with the line intensity decreasing rather regularly with increasing ion charge. Although the intensity gradation is affected by electronic structure in some light elements, the effect is not very distinctive, and was not recognized as a spectral characteristic until spark source mass spectra had been used analytically for several years.[28] The presence of small intensities of $CuFe^+$, Cu_2Fe^+, and similar binary ions in the mass spectra obtained by sparking an electrode of high-purity iron with one of high-purity copper demonstrates that the observation of molecular ions gives no assurance that the elemental constituents were at all intimately related in the solid before sparking. The spark-source spectra, however, do have one feature useful for quantitative evaluation of the data, the tendency of each atom to contribute about equally to the intensity of its particular isotopic line regardless of its chemical identity or state of binding.[29] This is an advantage in survey work; there is little chance that an element will go undetected because it has an exceptionally low sensitivity.

3.2.4.2. Interpretation of Spectra—Qualitative

The initial interpretation that must come before quantitative considerations is assignment of the spectral lines to a specific ion or ions. The basic tool for this task is a table of nuclidic masses, and a little division and combination of numbers will allow a valid prediction of all lines likely to be observed in the mass spectrum of a specimen containing only one or two elements. Computer-generated tables of the predicted lines for all elements are available, so the job need not be repeated by every individual scientist.[30] When many elements are present, a computer may also be used to make the line assignments, but with the simpler compositions encountered in many studies, the tables or even a visual inspection process may be satisfactory.

The problem of interference is present in spark-source mass spectrometry, particularly with elements for which the principal isotopes have even atomic masses. The very high resolution necessary to solve this problem completely is not available in most spark-source instruments; the usual resolution of 2000–5000, although high for an instrument analyzing ions with a wide energy spread, is not enough to avoid many of the possible interferences. The line found for the $^{56}Fe^{2+}$ ion in an iron of low silicon content will generally spread enough to completely mask the weak $^{28}Si^{+}$ line nearby, and the same situation will occur with the $^{28}Si_2{}^{+}$ dimer ion at m/e 56 where the $^{56}Fe^{+}$ line of a small iron impurity in the silicon should be observed.[31] These interferences occur as a result of isotopic mass relationships, and all isotopes of a single element may not be affected, leaving some elemental lines open to interference-free interpretation. The relative concentrations of the various elements in the specimen obviously also play a role in establishing the extent of the interference. Each interference must be considered in relation to the particular problem.

Carbon and oxygen have been troublesome in spark-source mass spectral work. Unless the source chamber is kept clean and evacuated by a high-capacity pump, carbon and oxygen lines extraneous to the true spectrum may frequently be observed. The source of these lines seems to be chemisorbed matter on the specimen and elsewhere on the inside of the source chamber. When this matter is bombarded by neutral sputtered matter from the spark, CO or some similar gas is released and travels about the spark chamber. On ionization of this CO at the spark, the ions produced appear at the detector in the same locations as the carbon and oxygen lines derived from sample material proper. This condition affects quantitative interpretation of carbon and oxygen lines as well as qualitative interpretation.

3.2.4.3. Interpretation of Spectra—Quantitative

The methods for calculating quantitative results from spark-source mass spectra are well established in the case of bulk analysis.[32] Similar methods are applied to the derivation of quantitative data for surface studies, but the need for special calibration and investigation of assumptions in the calculational method is greater with surfaces.

A first step in reducing line blackening data recorded on an ion-sensitive plate to quantitative results in terms of composition is conversion of the blackening data to their equivalents in terms of relative numbers of ions collected. The light not transmitted (100 − % transmittance) through a line on a processed plate developed under standardized conditions is related to the number of ions causing the line by a growth-type function with a saturation value. For the Ilford Q-2 plate, the growth function is very similar to the logarithmic–logistic relation

$$\log\left(\frac{1}{T_r} - 1\right) = R \log E \tag{2}$$

where E is the number of ions giving rise to the line, R is an experimental constant near unity, and T_r is relative transmittance. Here, T_r is measured on a linear scale between the level of the light transmitted through the plate at zero ion exposure T_0 and the level at an area of the plate blackened to the saturation level T_∞ by a very large exposure of the ion in question; T_r is given by

$$T_r = \frac{T - T_\infty}{T_0 - T_\infty} \tag{3}$$

There are a number of other mathematical relations that can be used effectively to represent the results of the ion-exposure process, and graphical methods for handling the situation have also been reported. The author has discussed the relation among these methods in a review of ion-sensitive-plate detectors.[33] Analogous methods can be used in conjunction with computer-controlled plate readers to obtain ion-intensity data from the plates. This conversion of light transmittance to ion intensity is avoided if electrical detection methods are used in lieu of the ion-sensitive plate.

Once ion intensities have been determined, it is usual to assume that the amount of ion detected is directly proportional to the amount of the progenitor isotopic atom in the volume or surface segment actually consumed by the spark. In bulk analysis by the spark-source method, it is found that the proportionality constants for the formation of singly charged ions are about the same (within a factor of 3) regardless of the element or its chemical or physical environment.[29] Regardless of the numerical similarity among the proportionality constants, it is necessary to do some calibration work to obtain accurate results. The field of surface analysis with spark-source mass spectrometers is less well developed than that of bulk analysis, and more calibration will be required to put surface studies on a quantitative basis. The penetration of the spark through the specimen surface is not uniform, and microscopic examination of spark craters on the specimen for depth contours is called for. Electron microscopic techniques have also been used for measurement of spark craters. If the investigation concerned the composition of a surface inclusion, then the relative amounts of inclusion, and matrix material consumed in producing the crater must be determined by microscopic techniques.

It is thus possible to calibrate and get elemental data for surface characterization with a spark-source mass spectrometer. These characterizations may be further studied by stable-isotope tracer techniques, because the mass spectrometer gives data primarily on an isotopic basis. But the very effectiveness of the spark source for elemental and isotopic investigations is directly connected with the general downfall of the technique as a direct means for obtaining molecular information. The energy available in the spark simply atomizes what the spark consumes, and the molecules do not stay intact.

3.2.4.4. Example Applications

Vidal, Galmard, and Lanusse used controlled-thickness electroplatings of nickel on a copper substrate, and then extended their investigations to thin layers of copper on iron and lead on iron.[34] They used a standard spark-source mass spectrometer to examine layers of nickel with thicknesses between 7 and 290 μ, and found the maximum ion yield that could be recorded on an ion-sensitive plate before the lines of the copper substrate appeared. From these data, they concluded that impurities in the electroplated layer could be determined at a level of 1000 parts per million atomic (ppma) in a 10-μ layer, 10 ppma in a 70-μ layer, and 0.1 ppma in a 400-μ layer if the standard sensitivity of the spark-source system for bulk analysis prevailed. Their analysis of a 200-μ electroplated lead coating indicated the presence of C, O, N, F, K, Ca, and Cu in the range of 10–100 ppma, Na, Mg, Al, Si, Cr, Fe, Co and other elements in the range of 1–10 ppma, and Ag, Cd, Sn, Sb, and other elements in the range of 0.1–1 ppma.

In a similar study on gold electroplated by various processes and containing different hardening elements, Vasile and Malm were able to detect the presence of the hardening element and C, N, O, and K.[35]

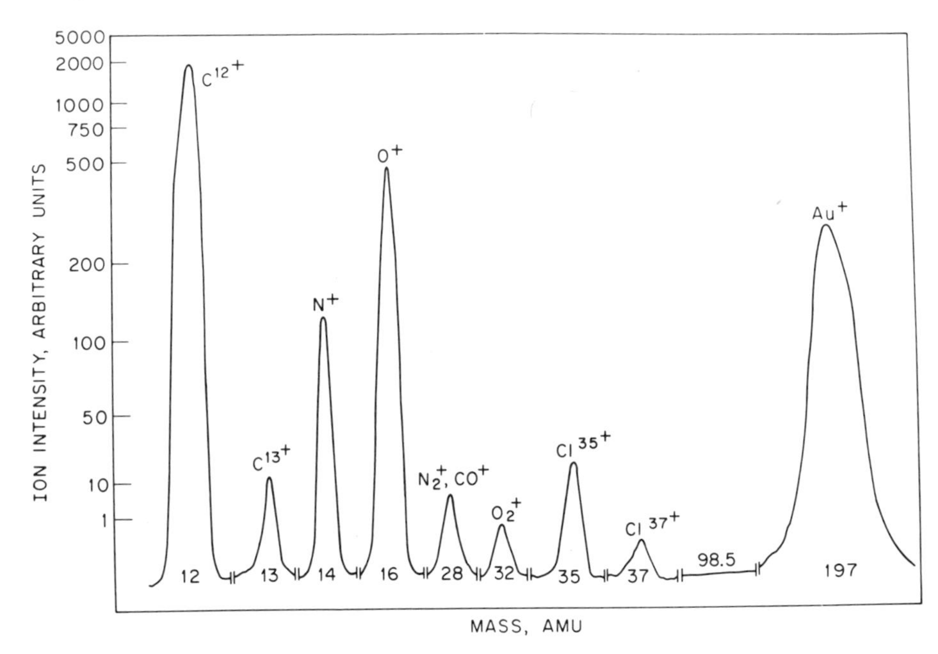

Fig. 21-4. Microphotometer trace of an ion-sensitive plate resulting from a surface analysis of a soft gold electroplate. (From Vasile and Malm,[35] © 1972, American Chemical Society, by permission of the copyright owner.)

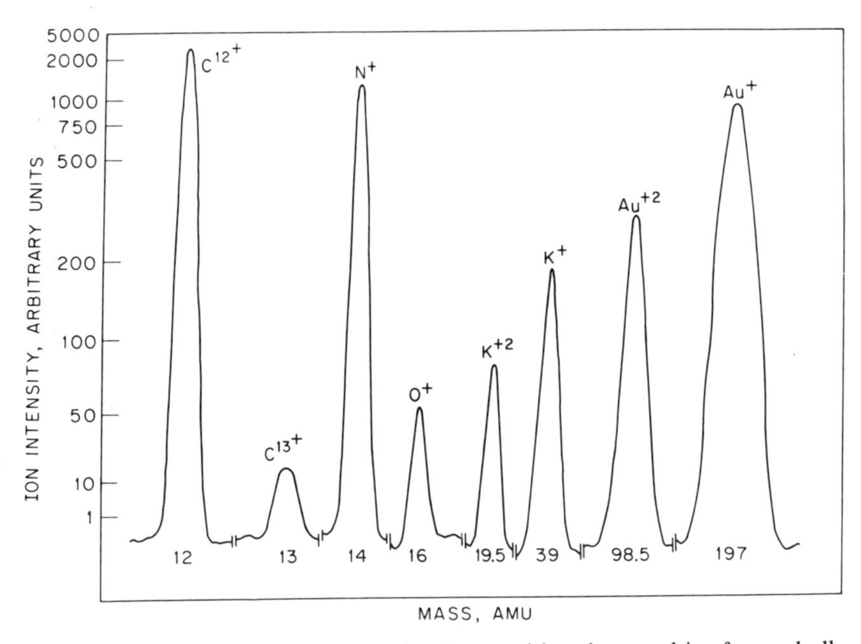

Fig. 21-5. Microphotometer trace of an ion-sensitive plate resulting from a bulk analysis of a nickel-hardened gold electroplate. (From Vasile and Malm,[35] © 1972, American Chemical Society, by permission of the copyright owner.)

The platings were at least 25 μ thick on tapered cylindrical substrates of iron–nickel alloy or high-purity copper, with a rounded tip of about 0.4-mm radius of curvature. The platings were deliberately made thick enough to prevent penetration of the discharge to substrate material. Figure 21-4 shows a spectrum obtained from a specimen without any presparking, and Fig. 21-5 shows a spectrum taken on another specimen after presparking. By use of data obtained from such spectra, they were able to obtain depth profiles of the various elements in terms of ionization fraction. Figure 21-6 gives the profiles obtained for gold and carbon in an indium-hardened gold electroplate.

Hickam and his associates have used the rotating-disk technique to examine known stearate and stearate-soap polylayers.[36] They were unable to detect the presence of stearic acid at 3 and 9 monolayers, probably due to volatilization of the acid prior to sparking. They did not detect barium from 1 monolayer of barium stearate, but detected barium when 2 and 9 layers were present. Evidence of organic fragment ions were noted in the case of the thicker layer, and also evidence of a variation of Ba content with *pH* of the layer-preparation process. In the same investigation, they detected Cu in a nine-layer copper stearate film but failed to find meaningful fragment ions of cholesterol or polyvinylbenzoate under similar conditions. In another investigation of extruded polyvinyl chloride containing a titanium dioxide filler, they were unable

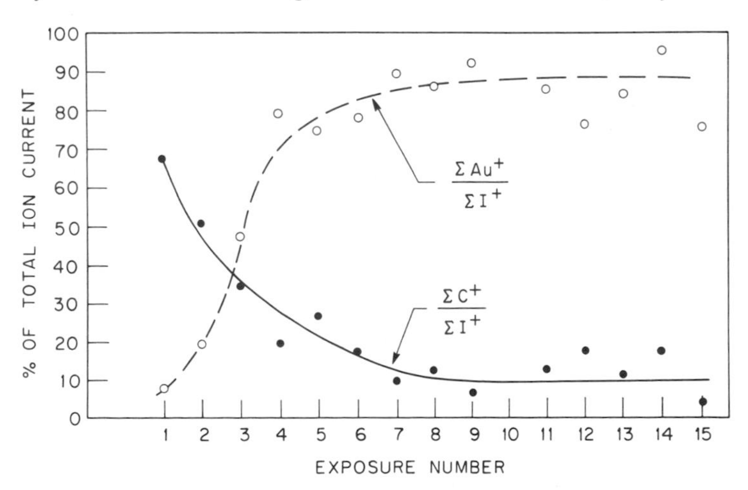

Fig. 21-6. Variation of relative ion yield with sampling depth of an indium-hardened gold electroplate. The quantities $\Sigma Au^+/\Sigma I^+$ and $\Sigma C^+/\Sigma I^+$ represent the percentages of the positive ions found as gold ions of various charge and as carbon ions of various charge. (From Vasile and Malm,[35] © 1972, American Chemical Society, by permission of the copyright owner.)

to detect Ti and concluded that the titanium particles were fully coated by the resin.[37] Meaningful ions were obtained from experiments with disks coated with polynuclear aromatics and coal tar pitch.[38] Clegg's group used the rotary spiral scanning principle in studies of GaAs on germanium, Al on SiO_2–Si, epitaxially grown Si, and also of a boron-doped epitaxial layer of Si on an antimony-doped Si substrate.[24] Their results on the Si on Sb-doped substrate specimens indicated that chlorine had been introduced during growth of the epitaxial layer. The substrate became detectable due to penetration of the spark after removal of about 3 μ of surface.

3.3. The Ion-Microprobe Mass Spectrometer

Herzog, who participated in the development of the ion optical system used for spark-source mass spectrometers, in collaboration with Viehböch, investigated secondary-ion mass analysis as a technique for characterizing surfaces.[39] Honig investigated the ion species sputtered from silver, germanium and germanium–silicon surfaces and reached the conclusion that sputtering combined with mass analysis would be a useful technique for surface characterization.[40] In the development of instruments for systematic study of the field, Liebl started in conjunction with Herzog and went on to develop more elaborate apparatus at other locations in the U.S.A. and Germany.[41–44] These instruments are functionally similar to electron microprobes, and have mechanically controlled stages where the specimens can be readily viewed under a microscope. The primary ion beam, and, alternatively in the case of the latest Liebl instrument, a primary electron beam, can be focused on the specimen where its impact may be quite visible as luminescence over the impacted area. The primary spot can be focused to a diameter as small as 1 μ, and can be positioned electrically over the specimen surface. Secondary ions are collected by a high-aperture double-focusing mass spectrometer, and detected by a highly sensitive ion detector. Only one specific mass can be detected at a time. It is possible to display the ion intensity as the brightness of a spot on an oscilloscope screen; then by synchronously sweeping the primary ion spot on the specimen and the spot representing the ion intensity, a picture of the concentration of that isotope on the surface of the specimen is produced. The picture or isotopic concentration map obtained is achieved at a sample-consumption rate that may be significant if several isotopes of several elements are to be mapped because the pictures must be obtained for each element in sequence. Other instruments that depend on the mass analysis of sputtered ions but that lack precise focusing characteristics for the primary beam will be discussed in Section 3.5.2.

3.3.1. Instrument Geometries

The ion-microprobe designs in essence are a blending of two mass spectrometer systems. The primary ion system consists of a high-efficiency ion source for gases, often a duoplasmatron, with a mass analyzing and focusing system for positioning the primary ion beam on the specimen surface. Secondary ions are gathered from the specimen by a second ion optical system that is double focusing and designed for high ion acceptance and transmission. The optical axes of the primary and secondary systems are usually inclined at about 45°, although a perpendicular arrangement has also been employed.[42] One might expect that angular distribution of the secondary ion production could have a great effect on instrument capabilities. A sensitive ion detector is used to determine the ions in the mass-resolved secondary ion beam because the ion current at this final stage is very low. A schematic diagram of an ion-microprobe system is shown in Fig. 21-7. These instruments are made commercially by Applied Research Laboratories, Sunland, Calif., and the GCA Corporation, Bedford, Massachusetts. AEI Scientific Apparatus is developing a microprobe attachment for their standard spark-source mass spectrometer.

3.3.2. Source Characteristics

In this section, the solid specimen will be considered as the significant ion source for the microprobe, although it is indeed a secondary, rather than a primary, generation of ions that takes place there. In the spark-source mass spectrometer, the basic instrument is intended as a general instrument for the analysis of solids, and solid surface studies are more conveniently handled by modification of the basic source design. The ion microprobe is intended primarily to characterize solid surfaces, and perhaps incidentally to build up information on bulk analysis. All microprobe sources must have positioning controls for locating the primary ion spot with respect to the specimen surface, and probably will also include a spot-size control, a primary ion energy control, and an adjustment for varying the primary ion type. The gas used to produce primary ions may vary, with some limitations as to corrosiveness to the source, at the researcher's will. Argon, helium, neon, and krypton are, in decreasing order of popularity, the inert gases generally used in the primary ion source. A tendency for the secondary ion yield to decrease more or less exponentially with time of bombardment is noted with inert-gas ions. Andersen and others have found that yields stable with time can be obtained if ions of oxygen, chlorine, or a similar oxidizing gas are used to produce the secondary ions.[45] This use of reactive-gas ions obviously is a chemical conditioning of the surface during the

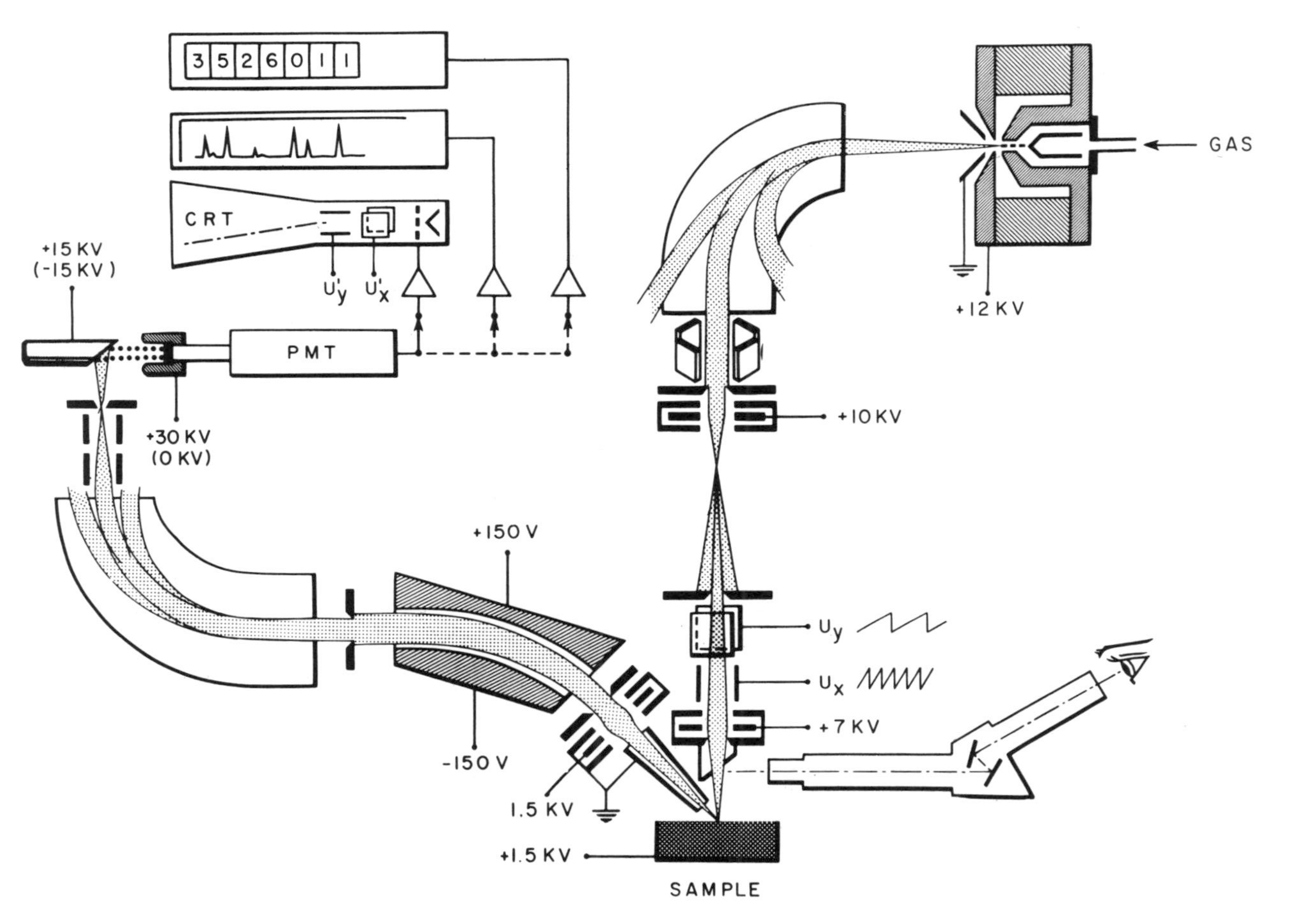

Fig. 21-7. A schematic diagram of an ion microprobe mass spectrometer. The *U*'s represent sweep voltages in *x* and *y* directions: CRT, a cathode ray tube; and PMT, a photomultiplier tube. (From Andersen and Hinthorne[48] © 1972 by the American Association for the Advancement of Science.)

experiment to enhance its secondary ion-emitting properties by maintaining an oxidized surface rather than an ion-cleaned surface. Indeed, Andersen interprets the drop of secondary ion intensity with time under bombardment with inert-gas ions as a result of cleaning the surface oxidation by the primary ions. The composition of the residual gas in the specimen region is also a matter of concern, and massive pumping systems are used to keep residual gas pressures very, very low.

Nonconducting solids pose a particular problem in ion microprobes, because the buildup of charge on the specimen as a result of impact of a positive ion beam causes large local voltages that defocus or deflect the primary ion beam. Conducting overcoats or grids on the specimen can overcome the charge buildup, but the danger of contamination is greatly increased. In the case of the grid there is obviously physical obstruction to examination of the entire specimen. One solution to the charging problem has been the use of a biased hot filament to emit low-energy electrons into the region around the specimen and neutralize the residual positive charge.[46] A second has been to use a primary ion beam of negative ions; this technique apparently works because the impact of the negative ion ejects enough low-energy electrons to maintain a steady-state condition with little net charge accumulation on the nonconducting specimen.[47]

The intensity and area of the primary ion beam can generally be controlled to achieve the removal of matter from the specimen surface at rates between 1 and 100 μ of depth per hour. As the removal proceeds, the area being studied gradually becomes rougher in contour, and the depth resolution of the resulting analytical data deteriorates. The surface characterization in conjunction with the ion planing for depth-profile information is the most significant operational characteristic of the ion-sputtering source. Although the ion microprobe may be used to characterize bulk material, the data are obtained as a set of surface analyses at sufficient depths to give a steady composition that may be considered as a bulk assay. In long-duration microprobe runs on a single region of a specimen, material from the bombarded region tends to deposit as a rim of debris about the bombardment crater, and makes in-depth studies of truly adjacent areas on a specimen difficult to interpret. It should be remembered that the spark craters in the spark-source mass spectrometer also have rims with subsurface debris.

3.3.3. Analytical Results

The characteristics of the ion-bombardment source just discussed have a determining role on the type of results obtainable with an ion microprobe. The analytical characterization is like that in the spark-source mass spectrometer, essentially destructive. The usefulness of chemical conditioning for achieving stable ion yields adds yet another step to the complexity of the interpretative process, as does selectivity in the sputtering. Some of the considerations in relating experimental data to surface composition follow.

3.3.3.1. Characteristics and Qualitative Interpretation of Spectra

The secondary ion yield as revealed by mass analysis is comprised mainly of singly and doubly charged atomic ions of the elements in the surface being subjected to ion bombardment. Triply and quadruply charged secondary ions have low relative intensities, and ions of still higher charge level are very rare. Ion molecules are rather more common than in spark-source mass spectra if the ion optics are operated to accept secondary ions of both low and high kinetic energy. If ions with energies less than 200 eV are rejected, then most of the molecular ions are rejected also.

The qualitative identification of the atomic ions in the secondary mass-resolved spectrum is made by the same methods as with spark-source mass spectra, with mass number and isotopic abundance playing important roles. The molecular ions encountered in the low-energy-ion spectra may cause some problems in identification; they are not typically found in the mass spectra of gases and vapors examined in an electron-bombardment source but represent fragments of the surface contamination material or other features of the surface layers.

3.3.3.2. Interpretation of Spectra—Quantitative

Instead of the simplistic situation observed in spark-source mass spectra where singly charged ion yields per isotopic atom in the specimen fall within a narrow range regardless of the element and surface composition, the yield of secondary ions is drastically dependent on elemental identity and the chemical nature of the surface. The reasons given for the use of oxygen to produce primary ions are convincing evidence of specificity in secondary ion yields. The variation of yield from element to element is reminiscent of properties related to the periodic table, and the scale of the variation is tremendous. The graph in Fig. 21-8 shows a variation over more than four cycles on a logarithmic scale.[48] Even casual consideration of these properties shows a need for conservatism and confirmatory calibration when the secondary ion sputtering process is involved. Along with the need for calibration comes the uneven level of detection present in the sputtering technique. For example, sputtering is an excellent way to look for aluminum, silicon, and titanium, and a poorer one to use in looking for cadmium, tellurium, and gold. Quanti-

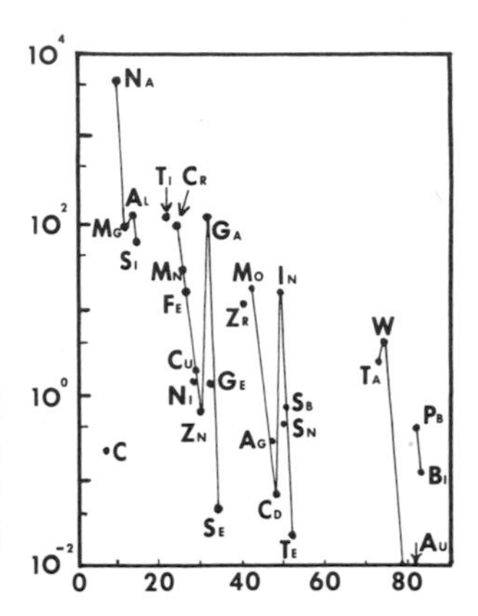

Fig. 21-8. Relative intensities of sputtered singly charged positive ions for pure elements bombarded with 11-kV $^{16}O^-$ ions (ordinate) *vs.* the atomic number (abscissa). The intensities have been corrected for natural isotopic abundances. (Adapted from Andersen.(50))

tative calculations can certainly be made, but the precision is only as good as the instrumental repeatability for that type of surface, and the accuracy is only as good as the relative ion-yield data used. Unfortunately, the ion microprobe is a destructive analysis, and true repetition of any phase of an analytical run is simply not possible.

3.3.3.3. Example Applications

Two reviews on the ion microprobe and its applications have appeared recently. Bayard's article is designed for general reading, but includes an excellent set of beam-scanning images for the mineral helvite—$(MnFe)_4Be_3Si_3O_{12}S$—and a discussion of applications to detecting problems in integrated-circuit devices and determining diffusion profiles in a gold–palladium—nickel contact device.(49) Andersen and Hinthorne present an in-depth review of the instrumentation and its applications to metallurgy and analysis of geological and lunar-soil samples.(48) Andersen has discussed his analytic methods in more detail elsewhere.(45,50) These methods have been applied to lunar samples obtained in the Apollo program.(51,52,53) Thin-film investigations by Satkiewicz have been presented in a project report.(54) Satkiewicz(55) has also applied the ion-microprobe technique to the study of air-borne particulates and their surfaces, and Evans(56) has calibrated the oxygen-in-copper system and used it to determine the variation of trace oxygen in copper in the vicinity of welds. Bayard has applied the technique to analysis of lead-bearing particulates in auto-exhaust samples and found that on the outside of the particles the bromide-to-chloride ratio is substantially greater than at greater depths.(57) Many of these studies have been on materials where the complexity of the system has made preparation of meaningful calibration standards virtually impossible.

Other thin-film studies have been conducted by Tamura, Kondo, and Doi,[58] and Evans and Pemsler.[59] The latter workers were able to confirm their microprobe data by the use of specially anodized tantalum specimens incorporating an ^{18}O tracer. They also followed the phosphorus gradients in specimens prepared under different anodizing conditions (0.9 M and 14 M H_3PO_4). Their results, such as those shown in Fig. 21-9, give firm justification for the use of the ion microprobe in depth-profile analysis. Such justification is highly desirable, because there have been reports of isotope segregations for the lead and lithium isotopes in some specimens.[60] Such reports are disturbing to anyone who has ever experienced the difficulty of fractionation of isotopes by any physical or chemical process that does not involve mass spectrometry, nuclear recoil experiments, or nuclear reactions. Hopefully, scientists encountering evidence of isotope segregation in any specimen will examine their data very carefully to confirm that such segregation is indeed possible on the basis of known processes and is not merely an artifact of the instrumentation or the result of some unrecognized interfering ion in the spectra obtained.

3.4. The Ion Microscope

After work on advanced design electron microscopes, Castaing and Slodzian turned their attention to the analogous instrument in which ion beams carry the image information.[61] The ion microscope in some respects is truly a microscope rather than a mass spectrometer, but its applications closely parallel those of the ion microprobes, and inclusion in a chapter on mass spectrometric methods is warranted on this basis. The developers' concept for the instrument was that of a microscope that displayed an image of the surface with intensities in the image related directly to the concentrations of a given isotope on the surface examined. The image is obtained simultaneously from all locations on the specimen surface. The distribution of additional isotopes must be obtained by a sequence of exposures. The image actually viewed or photographed is one produced by tertiary electrons on a fluorescent screen, but surface material is eroded by the primary ion beam and the image changes appearance as surface composition changes take place during a run.

3.4.1. Instrument Geometry

Either a radio-frequency or a duoplasmatron source may be used to generate the primary ions, which are often ions of inert gases. The ion optical path of this instrument is shown in Fig. 21-10. Secondary ions produced by a uniform incidence of primary ions on a selected region of specimen surface are mass selected and imaged into an ion-to-electron

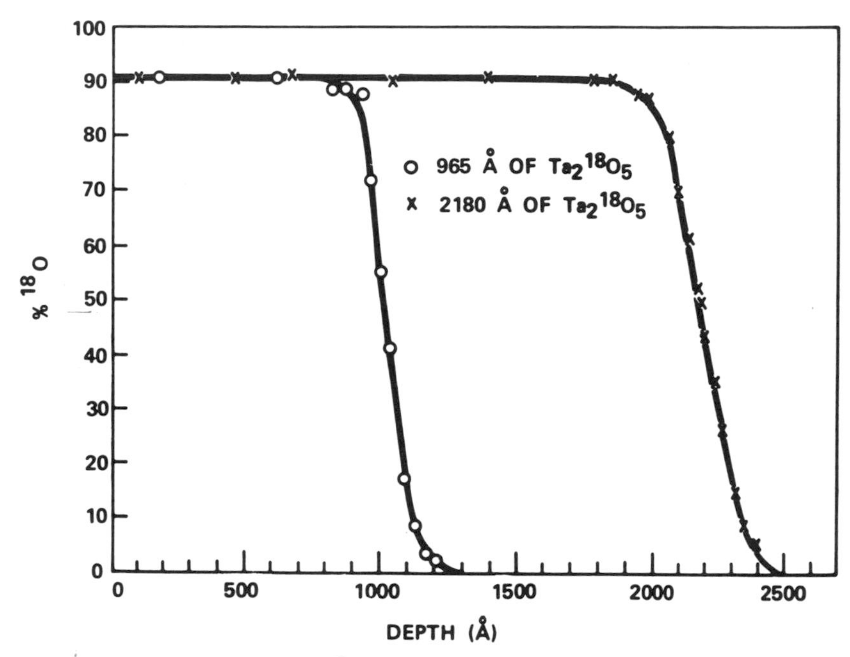

Fig. 21-9. Depth profiles for ^{18}O through a Ta_2O_3 layer anodized on Ta. The anodizing was done stepwise, first in $H_2{}^{16}O$ and then in $H_2{}^{18}O$. (From Evans and Pemsler,[59] © 1970, American Chemical Society, by permission of the copyright owner.)

converter. The tertiary electrons produced by the mass-selected secondary ions are in turn imaged on a sensitive fluorescent screen. The final image on the fluorescent screen may be viewed through a magnifier or photographed for record purposes. Mass resolution in recent instruments is adjustable and may be varied from 40 to 1000. The low resolutions are useful for imaging work with light elements, and the high values for resolving hydrocarbon interferences from elemental ions. The dimensions of the smallest area selectable for study are in the range of 1–5 μ. The rate of material removal for imaging purposes is high and may be as much as 1 μ per minute. A multiplier detector can be used to attain more sensitive ion detection, and a multichannel analyzer with digital storage can be used with the multiplier detector to aid in the recording of depth profiles or the progressive accumulation of data on a number of isotopes or elements. Lewis and Autier have recently described developments and applications of the Cameca Ion Analyzer, a commercial version of the ion microscope.[63]

3.4.2. Analytical Results

Socha has reviewed the applications of the ion microscope used as a microscope or a microprobe to the study of surfaces.[64] Updegrove,

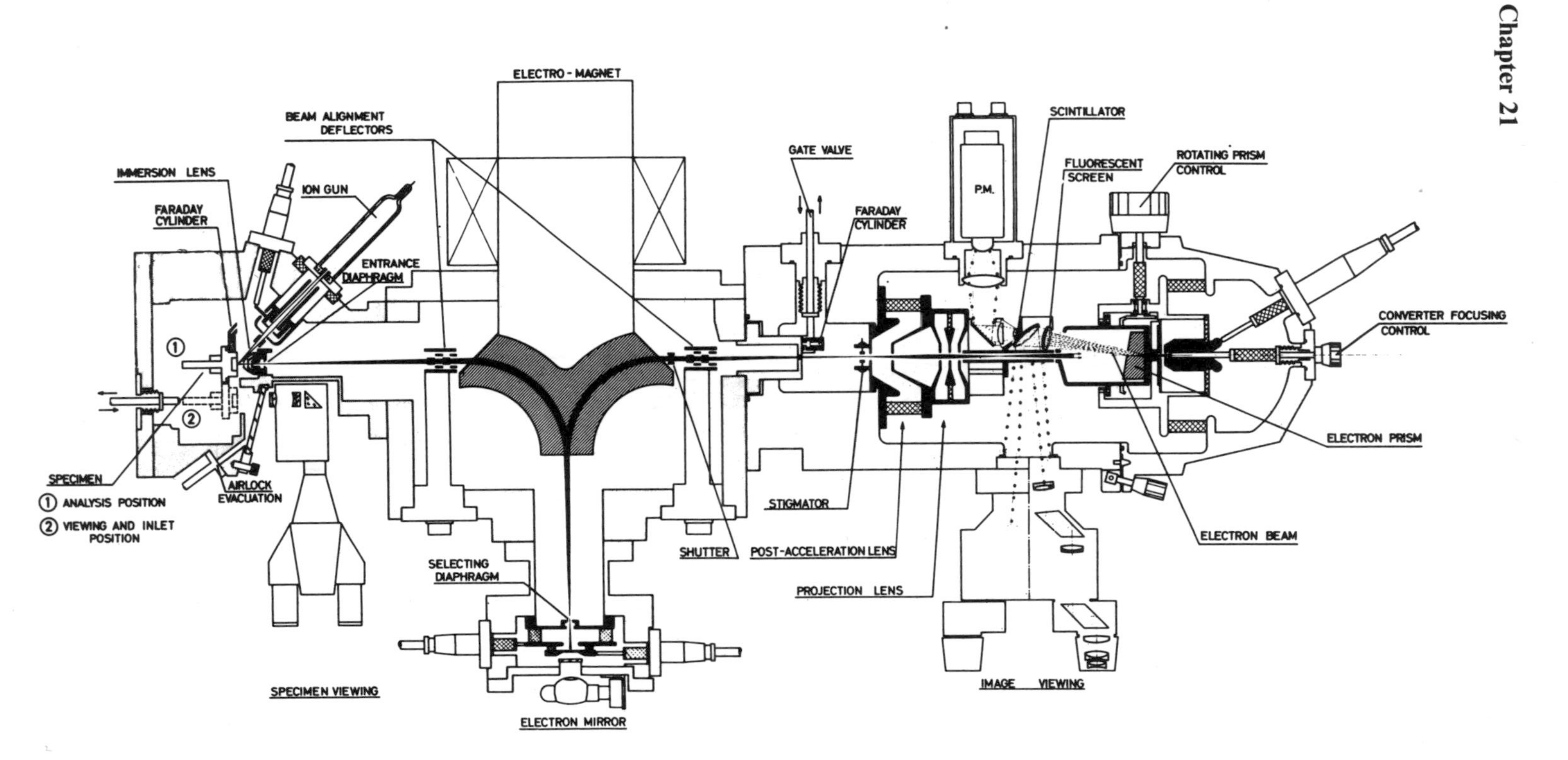

Fig. 21-10. Schematic diagram of an imaging ion microscope system. (From J.-M. Rouberol *et al.*[62])

Oro, and Lewis have successfully used the microscope and microprobe capabilities of the instrument to study meteorites and chondrites.[65] They found chromium segregation in the meteorite and pronounced calcium and potassium segregation in the chondrite. Phillips, Baxter, and Blosser[66] have discussed briefly the investigation of electronic materials, metal surfaces, sectioned teeth, and tissue sections from rabbit and rat aorta. They obtained depth profiles for thermally grown SiO_2 layers on Si, and found that specimens with implanted Al ions confirmed the general nature of the depth profiles. An unexpected and drastic thousandfold reduction in ion yield of all the Si species was noted near the oxide surface. The investigators have provisionally interpreted this as a result of a high defect concentration produced by the ion-implantation process. The work on teeth indicated the relation between the fluorine distribution and the external sources of the fluorine, and the tissue experiments were part of a physiological study on high lead exposure.

DiLorenzo and his co-workers have made a detailed study of homoepitaxial n on n^+ films of GaAs to discover possible reasons for the presence of the high-resistivity regions called the "*i* layer."[67–68] The cause of the high-resistivity layer has been traced to a high concentration of silicon at the interface and a subsequent isolation of the lightly doped layer from impurities segregated at the interface. The source of the silicon was considered to be contamination originating in the reactor hardware rather than a fractionation of impurities in the materials being processed. These DiLorenzo studies form one of the most comprehensive and systematic applications of the ion microscope yet reported.

3.5. Mass Spectrometers as Analyzers for Special Sampling Methods

The methods discussed here represent a more direct coupling of the specimen-sampling and mass-analysis processes than investigations mentioned in Section 2.4. In an attempt to classify the diverse and varying degrees of interrelation between the specimen and the mass analysis, thermal and desorption methods are being considered as sample-preparation methods for a subsequent mass spectrometric analysis. This classification has the awkward effect of classifying a mass spectrometer equipped with a Knudsen-cell source as part sample-preparation apparatus and part mass spectrometer. On the other hand, the classification serves to distinguish laser irradiation, field ion sampling of solids, and ion or electron sputtering as a related group of special ionization techniques.

3.5.1. Laser Techniques

The coherent light beams or pulses from lasers can be focused on a specimen in a mass spectrometer if the source chamber is merely provided with a suitably positioned window, and ion draw-out and focusing plates of the source are arranged so as not to obstruct the laser light. Light focused on small areas of the specimen surface gives a very high energy density, and surface material is decomposed and volatilized as charged and neutral molecular clusters and atoms. The actual proportion of ions produced directly by the laser light is not large, and an auxiliary electron beam, plasma discharge, or spark discharge is useful for increasing the ion yield.

Laser irradiation has some of the interpretational difficulties associated with a high-energy process such as the spark source. The specimens are most certainly thermally decomposed and disrupted. The relationship of the identities and relative intensities of the resulting ions to surface structure must be established by careful reasoning and supported by calibration experiments where necessary. Crater rims about the spots or irradiation also spoil adjacent surface regions on the specimen for further meaningful study.

Honig irradiated specimens in the source region of a slightly modified Mattauch–Herzog spark-source mass spectrometer and obtained spectra for both thermal ions and ionized neutrals from laser-pulse plumes.[69] The ions from single light pulses were detected as separate exposures on an ion-sensitive plate. The microvolumes of specimen sampled by the 1-J pulses were typically 20–100 μ in diameter and relatively very, very deep, with depths of up to 1000 μ. Estimated surface temperatures were 10,000°K for metals and 6000°K for semiconductors and insulators. The total ion yield per pulse was about 10^8 on collection of the thermal ions, and about 10^{10} with auxiliary ionization of the vapor plume. The thermal-ion yield was apparently very dependent on the surface work function as expected from the Langmuir-Saha relation, and the ions from ionization of the plume represented closely the elemental composition of the solid matrix with a proportion of multiply charged ions reminiscent of spark-source mass spectra. Mossotti and co-workers used a low-power laser to study solids, and apparently overcame the disadvantage of using a high-power pulsed laser for surface studies that Honig encountered with excessive penetration depths.[70] Sharkey and his associates have used a laser in two ways in their mass spectrometric investigations of coal, flue gas adsorbants, and air-borne particulate matter.[71] They found it desirable to use a low-power laser as a simple surface-heating device for preparing samples of gas for mass spectrometric analysis and gaining an insight into surface oxidation and surface absorption on adsorbants used in the purification of flue gases. In the study of air-borne particulates and the rapid pyrolysis of coals, they adopted the laser-source–time-of-flight mass analyzer system for use with a high-power ruby laser. The laser–time-of-flight system has also been used by Vastola and his associates in studies of fuels, biological materials, and other solids.[72]

Strasheim, Scott, and Jackson have redesigned the source of a spark-source mass spectrometer for laser-source operation at high pulse power and pulse repetition rate.[73–74] They also move the specimen slowly to present a fresh surface to the laser spot. The primary interest of these workers is bulk analysis of geological materials. Their apparatus gives rectangular line profile on the ion-sensitive plate. Honig reported difficulty in achieving good line profiles, and the reason for the different results is not obvious from a study of the published data.

Laser techniques will without doubt be studied further, and results more closely related to surface studies can be expected as means for selecting pulsing rate, spot size, energy density, and irradiation wavelength with flexibility are developed.

3.5.2. Sputtering-Source Techniques

Sputtering-source techniques with precise primary-beam control and imaging capabilities have already been discussed in Sections 3.3 and 3.4 as ion microprobes and ion microscopes. There have been a number of investigations with much simpler apparatus that utilize the sputtering mechanism of particle production followed by ionization and mass analysis of some of the particles. Lichtman and his co-workers have used electrons as the primary beam, and call their technique electron-probe surface mass spectrometry.[75] They studied hydrogen adsorption on the (100), (110), (111), and polycrystalline surfaces of nickel. An intense H^+ ion was observed whose source was shown to be the metal surface rather than the ionization of the gaseous-hydrogen backfill gas. A single-focusing magnetic-sector mass spectrometer was used for these experiments.

Honig's applications of ion bombardment in sources for 180° magnetic-sector secondary-ion mass spectrometry have yielded an interesting result for surfaces of Ge, Ag, Ge–Si, SiC, graphite, coal, and diamond.[3,40] A study of a stabilized germanium surface that had been cleaned and ethylated by a Grignard reaction yielded complex positive- and negative-ion spectra. The positive-ion spectrum revealed a series of alkylated Ge ions such as $GeC_2H_5{}^+$, $GeC_4H_9{}^+$, and similar ions with two and four less hydrogen atoms. The $GeC_3H_3{}^+$ was also noted.

In total, the mass spectrometric data provided strong evidence that the surface-treatment technique results in actual ethylation of the germanium surface. Kerr reported the construction of a trochoidal-orbit mass spectrometer with an ion-sputtering source. He used it for the study of various glasses.[76] The expected elemental ions were observed, but the data for the ions of silicon and oxygen were interesting. In the positive-ion spectra, Si^+, SiO^+, and $SiOH^+$ were prominent, while SiO_2, SiO_3, and their hydrate clusters were found with a negative charge. Oxygen was removed almost entirely as O^-. Benninghoven has pointed out the desirability of a slow controlled sputtering if there is to be any hope of drawing conclusions about the monolayer characteristics of the surface.[77]

In addition to the more complicated microprobe and microscope instrumentation, some workers with sputtering sources have found it desirable to use double-focusing instruments for the mass-analysis step. Hernandez, Lanusse, Slodzian, and Vidal have equipped a Mattauch-Herzog instrument with a sputtering source without secondary ionization of neutrals.[78–79] They have reported data on copper and aluminum surfaces, and the detection of an upper layer with hydrocarbon components, a middle layer consisting principally of oxides, and the final surface of the bulk matrix. McHugh reported the design and application of a sputtering source and three-stage double-focusing system with a sensitive multiplier detector.[80–82] The $CO_2{}^+$ ions obtained by electron bombardment of carbon dioxide are used as the primary ion beam, and the secondary ions produced are mass analyzed without attempt to ionize any neutrals sputtered. The source is said to yield a very stable beam of secondary ions with a relatively low energy spread. For ease of operation, the source incorporates a ferris-wheel-type specimen changer for loading several specimens each time the vacuum is broken for reloading. McHugh has calibrated this apparatus against mass of ^9Be placed on the surface of a tungsten-ribbon filament and has found that the overall instrument response is linearly related to sample mass over a range of 10^{-12}–10^{-9} g of ^9Be. The tungsten ribbon was electrically heated to about 900°C to prevent the adsorption of any extraneous material on the specimen that might overcoat the beryllium and interfere with its sputter ionization. This finding has an important bearing on the problem of obtaining quantitative data from the more elaborate ion microprobe instruments.

3.5.3. Field Ionization Source Techniques

Field ionization is now applied in at least two ways in mass spectrometry. Field ionization is obtained in gas-ionization sources where a large electric field in the vicinity of a point or edge produces an ionization of complex organic molecules. A much higher proportion of undecomposed or unfragmented molecular ions is achieved than is the case with the more usual electron-bombardment spectra. The surfaces of the solids at the points and edges of field-ionization sources can also participate directly in field ionization with the production of ions derived from the solid surface material. Beckey, who was an early investigator of field ionization for gaseous material, has recently participated in the development of a field-ionization source for both gaseous and solid sample material.[83] The field-forming emitter is dipped into a solution of the sample, and the resulting solid residue is desorbed and field ionized into the mass analyzer. Beckey has called the technique field-desorption mass spectrometry; it gives mass spectra that are indicative of even less molecular rearrangement than the gaseous field-ionization spectra. Müller has used the field ionization of a solid specimen to obtain a microscopic image revealing atomic structural details of the surface[84] (see Chapter 16 for added details on the field ion microscope). To obtain more information on adsorbed gases and some other atoms in alloys, Müller and his co-workers have coupled a field ion microscope with a time-of-flight mass analyzer. This mass spectrometric instrument with a field ion source is called an atom-probe field ion microscope.[85,86] This new apparatus has shown a strong tendency of many metals to form triply or doubly charged ions during the pulsed field-ionization process. Dimer ions and intermetallic ions are formed also, but with considerably lower abundances. The inert gases used as image gases in the field-ion microscope can also be studied with respect to ionization and adsorption on the surface.

4. STATUS, PROGNOSIS, AND INFORMATION SOURCES

From the examples just discussed, it may be seen that a number of mass spectrometric techniques may be usefully applied to the characterization or analysis of solid surfaces. Where the main interest is the variation of composition with depth rather than variation in the surface plane, the spiral-path source used in conjunction with a spark-source mass spectrometer would seem to be the preferred technique. An ion microprobe operated with a defocused primary beam or a simpler sputter-source mass spectrometer might have distinct advantages for the determination of certain elements or the examination of certain types of samples. Where the object of the investigation is composition variation on the surface itself, the less-penetrating ion microscope and ion microprobe have distinct advantages over a probe use of the spark-source mass spectrometer or laser-source instruments. The quantitative situation as regards secondary-ion emission from solids has been summarized by Castaing and Hennequin[87] as follows:

> "We may conclude that our understanding of the intricate physical processes involved in kinetic ion emission has reached the point where quantitative interpretations were possible, at least in the case of metallic alloys. A lot of work has still to be done before we can apply secondary ion microanalysis to the quantitative analysis of any type of chemical compounds.
>
> "Nevertheless, the technique has proved to be most valuable for semi-quantitative applications in the fields of metallurgy and mineralogy, of oxidation and catalytic processes, and for studying the local composition of meteorites. A wide range of applications seems to be opened in the field of biology."

The field-ionization technique appears to offer real opportunities to add more structural information to the knowledge of surfaces than do the spark-source and sputtering methods which are so well adapted to elemental analysis.

Substantial progress in the study of surfaces by mass spectrometric methods can be expected to follow on the advances in instrumentation made between 1960 and 1970. If the history of the application of spark-source mass spectrometry to bulk analysis is used as a guide, the reporting of data on a substantial scale from commercial instruments should follow the installation of the first eight or ten commercially produced instruments by about five years, and continue thereafter for an additional five years before a condition of maturity and a decrease in new applications sets in. This means that surface scientists can expect a continuing flow of new information on the composition and structure of surfaces until 1985, first from ion microprobes and microscopes and then from field-ionization devices. Lasers will play an increasing role on mass spectrometric study of surfaces if means to control the excessive depth penetration are discovered.

Where might the surface scientist not primarily a mass spectrometrist look to keep up to date with the trends of mass spectrometric developments? In the list of references to this chapter, the writer has indicated the number of citations made in most of the source papers. They should be of help in following any aspect of mass spectrometric applications or instrumentation in more detail. Future developments in the field as a whole and in most details of it will receive early coverage at the Annual Conferences on Mass Spectrometry and Allied Topics held in the United States and at the triennial International Mass Spectrometry Conferences held in Europe. Attendees at the American conference and members of The American Society for Mass Spectrometry receive a bound volume of conference papers and abstracts about four months after each conference, but this information is not otherwise distributed by the society or through commercial channels. Commercial publication of proceedings for the European conferences has been traditional, with the Institute of Petroleum acting as publisher of a number of the *Advances in Mass Spectrometry* volumes. Commercial developments in instrumentation and information on new applications of these instruments will be described at numerous conferences on analytical and surface chemistry. The annual Pittsburgh Conference on Analytical Chemistry and Applied Spectroscopy, held in late winter in Cleveland, Ohio or Pittsburgh, Pennsylvania, features a large exhibit of analytical and spectrographic instrumentation in addition to a great selection of papers on the subject. The *International Journal of Mass Spectrometry and Ion Physics* may be expected to carry as many descriptions of new apparatus and techniques for surface analysis and characterization as any other journal. Reviews on mass spectrometry carry considerable data on surface analysis, although the percentage of the references devoted or related to surface studies may be low. The 1972 *Analytical Chemistry* review has 1743 citations.[88] This author considers only about 50 of them to be significantly related to surface studies by mass spectrometric methods. Another 50 might be considered to be less directly related to surface studies, and a number of others concern mainly photoelectron and Auger spectroscopy. The applications to catalytic surfaces are well represented in the 50 core papers. This relatively low representation of surface studies in the entire field of mass spectrometry is likely to continue because of the great number of applications to organic chemistry and biochemistry. A review emphasizing physics rather than chemistry might be expected to show a greater proportion of applications to surface problems. Certainly many papers on the applications of mass spectrometry to surface characterization will appear now that suitable instrumentation is available in many laboratories.

5. ACKNOWLEDGMENTS

The author is indebted to many of his colleagues in the field of mass spectrometry for assistance in the form of reprints and for general comments on the relationship of mass spectrometry to surface characterization. In this connection he would like to thank Dr. Richard E. Honig, RCA Laboratories, Princeton, New Jersey; Dr. C. A. Andersen, Applied Research Laboratories, Inc., Goleta, California; Dr. Edgar Berkey and Mr. W. M. Hickam, Westinghouse Research Laboratories, Pittsburgh, Pennsylvania; Mr. Donald L. Malm, Bell Laboratories, Murray Hill, New Jersey; and Mr. A. G. Sharkey, Jr., U.S. Bureau of Mines, Pittsburgh, Pennsylvania. A special note of appreciation is due to Dr. Rodney K. Skogerboe, Colorado State University, Fort Collins, Colorado, for supplying an advance copy of his manuscript on surface analysis by mass spectrometry that appears as a chapter in Ahearn's book on solids mass spectrometry.[14]

Thanks are extended to the authors and the journals and publishers for permission to use the illustrative material for the figures in this chapter. An important factor in introducing the subject of mass spectrometry at the start of the chapter was the kind permission of the American Society for Testing and Materials to use the standardized definitions from ASTM practices *verbatim*.

6. REFERENCES

1. Recommended practice for evaluation of mass spectrometers for use in chemical analysis, Designation E-137, in *Book of ASTM Standards*, Part 30, pp. 280–283, American Society for Testing and Materials, Philadelphia (1971). No references.
2. Recommended practice for use and evaluation of mass spectrometers for mass spectrochemical analysis of solids, Designation E-304, in *Book of ASTM Standards*, Part 30, pp. 916–920, American Society for Testing and Materials, Philadelphia (1971). No references.
3. R. E. Honig, Mass spectrometric studies of solid surfaces, in *Advances in Mass Spectrometry* (R. M. Elliott, ed.) Vol. 2, pp. 25–37, Pergamon Press, London (1962). 17 references.
4. J. H. Beynon, *Mass Spectrometry and its Application to Organic Chemistry*, Elsevier, Amsterdam (1960). 2213 references.
5. R. E. Honig, Gas flow in the mass spectrometer, *J. Appl. Phys.* **16**, 646–654 (1945). 21 references.
6. R. E. Halstead and A. O. Nier, Gas flow through the mass spectrometer viscous leak, *Rev. Sci. Instc.* **21**, 1019–1021 (1950). 4 references.
7. S. Meyerson, Trace components by sorption and vaporization in mass spectrometry, *Anal. Chem.* **28**, 317–318 (1956). 15 references.
8. H. E. Lumpkin and T. Aczel, Identification and analysis of aromatic oxygenated compounds by mass spectrometry, paper 2, Analytical Division, presented at the National Meeting of the American Chemical Society, Washington, D.C. (1962).
9. J. M. McCrea, The mass spectrometer as an effusiometer, in *Isotopic and Cosmic Chemistry* (H. Craig, S. L. Miller, and G. J. Wasserburg, eds.) pp. 71–91, North-Holland, Amsterdam (1964). 20 references.
10. J. Roboz and R. A. Wallace, Mass spectrometric investigation of gas evolution from metals, paper 33, presented at the Tenth Annual Conference on Mass Spectrometry and Allied Topics, New Orleans (1962).
11. J. Roboz, *Introduction to Mass Spectrometry—Instrumentation and Techniques*, pp. 382–383, Interscience Publishers, New York (1968). 6 related references.
12. E. K. Gibson, Thermal analysis–gas release studies of selected samples via interfaced thermoanalyzer–mass spectrometer–computer system, paper 121, presented at the 23rd Pittsburgh Conference on Analytical Chemistry and Applied Spectroscopy, Cleveland (1972). 2 references.
13. I. M. Stewart, What price surfaces?, paper 153, presented at the 23rd Pittsburgh Conference on Analytical Chemistry and Applied Spectroscopy, Cleveland (1972).
14. A. J. Ahearn (ed.), *Mass Spectrometric Analysis of Solids*, Elsevier, Amsterdam (1966). 243 references.

15. R. K. Skogerboe, in *Trace Analysis by Mass Spectrometry* (A. J. Ahearn, ed.), pp. 401–422, Academic Press, New York (1972), 37 references.
16. A. J. Dempster, Ion source for mass spectroscopy, *Rev. Sci. Instr.* **7**, 46–52 (1936). 15 references.
17. J. Mattauch and R. Herzog, Über einen neuen massenspektrographen, *Z. Physik* **89**, 786–795 (1934). 6 references.
18. J. Mattauch, A double-focusing mass spectrograph and the masses of N^{15} and O^{18}, *Phys. Rev.* **50**, 617–623, and erratum, 1089 (1936). 22 references.
19. Ilford Ltd., New photographic emulsions of interest to physicists, *J. Sci. Instr.* **12**, 333–335 (1935). No references.
20. D. L. Malm, Bell Laboratories (private communication).
21. F. Aulinger, Erfahrungen mit drehelektroden in der massenspektropischen analyse von metallen, *Z. Anal. Chem.* **221**, 70–79 (1966). 10 references.
22. W. M. Hickam and Y. L. Sandler, Mass spectrographic methods for studying surfaces and reactions at surfaces, in *Surface Effects in Detection* (J. I. Bregman and A. Dravnieks, eds.) pp. 189–196, Spartan Books, Washington, D.C. (1965). 8 references.
23. D. L. Malm, Rf spark source mass spectrometry for the analysis of thin films, in *Physical Measurement and Analysis of Thin Films* (E. M. Murt and W. G. Goldner, eds.) pp. 148–167, Plenum Press, New York (1969). 10 references.
24. J. B. Clegg, E. J. Millett, and J. A. Roberts, Direct analysis of thin layers by spark source mass spectrography, *Anal. Chem.* **42**, 713–719 (1970). 11 references.
25. R. A. Meyer, A spiral sparking source for surface analysis, paper J6, presented at the 18th Annual Conference on Mass Spectrometry and Allied Topics, San Francisco, (1970).
26. K. G. Orzhonikidze, V. A. Melashvili, G. G. Sikharulidze, and V. M. Glinskikh, Mass spectrographic analysis of nonconducting solids, in *Recent Developments in Mass Spectroscopy* (K. Ogata and T. Hayakawa, eds.) pp. 362–366, University Park Press, Baltimore (1970); K. G. Orzhonikidze (private communication). 7 references.
27. A. J. Ahearn, Mass spectrographic detection of impurities in liquids, *J. Appl. Phys.* **32**, 1197–1201 (1961). 3 references.
28. J. M. McCrea, Electronic configurations and the spark-source mass spectra of some elements, *Int. J. Mass Spectry. Ion Phys.* **3**, 189–196 (1969). 10 references.
29. R. D. Craig, G. A. Errock, and J. D. Waldron, Determination of impurities in solids by spark-source mass spectrometry, in *Advances in Mass Spectrometry*, Vol. 1, pp. 136–156, Pergamon Press, London (1959). 8 references.
30. E. B. Owens and A. M. Sherman, Mass spectrographic lines of the elements, M. I. T. Lincoln Laboratory Report No. 265 (April 3, 1962).
31. J. M. McCrea, Spark-source mass spectrometry as an analytical technique for high-purity iron, *Appl. Spectry.* **23**, 55–62 (1969). 15 references.
32. A. J. Ahearn, Spark source mass spectrometric analysis of solids, in *Trace Characterization—Chemical and Physical*, National Bureau of Standards Monograph No. 100, (W. W. Meinke and B. F. Scribner, eds.) pp. 346–376, U.S. Government Printing Office, Washington, D.C. (1967). 59 references.
33. J. M. McCrea, Ion-sensitive plate detectors in mass spectrometry, *Appl. Spectry.* **25**, 246–252 (1971). 45 references.
34. G. Vidal, P. Galmard, and P. Lanusse, Dosage des impurités dans les dépots metalliques superficiels par spectrographie de mass a étincelles, *Int. J. Mass Spectry. Ion Phys.* **2**, 373–384 (1969). 22 references.

35. M. J. Vasile and D. L. Malm, Study of electroplated gold by spark-source mass spectrometry, *Anal. Chem.* **44**, 650–655 (1972). 10 references.
36. L. C. Scala, G. G. Sweeney, and W. M. Hickam, Analysis of metal-containing organic thin layers using the spinning electrode spark-source mass spectrograph, *J. Paint Technol.* **38**, 402–406 (1966). 22 references.
37. G. G. Sweeney, W. M. Hickam, and L. C. Crider, Spinning electrode spark-source mass spectrographic study of PVC containing TiO_2, paper 44, presented at the 14th Annual Conference on Mass Spectrometry and Allied Topics, Dallas (1966).
38. T. Kessler, A. G. Sharkey, Jr., W. M. Hickam, and G. G. Sweeney, Spark-source mass spectra of several aromatic hydrocarbons using a spinning electrode, *Appl. Spectry.* **21**, 81–85 (1967). 11 references.
39. R. F. K. Herzog and F. P. Viehböch, Ion source for mass spectrography, *Phys. Rev.* **76**, 855–856 (1949). No references.
40. R. E. Honig, Sputtering of surfaces by positive ion beams of low energy, *J. Appl. Phys.* **29**, 549–555 (1958). 21 references.
41. R. F. K. Herzog, W. P. Poschenrieder, H. J. Liebl, and A. E. Barrington, Solids mass spectrometer, NASA Contract NAS w-839, GCA Technical Report No. 65-7N47 (1965).
42. A. E. Barrington, R. F. K. Herzog, and W. P. Poschenrieder, Vacuum system of ion-microprobe spectrometer, *J. Vac. Sci. Technol.* **3**, 239–251 (1966). 9 references.
43. H. J. Liebl, Ion microprobe mass spectrometer, *J. Appl. Phys.* **38**, 5277–5283 (1967). 26 references.
44. H. J. Liebl, A combined ion and electron microprobe, in *Advances in Mass Spectrometry* (A. Quayle, ed.) Vol. 5, pp. 433–435, Institute of Petroleum, London (1971). 2 references.
45. C. A. Andersen, Progress in analytic methods for the ion microprobe mass analyzer, *Int. J. Mass Spectry. Ion Phys.* **2**, 61–74 (1969). 27 references.
46. R. L. Hines and R. Wallor, Sputtering of vitreous silica by 20 to 60-kev Xe^+ ions, *J. Appl. Phys.* **32**, 202–204 (1961). 9 references.
47. C. A. Andersen, H. J. Roden, and C. F. Robinson, Negative ion bombardment of insulators to alleviate surface charge-up, *J. Appl. Phys.* **40**, 3419–3420 (1969). 13 references.
48. C. A. Andersen and J. R. Hinthorne, Ion microprobe mass analyzer, *Science* **175**, 853–860 (1972). 32 references.
49. M. Bayard, The ion microprobe, *Amer. Lab.* **3**(4), 15–22 (1971). No references.
50. C. A. Andersen, Analytic methods for the ion microprobe mass analyzer, *Int. J. Mass Spectry. Ion Phys.* **3**, 413–428 (1970). 21 references and 8 additional literature sources.
51. K. Fredricksson, J. Nelen, W. G. Melson, E. P. Henderson, and C. A. Andersen, Lunar glasses and micro-breccias: Properties and origin, *Science* **167**, 664–666 (1970). 8 references.
52. C. A. Andersen, J. R. Hinthorne, and K. Fredricksson, Ion microprobe analysis of lunar material from Apollo 11, in *Proceedings of the Apollo 11 Lunar Science Conference* (A. A. Levinson, ed.) Vol. 1, pp. 159–167, Pergamon Press, New York (1970). 17 references.
53. K. Fredricksson, J. Nelen, A. Noonan, C. A. Andersen, and J. R. Hinthorne, Glasses and sialic components in mare procellarum soil, in *Proceedings of the Second Lunar Science Conference*, Vol. 1, pp. 727–735, M. I. T. Press, Cambridge, Massachusetts (1971).

54. F. G. Satkiewicz, Sputter-ion source mass spectrometer studies of thin films, Technical Report AFAL-TR-69-332, Air Force Avionics Laboratory, Wright-Patterson ABF, Ohio (1970).
55. F. G. Satkiewicz, Sputter-ion source mass spectrometer analysis of particulates, paper Q6, presented at the 19th Annual Conference on Mass Spectrometry and Allied Topics, Atlanta, Georgia (1971).
56. C. A. Evans, Jr., Ion microprobe mass spectrometric determination of oxygen in copper, *Anal. Chem.* **42**, 1130–1132 (1970). 10 references.
57. M. Bayard, The ion microprobe, paper 154, presented at the 23rd Annual Pittsburgh Conference on Analytical Chemistry and Applied Spectroscopy, Cleveland (1972).
58. H. Tamura, T. Kondo, and H. Doi, Analysis of thin films by ion microprobe mass analyzer, in *Advances in Mass Spectrometry* (A. Quayle, ed.) Vol. 5, pp. 441–443, Institute of Petroleum, London (1971). 1 reference.
59. C. A. Evans, Jr. and J. P. Pemsler, Analysis of thin films by ion microprobe mass spectrometry, *Anal. Chem.* **42**, 1060–1064 (1970). 19 references.
60. C. A. Andersen, H. J. Roden, and C. F. Robinson, Application of the ion microprobe mass analyzer, in *Recent Developments in Mass Spectroscopy* (K. Ogata and T. Hayakawa, eds.) pp. 214–224, University Park Press, Baltimore (1970). No references.
61. R. Castaing and G. Slodzian, Microanalyse par émission ionique secondaire, *J. Microscopie* **1**, 395–410 (1962). 9 references.
62. J.-M. Rouberd, J. Guernet, P. Deschamps, J.-P. Dagnot, and J.-M. Guyon de la Berge, Secondary Ion Emission Microanalyzer, paper 82, presented at the 16th Annual Conference of Mass Spectrometry and Allied Topics, Pittsburgh (1968). 4 references.
63. R. Lewis and B. Autier, Recent developments on the cameca ion analyzer and new applications, paper 155, presented at the 23rd Annual Pittsburgh Conference on Analytical Chemistry and Applied Spectroscopy, Cleveland (1972).
64. A. J. Socha, Analysis of surfaces utilizing sputter ion source instruments, *Surface Sci.* **25**, 147–170 (1971). 23 references.
65. W. S. Updegrove, J. Oro, and R. Lewis, Direct imaging spatial mass spectrometric analysis of canyon diablo meteorite, pueblito de allende chondrite and Apollo 11 returned lunar material, paper J3, presented at the 18th Annual Conference on Mass Spectrometry and Allied Topics, San Francisco (1970).
66. B. F. Phillips, R. D. Baxter, and E. R. Blosser, Recent analytical data obtained using an ion microanalyzer, paper 156, presented at the 23rd Annual Pittsburgh Conference on Analytical Chemistry and Applied Spectroscopy, Cleveland (1972).
67. J. V. DiLorenzo, R. B. Marcus, and R. Lewis, Analysis of impurity distribution in homoepitaxial n on n^+ films of GaAs which contain high-resistivity regions, *J. Appl. Phys.* **42**, 729–739 (1971). 19 references.
68. J. V. DiLorenzo, Analysis of impurity distribution in homoepitaxial n on n^+ films of GaAs. II, *J. Electrochem. Soc.* **118**, 1645–1649 (1971). 9 references.
69. R. E. Honig and J. R. Woolston, Laser-induced emission of electrons, ions and neutrals from solid surfaces, *Appl. Phys. Letters* **2**, 138–139 (1963). 2 references.
70. V. G. Mossotti, D. W. Golightly, and W. C. Phillips, Laser-assisted spark source mass spectrographic trace analysis of semiconducting and nonconducting materials, paper presented at XV Colloquium Spectroscopicum Internationale, Madrid (1969).
71. A. G. Sharkey, Jr., A. F. Logar, and R. G. Lett, Laser time-of-flight mass spectrommeter studies of sorbed species and surface reactions, paper S4, presented at the 19th Annual Conference on Mass Spectrometry and Allied Topics, Atlanta (1971).

72. F. J. Vastola, A. J. Pirone, and R. O. Mumma, Analysis of biological related organic salts by laser irradiation, paper 105, presented at the 16th Annual Conference on Mass Spectrometry and Allied Topics, Pittsburgh (1968).
73. R. H. Scott, P. F. S. Jackson, and A. Strasheim, Application of laser source mass spectroscopy to analysis of geological material, *Nature* **232**, 623–624 (1971). 3 references.
74. A. Strasheim, R. H. Scott, and P. F. S. Jackson, Laser Source Mass Spectrography, paper 90, presented at the 10th National Meeting of the Society for Applied Spectroscopy, St. Louis (1971).
75. D. Lichtman, F. N. Simon, and T. R. Kirst, Electron probe surface mass spectrometry study of the hydrogen–100 nickel system, *Surface Sci.* **9**, 325–346 (1968). 25 references.
76. J. T. Kerr, The observation of ion and neutral species sputtered from glass surfaces, paper 95, presented at the 15th Annual Conference on Mass Spectrometry and Allied Topics, Denver (1967).
77. A. Benninghoven, Mass spectrometric analysis of monomolecular layers of solids by secondary ion emission, in *Advances in Mass Spectrometry* (A. Quayle, ed.) Vol. 5, pp. 444–447, Institute of Petroleum, London (1971). 7 references.
78. R. Hernandez, P. Lanusse, G. Slodzian, and G. Vidal, Premiers résultats sur l'émission ionique secondaire des couches superficielles d'une cible d'aluminium, *Meth. Phys. Anal. (GAMS)* **6**, 411–413 (1970). 8 references.
79. R. Hernandez, P. Lanusse, and G. Slodzian, Application de l'émission ionique secondaire a l'analyse des couches superficielles, *Comp. Rend.* **271**, 1033–1036 (1970). 8 references, same citations as Ref. 78.
80. J. A. McHugh, A low current density sputtering source and its application to the isotopic and quantitative analysis of nanogram samples, paper 94, presented at the 15th Annual Conference on Mass Spectrometry and Allied Topics, Denver (1967).
81. J. A. McHugh and J. C. Sheffield, Mass analysis of subnanogram quantities of iodine, *Anal. Chem.* **37**, 1099–1101 (1965). 3 references.
82. J. A. McHugh and J. C. Sheffield, Mass spectrometric determination of beryllium at the subnanogram level, *Anal. Chem.* **39**, 377–378 (1967). 4 references.
83. H. D. Beckey, G. Hoffmann, K. H. Maurer, and H. U. Winkler, High and low resolution field desorption mass spectra of organic molecules, in *Advances in Mass Spectrometry* (A. Quayle, ed.) Vol. 5, pp. 626–631, Institute of Petroleum, London (1971). 6 references.
84. E. W. Müller and T. T. Tsong, *Field Ion Microscopy, Principles and Applications*, Elsevier, Amsterdam (1969). 561 references.
85. E. W. Müller, J. A. Panitz, and S. B. McLane, The atom-probe field ion microscope, *Rev. Sci. Instr.* **39**, 83–89 (1968). 11 references.
86. E. W. Müller, The atom-probe field ion microscope, in *Advances in Mass Spectrometry* (A. Quayle, ed.) Vol. 5, pp. 427–432, Institute of Petroleum, London (1971). 15 references.
87. R. Castaing and J.-F. Hennequin, Secondary ion emission from solid surfaces, in *Advances in Mass Spectrometry* (A. Quayle, ed.) Vol. 5, pp. 419–426, Institute of Petroleum, London (1971). 26 references.
88. A. L. Burlingame and G. A. Johanson, Mass spectrometry, *Anal. Chem.* **44**, 337R–378R (1972). 1734 references.

22

IMPURITY-MOVEMENT PROBLEMS IN ANALYSIS METHODS USING PARTICLE BOMBARDMENT

Daniel V. McCaughan and R. A. Kushner

Bell Laboratories
Murray Hill, New Jersey

1. INTRODUCTION

Most of the analysis methods described in this book rely upon the use of charged-particle beams of either ions or electrons to produce secondary species or radiation which on detection reveals the constituents of the materials under analysis. Two generic types of analysis may be distinguished: (1) those in which the bulk is analyzed directly, and (2) those in which the surface is analyzed directly and the bulk characteristics are determined by profiling either by the use of the analyzing beam or by the use of an auxiliary sputtering beam. Typical of the first type would be electron probe microanalysis (see Chapter 18), in which an electron beam of energy of the order of 20 keV is used to excite fluorescent x rays from the solid under analysis. The range of such electrons is of the order of a few micrometers. Analysis methods of the second type include ion microprobe microanalysis (see Chapter 21), ion-scattering spectrometry,[1] Auger electron spectroscopy (see Chapter 20), and SCANIIR (see Chapter 23). The assumption made in many of these analysis methods is that the beam does not disturb the distribution of elements in the solid which is being analyzed. For many methods which use charged-particle beams, this is not, in fact, the case. In this chapter we will discuss the ion-migration problems which may occur in all these methods of surface analysis.

2. ION MIGRATION IN INSULATORS INDUCED BY ELECTRON BOMBARDMENT

Electron probe microanalysis is a powerful technique for the study of solids. Here an energetic electron beam (typically 20 keV) is used to irradiate a specimen. The characteristic x rays from sample constituents are then used to determine the concentration of impurities or bulk elements.

The assumption inherent in such an analysis is that during the period of analysis no change occurs in the sample constituents. That

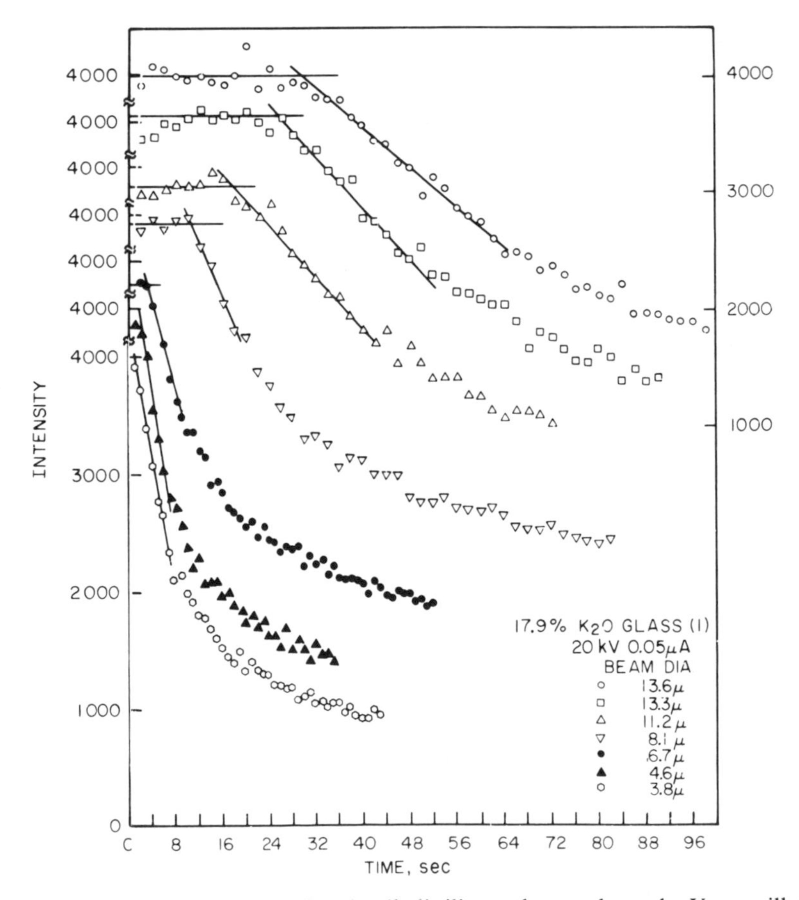

Fig. 22-1. Migration of potassium in alkali silicate glass as shown by Vassamillet and Caldwell.[2] Here the x-ray intensity on 20-keV electron bombardment is plotted as a function of time; the intensity falls with time indicating that alkali atoms are drifting out of the region being analyzed.

this may indeed not be so was shown by Vassamillet and Caldwell.[2] They used a 20-keV electron beam to analyze alkali silicate glasses for their alkali content. Figure 22-1 shows a typical result obtained by them. Here they plotted x-ray intensity *vs.* time and flux. For each flux, after a short "waiting period," measured in seconds, the intensity of characteristic x radiation, from sodium, in this case, fell very rapidly. Similar results were found for potassium. They interpreted their results solely on the basis of temperature rise and diffusion in the samples; this interpretation is probably not fully correct. Electron irradiation at such energy can cause the breaking of bonds and thus the formation of ion–electron pairs. Since the secondary emission coefficient for electrons is greater than 1, the surface of the glass may become positively charged, leading to field-enhanced diffusion of the alkali ions released by the irradiation. That such effects do occur is shown by the work of Aubouchon[3] in which it was demonstrated that alkali-ion migration contributed to the sensitivity to ionizing radiation of transistor devices containing insulating layers.

This type of ion-migration problem will be most troublesome in the analysis of insulators, particularly glasses, containing monovalent impurities.

3. ION MIGRATION IN INSULATORS INDUCED BY ION BOMBARDMENT

The methods of ion microprobe mass analysis,[4] ion scattering spectrometry,[1] and Auger electron spectroscopy[5] all use ion beams to analyze or profile solids. In the first two, the ion beam is used for both purposes.

In ion microprobe mass analysis the first problem which occurs in insulator analysis is that of charge-up of the insulator surface. The method used for reduction of the problem is the deposition of a thin film or grid of gold, say, over the surface. This, however, is only a partial solution; as soon as the ion beam sputters away the metal, the beam ions interact directly with the insulator surface. The theory of this interaction will be discussed in Section 6. One result of the ion–insulator interaction is, however, the transfer of charge from the ions to the surface of the insulator. A potential is therefore developed across the insulator between front surface and substrate, with the surface positive.[6] Ion bombardment of insulators, in the form of thin films at least, can cause a further effect; i.e., the migration of ions away from the surface of the insulator.

The magnitude of this problem has been demonstrated by McCaughan and Kushner.[7–9] They bombarded thin films (up to 5000 Å thick) of SiO_2 on silicon, onto the surface of which had been evaporated up to 2×10^{13} Na atoms/cm^2 (as sodium chloride) which contained ^{22}Na as a radiotracer. Through counting the coincident γ rays from the decay of the ^{22}Na, the profile of ^{22}Na in the SiO_2 film could be uniquely determined by planar etching, with a sensitivity of 2×10^9 Na atoms/cm^2. Samples not subjected to ion bombardment retained all sodium on the outside surface, with no penetration below the first monolayer or so. Figure 22-2 shows the profile of sodium in an identical sample after bombardment by 1×10^{15} Ar^+ ions/cm^2 at 500-eV energy. *Almost 50% of the sodium, initially all on the surface, has left the surface and moved to the SiO_2–Si interface.*

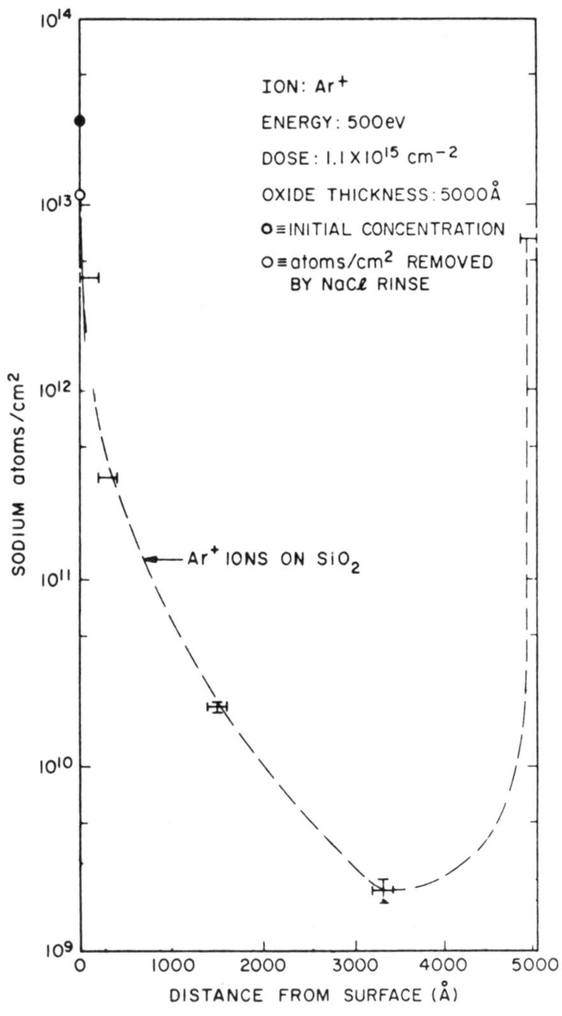

Fig. 22-2. Profile of radiotracer ^{22}Na in ion-bombarded SiO_2, initially all at the SiO_2–air surface, but moved to the SiO_2–Si interface by the ion beam.

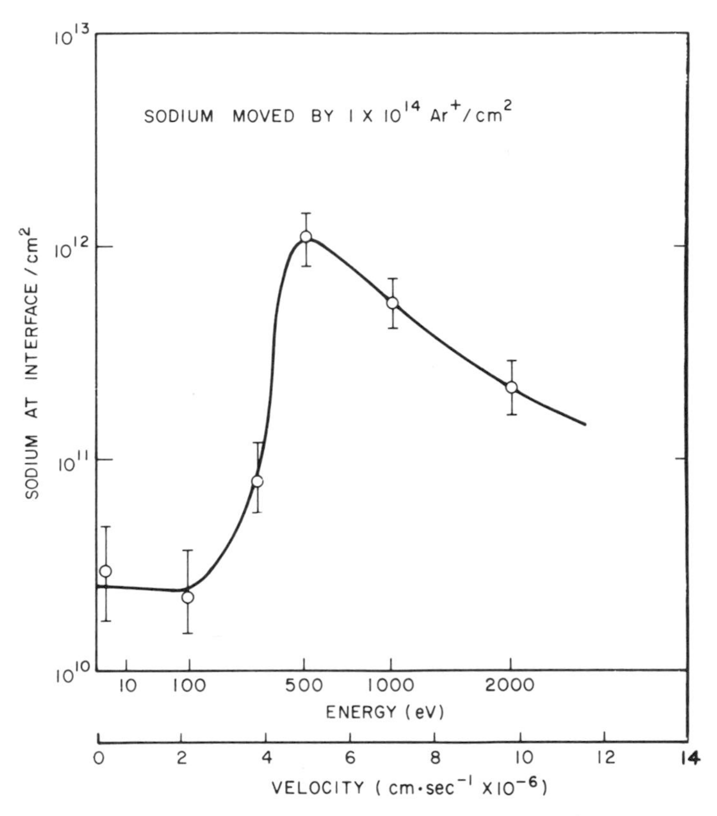

Fig. 22-3. Amount of sodium moved in SiO_2 by Ar^+ ions as a function of energy at constant dose.[9]

Such behavior is also found after bombardment by ions of $N_2{}^+$, N^+, Ne^+, He^+, O^+, and $H_2{}^+$. The effect of Ar^+-ion energy on the amount of sodium transferred across the oxide film is shown in Fig. 22-3.[9] It may be noted that some sodium moves across even at bombarding energies as low as 5 eV and up to 2 keV or more. Oxygen ions, O^+, move sodium similarly at energies up to 20 keV. The effect of ion dose is shown in Fig. 22-4.[9] The amount of sodium moved across rises linearly with dose. The saturation effect appears to be chiefly related to depletion of available sodium from the insulator surface. Efficiency before saturation is of the order of 1 %.

That this may be a major problem in analysis of some insulators by ion microprobe analysis may be understood by a simple dose calculation. Doses of 1–5 mA/cm^2 are common in such machines; 1 mA/cm^2 is equivalent to 6.25×10^{15} ions/sec. Typical sputtering rates are of the order of 1 monolayer/sec, i.e., 2×10^{15} atoms/sec. Thus, before even one monolayer is removed, an ion dose sufficient to

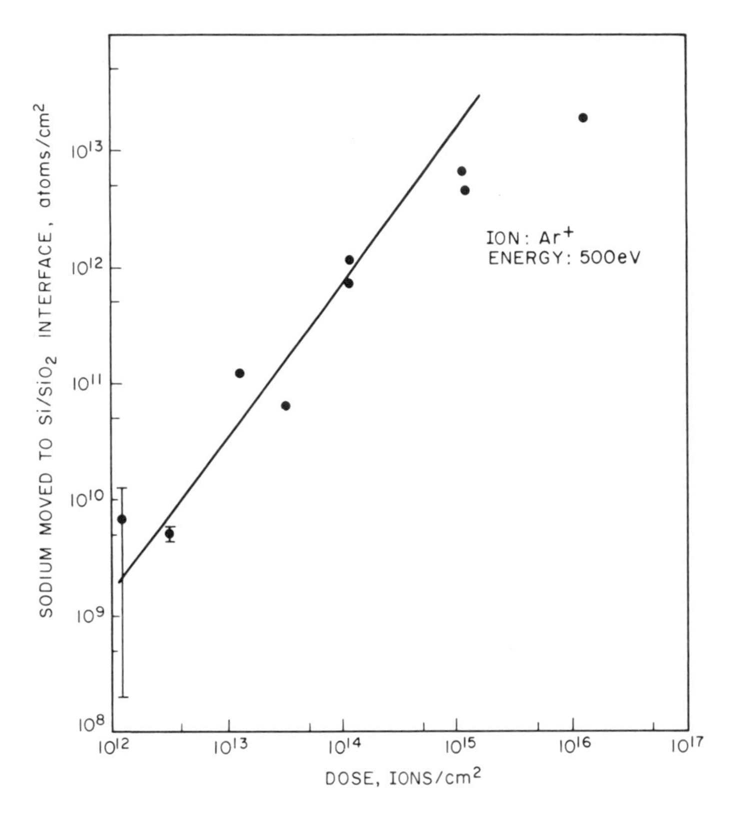

Fig. 22-4. Amount of sodium moved in SiO_2 by Ar^+ ions at 500 eV as a function of dose.[9]

move approximately 6×10^{13} Na atoms/cm^2 away from the surface into the insulator bulk has been given to the sample. We may conclude therefore that profiles of monovalent ions in some insulators obtained by ion microprobe analysis may be suspect.

An example of such problems is given by the work of Baxter.[10] He gives profiles of sodium, aluminum, and potassium in thermally grown SiO_2 films. The K and Na are shown as being piled up at the SiO_2 interface, as shown in Fig. 22-5; the similarity to Fig. 22-2 is immediately apparent. Hughes, Baxter, and Phillips[11] later reported analyses of SiO_2 films which were intentionally implanted with sodium. Typical results obtained with O^+ and O^- analyzing beams are shown in Fig. 22-6. The O^+ caused pile-up of sodium at the SiO_2–Si interface; the O^- did not. Furthermore, in the work of Lorenze,[12] the migration and clustering of alkalies and protons in germanium silicate glasses has been recently demonstrated in ESR experiments.

We conclude then, that depth profiles obtained by ion microprobe microanalysis of some elements in some insulators, in particular, alkalies in SiO_2, may be distorted due to ion-migration effects.

Ion-scattering spectrometry[1,13] is another technique which uses low-energy ion impact on surfaces for analysis purposes. Here ions of noble gases, He^+ and Ar^+, are accelerated to an energy in the range of 500 eV–3 keV and impacted (at 45° to the normal to the surface) on the surface to be analyzed. Those ions reflected at 90° are energy analyzed by an electrostatic analyzer using the formula

$$E_R = \frac{M_2 - M_1}{M_2 + M_1} E_P \tag{1}$$

where E_R is the energy with which the reflected ion leaves the surface, E_P is the energy at which the primary ion reaches the surface, and

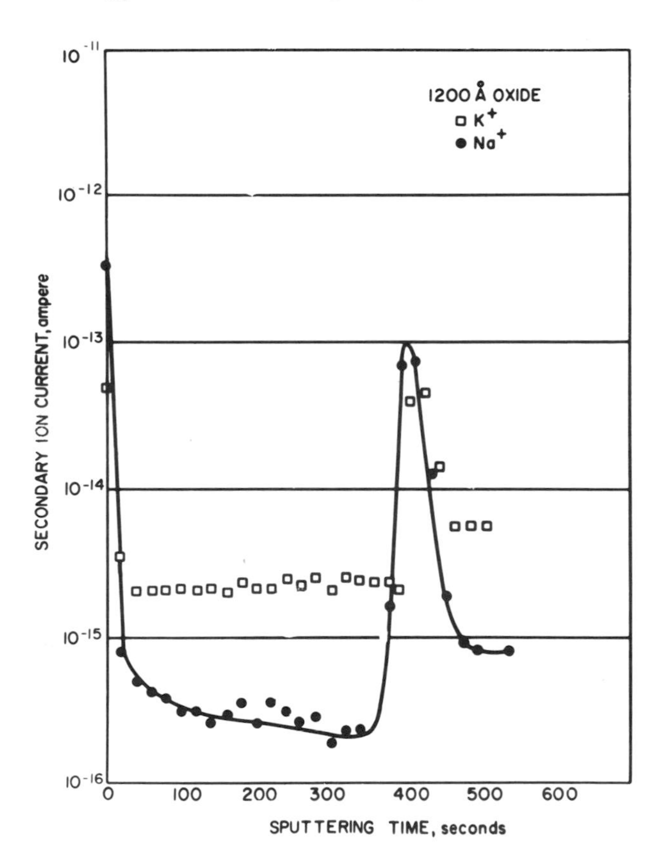

Fig. 22-5. Profile of sodium and potassium in SiO_2 as measured by Baxter[10] by ion microanalyzer. Note the similarity to Fig. 22-2.

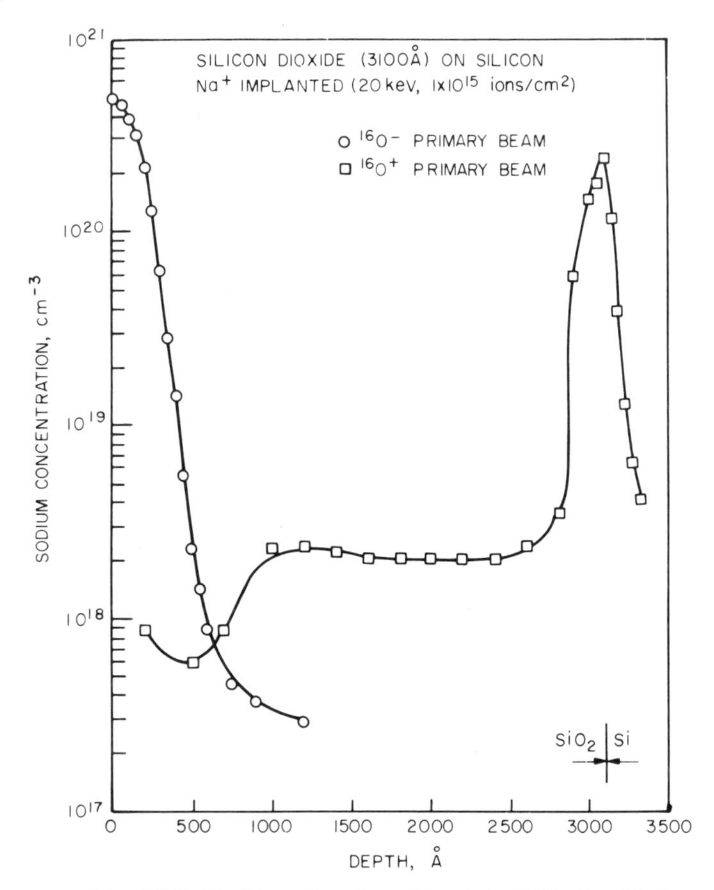

Fig. 22-6. Profiles of sodium found in SiO_2 by Hughes *et al.*[11] by ion microanalyzer. Note the difference in profiles found by using O^+ and O^- primary beams, due to effects caused by ion-neutralization processes.

M_1 and M_2 are the masses of the bombarding ion and target atom, respectively. It is directly assumed that surface binding effects are negligible.

To expedite the analysis of insulators, this method of analysis uses a hot filament placed near the surface to flood the surface with low-energy electrons. This technique is supposed to "neutralize" the surface, and thus, by implication, suppress the processes described above. That this may not be the case is shown in the next section.

4. EFFECTS OF ELECTRON FLOODING DURING ION BOMBARDMENT OF INSULATORS

In some analysis methods using ion bombardment, electrons are caused to flood the insulator surface during the ion bombardment.

For example, in ion-scattering spectrometry[1,13] thermal electrons from a filament are used to flood insulator surfaces during bombardment by Ar^+ or He^+ ions at energies up to 3 keV. In Auger electron spectroscopy, while sputtering is being performed, higher-energy electrons (typically 1 keV) flood the substrate.

The effect of such electron flooding is to reduce the effective field across the insulator during ion bombardment. It does not change the basic atomic processes occurring at the surface, and therefore ion-migration problems can still take place. Recent experiments[14] have demonstrated the effects of simultaneous ion bombardment and thermal-electron flooding. Radiotracer techniques similar to those described in Section 3 were used to monitor sodium profiles during such experiments. In a typical experiment, the effects on sodium migration of bombardment by ions, ions in the presence of thermal electrons, and neutral atoms were investigated. Results of such an experiment are shown in Fig. 22-7.[14] The main points to note are (1) the large amount of sodium moved by ion bombardment, (2) the intermediate (but non-negligible) amount moved by the ion beam in the presence of thermal electrons, and (3) the negligible amount moved by the neutral atom bombardment at the doses used here. It should be noted, of course, that in a typical *profiling* experiment a much greater dose than that used for these experiments will be given to the samples, with a corresponding increase in the amount of sodium moved. It should also be noted that in experiments both with ion bombardment and with ion bombardment with thermal-electron flooding, a spurious peak will be measured at the interface and an erroneous *surface* concentration found.

Recent experiments by M. J. Vasile[15] in which electron flooding was found to be sometimes ineffectual in "neutralizing" the surface of polyethylene lend credence to the picture of charge migration away from the insulator surface during ion bombardment despite electron flooding.

We conclude here that the effect of electron flooding is to decrease the macroscopic field across bombarded insulators. The migration problems associated with ion bombardment are thereby reduced, but not eliminated.

5. ELIMINATION OF ION MIGRATION BY USE OF NEUTRAL ATOM BOMBARDMENT

All the ion-migration problems described in the previous two sections may be avoided by the use of neutral atom bombardment instead of ion bombardment in analysis methods. Figure 22-7 illustrates that neutral atom bombardment gives less sodium movement than ion

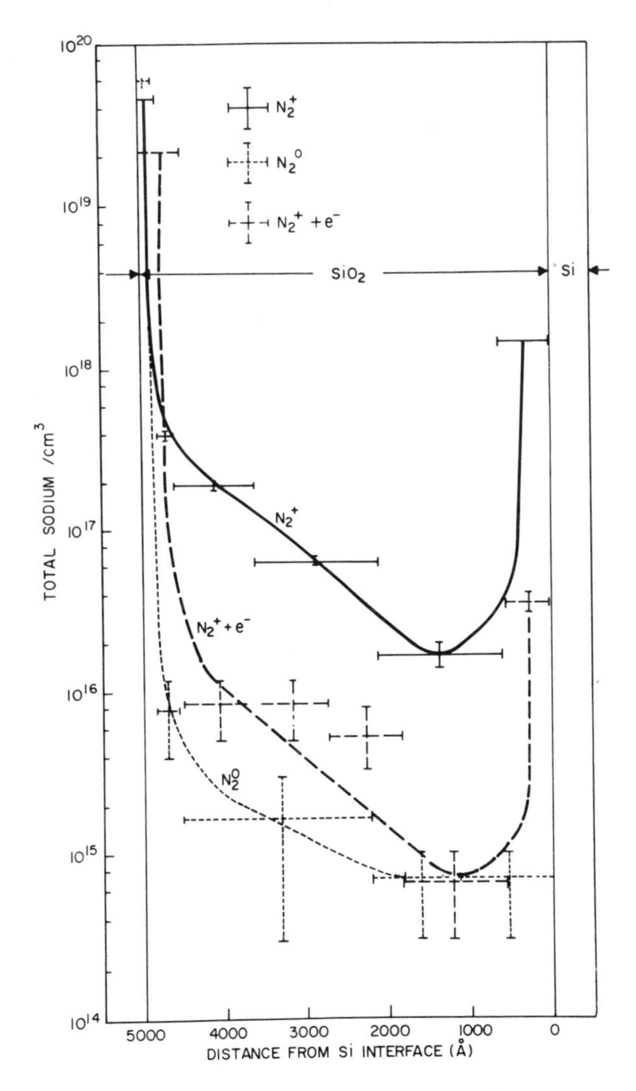

Fig. 22-7. Sodium moved in SiO_2 by N_2^0, N_2^+ and N_2^+ in presence of thermal electrons. Note how (1) much sodium is moved by ions, (2) reduced amount is moved when electrons flood the surface, and (3) no sodium is moved by the neutral atom beam.[14]

bombardment by at least four orders of magnitude, and less movement than with the electron-flooding method by at least two orders of magnitude. The basic reason for this is the absence of the surface charge-exchange process characteristic of the other methods. This has the additional bonus that no surface charge-up problems (other than those caused by secondary electron emission) exist in insulators under analysis by neutral-beam methods. An example of such a method is SCANIIR[20] (see Chapter 23).

6. THEORY OF THE ION–INSULATOR INTERACTION

The theory of the ion–insulator interactions causing ion migration has been discussed by McCaughan and Murphy,[8] and McCaughan, Kushner, and Murphy.[7,9,16] The theory covers the interaction of ions, neutral atoms, and ions in the presence of thermal electrons with an insulator surface. We will primarily discuss here ions of ionization potential greater than about 10 eV.

Consider an insulator such as SiO_2. We may discuss the processes of ion neutralization at the surface in the following series of steps: (1) what happens to the ion as it approaches the surface as a result of dielectric polarization of the surface, (2) what are the neutralization processes at the surface, and (3) what are the subsequent charge-transfer processes. As an example, we have chosen SiO_2 for the insulator since its band structure has been discussed in some detail.[17] Figure 22-8 illustrates the band structure of SiO_2 and the neutralization processes; the band gap and valence-band structure are according to DiStefano and Eastman,[17] though the neutralization mechanisms discussed here are not dependent on the exact valence-band structure for their validity.

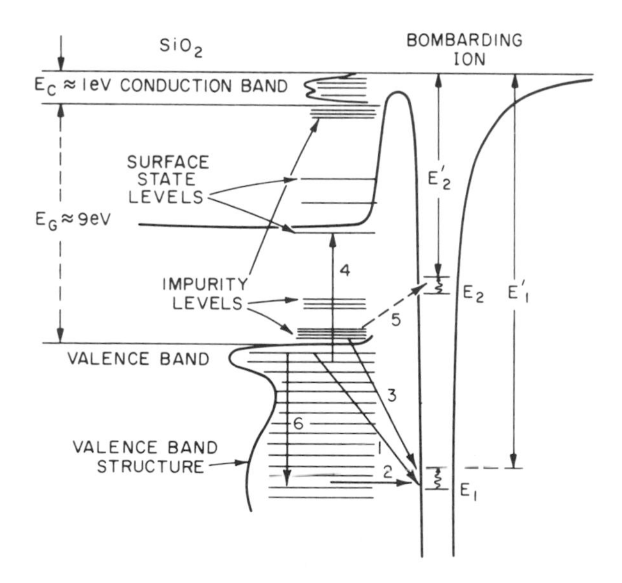

Fig. 22-8. Neutralization processes undergone by ions at an insulator surface. See text for details of processes shown.[16]

As a positive ion, such as Ar^+, approaches the surface of an insulator, its ionization potential is altered by the presence of the surface through image-force effects. The shift in ionization potential may be written[16]

$$\Delta E = \frac{3.6Q^2(\varepsilon - 1)}{D(\varepsilon + 1)} \tag{2}$$

where Q is the ion's charge in units of electronic charge, ε is the insulator's dielectric constant, D is the distance from the surface in angstroms, and ΔE the consequent shift in units of electron volts. This shift in the ionization potential is illustrated diagrammatically in Fig. 22-8.

Impurity levels are shown in the insulator band gap. The exact location of impurity levels at the surface of an insulator such as SiO_2 is not known. The location of those levels shown here, and corresponding to sodium in SiO_2, is inferred from the work of Segal[18] on uv absorption in glasses. As the ion approaches within 3 Å of the surface, the probability of its neutralization reaches a maximum. The type of neutralization processes which it may undergo are illustrated in Fig. 22-8. These are resonance processes and Auger processes, and are only possible if the ionization potential of the oncoming ion is greater than $(E_G + E_c)$, i.e., if it is deeper than the valence band edge.[16] In these processes, an electron from the surface neutralizes the oncoming ion before it actually impacts the surface. The neutralizing electron may come (as in transitions 1 and 2) from a bulk level in the solid or (as in transition 4) from an impurity level. The "up" Auger electron will be raised to a surface-trap level.[19] Transitions from "bulk" levels (1 or 2) give rise to a positively charged "hole" which, depending on the insulator, may or may not be mobile. Transitions such as 4, however, give rise to positive ions of the impurity. An analogous way of looking at the process (say for Ar^+ ions and sodium impurity levels) is by writing the process

$$\equiv\text{Si-O}^-\text{Na}^+ + \text{Ar}^+ \rightarrow \text{Ar}^0 + \equiv\text{Si-O} + \text{Na}^+ \tag{3}$$

The Na^+, previously bound coulombically, is now free to move in the insulator away from the surface.

The effects of increasing ion velocity are to change the ionization potential more than in the zero-bias case and, for high enough primary ion energy, to reduce the transition rates considerably, to the point where the probability of transitions such as those described is considerably reduced at high ion velocity (e.g., for Ar^+ ions $>$ 20 keV).

It will be immediately obvious that these processes can only be undergone by positive ions, the requisite levels not being available in the cases of neutral atoms and negative ions. In addition, fields of the appropriate polarity would not be available when negative ions are used, and reduced fields (caused by secondary electron emission) would occur on neutral atom bombardment. Thus neutral atoms will certainly not cause impurity-migration effects to the same degree as positive ions; this has been shown by White *et al.*[14] for alkali atoms in SiO_2. The case of negative ions is somewhat more complex, in that the neutralization of negative ions on the insulator surface may give very high internal fields in the bombarded insulator, thus causing some field-assisted drift of positive impurity ions toward the insulator surface. We see an excellent example of the effects of these neutralization and migration processes in the work of Hughes, Baxter, and Phillips,[11] where O^+ and O^- ions were used in the ion microprobe (Fig. 22-6). The sodium profile found using O^+ ions was totally distorted by the action of the beam; the profile obtained using O^- ions, however, was also not consistent with the implanted sodium profile, implying that some field-assisted sodium drift toward the surface may have occurred.

The effects of using electrons to flood the surface are simply to fill some of the available surface-trap levels with electrons. There is a limit to the number of electrons ($\approx 10^{13}$ electrons/cm^2) which may be accommodated in such levels, however, due to the large fields and surface-potential build-up which occur.[6] Flooding an insulator surface with thermal electrons does not, of course, neutralize the ions outside the surface (due to space-charge limitations on electron density and the small cross section for such processes). At the surface itself, there are of the order of 10^{17} electrons/cm^2 available for neutralizing oncoming ions from the valence band, while only $\approx 10^{13}$ electrons/cm^2 are present in the surface traps. Thus, the neutralization processes described above for ions continue almost as before in the presence of flooding thermal electrons. The main effect of the electrons is to reduce the *macroscopic* field across the insulator, and thus reduce, but not eliminate, the impurity- and charge-migration effects earlier discussed. White *et al.*[14] have shown that impurity-migration effects occur when N_2^+ ions are used to bombard SiO_2, even in the presence of electron flooding.

7. SUMMARY

Both experiment and theory therefore show that ion bombardment of some insulators may cause impurity-migration effects and consequent profile distortion, even in the presence of flooding electrons at thermal energies. These effects may occur in ion microprobe mass analysis, ion-scattering spectrometry, and in Auger electron spectroscopy when ion beams are used for sputter profiling; electron flooding is only a partial answer to these problems. Neutral atom beams are probably preferable to ion beams in all these applications. Electron bombardment as in electron probe microanalysis can also lead to impurity-migration problems which must be guarded against.

8. REFERENCES

1. D. P. Smith, *J. Appl. Phys.* **38**, 340 (1967).
2. L. F. Vassamillet and V. E. Caldwell, *J. Appl. Phys.* **40**, 1637 (1969).
3. K. Aubouchon, *IEEE Trans. Nucl. Sci.* **NS-18 117**, (1971).
4. R. Castaing and G. Slodzian, *Compt. Rend.* **225**, 1893 (1962).
5. C. C. Chang, in *Characterization of Solid Surfaces* (P. F. Kane and G. B. Larrabee, eds.) Chapter 20, Plenum Press, New York (1973).
6. D. V. McCaughan and V. T. Murphy, *J. Appl. Phys.* **44**, 3182 (1973).
7. D. V. McCaughan, R. A. Kushner, and V. T. Murphy, Abstracts 28 and 29, presented at the Seventh National Electron Microprobe Conference, San Francisco (1972).
8. D. V. McCaughan and V. T. Murphy, *IEEE Trans. Nucl. Sci.* **NS-19**, 249 (1972).
9. R. A. Kushner, D. V. McCaughan, and V. T. Murphy, *Phys. Rev. B* (to be published).
10. R. D. Baxter, Investigation of Thermally Grown SiO_2 films by Ion Microanalysis, Final Report, Batelle Memorial Institute, (December 1971).
11. H. Hughes, R. D. Baxter, and B. Phillips, *IEEE Trans. Nucl. Sci.* **NS-19**, 256 (1972).
12. R. V. Lorenze, Ph.D Thesis, Lehigh Univ. (1972).
13. R. F. Goff, *J. Vac. Sci. Tech.* **9**, 154 (1972); D. J. Ball, T. M. Buck, D. MacNair, and G. H. Wheatley, *Surface Sci.* **30**, 69 (1972).
14. C. W. White, R. A. Kushner, D. V. McCaughan, N. H. Tolk, and D. L. Simms (unpublished data).
15. M. J. Vasile (private communication).
16. D. V. McCaughan, R. A. Kushner, and V. T. Murphy, *Phys. Rev. Letters* **30**, 614 (1973).
17. T. H. DiStefano and D. E. Eastman, *Phys. Rev. Letters* **27**, 1560 (1971).
18. G. Segal, *J. Phys. Chem. Solids* **32**, 2373 (1971).
19. D. Vance, *J. Appl. Phys.* **42**, 5430 (1971).
20. C. W. White, D. L. Simms, and N. H. Tolk, *Science* **177**, 481 (1972).

23

SURFACE COMPOSITION BY ANALYSIS OF NEUTRAL AND ION IMPACT RADIATION

C. W. White, D. L. Simms, and N. H. Tolk

Bell Laboratories
Murray Hill, New Jersey

1. INTRODUCTION

A recently developed method for surface composition analysis is the SCANIIR (Surface Composition by Analysis of Neutral and Ion Impact Radiation) surface analysis technique.[1] This method has evolved from recent experiments[1–6] which show that visible, ultraviolet, and infrared radiation is produced when beams of low-energy ions or neutral particles impact on a solid surface.* Surface constituents are determined by identification and analysis of optical lines and bands that are produced in the collision process.

Because low bombarding energies are used, the SCANIIR technique is a very sensitive probe of the surface region since the range of the incident projectile in the solid is limited to, at most, a few monolayers. This technique is reasonably nondestructive, and damage to the sample can be minimized by using low bombarding energies and low current densities. However, since surface atoms are being continuously removed by the impinging beam, measurements made as a function of bombardment time can be used to provide depth-profile information.

* The production of optical radiation has also been reported in ion–solid collisions at bombarding energies greater than 10 keV.[7–10]

Subsequent sections of this chapter will discuss the following aspects of the SCANIIR technique: In Section 2, a brief description of experimental apparatus is given. In Section 3, several mechanisms for the production of optical radiation in particle–solid collisions are indicated. Radiation arising from excited states of sputtered atoms is discussed in Section 4. Effects of nonradiative de-excitation processes are indicated in Section 5. Examples of radiation from complex samples are presented in Section 6. Detection sensitivities for various impurities in an SiO_2 matrix are presented in Section 7. Preliminary depth-profiling results are discussed in Section 8, followed by a brief discussion of radiation produced by competing mechanisms in Section 9. Section 10 contains a summary and conclusions.

2. EXPERIMENTAL APPARATUS AND TECHNIQUE

Low-energy experimental investigations of optical radiation induced by heavy-particle impact requires an apparatus which can produce monoenergetic beams of ions and neutrals in the general range of 10 eV to 10 keV. An apparatus which satisfies this criterion is shown in Fig. 23-1. Neutral beams are useful both for comparison with corresponding ion-impact measurements as well as to avoid problems associated with charge accumulation and ion-induced impurity transport on nonconducting surfaces.*

The first step in the use of either ions or neutrals as the bombarding projectile is the creation of ions in the source region by means of electron-bombardment ionization, usually of nitrogen molecules or one of the rare gases. Ions are then withdrawn from the source region, accelerated to the desired energy, and focused into a charge-exchange chamber where fast neutrals can be created from the ions by means of resonant charge transfer. This process involves little momentum transfer, and consequently the resulting neutrals have essentially the same energy and direction as the initial ions.[11] In general, the gas pressure in the charge-exchange chamber is adjusted to produce approximately 30% neutralization of the incident ion beam. After passage through the charge-exchange chamber, the remaining ions can be deflected out of the beam, leaving only neutrals to enter the target chamber. Alternatively, to use ions as bombarding projectiles, gas is removed from the charge-exchange chamber and the ion-deflecting voltage is reduced to zero. The solid to be bombarded is oriented in the target chamber with the

* Problems related to impurity transport in insulators subjected to ion bombardment are discussed in Chapter 22. Included in Chapter 22 is a discussion of the drastic reduction in impurity transport which has been observed when neutrals rather than ions are used for bombardment.

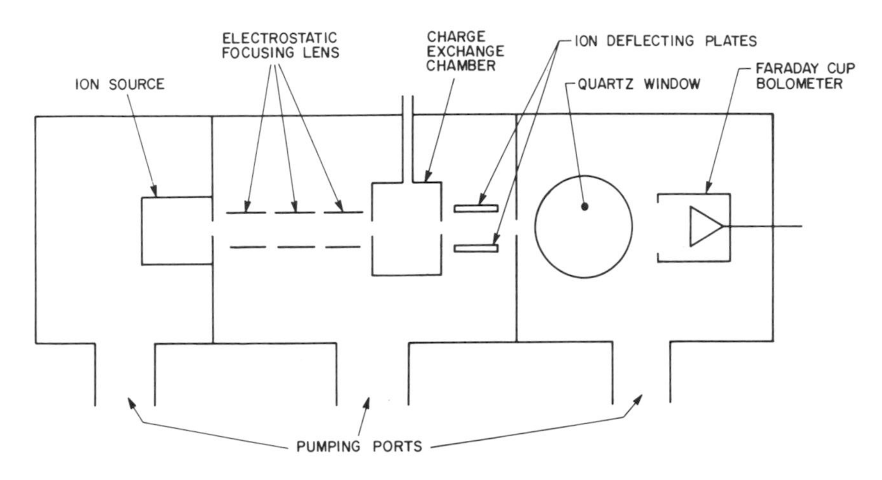

Fig. 23-1. Schematic diagram of apparatus to produce low-energy ion and neutral beams.

surface normal at an angle of approximately 45–60° with respect to the direction of the incident beam. The target chamber is equipped with a quartz window for viewing the interaction region, and photons which are produced as a result of the particle–solid collision are focused by a quartz lens into an $f/5$ 0.3-m monochromator. The monochromator and a cooled photomultiplier are used to record the spectral distribution of radiation produced in the collision process. Single-photon counting techniques are used to enhance the sensitivity of the system for the detection of radiation. The flux of the neutral beam in the target chamber is measured by a bolometer which also can be operated as a Faraday cup to measure ion current.[12] The neutral beam intensity varies as a function of beam energy from 1×10^{-8} A (equivalent neutral current) at 30 eV to 1×10^{-6} A at 4 keV. The beam diameter ranges from 2 to 3 mm. For some experiments, particularly those involved with profiling, it is desirable to use higher beam current density, and in that case we find it advantageous to replace the electron-bombardment source with an rf-discharge ion source.

The aperture separating the intermediate chamber (the charge-exchange chamber is contained in the intermediate chamber) and the target chamber is 3 mm in diameter. Each chamber is evacuated with a mercury diffusion pump having a liquid-nitrogen trap in order to minimize excessive hydrocarbon contamination which might possibly be incurred if oil diffusion pumps were used. A pressure of 3×10^{-7} Torr is achieved in the target chamber with a gas pressure of 10^{-3} Torr in the source and charge-exchange chambers.

In this work, two different kinds of measurements are taken. The first of these is spectral analysis of the emitted radiation obtained by scanning the monochromator through the accessible wavelength range, which for this work is 2000–8500 Å. The second type of measurement involves observation of the intensity of a single line or of a limited segment of the spectrum as a function of beam energy or as a function of bombardment time at fixed energy using either a monochromator or an interference filter. As will be described below, a wide variety of beam species and target materials have been used in these measurements.

3. MECHANISMS FOR PRODUCING OPTICAL RADIATION IN PARTICLE–SOLID COLLISIONS

In the low- and intermediate-energy particle–solid collisions, three distinct kinds of collision-induced optical radiation have been identified. These are:

1. Radiation from excited states of sputtered surface constituents.
2. Radiation from excited states of backscattered beam particles.
3. Radiation resulting from the excitation of electrons in the solid.

It is the first mechanism which provides the basis for the SCANIIR technique, a sensitive technique to identify the constituents of the surface. The other two types of radiation provide important information necessary for understanding fundamental particle–solid interactions and will be discussed in Section 9.

4. RADIATION FROM EXCITED STATES OF SPUTTERED SURFACE CONSTITUENTS

Recent experiments[1–6] have shown that the interaction of a low-energy ion or neutral beam with the surface results in the emission of optical radiation accompanying the ejection of neutral atoms, neutral molecules, and ions from the surface through a process known as sputtering, with simultaneous electronic excitation. A significant portion of the sputtered fragments leave the surface in excited electronic states, subsequently giving rise to infrared, visible, and ultraviolet radiation. One can determine the surface constituents by analyzing the spectral distribution of radiation and identifying prominent optical lines and bands. The intensities of the individual spectral lines are proportional to the absolute concentrations of specific surface constituents within the penetration depth of the incident projectile.

Examples of radiation arising from excited states of sputtered surface constituents in low-energy ion and neutral impact on solid surfaces are illustrated in Figs. 23-2 and 23-3. The spectral distribution

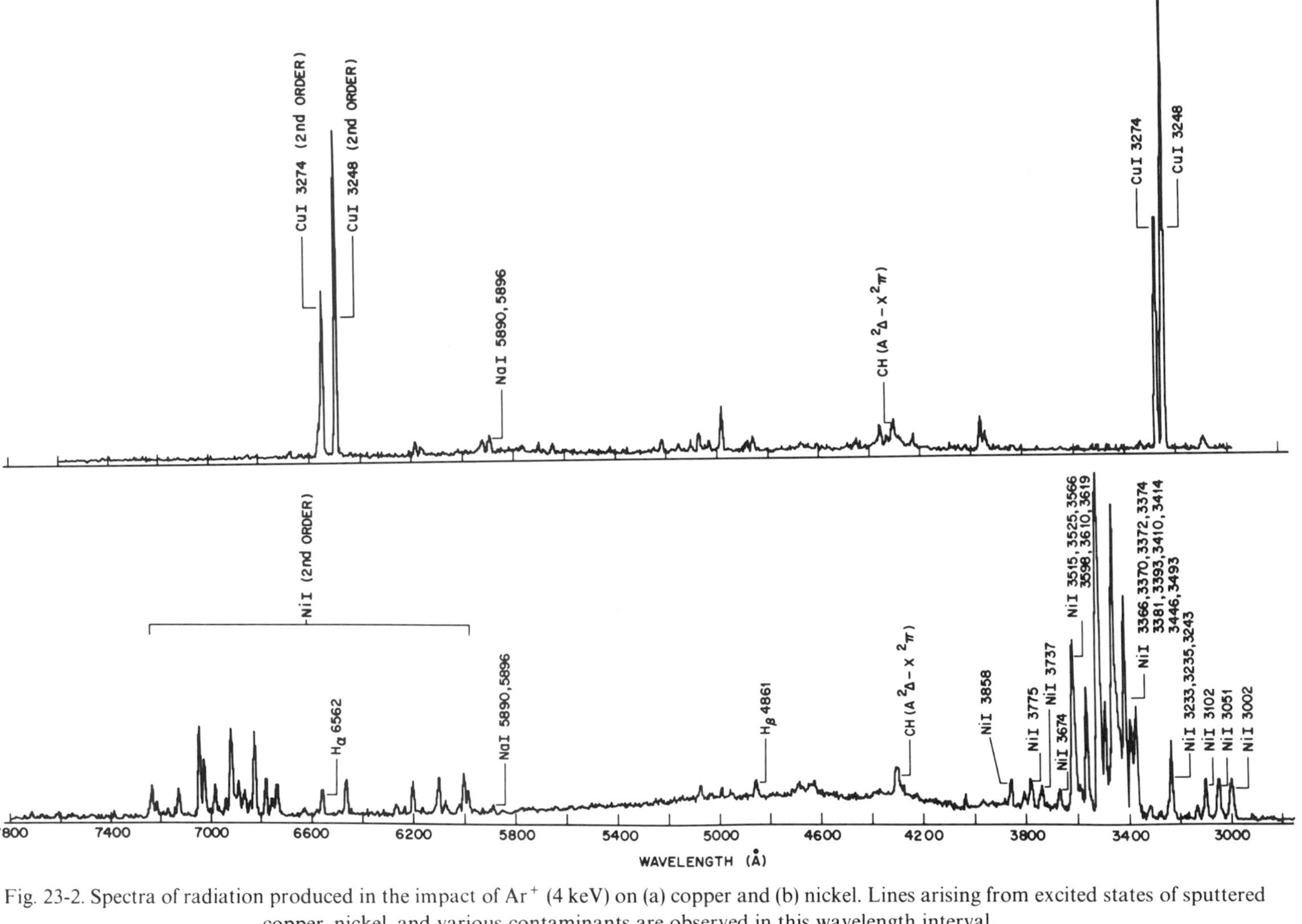

Fig. 23-2. Spectra of radiation produced in the impact of Ar^+ (4 keV) on (a) copper and (b) nickel. Lines arising from excited states of sputtered copper, nickel, and various contaminants are observed in this wavelength interval.

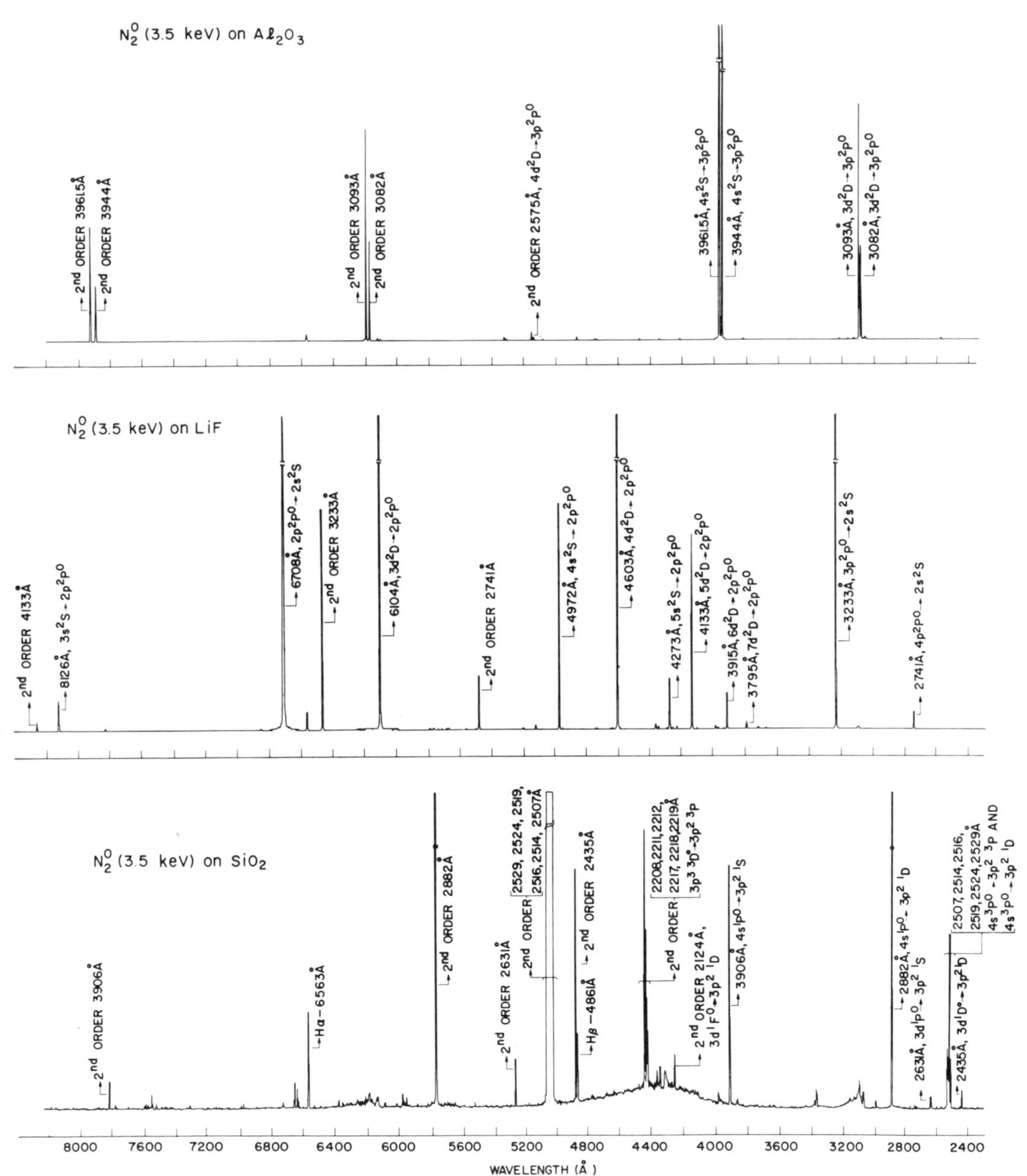

Fig. 23-3. Spectra of radiation produced in the impact of $N_2{}^0$ (3.5 keV) on (a) Al_2O_3, (b) LiF, and (c) SiO_2. Lines arising from excited states of neutral aluminum, lithium, and silicon are observed in this wavelength interval. The wavelength and electronic transition are indicated beside each line. Two Balmer lines of neutral hydrogen, H_α and H_β, are also observed on the SiO_2 scan.

of radiation produced in the impact of Ar^+ (4 keV) on nickel and copper is shown in Fig. 23-2. Most of the prominent lines in these spectral scans have been identified as arising from low-lying energy levels of neutral Ni and Cu, sputtered off the surface in excited states by the incident ion beam. In addition to spectral lines from the surface target materials, radiation is also often observed which is characteristic of surface contaminants. The molecular radiation centered at 4300 Å and 3900 Å has been identified as arising from the $A^2\Delta \rightarrow X^2\Pi$ and $B^2\Sigma \rightarrow X^2\Pi$ electronic transitions of the CH molecule. The origin of this radiation is believed to be collisional excitation of adsorbed hydrocarbon surface contamination. The prominent line which is often observed at approximately 5900 Å is the sodium *D* line doublet which is assumed to arise from sodium contamination deposited on the surface. By prolonged exposure (approximately 20 min) to a 4-keV ion beam or by heating the target, the surface can be cleaned to such an extent that the contaminant radiation is negligible.

Figure 23-3 shows the spectral distribution of radiation produced by the impact of neutral nitrogen (N_2^0) on three insulating targets: Al_2O_3, LiF, and SiO_2. Optical scans taken using neutral beams of neon, argon, and other heavy particles give similar results. The intense lines in Fig. 23-3 have been identified as arising from the decay of excited states of neutral aluminum, lithium, silicon, and hydrogen. Because the line widths are found in these experiments to be equal to the instrumental resolution (about 1 Å), we may assume that the radiation originates from individual atoms and molecules which have been sputtered off the surface in excited states and subsequently decay by photon emission. The Balmer lines of neutral hydrogen are believed to arise from the sputtering of surface contaminants.

Neutral beams rather than ions are used for bombardment of these insulators in order to avoid ion-beam defocusing and energy decrease due to charge buildup on the insulator surface. Ion bombardment may also be used if the insulator surface is bathed in electrons emitted from a nearby heated filament. However, if electrons are used to neutralize the insulating surface during ion bombardment, there is evidence which suggests that impurity transport in the insulator is substantially greater than when neutrals are used as bombarding projectiles. This has been discussed more fully in Chapter 22.

For all cases studied, using metals, semiconductors, and insulators, optical radiation has been observed due to sputtering with simultaneous excitation. If a metal or semiconducting target is used, neutrals and ions (of the same species) are found to produce photons with equal efficiency.[1,3] The reason for this is that at low energies ions impinging on a metallic target are neutralized by nonradiative processes several angstroms in front of the surface[13] before the close-in sputtering encounter occurs. Photons result from the sputtering encounter, and thus the photon-production efficiency from a metallic target should be the same for both ions and neutrals at the same energy; this prediction has been confirmed by experiment. This result can be used as a means for the measurement of the neutral-beam flux in the low-energy region. The ratio of the intensity of the optical line radiation produced when a metallic target is bombarded with neutrals to that produced when a metallic target is bombarded with ions is equal to the ratio of the neutral "current" to the easily measured ion current.

The photon-production efficiency has been found to vary substantially depending on the nature of the bombarded target. For the case of metals at a bombarding energy of ~2–4 keV, typical prominent lines have been measured to have excitation efficiencies of the order of 10^{-4} photons per incident projectile; while in the case of insulators at the same energy, the excitation efficiencies are measured to be two or three orders of magnitude higher. Although the total sputter yields of these two types of solids are different, in general they do not differ by two to three orders of magnitude. A more plausible explanation for the large difference in excitation efficiencies can be found by considering the effects of nonradiative de-excitation processes which are discussed in the next section.

5. NONRADIATIVE DE-EXCITATION

When an excited atom is in the vicinity of a solid surface, there are nonradiative processes resulting from the interaction of the excited atom with the solid which can very efficiently compete with radiative de-excitation of the excited atomic state.[13–15] These nonradiative processes are of two general types; one-electron resonance processes and two-electron Auger processes. When an excited atom is within a few angstroms of the solid, these nonradiative processes are, if energetically allowed, significantly more efficient than radiative decay; this can result in a substantial decrease in the intensity of optical radiation produced in particle–solid collisions.

A vivid example of the effect of nonradiative de-excitation is shown in Fig. 23-4 where we have profiled a composite structure consisting of a thin SiO_2 film (1200 Å thick) on silicon by measuring the intensity of the neutral Si I–2882 Å emission line as a function of bombardment time. In this figure, the normalized intensity of the Si I–2882 Å optical emission line is plotted on a logarithmic scale as a function of the bombardment time. Since atoms are continuously sputtered away by

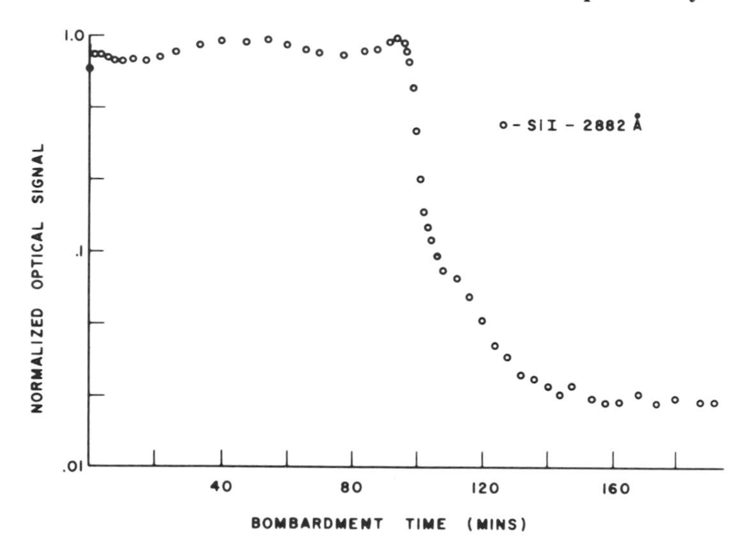

Fig. 23-4. Intensity of Si I-2882 Å ($4s\,^1P^0 \rightarrow 3p^2\,^1D$) measured as a function of bombardment time. The sample used was a thin SiO_2 film (1200 Å thick) on silicon. The location of the SiO_2–Si interface is indicated by the rapid decrease in the intensity of the optical line.

the impinging beam, bombardment time is related to the depth into the SiO_2 film. In Fig. 23-4, the location of the SiO_2–Si interface is determined by the time (or depth) at which the rapid decrease in the intensity of Si I–2882 Å line occurs. These results show that the intensity of the Si I–2882 Å line is reduced by a factor of 50 in sputtering from SiO_2 into silicon. As indicated in Fig. 23-4, the measured width of the transition region, arbitrarily defined as the time or depth required to reduce the optical line intensity from 90% to 10%, is $\sim$120 Å. However, the measured width is a convolution of the true transition width and the depth resolution of the SCANIIR technique. The depth resolution of the SCANIIR technique has not been experimentally determined.

The results of Fig. 23-4 are consistent with the hypothesis that a very efficient mechanism for de-exciting an excited Si atom in the immediate vicinity of a silicon surface involves nonradiative de-excitation. Possible electronic processes which can result in nonradiative de-excitation of an excited Si atom through an interaction with the silicon surface are illustrated in Fig. 23-5. Neglecting broadening of atomic levels, the excited Si atomic level lies above the conduction-band edge and either the one-electron resonance ionization process or the two-electron Auger de-excitation process is energetically allowed. Which of these is the dominant process cannot be determined from this

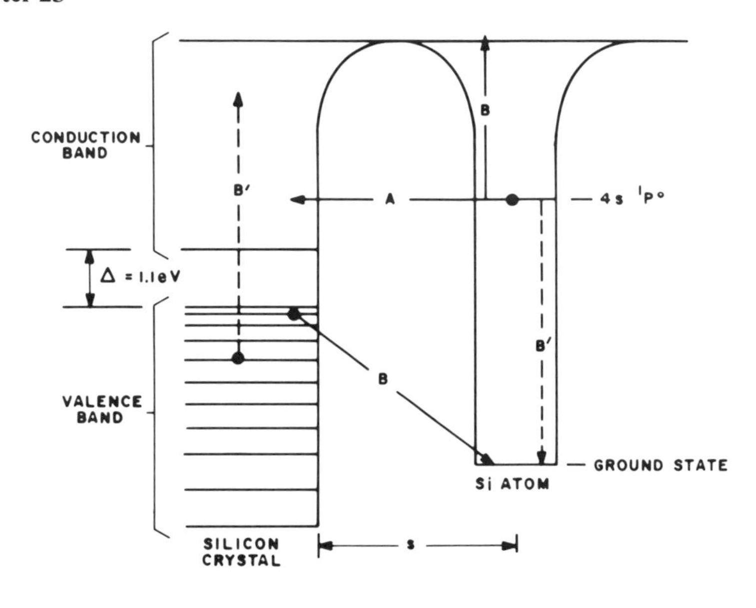

Fig. 23-5. Nonradiative electronic processes resulting from the interaction of an excited Si atom ($4s\ ^1P^0$) and a silicon crystal.

experiment. The nonradiative de-excitation rate $P(s)$ for either of these processes is a very strong function of distance(s) from the surface and has been approximated by[2,13,16]

$$P(s) = A\,e^{-as} \tag{1}$$

where A is the nonradiative transition rate at the surface ($A \sim 10^{14}$–10^{15}/sec, and $a \sim 2 \times 10^8$ cm).[13,16] From Eq. (1), it can be shown that the probability R that an excited atom will escape from the surface and subsequently radiate a photon is given by

$$R = \exp(-A/av_{\perp}) \tag{2}$$

where $v_{\perp}$ is the velocity component of the sputtered atom normal to the surface.

From Eq. (2), the probability of escape from the surface as an excited atom increases exponentially as the velocity component of the sputtered excited atom normal to the surface increases, and therefore the fast, excited sputtered Si atoms are expected to contribute most efficiently to the observed radiation from the silicon substrate. This then accounts for the reduced optical intensity from the silicon substrate. Excited Si atoms which come off the silicon surface with low velocities (the great majority of excited sputtered Si atoms) are very efficiently de-excited by nonradiative processes, leaving only those which came off the surface with high velocity to contribute to the detected radiation.

The very large forbidden band gap ($\sim$8 eV) of SiO_2 results in the enhanced optical intensity observed from SiO_2. In the case of SiO_2, the excited Si atomic level lies below the conduction-band edge and consequently resonance ionization is prohibited due to the lack of available levels. Similarly, if transfer of excitation energy to an electron in the valence band were to occur (Auger de-excitation), it would require lifting the electron from the valence band into the band gap. This is prohibited due to the paucity of available electronic levels. Under these conditions, the alternative channel for de-excitation is radiative decay and thus, from SiO_2, all excited atoms, even those with low velocity, are expected to contribute to the detected radiation.

To summarize, the large forbidden band gap of SiO_2 serves as a barrier to prevent nonradiative de-excitation of excited sputtered Si atoms. In the case of the silicon substrate, nonradiative processes very efficiently de-excite those sputtered excited atoms which come off the surface at low velocities, leaving only the small fraction which are ejected at high velocities to contribute to the radiation. The drastic decrease in the intensity of optical radiation therefore can be used as a means for determining the location of the SiO_2–Si interface in a back-sputtering or ion-milling operation.

The above explanation predicts that the velocity distribution of radiating atoms from silicon and SiO_2 should be quite different, and consequently one might expect to see a significant difference in the width of the optical emission lines from the two different materials. This difference is obvious in Fig. 23-6, which compares second-order optical emission line profiles of the Si I–2882 Å line from SiO_2 and silicon at higher bombarding energies (80 keV). Measurements were made at high bombarding energies and in second order to enhance possible Doppler broadening and shifts of the emission lines. These line profiles were obtained under the same experimental conditions (40-μ slit settings, surface normal at 60° with respect to the beam direction) with samples similar to those used in obtaining the intensity profile of Fig. 23-4. The line profile measured from the SiO_2 film is approximately Gaussian with a full width at half maximum of $\sim$1.1 Å which corresponds within experimental uncertainty to the linear dispersion of the monochromator; this width is therefore assumed to be the instrumental resolution at these slit settings. The measured line profile of the same optical line arising from the silicon substrate is also shown in Fig. 23-6, and the measured width ($\sim$5 Å) is significantly greater than

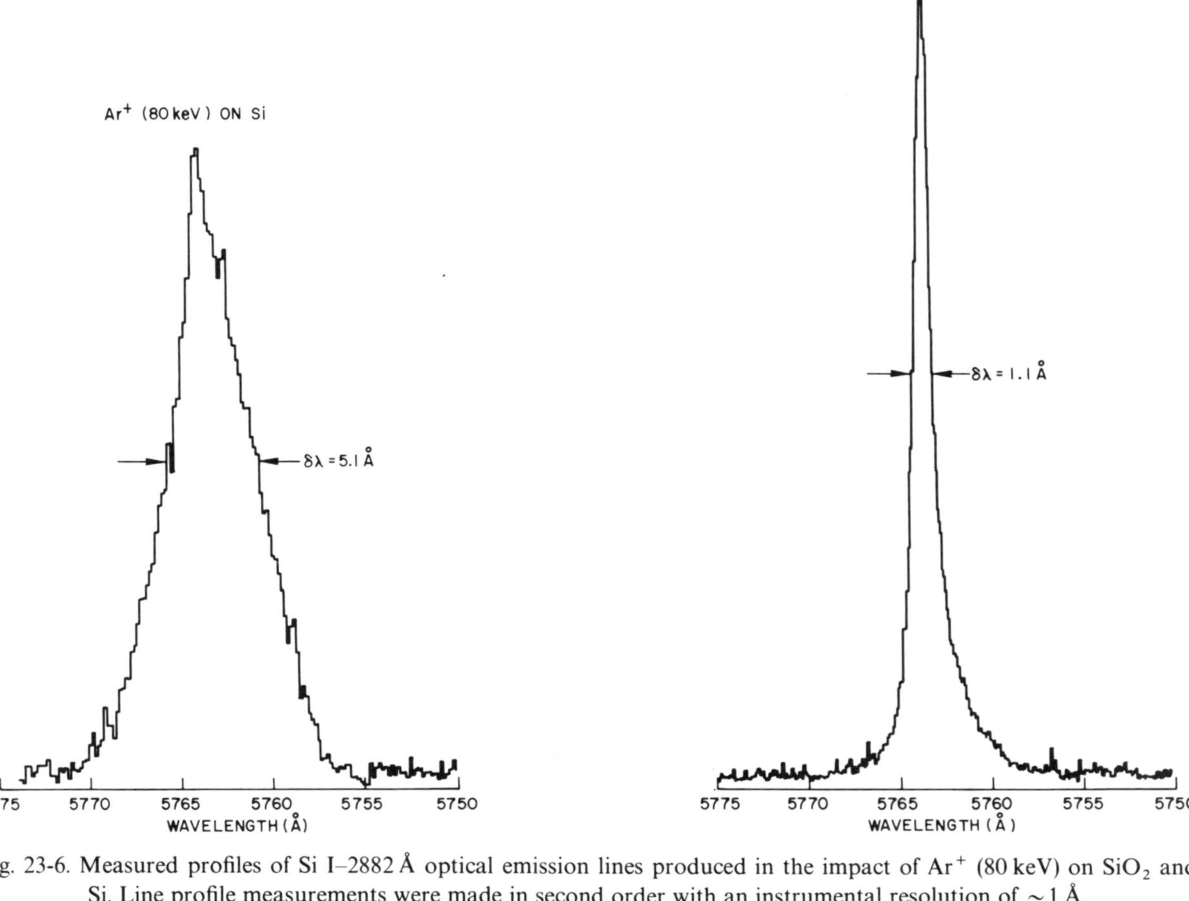

Fig. 23-6. Measured profiles of Si I–2882 Å optical emission lines produced in the impact of Ar^+ (80 keV) on SiO_2 and Si. Line profile measurements were made in second order with an instrumental resolution of ~ 1 Å.

that from SiO_2. If an instrumental resolution of ~ 1 Å is assumed, then most of the width from the silicon surface is real. These line-profile results are consistent with the interpretation that the silicon atoms contributing to the observed radiation in ion bombardment of silicon are moving with significantly higher velocities (which leads to a greater Doppler shift) than in the case of ion bombardment of SiO_2. Nonradiative processes have very efficiently de-excited the excited states of low-velocity Si atoms sputtered from the silicon substrate, leaving only the small fraction ejected at high velocities (which can result in substantial Doppler shifts) to contribute to optical emission. Integration by the spectrometer over the angular distribution of high-velocity radiating Si atoms then results in the enhanced width of the optical emission line.

The effects of nonradiative de-excitation processes and the variation of sputtering yield as a function of the matrix require calibration of the SCANIIR technique in order to obtain reliable quantitative information. This calibration can be achieved using samples containing impurities of interest distributed homogeneously at known concentrations in the matrix, or by depositing a submonolayer of an impurity at known concentration on the surface of the sample. Application of the SCANIIR technique to samples containing impurities at known concentration will be discussed in Section 7. In addition, a method by which the sputtering rate of the matrix can be experimentally measured will be indicated in Section 8 which discusses depth profiling.

Effects due to the variation of sputtering yield and nonradiative processes are commonly encountered in other surface-analysis techniques. For example, in secondary-ion mass spectroscopy,[17] detection of secondary ions emitted from the bombarded surface provides the basis for the measurement technique, while in low-energy ion scattering,[18] energy analysis of backscattered ions is used to identify surface constituents. Both the yield of secondary ions and the yield of backscattered ions are affected by nonradiative ion-neutralization processes similar to the nonradiative de-excitation processes discussed in this section. However, the very preliminary work that we have done indicates that effects due to nonradiative de-excitation processes can be reduced or possibly eliminated by using reactive beams (oxygen).

6. RADIATION FROM COMPLEX SAMPLES

The SCANIIR technique has been applied to a number of complex unknown solids.[1] As examples of this type of application, we present results obtained from the use of samples of volcanic glass and a tektite. Both of these petrological samples were washed in xylene, acetone, and

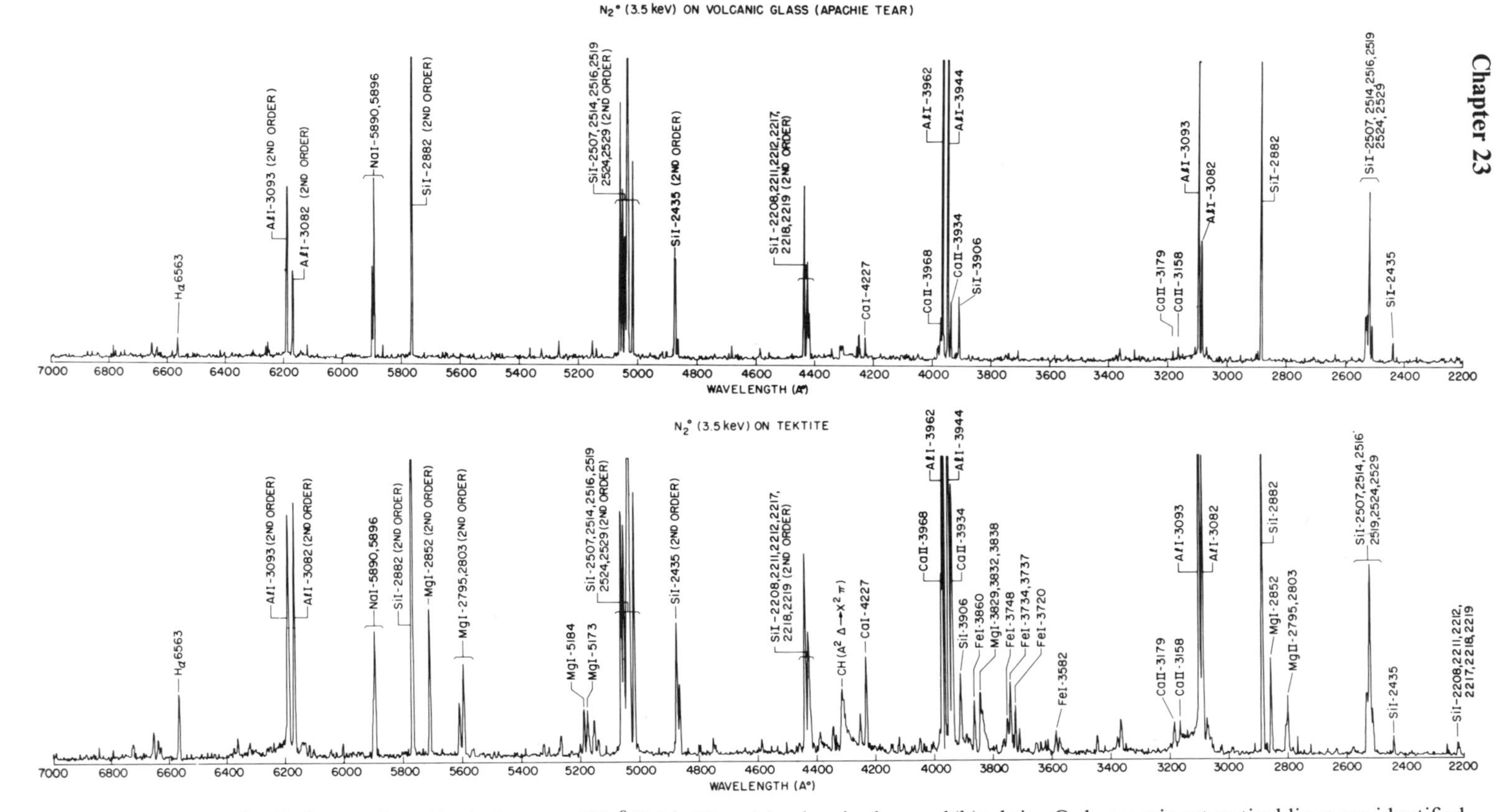

Fig. 23.7. Spectra of radiation produced in the impact of $N_2{}^0$ (3.5 keV) on (a) volcanic glass and (b) tektite. Only prominent optical lines are identified in the scans, and these are labeled by element and wavelength. Roman numerals I and II refer to neutral and singly ionized atoms, respectively. Slightly different slit settings were used for these two scans, 125 μ for volcanic glass and 160 μ for the tektite.

deionized water prior to insertion into the apparatus to remove dust and other contaminants.

The spectral distributions of radiation produced by the impact of N_2^0 (energy of 3.5 keV) on the surfaces of the volcanic glass and the tektite are shown in Fig. 23-7. These spectral scans were acquired in a time period of 30 minutes each, and in each case an equivalent neutral beam current of 3×10^{-7} A was used. It is possible to assign the prominent optical lines to specific elements by comparing the spectrum from the unknown with previously determined spectra characteristic of pure materials. Radiation from common elements such as silicon, aluminum, sodium, and calcium is evident in both scans. In addition, the tektite contains iron and magnesium. There are a number of lines in both scans which have not as yet been assigned to specific elements.

The SCANIIR technique also has been applied to the qualitative analysis of the composition of very small quantities of liquid residues which contain both organic and inorganic material. The experimental procedure we have used is to deposit the liquid on a clean quartz substrate and allow the water to evaporate. Neutral argon or nitrogen is used to bombard the substrate, and the collision-induced spectrum of radiation is recorded and analyzed. In this manner, results have been obtained using very small quantities (0.1–10 μl) of human blood.[1]

7. DETECTION SENSITIVITIES

Detection sensitivities for some elements in SiO_2 have been established by using samples containing known impurities distributed homogeneously at known concentration in the SiO_2 matrix. These samples were impacted by neutral nitrogen (N_2^0) at an energy of 3.5 keV, and the collision-induced spectrum of radiation was recorded and analyzed. From these measurements, signal-to-noise ratios (S/N) were determined for the most prominent optical line arising from each impurity in the sample. Detection sensitivities for each impurity, defined as the impurity concentration necessary for observation with a S/N ratio of 1:1, were then determined from the known impurity concentration and the experimentally determined S/N ratio.

Table 23-1 summarizes results which have been obtained to date using several SiO_2 samples with oxide impurities distributed homogeneously in the matrix. Elements distributed as oxide impurities in these samples are listed in the first column. The detection sensitivity for each element is listed in the second column. These detection sensitivities refer to the weight fraction of the oxide impurity necessary to be seen with a S/N ratio of 1:1. These measurements were obtained under the following experimental conditions: an equivalent neutral current of

TABLE 23-1. SCANIIR Detection Sensitivities in SiO_2

Element	Sensitivity
Al	7×10^{-5}
Fe	8×10^{-4}
Ca	2×10^{-5}
Mg	1×10^{-4}
K	8×10^{-4}
Na	3×10^{-5}
Rb	7×10^{-5}
Li	1×10^{-5}
Sr	7×10^{-5}
Ba	8×10^{-5}
Mn	1×10^{-4}
Ti	1×10^{-3}
Be	4×10^{-5}

3×10^{-7} A, 150-μ monochromator slits (about 4-Å resolution), grating blazed at 5000 Å, a 10-sec photon integration time, and a photon-counting efficiency (photons counted per emitted photon) of ~ 1 part in 10^5 at 5000 Å. Under the outlined experimental conditions, the detection sensitivities in Table 23-1 vary from 1 part in 10^5 to 1 part in 10^3.

The detection sensitivities listed in Table 23-1 can be improved in a number of ways. One way is to use narrow-band interference filters in place of the monochromator to isolate prominent optical lines emitted by the sputtered impurity atoms. Measurements using an interference filter to isolate the Na D_2 line indicate that a factor of 40 improvement in the S/N ratio can be achieved. An improvement of a factor of 40 results in a detection sensitivity of 6 parts in 10^7 for sodium in this homogeneous sample. Further improvement can be achieved by increasing the neutral-beam intensity, by increasing the photon-collection efficiency, and by increasing the photon integration time. With these improvements, detection sensitivities of better than 1 part in 10^8 might be achieved in a homogeneous sample.

8. PROFILING

The SCANIIR technique can also be used to profile thin films or layered structures. One way that this can be done is to measure the intensity of prominent optical lines as a function of bombardment time. Bombardment time can then be related to depth into the sample if the sputtering efficiency for the matrix material is known. As an example of this type of application, we present results which were obtained using a composite structure consisting of SiO_2 (2000 Å thick) on Al_2O_3 (500 Å thick) on SiO_2 (1000 Å thick) on Si (∞). Results which were obtained using this sample are shown in Fig. 23-8 where we have plotted the intensity (normalized to unity) of prominent silicon (Si I–2882 Å) and aluminum (Al I–3962 Å) optical lines on a logarithmic scale as a function of bombardment time. For these measurements, 5-keV argon ions were used as bombarding projectiles, and gross charging of the insulator surface was prevented by flooding the surface with electrons from a nearby heated filament during bombardment. Ion rather than neutral bombardment was used to facilitate rastering of the beam which is necessary to reduce beam cratering phenomena and edge effects. In addition, photons were detected from a 0.1-cm^2 area in the center of the bombarded region (0.4 cm^2) through an optical collection system with a 5° field of view. Decreasing the field of view of the optical collection

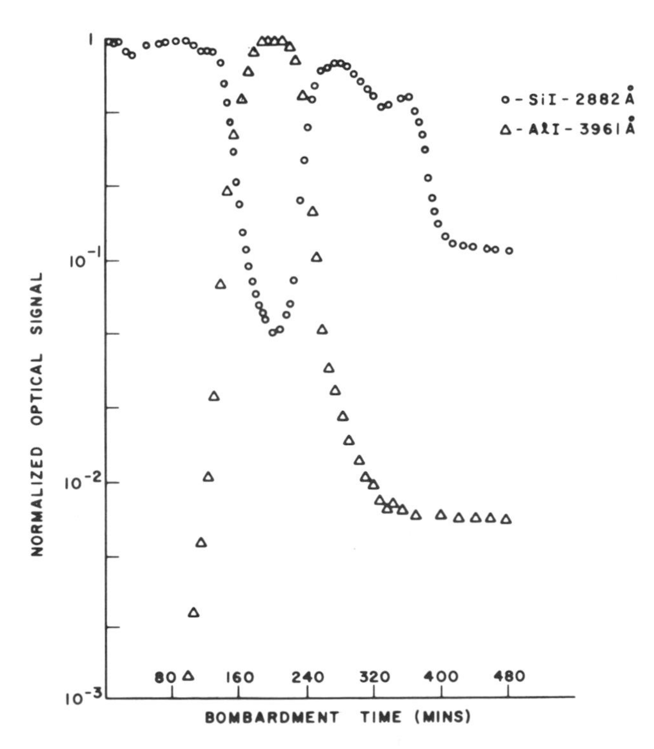

Fig. 23-8. SCANIIR profile measurements of thin film. The sample consisted of SiO_2 (2000 Å thick) on Al_2O_3 (500 Å thick) on SiO_2 (1000 Å thick) on Si. The intensity of silicon and aluminum optical lines is plotted as a function of bombardment time.

system reduces sensitivity, but is necessary to reduce edge effects and cratering phenomena.

The location of the three interfaces is clearly evident in Fig. 23-8, but there is evidence that edge effects or surface nonuniformity are still contributing to systematic errors. The intensity of the Si I–2882 Å decreases to only 5% of its maximum value in the center of the Al_2O_3 layer, and the intensity of the same line decreases by only a factor of 8 in going from the inner SiO_2 film into the silicon substrate. These effects are due either to surface nonuniformity or to edge effects (photons emitted from the edges of the bombarded region being scattered into the optical collection system). Later profiling results, such as those of Fig. 23-4, which were obtained by detecting photons from a smaller surface area ($\sim 1.6 \times 10^{-2}$ cm^2), indicate that further work is needed to eliminate these systematic errors.

Results such as those shown in Figs. 23-8 and 23-4 suggest several applications for the SCANIIR technique. The location of the interface can be determined by detecting optical radiation produced as a result of particle–solid collisions in profiling a thin film using backsputtering or ion-milling techniques. In addition, sputtering efficiencies or sputtering rates for various thin films can be determined by measuring the time required to sputter through a film of known thickness. For example, measurements in Fig. 23-8 indicate that the sputtering rate of plasma-deposited SiO_2 (the outer layer) is almost a factor of 2 greater than that of thermally grown SiO_2 (the inner layer). Finally, detection of optical radiation can be used as a means to monitor the extent of surface cleanliness in a typical backsputtering operation.

9. RADIATION RESULTING FROM OTHER MECHANISMS

In addition to radiation from excited states of sputtered surface constituents, radiation is also produced by other mechanisms in particle–solid collisions, as was indicated in Section 3. Line radiation is observed arising from excited states of backscattered beam particles which have escaped the surface after having experienced violent collisions in the bulk material. Studies of backscattered radiation provide more complete information on final states than do conventional charge-state measurements. Although the processes which create excited states in backscattered particles are likely to be different from those responsible for excited sputtered particles, upon leaving the surface, both classes of particles may be treated identically in terms of nonradiative de-excitation processes which compete with radiative decay.

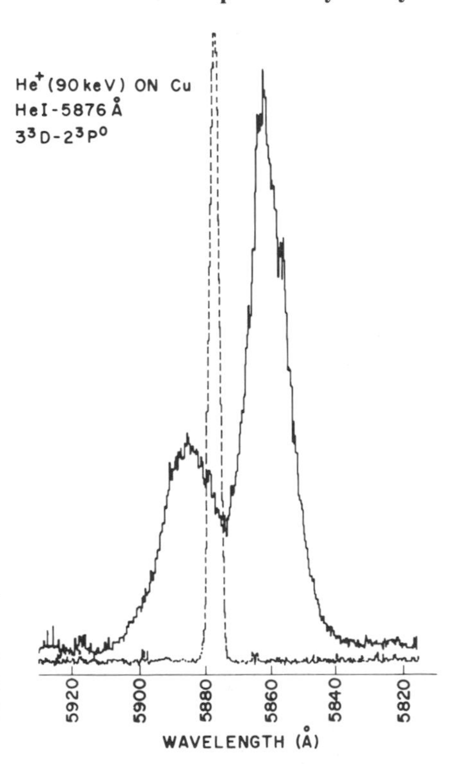

Fig. 23-9. Measured line profile of He I–5876 Å produced in the impact of He^+ (90 keV) on copper. The measured line profile is indicated by the solid line. The position of the unshifted line and the instrumental resolution is indicated by the dotted line.

Radiation arising from this mechanism is illustrated in Fig. 23-9, which shows the measured line profile of the neutral He I–5876 Å produced in the impact of He^+ (at 90 keV) on copper. This emission line is observed to be substantially Doppler-shifted and broadened, reflecting the wide range of energies and directions of the backscattered (neutralized) projectiles as well as effects due to nonradiative de-excitation. The position of the unshifted line and the instrumental resolution is indicated by the dotted line in Fig. 23-9. The shoulder toward the long-wavelength end of the spectrum in Fig. 23-9 arises in part from photons reflected off the copper target into the spectrometer. Pioneering experiments using light projectiles (H^+, $H_2{}^+$, and He^+) on a variety of metal surfaces have been previously reported.[19–21] Future studies should provide important information on the relative role of the bulk and the surface in determining excited-state distributions of backscattered particles.

Intense radiative continua arising from the third mechanism indicated in Section 3 are commonly observed when low-Z projectiles bombard insulating targets.[3] An example is given in Fig. 23-10 which shows the spectral distribution of radiation produced in the impact of He^0 (5 keV) on a crystal of CaF_2. The impact of neutral helium results

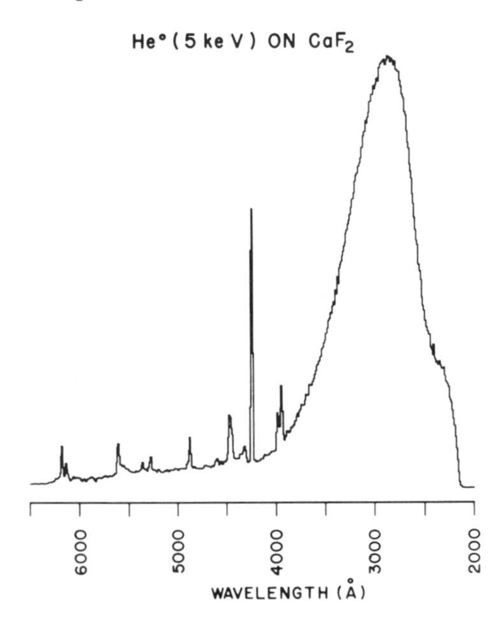

Fig. 23-10. Spectrum of radiation produced in the impact of He^0 (5 keV) on CaF_2.

in the production of an intense continuum of radiation extending over ~1800 Å with a single pronounced peak at ~2800 Å. Prominent line radiation in Fig. 23-7 arises from excited states of Ca and Ca^+. We find the same continuum radiation is produced when low-energy electrons (200 eV to 3 keV) impact on this crystal, and similar results[(22)] have been reported in high-energy electron-bombardment experiments (20 keV). This continuum of radiation is observed when neutral He, H, and H_2 are used as projectiles; when heavier projectiles such as Ne, Ar, and N_2 are used at the same energy, the intensity of the continuum is decreased by two to three orders of magnitude and the spectral distribution is dominated entirely by radiation from excited states of sputtered atoms and ions.

In the case of CaF_2, Hayes *et al.*[(23,24)] have provided convincing evidence that the continuum arises from radiative recombination of electrons with the self-trapped v_k center (the F_2^- molecule) of the CaF_2 crystal. We believe that in the case of low-Z neutral-particle bombardment, the production of the v_k centers results from the inelastic energy transfer of beam projectile energy to bound electrons in the solid (inelastic energy loss in the solid). The continuum then results from the radiative recombination of the self-trapped hole and the mobile electrons created pairwise with holes in the inelastic interaction of the projectile with electrons in the crystal. The absence of a significant continuum in high-Z neutral-particle bombardment at the same energy is consistent with this model since in that case elastic (or nuclear) collisions with target nuclei is the dominant energy-loss mechanism for the projectile moving in the solid.

The production of a radiative continuum in collisions of ions with metal targets also has been reported recently in collisions of heavy ions with Ni, Ta, and Mo.[25] The continuum observed in this experiment was uniquely characterized by the following: (1) No fine structure was observed down to a spectrometer resolution of 2.6 Å. (2) The spectra varied in intensity and shape as a function primarily of the species of bombarded metal. (3) At a given bombardment energy, the radiation intensity was measured to increase monotonically with increasing mass of the bombarding ion. Since the radiative efficiency was observed to increase with increasing mass (at fixed energy), it is doubtful whether the same mechanism that is responsible for producing continuum radiation from insulator targets was also responsible for continuum radiation in the ion–metal collision experiments. Further experimentation in this area is necessary to clarify the detailed nature of the mechanism involved.

10. CONCLUSIONS

Studies of the optical radiation produced in particle–solid collisions constitutes a powerful method for studying fundamental particle–solid interactions. Bombarding solids with low-energy ions and neutrals results in the ejection of a significant fraction of sputtered atoms and molecules from the surface in excited electronic states. Optical emission lines, which result from the decay of these excited states, provide the basis for a simple and sensitive technique for the identification of surface species. Experimental results presented here demonstrate that with the SCANIIR technique, which is based on the excited-state sputtering phenomena, one is able to detect a large variety of surface constituents and contaminants by the identification of the characteristic optical radiation produced when low-energy projectiles bring about the sputtering of surface atoms and molecules. Detection limits for trace impurities in SiO_2 in these experiments range from 1×10^{-3} to 6×10^{-7}. Preliminary evidence indicates that in a homogeneous bulk sample, detection limits for trace impurities can be lowered to 10 parts per 10^9 by increasing the photon-collection efficiency and by increasing the particle flux. Measurements of the optical radiation as a function of bombardment time can be used to obtain depth-profile information in the first few thousand angstroms of the solid. Because of its sensitivity as a surface probe, and because only very small quantities of material are required for analysis, it seems likely that the SCANIIR technique will become an important tool for surface analysis with application to scientific, technological, and environmental problems.

11. ACKNOWLEDGMENTS

We would like to thank R. Preston Watts of Huntsville, Alabama for supplying the samples of volcanic glass and tektite discussed in Section 6. We would also like to thank Dr. L. S. Walter of the Geochemistry Laboratory and Laboratory for Theoretical Studies, Goddard Space Flight Center, Greenbelt, Maryland for supplying the SiO_2 samples which were used to obtain the data summarized in Table 23-1.

12. REFERENCES

1. C. W. White, D. L. Simms, and N. H. Tolk, *Science* **177**, 481 (1972).
2. C. W. White and N. H. Tolk, *Phys. Rev. Letters* **26**, 486 (1971).
3. N. H. Tolk, D. L. Simms, E. B. Foley, and C. W. White, *Radiation Effects* **18**, 221 (1973).
4. J. P. Meriaux, J. M. Gutierrez, Ch. Schneider, R. Goutte, and Cl. Guilland, *Nouv. Rev. Opt. Appl.* **2**, 81 (1971).
5. I. S. T. Tsong, *Phys. Status Solidi* (*A*) **7**, 451 (1971).
6. H. Kerkow, *Phys. Status Solidi* (*A*) **10**, 501 (1972).
7. J. M. Fluint, L. Friedman, J. Van Eck, C. Snoek, and J. Kistemaker, *Proceedings of the Fifth International Conference on Ionization Phenomena in Gases, Munich, Germany, 1961* (H. Maecker, ed.) p. 131, North-Holland, Amsterdam (1962); I. Terzic and B. Perovic, *Surface Sci.* **21**, 86 (1970).
8. C. Snoek, W. F. van der Weg, and P. K. Rol, *Physica* **30**, 341 (1964).
9. W. F. van der Weg and D. J. Bierman, *Physica* **44**, 206 (1969).
10. I. Terzic and B. Perovic, *Surface Sci.* **21**, 86 (1970).
11. H. S. W. Massey and E. H. S. Burhop, *Electronic and Ionic Impact Phenomena*, Clarendon Press, Oxford (1952).
12. C. A. van de Runstraat, R. Wijnanendts van Resandt, and J. Los, *J. Sci. Instr.* **3**, 575 (1970).
13. H. D. Hagstrum, *Phys. Rev.* **123**, 758 (1961); *Phys. Rev.* **96**, 336 (1954).
14. S. S. Shekhter, *Zh. Eksp. Theor. Fiz.* **7**, 750 (1937).
15. A. Cobas and W. E. Lamb, *Phys. Rev.* **65**, 327 (1944).
16. W. F. van der Weg and D. J. Bierman, *Physica* **44**, 206 (1969).
17. R. Castaing and G. Slodzian, *Compt. Rend.* **255**, 1893 (1962).
18. D. P. Smith, *Surface Sci.* **25**, 171 (1971).
19. V. V. Gritsyna, T. S. Kujan, A. G. Koval, and Ya. M. Fogel, *Sov. Phys.—JETP* **31**, 796 (1970).
20. G. M. McCracken and S. K. Erents, *Physics Letters* **31A**, 429 (1970).
21. C. Kerkdijk and E. W. Thomas, *Physica* **63**, 577 (1973).
22. D. R. Rao and H. N. Bose, *Physica* **52**, 371 (1971).
23. W. Hayes, D. L. Kirk, and G. P. Summers, *Solid State Commun.* **7**, 1061 (1969).
24. J. H. Beaumont, W. Hayes, D. L. Kirk, and G. P. Summers, *Proc. Roy. Soc.* (*London*) **315**, 69 (1970).
25. N. H. Tolk, C. W. White, and P. Sigmund, *Bull. Am. Phys. Soc.* **18**, 686 (1973).

INDEX